Brief Contents

Preface viii
About the Authors xvi

CHAPTER 1 Introduction: Using Statistics to Study the Social World 1
CHAPTER 2 Getting to Know Your Data: Frequency Distributions and Visual Representations of Data 55
CHAPTER 3 Examining Relationships between Two Variables: Cross-Tabulations 117
CHAPTER 4 Typical Values in a Group: Measures of Central Tendency 153
CHAPTER 5 The Diversity of Values in a Group: Measures of Variability 196
CHAPTER 6 Probability and the Normal Distribution 233
CHAPTER 7 From Sample to Population: Sampling Distributions and the Central Limit Theorem 271
CHAPTER 8 Estimating Population Parameters: Confidence Intervals 308
CHAPTER 9 Differences between Samples and Populations: One-Sample Hypothesis Tests 350
CHAPTER 10 Comparing Groups: Two-Sample Hypothesis Tests 392
CHAPTER 11 Testing Mean Differences among Multiple Groups: Analysis of Variance 428
CHAPTER 12 Testing the Statistical Significance of Relationships in Cross-Tabulations: The Chi-Square Test 459
CHAPTER 13 Ruling Out Competing Explanations for Relationships between Variables: Control Variables and the Logic of Causality 500
CHAPTER 14 Describing Linear Relationships between Variables: Correlation and Regression 543

SOLUTIONS TO ODD-NUMBERED PRACTICE PROBLEMS 600
GLOSSARY 657
APPENDIX A Normal Table 665
APPENDIX B Table of t-Values 667
APPENDIX C F Table, for Alpha = .05 669
APPENDIX D Chi-Square Table 671
APPENDIX E Selected List of Formulas 673
APPENDIX F Choosing Tests for Bivariate Relationships 677
Index 679

Contents

Preface   viii
About the Authors   xvi

CHAPTER 1 Introduction   1

Why Study Statistics?   1
Research Questions, Variables, and the Research Process   3
Pinning Things Down: Variables and Measurement   4
Units of Analysis   5
Measurement Error: Validity and Reliability   6
Levels of Measurement   8
   Interval-Ratio-Level Variables   9
   Ordinal-Level Variables   9
   Nominal-Level Variables   9
   What about Scales?   10
Causation: Independent and Dependent Variables   10
Getting the Data: Sampling and Generalizing   11
   Sampling Methods   12
Sources of Secondary Data: Existing Data Sets, Reports, and "Big Data"   14
   Big Data   15
Growth Mindset and Math Anxiety   17
Using This Book   18
Statistical Software   19
Chapter Summary   21
Using Stata   23
Using SPSS   31
Practice Problems   43
Notes   53

CHAPTER 2 Getting to Know Your Data   55

Frequency Distributions   56
Percentages and Proportions   58
Cumulative Percentage and Percentile   60
Percent Change   63
Rates and Ratios   64
   Rates   64
   Ratios   65
Working with Frequency Distribution Tables   66
   Missing Values   66
   Simplifying Tables by Collapsing Categories   67
Graphical Displays of a Single Variable: Bar Graphs, Pie Charts, and Histograms   70
   Bar Graphs and Pie Charts   70
   Histograms   72
Time Series Charts   74
Comparing Two Groups on the Same Variable Using Tables, Graphs, and Charts   75
Chapter Summary   80
Using Stata   81
Using SPSS   90
Practice Problems   102
Notes   116

CHAPTER 3 Examining Relationships between Two Variables   117

Cross-Tabulations and Relationships between Variables   118
   Independent and Dependent Variables   119
   Calculating Column Percentages   123
   Row Percentages   126
   Comparing Column and Row Percentages   127
   Total Percentages   127
   Comparing Column, Row, and Total Percentages   128

STATISTICS for SOCIAL UNDERSTANDING

A Problems-Based Approach

Second Edition

NANCY WHITTIER
Smith College

TINA WILDHAGEN
Smith College

HOWARD J. GOLD
Smith College

ROWMAN & LITTLEFIELD
Lanham • Boulder • New York • London

Published by Rowman & Littlefield
An imprint of The Rowman & Littlefield Publishing Group, Inc.
4501 Forbes Boulevard, Suite 200, Lanham, Maryland 20706
www.rowman.com

86-90 Paul Street, London EC2A 4NE

Copyright © 2025 by The Rowman & Littlefield Publishing Group, Inc.

All rights reserved. No part of this book may be reproduced in any form or by any electronic or mechanical means, including information storage and retrieval systems, without written permission from the publisher, except by a reviewer who may quote passages in a review.

British Library Cataloguing in Publication Information Available

Library of Congress Cataloging-in-Publication Data

Names: Whittier, Nancy, 1966– author. | Wildhagen, Tina, 1980– author. | Gold, Howard J., 1958– author.
Title: Statistics for social understanding : a problems-based approach / Nancy E. Whittier, Smith College, Tina Wildhagen, Smith College, Howard J. Gold, Smith College.
Description: Second edition. | [Lanham] : [Rowman & Littlefield], [2024] | Includes bibliographical references and index.
Identifiers: LCCN 2023059140 (print) | LCCN 2023059141 (ebook) | ISBN 9781538175927 (cloth) | ISBN 9781538175934 (paperback) | ISBN 9781538175941 (epub)
Subjects: LCSH: Statistics. | Social sciences—Statistical methods. | Stata. | BISAC: SOCIAL SCIENCE / Statistics
Classification: LCC QA276.12 .W5375 2024 (print) | LCC QA276.12 (ebook) | DDC 519.5—dc23/eng/20240119
LC record available at https://lccn.loc.gov/2023059140
LC ebook record available at https://lccn.loc.gov/2023059141

∞™ The paper used in this publication meets the minimum requirements of American National Standard for Information Sciences—Permanence of Paper for Printed Library Materials, ANSI/NISO Z39.48-1992.

Interpreting the Strength of Relationships   130
Interpreting the Direction of Relationships   132
Putting the Results into Words   134
Graphical Representations of Bivariate Relationships   135
Chapter Summary   138
Using Stata   139
Using SPSS   141
Practice Problems   145
Notes   152

CHAPTER 4 Typical Values in a Group   153

What Does It Mean to Describe What Is Typical?   154
Mean   155
Median   159
Mode   163
Finding the Mode, Median, and Mean in Frequency Distributions   165
   Finding the Mode in a Frequency Distribution   165
   Finding the Median in a Frequency Distribution   165
   Finding the Mean in a Frequency Distribution   166
Choosing the Appropriate Measure of Central Tendency   167
Median versus Mean Income   171
Chapter Summary   173
Using Stata   174
Using SPSS   179
Practice Problems   184
Notes   195

CHAPTER 5 The Diversity of Values in a Group   196

Range   197
Interquartile Range   198
   Finding the Interquartile Range in a Frequency Distribution   199
   Visualizing the Interquartile Range with Box Plots   201

Standard Deviation   202
   Calculating Standard Deviation   204
Using the Standard Deviation to Compare Distributions   205
Comparing Apples and Oranges   206
Skewed versus Symmetric Distributions   211
Chapter Summary   213
Using Stata   214
Using SPSS   217
Practice Problems   223
Notes   232

CHAPTER 6 Probability and the Normal Distribution   233

The Rules of Probability   234
   The Addition Rule   237
   The Complement Rule   238
   The Multiplication Rule with Independence   240
   The Multiplication Rule without Independence   241
Applying the Multiplication Rule with Independence to the "Linda" and "Birth-Order" Probability Problems   242
Probability Distributions   244
The Normal Distribution   246
   The Empirical Rule   247
Standardizing Variables and Calculating z-Scores   249
Chapter Summary   257
Using Stata   258
Using SPSS   261
Practice Problems   263
Notes   270

CHAPTER 7 From Sample to Population   271

Repeated Sampling, Sample Statistics, and the Population Parameter   272
Sampling Distributions   275
Finding the Probability of Obtaining a Specific Sample Statistic   278
Estimating the Standard Error from a Known Population Standard Deviation   279

Finding and Interpreting the z-Score for Sample Means 279

Finding and Interpreting the z-Score for Sample Proportions 283

The Impact of Sample Size on the Standard Error 284

Chapter Summary 286

Using Stata 287

Using SPSS 292

Practice Problems 297

Note 307

CHAPTER 8 Estimating Population Parameters 308

Inferential Statistics and the Estimation of Population Parameters 309

Confidence Intervals Manage Uncertainty through Margins of Error 311

Certainty and Precision of Confidence Intervals 311

Confidence Intervals for Proportions 312

Constructing a Confidence Interval for Proportions: Examples 316

Confidence Intervals for Means 319

The t-Distribution 320

Calculating Confidence Intervals for Means: Examples 322

The Relationship between Sample Size and Confidence Interval Range 326

The Relationship between Confidence Level and Confidence Interval Range 330

Interpreting Confidence Intervals 331

How Big a Sample? 332

Assumptions for Confidence Intervals 335

Chapter Summary 336

Using Stata 338

Using SPSS 340

Practice Problems 342

Notes 348

CHAPTER 9 Differences between Samples and Populations 350

The Logic of Hypothesis Testing 352

Null Hypotheses (H_0) and Alternative Hypotheses (H_a) 353

One-Tailed and Two-Tailed Tests 354

Hypothesis Tests for Proportions 354

The Steps of the Hypothesis Test 359

One-Tailed and Two-Tailed Tests in Practice 360

Hypothesis Tests for Means 362

Example: Testing a Claim about a Population Mean 367

Error and Limitations: How Do We Know We Are Correct? 370

Type I and Type II Errors 371

What Does Statistical Significance Really Tell Us? Statistical and Practical Significance 374

Chapter Summary 376

Using Stata 377

Using SPSS 380

Practice Problems 384

Notes 390

CHAPTER 10 Comparing Groups 392

Two-Sample Hypothesis Tests 393

The Logic of the Null and Alternative Hypotheses in Two-Sample Tests 393

Notation for Two-Sample Tests 395

The Sampling Distribution for Two-Sample Tests 395

Hypothesis Tests for Differences between Means 396

Confidence Intervals for Differences between Means 403

Hypothesis Tests for Differences between Proportions 404

Confidence Intervals for Differences between Proportions 408

Statistical and Practical Significance in Two-Sample Tests 410

Chapter Summary 411

Using Stata 412

Using SPSS 416

Practice Problems 421

Notes 427

CHAPTER 11 Testing Mean Differences among Multiple Groups 428

Comparing Variation within and between Groups 429
Hypothesis Testing Using ANOVA 431
Analysis of Variance Assumptions 432
The Steps of an ANOVA Test 432
Determining Which Means Are Different: Post Hoc Tests 439
ANOVA Compared to Repeated t-Tests 440
Chapter Summary 441
Using Stata 442
Using SPSS 444
Practice Problems 447
Notes 458

CHAPTER 12 Testing the Statistical Significance of Relationships in Cross-Tabulations 459

The Logic of Hypothesis Testing with Chi-Square 462
The Steps of a Chi-Square Test 465
Size and Direction of Effects: Analysis of Residuals 471
Example: Gender and Perceptions of Health 472
Assumptions of Chi-Square 477
Statistical Significance and Sample Size 477
Chapter Summary 482
Using Stata 483
Using SPSS 485
Practice Problems 488
Notes 499

CHAPTER 13 Ruling Out Competing Explanations for Relationships between Variables 500

Criteria for Causal Relationships 505
Modeling Spurious Relationships 507
Modeling Non-Spurious Relationships 512
Chapter Summary 518
Using Stata 519

Using SPSS 524
Practice Problems 530
Notes 542

CHAPTER 14 Describing Linear Relationships between Variables 543

Correlation Coefficients 544
   Calculating Correlation Coefficients 545
Scatterplots: Visualizing Correlations 547
Regression: Fitting a Line to a Scatterplot 551
The "Best-Fitting" Line 552
Slope and Intercept 553
   Calculating the Slope and Intercept 556
Goodness-of-Fit Measures 556
   R-Squared (r^2) 557
   Standard Error of the Estimate 558
Dichotomous ("Dummy") Independent Variables 558
Multiple Regression 562
Statistical Inference for Regression 565
   The F-Statistic 566
Standard Error of the Slope 568
Assumptions of Regression 570
Chapter Summary 573
Using Stata 575
Using SPSS 581
Practice Problems 588
Notes 598

SOLUTIONS TO ODD-NUMBERED PRACTICE PROBLEMS 600
GLOSSARY 657
APPENDIX A Normal Table 665
APPENDIX B Table of t-Values 667
APPENDIX C F Table, for Alpha = .05 669
APPENDIX D Chi-Square Table 671
APPENDIX E Selected List of Formulas 673
APPENDIX F Choosing Tests for Bivariate Relationships 677

Index 679

Preface

The idea for *Statistics for Social Understanding: A Problems-Based Approach* began with our desire to offer a different kind of book to our statistics students. We wanted a book that would introduce students to the way statistics are actually used in the social sciences: as a tool for advancing understanding of the social world. We wanted thorough coverage of statistical topics, with a balanced approach to calculation and the use of statistical software, and we wanted the textbook to cover the use of software as a way to explore data and answer exciting questions. We wanted a textbook that incorporated Stata, which is widely used in graduate programs and is increasingly used in undergraduate classes, as well as Statistical Package for Social Sciences (SPSS), which remains widespread. We wanted a book designed for introductory students in the social sciences, including those with little quantitative background, but one that did not talk down to students and that covered the conceptual aspects of statistics in detail even when the mathematical details were minimized. We wanted a clearly written, engaging book, with plenty of practice problems of every type and easily available data sets for classroom use.

We are excited to introduce the second edition of *Statistics for Social Understanding* to students and instructors. We are three experienced instructors of statistics, two sociologists and a political scientist, with more than seventy combined years of teaching experience in this area. We drew on our teaching experience and on research about the teaching and learning of statistics to write what we think will be a more effective textbook for fostering student learning.

The second edition of this book includes all new examples from recent research and updated examples based on the most recent available datasets throughout each chapter. In revising, we have paid special attention to clear writing, engaging examples, and thorough explanations. We focus on showing students not just *how* to conduct statistical analysis but *why* to do so: to answer important research questions. And we emphasize interpretation as much as we do calculation.

We also have drawn on our experiences teaching statistics to social science students by authoring this book's ancillary materials, which include practice problems, test banks, data sets, suggested class exercises, PowerPoint slides, assignments, and lecture notes.

Distinguishing Features

1. *Incorporates real-world research and a real-world approach to the use of statistics*

   Each chapter begins with an engaging real-world social science question that can be addressed using a given statistical technique. The chapter then explains that technique, provides examples using the same question and other engaging questions, and shows how related questions can also be addressed using Stata or SPSS. Every chapter asks questions about relationships between variables or comparisons across groups, showing students the power of statistics to answer important real-world questions.

   Applied questions draw from sociology, political science, criminology, education, and related fields. We use data sets that are widely used in the social sciences, including the General Social Survey, American National Election Study, World Values Survey, and Police Public Contact Survey.

   Throughout this book, we show students how to address issues that working social scientists routinely confront, such as how to use missing data, recode variables (including conceptual and statistical considerations), combine variables into new measures, think about outliers or atypical cases, choose appropriate measures, weigh considerations of causation, and interpret results.

2. *Balances calculation and concepts*

   This book frames learning statistics as a way to analyze important social science questions. We believe students need to be able to reason statistically, not simply use software to produce results, but we recognize that most working researchers rely on statistical software, and we strike a balance among these skills. We include some formulas and hand calculation to help students understand where the numbers come from. We spend more time on conceptual understanding, including topics relating to causality, and we include topics often omitted from other texts such as the use of confidence intervals as a follow-up to a hypothesis test. A lighter focus on hand calculation opens up time in the semester for topics that are most important to understanding statistical social sciences. Our aim is to give students the tools they might use as working researchers in a variety of professions (from jobs in small organizations where they might be reading and writing up external data or doing program evaluation, to research or data analysis jobs) and prepare them for higher-level statistics classes, if they choose to take them.

3. *Uses accessible, non-condescending approach and tone*

   We have written a text that is student-friendly but not condescending. In an effort to assuage students' anxiety about statistics, some texts strike a tone that communicates the *expectation* that students lack confidence in their abilities. We are conscious of the possibility that addressing students with the assumption that they hate or are intimidated by statistics could activate stereotype threat—the well-established fact that, when students feel that they are expected to perform poorly, their anxiety over disproving that stereotype makes their performance worse than it otherwise would be.[1] In selecting examples, we have remained alert to the risk of stereotype threat, choosing examples that do not activate (or even challenge) gender or racial stereotypes about academic performance.

4. *Draws on teaching and learning research*

Our approach is informed by research on teaching and learning in math and statistics and takes a universal design approach to accommodate multiple learning styles. We take the following research-based approaches:

- Research on teaching math shows that students learn better when teachers use multiple examples and explanations of topics.[2] This book explains topics in multiple ways, using both alternative verbal explanations and visual representations. As experienced instructors, we know the topics that students frequently stumble over and give special attention to explaining these areas in multiple ways.
- Some chapter examples and practice problems lead students through the process of addressing a problem by acknowledging commonly held misconceptions before presenting the proper solution. This approach is based on research that shows that simply presenting students with information that corrects their statistical misconceptions is not enough to change these "strong and resilient" misconceptions.[3] Students need to be able to examine the differences in the reasoning underlying incorrect and correct approaches.
- Each chapter provides numerous, carefully proofread, practice problems. Students learn best by doing, and this book provides numerous opportunities for problem-solving.

- This book avoids the "busy" layout used by some textbooks, which can distract students' attention from the content, particularly those with learning differences. Drawing on the principles of universal design, our book utilizes a clean, streamlined layout that allows all students to focus on the content without unnecessary distractions.[4]

5. *Instructors can choose either Stata or SPSS*

While most existing textbooks use only SPSS, or assume that students will purchase an additional, costly, supplemental text for Stata, this book can be used with either Stata or SPSS. We include parallel sections for both SPSS and Stata at the end of every chapter. These sections are written to ensure that students understand that software is a tool to be used to improve their own statistical reasoning, not a replacement for it.[5] This book walks students through how to use Stata and SPSS to analyze interesting and relevant research questions. We not only provide students with the syntax or menu selections that they will use to carry out these commands but also carefully explain the statistical procedures that the commands are telling Stata or SPSS to perform. In this way, we encourage students to engage in statistical reasoning as they use software, not to think of Stata or SPSS as doing the statistical reasoning for them. For Stata, we teach students the basic underlying structure of Stata code. This approach facilitates a more intuitive understanding of how the program works, promoting greater confidence and competence among students. For SPSS, we teach students to navigate the menus fluently.

Tools for Learning

1. **"In Depth" boxes** encourage critical thinking by tackling tricky statistical queries and **"Application" boxes** provide additional examples.
2. **Bulleted chapter summaries** recap important concepts introduced in the chapter.
3. **Chapter key terms and a glossary** build students' terminology for discussing statistical methods and techniques.
4. **"Using Stata"** and **"Using SPSS"** features teach students to use statistical software.
5. **Practice Problems, Stata Problems,** and **SPSS Problems** at the end of every chapter allow students to test their knowledge. **Solutions to Odd-Numbered Questions** are answered in the back of this book. Practice problems are designed so that instructors can cover all topics of the chapter by assigning either odd- or even-numbered questions. For example, students could work odd-numbered problems as practice, checking their answers in this book, and complete even-numbered problems as homework.

For Instructors

Organization of this Text

This textbook begins with descriptive statistics in chapters 2 through 5. One key difference from many introductory statistics texts is that we introduce cross-tabulations early, after frequency distributions and before central tendency and variability. In our experience as instructors, we have noticed that students often begin thinking about relationships between variables at the very beginning of the class, asking questions about how groups differ in their frequency distributions of some variable, for example. Cross-tabulations follow naturally at this point in the class and allow students to engage in real-world data analysis and investigate questions of causality relatively early in the course. Chapters 6 and 7 lay the foundation for inferential statistics, covering probability, the normal distribution, and sampling distributions. We cover elementary probability in the context of the normal distribution, with a focus on the logic of probability and probabilistic reasoning in order to lay the groundwork for an understanding of inferential statistics. Chapters 8 through 12 cover the basics of inferential statistics, including confidence intervals, hypothesis testing, z- and t-tests, analysis of variance, and chi-square. Chapter 13, unusual among introductory statistics texts, focuses on the logic of causality and control variables. Most existing texts address this topic more briefly (or not at all), but, in our experience, it is an important topic that we all supplement in lecture. Finally, chapter 14 covers correlation and regression. While that chapter is pitched to an introductory level, we pay more attention to multiple regression than do many texts, because it is so widely used, and we have a box on logistic regression to introduce students to the range of models that working social scientists employ.

Alternative Options for Organization

Instructors who wish to cover chapters in a different order—for example, delaying cross-tabulations until later in the semester—can readily do so. Instructors teaching on a quarter system may choose to omit some chapters or sections, such as those on probability, analysis of variance, or the final chapters of this book. Courses moving at a faster pace might cover both chapters 2 and 3 and/or 4 and 5 in one week; courses moving at a slower pace might omit parts of some chapters and end the course before chapters 13 and 14. The ancillary materials make it easy for instructors to adapt to their own organization, either using the materials in full and in order, or selecting from them.

For Students

In a course evaluation, one of our students offered advice to future students:

> Use the textbook! it is incredibly specific and helpful.

We agree, and not just because we wrote it! We suggest reading the assigned section of the chapter before class and working the example problems, pencil in hand, as you read. Make a note of anything you don't understand and ask questions or attend especially to that material in class. After class, look back at the chapter summary and work the practice problems to consolidate your understanding. If you find a chapter especially difficult on your first pass through, try to reread it after you have covered the material in class. This may seem time-consuming, but you not only will improve your understanding (and your grade) but will save time when it comes to studying for midterm and final exams or completing class projects. As another student explained:

> The textbook format let me go through the material from class at a slower pace and I could turn to it for step-by-step help in doing the assignments.

Similarly, you should look through the software sections before you conduct these exercises in class or lab. You do not need to try to memorize the SPSS or Stata commands but familiarize yourself with the procedures and the reasons for them. As with the rest of the chapter, hands-on practice is key here, too.

Remember, you are taking this class because you want to understand the social world.

As another student wrote:

> If you are not too familiar with working with numbers, that is just fine! This course is designed as an analytical course which means that you will be focusing more so on the meaning behind numbers and statistics rather than just focusing on finding "correct" answers.

The companion website contains more study materials and gives you access to the data sets used for the software sections in this textbook. You can use these data sets and your newfound skill in SPSS or Stata to investigate questions you are interested in, beyond those we cover.

This book does not assume or require any mathematical skills beyond being able to plug numbers into formulas and is designed to be engaging and clear for students with a lot or a little math background. Chapter 1 contains more tips on studying and learning as well as overcoming math anxiety.

Student and Instructor Resources

This book is accompanied by a learning package, written by the authors, that is designed to enhance the experience of both instructors and students. All practice problems and test bank items have been accuracy-checked by a panel of outside reviewers.

For Instructors

Instructor's manual with solutions. This valuable resource includes a sample course syllabus and links to the publicly available data sets used in the Stata and SPSS sections of this text. For each chapter, it includes lecture notes, suggested classroom activities, discussion questions, and the solutions to the practice problems. The *Instructor's Manual with Solutions* is available to adopters for download on the text's catalog page at www.rowman.com/ISBN/9781538175927.

Test bank. The *Test Bank* includes both short-answer and multiple-choice items and is available in Word. In either format, the *Test Bank* can be fully edited and customized to best meet your needs.

PowerPoint® slides. The PowerPoint presentation provides lecture slides for every chapter. In addition, multiple-choice review slides for classroom use are available for each chapter.

For Students

Companion website. Accompanying the text is an open-access companion website designed to reinforce key topics and concepts. For each chapter, students will have access to:

Publicly available data sets used in the Stata and SPSS sections
Flashcards of key concepts

Students can access the companion website from their computers or mobile devices at www.textbooks.rowman.com/whittier2e.

Acknowledgments

We are grateful to many manuscript reviewers, both those who are identified here and those who chose to remain anonymous, for their in-depth and thoughtful comments as we developed this text. We are fortunate to have benefited from their knowledgeable and helpful input. We thank the following reviewers who offered comments for the **second edition** of this text:

Paul Babbitt, *Southern Arkansas University*
Michelle Bannoura, *Hudson Valley Community College*
Raymond Barranco, *Mississippi State University*
Mandi Barringer, *University of North Florida*
Maybeth Beller, *Marshall University*
Hanna Brant, *SUNY Geneseo*
Ross Burkhart, *Boise State University*
Robert Carini, *University of Louisville*
Elise Cohen, *Loyola University Chicago*
Benjamin Cole, *Simmons University*
Ed Collom, *CSU Fullerton*
Bridget Costello, *Lonestar College-Kingwood*
Cynthia Cready, *University of North Texas*
Todd Curry, *University of Texas at El Paso*
G. Doug Davis, *Troy University*
John Doces, *Bucknell University*
Rebecca Eissler, *San Francisco State University*
Todd Ferguson, *University of Mary Hardin-Baylor*
Tracy Freudenthaler, *Northeastern State University*
Luis Garcia-Fierro, *Florida Gulf Coast University*
Larry Gibbs, *Southern Oregon University*
Matthew Green, *College of DuPage*
William Haller, *Clemson University*
Chris Heasley, *Saint Joseph's University*
Patrick Homan, *Dominican University*
Jonathan Hoover, *Regent University*
Ting Jiang, *Metropolitan State University of Denver*
Nicole Kalaf-Hughes, *Bowling Green State University*
Esther Kim, *University of Southern California*
Athena King, *Virginia State University*
Kyle Knight, *University of Alabama in Huntsville*

Adriana Lopez-Ramirez, *University of Arkansas at Little Rock*
Luis Felipe Mantilla, *University of South Florida*
Fred Markowitz, *Northern Illinois University*
Kyle Mattes, *Florida International University*
Ronald McGauvran, *Tennessee Tech University*
Karyn Miller, *Texas A&M University-Commerce*
Nikola Mirilovic, *University of Central Florida*
Melanie Mogavero, *Georgian Court University*
Michael Morgan, *Marietta College*
David Nalbone, *Purdue University*
Ingrid Nelson, *Bowdoin College*
Boniface Noyongoyo, *Marshall University*
Emily Pakhtigian, *Penn State University*
Hyung Sam Park, *University of Central Florida*
B. Mitchell Peck, *University of Oklahoma*
Margarita Pivovarova, *Arizona State University*
Etienne Poliquin, *College of William and Mary*
Chad Posick, *Georgia Southern University*
Allison Quatrini, *Eckerd College*
John Rausch, *West Texas A&M University*
Karen Robinson, *CSU San Bernadino*
Cristabel Rogalin, *Purdue University Northwest*
Kurt Schock, *Rutgers*
Richard Seltzer, *Howard University*
Brion Sever, *Florida Gulf Coast University*
Amanda Simpfenderfer, *College of William and Mary*
Brian William Smith, *Saint Edward's University*
Jess Smith, *Virginia Commonwealth University*
Michelle Steimer, *Waynesburg University*
Jeffrey Stone, *CSU Los Angeles*
Jennifer Strangfeld, *CSU Stanilaus*
Xiaoli Su, *Jackson State University*
Kirk Swortzel, *Mississippi State University*
May Takeuchi, *University of North Alabama*
Jessica Terman, *George Mason University*
Victor Thompson, *Ride University*
Jeffrey Timberlake, *University of Cincinnati*
Elizabeth Trudeau, *University of Notre Dame*
Carol Walther, *Northern Illinois University*
Nicole Weller, *Indiana University Kokomo*
Joe Wert, *Indiana University Southeast*
Philip Yang, *Texas Woman's University*
Huijun Yi, *Troy University*
Ryan Yonk, *Utah State University*

The following instructors and researchers were kind enough to accuracy-check the practice problems for the book:

Raymond Barranco, *Mississippi State University*
Amy L. Benton, *Samford University*
Jeremy Brenner-Levoy, *University of Cincinnati*
Ross E. Burkhart, *Boise State University*
Cynthia M. Cready, *University of North Texas*
Qingwen Dong, *University of the Pacific*
Lisa Jo Elliott, *Penn State*
William Haller, *Clemson University*
Adriana López Ramírez, *University of Arkansas at Little Rock*
Hunter Martaindale, *Texas State University*
Michael A. Morgan, *Marietta College*
Margaret Ralston, *Mississippi State University*
Judy Reed Sandlin, *Texas A&M University*
Mark B. Schure, *Montana State University*
Brad Seaberg, *Sam Houston State University*
Sara Skiles, *University of Notre Dame*
Steve Song, *Concordia University Chicago*
Jeffrey D. Stone, *California State University Los Angeles*

We thank Alessandra Bazo Vienrich, Georgiana Bostean, and John Doces for their helpful comments on the first edition. We also thank our research assistants at Smith College who helped with the first edition. Sarah Feldman helped with generating clear figures and practice problems and gave feedback on the text early on, Elaona Lemoto assisted with the final stages, and Sydney Pine helped with the ancillary materials. Dan Bennet, from the Smith College Information Technology Media Production Department, helped us figure out how to generate high-quality screenshots for the SPSS and Stata sections. Leslie King offered helpful feedback on early drafts of some chapters and Bobby Innes-Gold read and commented on some chapters.

At Rowman & Littlefield, we are grateful to Alyssa Palazzo and Samantha Delwarte for their help as we wrote the second edition and to Alden Perkins for shepherding the book through production. We are grateful to Bill Rising of Stata's author support program for his detailed comments on the accuracy of the text and the Stata code in the first edition.

Finally, our greatest thanks go to our students. Their questions, points of confusion, and enthusiasm for learning helped us craft this text and inspire us in our teaching. This book is dedicated to them.

Notes

[1] S. J. Spencer, C. Logel, and P. G. Davies. 2016. "Stereotype Threat." *Annual Review of Psychology* 67: 415–437.

[2] J. R. Star. 2016. "Small Steps Forward: Improving Mathematics Instruction Incrementally." *Phi Delta Kappan* 97: 58–62.

[3] J. Garfield and D. Ben-Zvi. 2007. "How Students Learn Statistics Revisited: A Current Review of Research on Teaching and Learning Statistics." *International Statistical Review* 75: 372–396.

[4] S. E. Burgstahler. 2015. *Universal Design in Higher Education: From Principles to Practice*. Cambridge, MA: Harvard Education Press.

[5] S. Friel. 2007. "The Research Frontier: Where Technology Interacts with the Teaching and Learning of Data Analysis." In M. K. Heid and G. W. Blume (eds.), *Research on Technology and the Teaching and Learning of Mathematics: Syntheses and Perspectives*, vol. 2 (279–331). Greenwich: Information Age Publishing, Inc.

About the Authors

Nancy Whittier is Sophia Smith Professor of Sociology at Smith College. She has taught statistics and research methods for over 30 years and also teaches classes on gender, sexuality, and social movements. She is the author of *Frenemies: Feminists, Conservatives, and Sexual Violence* (2018), *The Politics of Child Sexual Abuse: Emotions, Social Movements, and the State* (2009), *Feminist Generations* (1995), and numerous articles on social movements, gender, and sexual violence, and is coeditor of *Feminist Frontiers* (tenth edition, 2020, with Verta Taylor and Leila Rupp).

Tina Wildhagen is associate professor of sociology at Smith College. She has taught statistics and quantitative research methods for 15 years and also teaches courses on privilege and power in American education and inequality in higher education. Her research and teaching focus on social inequality in the American education system and on first-generation college students. Her work appears in various scholarly journals, including *The Sociological Quarterly*, *Sociological Perspectives*, *The Teachers College Record*, *The Journal of Negro Education*, and *Sociology Compass*.

Howard J. Gold is professor of government at Smith College. He has taught statistics for over 30 years and also teaches courses on American elections, public opinion and the media, and political behavior. His research focuses on voting behavior, partisanship, and third parties. He is the coauthor (with Donald Baumer) of *Parties, Polarization and Democracy in the United States* and author of *Hollow Mandates: American Public Opinion and the Conservative Shift*. His work has also appeared in *American Politics Quarterly*, *Political Research Quarterly*, *Polity*, *Public Opinion Quarterly*, and the *Social Science Journal*.

Chapter 1

Introduction
Using Statistics to Study the Social World

Why Study Statistics?

We all live in social situations. We observe our surroundings, are socialized into our cultures, navigate social norms, make political judgments and decisions, and participate in social institutions. But social sciences assume that what we see as individuals does not tell the whole story of our social world. Large-scale political and social institutions and processes are difficult to see and understand without systematic research. For many students in a social science statistics class, this basic insight is part of what drives their interest in this field. Maybe you want to understand political processes more thoroughly, understand how inequalities are produced, or understand the operation of the criminal justice system.

Many students reading this book are taking a statistics class because it is required for their major. Some of you are passionate about statistics, but most of you are probably mainly interested in sociology, political science, criminology, anthropology, education, or whatever your specific major is. Whatever your specific interest, statistics can deepen your understanding and build your toolkit for communicating social science insights to diverse audiences. You may think of statistics as a form of math, but, in fact, statistics are more about thinking with numbers than they are about computation. Although we do cover some simple computation in this book, our emphasis is on understanding the logic and application of statistics and interpreting their meaning for concrete topics in the social sciences. There is a good reason that statistics are required for many social science majors: Statistical methods can tell us a lot about the most interesting and important questions that social scientists study. Statistics also can tell you a lot about the questions that motivated your own interest in social sciences.

Statistics and quantitative data are important tools for understanding large-scale social and political processes and institutions and how these structures shape individual lives. They help us to comprehend trends and patterns that are too large for us to see in other ways. Statistics do this in three main ways. First, they help us to describe large-scale patterns. For example, what is the average income of residents in a given state?

Second, statistics help us determine the factors that shape these patterns. This includes simple comparisons, such as how income varies by gender or by age. It also includes more complicated mathematical models that can show how multiple forces shape a given outcome. How do gender, age, race, and education interact to shape income, for example?

Third, statistics help us understand how and whether we can generalize from data gathered from only some members of a group to draw conclusions about all members of that group. This aspect of statistics, called inferential statistics, uses ideas about probability to determine what kinds of generalizations we can make. It is what allows researchers to draw meaningful conclusions from data about relatively small numbers of people.

In this book, we emphasize what we can *do* with statistics, focusing on real social science research and analyzing real data. Readers will develop a strong sense of how quantitative social scientists conduct their research and will get plenty of practice in analyzing social science data. Not all of this book's readers will pursue careers as researchers, but many of you will have careers that include analyzing and presenting information. And all of you face the task of making sense of mountains of information, including social science research findings, communicated by various media. This book provides essential tools for doing so.

Recently, some commentators have noted that we have entered a "post-fact," or "post-truth," era. People mean different things by this, but one meaning is that the sheer volume of people and agencies producing information has multiplied to the point that an expert can be found to attest to the accuracy of just about any claim.[1] Just think of the amount of information that you are exposed to on a weekly basis from various social media platforms, websites, television programs, and other forms of media. How do you make sense of it? How do you decide whether a claim you read online is true or false? Statistics can powerfully influence opinion because they use numerical data, which American culture assumes are objective and legitimate. But not all claims are factual, even those that appear to be backed up by statistics. This book will equip you with an understanding of how statistics work so that you can evaluate the meaning and credibility of statistical data for yourself.

When quantitative research is carefully conceived and conducted, the results of statistical analyses can yield valuable information not only about how the social world works but also about how to effectively address social problems. For example, sociologist Natasha Quadlin examined how college grades affected men's and women's employment prospects after graduation.[2] Quadlin wanted to understand whether students with higher GPAs were more likely to be called back for an interview, and whether this differs for women and men. Quadlin conducted a resume audit study, in which she submitted 2,106 fictional job applications to actual job openings. Resumes had high, moderate, or low GPAs and men's or women's names. Did resumes with higher GPAs fare better in the job search process?

It turned out that GPA made no real difference for men: those with low GPAs were called back for interviews about as often as those with high GPAs. In contrast, women with GPAs in the B range were called back significantly more often than those with the lowest GPAs, but women with GPAs in the A range were called back *less* often. In other words, moderately strong grades helped women while higher achievement actually hurt them. In subsequent research, Quadlin found that employers viewed moderate-achieving women as more likable than high-achieving women, relying on this stereotype even though they had not met any of the (fictional) applicants.

By varying the applicant's gender and GPA on otherwise identical resumes, Quadlin controlled for alternative explanations, such as specific skills or work histories listed on resumes. In other words, employers were reacting to the gender and GPA themselves.

Quadlin's study contains many of the key elements of statistical analysis that we discuss in this book: assessment of the relationship between variables (GPA, gender, and employer callbacks); a careful investigation of whether one of the variables (GPA) has a causal impact on the other (employer callback) and, if so, whether that causal impact varies by another factor (gender); and examination of the generalizability of the results.

Research Questions, Variables, and the Research Process

Most research starts with a **research question**, which asks how two or more variables are related. A **variable** is any characteristic that has more than one category or value. In the social sciences, we must be able to answer our research questions using data. In many cases, these questions may be fairly general. For example, a criminologist may begin by wanting to know what factors make people convicted of felonies more or less likely to register to vote upon release from prison. In this example, the key variable is whether or not people register to vote. It has two values, registering or not registering to vote. Felony conviction is *not* a variable, because all the people the researcher is studying have felony convictions. Having a felony conviction was a selection criterion for someone to be included in the research.

In other cases, a question may expand on prior research. For example, research has shown that Internet skills vary by class, race, and age.[3] Do these factors affect how often Internet users post on Twitter or contribute to TikTok? The variables in this example are class, race, age, and frequency of posting on Twitter or contributing to TikTok. Or, if we know that children tend to generally share their parents' political viewpoints, does this hold true in votes for candidates in primaries? The variables here are parents' votes and children's votes.

Some research begins with a **hypothesis**, a specific prediction about how variables are related. For example, a researcher studying political protest might hypothesize that larger protests produce more news media coverage. (The variables are size of protest and amount of media coverage.) Other research begins at a more exploratory level. For example, the same researcher might collect data on several possible variables about protests, such as the issue they focus on, the organizations that sponsor them, whether they include violence, as well as their size, in order to explore what shapes media coverage. Statistical methods can support both approaches to research.

This book focuses on **quantitative analysis**—that is, analyses that use statistical techniques to analyze numerical data. Many social scientists also use qualitative methods. **Qualitative methods** use data that are not numerical, such as the text of documents, interviews, or field observations. Qualitative data analysis often focuses on meanings, processes, and interactions, which can be difficult to analyze through quantitative data. Like quantitative research, qualitative research may test hypotheses or be more exploratory in nature. Qualitative research analysis often uses specialized software programs. Many researchers use

mixed methods, which employ both qualitative and quantitative data and analysis. While this book focuses on quantitative analysis, combining both methods can yield a richer and more accurate understanding of social phenomena than either approach alone.

Pinning Things Down: Variables and Measurement

Answering any kind of social science research question entails gathering data. Gathering useful data requires formulating the research question as precisely as possible. Quantitative researchers first identify and define the question's key concepts. **Concepts** are the abstract factors or ideas, not always directly observable, that the researcher wants to study. Many concepts have multiple dimensions. For example, a researcher interested in how people's social class affects their sense of well-being must define what social class and well-being mean before examining whether they are related. Using existing research and theory, the researcher might define a social class as a segment of the population with similar levels of financial, social, and cultural resources. She might decide that well-being is one's sense of overall health, satisfaction, and comfort in life. Stating clear definitions of concepts ensures that the researcher and her audience understand what is meant by those concepts in the particular project at hand.

Once researchers specify, or define, their concepts, they must decide how to **measure** these concepts. Deciding how to measure a concept is also referred to as **operationalizing** a concept, or **operationalization**. Operationalization is the process of transforming concepts into variables that can be measured. It is how the researcher will observe concepts using empirical data. Staying with the example of social class and well-being, how would we place people into different class categories? Using the conceptual definition described above, the researcher might decide to use people's income, wealth, highest level of education, and occupation to measure their social class. All of these are empirical indicators of financial, social, and cultural resources. To operationalize well-being, the researcher might decide to measure an array of behaviors (e.g., number of times per week that one exercises) and attitudes (e.g., overall sense of satisfaction with one's life).

This process of conceptualization and measurement, or operationalization, is how concepts become variables in quantitative research. Figure 1.1 shows this process for the concept of well-being.

Starting from the top of the figure and moving down, Figure 1.1 shows how researchers move from defining a key concept to specifying how that concept will be empirically measured and transformed into variables. First, the concept of well-being is defined. Next, the dimensions of the concept (physical, mental, and spiritual) are specified. Finally, the researcher establishes empirical measures for each dimension (e.g., frequency of exercise as an indicator of physical well-being). These empirical measures are variables. The vertical arrow on the right side of Figure 1.1 shows how moving from defining concepts to measuring them shifts from the theoretical or abstract to the empirical realm, where variables can be measured. Studying relationships among variables is the central focus of quantitative social science research. A variable, remember, is any single factor that has more than one category or value. For example, gender is a variable with multiple categories

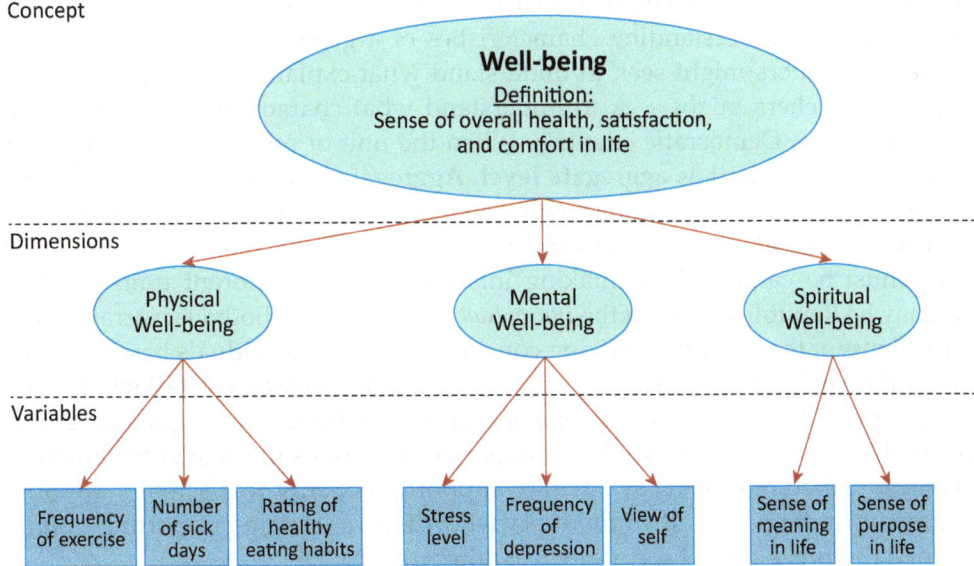

Figure 1.1 Conceptualization and Measurement of a Key Concept

(e.g., man, woman, gender non-binary, etc.). For some variables, such as number of sick days, it is straightforward to determine the value of the variable for different individuals. For variables that lack a clear measurement standard, such as sense of purpose in life, researchers must establish their categories and methods of measurement, usually guided by existing research. In quantitative social science research, the survey item—that is, survey question—is among the most common tools used to operationalize concepts. Survey items have either closed- or open-ended response options. **Closed-ended survey items** provide survey respondents with predefined response categories—or "answers" to the question. The number of categories can range from as few as two (e.g., yes or no) to very many (e.g., a feeling thermometer that asks respondents to rate their feeling about something on a scale from 0 to 100 degrees). With closed-ended survey items, researchers decide on the measurement of the concept before administering the survey. **Open-ended survey items** do not provide response categories. For example, an item might ask respondents to name the issue that is most important to them in casting a vote for a candidate. Open-ended items give respondents more leeway in answering questions. Once researchers have all responses to an open-ended item, they often devise response categories informed by the responses themselves and then assign respondents to those categories based on their responses. For example, with an open-ended question about which issues are important to voters, a researcher might combine various responses having to do with jobs or the economy into one category.

Units of Analysis

Who or what is being studied is the **unit of analysis**. In the social sciences, researchers are interested in studying the characteristics of individuals and the characteristics of groups. A study of people's voting patterns and political party affiliation focuses on understanding individuals. But a study of counties that voted for a Republican versus Democratic candidate focuses on understanding characteristics of a group, in this case counties. In the first case, researchers might seek to understand what explains people's votes; in the second case, researchers might seek to understand what characteristics are associated with Republican versus Democratic counties. When the unit of measurement is the group, we sometimes also refer to it as **aggregate level**. Aggregate level units that researchers might be interested in include geographic areas, organizations, religious congregations, families, sports teams, musical groups, or businesses.

One must be careful about making inferences across different units of analysis. A *county* may be Republican, but at the *individual* level, there are both Democratic and Republican residents of that county. Drawing conclusions about individuals based on the groups to which they belong is an error in logic known as the **ecological fallacy**. Similarly, the death toll from COVID, controlled for population, has been much higher in counties that voted for Trump in the 2020 election, compared to counties that voted for Biden.[4] But not everyone who died of COVID in a Trump-supporting county was actually a Trump voter. Wrongly concluding that all individuals who died in Trump-supporting counties had voted for Trump is a case of the ecological fallacy.

Measurement Error: Validity and Reliability

Most variables in the social sciences include some amount of error, which means that the values recorded for a variable are to some degree inaccurate. Even many variables that one might suspect would be simple to measure accurately, such as income, contain error. How much money did you receive as income in the last calendar year? Some readers may know the exact figure. But others would have to offer an estimate, maybe because they cannot recall or because they worked multiple jobs and have trouble keeping track of the income produced by each of them. Still others might purposefully report a number that is higher or lower than their actual income. Researchers never know for sure how much error their variables contain, but we can evaluate and minimize error in measurement by assessing the validity and reliability of our variables.

Validity is the extent to which variables actually measure what they claim to measure. When measures have a high degree of validity, this means that there is a strong connection between the measurement of a concept and its conceptual definition. In other words, valid measures are accurate indicators of the underlying concept. Imagine a researcher who claims that he has found that happiness declines as people exercise more. How is that researcher measuring happiness? It turns out that he has operationalized happiness through responses to two survey questions: "How much energy do you feel you have?" and "How much do you look forward to participating in family activities?" Do you think answers to these questions are good measures of happiness? They may get at elements of happiness—happier people may have more energy or look forward to participating in activities more. But they are not direct measures of happiness, and we could argue that they measure other things instead (such as how busy people are, the quality of their family relationships, or their health). What about a researcher who wants to measure the prevalence of food insecurity, in which people do not have consistent access to sufficient food? This could be operationalized in a survey question such as, "How often do you have insufficient food for yourself and your family?" or "How often do you go hungry because of inability to get sufficient food for yourself or your family?" It could also be operationalized by the number and size of food pantries per capita or food stamp usage. Which way of operationalizing food insecurity is more accurate? The survey questions have greater validity because both food pantries and food stamp usage are affected by forces other than food insecurity (urban areas may have more food pantries per capita than rural areas, not all people eligible for food stamps use them, and so forth). If the researcher were interested instead in social services to reduce food insecurity, looking at food pantries and food stamps would be a valid measure. Even if a measure is valid, it may not yield consistent answers. This is the question of **reliability**. Reliable measures are those whose values are unaffected by the measurement process or the measurement instrument itself (e.g., the survey). Imagine asking the same group of college students to rate how often in a typical week they spend time with friends, with the following response choices: "often," "a few times," "occasionally," and "rarely." These response choices are likely to lead to problems with reliability because they are not precise. A student who gets together with friends about four times a week might choose "often" or "a few times," and if you asked her the question again a week later, she might choose the other option, even if her underlying estimate of how often she spent time with friends was unchanged. In other words, the same students may give different, or inconsistent, responses if asked the question repeatedly. Measures also tend not to be reliable when they ask questions that respondents may not have detailed understanding or information about. For example, a survey might ask how many minutes a week people spend doing housework, or a survey of Americans might ask their opinion of Britain's foreign policy toward Chile. Because people do not generally precisely track minutes spent doing housework, and Americans are unlikely to know much about British foreign policy, their responses to such questions will be inconsistent. Reliability and validity do not necessarily coincide. For example, the time shown on a clock may be reliable without being valid. Some households may deliberately set their clocks to be a few minutes fast, ensuring that when the alarm goes off at what the clock says is 6:45, the actual time is 6:30. In this case, the clock consistently—that is, reliably—tells time, but that time is always wrong (or invalid). Figure 1.2 uses a feeling thermometer, which asks people to rate their feeling about something on a scale from 0 to 100 degrees, to illustrate how reliability and validity can coincide or not. Imagine each dot is an individual's response to the same feeling thermometer item asked five separate times. The true value of the person's feeling is 42 degrees. In scenario A, the responses have a high degree of validity, or accuracy, because they are all near 42 degrees, the accurate value. There is also a high degree of reliability because the responses are consistent. Researchers strive to attain scenario A by using accurate and consistent measures. In scenario B, there is still a high degree of consistency, and therefore reliability, in the measure. However, validity is low because the responses are far from the true value of 42 degrees. Finally, scenario C reflects both low reliability and low validity. The responses are inconsistent, or scattered across the range of the temperature scale, and many fall far from 42 degrees. Notice that there is no scenario D, in which reliability is low and validity is high. This is because the overall accuracy of a measure requires that it be reliably measured.

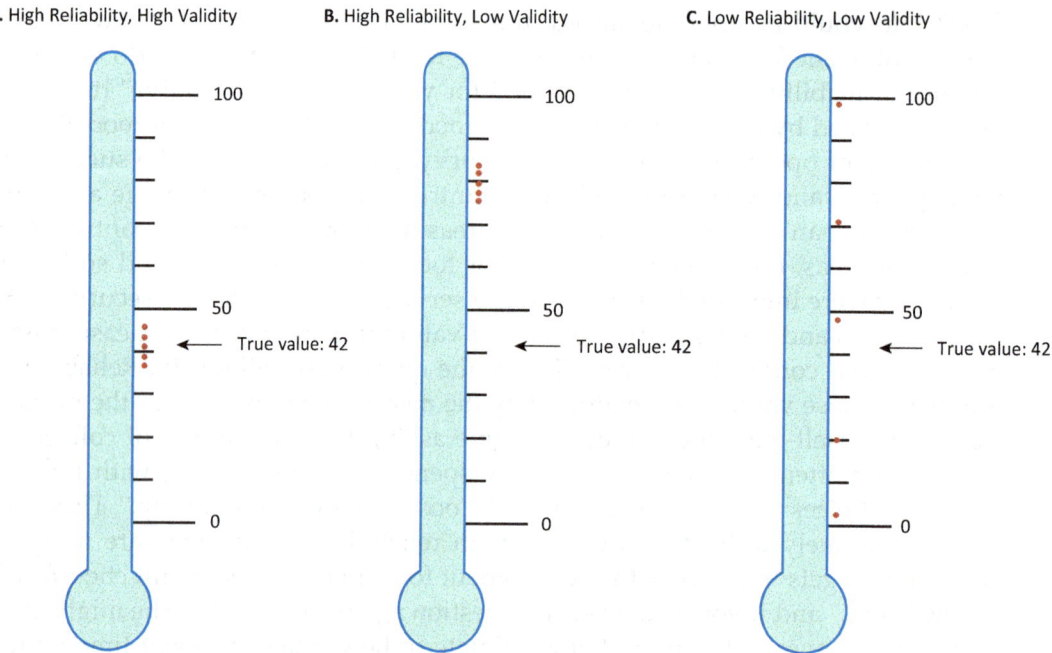

Figure 1.2 Visualizing Reliability and Validity

Levels of Measurement

There is another consideration about how to measure variables—whether they will be measured in a way that will yield data that are numerical. This is very important for statistical analysis because it determines what statistics and graphics can be employed, as we will explain below. Consider a variable measuring employment status. A survey question could ask respondents how many hours they worked in the preceding week. The answers would all be *numbers*, such as 35 hours, 12 hours, and so forth. Alternatively, a survey question could ask whether respondents are employed full-time, part-time, or not at all. The answers to this question are not numbers, although they can be placed in *rank order*, since those who are employed full-time are working more than those who are employed part-time. Variables also can be measured in ways that are neither numerical nor rank-able. For example, a question about employment might ask what type of job the respondents hold and provide response categories such as "officials and managers," "professionals," "technicians," "sales," "clerical," "skilled trades," and so forth.* These answers are *categories*, but they do not have any quantitative meaning because none of them can be considered to have a greater value than others. A variable's **level of measurement** refers to whether the "answers," or possible values of the variable, are numerical; rank-able but not numerical; or categorical.

---

* All federal agencies in the United States use the Standard Occupational Classification system, which classifies all workers into 867 detailed occupations. A full list of these occupations can be found in the 2018 *Standard Occupational Classification Manual* at https://www.bls.gov/soc/2018/soc_2018_manual.pdf.

Interval-Ratio-Level Variables

Variables with values that are numerical, or quantitative, are called **interval** or **ratio** level. For these variables, the distance between each consecutive value of the variable is identical. For example, in the variable number of hours worked, the distance between twenty hours and twenty-one hours (one hour) is the same as the distance between twenty-one hours and twenty-two hours or between any other adjacent values. Ratio level variables have a meaningful 0 value that represents a true value of 0 for the variable being measured (such as 0 hours of work or 0 dollars). Interval-level variables do not have a true 0 value. For example, temperature is an interval-level variable because a value of 0 on any temperature scale does not mean the "absence" of temperature. For our purposes, interval- and ratio-level variables are treated in the same way, and we will refer to them as **interval-ratio** variables. Examples of interval-ratio variables include scores on exams, hours or minutes spent on any activity (e.g., hours spent watching videos or doing housework), number of times participating in an activity (e.g., number of times per month attending religious services or exercising), number of sexual partners, family members, or children, and many more. Interval-ratio variables can be continuous or discrete. **Discrete variables** are measured in whole numbers and cannot be broken down further. For example, number of children is a discrete variable because the values of that variable (the number of children) only can be whole numbers. One cannot have 2.5 children. **Continuous variables** have values that can be continually subdivided. Savings measured in dollars, length of employment measured in years, and length of commute measured in miles are all examples of continuous variables.* Although we may round these variables (to dollars, days, or half miles), in theory these units can be subdivided further and further.

Ordinal-Level Variables

Variables with values that can be rank ordered, but which are not numerical and where the distance between each value of the variable is not identical, are **ordinal** level. For example, in the variable employment status, "full-time" represents a greater amount of employment than "part-time," but the difference between the two categories cannot be expressed in a specific numerical amount. Social science variables that are ordinal level also include questions in which the response categories are not equal in size. For example, when measuring frequency of exercise, a variable could include response categories such as "daily," "several times a week," "weekly," "two or three times a month," and "monthly or less." While these categories can clearly be ranked in order of frequency, the difference between exercising daily and exercising several times a week (or between any other two categories) is not numerically precise. Other examples include variables like "How happy are you?" or "How satisfied are you with your job?" that have response categories like "very," "somewhat," "little," or "not at all."

Nominal-Level Variables

Finally, variables that are not numerical and also cannot be rank ordered are **nominal** level. The response categories for nominal-level variables are simply categories, without

---

* "Dollars" is technically a discrete variable because its units cannot be subdivided below one cent. However, when dealing with large quantities (e.g., hundreds or thousands), dollars can be treated as a continuous variable.

any quantitative meaning. As a result, nominal variables are sometimes also called "categorical" variables. Many variables that social scientists use are nominal level. These are variables such as race, gender, religious affiliation, region of residence, marital status, occupation, or political party affiliation. For example, if the categories of political party affiliation are "Democrat," "Republican," "Independent," and "other," we cannot rank these categories; they are simply names for the different affiliations. Questions that are answered with "yes" or "no" are also usually considered nominal.[5]

What About Scales?

There is one more important piece of information about levels of measurement. There are many variables in social science research that are scales ranging from "strongly agree" to "strongly disagree." They are often questions about opinions. These are ordinal variables since the distance between each pair of categories is not numerically precise. However, in practice, researchers generally treat them as interval-ratio level if they have at least five categories. That means, for example, that a researcher might calculate an average for such a variable, saying, for example, that "On a scale of 1 to 10, average support for measures to reduce climate change was 8.2." Why does a variable's level of measurement matter? It determines what kind of statistical calculations can be performed. Many statistics can be calculated only for interval-ratio variables. Consider the mean, or average. You may know that calculating an average requires adding up the values of the variable for all the cases and then dividing by the total number of cases. But you only can add values that are actually numbers, such as hours spent online. You can't add values for nominal variables. (How would you add "Protestant" + "Catholic," for example?) You also can't add values for ordinal variables. (How would you add "Very much" + "Somewhat"?) We will cover this in much more detail in the chapters that follow. For now, remember that determining the level of measurement of a variable is the first important task in statistical analysis.

Causation: Independent and Dependent Variables

A major purpose of statistics in the social sciences is to study relationships among variables. Many social scientists are interested in studying a specific kind of relationship: causal relationships. In a causal relationship, one variable, called the **independent variable**, causes changes in another variable, called the **dependent variable**. For example, a criminologist might be interested in studying the effects of rehabilitation programs offered in prison (such as job training) on recidivism, the likelihood of being re-arrested. Does participation in such programs have a causal impact on the likelihood of reoffending? The independent variable in this example is participation in the program and the dependent variable is being re-arrested.

As we will see in chapter 13, determining whether one variable causes changes in another is no simple task. One might observe, for example, that former offenders who participated in rehabilitation programs have an overall lower rate of recidivism than do those who did not participate in those programs. But to establish that this relationship is causal—that it is the programs themselves that actually deter former offenders from reoffending—the researcher must rule out alternative explanations. For example, it could be that rehabilitative programs are more likely to exist in states that also have higher expenditures on social service programs. The researcher would hold constant or "control" for this third variable—state expenditures on social service programs—to see if the relationship between rehabilitative programs and reoffending were still present. If there were no longer a relationship after holding constant state expenditures on social service programs, this could indicate that lower recidivism rates among those who participate in rehabilitative programs are caused not by those programs but by higher spending on social services in general, which also happens to be correlated with the number of rehabilitative programs that states offer. For example, if people released from prison have access to job training, housing support, or supplemental nutrition programs, they might be less likely to reoffend. There are two basic ways of controlling for alternative causal explanations. Researchers using experimental research designs employ **experimental control** by randomly assigning research participants to treatment and control groups to ensure that participants in one group are not systematically different from those in the other group. Participants in the treatment group receive the "treatment" (e.g., participate in a rehabilitative program), while those in the control group do not. We would assume that any difference in the outcome (i.e., the dependent variable) between the groups was caused by the treatment because of the random assignment of participants to the two groups. Because experimental designs are often impractical, most social scientists must employ the other method of ruling out alternative explanations: statistical control. **Statistical control** is employed in a variety of ways in the data analysis process to ensure that a third variable does not account for the relationship between the independent and dependent variables. You will learn much more about how to do this later in this book.

Getting the Data: Sampling and Generalizing

During presidential election campaigns, we are inundated with surveys about the candidates' relative standing. These surveys are meant to give us a sense of who is ahead, who is behind, and by how much. For example, on October 5, 2020, about one month before the presidential election, an IBD/TIPP poll reported that 48.6% of likely voters expressed support for Joe Biden, compared to for 45.9% Donald Trump.[6] But for obvious reasons, this poll, and every other poll, interviewed a relatively small number of people—it was based on interviews with a sample of 1,221 registered voters. If truth be told, we would not be all that interested in the views of these 1,221 people if they were not representative of the full population of US voters. But they were. Each person in the sample was randomly selected to participate in the survey. This random selection gives us a high degree of confidence that our sample results—Biden 48.6%, Trump 45.9%—are close to what we would have obtained had we somehow managed to interview all 159 million voters.

Inferring from a smaller sample to a larger population is one of the central goals of statistics. A **population** includes every individual or **case** in a category of interest, such as voters. A **sample** is made up of a small group of individuals or cases drawn from the larger population of interest. Most of the time, it is not practical to study all the members of a population directly—unless that population is relatively small and well defined. For example, we could imagine drawing up a full list of every county in the United States or every student at your school in order to study them directly. When we are able to study all members of a population, we use a variety of statistical tools to describe variables and their relationships within this population. There is no need to make inferences about the population because we have actual, direct data about the full population. But most of the time, this is not possible. Instead, researchers draw random samples from populations in order to make inferences about the population based on the characteristics of the sample. If a researcher wishes to generalize from a sample to the population, then that sample must be randomly selected from the population. Chapters 2 through 5 focus on **descriptive statistics**, statistical techniques for describing the patterns found in a set of data, whether those data are based on a full population or a sample. In chapters 6 through 14, we focus on **inferential** statistics, which examine whether information from a sample can be generalized to a population. Those chapters explain the idea of "inference" and the various statistics researchers employ to determine whether and how the results they find in a sample can be generalized. (Chapter 14 also covers some descriptive statistics for examining relationships between variables.) The ability to infer from a sample to a population is based on the idea of randomness. Randomness is at the core of "probability samples." In a **probability sample**, every member of the population must have an equal probability of being selected for the sample, and the selection of cases from the population must be made randomly. Most election polls reported by the media employ probability samples. On the other hand, you may have come across Internet polls or call-in polls on the local news. These are **non-probability samples**. In such instances, members of the sample are self-selected, they are not drawn randomly, and most of the time there are biases associated with who chooses to participate and who doesn't. Although the results of such polls may be interesting, they cannot tell us about a larger population beyond those who responded.

Sampling Methods

There are a variety of methods for drawing a probability sample that allow for inference to a larger population. The most basic method is known as **simple random sampling**. Here, we make a list of all the members of a population and randomly draw our desired number of cases from that population into the sample. We must be able to make a full list of all the members of the population so that we can randomly draw from that list. The list that we draw our sample from is called a **sampling frame**. For example, we could list all 2,600 students enrolled at Smith College, the school where the authors of this book teach, and then randomly draw a sample of 200 of them. Mechanically, these are the steps we might follow to draw this sample:

1. Obtain a list of all 2,600 students at Smith College.
2. Assign every Smith College student a number between 1 and 2,600.
3. Use a random number generator to select 200 numbers between 1 and 2,600.
4. Match each selected number with the student assigned to that number.

We would now have a randomly selected sample of 200 Smith College students.

Because simple random samples require a list of every member of the population, they are practical to use only with fairly small and well-defined populations, such as the students at a small school or all the counties in the state of California. On the other hand, large or constantly changing populations should not be sampled using this method. For example, it would not be possible to list the names of all 159 million voters in the United States.

Stratified random sampling is a variation of simple random sampling. A **stratified random sample** allows the researcher to randomly sample from subgroups in a population to ensure that the sample is representative of population subgroups that are of interest to the researcher, such as students from different class years or residents of rural and urban counties.

Assembling a sampling frame can be harder than it sounds. Sometimes, lists of all members of a population are available through, for example, records of students enrolled at a school, voter registration rolls, telephone directories, or lists of mailing addresses. But these lists are not always publicly available, and the lists themselves can have errors. Sometimes random samples are drawn by randomly dialed telephone numbers (through a computer program that begins with area codes and the three-digit prefixes associated with that area code and then randomly selects the final four digits of a phone number). Of course, not everyone has a telephone; cell phone numbers are not listed in directories; and some numbers produced by randomly generated digits will not be working numbers, and others will be assigned to businesses. For paper or face-to-face surveys, researchers can purchase address lists for many areas from the US Postal Service.[7] In many countries other than the United States, similar procedures are available. Nevertheless, for large populations, these procedures are cumbersome.

There are methods of probability sampling that do not require a full listing of the target population. The most common is **cluster sampling**, where we randomly sample clusters of cases instead of individuals and then randomly sample individuals from within these clusters. For example, we might not be able to put together a complete list of individuals in a large metropolitan area, but we can assemble a full list of census tracts or city blocks. A cluster sample might start by the researcher putting together a complete list of city blocks, randomly selecting a number of them, assembling a list of households on those city blocks, randomly selecting a number of those households, and then randomly selecting one individual from each household. This method is sometimes called **multistage cluster sampling**. Its main advantage is that it allows the researcher to put together a random sample of individuals from a large population without a complete list of individuals in that population.

Even proper probability sampling techniques can yield a sample that is not representative of a population of interest. This is because of **nonresponse bias**, which occurs when individuals who are invited to take a survey vary systematically in the likelihood that they will complete the survey (or particular survey items). For example, if a survey begins with a question about citizenship status, undocumented immigrants may be less likely to respond to the survey than citizens. Or if a survey is administered during the day, it may be more difficult to reach people who are at work. In these cases, the sample data would not be generalizable to the population because one group of intended respondents was much less likely to answer the survey than others and is, therefore, underrepresented in the sample. Regardless of the sampling method employed, it is important not to lose sight of our central objectives. We use samples because they shed light on a larger population. When we study data from samples, we generate statistics that help us describe characteristics of the sample. We use these statistics to make educated guesses about the value of an unknown population characteristic in which we are interested. For example, we measure the percentage of our sample who state they will support candidate A because that tells us approximately how much support candidate A has in the population. We measure the average income in a sample because that tells us approximately what the average population income is. A lot of what we do in the chapters that follow is based on this simple notion: we use statistics to describe a sample and then to infer from that sample to the population.

Sources of Secondary Data: Existing Data Sets, Reports, and "Big Data"

In addition to collecting their own data to address research questions, social scientists often use **secondary data**, or data that have been collected previously, usually by someone else and often for a purpose that might differ from an individual researcher's. In these cases, the researcher is usually not involved in the sampling process, but it is still very important to understand the sampling strategies used to collect any source of secondary data. If the goal of a study is to yield results that can be generalized to a population, only secondary data collected through probability sampling is appropriate.

Fortunately, there are many sources of high-quality secondary data available to social scientists that are collected with generalizability as a primary goal. These data sources are usually the product of large-scale surveys conducted by university researchers with support from various private and public agencies. Most secondary data sets follow a general theme (e.g., political beliefs) yet still ask questions about a wide enough range of topics that researchers can use the data to address a variety of research questions. Throughout this book, we work with a number of publicly available secondary data sets, all collected using probability sampling. Many of these data sets are available for download on this book's website, including the following:

1. General Social Survey (GSS)
2. American National Election Study (ANES)
3. World Values Survey (WVS)
4. Police Public Contact Survey (PPCS)[8]

These data sets allow us to address a range of interesting social science topics. The WVS is a cross-national survey with probability samples of nearly 100,000 respondents from sixty countries. The rest of the data sets employ probability samples of respondents from the United States. The unit of analysis for the GSS, ANES, WVS, and PPCS is the individual. These surveys ask individuals about a range of topics such as their social backgrounds, financial resources, activities, families, opinions, and political beliefs. Along with the data sets themselves, users can download the **codebooks** for the data sets. Codebooks are so named because they provide the "code" necessary for interpreting the meaning of each variable. When a data set is created, variables are given names, and numbers are assigned to the categories of the variables. Codebooks contain the following essential information about the variables in a data set:

- the name and description of each variable;
- descriptions of each category of every variable; and
- the numerical value assigned to each category of every variable.

Figure 1.3 shows an excerpt from the PPCS codebook, for a variable called *V344*.

V344 - TIME OF DAY OF CONTACT

Location: 781-782 (width: 2; decimal: 0)

Variable Type: numeric

Question:	
About what time of day did this contact occur?	
Value	Label
01	After 6 a.m. - 12 noon
02	After 12 noon - 6 p.m.
03	Don't know what time of day
04	After 6 p.m. - 12 midnight
05	After 12 midnight - 6 a.m.
06	Don't know what time of night
07	Don't know whether day or night
98	Refused
-9 (M)	Out of universe/missing

Figure 1.3 Codebook Excerpt from Police Public Contact Survey (PPCS)

The codebook tells us that the variable called *V344* measures what time of day the respondent's most recent contact with a police officer occurred. It also tells us that this variable has eight categories: (1) after 6 a.m. to 12 noon, (2) after 12 noon to 6 p.m., (3) don't know what time of day, (4) after 6 p.m. to 12 midnight, (5) after 12 midnight to 6 a.m., (6) don't know what time of night, (7) don't know whether day or night, and (98) refused. The last category listed, (-9) out of universe, represents missing data. Notice that the numbers assigned to each category are only labels for the categories and are not meaningful as numbers. Category 1 does not mean that the respondent had contact with a police officer at 1:00, for example; it means that the contact occurred between 6 a.m. and noon. When researchers use secondary data, they can decide whether to use the original code for any given variable or recode the variable in some other way. For example, a researcher might use *V344* to create a new variable that measures whether the respondent had contact with the police officer during the day, evening, or night.

Big Data

By now, most people have heard the term "big data," but what does it mean, and how is it related to statistics? There is a key distinction between "big data" and data collected through traditional survey methods. Whereas traditional survey methods collect data for a specific purpose, **big data** emerge as a by-product of the electronic tracking of people's behavior online and in the real world. Big data emanate from various sources, such as administrative information (e.g., electronic medical records), social media, geolocation tracking such as by apps on cell phones, and records of online searches. One way of thinking about big data is to imagine individuals' actions, especially their online actions, as leaving an invisible residue, or digital trace. This residue constantly adds to the ever-growing store of big data. Big data are collected by corporations (tracking purchasing and search information, for example), by technology companies such as Google and Meta (which owns Facebook and Instagram), and by other entities. Some big data are proprietary, owned and accessible only by those who collect them, but many big data records can be obtained by independent researchers.

Whereas in survey research, researchers determine the questions and their possible answers by constructing variables and their response categories, big data directly reflect people's actions without categories imposed by a researcher. As sociologist Amir Goldberg notes, with big data, the approach to data analysis is more open-ended. Big data researchers are less likely to approach their analyses with pre-formulated hypotheses and more likely to "let the data speak," opening up possibilities for finding unanticipated patterns in the data.[9] For example, a team of researchers in Wisconsin used linked administrative records from social service agencies in the state to study patterns of disconnection from sources of public assistance for those who are in need of them.[10] One of the key findings is that the traditional notion of what it means for a family to be "disconnected" from public financial assistance—when a family is eligible and in need of financial assistance but no longer receives it—misses a number of other classes of "disconnection" uncovered in the data, such as families who receive food assistance through the Supplemental Nutrition Assistance Program (SNAP) but not financial assistance. If the researchers had relied on a predetermined measure of disconnection, as survey research might have, they would have missed these other ways of thinking about disconnection. Big data researchers have used online wedding photos to analyze interracial friendship, studied Google search data to predict fertility rates in areas with different socioeconomic characteristics, used geolocation data from cell phones or Tweets to track migration, and used images from Google Street View to characterize demographic characteristics and voting patterns in different neighborhoods. Because of the vast amount of data, such researchers often use machine learning, techniques that train algorithms to recognize and characterize patterns.[11] But where big data enthusiasts see possibility, critics argue that its push toward more open-ended approaches to data analysis—letting the data speak—will pull the social sciences away from building theoretically informed explanations for social phenomena and toward simplistic descriptions of social behaviors and attitudes. For example, danah boyd and Kate Crawford point out that cell phone data might show that cell phone users have more social media and text communications with their work colleagues than with their spouses. Without applying the theoretical tools of the social sciences, we might conclude that coworkers are more important to people than are their spouses. However, it is more likely that text and social media communications reflect what sociologists call "weak ties" but are poor indicators of "strong ties," or close interpersonal relationships marked by emotional connection.[12] Big data also must grapple with the same considerations about sampling frame, the list of all members of the population, that researchers using probability samples must consider. Namely, is the sampling frame biased? Does it actually contain all members of the population of interest? As many observers have noted, big data from social network sites, such as X (formerly Twitter) and Facebook, represent biased sampling frames because social background and demographic characteristics, such as race and age, are related to whether people use social media sites.[13] Thus, inferences about the general population should not be drawn from big data derived from social media.

One final major concern about big data is ethical and privacy implications. All research involving human subjects must ensure that the safety and privacy of the research participants will not be compromised by participating in the study. Researchers must ensure that all participants give their informed consent to participate in the study. Because big data are made up of the digital traces people leave behind, it is impossible for researchers to obtain the consent of the people whose behaviors left the traces. In addition, for some sources of big data, anonymity cannot always be maintained. For example, using data from credit card transactions for 1.1 million users that did not contain identifiable information (i.e., no names or account numbers), researchers were able to "re-identify" many of the 1.1 million users using limited pieces of information available in the data, such as the price of the transaction.[14]

In sum, big data offer new and exciting possibilities for researchers interested in social behavior. There is no question that research using big data will contribute mightily to social science. However, there remains an important place for traditional statistical methods in the social sciences. The findings from research using traditional, theoretically informed, and ethical statistical methods can provide the context necessary for making sense of the findings yielded by big data.

Growth Mindset and Math Anxiety

"I'm not a math person." At some point, you likely have heard someone utter this statement, or maybe you have said it yourself. Underneath this statement lies a potentially harmful view of math and one's relationship to it. In general, this statement communicates a view of one's mathematical capabilities as fixed and impervious to growth. Saying that one is not a math person also can indicate some level of anxiety about the material itself, perhaps tied to previous difficulties with math. In this section, we discuss how adopting a growth mindset can help all students do better in statistics. For those who have some level of anxiety about studying a subject that does utilize math, a growth mindset can be a particularly valuable ingredient for success in statistics.

Researcher Carol Dweck has written extensively about the benefits of what she calls a growth mindset approach to learning. As opposed to a fixed mindset, which views intelligence as a fixed and essential characteristic of individuals, a **growth mindset** views intelligence as something that develops over time through hard work and effort.[15] Research in neuroscience has demonstrated the human brain's ability to become smarter in response to targeted effort, indicating that the human brain works much more like the vision of the growth mindset than the fixed mindset. So when we hear that someone is not a math person, we know that neuroscience tells us otherwise. To be sure, individuals differ in their intellectual interests and talents, but most people's intellectual skills can improve through effort and engagement. In fact, a number of experiments have shown that students who are explicitly taught to adopt the view that intelligence is not fixed, but develops through work and effort, experience greater gains in mathematics learning than control groups.[16] In other words, evidence suggests that adopting a growth mindset when it comes to statistics can go a long way toward actually helping people to do well in statistics. Believing that competence can improve in an area, such as statistics, is just one element of a growth mindset. The other element, equally important, is understanding that this competence is the outcome of applied effort.

Sometimes, adopting a growth mindset when it comes to learning statistics may not be enough to overcome math anxiety, which is "an adverse emotional reaction to math or the prospect of doing math."[17] With about 17% of the US population having math anxiety,[18] this is no small issue. Fortunately, when it comes to the study of statistics, and particularly the approach taken by this book, there are ways to combat the potentially disruptive effects of math anxiety on learning statistics. The first way to lessen the effect of math anxiety on your performance in your statistics course is to recognize that, while statistics does depend on basic math skills, most statistics courses taught from a social science perspective draw more upon verbal and inductive reasoning than math skills themselves.[19] The focus of this book is much more on statistical reasoning than the math underlying the statistics. Thus, even students who have some level of anxiety about math can be reassured that this book presents statistics as a tool for understanding social phenomena, requiring students to draw upon only basic math skills.

For students who still have some anxiety about studying statistics stemming from anxiety about their math abilities, research suggests a simple way to counteract that anxiety. A team of psychologists asked college students with high and low levels of math anxiety to complete a math test. They wondered if completing an expressive writing task, in which students were asked to write for seven minutes "as openly as possible about [their] thoughts and feelings regarding the math problems [they were] about to perform," would lead to smaller differences in performance on the test between students with high and low levels of math anxiety. In fact, there was a dramatically smaller gap in performance between high- and low-anxiety students in the expressive writing task group than in the control group in which students were simply given the test.[20] Take a moment to reflect on this: the math performance of math-anxious students improved dramatically when they wrote openly about their math anxieties *without any effort to improve their math abilities*. These results suggest that the threat of math anxiety is not primarily a tale of those with high anxiety having worse math skills. As the researchers speculate, it is likely much more a story of how math anxiety distracts one's cognitive abilities from the task at hand. This study measured the positive effects of expressive writing on performance on a brief math test, but it is plausible to think that there may be positive effects of acknowledging one's math anxiety on one's performance in a statistics course. It is worth trying an expressive writing exercise similar to the one in the experiment, in which you openly express your thoughts and feelings about the material in your statistics course.

To recap, our recommendations for counteracting the negative effects of math anxiety on statistics performance include, first, adopting a growth mindset when it comes to mastery of statistics and, second, openly acknowledging one's math anxiety regularly throughout the course. This advice suggests neither that math anxiety can be easily eradicated nor that it should be completely eradicated. In fact, a frequently replicated empirical finding indicates that both high *and* low levels of anxiety in a given domain can hurt performance in that domain. The finding has been replicated so many times that the phenomenon has a name: the Yerkes–Dodson Law. Using a sample of students from a university's Introduction to Statistics course, researchers found that the Yerkes–Dodson Law applied to students' statistics performance. Students with very high and low levels of statistics anxiety performed worse than students who reported a medium level of anxiety.[21] This research suggests that there is an optimal level of anxiety that motivates students to seek to improve, as a growth mindset would call upon students to do but does not monopolize students' cognitive resources in a damaging way.

⊖ Using This Book

This book is designed to be used with a growth mindset approach to statistics. This means that we encourage readers to use this book as a tool to help them actively develop and sharpen their understanding of statistics. As with most kinds of knowledge, developing statistical knowledge is not a linear process. Just when you think you understand something, you might find that you're confused about the concept all over again. This is quite typical with statistics, and you are not alone. Even seasoned researchers can benefit from returning to core statistical concepts to refresh their memories. This means that you should expect to work with and return to various concepts throughout this book many times.

Throughout this book, we offer readers a number of ways to develop and practice their skills and check their understanding of the material. First, each chapter includes rich examples of how to use statistical tools to answer interesting social science questions. In addition to reading these examples, we encourage you to work through each of the examples on your own. Second, the chapters include two kinds of boxes separate from the main text. "In Depth" boxes go into more depth about a topic that is covered in the main text, and "Application" boxes walk readers through additional examples employing the relevant statistical method. We know that the temptation is strong to focus on the main text in the name of efficiency, but these boxes provide readers the opportunity to practice and dig deeper into some of the key topics covered in the text. Third, each chapter ends with sections on using two common statistical software programs, Stata and Statistical Package for Social Sciences (SPSS). (You will use the section for the program that your class is using.) These sections provide an opportunity to apply the tools learned in a given chapter to real social science data using statistical software. After reading the "Using Stata" or "Using SPSS" section, you can complete the exercises on your own or in a lab associated with your statistics class. For an extra challenge, we encourage you to apply the same techniques used in the software sections to variables that are not used in the examples. Fourth, each chapter includes a set of practice problems, designed to help you check your understanding of the material and provide opportunities to challenge yourself. Finally, the companion website to this textbook contains many resources for students to practice their skills. We believe that "statistical intelligence" is something that everyone can obtain through work and effort, and we encourage readers to use this book to challenge themselves to develop this intelligence.

Statistical Software

Statistical software programs can analyze patterns in data sets that include large numbers of cases. Throughout this book, as we explain statistical techniques, we often show you how to calculate a result by hand, but these calculations are very time consuming when data sets are large. Almost all statistical research now relies on computers to do calculations. Statistical software programs ease the computational burden on the user and allow for the analysis of data sets that are too large for the human brain to analyze in a reasonable amount of time.

The first statistical software program was developed in 1957, and since then scientists have developed many more programs.[22] Today, analysts are faced with a dizzying array of these programs, ranging from those designed for general use to those designed for the use of highly specialized statistics.

> **BOX 1.1: IN DEPTH**
>
> ### Punched Cards and Data Analysis before the Digital Era
>
> Before the technology existed for the electronic storage and analysis of large data sets, data were stored on small "punched cards." Machines punched small circles or rectangles at specific locations on the card to indicate the value for a specific variable, with the presence or absence of a hole indicating the case's value for that variable. The US Census Bureau commissioned the inventor Herman Hollerith to develop this "punched card" technology to aid in the collection and analysis of information about the US population. Figure 1.4 shows an image of a census worker punching a card for the 1920 Census.
>
>
>
> **Figure 1.4** A Census Worker Punches a Card from the 1920 Census

In this book, we will use Stata and SPSS, two programs that enjoy wide popularity among social scientists.* Most students will be using only one of these programs, depending on what is available on your campus. You should read only the section of each chapter pertaining to the program you are using in your class. These sections give you the opportunity to use Stata or SPSS to find answers to interesting social science questions using real social science data. At the end of this chapter, we present a general introduction to each program.

---

* According to Stata's FAQs, the name "Stata"—pronounced with a long "a" (stayta)—is not an acronym but an invented word stemming from the combination of "statistics" and "data," and, as such, only the first letter is capitalized. SPSS was founded in 1968 by three individuals affiliated with Stanford University. It stands for Statistical Package for the Social Sciences. For an overview of the corporate history of SPSS, see http://www.spss.com.hk/corpinfo/history.htm.

Chapter Summary

This chapter covered the key parts of the process of conducting social science research with quantitative data that precede data analysis. We also discussed available sources of quantitative data and how best to approach learning statistics from a social science perspective. Below, we review key terms.

- The research process proceeds in four major steps:
    1. A social science **research question** asks how two or more variables are related and must be able to be answered using data.
    2. Defining **concepts** and their dimensions. Concepts are the abstract factors or ideas that the researcher wants to study. Concepts may have multiple dimensions.
    3. **Measurement** or **operationalization** is the process of transforming concepts into observable data, or variables. It includes specifying the dimensions of each concept and establishing the variables that are empirical measures of each dimension. Operationalization determines how the researcher will observe concepts using empirical data.
    4. **Sampling** is the process of choosing cases from the population to study.
- A **hypothesis** is a specific prediction about how variables are related. Research questions may specify hypotheses or be more exploratory.
- **Quantitative analysis** uses statistical techniques to analyze numerical data.
- **Qualitative methods** start with data that are not numerical, such as the text of documents, interviews, or field observations. Qualitative data analysis often focuses on meanings, processes, and interactions; like quantitative research, it may test hypotheses or be more exploratory in nature.
- **Mixed methods** employ both qualitative and quantitative data and analysis.
- An **independent variable** is the cause of changes in another variable.
- A **dependent variable** is affected by another variable.
- **Descriptive statistics** are statistical techniques for describing the patterns found in a set of data.
- **Statistical control** controls for alternative causal explanations by using statistical techniques.
- Key terms involving variables and measurement:
    - A **variable** is any characteristic that has more than one category or value.
    - **Level of measurement** refers to whether variables are nominal, ordinal, or interval-ratio. It determines what statistical techniques can be applied to variables.
    - **Ratio-level variables** have numerical values, with identical distances between each value, and a meaningful 0 value that represents a true value of 0 for the variable being measured.

- **Interval-level-variables** have numerical values, with identical distances between each value, and no true 0 value.
- **Interval-ratio** refers to both interval and ratio-level variables.
- **Ordinal-level variables** have non-numerical values that can be rank-ordered; the distance between each value of the variable is not identical.
- **Nominal-level variables** are not numerical and cannot be rank-ordered.
- **Scales**, such as those ranging from "strongly agree" to "strongly disagree," are ordinal-level variables that can be treated as interval-ratio in practice if they have at least five categories.
- **Discrete variables** are measured in whole numbers and cannot be broken down further.
- **Continuous variables** have values that can be continually subdivided.
- **Validity** is the extent to which variables actually measure what they claim to measure; accurate responses.
- **Reliability** is the extent to which responses are consistent and unaffected by the measurement process.
- **Closed-ended** survey items provide survey respondents with predefined response categories.
- **Open-ended** survey items do not provide response categories; respondents supply their own answers.

- Key terms involving sampling and generalizing:
  - The **unit of analysis** is the object of study, either individuals or groups.
  - When the unit of measurement is the group, we sometimes also refer to it as **aggregate level**.
  - The **ecological fallacy** is drawing conclusions about individuals based on groups.
  - A **sample** is a small group of cases drawn from a larger population of interest.
  - A **population** is every case in the group of interest.
  - A **sampling frame** is a complete list of all members of a population.
  - A **probability sample** is drawn randomly from a population in which every member has an equal probability of being selected for the sample. Simple random samples, stratified random samples, and cluster samples are all probability samples.
  - A **simple random sample** is a random selection of members of a population from a complete list of all members.
  - A **stratified random sample** is derived from random samples of subgroups of interest in a population.
  - A **cluster sample** is a multistage sample where the researcher randomly samples clusters of cases instead of individuals, followed by randomly sampling individual cases from the sampled clusters. It allows the researcher to put together a random sample of individuals from a large population without having a complete list of individuals in that population.
  - A **nonprobability sample** is one in which members of the sample are self-selected or are not selected randomly.

- **Inferential statistics** examine whether information from a sample can be generalized to a population. They can be used with probability samples only.
- Sources of data not collected directly by the researcher:
  - **Secondary data** refers to data that are analyzed but not collected by the researcher.
  - **Big data** are data not collected through traditional sampling methods; the by-product of the electronic tracking of people's behavior online and in the real world.
- How to use this book:
  - Adopt a **growth mindset**, the view that ability increases over time through hard work and effort and is not a fixed characteristic of individuals.
  - Address math anxiety by openly expressing your thoughts and feelings about doing math in writing.
  - Practice working the problems in this book and on the companion website.

Using Stata

Stata is one of the most commonly used statistical software programs. With Stata, users have a choice between "point-and-click" and command-driven use. With the point-and-click method, the analyst uses Stata's graphical user interface (GUI), or series of drop-down menus, to select the actions for Stata to perform on the data. With the command approach, analysts use code to write their own commands that tell Stata which analyses to perform. While both methods will yield the same results if done properly, we focus on the command-driven approach to Stata in this book. Using the GUI is a perfectly legitimate way to use Stata, but most Stata users prefer to write their own code. This section was developed using Stata for Windows Version 17. If you are using a different version of the software, you may find slight variations in the commands, but, mostly, they will be very similar.

How Stata Looks

There are two ways to launch Stata. You can launch the program just as you would launch any program (such as your word processing software) on your computer. You then click on "File" in the upper-left corner, choose "Open," and navigate to the location of the saved data file. Alternatively, if you have a saved Stata data set on your computer, you can click on that. This will automatically open Stata, and you can begin analyzing this data set. Stata data files always end with the extension ".dta" (e.g., `filename.dta`). Here, we will use a small example data set called "`ExampleStataData.dta`," with information about fictitious college students after their first year of college. Figure 1.5 shows how Stata will appear once you open the data set.

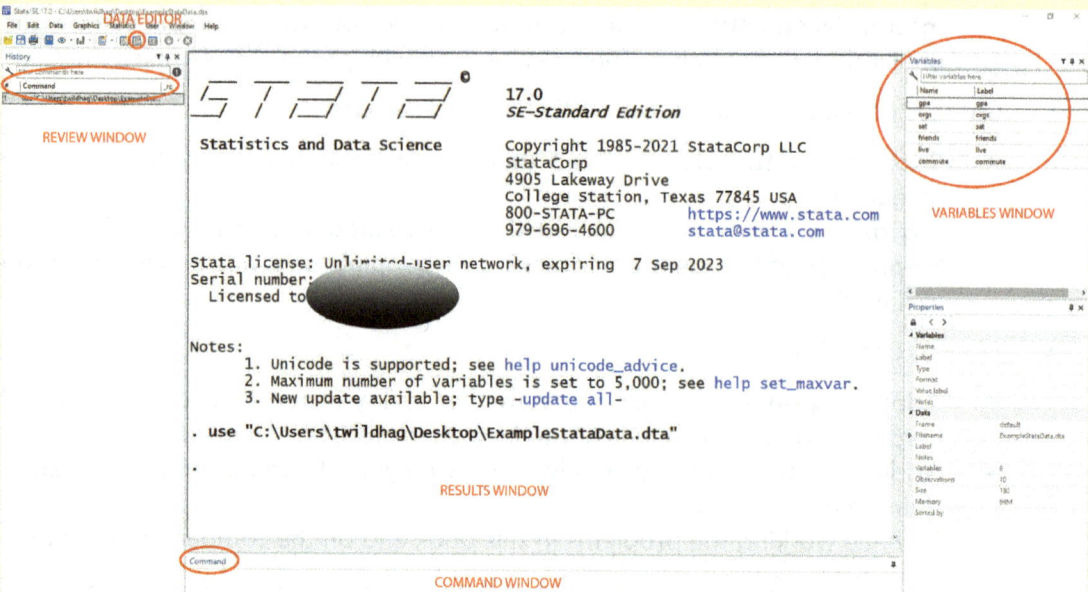

Figure 1.5

The large white space in the middle of the screen is the Results Window, where Stata will display the results of the statistical analyses that you tell it to conduct. The Command Window, circled at the bottom of the screen, is the place where you can type your commands. The long, narrow Review Window on the left side of the screen is the place where your commands are archived. Each time you enter a command into the Command Window, Stata automatically also places that command in the Review Window on the left. You can copy a command from the Review Window back into the Command Window by clicking on it at any time. (Note that Stata does not save this archive of commands. If you want to save a record of your commands, you will need to enter your code in a separate file, which we discuss below.) On the right side of the screen, circled, is the Variables Window, which lists all of the variables in the data set. We see that in this example data set, there are six variables.

We can view the data by clicking on the Data Editor icon at the top of the screen, circled. (Note that there are two versions of this icon, one that allows the user to edit the data and one that does not allow editing. We are using the icon that allows only for browsing the data, not editing it.) You can also open the Data Editor by selecting "Data" in the menu in the top taskbar and then selecting "Data Editor" → "Data Editor (Browse)." Once you click on the "Data Editor" icon, you will see the image shown in Figure 1.6, which displays the data in this data set.

In this data set, the unit of analysis is college student, so each row reports information for a unique college student in the data set.

We can see that this data set is small, including only ten college students. Each column shows the values for a unique variable, in this case six of them. (The data sets that we will

	gpa	orgs	sat	friends	live	commute
1	2.8	5	9	12	1	.
2	3	1	6	5	1	.
3	3.2	0	3	0	0	30
4	3.1	1	8	5	1	.
5	2.95	0	5	1	0	10
6	3.5	3	8	4	1	.
7	2.75	2	4	1	1	.
8	3.75	0	5	0	0	45
9	3.4	3	8	7	1	.
10	3	1	8	4	1	.

Figure 1.6

use in later chapters have many more cases and variables.) The six variables in this data set include:

- *gpa* (students' GPAs after the first year of college);
- *orgs* (the number of student organizations that students have joined);
- *sat* (students' satisfaction with their social life on campus, ranked from 1 to 10);
- *friends* (the number of close friends students have on campus);
- *live* (whether the student lives on or off campus; 1 = on campus; 0 = off campus); and
- *commute* (the daily commute time for students living off campus, in minutes).

We can see that the first student in the data set had a GPA of 2.8, belonged to five student organizations, rated campus social life as a 9, had twelve close friends on campus, and lived on campus. Why do we see a period in the cell for the *commute* variable for this student? The period means that there is no value reported for this variable for this student. We refer to missing information on a variable as "missing data." We could observe missing data for a variable for a number of reasons. For example, a respondent may have refused to respond to that particular question, a researcher may have failed to record the respondent's value, or that variable may not be applicable to that particular case in the data set. In this case, Student 1 has no value for the *commute* variable because the student did not live off campus. Only students who lived off campus were asked to report their commute time to campus. Once you have reviewed the data in the Data Editor, you can close it, and Stata will return you to the initial interface that you saw when you first opened the program. When we walk you through the analyses in the "Using Stata" sections at the end of each chapter, you will type those commands into the Command Window at the bottom of the window. Once you have finished typing a command, hit "Enter," and Stata will display the results of the analysis in the Results Window. The only time that the result will not be displayed in the Results Window is when you ask Stata to generate a graph or chart, which will always open in a separate window. You will be impressed to find that Stata can produce results, even with massive data sets, in fractions of a second. No matter how quickly you can do computations by hand, Stata will beat you every time!

Basic Logic of Stata Code

For many people, the most intimidating part of learning a new statistics software program is learning how the code works. If you have studied any language, you know that every language has its own grammar that lays out its structural rules. As a language, Stata code is no different. All Stata code commands follow the same basic structure:

```
command variable name(s), options
```

Commands can be customized in various ways, as we will see in examples throughout this book, but the basic rule is to state the specific analysis that you want Stata to conduct (the "command") followed by the variable(s) on which it should conduct this analysis, followed by any special options for that analysis.

Three Categories of Commands: Create New Variables, Transform Existing Variables, and Analyze Existing Variables

There are three basic kinds of commands that we will use in this book: (1) those that create new variables, (2) those that transform existing variables, and (3) those that conduct statistical analyses on variables that already exist in the data set. We offer simple versions of each category of command here as a preview. You can try them using the example data set, available on the website.

1. *Create a New Variable*: Here is an example of a command that creates a new variable in Stata:

   ```
   generate commute2 = commute/60
   ```

   The "generate" command asks Stata to create a new variable called *commute2*. For each student in the example data set, the value of this new variable is equal to the value of the *commute* variable divided by 60. Because the *commute* variable is measured in minutes, dividing the variable by 60 converts the variable into hours. After running this command, if you click on the Data Editor, you can see that the value for *commute2* is missing if there was also missing information for *commute*. For the three cases that had values for *commute*, the values for *commute2* are now measured in hours. For example, for Student 3, the value of *commute* is 30, and the value of *commute2* is .5.

   We could have chosen any name for the new *commute2* variable as long as it conformed to the rules for Stata variable names. Variable names in Stata can be up to thirty-two characters, including capital or lowercase letters, numbers, and underscores. When you create new variables in Stata, we recommend keeping the variable name to around eight characters. Variable names that are longer than twelve characters will be truncated in most Stata output. Variable names are case sensitive, so be careful about specifying capital or lowercase letters. We recommend sticking to lowercase letters to simplify matters.

2. *Transform an Existing Variable*: Here is a series of commands that transforms an existing variable, in this case *orgs*, which measures the number of student organizations that students have joined:

```
generate orgs2 = orgs
replace orgs2 = 1 if orgs>=1
replace orgs2 = 0 if orgs==0
```

This series of commands follows a convention that we adopt throughout the book: *whenever transforming an existing variable, create a duplicate version of that variable first, and transform only the duplicate version.* This preserves the original variable and allows you to check that the transformation proceeded properly by comparing values of the new variable to the original one.

With these three commands, we create a new variable, called *orgs2*, which indicates whether students joined any organization during their first year of college. In the new variable, all students who joined one or more organizations are combined into one category. The first "generate" command creates the new *orgs2* variable. By writing "= orgs," we are telling Stata to set the values for the new variable equal to the values for the existing *orgs* variable (e.g., all cases will have the same value for *orgs2* that they have for *orgs*). The second two "replace" commands change, or recode, the values of the *orgs2* variable. The first "replace" command assigns *orgs2* a value of 1 if the student's value for *orgs* is greater than or equal to 1. The second "replace" command assigns a value of 0 to *orgs2* if the value for *orgs* for that student is equal to 0. (Note that this command is redundant because *orgs2* was already equal to 0 if the student's value for *orgs* was 0, but we show the command for consistency.) Also notice that we have used two equals signs after "if" in the second "replace" command. This is not an error. It is a rule in Stata syntax that you must use two equals signs after "if" in a command. As we see in the "replace" commands, if an equals sign precedes "if" (or if there is no "if" used in the command), then we use only one equals sign.

It is essential to check your work every time you transform an existing variable. Even seasoned data analysts make coding mistakes, and one small error in a line of code can have profound consequences for the results of a statistical analysis. Remember to check your work and check it again! As long as you do not violate any of Stata's syntax, it will run your command, even if the code contains an error for your purposes. While Stata's brain may be much faster than the human brain, in this way, the human brain is smarter. In this case, we can check our work by making sure that the number of cases with a value of 1 for *orgs2* is equal to the number of cases with value of 1 or higher for *orgs*. (Since we are dealing with a small data set, we can check this manually in the Data Editor.)

3. *Analyze an Existing Variable*: Here is an example of a command that asks for a simple analysis of an existing variable:

`summarize friends`

We will see this "summarize" command many times throughout this book, as it is one of the most commonly used commands in Stata. Here, we have asked Stata to summarize the *friends* variable. The output from this command, as shown in Stata's Results Window, is shown in Figure 1.7.

```
. summarize friends
```

Variable	Obs	Mean	Std. dev.	Min	Max
friends	10	3.9	3.725289	0	12

Figure 1.7

In response to the "summarize" command, Stata produces a table reporting the following information about the *friends* variable, from left to right: the number of observations for which there are available data (10), the mean (3.9), the standard deviation (3.7), the minimum value (0), and the maximum value (12). (We will cover all of these statistics in detail in chapters 4 and 5.)

As you start to become comfortable with Stata, you might discover that different commands can often be used to generate the same result. If you find yourself developing your own ideas for how you might ask Stata to accomplish a task that vary from the examples that we offer in this book, this is a good sign that you are internalizing the language of Stata.

Error Messages in Stata

It is not uncommon for Stata to return errors instead of output in the Results Window. Whenever you make a mistake that violates one of Stata's syntax rules, you will see an error message. To show a simple example, if we had misspelled the word "summarize" in the command that we used above as "sumarize," we would have received the following error message:

`. sumarize friends`

`command sumarize is unrecognized`

`r(199);`

All error messages give a brief explanation for the error (in red) and provide a link to the error code (in blue). Here, Stata is telling us that the command "sumarize" is not recognized by the program. That is simply because we mistyped the command. If you click on "r(199)," Stata will take you to a brief explanation of the error code, in this case 199. The description of the error code tells us that there is probably a typographical error in our code.

This simple error message illustrates the important role that error messages can play in learning Stata code. If you read the error message carefully and take the time to click on the error code, you will often identify the source of the problem in your code. If you can resist the urge to panic when you see an error message, which can be strong for new Stata users, then error messages can become one of your most valuable Stata teachers.

Operators in Stata Code

As you become familiar with Stata code, it will be important for you to know the basic operators in Stata syntax.[23] For example, as we saw in one of the above commands, if we want to divide the values of a variable by something, we do not write out the word "divide"; we use the "/" sign. The commonly used arithmetic, relational, and logical operators in Stata are shown in Table 1.1.

Table 1.1

Arithmetic Operators	Relational Operators	Logical Operators
Addition: +	Greater than: >	And: &
Subtraction: −	Less than: <	Or: |
Multiplication: *	Greater than or equal to: >=	Not: ! (or) ~
Division: /	Less than or equal to: <=	Through: /
Raise to a power: ^	Equal to: ==	
	Not equal to: ! (or) ~=	

Data Files, Do-Files, and Saving Your Work

As we work with Stata throughout the book, we will use a number of data files, which contain the actual data that you will ask Stata to analyze. As we noted above, the common feature of all Stata data files is that they end in the extension ".dta."

Do-files are another kind of Stata file, and they end in the extension ".do." Do-files are where you can write and save commands for future use. While you need not use a do-file in order to follow along with any of the Stata exercises in this book (all commands can be typed directly into the Command Window), we recommend starting a do-file if you are working on projects outside of the exercises in this book. To open a do-file, simply click on the do-file icon in Stata, circled in Figure 1.8.

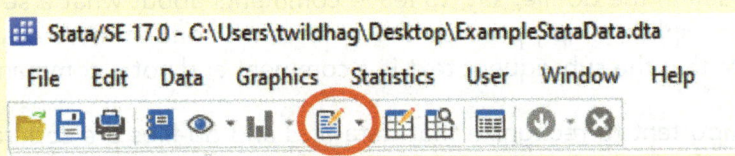

Figure 1.8

Clicking on the do-file icon, as shown above, will open a do-file, as shown in Figure 1.9.

Figure 1.9

Instead of typing single commands into the Command Window, you can collect all of your code in one savable do-file. To use the do-file, simply write your commands directly in the file, with each command on a separate line. (You can also copy and paste commands that you have run from the Command Window into a do-file.) You can either run one command at a time, as you would in the Command Window, or select chunks of syntax to run all at once. The easiest way to do this is to highlight the command(s) that you want to run, select the "Tools" menu, and then "Execute Selection." Once you have run the commands, you can toggle to the Results Window to review the output. You will see that the commands you just ran now also appear in the Review Window. If you want to make comments to yourself in the do-file, say, to leave comments about what a set of commands is meant to accomplish, simply precede those notes with one or more asterisks (*), which signals to Stata that the subsequent text is a comment and not a command.

Saving: It is important to recognize that data files and do-files are separate files, saved independently of one another. Saving a data file does not mean that you have saved the do-file that you have been using, and vice versa.

Sources of Stata Help

The Stata *User's Guide*, Stata *Base Reference Manual*, and Stata help files provide useful sources of extra help and support as you learn Stata. The most recent versions of the Stata *User's Guide* and *Base Reference Manual* can be downloaded for free at http://www.stata.com/bookstore/users-guide/ and https://www.stata.com/manuals/r.pdf, respectively. The Stata *User's Guide* gives an overview of how Stata works, and the *Base Reference Manual* offers detailed descriptions and examples of all Stata commands. Chapter 3 of the *User's Guide* lists a number of sources of information about Stata, including the Stata website, the Stata YouTube channel, and the Stata blog. Typing "help" followed by any Stata command (e.g., "summarize") into the Command Window will open the help file associated with that command, which describes how to use the commands and provides links to the place in the *User's Guide* where that command is described.

Using SPSS

SPSS is a statistical software program that enjoys wide popularity among social scientists. SPSS can be used in two modes: "point-and-click" and command-driven use. With the point-and-click method, the analyst uses SPSS's graphical user interface (GUI), or series of drop-down menus, to select the statistical analyses for SPSS to perform on the data. With the command approach, analysts use a special language called syntax to write their own commands that tell SPSS which analyses to perform.[24] While both methods will yield the same results, if done properly, in this book, we focus on the point-and-click approach to SPSS, which is the more widely used method. The SPSS exercises that follow (here and in subsequent chapters) were developed using SPSS Version 27 for the Mac. If you are using a different version of SPSS, there may be very slight differences in the appearance of commands and procedures.

How SPSS Looks

There are two ways to launch SPSS. You can launch the program just as you would launch any program (such as your word processing software) on your computer. You then navigate to "File" → "Open Data" and select the SPSS data set you wish to open. Alternatively, if you have a saved SPSS data set on your computer, you can click on that. This will automatically open SPSS, and you can begin analyzing this data set. SPSS data files always end with the extension ".sav" (e.g., filename.sav).

Regardless of how you launch the software, SPSS operates with two windows, and you will navigate back and forth between them. If you launch the program by clicking on the SPSS icon, you will see the SPSS "Data Editor" window, but it will be empty (since you have not yet indicated what data set you are using). If you launch SPSS by clicking on the saved SPSS data file, then it will open the "Data Editor" window, which will be populated with information about the variables in the data set and the actual data. When you launch

SPSS, it will also open an "output" window at the same time it opens the "Data Editor" window. The output window is where the results of your statistical analyses will appear, but it will be empty when you first open the program. For illustration purposes, we will use a small SPSS data set, called `ExampleSPSSData.sav`, with information about fictitious college students after their first year of college. You can try the procedures we show here, using the example data set, available on the website. We launch SPSS by clicking on this file. This opens the SPSS "Data Editor" window, shown in Figure 1.10.

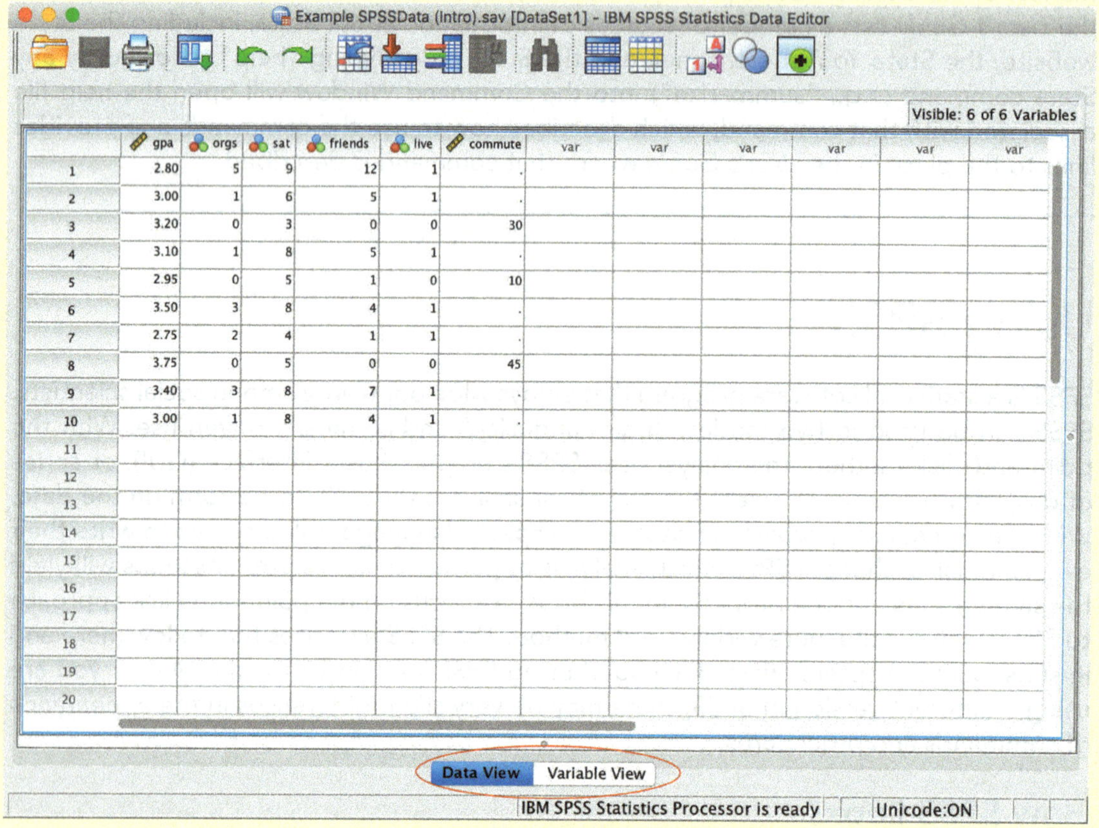

Figure 1.10

The "Data Editor" window has two options, "Data View" and "Variable View," circled at the bottom of Figure 1.10. When the "Data View" window is active, as shown in Figure 1.10, we see the data. Each row corresponds to a case in our data set. In this data set, the unit of analysis is the student, so each row reports information for a unique college student in the data set. We can see that the data set is small, with data for only ten college students. Each column shows the values of a different variable. The names of the variables are shown at the top of each column. The six variables in this data set are:

- *gpa* (students' GPA after the first year of college);
- *orgs* (the number of student organizations that students have joined);

- *sat* (students' satisfaction with their social life on campus, ranked from 1 to 10);
- *friends* (the number of close friends students have on campus);
- *live* (whether the student lives on or off campus; 1 = on campus; 0 = off campus); and
- *commute* (the daily commute time for students living off campus).

In the illustration above, we can see that Respondent 1 has a GPA of 2.80, belonged to five student organizations, rated campus social life as a 9, had twelve close friends on campus, and lived on campus. If you look at the *commute* variable on the right, you will note that we have data for only three of the ten respondents; for seven respondents, the data are missing. SPSS puts in a period to represent data that are missing. We could observe missing data for a variable for a number of reasons. For example, a respondent may have refused to respond to that particular question, a researcher may have failed to record the respondent's value, or that variable may not be applicable to that particular case in the data set. In this case, Respondent 1 has no value for the *commute* variable because the student did not live off campus. Only students who lived off campus were asked to report their commute time to campus.

There is an alternative way of examining the information in this data set. Instead of "Data View," we can select "Variable View" by clicking on "Variable View" at the bottom of the window (circled in Figure 1.10). Instead of seeing the raw data for each respondent, we see summary information about the variables. The "Variable View" window is shown in Figure 1.11.

SPSS now shows us the variable name, the type of variable (numeric or text), the width of the variable as it is recorded in the data file, the number of decimal places, the variable's label, and the labels associated with each value of the variable. We can also use this window to add or edit certain information about the variables. We will return to this point shortly.

Before you begin to analyze your data, we always recommend getting a sense of your data set by looking at both the "Data View" and the "Variable View" windows. Once you begin analyzing data, SPSS will deliver the output in a different window, called the "Output" window. One can use the SPSS dropdown menus when either window—"Data Editor" or "Output"—is active. Any time you ask SPSS to run a procedure, it will append the output to the bottom of the "Output" window. You will be impressed to find that SPSS can produce results, even with massive data sets, in fractions of a second. No matter how quickly you can do computations in your head, SPSS will beat you every time!

Three Categories of Commands: Create New Variables, Transform Existing Variables, and Analyze Existing Variables

There are three basic kinds of commands that we will use in this book: (1) those that create new variables, (2) those that transform existing variables, and (3) those that conduct

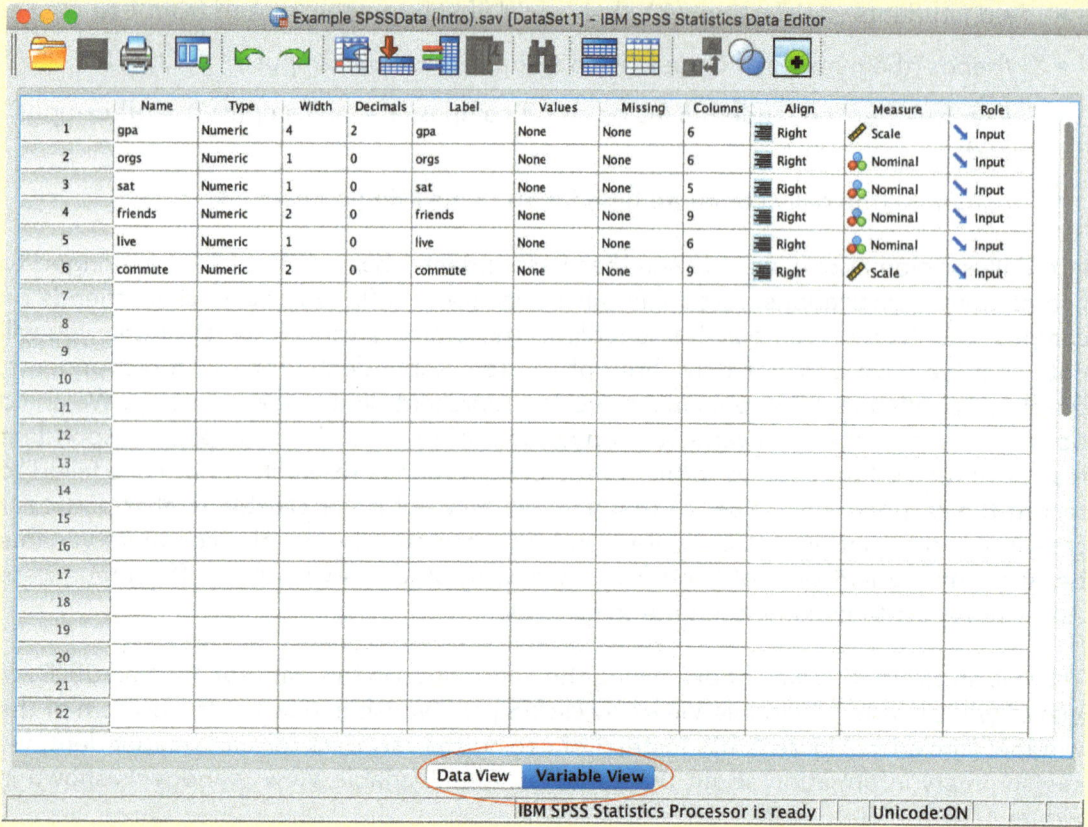

Figure 1.11

statistical analyses on variables that already exist in the data set. We offer simple versions of each category of command here as a preview.

1. *Creating a New Variable*: One of the variables in the ExampleSPSSDATA.sav data set is called *commute*. It represents the daily commute time for students living off campus, measured in minutes. We are going to tell SPSS to create a new variable called *commute2* by taking each respondent's score on the original *commute* variable and dividing it by 60. Thus, the new variable, *commute2*, will represent respondents' commute time, measured in hours.

   To create a new variable, we use the SPSS drop-down menu called "Transform"; we click on it and then click on "Compute Variable," as shown in Figure 1.12.

   `Transform → Compute Variable`

   This sequence opens the "Compute Variable" dialog box, as shown in Figure 1.13.

   We will type the new variable's name, *commute2*, into the "Target Variable" box, and we will define the construction of the new variable by putting the appropriate

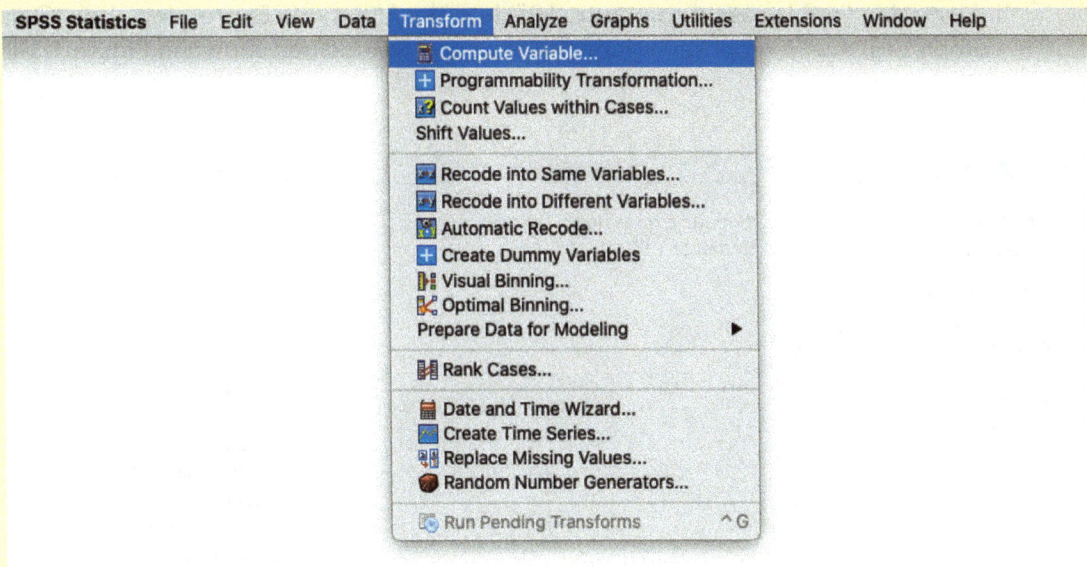

Figure 1.12

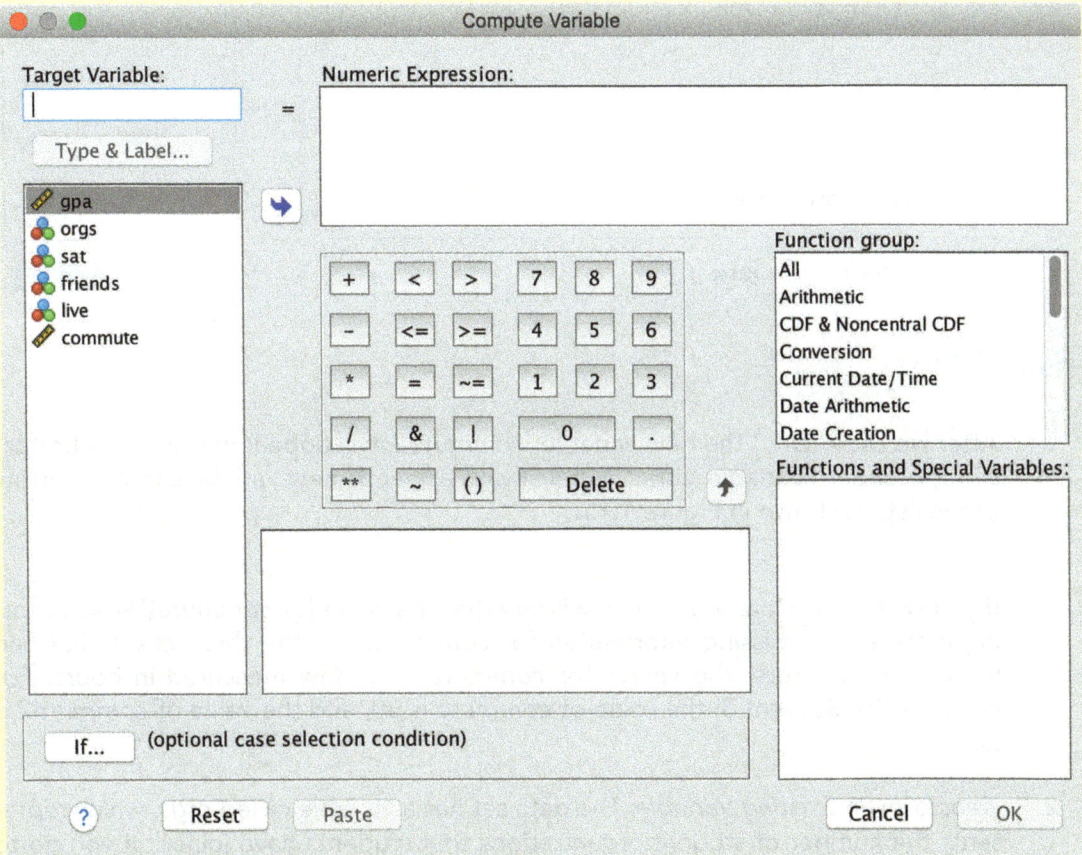

Figure 1.13

information (in this case, "commute/60") into the "Numeric Expression" box. Once completed, the "Compute Variable" dialog box will look as shown in Figure 1.14.

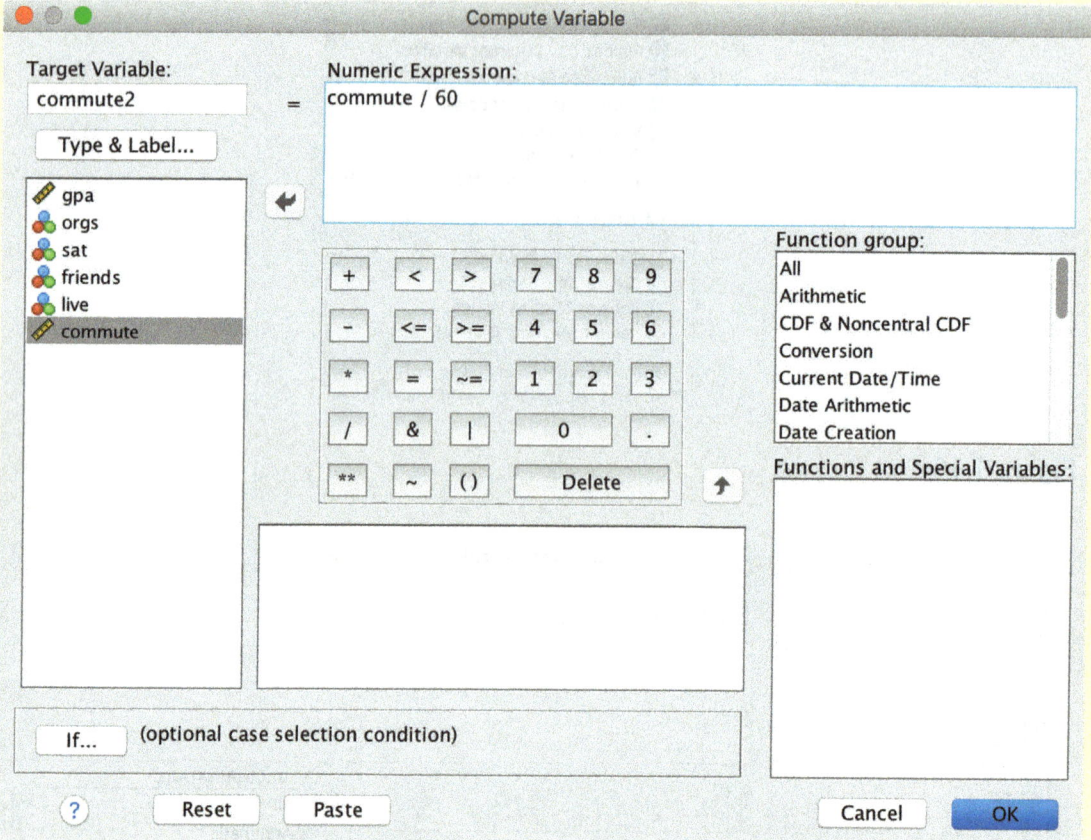

Figure 1.14

After we click "OK," the new variable, *commute2*, will appear in the "Data Editor" window under "Variable View." SPSS always appends new variables to the bottom of the list, as shown in Figure 1.15.

If you switch to "Data View," you will see that the value for *commute2* is also missing if there was missing information for *commute*. For the three cases that had values for *commute*, the values for *commute2* are now measured in hours. For example, for Student 3, the value of *commute* is 30, and the value of *commute2* is .5.

2. *Transform an Existing Variable*: The data set has a variable called *orgs*, which represents the number of student organizations that students have joined. If you go to the "Data Editor" window and switch to "Data View," you will see that the values

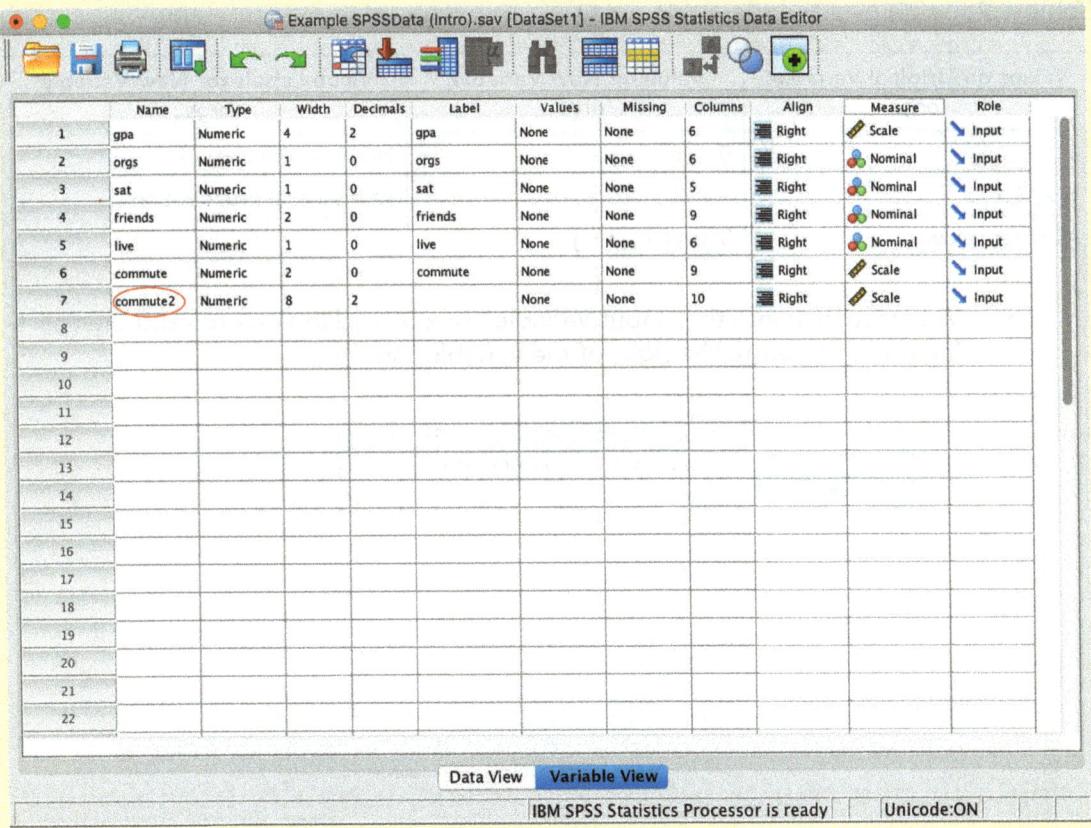

Figure 1.15

range from 0 to 5. We will "recode" this variable so that there are only two categories: 0 (for students who do not belong to any organizations)

and 1 (for students who belong to *at least one* organization). SPSS offers us two methods of recoding: "Recode into Same Variables" and "Recode into Different Variables." If we choose the first method, then SPSS will overwrite the existing *orgs* variable, and we will lose the original information. As a result, it is always a good idea to *choose the second method*: "Recode into Different Variables." When we use this method, SPSS maintains the original variable and creates a new variable according to whatever specifications we provide. This way we do not lose any information.

We are going to tell SPSS to recode *orgs* into *orgs2* (while maintaining the original variable), using the specifications shown in Table 1.2.

To recode a variable, we use the following SPSS sequence of menu commands:

`Transform → Recode into Different Variables`

This opens the "Recode into Different Variables" dialog box, shown in Figure 1.16.

Table 1.2

Values on Original Variable, *orgs*	Value on New Variable, *orgs2*	Value Labels for New Variable, *orgs2*
0	0	No organizations
1, 2, 3, 4, 5	1	At least one organization
All other values	System-missing	

- We move *orgs* into the "Input Variable" box by clicking on *orgs* and then clicking on the arrow to the right of the variable list.

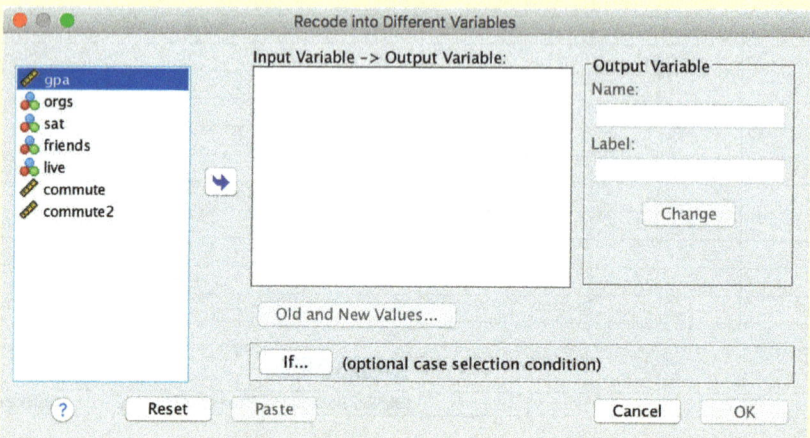

Figure 1.16

- We type the new variable name, *orgs2*, into the "Output Variable: Name" box. We then click on "Change."
- We click on "Old and New Values." This opens another dialog box, shown in Figure 1.17, where we tell SPSS how to recode the variable, using the specifications presented in the table above.

We enter these recode commands one at a time. After we put the appropriate information into the "Old Value" and "New Value" boxes, we must click on "Add" so that SPSS adds them to the "Old → New" window. Note as well that SPSS offers us a series of shortcuts. For example, instead of entering the following:

```
old value 0 = new value 0,
old value 1 = new value 1,
old value 2 = new value 1,
old value 3 = new value 1,
old value 4 = new value 1,
old value 5 = new value 1,
```

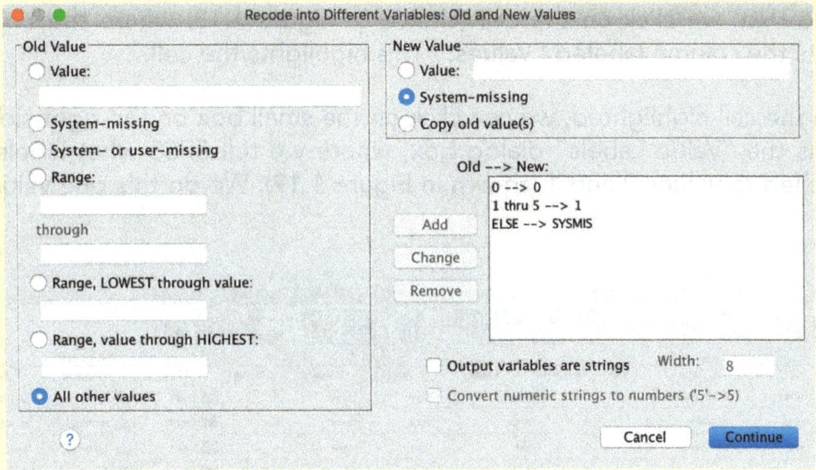

[Figure 1.17 — screenshot of SPSS "Recode into Different Variables: Old and New Values" dialog]

Figure 1.17

SPSS allows us to perform this recode using a single entry with a range of values (e.g., 1 through 5 = 1) by selecting "Range" under the "Old Value" options.

If we have a large data set and are not sure of the lowest and highest values, SPSS gives us lowest and highest value shortcuts as well.

- We have told SPSS that:"Old Value" 0 (on the *orgs* variable) will be 0 on the new variable (*orgs2*).
- Any respondent with a score between 1 and 5 on the original *orgs* variable will be a "1" on the new *orgs2* variable.
- Just in case we missed anything, "All other values" on the original variable will be "System missing" on the new variable.

Once we have finished entering the recode specifications, we click on "Continue," then "OK," and SPSS will create the new, recoded variable *orgs2*. Once again, to be certain, look at the "Data Editor" window and make sure "Variable View" is active, and you will see *orgs2* appended to the bottom of the list. When we create a new variable using the "Recode" or "Compute" procedure, we have the option of adding information about the newly created variable. For example, say we want to tell SPSS that when it produces output involving the new *orgs2* variable, it should include the labels that are associated with each of the two categories. We need to tell SPSS that on the new *orgs2* variable, 0 is equal to no organizations, and 1 is equal to at least one organization. We can easily add this information when the "Variable View" is active in the "Data Editor" window.

To do that, we click on the cell (circled in Figure 1.18) where the *orgs2* variable meets the column labeled "Values." This highlights the cell.

With the cell highlighted, we can click on the small box on the right side of it. This opens the "Value Labels" dialog box, where we tell SPSS what labels should be attached to values 0 and 1 (shown in Figure 1.19). We do this one value at a time,

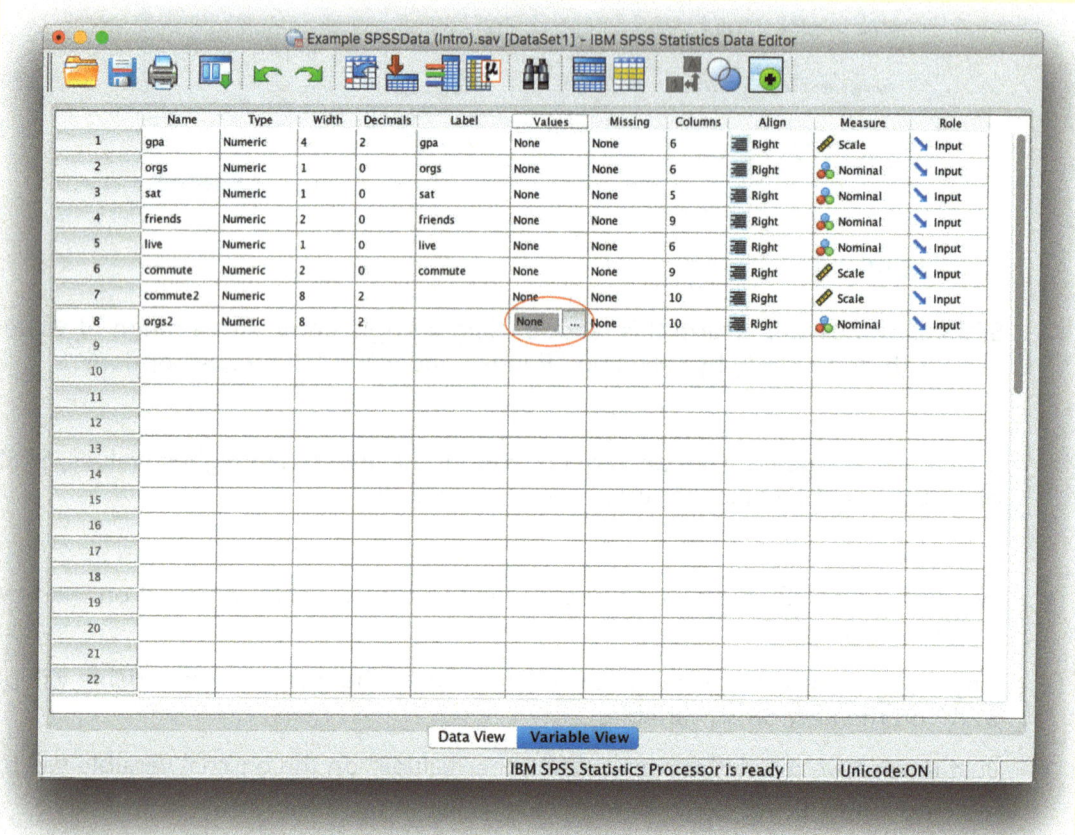

Figure 1.18

and we click "Add" after we have entered each value and label into the appropriate boxes. When this is complete, we click "OK," and SPSS will store these value labels. We can also use the "Variable View" window to add a full label to the variable name itself (for example, we might call it "Recoded Organizations"). To do that, we click on the cell where the *orgs2* variable meets the column headed by "Label." Similarly, we can tell SPSS to treat certain values as "missing" for any variable by clicking on the appropriate cell in the "Missing" column and telling SPSS which values should be considered "missing."

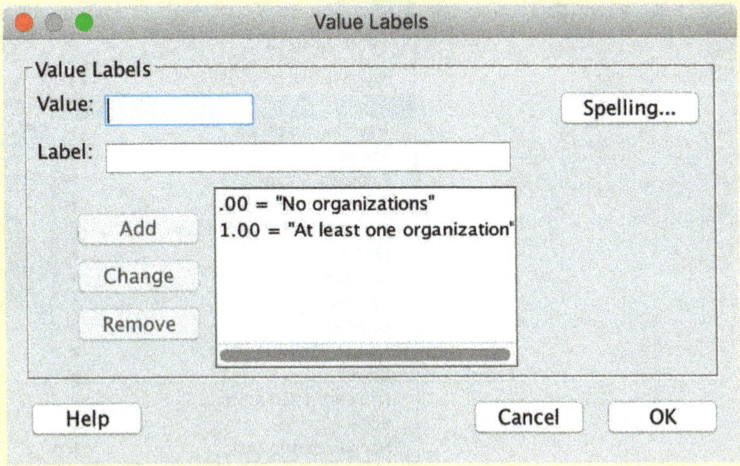

Figure 1.19

3. *Analyze an Existing Variable*: Here is an example of a sequence of menu commands that asks for a simple analysis of an existing variable.

   `Analyze → Descriptive Statistics → Descriptives`

   This sequence will open the "Descriptives" dialog box, shown in Figure 1.20. We are going to ask SPSS to calculate some basic statistics for a variable called *friends* (number of close friends that students have on campus). To do that, we move *friends* into the "Variable(s)" box and click "OK," as shown in Figure 1.21.

   We will see this "Descriptives" procedure many times throughout the book, as it generates some commonly used statistics. Here is the output from this procedure, which you will find in the SPSS "Output" window, as shown in Figure 1.22.

   SPSS has produced a table reporting the following information about the *friends* variable, from left to right: the number of cases for which there are available data (10), the minimum value (0), the maximum value (12), the mean (3.9), and the standard deviation (3.73). (We will cover all of these statistics in detail in chapters 4 and 5.)

Saving Your Work

When you finish a session in SPSS, you can save the data file, save the output file, or both. In either case, you simply click on the save icon or navigate to "File" → "Save." You might save a data file if you have created new variables that you want to be able to use later. You might save an output file if you want to be able to revisit the output or use it in a report later.

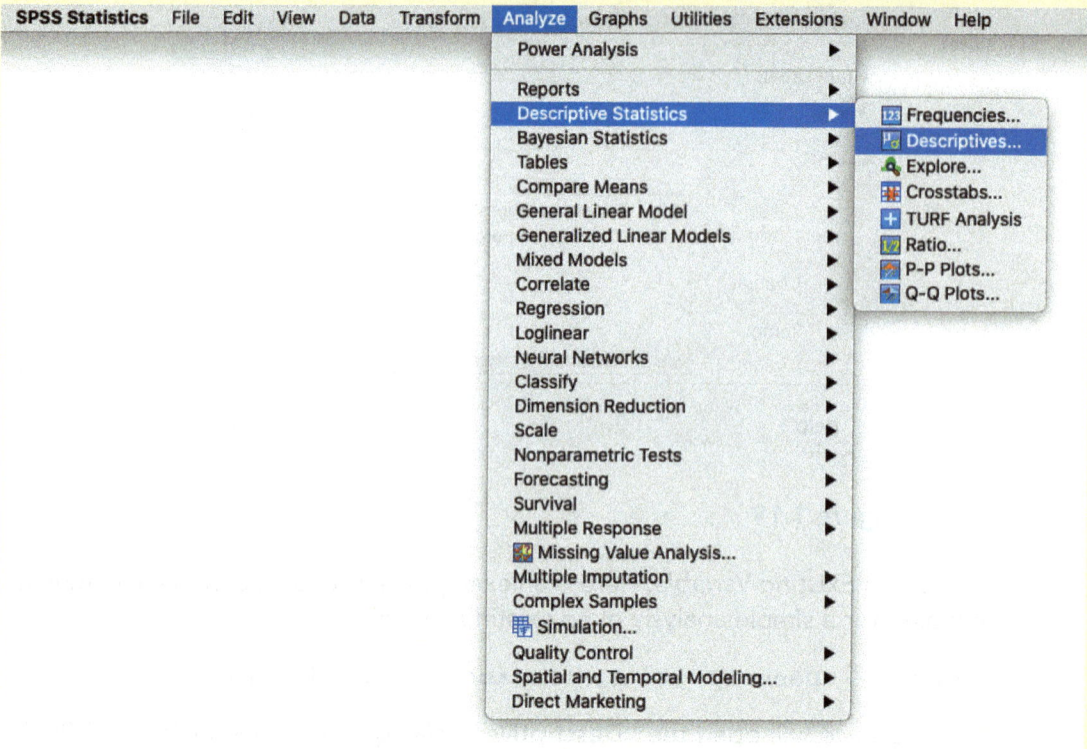

Figure 1.20

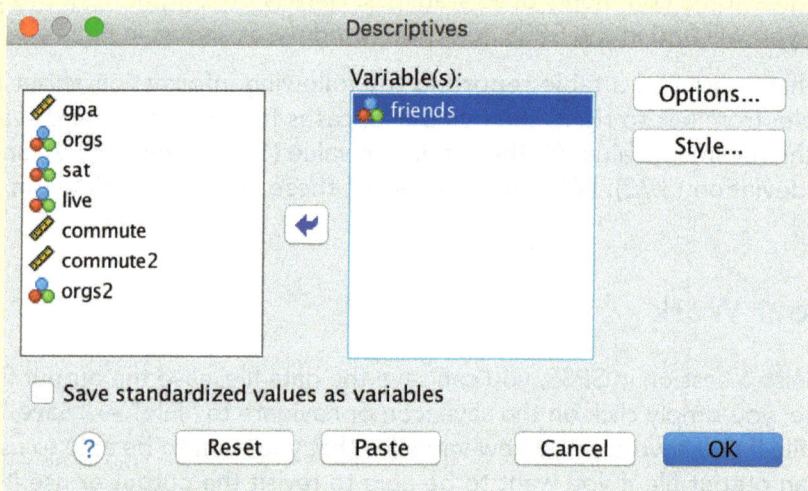

Figure 1.21

Descriptive Statistics					
	N	Minimum	Maximum	Mean	Std. Deviation
friends	10	0	12	3.90	3.725
Valid N (listwise)	10				

Figure 1.22

Practice Problems

1. A team of researchers is studying the effect of inequality in metropolitan areas on aggregate levels of trust. The researchers define inequality as differences in access to resources between groups of people in the area. This variable is measured by an index of inequality for each metropolitan area, ranging from 0 (no difference in access to resources between groups) to 100 (some groups have access and others have none). The other variable is the average level of trust in metropolitan areas.

    a. What is the independent variable in this study?
    b. What is the dependent variable?
    c. What is the unit of analysis?
    d. What is the sampling strategy?

2. A researcher wants to know if autonomy increases employees' job satisfaction. The researcher measures autonomy by rating the extent to which employees make their own choices about how to complete work tasks on a scale from 0 (no control over how to complete tasks) to 10 (complete control over how to complete tasks). Job satisfaction is measured with a survey item asking employees the rate their agreement with the statement: I am satisfied with my job.

    a. What is the independent variable in this study?
    b. What is the dependent variable?
    c. What is the unit of analysis?
    d. What is the sampling strategy?

3. Consider the following item on a survey about how people use social media: *In a typical week, how many times do you post content on a social media site? Please choose the response that fits best.*

    - Never
    - Rarely
    - Often
    - All the time

a. Is this a closed- or open-ended survey item? Explain why.
b. What is the variable being measured by this survey item?
c. What are the categories of this variable?
d. What level of measurement is this variable? Explain how you know.

4. A survey asks a sample of recent college graduates the following question: *What was your major in college?*

   a. Is this a closed- or open-ended survey item? Explain why.
   b. What is the variable being measured by this survey item?
   c. What are the categories of this variable?
   d. What level of measurement is this variable? Explain how you know.

5. Before administering a survey about social media use to a sample of college students, a researcher pilots the survey on a small group of respondents, asking them to answer the question every Monday for ten weeks. After the ten weeks have passed, he assesses the reliability and validity of a survey item asking about the frequency of social media use. He notices variation in many of the individuals' responses. One respondent gave the following responses over the ten weeks:

Table 1.3 One Individual's Reported Frequency of Social Media Postings, Over Ten Weeks

	Frequency of Responses
Never	2
Rarely	2
Often	4
All the time	2

   a. Based on this information, how would you characterize the reliability of this measure?
   b. What changes could be made to the response categories to increase the reliability of the measure?

6. A study of the long-term effects of exercise on people's lives asks participants to record how frequently they exercised each week for two months. Participants recorded their exercise in a time-use diary at the end of each week. One respondent exercised two or three times per week throughout the study. However, on a weekly survey asking about the frequency of exercise, this same respondent chose the response "almost never exercised" in all ten of the survey weeks.

a. For this pilot respondent, is reliability high or low for this survey item?

b. For this pilot respondent, is validity high or low for this survey item? Explain your answer.

7. A researcher wants to study how people cope with the stress of being homeless. Her goal is to make inferences about the total homeless population in the city from one probability sample. She decides to devise her sampling frame by asking all of the homeless shelters in the city for lists of the people who stayed at their shelter for at least one night during the last thirty days. She plans to draw a random sample from the lists of residents. It is the beginning of the summer, and the researcher wants to finish collecting data by the end of the summer.

   a. What is the researcher's sampling strategy?

   b. Do you think that this sampling frame will yield a sample that is representative of the city's entire homeless population? Explain your answer.

   c. Does collecting data over the summer months present any threats to the generalizability of the data? Explain your answer.

8. A study examining the experiences of high school students in the United States first draws a random sample of high schools from cities and towns across the country. Second, students are selected randomly from each of the chosen schools. What kind of sampling design is this?

9. A popular website called Over the Top is known for publishing popular top ten lists. One of its recent lists, "The Best New Ice Cream Flavors of the Summer," was shared widely on social media. The data for this list come from a survey of thirty ice cream shops around the country. Managers of each shop were asked whether they would be selling the following ten flavors during the coming summer:

   - chocolate mocha
   - apricot
   - green tea
   - black raspberry
   - burnt sugar
   - chocolate swirl
   - ginger lemongrass
   - chocolate
   - vanilla peanut swirl
   - tamarind coconut pineapple

   a. What is the variable being measured and what is its level of measurement?

   b. For their list, Over the Top ranked each flavor according to how many shops would be serving that flavor. Do you agree with Over the Top's decision to call the list "The Best New Ice Cream Flavors this Summer"? Explain your answer.

   c. Propose a revised title for the list, and explain why it works better than Over the Top's title.

10. A community recreation center disseminates a survey to its members to learn which of its programs were most popular among its members. One question asks members which team sports they participated in through the recreation center. The recreation center offers the following team sports:

　　1. baseball
　　2. basketball
　　3. softball
　　4. football
　　5. soccer
　　6. volleyball

　a. What is the level of measurement for this variable?

　b. The recreation center finds that more people participate in soccer than any other team sport. They send an email to all members with this subject line: "Soccer is Our Community's Favorite Team Sport! Sign Up Today!" Is the subject line of this email accurate?

　c. Propose a revised subject line for this email and explain why it works better than the one written by the recreation center.

11. Employees of a humanitarian organization are concerned about the separation of migrant children from their parents when families try to cross the US border. They want to know how they can use their organization's resources to decrease the number of children separated from their families. One group thinks that investing their resources in as many social media campaigns as possible will be more effective at reducing the numbers, while another group thinks that focusing on staging as many protest events as possible will work better.

　a. Turn the two proposed strategies into two separate social science research questions. Be sure to pose each as a question.

　b. For each research question, identify the variables. Which variables are the independent variables, and which are the dependent variables for each question?

　c. What level of measurement are each of the variables from each research question?

12. A group of community activists wants to increase participation in volunteer activities among residents of their town. The group has two ideas for how to increase volunteerism. The first idea is to raise awareness of the need for volunteer labor through a social media campaign. The second idea is to hold parties for volunteers following volunteer events.

　a. Turn the two proposed strategies into two separate social science research questions. Be sure to pose each as a question.

　b. For each research question, identify the variables. Which variables are the independent variables, and which are the dependent variables for each question?

c. What level of measurement are each of the variables from each research question?

13. A hypothetical study found in a probability sample of children that children's average score on a self-discipline scale, ranging from 0 to 100, was 50. A researcher conducting a follow-up study decides to measure the self-discipline of children in her small sample *relative* to the average self-discipline score from the previous study, 50. To calculate each child's *relative self-discipline*, she will subtract the average score from previous research (50) from each child's score in her sample. Figure 1.23 shows the scores for the first five children in her sample.

	Child's Self-Discipline Score	Child's Self-Discipline Score Minus Average Score	Child's Relative Self-Discipline Score
Child 1	10	10 - 50	-40
Child 2	60	60 - 50	10
Child 3	35	35 - 50	-15
Child 4	75	75 - 50	25
Child 5	50	50 - 50	0

Figure 1.23 Self-Discipline Scores for First Five Children in the Sample

a. The last column in Figure 1.23, circled, shows the relative measure of self-discipline for each child (the child's raw score minus the average self-discipline score from the previous study). Is the variable *relative self-discipline* an interval- or ratio-level variable? Explain how you know.

b. The researcher drew a larger sample of children and decided to transform the *relative self-discipline* variable into a new variable. This new variable has three categories: "low" (scores below the average score of 50), "medium" (scores equal to the average score of 50), and "high" (scores above the average score of 50). Assign each child in Figure 1.23 to the proper category of this new variable.

c. What level of measurement is the new variable? Explain your answer.

14. An open-ended survey item asked residents of a neighborhood to explain what they liked least about their neighborhood. Here are six of the responses:

- *None of my neighbors talk to each other.*
- *When I walk around, I don't feel like I can trust people I pass on the street.*
- *I have to drive too far to get to a grocery store.*
- *If I want to walk anywhere, I have to leave my house an hour early.*
- *I wish we still had block parties, but people stopped organizing them a few years ago.*
- *There isn't enough public transportation.*

a. Assign each of the six responses to one of the two categories of a new variable: (1) Transportation or (2) Social Relationships.
b. Propose a name for this new variable.
c. What level of measurement is this new variable? Explain how you know.

15. Figure 1.24 shows an excerpt from the GSS codebook for one of its survey items. It shows the distribution of responses to the question over time, since the GSS first asked it in 1988.

1822. Have your sex partners in the last five years been...

[VAR: SEXSEX5]

RESPONSE	PUNCH	1972-82	1982B	1983-87	1987B	1988-91	1993-98	2000	2002	2004	2006	2008	2010	2012	2014	2016	ALL
Exclusively male	1	0	0	0	0	601	4147	1021	968	1022	1059	759	800	758	1023	769	12927
Both male and female	2	0	0	0	0	14	118	43	34	14	40	25	25	36	40	47	436
Exclusively female	3	0	0	0	0	455	3475	882	879	882	893	703	707	654	913	687	11130
Don't know	8	0	0	0	0	0	0	0	9	3	0	4	2	0	2	1	21
No answer	9	0	0	0	0	45	435	239	78	54	71	31	45	43	61	29	1131

Figure 1.24 Excerpt of GSS Codebook for One Variable

a. What is this variable measuring?
b. How many categories does this variable have, and what are the labels for each of them? What are the numbers assigned to each category?
c. Between 1988 and 2016, how many respondents said that their sex partners had been both male and female?
d. What level of measurement is this variable? Explain how you know.
e. Which categories should be defined as "missing data"? Explain why.

16. Figure 1.25 (see facing page) shows an excerpt from the 2021 GSS codebook for one of its survey items. It shows the distribution of responses to the question.

a. What is this variable measuring?
b. How many categories does this variable have, and what are the labels for each of them? What are the numbers assigned to each category?
c. How many respondents reported using the Internet for ten or more hours per week? How many respondents reported using the Internet for fewer than ten hours per week?
d. What level of measurement is this variable? Explain how you know.

17. State the level of measurement for each variable listed below:

a. Party identification, where 1 = Democrat; 2 = Republican; 3 = Independent; 4 = Other

```
Variable: WWWHR    Type: Numeric

Label: Not counting e-mail,
about how many minutes or hours per week do you use the
Web?
(Include time you spend visiting regular web sites and time
spent using interactive Internet services like social media,
streaming services, chat rooms,
online conferencing services, discussion boards or forums,
and the like.)

Notes:
```

LABEL	VALUE	COUNT	PCT	PCT Excl. Reserve Codes
ZERO HOURS	0	76	1.9%	3.1%
1 HOUR	1	168	4.2%	6.8%
2 HOURS	2	218	5.4%	8.8%
3 HOURS	3	172	4.3%	7.0%
4 HOURS	4	143	3.5%	5.8%
5-9 HOURS	5	460	11.4%	18.7%
10-14 HOURS	10	371	9.2%	15.0%
15-19 HOURS	15	116	2.9%	4.7%
20-29 HOURS	20	342	8.5%	13.9%
30-39 HOURS	30	146	3.6%	5.9%
40-49 HOURS	40	114	2.8%	4.6%
50+ HOURS	50	140	3.5%	5.7%

Figure 1.25 Implicit Association Test for Implicit LGBTQ Bias

Source: Adapted from Project Implicit, https://implicit.harvard.edu/implicit/selectatest.html.

b. Party identification, where 1 = Democrat; 2 = Independent; 3 = Republican

c. Income, measured in dollars

d. Income, where 1 = low; 2 = middle; 3 = high

e. Feeling thermometer rating of the military, 0 to 100 degrees

f. Feeling thermometer rating of the military, where 0 to 25 = very cold; 26 to 49 = cold; 50 = neutral; 51 to 74 = warm; 75 to 100 = very warm

g. How many days exercised in a week, 1 to 7

h. Gender, where 1 = girl and 2 = boy

i. Girl, where 0 = not a girl and 1 = is a girl

18. State the level of measurement for each variable listed below:

    a. College major, where 1 = social sciences; 2 = sciences; 3 = humanities; 4 = other
    b. Time spent on homework, measured in hours
    c. Time spent on homework, where 1 = less than one hour; 2 = one to two hours; 3 = more than two hours
    d. Feeling thermometer rating of higher education, 0 to 100 degrees
    e. Feeling thermometer rating of higher education, where 0 to 25 = very cold; 26 to 49 = cold; 50 = neutral; 51 to 74 = warm; 75 to 100 = very warm
    f. How many days posted on social media in a week, 1 to 7
    g. Work schedule, where 1 = works days and 2 = works nights
    h. Works days, where 0 = does not work days and 1 = works days

19. Table 1.4 describes the research question and data collection strategy for five studies. For each study:

    a. Identify the unit of analysis.
    b. Identify whether original data is being collected or if secondary data is being used. If original data is being collected, identify the sampling strategy: non-probability sampling, simple random sampling, stratified sampling, or cluster sampling.

Table 1.4 Research Questions and Sampling Design for Five Studies

	Research Project	Unit of Analysis	Sampling Method
1	*Research Question*: Are college Republicans or Democrats more likely to engage in campus activism? *Data Collection*: Make a list of all members of the Republican and Democrat student organizations at a large university. Use a random number generator to draw a sample from the list.	_____	_____
2	*Research Question*: Is gender related to the number of close friends that high school students have? *Data Collection*: Make a list of all high schools in Texas and draw a random sample of schools. Draw random samples of students from each of the sampled schools.	_____	_____

Table 1.4 Continued

Research Project	Unit of Analysis	Sampling Method
3 **Research Question**: Are companies' parental leave policies related to the gender composition of their upper-level staff? **Data Collection**: Find ten companies that will allow the researcher to collect information about their organizational policies and staff.	_____	_____
4 **Research Question**: Is the frequency of a city's misdemeanor arrests related to the frequency of felonies committed in the city? **Data Collection**: Use publicly available data from the Bureau of Justice Statistics.	_____	_____
5 **Research Question**: Who has the highest frequency of contact with the police: African Americans, whites, Asians, or Hispanics? **Data Collection**: Create separate sampling frames of all African American, Asian, white, and Hispanic residents of the same geographic area. Draw random samples from each of the four sampling frames.	_____	_____

20. Table 1.5 describes the research question and data collection strategy for five studies. For each study:

    a. Identify the unit of analysis.

    b. Identify whether original data is being collected or if secondary data is being used. If original data is being collected, identify the sampling strategy: non-probability sampling, simple random sampling, stratified sampling, or cluster sampling.

Table 1.5 Research Questions and Sampling Design for Five Studies

Research Project	Unit of Analysis	Sampling Method
1 **Research Question**: Are social science majors more likely to volunteer during college? **Data Collection**: Obtain a list of all registered students with declared majors at a large university. Use a random number generator to draw a sample from the list.	_____	_____
2 **Research Question**: Is gender related to job satisfaction among doctors who work in hospitals? **Data Collection**: Make a list of all hospitals in New York and draw a random sample of hospitals. Draw random samples of doctors from each of the sampled hospitals.	_____	_____

Table 1.5 Continued

	Research Project	Unit of Analysis	Sampling Method
3	*Research Question*: Are social media companies' privacy policies affected by the political opinions of the CEO? *Data Collection*: Find five social media companies that will allow the researcher to collect information about their organizational policies and staff.	_____	_____
4	*Research Question*: Are schools' extracurricular offerings related to the socioeconomic status of their students? *Data Collection*: Use publicly available data from the National Center for Education Statistics.	_____	_____
5	*Research Question*: Who reports the highest level of satisfaction with their lives: people under 30, people between 30 and 40, people between 40 and 50, or people over 50? *Data Collection*: Create separate sampling frames of all residents of the same geographic area by age group. Draw random samples from each of the age groups.	_____	_____

Stata Problems

Open the `GSS2021.dta`. Here, we will focus on the variable *agekdbrn*, which measures the age of respondents at the time their first children were born.

1. Click on the Data Browser to visually scan the data. You will notice that the values for some variables are presented as "value labels," while others show the numerical values assigned to the variable's categories. Find the variable called *agekdbrn*. Are the categories of the variable presented as numbers or labels?

2. Use the summarize command to generate summary information about *agekdbrn*. How many people answered this question? What is the youngest age, and oldest age, at which first children were born to this sample of respondents? What is the mean age at which first children were born?

3. Use the generate command to create a new version of *agekdbrn*, and name it *agekdbrn2*.

4. Use the recode command to transform *agekdbrn2* into a variable with three categories: "Under 30," "30 to 40," and "Over 40." Assign 1 to "Under 30," 2 to "30 to 40," and 3 to "Over 40."

5. Click on the Data Browser again and scroll to the last column in the spreadsheet. Scroll down the "*agekdbrn2*" column. How can you tell that the recoding occurred?

SPSS Problems

Open the `GSS2021.sav`. Here, we will focus on the variable *agekdbrn*, which measures the age of respondents at the time their first children were born.

1. Navigate to the "Data View" window to visually scan the data. You will notice that the values for some variables are presented as "value labels," while others show the numerical values assigned to the variable's categories. Find the variable called *agekdbrn*. Are the categories of the variable presented as numbers or labels?

2. Use the "Descriptives" dialog box to generate summary information about *agekdbrn*. How many people answered this question? What is the youngest age, and oldest age, at which first children were born to this sample of respondents? What is the mean age at which first children were born?

3. Use the "Recode into Different Variables" dialog box to create a new version of *agekdbrn* called *agekdbrn2*, with three categories: "Under 30," "30 to 40," and "Over 40." Assign 1 to "Under 30," 2 to "30 to 40," and 3 to "Over 40."

4. Navigate to the "Data View" window and scroll to the last column in the spreadsheet. Scroll down the *agekdbrn2* column. How can you tell that the recoding occurred?

Notes

[1] William Davies. 2016. "The Age of Post-Truth Politics." *New York Times*. August 24. Retrieved from https://www.nytimes.com/2016/08/24/opinion/campaign-stops/the-age-of-post-truth-politics.html.

[2] Natasha Quadlin. 2018. "The Mark of a Woman's Record: Gender and Academic Performance in Hiring." *American Sociological Review* 83(2): 331–360. The resumes also varied by college major.

[3] Eszter Hargittai. 2010. "Digital Na(t)ives? Variation in Internet Skills and Uses among Members of the 'Net Generation.'" *Sociological Inquiry* 80(1): 92–113.

[4] Philip Bump. 2022. "Even Moved to the Background, Pandemic Deaths Are Shaped by Politics." *Washington Post*. September 8.

[5] Some analysts treat yes-no variables as ordinal, with the idea that a "yes" means presence of a property and "no" means its absence.

[6] "IBD/TIPP Election Poll Shows Trump Closes Gap in Presidential Race; Biden Leads by Just 2.7 Points." *AP News*, October 5. Retrieved from https://apnews.com/press-release/business-wire/election-2020-joe-biden-public-opinion-campaign-2016-social-affairs-1afa1897618f4d1f9ee9432510bba9e2.

[7] On random-digit dialing and telephone sampling methods, see J. Michael Brick and Clyde Tucker. 2007. "Mifosky-Waksberg: Learning from the Past." *Public Opinion Quarterly* 71(5): 703–716. On address-based sampling frames, see American Association for Public Opinion Research. 2016. *AAPOR Report: Address-Based Sampling*. Retrieved from http://www.aapor.org/EducationResources/Reports/AddressbasedSampling.aspx.

[8] We use several waves of the GSS data, each of which can be downloaded here: www.gss.norc.org/Get-The-Data. We use several waves of the ANES data, which can be downloaded here: www.electionstudies.org/data-center/. We use Wave 7 of the WVS data, which can be downloaded here: www.worldvaluessurvey.org/WVSDocumentationWV7.jsp. We use the 2020 wave of the PPCS data, which can be downloaded here: https://www.icpsr.umich.edu/web/NACJD/studies/38320.

[9] Amir Goldberg. 2015. "In Defense of Forensic Social Science." *Big Data & Society* 2(2): 1–3.

10. Maria Cancian, Eunhee han, and Jennifer L. Noyes. 2014. "From Multiple Program Participation to Disconnection: Changing Trajectories of TANF and SNAP Beneficiaries in Wisconsin." *Children and Youth Services Review* 42: 91–102.

11. On interracial friendship see: Brent Berry. 2006. "Friends for Better or for Worse: Interracial Friendship in the United States as Seen through Wedding Party Photos." *Demography* 43(3): 491–510. On fertility rates see J. Ojala, E. Zagheni, F. C. Billari, and I. Weber. 2017. "Fertility and Its Meaning: Evidence from Search Behavior. ArXiv:1703.03935. On migration see Cesare Nina, Lee Hedwig, McCormick Tyler, Spiro Emma, and Zagheni Emilio. 2018. "Promises and Pitfalls of Using Digital Traces for Demographic Research." *Demography* 55(5): 1979–1999. On Google Street View see Gebru Timnit, Krause Jonathan, Wang Yilun, Chen Duyun, Deng Jia, Aiden Erez L, and Fei-Fei Li. 2017. Using Deep Learning and Google Street View to Estimate the Demographic Makeup of Neighborhoods across the United States." *Proceedings of the National Academy of Sciences* (*PNAS*) 114(50): 13108–13113. On machine learning and for an overview of the field see Achim Edelmann, Tom Wolff, Danielle Montagne, and Christopher A. Bail. 2020. "Computational Social Science and Sociology." *Annual Review of Sociology* 46(1): 6181.

12. danah boyd and Kate Crawford. 2012. "Critical Questions for Big Data: Provocations for a Cultural, Technological, and Scholarly Phenomenon." *Information, Communication, & Society* 15: 662–679.

13. Ibid., and Eszter Hargittai. 2015. "Is Bigger Always Better? Potential Biases of Big Data Derived from Social Network Sites." *The Annals of the American Academy of Political and Social Science* 659(1): 63–76.

14. Yves-Alexandre de Montjoye, Laura Radaelli, Vivek Kumar Singh, and Alex Pentland. 2015. "Unique in the Shopping Mall: On the Reidentifiability of Credit Card Metadata." *Science* 347: 536–539.

15. C. S. Dweck. 2010. "Even Geniuses Work Hard." *Educational Leadership* 68(1): 16–20.

16. Jo Boaler. 2013. "Ability and Mathematics: The Mindset Revolution That Is Reshaping Education." *Forum* 55(1): 143–152.

17. E. A. Maloney and S. L. Beilock. 2012. "Math Anxiety: Who Has It, Why It Develops, and How to Guard against It." *Trends in Cognitive Sciences* 16(8): 404–406 (404).

18. M. H. Ashcraft and A. M. Moore. 2009. "Mathematics Anxiety and the Affective Drop in Performance." *Journal of Psychoeducational Assessment* 27(3): 197–205.

19. M. Paechter, D. Macher, K. Martskvishvili, S. Wimmer, and I. Papousek. 2017. "Mathematics Anxiety and Statistics Anxiety: Shared but Also Unshared Components and Antagonistic Contributions to Performance in Statistics." *Frontiers in Psychology* 8: 1–13.

20. D. Park, G. Ramirez, and S. L. Beilock. 2014. "The Role of Expressive Writing in Math Anxiety." *Journal of Experimental Psychology: Applied* 20(2): 103–111.

21. J. Keeley, R. Zayac, and C. Correia. 2008. "Curvilinear Relationships between Statistics Anxiety and Performance among Undergraduate Students: Evidence for Optimal Anxiety." *Statistics Education Research Journal* 7(1): 4–15.

22. J. De Leeuw. 2011. "Statistical Software: An Overview." In M. Lovric (ed.), *International Encyclopedia of Statistical Science* (1470–1473). Heidelberg: Springer Berlin Heidelberg.

23. See chapter 13 in the Stata *User's Manual* for more on operators.

24. For more information about SPSS syntax, see "Introduction: A Guide to Command Syntax." Retrieved from https://www.ibm.com/docs/en/spss-statistics/26.0.0?topic=reference-introduction-guide-command-syntax.

Chapter 2

Getting to Know Your Data

Frequency Distributions and Visual Representations of Data

Younger generations are often faster to embrace cultural change than are older generations. After all, young people often drive these cultural changes. Such generational differences in positions on social issues can lead to difficulties navigating personal relationships across generations, and at times they have contributed to large-scale political conflict. The popular "OK, Boomer" videos on TikTok, in which younger people lampoon the supposedly out-of-touch views of people from the baby boom generation, highlight the cultural assumption that older and younger people are socially disconnected. How real is this generational clash? What do data tell us about differences between older and younger people on contemporary social issues?

Surveys of teenagers' and adults' attitudes about various social and political issues offer one way to investigate this question. For instance, in 2022, the Pew Research Center asked US teens and adults how they felt about Black Lives Matter and societal acceptance of transgender people. Pew found that 70% of teenagers aged thirteen to seventeen, and 66% of eighteen- to twenty-nine-year-olds, expressed support for the Black Lives Matter movement. Support was not as strong among those aged thirty to forty-nine (54%) and fifty to sixty-four (54%).[1] Another Pew survey about attitudes toward LGBTQ people showed that 47% of eighteen- to twenty-nine-year-olds thought that society had not gone far enough in accepting transgender people, compared to 31% of fifty- to sixty-four-year-olds.[2] These percentages are consistent with the claim that younger people tend to be more socially liberal than older people.

Before drawing firm conclusions about how groups differ, however, it is important to "get to know" the data. That means familiarizing oneself with each of the variables that one is asking about in a research question. Getting to know the data includes examining the distribution of respondents across all possible response categories of one's variables. For example, the question about how far society has gone toward accepting transgender people actually has three response categories: "society has gone too far," "society has not gone far enough," and "society's response has been about right." How are all of the responses to the survey distributed across all of the categories of this variable? Among all people surveyed, 38% said that society had gone too far in accepting transgender people, 36% said society had not gone far enough, and 23% said society had been about right. We see that there is no overall consensus about how society is doing when it comes to accepting transgender people.

Getting to know your data in this way, before examining relationships between variables (e.g., age and social attitudes), provides context for understanding those relationships. As we mentioned above, 47% of young people said that society has more work to do in accepting transgender people. Is that a large percentage? Part of answering that question lies in our examination of how all people responded to that question.

One major task of the social sciences is to explain these kinds of patterns among groups of people. In this example, we are looking at patterns in attitudes about social and political issues for different generations. Before explaining attitudinal differences across groups, we begin by simply describing the patterns numerically. At the most basic level, this means counting the numbers of people who fit into different categories. These counts are called frequencies. For example, the researchers counted the number of people who said that society had not gone far enough in accepting transgender people. They compared that number to the total number of people studied to calculate the percentage of people who held this opinion. Comparing these percentages across generations can tell us how the generations are different.

This chapter will show you how to get to know your data by examining patterns using frequencies and percentages, how to represent these patterns visually in tables and graphs, and how to begin comparing patterns for different groups through rates, ratios, and the comparison of percentages. The next several chapters will cover more types of descriptive statistics, which are used for summarizing and describing data.

● Frequency Distributions

Frequency distributions simply give the number, percentage, or proportion of those in a sample who fall into each category of a variable. This might give the breakdown for events like marriage, answers on an opinion poll, or votes for candidates. The unit of analysis may be the individual person, as in the preceding examples. The unit of analysis might also be groups, as with aggregate data. For example, a frequency distribution might show carbon emissions for countries or high school graduation rates for states. In other words, **frequency** distributions show the number of cases that fall into each category of a variable.

Remember that a variable measures one concept and has different values. Each value is a category. For example, if the variable is full-time employment, the values could be "employed full-time" and "not employed full-time." Frequency distributions describe one variable at a time. That is, they are **univariate statistics**.

Let's start with the simplest frequency distribution, which shows the number, or **raw frequency** of cases in each category. We will look first at variables with a relatively small number of response categories. These include most nominal and ordinal variables and some interval-ratio variables.

How did Americans vote in the 2020 presidential election? This is a nominal variable. (Why? The three response categories [Trump, Biden, and other] are different in kind, not amount.) If we look at votes for the two leading candidates, Republican Donald Trump and Democrat Joe Biden, we find that 74,225,845 people voted for Trump and 81,286,365 voted for Biden.[3] We also find that more than 3 million voters cast a ballot for a minor candidate. We could present these numbers in a table, as shown in Table 2.1.

Table 2.1   Frequency of Votes for Presidential Candidates, 2020

Candidate	Total Votes
Donald Trump	74,225,845
Joe Biden	81,286,365
Other	3,078,684
Total	158,590,984

Source: Dave Leip's *Atlas of US Presidential Elections*, https://uselectionatlas.org/.

Note that, in addition to showing the number of people who voted for each candidate, Table 2.1 also shows the total number of votes. Note also that the table includes a title that describes its contents, and each column is labeled. Frequency distribution tables should always include a title and column labels and show the total at the bottom of the table. They should also include a source.

Looking at another example, how exciting do you find your life? Do you think you are typical in your response? The General Social Survey (GSS) has asked this question regularly: "In general, do you find life exciting, pretty routine, or dull?" Of the 2,669 people who answered in 2021, 136 said they found life generally dull; 1,571 said they found life routine; and 962 said they found life exciting. We can present these results in a table, as shown in Table 2.2.

Table 2.2   Number of Americans Who Say Life Is Exciting, Routine, or Dull

Rating of Life	Frequency
Dull	136
Routine	1,571
Exciting	962
Total	2,669

Source: 2021 General Social Survey.

Which response is the most common, according to Table 2.2? Which is the least common? How big are the differences between the numbers of people who give each response? We can see that relatively few people find their lives dull. The largest number of people find their lives routine, but many say their lives are exciting.

This is an ordinal variable. While these response categories do not have precise numerical values, they can be rank-ordered. Someone who says their life is "exciting" finds life more exciting than someone who says life is "routine." It is not an interval-ratio variable, because we cannot assign numerical values to the categories. We cannot say that those who choose one category find life twice as exciting or ten "excitement units" more exciting. We construct frequency tables in the same way for interval-ratio variables. For example, how many children do people have? The results from the 2021 GSS are presented in Table 2.3.

Table 2.3 Number of Children

Number of Children	Frequency
0	1,163
1	646
2	1,152
3	578
4	277
5	79
6	52
7	17
8 or more*	19
Total	3,983

Source: 2021 General Social Survey.
* The top category is open-ended; in practice such variables are often treated as interval-ratio.

Frequency distributions like these, showing the number of responses in a given category, give us a first look at the data. But as the number of response categories grows, it is harder for us to grasp the pattern in the data. We can see pretty quickly that the largest numbers of people have no children or two children. We also can see that fewer people have four or more children. It is more difficult to assess differences in size among the categories precisely.

Percentages and Proportions

It is often easier for people to interpret results when they are standardized as percentages or proportions. Percentages or proportions are relative frequencies—that is, they show the size of each response category relative to the overall number of cases. (Rates, discussed later in this chapter, are another means of standardizing results.)

Calculating a percentage entails comparing the number of responses in one category to the total number of responses. (In mathematical terms, "comparing" generally means "dividing.") To calculate a **percentage**, divide the number of responses in one category by the total number of responses. This will give you the **proportion**, which is a decimal between 0 and 1.0. To convert proportion to percentage, multiply by 100, or move the decimal two places to the right. The formula looks like this:

$$Percentage = \frac{f}{N}(100)$$

In the formula, f is frequency, the number of responses in one category. N is the total number of cases. In statistics, lowercase "f" refers to frequency and uppercase "N" to the total number of cases.

Looking again at the frequency distribution for number of children, we see that 3,983 people responded to the question. Of these, 1,163 had zero children. Using the formula:

$$Percentage = \frac{1163}{3983}(100)$$

$$= 0.292\,(100)$$

$$= 29.2$$

Of respondents, 29.2% have no children. Take a moment to calculate the percentage for the other response categories. What percentage of respondents have one child? Two children? Table 2.4 includes the percentages in the last column.

Table 2.4 Number of Children, Including Percentage

Number of Children	Frequency	Percent
0	1,163	29.2
1	646	16.2
2	1,152	28.9
3	578	14.5
4	277	7.0
5	79	2.0
6	52	1.3
7	17	0.4
8 or more	19	0.5
Total	3,983	100

Source: 2021 General Social Survey.
Total percent may not sum to 100 due to rounding.

BOX 2.1: IN DEPTH

Assessing Relative Size Using Percentages and Frequencies

Research shows that almost everyone has trouble accurately grasping the relative sizes of groups or the relative likelihood of events, particularly very large or very small groups or likelihoods. This is even true of professional statisticians, researchers, and statistics teachers![4] Knowing this, we should be especially careful to look at both percentages and raw frequencies to check our understanding. If you go on to work with statistical information in other courses or employment, you should present data in a variety of ways to help your audience understand it.

Do you find it easier to see the overall patterns with percentages included in the table? Which is easier to comprehend, the fact that seventy-nine out of 3,983 people had five children or the fact that 2.0% of people had five children? To put the percentage in different terms, two people out of every one hundred people would have five children.

For extra practice, go back to the frequency tables showing the candidates people voted for and whether they find life exciting or dull (Tables 2.1 and 2.2, respectively). Calculate the percentages for these tables. They are presented below (Tables 2.5 and 2.6), with percentages filled in.

Table 2.5 Votes for 2020 Presidential Candidates

Candidate	Frequency	Percent
Donald Trump	74,225,845	46.8
Joe Biden	81,286,365	51.3
Other	3,078,684	1.9
Total	158,590,984	100

Source: Dave Leip's *Atlas of US Presidential Elections*, https://uselectionatlas.org/.

Table 2.6 Americans Who Say Life Is Exciting, Routine, or Dull

Rating of Life	Frequency	Percent
Dull	136	5.1
Routine	1,571	58.9
Exciting	962	36.0
Total	2,669	100

Source: 2021 General Social Survey.
Total percent may not sum to 100 due to rounding.

How does seeing the percentage change your understanding of the final votes in the 2020 presidential election? How does seeing the percentage change your understanding of how Americans view their lives? If you are like most people, the percentages probably help you assess the relative size of the groups who voted for Donald Trump and Joe Biden, because it is hard for most people to intuitively grasp the size of very large numbers like these hundreds of millions. But the percentages may or may not improve your understanding of the relative frequency of assessing one's life as exciting or dull.

Cumulative Percentage and Percentile

Another important, basic number for thinking about the distribution of ordinal- or interval-ratio-level variables is cumulative percentage. **Cumulative percentage** is the total percentage of cases with a given value or below. For example, cumulative percentage could tell you what percentage of people have two or fewer children, or what percentage say their lives are routine or worse. Why is cumulative percentage used only with ordinal- or interval-level variables and never with nominal variables? Because it relies on the idea that the values, the "answers to the questions," can be rank-ordered from least to most. Nominal variables, by definition, cannot be rank-ordered.

For cumulative percentage to be calculated, the categories must first be rank-ordered. Usually, they are listed from the lowest to the highest value. To calculate the cumulative percentage at each value, add the percentage of cases that have that value to the percentage of cases that have lower values. You can also start with a cumulative frequency, by adding the total number of cases with a given value or below. Then convert the cumulative frequency to a percentage by dividing it by the total N and multiplying by 100. By formula:

$$Cum\% = \frac{cumf}{N}(100)$$

"Cum%" stands for cumulative percentage, and "cumf" stands for cumulative frequency. As before, "N" stands for the total number of cases.

For example, as Table 2.6 shows, 5.1% of respondents said their life was "dull," and 58.9% said their life was "routine." The cumulative percentage for routine will tell us what percentage of respondents said life was routine or worse. To calculate the cumulative percentage, you can add the percentages of respondents who say life is dull or routine (5.1 + 58.9 = 64). Or, you can add the frequencies of respondents who say life is dull or routine and then convert to percentage, using the above formula. (*Cumf* = 136 + 1571 = 1707. Cum% = (1707/2669)(100) = 64.0.) Table 2.7 shows the data for the excitement for life variable, including cumulative percent in the last column.

Table 2.7 Americans Who Say Life Is Exciting, Routine, or Dull (Frequency, Percentage, and Cumulative Percentage)

Rating of Life	Frequency	Percent	Cumulative Percent
Dull	136	5.1	5.1
Routine	1,571	58.9	64.0
Exciting	962	36.0	100
Total	2,669	100	

Source: 2021 General Social Survey.
Total percent may not sum to 100 due to rounding.

Note that cumulative percentages can tell you what percentage of cases have a given value *or more*, or a given value *or less*. This depends on the order in which the variable values are listed. Look at the frequency distribution for number of children (Table 2.4). Because number of children is listed from lowest to highest, cumulative percentages tell us the percentage who have that many children or fewer. For example, 29.2% of respondents had zero children, and 16.2% had one child. The cumulative percentage will tell us what percentage of respondents had one or fewer children. As before, to calculate the cumulative percentage, you can add the percentages of respondents with zero and one child (29.2 + 16.2 = 45.4). Or, you can add the frequencies of respondents with zero or one child and then convert to percentage, using the above formula. (Cumf = 1163 + 646 = 1809. Cum% = (1809/3983)(100) = 45.4.).

Take a moment to calculate the cumulative percentages for the rest of the table. Table 2.8 reflects those calculations.

Table 2.8 Number of Children, Including Cumulative Percentage

Number of Children	Frequency	Percent	Cumulative Percent
0	1,163	29.2	29.2
1	646	16.2	45.4
2	1,152	28.9	74.3
3	578	14.5	88.9
4	277	7.0	95.8
5	79	2.0	97.8
6	52	1.3	99.1
7	17	0.4	99.5
8 or more	19	0.5	100
Total	3,983	100	

Source: 2021 General Social Survey.
Total percent may not sum to 100 due to rounding.

What percentage of respondents had three or fewer children? If you want to know what percentage have a certain number of children or more, you have two options. You can reverse the order that the values are presented in the table, with "8 or more" children at the top, or you can calculate the cumulative percentages starting at the bottom of the table. Table 2.9 presents the latter option.

Table 2.9 Number of Children, Including Cumulative Percentage (Reverse Order)

Number of Children	Frequency	Percent	Cumulative Percent
0	1,163	29.2	100
1	646	16.2	70.8
2	1,152	28.9	54.6
3	578	14.5	25.7
4	277	7.0	11.2
5	79	2.0	4.2
6	52	1.3	2.2
7	17	0.4	0.9
8 or more	19	0.5	0.5
Total	3,983	100	

Source: 2021 General Social Survey.
Total percent may not sum to 100 due to rounding.

Looking at Table 2.9, you can see that 54.6% of people have two or more children, 70.8% have one or more children, but only 25.7% have three or more children. Which way should you present a table or calculate a cumulative percentage? It depends on what you want to know.

A similar idea to cumulative percentage is percentile. **Percentile** indicates the position of any given case or value in the overall distribution. Most of you have encountered percentiles in your daily life. For example, as children grow, their height and weight are measured in terms of percentile. A child who is in the 50th percentile for height has a height for which the cumulative percentage would be 50. Of all children, 50% are below that height. A shorter child, in the 10th percentile for height, would have a height for which the cumulative percentage is 10. Of all children, 10% would be below that height. Most standardized tests report results in percentile as well as the actual score. For example, you may have taken the SAT or ACT for college admissions. If so, you received a numerical score (such as 465 or 28) and a percentile rank (such as 60th percentile), which is also the cumulative percentage of test-takers who scored below that score.

Percent Change

Another important concept is percent change. For example, if a mentoring program for high school students improved the graduation rate at a given school, we would want to know *how much* the graduation rate improved, relative to the size of the initial graduation rate. Imagine that, before the mentoring program, 75% of students who entered ninth grade graduated from high school four years later. After the mentoring program was implemented, 90% of students who entered ninth grade completed high school on time. How much improvement is this? Be careful! Many people will say that the program increased the graduation rate by 15%, because the difference between the graduation rates after the program (90%) and before the program (75%) is 15 percentage points.

This is incorrect because it does not account for the *size* of the initial percent. **Percent change** tells us the size of the change in comparison to the initial percent or frequency. By formula:

$$\text{Percent change} = \frac{(\text{percent at time 2} - \text{percent at time 1})}{\text{percent at time 1}}(100)$$

If you are calculating percent change with frequencies instead of percentages, the formula is:

$$\text{Percent change} = \frac{f \text{ at time 2} - f \text{ at time 1}}{f \text{ at time 1}}(100)$$

With our example, what is the percent change in graduation rate at the school after the implementation of the mentoring program?

$$\text{Percent change} = \frac{90-75}{75}(100) = 20$$

We can see that the graduation rate increased by 20% after the program's implementation.

Rates and Ratios

There are two other simple mathematical calculations that help us measure and describe phenomena in standardized terms: rates and ratios.

Rates

Rates indicate the prevalence of outcomes or events of interest in a given population. A **rate** tells us about the frequency of an occurrence relative to the number of times that the event could have occurred in a given group. The formula for calculating a rate is:

$$Rate = \frac{f \text{ actual occurrences}}{f \text{ possible occurrences}}$$

Demographers often use rates to track changes in a population over time. Birth and death rates, marriage and divorce rates, and dropout and graduation rates are examples of rates studied in depth by social scientists across many disciplines. Popular and academic interest in these rates stems from the fact that changes in rates over time can indicate cultural and structural shifts in a given population. For instance, changes in cultural attitudes about childbearing and marriage are reflected in the rising birth rate among unmarried women in the United States, rising from 29.4 in 1980 to 39.9 in 2019.[5] Note that the birth rate here is expressed as the number of births to unmarried women per 1,000 unmarried women in the population.

While calculating a rate is simple, choosing the appropriate number of occurrences for a rate's numerator and denominator can be a more complicated task. Consider high school graduation rates as an example. These rates are used by different audiences to make different kinds of assessments and decisions about students and schools. While parents might compare schools' graduation rates as they choose a school for their child, a researcher might track changes in national graduation rates over time to see whether the population has become more educated.

If a researcher wants to know about the rate of high school completion in a given country, which figures should they choose for the numerator and denominator? We know that the numerator for calculating a rate is the number of actual occurrences, and the denominator is the number of possible occurrences. The denominator could include the entire population of the country, but that would overcount the number of possible occurrences by including people who are too young to have completed high school. What should the age cut-off be for inclusion in the denominator of this high school graduation rate? For people following the traditional route through the formal education system—beginning kindergarten at age five and advancing to the next grade every year until high school completion—we expect that most would finish high school by age eighteen.

We can use data from the 2021 Current Population Survey, a representative sample of the population of the United States to calculate this rate.[6] In this sample, there are 253,417 people aged eighteen or older, with 229,083 of them having completed high school. By formula, the graduation rate for people eighteen and older is:

$$\text{Graduation rate} = \frac{229{,}083}{253{,}417} = 0.90$$

That is, 90% of people eighteen and older have completed high school.

But say the researcher wanted to focus on the graduation rate not among all eligible adults, but for younger adults, in particular. The researcher decides to limit the figure in the denominator, all possible occurrences of high school completion, to people aged eighteen to twenty-four. There are 28,837 people in the sample aged eighteen to twenty-four, and 24,555 of them had completed high school. Among this group, the high school graduation rate is:

$$\frac{24{,}555}{28{,}837} = 0.85$$

The graduation rate among eighteen- to twenty-four-year-olds (0.85, or 85%) is lower than among all adults eighteen and older (0.90, or 90%) because some proportion of younger adults (probably around 5% of them) have not finished the requirements for high school completion but will do so later in life. These rates tell us something about the overall level of education in the population.

But these rates tell us nothing about how well a particular school district is doing at successfully shepherding its students through high school to graduation within four years. For this high school graduation rate, we need information about students over their four years of high school. Here, the number of possible occurrences for the denominator would be the number of students who began ninth grade in a particular school district. The number of actual occurrences, to be placed in the rate's numerator, is the number of cases who belonged to that entering class of students and graduated from a school in the district four years later. According to the Massachusetts Department of Education, 4,347 students started high school in Boston Public Schools in the fall of 2019.[7] Four years later, in the spring of 2023, 3,182 of those students graduated from a high school in the district. For Boston Public Schools, the four-year graduation rate (also called a cohort graduation rate) for the Class of 2021 was 0.732, or 73.2%. By formula:

$$\frac{3182}{4347} = 0.732$$

Choosing which numbers to use in a rate's numerator and denominator depend on what you want to know. Our first example, rate of high school completion among all adults, indicates the overall level of basic education in the population. Our second example, the four-year graduation rate for Boston Public Schools, indicates the fraction of high school students who graduate from high school in that district within four years.

Ratios

While a rate examines the frequency of an event given the number of times the event could have occurred, a ratio examines the size of one category relative to another. The formula for a ratio simply divides the frequency of one category (f_1) by that of another (f_2):

$$\text{Ratio} = \frac{f_1}{f_2}$$

While a divorce rate tells us about the prevalence of divorce relative to the number of divorces that could have happened (i.e., number of marriages in a given population), a ratio can tell us about the relative sizes of the divorced and married populations. In 2021, the number of married people in the United States was 266,140,000, and the number of people older than eighteen who reported being divorced was 25,910,000.[8] Thus, the ratio of divorced to married adults in the United States was 0.10 (25,910,000/266,140,000). In other words, for every married adult, there were 0.10 divorced adults. We can multiply the ratio by a multiple of ten to avoid talking about a fraction of a person. Multiplying the ratio by 1,000 tells us that for every 1,000 married adults, there were 100 divorced adults.

Working with Frequency Distribution Tables

The frequency distribution table is the most common device for displaying information about a variable. Although the presentation may differ a little, pretty much any statistical software package you use will display, in a frequency distribution table, the elements that we have examined: number of cases associated with each category, percentage of cases associated with each category, and cumulative percentage. When analyzing and presenting frequency data, you have to consider two other issues: missing values and whether to combine categories of the variable.

Let's look at another example of a frequency distribution to explore these issues. The 2020 American National Election Study (ANES) asked respondents to place themselves on a 7-point ideological scale, running from strong liberal to strong conservative. Table 2.10 displays a frequency distribution of their responses.

Missing Values

As we have seen, the "Frequency" column in Table 2.10 reports the number of respondents that chose each category. Overall, this survey interviewed 8,280 respondents. Among these respondents, 369 chose the label "extremely liberal." If we divide the number of extremely liberal respondents (369) by the total number of respondents (8,280), we derive the percentage of all respondents—4.5%—who are "extremely liberal." Those percentages are displayed in the "Percent" column. At the other end of the scale, 428 respondents described themselves as "extremely conservative." That represents 5.2% of all respondents.

Notice that Table 2.10 distinguishes between "percent" and "valid percent." The "Valid Percent" column computes the percentage of each category from a smaller denominator. That denominator excludes respondents who refused to answer the question, who said that they did not know what their ideological placement is, or who admitted that they haven't thought much about this question. Overall, 1,224 respondents did not select one of the seven ideological categories. This is called missing data. But is it really missing?

The answer depends on the point of view of the researcher. We know that 8,280 individuals were queried about their ideological identification, and we also know that 1,224 of

Table 2.10 Self-Placement on Ideological Scale

Self-Placement	Frequency	Percent	Valid Percent	Cumulative Percent
1: Extremely liberal	369	4.5	5.2	5.2
2: Liberal	1,210	14.6	17.1	22.3
3: Slightly liberal	918	11.1	13.0	35.4
4: Moderate	1,818	22.0	25.8	61.2
5: Slightly conservative	821	9.9	11.6	72.8
6: Conservative	1,492	18.0	21.1	93.9
7: Extremely conservative	428	5.2	6.1	100
Total, 1–7	7,056	85.2	100	
Refused	21	0.3		
Don't know	2	0.0		
Haven't thought much about this	1,201	14.5		
Total, missing	1,224	14.8		
Total	8,280	100		

Source: 2020 American National Election Study.
Total percent may not sum to 100 due to rounding.

these individuals did not choose an ideological category. The "Percent" column includes these 1,224 respondents in the denominator, and, thus, one might say that 4.5% of all respondents identified as extremely liberal. The "Valid Percent" column excludes those 1,224 respondents (the denominator for the "Valid Percent" column is 7,056), and, thus, one might say that 5.2% of all those who chose an ideological label identified as extremely liberal. Regardless of which figure one chooses to report, it is noteworthy that almost 15% of respondents admitted that they don't know or haven't thought much about where to place themselves on an ideological scale.

The last column of Table 2.10 displays the "Cumulative Percent." As we have already seen, one simply adds the valid percent as one moves down the categories of the variable.

So, 5.2% of respondents are extremely liberal, 22.3% (5.2 + 17.1) are extremely liberal or liberal, and so on. Cumulative percent cannot include missing values.

Simplifying Tables by Collapsing Categories

The ideological identification variable has seven categories, and although this contains precise information, it may be more than the researcher needs. The researcher may simply want to distinguish among the three main labels (liberal, moderate, conservative) without considering the finer distinctions (how conservative or how liberal). Toward that end, the researcher may choose to collapse categories of the variable and present a scaled-down version of the variable. When you are working with data using statistical software, combining or collapsing categories is done by **recoding**, as you will see at the end of this chapter.

How to collapse categories is a subjective decision. For a variable such as ideological identification, the recoding scheme is fairly self-evident. For other variables, it may be more complicated.

Table 2.11 presents a collapsed version of the ideological identification variable that was displayed in Table 2.10. The seven categories of the original variable have been collapsed into three categories. In this version of the variable, respondents who refused, didn't know, or hadn't thought about it have been excluded. The percentages are all based on the smaller denominator (the one that excludes "missing data"). For this reason, the percentages shown in the "Percent" and "Valid Percent" columns are identical.

Table 2.11   Self-Placement on Ideological Scale, Collapsed Version

Self-Placement	Frequency	Percent	Valid Percent	Cumulative Percent
Liberal	2,497	35.4	35.4	35.4
Moderate	1,818	25.8	25.8	61.2
Conservative	2,741	38.8	38.8	100
Total	7,056	100	100	

Source: 2020 American National Election Study.

Although we have less precise information than we did in Table 2.10, the collapsed variable displayed in Table 2.11 allows us to quickly see the ideological distribution of Americans. Conservative is the most common category, liberals are in second place, and moderates constitute just more than one-quarter of the sample.

As we stated, the recoding scheme for this variable was fairly obvious. But there are many examples where the researcher's subjective coding scheme will shape the description of empirical reality that emerges.

Consider the following example. An NBC News/*Wall Street Journal* poll in August 2022 found this breakdown of party identification among Americans (see Table 2.12).

Table 2.12   Party Identification, August 2022

Party Identification	Percent
Democrat	34
Independent who leans Democrat	8
Strictly Independent	12
Independent who leans Republican	9
Republican	31
Other	3
Don't Know	3
Total	100

Source: NBC News/*Wall Street Journal* poll.

Party identification has five categories (in addition to the two categories of respondents who chose "other" or "don't know"). Again, these fine distinctions may be more than the researcher wants or needs. If the researcher is mainly interested in knowing what percentage of people are Republicans, Democrats, and Independents, how should she recode this variable? There are at least two possible recoding schemes, displayed in Tables 2.13 and 2.14.

Table 2.13   Party Identification, August 2022, Version 1

Party Identification	Percent
Democrat (includes leaners)	42
Independent	12
Republican (includes leaners)	40
Other/don't know	6
Total	100

Source: NBC News/*Wall Street Journal* Poll.

The first scheme, in Table 2.13, treats respondents who said they are Independent but lean toward one of the parties as partisans. So a respondent who says that she is an Independent who leans toward the Democrats is treated as a Democrat; and similarly, a respondent who says that she is an Independent who leans toward the Republicans is treated as a Republican. The partisan portrait that emerges from Table 2.13 is of an electorate that is quite polarized, with relatively few people (only 12%) identifying as Independent.

The second scheme, in Table 2.14, treats leaners as Independents. If someone reports that he or she is an Independent who leans toward one of the parties, then that respondent is categorized as an Independent.

Table 2.14   Party Identification, August 2022, Version 2

Party Identification	Percent
Democrat	34
Independent (includes leaners)	29
Republican	31
Other/don't know	6
Total	100

Source: NBC News/*Wall Street Journal* poll.

As we see in Table 2.14, this produces a picture of an electorate that has a much larger middle—29% of respondents are Independents. Neither of these schemes is incorrect. The point is that the subjective decisions that the researcher makes about how to collapse categories or recode a variable are critical in shaping the image of reality that ensues.

Graphical Displays of a Single Variable: Bar Graphs, Pie Charts, and Histograms

The frequency distribution is a tabular representation of one variable. Alternatively, analysts can display variables in graphic form. In this section, we cover bar graphs and pie charts for displaying nominal and ordinal variables as well as **histograms** for displaying interval-ratio variables.

Bar Graphs and Pie Charts

The most common devices for the graphic representation of a nominal or ordinal variable are the **bar graph** and the **pie chart**. Figure 2.1 displays a bar graph for marital status among Americans from the 2021 GSS.

This is a nominal variable, with five categories. Each bar represents a category of the marital status variable. The vertical axis displays the percentage of cases. For example, 49.7% of respondents reported that they were married, and 16.3% reported that they were divorced. The space between the bars reinforces the point that this is a nominal variable and that one could display the bars in any order—there is after all no rank-ordering among the categories of marital status.

Some bar graphs report the number of cases along the vertical axis (instead of the percentage of cases). This is most appropriate when the number of cases is small. In general, bar graphs are an especially effective graphic device when the number of categories is relatively large.

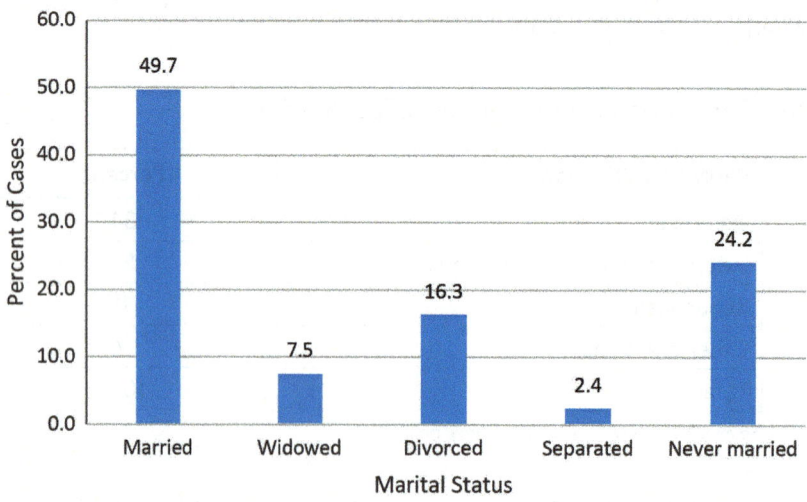

Figure 2.1 Marital Status among Americans, Bar Graph
Source: 2021 General Social Survey.

Alternatively, one could present the same variable by means of a pie chart, shown in Figure 2.2. Each slice of the pie represents the percentage of cases associated with each category. Pie charts are especially effective at conveying a quick visual sense of the relative

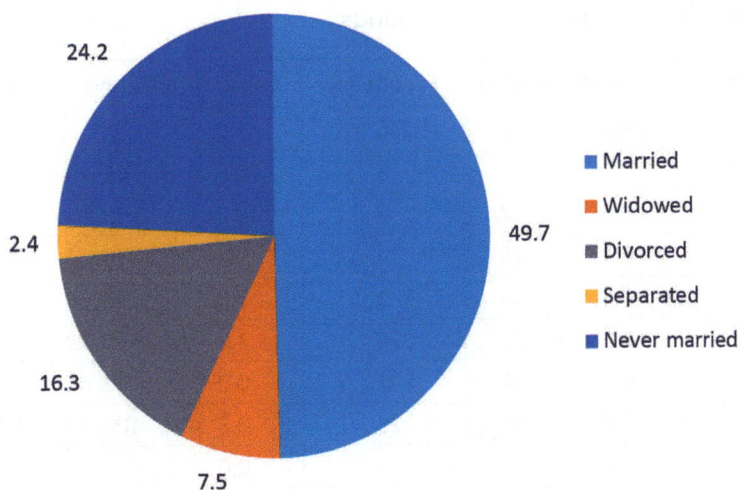

Figure 2.2 Marital Status among Americans, Pie Chart
Source: 2021 General Social Survey.

size of each category. The pie chart in Figure 2.2 conveys a quick and correct visual impression that married people constitute the largest group of respondents.

Interval-ratio-level variables present a different set of challenges. Typically, they have a much larger range of categories—think of percentage scores in an exam in your class, the ages of a random sample of Americans, or the family incomes of people in your county. Frequency distributions are unwieldy when a variable has so many categories.

Here is an example. In 2020, the ANES asked respondents how Donald Trump made them feel on a 100-degree feeling thermometer, where 0 represented negativity or coldness, 50 represented neutrality, and 100 represented positivity or warmth. Table 2.15 displays the frequency distribution for all the respondents in the survey.

As you can see, there are an awful lot of categories—101 possible ratings on this scale. As such, the table is hard to read, with very small numbers of respondents in most of the cells. To summarize this variable graphically, there is no question that we need to recode it. (If we produced a bar graph without recoding, it would have up to 101 bars. Similarly, a pie chart would have up to 101 slices.)

Once again, the subjective decisions of the researcher will influence the story that this variable tells. When we recode, we must decide on the width of the intervals. If the intervals are too wide (e.g., three intervals, as follows: 0–49 degrees, 50 degrees, 51–100 degrees), then we lose too much information. If the intervals are too narrow (e.g., twenty intervals, each 5 degrees of width), we have not really gained much efficiency relative to the original interval-ratio variable. The challenge is to strike a reasonable balance between the number and width of the intervals.

Here is one possible scheme for recoding the feeling thermometer variable:

- Very cold: 0 to 20 degrees
- Cold: 21 to 40 degrees
- Slightly cold to slightly warm: 41 to 60 degrees
- Warm: 61 to 80 degrees
- Very warm: 81 to 100 degrees

Table 2.15 Trump Feeling Thermometer Ratings

Rating	Frequency	Percent	Rating	Frequency	Percent	Rating	Frequency	Percent
0	3,189	39.6	50	245	3.0	90	110	1.4
1	15	0.2	51	1	0.0	91	1	0.0
2	11	0.1	52	2	0.0	92	1	0.0
3	6	0.1	55	14	0.2	94	1	0.0
5	29	0.4	56	1	0.0	95	50	0.6
6	2	0.0	60	398	4.9	98	6	0.1
7	1	0.0	64	1	0.0	99	11	0.1
8	3	0.0	65	28	0.3	100	1,212	15.1
9	2	0.0	66	1	0.0			
10	64	0.8	68	3	0.0			
12	2	0.0	69	1	0.0			
15	531	6.6	70	523	6.5			
18	1	0.0	72	1	0.0			
20	23	0.3	74	1	0.0			
24	1	0.0	75	61	0.8			
25	10	0.1	77	1	0.0			
29	1	0.0	80	43	0.5			
30	302	3.8	82	1	0.0			
34	1	0.0	85	864	10.7			
35	14	0.2	86	3	0.0			
40	241	3.0	87	2	0.0			
45	9	0.1	88	1	0.0			
49	1	0.0	89	1	0.0			
						Total	8,048	100

Source: 2020 American National Election Study.
Total percent may not sum to 100 due to rounding.

There are five intervals, and they are all approximately equal in size. With this scheme as the basis, we can construct both a frequency distribution and a graphical display, known as a histogram.

Histograms

The frequency distribution in Table 2.16 is similar to the frequency distributions we looked at earlier, except each category of the variable now represents a range of values. The histogram is the most common graphical display of an interval-ratio variable. Figure 2.3 shows the histogram for the same data that are shown in Table 2.16.

Table 2.16 Trump Feeling Thermometer Ratings, Recoded

Rating Interval	Frequency	Percent	Cumulative Percent
Very cold (0–20)	3,879	48.2	48.2
Cold (21–40)	570	7.1	55.3
Slightly cold to slightly warm (41–60)	671	8.3	63.6
Warm (61–80)	664	8.3	71.9
Very warm (81–100)	2,264	28.1	100
Total	8,048	100	

Source: 2020 American National Election Study.

In this example, each interval, or "bin," is 20 degrees wide. The height of each bar corresponds to the percentage of cases, which is displayed on the vertical axis. As with bar graphs, one can alternatively choose to display the number of cases on the vertical axis. Unlike bar graphs, there is no space between the bars because this is a continuous variable. The histogram in Figure 2.3 tells us a story: About half of people were very cold (0–20 degrees) toward Donald Trump, and about 28% were very warm (81–100 degrees). The heights of the outside bars are higher than the bars in the middle.

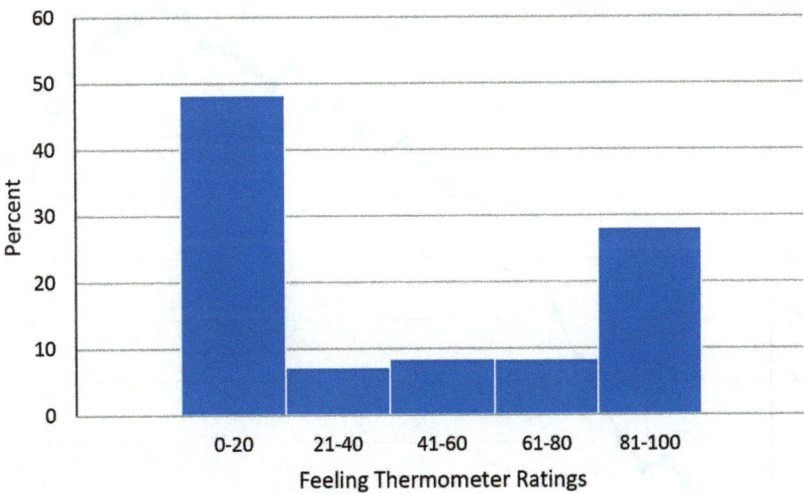

Figure 2.3 Histogram for Trump Feeling Thermometer, Recoded
Source: 2020 American National Election Study.

Because of their similar appearance, histograms and bar graphs are sometimes confused with each other. Histograms usually are used to display interval-ratio variables, often ones that have been collapsed into categories. Bar graphs, on the other hand, are typically used with nominal or ordinal-level variables that have relatively few categories.

Time Series Charts

Time series charts are particularly useful for showing how some statistic for a variable has changed over time. Time series charts draw a line to connect points, which are plotted at values of the variable of interest at different points in time. Thus, while it may seem that time series charts are reporting information about a single variable, they are actually reporting information about the relationship between two variables, the focal variable (represented on the y axis) and time (represented on the x axis).

For example, criminologists have paid close attention to the rising incarceration rate in the United States. According to the Bureau of Justice Statistics, the number of new defendants sentenced to at least one year in prison increased from 152,039 in 1978 to 608,318 in 2015. Adjusting for increases in the size of the general population, this means an increase in the rate of new sentences per million people from 683 to 1,900, a striking 178% increase. (Take a moment to check the math here and calculate the percent change for yourself.) But was the increase in the rate of new prison sentences over this time period a steady climb, or were there peaks and valleys along the way? We can evaluate the overall trend by examining a time series chart, as shown in Figure 2.4.

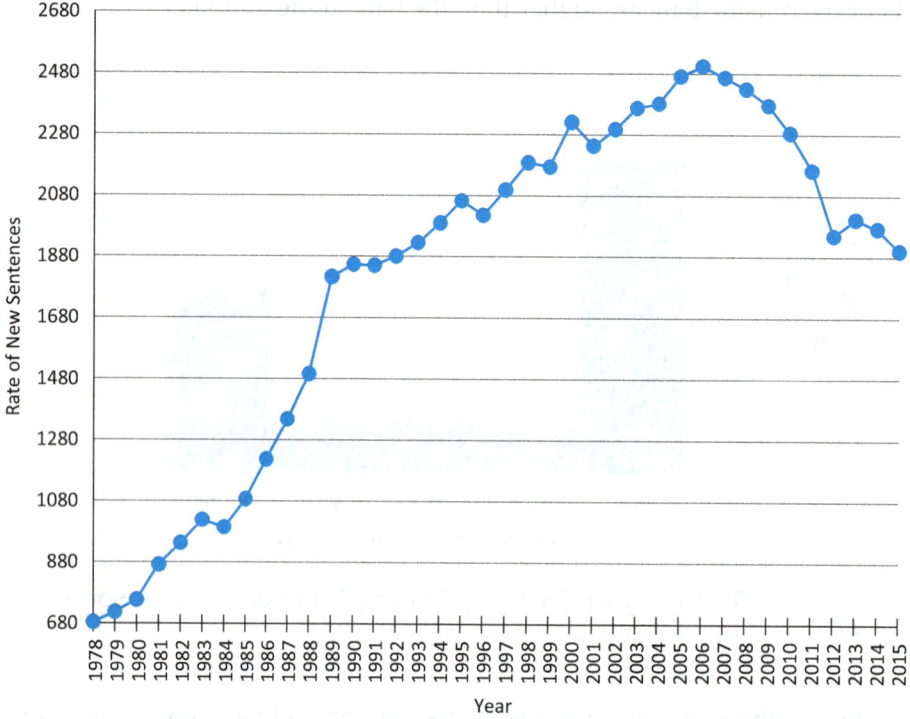

Figure 2.4 Rates of New Prison Sentences (per Million), 1978–2015
Source: Bureau of Justice Statistics.

Figure 2.4 shows the rate of new prison sentences each year between 1978 and 2015. While there has been a general steep upward trend, the rate of new prison sentences has declined in recent years. Still, the magnitude of these recent reductions has not matched the magnitude of the increases occurring throughout the 1980s and 1990s. Again, while it may seem that this time series chart is displaying information for only one variable, new prison sentence rates, it is actually displaying the relationship between two variables: new prison sentence rates and time.

Comparing Two Groups on the Same Variable Using Tables, Graphs, and Charts

So far in this chapter, we have examined different ways to describe and display data for a single variable. In later chapters, we will examine **associations** between two variables in detail. Here, we take a first look at how variables are distributed differently between subgroups. Descriptions of single variables offer a quick overview of the group as a whole, but general overviews can mask important differences across groups. For example, drawing conclusions about attitudes toward gay marriage in the United States by examining the distribution of the entire adult population across the categories of that variable masks important differences across groups in attitudes. According to the data from the General Social Survey, 79% of adults in the United States supported gay marriage in 2021. While the majority of respondents supported gay marriage in general, the story is more complicated when we examine support for gay marriage by subgroup.

Table 2.17 presents frequency distributions for gay marriage support by political affiliation and level of education. About 58% of Republicans either agreed or strongly agreed that gay people should be allowed to legally marry, compared to 92% of Democrats. Similarly,

Table 2.17 Gay Marriage Support, by Political Party and Level of Education

	Republican			Democrat	
	Frequency	Percent		Frequency	Percent
Strongly agree	187	29.2	Strongly agree	817	73.1
Agree	183	28.6	Agree	206	18.4
Disagree	99	15.4	Disagree	44	3.9
Strongly disagree	172	26.8	Strongly disagree	51	4.6
Total	187	100	Total	1,118	100
	High School or Less			Bachelor's Degree or Higher	
	Frequency	Percent		Frequency	Percent
Strongly agree	452	47.0	Strongly agree	811	64.0
Agree	240	25.0	Agree	256	20.2
Disagree	103	10.7	Disagree	75	5.9
Strongly disagree	166	17.3	Strongly disagree	126	9.9
Total	961	100	Total	1,268	100

Source: 2021 General Social Survey.

72% of those whose highest level of education was a high school diploma or less either agreed or strongly agreed that gay people should be allowed to legally marry, compared to 84% of those who had completed a bachelor's degree or higher. Thus, while the majority of Americans support gay marriage, these data show how that generalization masks variation along the lines of political affiliation and level of education. Saying that 79% of Americans support gay marriage overstates support among Republicans and those who have a high school diploma or less and understates support among Democrats and those who have attained a bachelor's degree or higher.

The bar graphs in Figure 2.5 present another option for depicting the same information shown in the frequency distributions in Table 2.17. The advantage of the bar graph is that it is able to communicate more information in a single figure, making it easier for viewers to visually recognize patterns in the data.

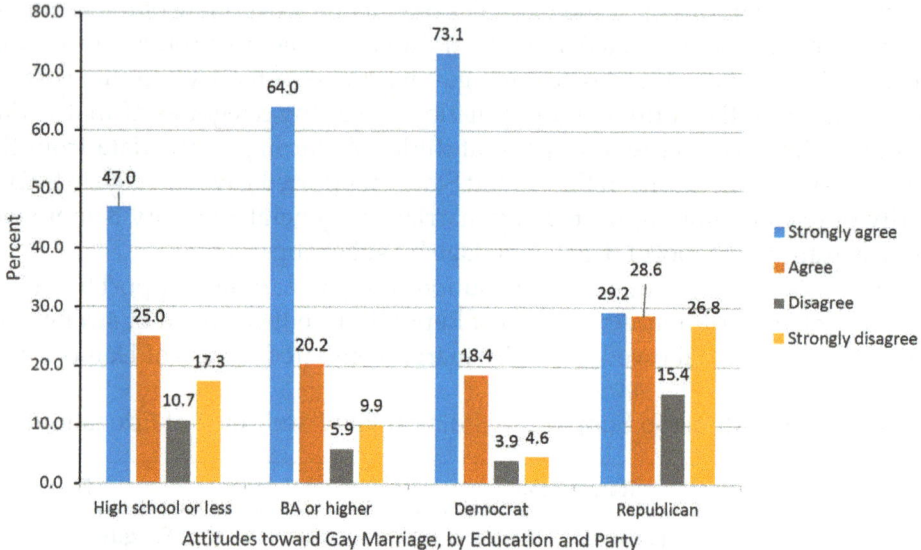

Figure 2.5 Bar Graphs of Attitudes toward Legalization of Gay Marriage, by Education and Political Affiliation
Source: 2021 General Social Survey.

Alternatively, one can use side-by-side pie charts to describe the same information. Figure 2.6 presents two sets of pie charts—analogous to the data presented in Figure 2.5's bar graph. The top set compares Republicans to Democrats. The bottom set examines attitudes toward gay marriage broken down by level of education. One advantage of pie charts is that, because they are fairly small, one can easily compare the distribution of a variable between two groups. A quick glance at Figure 2.6 reveals marked differences in the sizes of the pie slices, indicating differences in attitudes toward gay marriage based upon both education and partisanship.

For an interval-ratio variable, one can compare histograms across two groups to see whether the distribution of the variable varies across groups. Using the data on attitudes toward gay marriage from the General Social Survey, Figures 2.7a and 2.7b present separate histograms for respondent age for those who support the legalization of gay marriage and for those who oppose it.

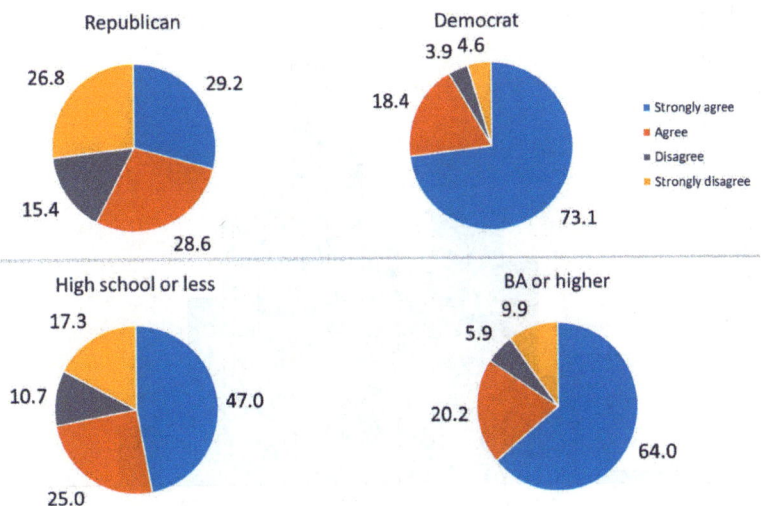

Figure 2.6 Pie Charts of Attitudes toward Legalization of Gay Marriage, by Education and Political Affiliation
Source: 2021 General Social Survey.

We see higher concentrations of younger people among supporters of gay marriage (Figure 2.7a) than among its opponents (Figure 2.7b). The distribution for opponents skews left, with opponents concentrated among older survey respondents.

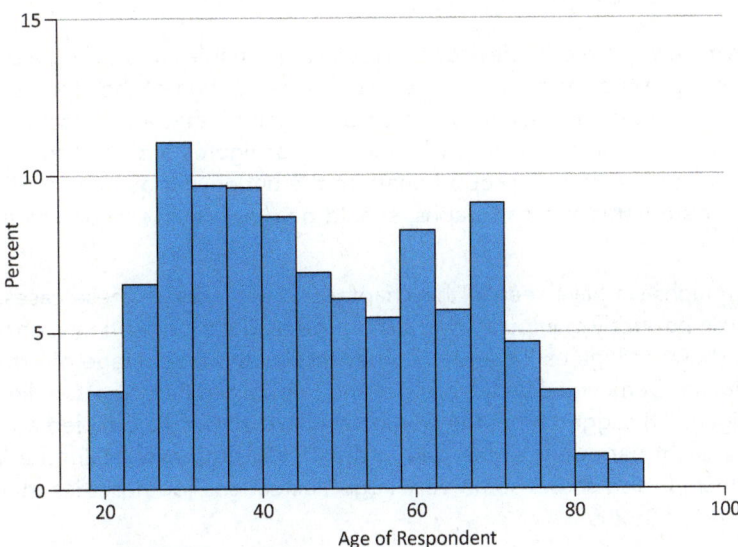

Figure 2.7a Age Distribution for Gay Marriage Supporters
Source: 2021 General Social Survey.

Uncovering the exact nature of the potential relationships among a set of variables will be the focus of later chapters. For example, social scientists have studied the question of whether people with more liberal social attitudes are more likely to pursue higher levels of education or whether staying in school longer leads one to adopt more liberal social attitudes. These data cannot answer that question. They do indicate, however, that education, political affiliation, and social attitudes are related.

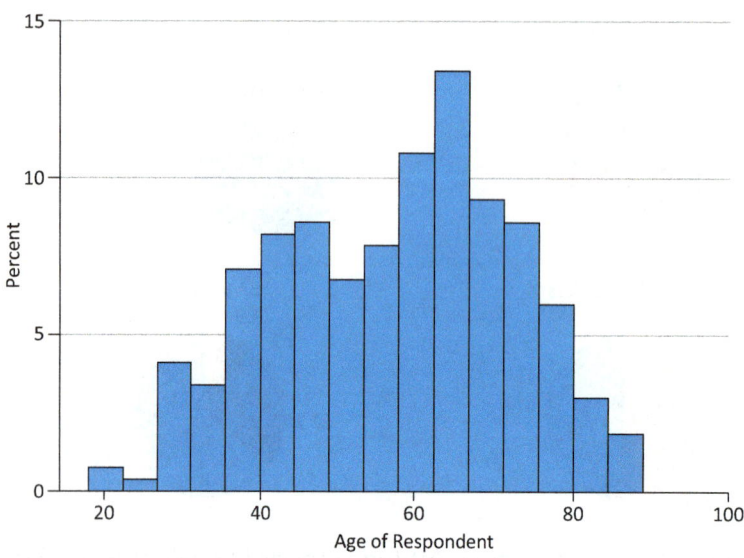

Figure 2.7b Age Distribution for Gay Marriage Opponents
Source: 2021 General Social Survey.

BOX 2.2: IN DEPTH

Misleading Graphics

Graphs are simple yet powerful devices for displaying variables and telling a story. But it is easy to manipulate a graph so as to distort the reader's perception of the story it is telling. Graphical honesty is important. Researchers should draw graphs that are reflective of the data they represent. The basic principle is that when images or figures are used to represent data, the size of those images should be proportionate to the underlying quantities. Slices of pie charts, or the bars in histograms and bar graphs, should be drawn in proportion to the data that they represent.

Many of the graphs we have seen in this chapter use two axes. In these cases, the scale of the vertical axis can be subtly manipulated to distort the reader's perception. Consider the two time series charts shown in Figures 2.8 and 2.9. They display the percentage of Americans who have identified with the Democratic Party (top line) and Republican Party (bottom line), between 1952 and 2016. Figure 2.8 suggests that the Democrats have enjoyed a decided advantage since the 1950s, with a slight narrowing of the gap in the 1980s, but have retained a large Democratic edge since then. It also shows some very large fluctuations (both parties in 1964 and a large Republican decline in 2008).

Figure 2.9 displays the exact same data, but the story it tells is much less dramatic. The gaps between parties look much narrower, and the year-to-year fluctuations seem much smaller. Whereas Figure 2.8 tells a story of rapidly shifting partisan identifications, Figure 2.9 displays stability more than change.

What's the difference between the two charts? In Figure 2.8, the vertical axis runs from about 30% to 60%. We call this a "compressed axis," because it does not show the full range of values of the variable. A compressed axis exaggerates changes between points on the horizontal axis.

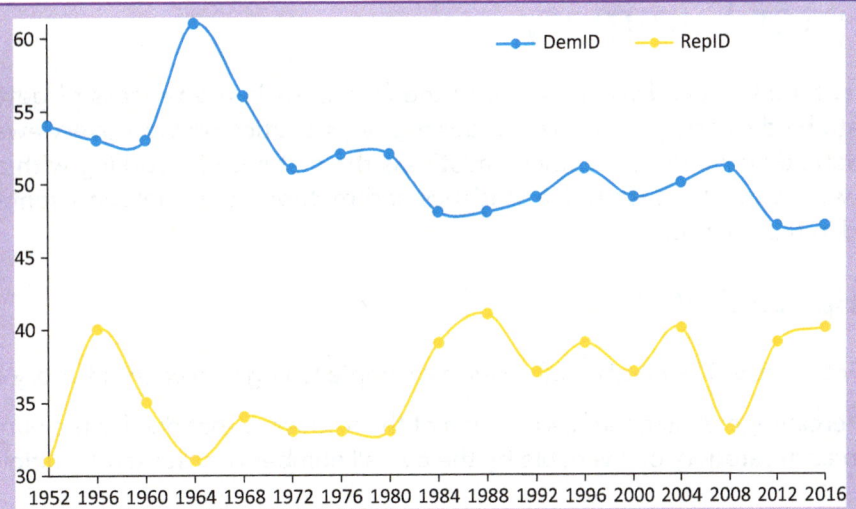

Figure 2.8 Party Identification in the United States, 1952–2016, Compressed Vertical Axis

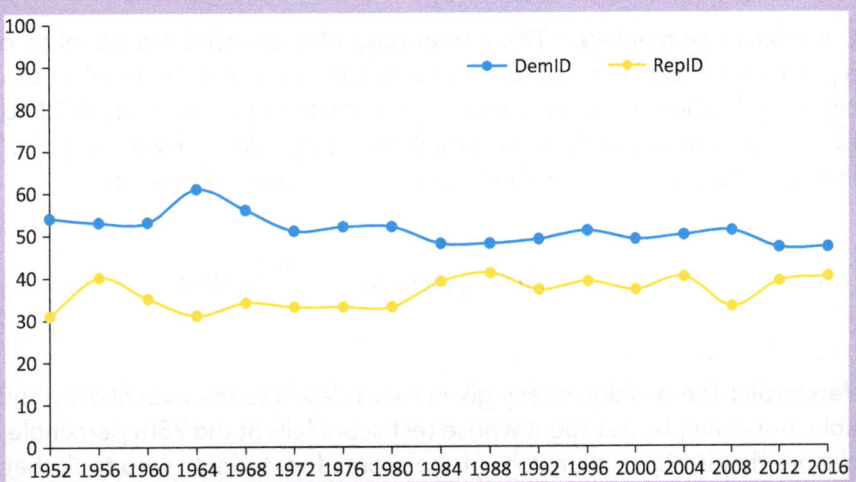

Figure 2.9 Party Identification in the United States, 1952–2016, Full Vertical Axis

The vertical axis in Figure 2.9 runs from 0 to 100%, representing the full range of possible values for party identification. Figure 2.9 is a more honest depiction of partisan change in the United States since the 1950s.

Chapter Summary

This chapter introduced you to some of the first steps in the process of getting to know your data by describing them. These descriptive statistics provide overviews of patterns in the data. Examining descriptive statistics is the first step in working with any data set. Below, we review the descriptive statistics and methods of visual representation of data covered in this chapter.

Descriptive Statistics

- **Frequency**: The number of cases in a sample falling into each category of a variable.
- **Percentage**: A standardized version of the frequency that divides the number of cases in each category of a variable by the overall number of cases (and multiplies by 100).

$$Percent = \frac{f}{N}(100)$$

Percentages are particularly useful when overall sample sizes are very large.

- **Cumulative percentage**: The percentage of cases that are equal to or lower than a particular value for a variable. Calculate by summing the number of cases in each category leading up to and including the category of interest, dividing by the number of cases in the sample, and multiplying by 100, or by adding the percentages in each category below and including the category of interest.

$$Cumulative\ percentage = \frac{cumf}{N}(100)$$

- **Percentile**: The position of any given case relative to the overall distribution for a variable. For example, a student whose test score falls at the 75th percentile has achieved a score that is higher than 75% of the scores for students who took that test.
- **Percent change**: A way of understanding the magnitude of a change in percent over time.

$$Percent\ change = \frac{(percent\ at\ time\ 2 - percent\ at\ time\ 1)}{percent\ at\ time\ 1}(100)$$

- **Rate**: The frequency of an event or outcome relative to the number of times that the event or outcome could have occurred in a given group.

$$Rate = \frac{f\ actual\ occurrences}{f\ possible\ occurrences}$$

- **Ratio**: The size of one category relative to another.

$$Ratio = \frac{f_1}{f_2}$$

Visual Representations of Data

- **Frequency distribution**: A tabular representation of frequencies, percentages, and cumulative percentages for each category of a variable. This may require collapsing, or recoding, data.
- **Bar graph**: A graphical representation of frequencies or percentages for each category of a nominal or ordinal-level variable. The length of each bar associated with each category is determined by the frequency or percentage of cases in that category relative to the overall sample. Bar graphs work well for variables that have many categories.
- **Pie chart**: A graphical representation of frequencies or percentages for each category of a nominal or ordinal-level variable. The size of the pie slice associated with each category is determined by the frequency or percentage of cases in that category relative to the overall sample. Pie charts should be used only for variables with relatively small numbers of categories.
- **Histogram**: A graphical representation of the distribution of an interval-ratio-level variable. Each bar in a histogram represents a range of values.
- **Time series chart**: A graphical representation of the change in a variable over time. A line connects some value for a variable at specified time intervals (e.g., years, months).

Using Stata

Chapter 1 gave you a general introduction to Stata and how it works. Now we will learn how to use Stata to generate the statistics and graphs that have been covered in this chapter. For our examples, we will use data from the 2021 GSS. In these examples, we will use two variables:

- *Degree* is an ordinal variable measuring the highest educational degree earned by the respondent. The five categories are 0 = less than high school; 1 = high school; 2 = community college (referred to by the GSS as "associate/junior college"); 3 = bachelor's degree; and 4 = graduate degree.
- *Age* is an interval-ratio variable measuring the respondent's current age.

Creating Frequency Distributions for a Variable

Frequency distributions are generated in Stata by using the "tabulate" command. Recall from chapter 1 that a general rule for Stata syntax is to type the kind of analysis that you want Stata to conduct first, followed by the name of the variable on which you want Stata to conduct that analysis. Open `GSS2021.dta` and type the following command into Stata:

`tabulate degree`

The "tabulate" command tells Stata to generate a frequency distribution for the *degree* variable in the Results Window (Figure 2.10). Recall that these results are referred to as output.

```
. tabulate degree

   r's highest degree |      Freq.     Percent        Cum.
----------------------+-----------------------------------
less than high school |        246        6.14        6.14
          high school |      1,597       39.84       45.97
associate/junior college |      370        9.23       55.20
           bachelor's |      1,036       25.84       81.04
             graduate |        760       18.96      100.00
----------------------+-----------------------------------
                Total |      4,009      100.00
```

Figure 2.10

The table in the output includes the number of cases (freq.), percentage (percent), and cumulative percentage (cum.) for each category of the variable. Of all the respondents included in the table (4,009), 1,036 reported that their highest educational degree is a bachelor's degree, accounting for 25.8% of the respondents who answered that question. Looking at the column for cumulative percentages, we can see that about 81% of the respondents had completed a bachelor's degree or less. Do you notice anything missing in this table? Here's a hint: with such a large sample, it is highly unlikely that every single person would provide a response for every single survey item. But this table suggests that we know every single respondent's value for this variable. Where is the information about missing data? We must tell Stata if we want to include missing data in the frequency distribution for a variable. In order to include those respondents for whom we do not have data on the *degree* variable, we need to add the "missing" option to our "tabulate" command:

`tabulate degree, missing`

The output is shown in Figure 2.11.

Stata uses a period to denote missing data. According to the output, data are missing on the *degree* variable for twenty-three respondents in the sample, accounting for only 0.6% of respondents. Including these missing cases in the denominator of the percentage for each category affects the percentage of respondents who fall into each. For example, including missing data in the denominator changes the percentage of respondents with a bachelor's degrees from 25.8% to 25.7%. This is a tiny change because the proportion of cases with missing data is very low for this variable, only 0.6%. The percentages displayed in the first frequency distribution give the valid percentage for each category because the denominator does not include those with missing data on the variable.

```
. tabulate degree, missing

        r's highest degree │    Freq.     Percent        Cum.
       ─────────────────────┼──────────────────────────────────
         less than high school      246       6.10        6.10
                  high school     1,597      39.61       45.71
       associate/junior college      370       9.18       54.89
                  bachelor's     1,036      25.69       80.58
                    graduate       760      18.85       99.43
                           .        23       0.57      100.00
       ─────────────────────┼──────────────────────────────────
                      Total     4,032     100.00
```

Figure 2.11

Recoding Variables

Recall that for various reasons researchers might choose to recode a variable by combining categories for that variable. In statistical software programs, each category of a variable is assigned a number. This is true even for nominal variables, in which categories cannot be rank-ordered. In order to recode a variable, the researcher must know the number assigned to each category of that variable. Most publicly available data sets have codebooks that describe each variable, including the number assigned to each category. Researchers can get a quick view of the numerical values assigned to each category without paging through the codebook by generating a frequency distribution without value labels. To do this, we add the "nolabel" option to the "tabulate" command:

tabulate degree, missing nolabel

```
. tabulate degree, missing nolabel

       r's highest │
            degree │    Freq.     Percent        Cum.
       ────────────┼──────────────────────────────────
                 0 │      246       6.10        6.10
                 1 │    1,597      39.61       45.71
                 2 │      370       9.18       54.89
                 3 │    1,036      25.69       80.58
                 4 │      760      18.85       99.43
                 . │       23       0.57      100.00
       ────────────┼──────────────────────────────────
             Total │    4,032     100.00
```

Figure 2.12

The output is shown in Figure 2.12.

The values in each cell of the table in Figure 2.12 are exactly the same as in Figure 2.11. The only difference is that instead of seeing the verbal descriptions of each category for the *degree* variable (e.g., "bachelor's"), we see the number assigned to each of those categories in Figure 2.12. We can see, for example, that the "bachelor's" category is assigned a value of 3. When recoding variables, we work with the numbers assigned to categories instead of their verbal descriptions.

With five categories for degree, it might make sense for some researchers to consider combining some of them. For example, for some research purposes it is probably not very meaningful to distinguish between parents who who did not finish high school and those whose highest level of education is a high school diploma. Here, we will recode *degree* such that there are only three categories: (1) highest level of education is high school or less, (2) respondents have some education beyond high school but did not complete a bachelor's degree, and (3) highest level of education is a bachelor's degree or higher.

Before recoding the variable, remember from chapter 1 that it is a good idea to create a new version of the original variable and recode the new version so that we leave the original variable unchanged. In order to create a new variable, we use the "generate" command. Type the following command into Stata:

```
generate degreerecode=degree
```

With this command Stata creates a new variable called *degreerecode* that is exactly the same as the original *degree* variable. Every respondent in the data set now has exactly the same value for the *degreerecode* variable as they have for the original *degree* variable. (You can confirm this by comparing the variables in the Data Browser or examining frequency distributions for each.)

Now we are ready to recode. Type the following command into Stata:

```
recode degreerecode 1=0 2=1 3 4=2
```

The "recode" command tells Stata that it should recode the *degreerecode* variable such that category 1 is reassigned a value of 0, category 2 is reassigned to 1, and 3 and 4 are reassigned to 2. Notice that listing 3 and 4 with a space between them followed by "=2" tells Stata to recode both 3 and 4 into a value of 2; it is not necessary to put a comma or "&" between 3 and 4. Notice also that we have not specified where to place the original category 0, which indicated that the respondent had not finished high school. This is because we want the original category 0 to remain in category 0, respondents whose highest level of education was high school completion or less. The recoded version of the variable should have three categories. Once you have recoded the variable, generate a frequency distribution of the new version. Be sure to ask Stata to include cases with missing data. The output should look like Figure 2.13.

```
. tabulate degreerecode, missing

degreerecod
         e │      Freq.     Percent        Cum.
───────────┼───────────────────────────────────
         0 │      1,843       45.71       45.71
         1 │        370        9.18       54.89
         2 │      1,796       44.54       99.43
         . │         23        0.57      100.00
───────────┼───────────────────────────────────
     Total │      4,032      100.00
```

Figure 2.13

You can check to make sure that there were no mistakes by confirming that the summed frequencies for cells in the previous frequency distribution match the frequencies of the combined cells shown in the distribution for the recoded *degreerecode* variable.

For example, the new category 0 includes 1,843 respondents. Looking at the previous frequency distribution, we see that 246 respondents reported that they had not finished high school and 1,597 reported that they had finished high school but had no further schooling. Adding those frequencies (246 + 1,597) yields a sum of 1,843, which matches the frequency for the recoded category 0.

Assigning Value Labels to Recoded Variables

Our new variable, *degreerecode*, is now recoded according to our three categories, but there are no value labels attached to those categories. The value labels are useful because they allow us to see descriptions of the categories without referring to the codebook. We can create the labels for *degreerecode* by following these two steps:

Step 1: Create a set of value labels and name it. Here we will name the set "degree." The only rule about naming a set of value labels is that it must be fewer than eighty characters. A set of value labels is not attached to a variable until we tell Stata to attach it (which we will do in Step 2). Type the following syntax:

```
label define degree 0 "High school or less" 1 "Some college" 2 "Bachelor's degree or higher"
```

The "label define" command tells Stata that we are creating a new set of value labels. We are telling Stata what we want each of the three categories to be called. For instance, category 0 will be called "High school or less."

Step 2: Attach the set of value labels to our recoded degree variable, *degreerecode*:

```
label values degreerecode degree
```

The "label values" command tells Stata that it should attach the values from the "degree" value set to the *degreerecode* variable.

This two-step procedure can be confusing when you are first learning it. It may be helpful to think about physical, hard-copy labels for, say, holiday gifts. The "label define" command is analogous to writing the names of the recipients on sticky labels, while the "label values" command is analogous to actually sticking those labels on the gifts themselves.

```
. tabulate degreerecode

       degreerecode |      Freq.     Percent        Cum.
--------------------+-----------------------------------
High school or less |      1,843       45.97       45.97
       Some college |        370        9.23       55.20
Bachelor's degree or higher |  1,796   44.80      100.00
--------------------+-----------------------------------
              Total |      4,009      100.00
```

Figure 2.14

A frequency distribution for *degreerecode*, shown in Figure 2.14, displays our value labels instead of the numbers attached to each category of the variable.

Generating Graphical Representations of Data

The general form of the "graph" command is to type "graph," followed by the type of graph, and then specify the variable name. In general, the "graph" commands in Stata are less intuitive than most other commands. Here, we will cover the syntax commands for generating pie charts, bar graphs, and histograms. Stata's drop-down "Graphics" menu for making figures allows you to customize figures in a number of ways, such as colors and line width.

To generate a pie chart of respondents' educational attainment, type the following command into Stata:

```
graph pie, over(degreerecode) plabel(_all percent)
```

Because the default is to provide no labels for the pie slices, we type "plabel(_all percent)" to tell Stata to show the percentage for each pie slice. The graph is shown in Figure 2.15.

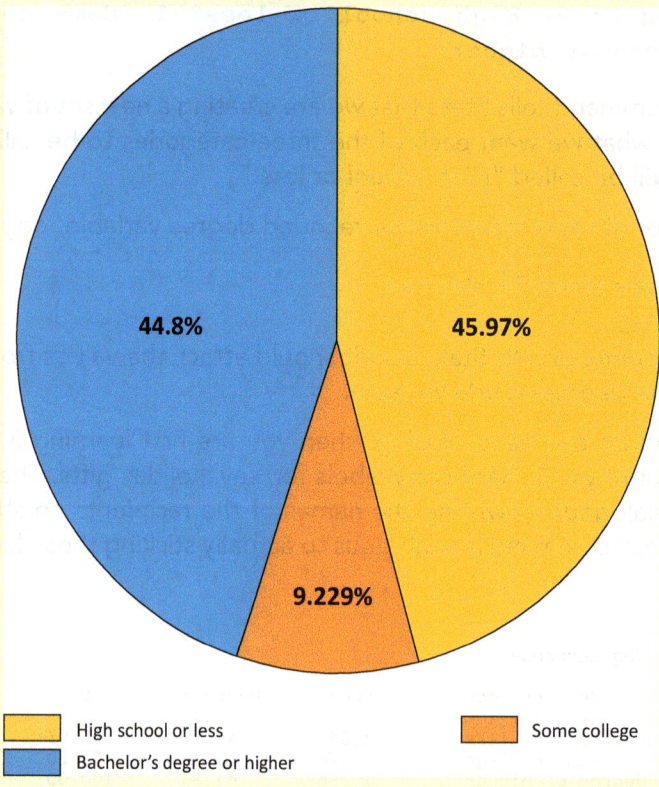

Figure 2.15

Notice that the percentages for each category shown in the pie chart match the percentages shown for the most recent frequency distribution for the recoded version of the variable, shown above, but not for the previous frequency distribution, in which missing data are included in the table. This is because percentages shown in the pie chart are the valid percentages, in which missing data are excluded from the denominator.

Type the following command to generate a bar chart for *degreerecode*:

```
graph bar, over(degreerecode)
```

The chart is shown in Figure 2.16.

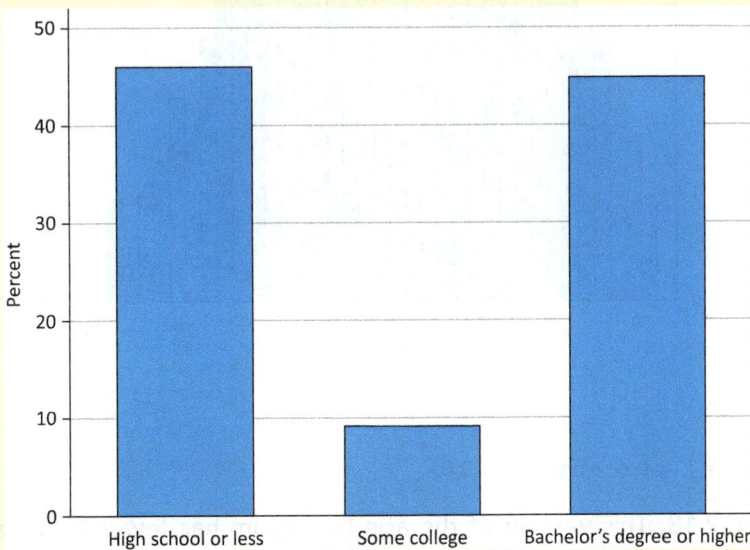

Figure 2.16

Which display of the distribution of cases across the categories of the degree variable do you prefer, and why?

Finally, we will learn how to generate a histogram in Stata. Since histograms are used with interval-ratio variables, we should not use the educational attainment variable, which is measured at the ordinal level. We can use the *age* variable, which reports respondents' ages, ranging from eighteen to eighty-nine years or older. The following command will generate a histogram:

```
histogram age, percent
```

The "histogram" command tells Stata to generate a histogram for the *age* variable. The "percent" option requests the histogram to report percentages for each value (Figure 2.17).

Our "histogram" command specified only one option—that Stata report the percentage of the sample who fell into each age range. We did not tell Stata how many intervals it should divide age into, nor the range of those intervals. Without our specifying those options, Stata makes its own decisions about the number of and range of intervals. In this case, Stata divided age into thirty-five "bins," each covering a range of about two years. To smooth the distribution, we could specify a smaller number of bins that cover a larger range of ages. Let's ask Stata to generate a histogram for age that shows ten bins. This time, our command will specify the number of bins:

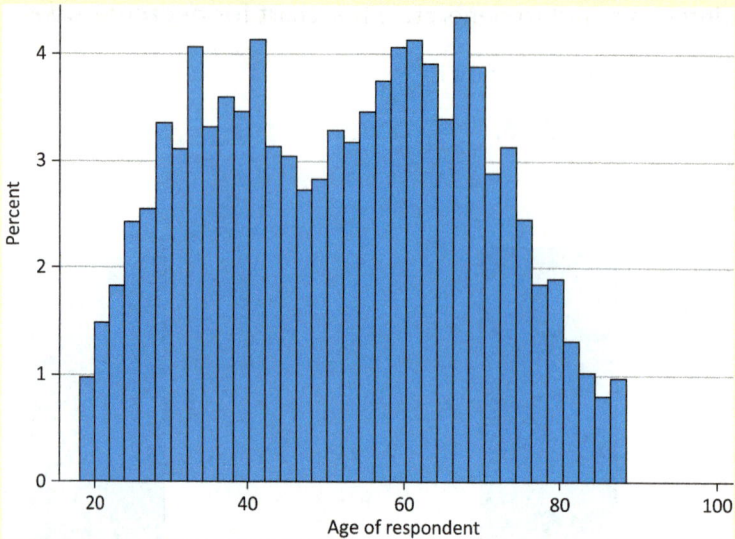

Figure 2.17

```
histogram age, percent bin(10)
```

Look at Figure 2.18. This version of the age histogram has fewer bins, each covering a larger range, about seven years. Notice that this histogram has fewer jagged edges than does the original histogram with thirty-five bins. Its appearance is "smoother," which can help us to see the overall shape of the distribution more easily than when a histogram is marked by many jagged edges.

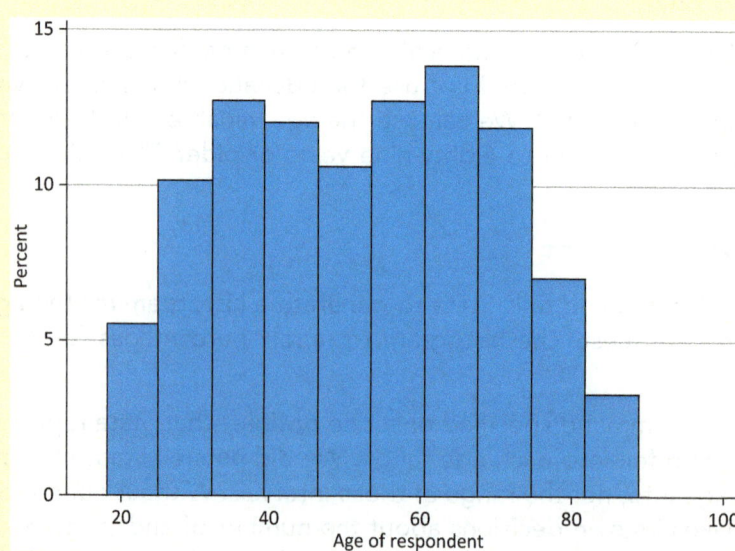

Figure 2.18

As a reminder, to save your work, copy the commands from the Command Window into a do-file and save that. If you wish, you can also save the data file with your new variables.

Review of Stata Commands

- Generate a frequency distribution excluding missing data

    `tabulate variable name`

- Generate a frequency distribution including missing data

    `tabulate variable name, missing`

- Create a new variable from an existing variable

    `generate new variable name=existing variable`

- Recode a variable

    `recode variable name [numbers assigned to existing categories to be collapsed into same new category]=number assigned to new category`

    (*Note*: Brackets shown here are not part of the Stata syntax. Do not include them in your syntax.)

- Creating and attaching labels to the categories of a variable

    Step 1: `label define name of value labels set [number assigned to first category] ["verbal label assigned to first category"] [number assigned to second category] ["verbal label assigned to second category"]` (for as many categories as the variable has)

    (*Note*: Brackets shown here are not part of the Stata syntax. Do not include them in your syntax.)

    Step 2: `label values variable name [name of value labels set]`

    (*Note*: Brackets shown here are not part of the Stata syntax. Do not include them in your syntax.)

- Generating pie charts showing percentages for each pie slice

    `graph pie, over(variable name) plabel (_allpercent)`

- Generating bar graphs

    `graph bar, over(variable name)`

    (Percentages are the default presentation for bar graphs, so there is no need to specify as we do for pie charts.)

- Generating histograms showing percentages

    `histogram variable name, percent`

Using SPSS

Chapter 1 gave you a general introduction to SPSS and how it works. Now we will learn how to use SPSS to generate the statistics and graphs that have been covered in this chapter. Remember, you may be using a different version of SPSS, so the menus for some procedures may look slightly different from those we show here.

For our examples, we will use data from the 2021 General Social Survey (GSS). In these examples, we will use two variables:

- *Degree* is an ordinal variable measuring the highest educational degree earned by the respondent. The five categories are 0 = less than high school; 1 = high school; 2 = community college (referred to by the GSS as "junior college"); 3 = bachelor's degree; 4 = graduate degree.
- *Age* is an interval-ratio variable measuring the respondent's current age.

Creating Frequency Distributions for a Variable

Open `GSS2021.sav`. We will use the drop-down menus to tell SPSS to generate a frequency distribution for the variable that measures the highest educational degree earned by the respondent (degree). The sequence of commands we use is:

`Analyze → Descriptive Statistics → Frequencies`

The image in Figure 2.19 displays this sequence of menu choices:

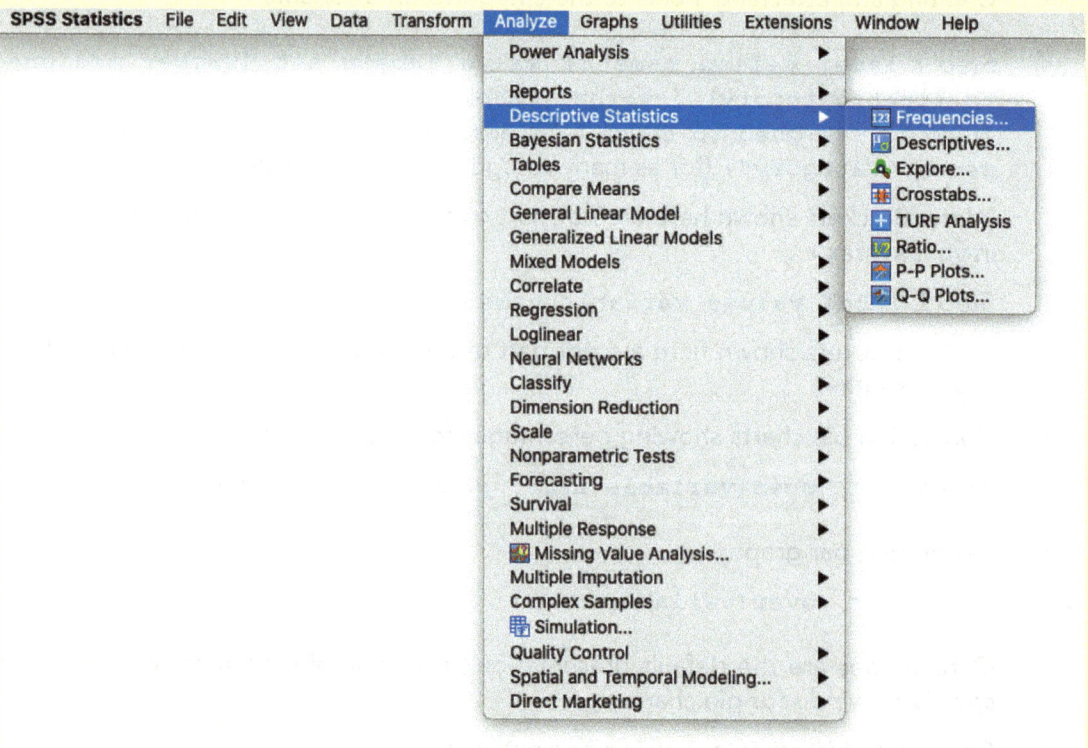

Figure 2.19

This sequence opens the "Frequencies" dialog box (Figure 2.20).

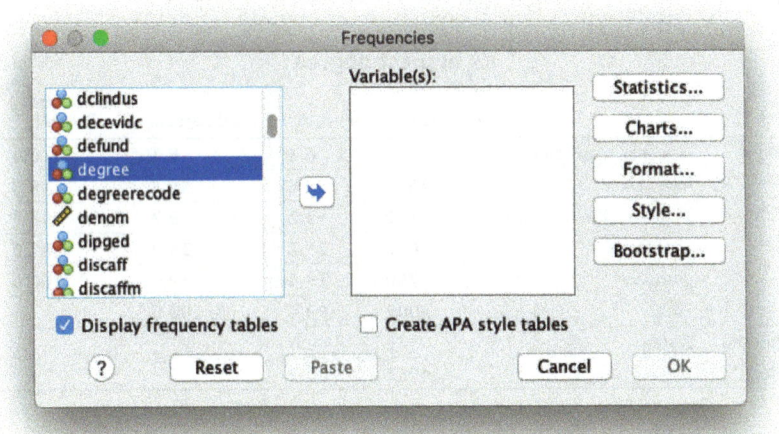

Figure 2.20

All the variables in the data set are arrayed in the box on the left. Since we want SPSS to generate a frequency distribution for the *degree* variable, we will move that variable from the variable list on the left into the "Variable(s)" box by clicking on *degree* and then clicking the arrow in the middle of the box, as displayed in Figure 2.21.

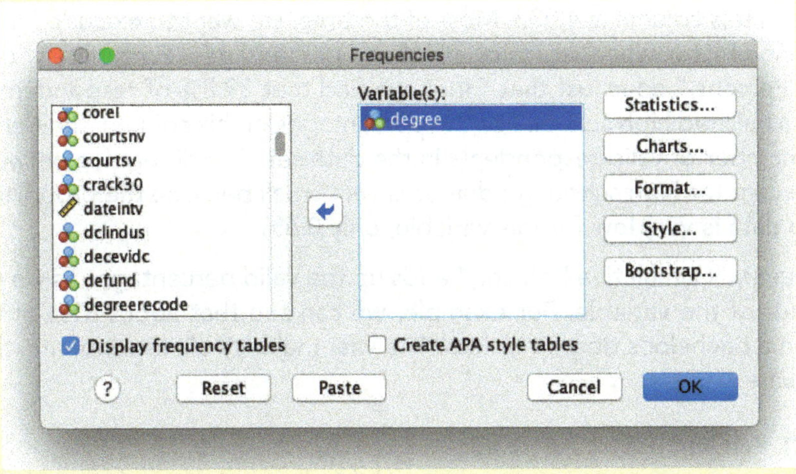

Figure 2.21

After we move at least one variable into the "Variables(s)" box, the "OK" button becomes active, and when we click on it, SPSS produces the desired frequency distribution, which will be found in the "Output" window, as displayed in Figure 2.22.

SPSS produces two tables. The first table (not displayed) is a summary of the number of valid responses and missing responses associated with this variable. The second, larger table, shown in Figure 2.22, also contains this information. As we can see, we have 4,009 respondents with information about their highest degree attained. There are an additional 23 respondents for whom we do not have this information.

r's highest degree

		Frequency	Percent	Valid Percent	Cumulative Percent
Valid	less than high school	246	6.1	6.1	6.1
	high school	1597	39.6	39.8	46.0
	associate/junior college	370	9.2	9.2	55.2
	bachelor's	1036	25.7	25.8	81.0
	graduate	760	18.8	19.0	100.0
	Total	4009	99.4	100.0	
Missing	System	23	.6		
Total		4032	100.0		

Figure 2.22

This table also displays the frequency distribution. The "Frequency" column displays the raw number of respondents associated with each educational category. For example, there are 1,597 respondents whose highest degree was a high school diploma. The next column, "Percent," tells us the percentage of all respondents associated with each educational level. We see that 39.6% of all respondents completed high school. It is important to note that the percentages in this column are calculated on the basis of all respondents, including those for whom we do not have data. In other words, the denominator for the percentages in this column is 4,032. Most of the time, we want to exclude "missing data" from the denominator when we calculate these percentages. Fortunately, the next column, "Valid Percent," does just that. Here, we find that 39.8% of respondents for whom we have data finished high school. For the percentages in this column, the denominator is 4,009—the number of valid respondents in the data set. The difference between percent and valid percent for high school graduates is very small because the proportion of cases with missing data is very low for this variable, only 0.6%.

The final column, "Cumulative Percent," adds up the valid percentages as we move down the categories of the variable. For example, we can see that about 81% of the respondents earned a bachelor's degree or less. The vast majority of respondents, then, do not have a graduate degree.

Recoding Variables

Recall that for various reasons researchers might choose to recode a variable by combining categories for that variable. In statistical software programs, each category of a variable is assigned a number. This is true even for nominal variables, in which categories cannot be rank-ordered. In order to recode a variable, the researcher must know the number assigned to each category of that variable. Most publicly available data sets have a codebook that describes each variable, including the number assigned to each category.

Researchers can get a quick view of the numerical values assigned to each category without paging through the codebook by using a simple SPSS menu sequence:

Utilities → Variables

This sequence opens the "Variables" dialog box. The user highlights the variable of interest, and SPSS will display information about the variable, including the numbers associated with each category. Let's try this procedure for the *degree* variable. The information that SPSS will provide to us is shown in Figure 2.23.

SPSS has provided us with summary information about this variable. We see, for example, that there are five categories of educational attainment, and we also see the number

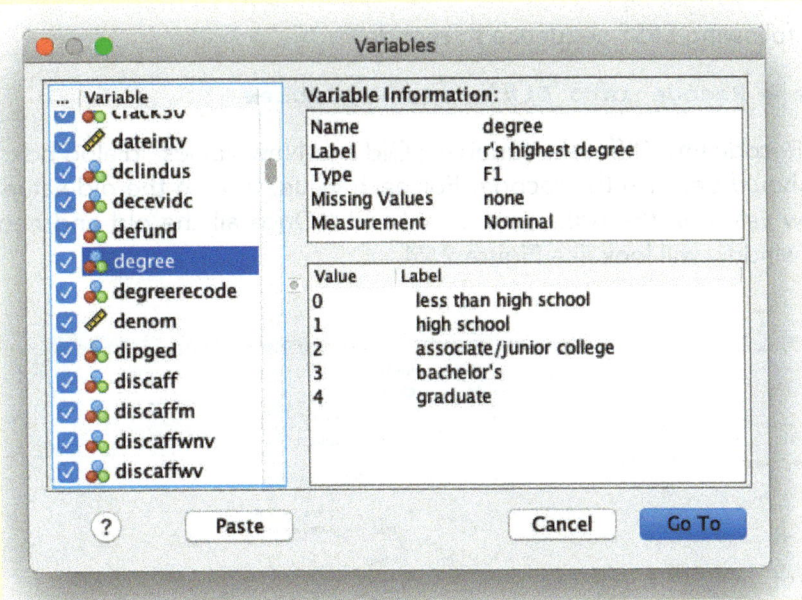

Figure 2.23

that is assigned to each category (0 is less than high school, 1 is high school, etc.). These numbers are important because, when recoding variables, we work with the numbers assigned to categories instead of their verbal descriptions.

With five categories for *degree*, it might make sense for some researchers to consider combining some of them. For example, for some research purposes it is probably not very meaningful to distinguish between respondents who did not finish high school and those whose highest level of education is a high school diploma.

Here, we will recode *degree* such that there are only three categories: (1) respondents whose highest level of education is high school or less, (2) respondents who have some education beyond high school but did not complete a bachelor's degree, and (3) respondents whose highest level of education is a bachelor's degree or higher.

We noted in chapter 1 that it is always a good idea to recode a variable into a different variable so that SPSS retains the original variable. We will tell SPSS to recode the existing variable, *degree*, into a new variable called *degreerecode* using the specifications listed in Table 2.18.

Table 2.18

Values on Original variable, *degree*	Value on New variable, *degreerecode*	Value Labels for new variable, *degreerecode*
0, 1	0	High school diploma or less
2	1	Some college
3, 4	2	Bachelor's degree or higher
All other values	System-Missing	

We use the following SPSS sequence to recode:

`Transform → Recode into Different Variables`

Using the "Recode into Different Variables: Old and New Values" dialog box, we indicate how SPSS should perform the recode. For each value, type in the old value on the left, add the new value on the right, and click "add." Once all the old and new values are added, the window will look like Figure 2.24.

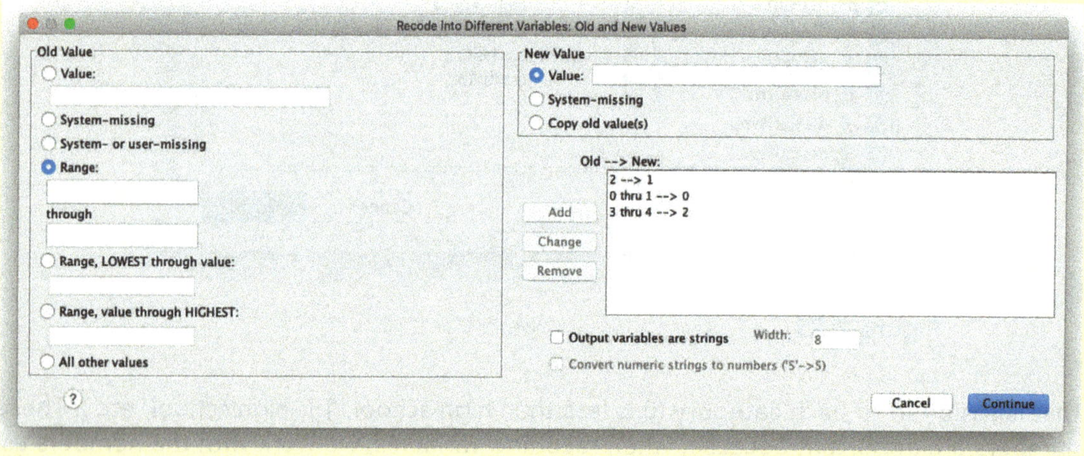

Figure 2.24

We click "Continue," then "OK," and SPSS performs the recode. To be certain, one can go to the "Data Editor" window and make "Variable View" active. The new variable, *degreerecode*, will be appended to the bottom of the variable list. Once you have recoded the variable, you can generate a frequency distribution of the new variable. Here is the output (Figure 2.25).

You can check to make sure that there were no mistakes by confirming that the summed frequencies for cells in the previous frequency distribution match the frequencies of the combined cells shown in the distribution for the recoded *degreerecode* variable. For example, the new category 0 includes 1,843 respondents. Looking at the frequency distribution for the original degree variable, we see that 246 respondents reported that they had not finished high school and 1,597 reported that they had finished high school but

degreerecode

		Frequency	Percent	Valid Percent	Cumulative Percent
Valid	.00	1843	45.7	46.0	46.0
	1.00	370	9.2	9.2	55.2
	2.00	1796	44.5	44.8	100.0
	Total	4009	99.4	100.0	
Missing	System	23	.6		
Total		4032	100.0		

Figure 2.25

had no further schooling. Adding those frequencies (246 + 1,597) yields a sum of 1,843, which matches the frequency for the recoded category 0.

Assigning Value Labels to Recoded Variables

Our new variable, *degreerecode*, is now recoded according to our three categories, but there are no value labels attached to those categories. The value labels are useful because they allow us to see descriptions of the categories without referring to the codebook. As we saw in chapter 1, we can easily add value labels through the "Data Editor" window. Recall that "Variable View" must be active.

1. Highlight the cell where the variable we are working on, *degreerecode*, meets the "Values" column.

2. Click on the highlighted box at the right of the cell. This opens the "Value Labels" dialog box.

3. Enter each value and its corresponding label. For example, put 0 into the "Value" box and "High school or less" into the "Label" box. Click on "Add." Repeat this step for each value and its associated label.

4. Click on "OK." SPSS will now attach labels to each value of the new variable, *degreerecode*.

A frequency distribution for *degreerecode*, shown in Figure 2.26, now displays our value labels instead of the numbers attached to each category of the variable.

degreerecode

		Frequency	Percent	Valid Percent	Cumulative Percent
Valid	High school or less	1843	45.7	46.0	46.0
	Some college	370	9.2	9.2	55.2
	Bachelor's degree or higher	1796	44.5	44.8	100.0
	Total	4009	99.4	100.0	
Missing	System	23	.6		
Total		4032	100.0		

Figure 2.26

Generating Graphical Representations of Data

In this section, we will demonstrate how to generate pie charts, bar charts, and histograms. Regardless of the graph desired, the sequence of menu commands we use is:

`Graphs → Legacy Dialogs → [desired graph]`

To generate a pie chart of respondents' educational attainment, use the following sequence:

`Graphs → Legacy Dialogs → Pie . . .`

SPSS will then offer the user three options for organizing the data. The default option, "Data in Chart Are: Summaries for Groups of Cases," is the method we use to produce a pie chart, so we click on "Define." This opens the "Define Pie: Summaries for Groups of Cases" dialog box. This is where we tell SPSS what variable to use to generate the pie chart. In this case, we will tell SPSS to produce a pie chart that displays the distribution of the variable *degreerecode*. In the dialog box, we move the variable *degreerecode* from the variable list into the "Define Slices By:" box. We then click "OK," and SPSS generates a pie chart, as shown in Figure 2.27.

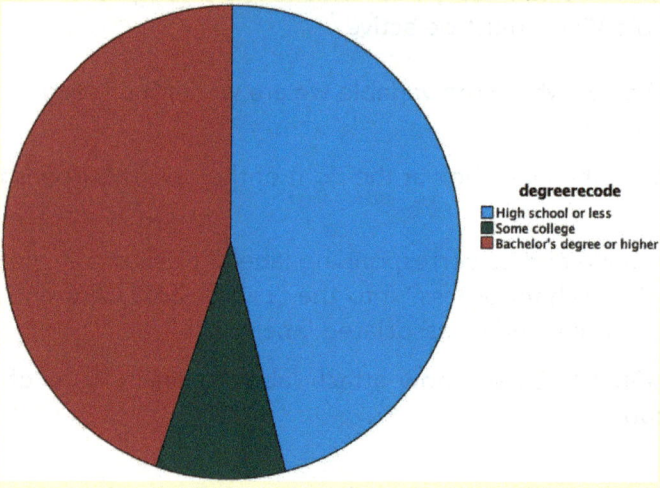

Figure 2.27

The default is to provide no labels on the pie slices. To add labels:

1. Double-click on the pie chart in the "Output" window. This opens the SPSS "Chart Editor."
2. Right-click on the chart and select "Show Data Labels." SPSS will automatically place the raw number of cases for each segment of the pie. It is simple to include percentages as well. In the "Properties" box, under "Data Value Labels," the user can move "Percent" from the "Not Displayed" box into the "Displayed" box.

3. Click on "Apply," and SPSS will add both frequencies and percentages to each segment of the pie, as displayed in Figure 2.28.

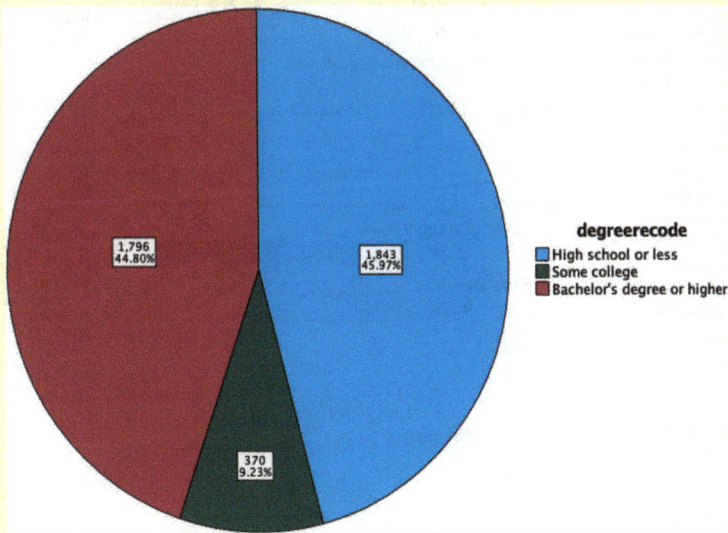

Figure 2.28

Let's use educational attainment (*degreerecode*) to make a bar chart. We use the same sequence, replacing "pie" with "bar":

`Graphs → Legacy Dialogs → Bar . . .`

SPSS gives us a variety of options. For a simple bar chart, select "Simple" and make sure that "Data in Chart Are: Summaries for groups of cases" is selected (these are the default choices). Click on "Define," and SPSS opens the "Define Simple Bar: Summaries for Groups of Cases" dialog box. We move our desired variable (*degreerecode*) into the "Category Axis" box. We can also tell SPSS if the bars should represent the number of cases or the percentage of cases. Let's select percentage of cases. These specifications are shown in the dialog box in Figure 2.29.

We click "OK." The bar chart is shown in Figure 2.30.

Which display of the distribution of cases across the categories of the degree variable do you prefer, and why?

Finally, we will learn how to generate a histogram. Since histograms are used with interval-ratio variables, we should not use the educational attainment variables, which are measured at the ordinal level. We can use the *age* variable, which reports respondents' ages, ranging from eighteen to eighty-nine years or older. We use the following sequence of commands:

`Graphs → Legacy Dialogs → Histogram . . .`

This opens the "Histogram" dialog box. Here, we move the variable *age* from the variable list on the left into the "Variable" box. We click "OK," and SPSS generates the following histogram (Figure 2.31).

Figure 2.29

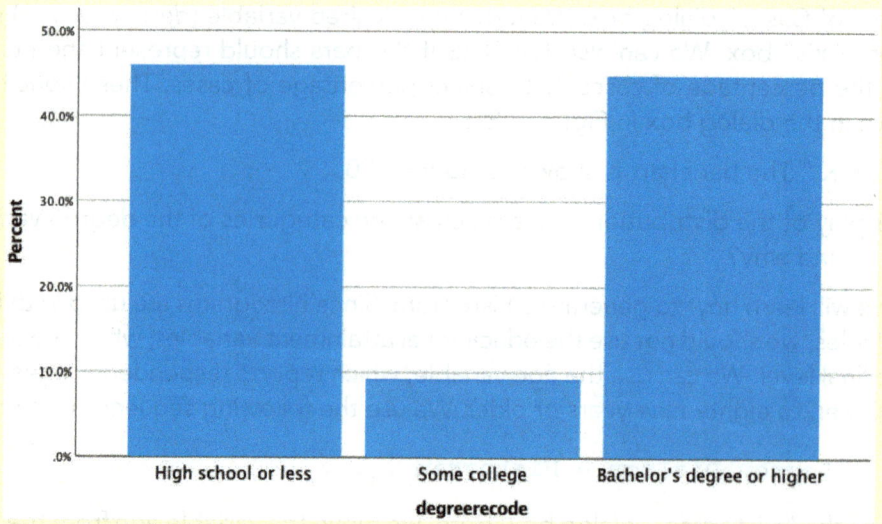

Figure 2.30

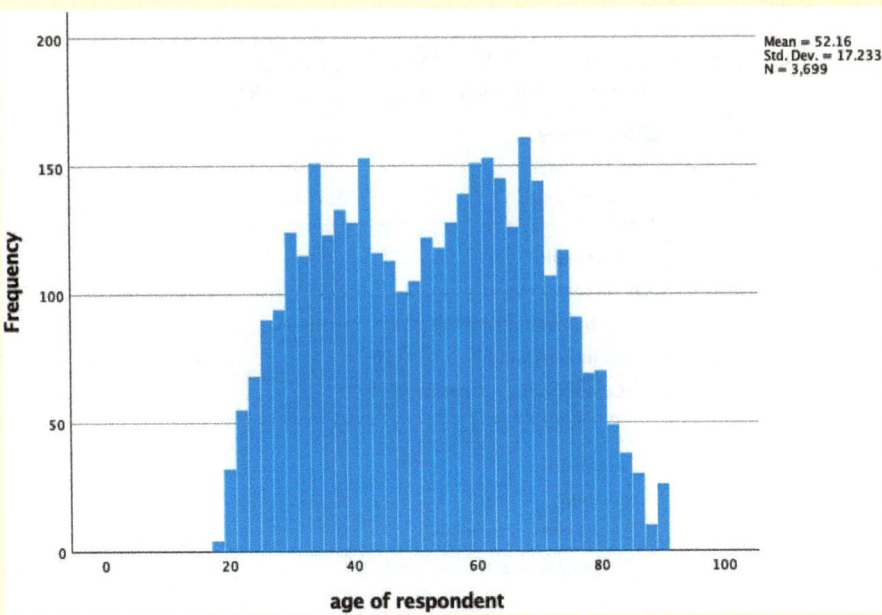

Figure 2.31

By default, SPSS places the frequency on the vertical axis. It also makes its own decisions about the number and range of intervals. In this case, SPSS divided age into thirty-seven "bins," each covering a range of about two years. To smooth the distribution, we could specify a smaller number of bins that cover a larger range of ages.

Let's ask SPSS to generate a histogram for *age* that shows eight bins, each one representing about a ten-year interval. Here are the steps:

1. In the "Output" window, double-click on the histogram. This opens the "Chart Editor."

2. In the "Chart Editor," double-click again on the histogram (making sure it is highlighted). This opens the "Properties" dialog box.

3. Click on the "Binning" tab. This is where we tell SPSS how wide the intervals should be when it constructs the histogram. First, select "custom" rather than "automatic" under the x axis. Next, we will tell SPSS to change the histogram so that each bin represents a ten-year interval (thus producing eight bins). These specifications are displayed in Figure 2.32.

We click on "Apply," and the histogram is redrawn with eight bins, each of which represents a range of ten years, as shown in Figure 2.33. (For example, the second bar represents the number of respondents aged twenty to thirty.)

This version of the *age* histogram has fewer bins, each covering a larger range than the original histogram. Notice that this histogram has fewer jagged edges than does the original histogram with thirty-seven bins. Its appearance is "smoother," which can help us to see the overall shape of the distribution more easily.

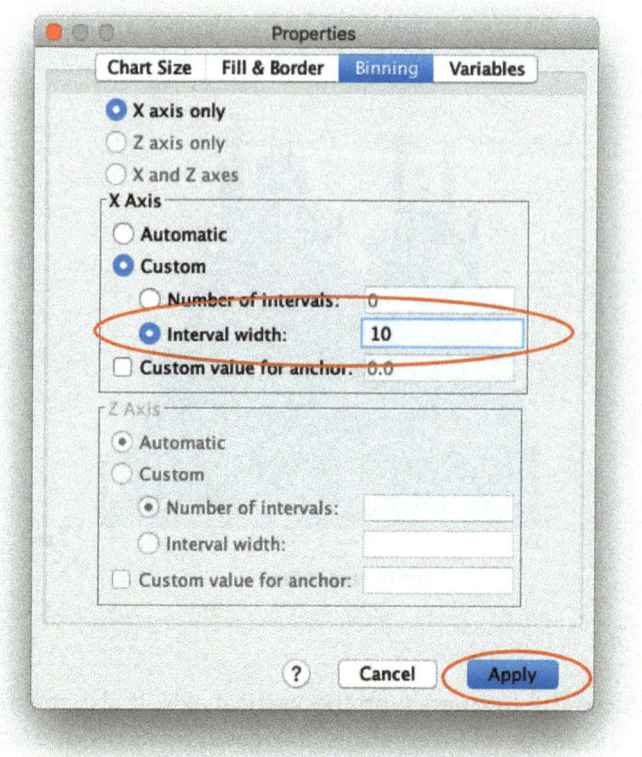

Figure 2.32

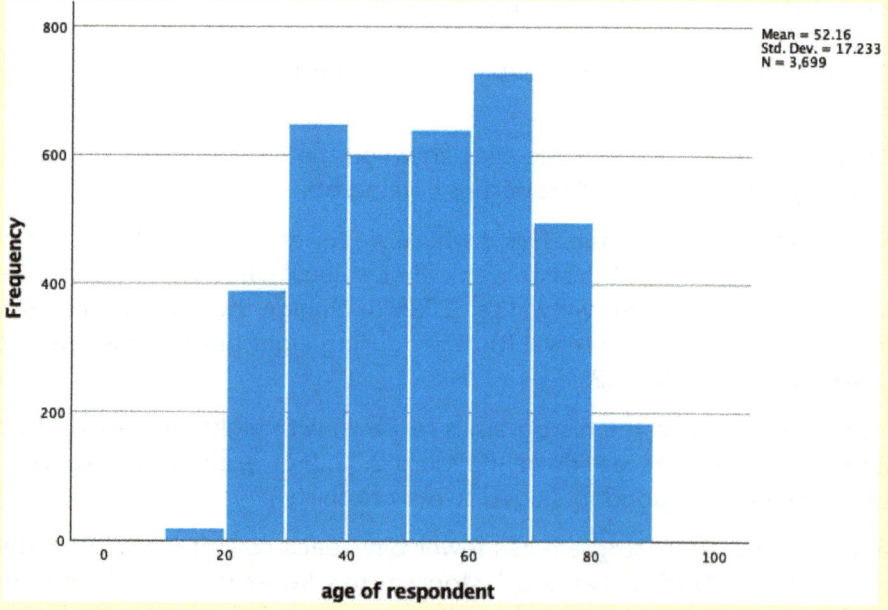

Figure 2.33

By default, SPSS generates a frequency histogram. To change the y axis into percentages instead of raw frequencies, we must use a different sequence of SPSS commands to generate the histogram:

`Graphs → Chart Builder`

If SPSS shows you a box giving you the option of clicking "OK" or "Define Variable Properties," click "OK." This opens the "Chart Builder" dialog box, which offers an alternative method of generating graphs, one that gives the user more control over many of the graph's elements. To construct a histogram using this sequence, we follow these steps:

1. Select histogram from the list of graphical options in the dialog box.
2. Drag the image of the histogram from the "Gallery" into the empty box above it.
3. Drag the desired variable (in this case, *age*) into the box for the x axis.
4. On the right, in the "Statistic" box, select "Histogram Percent."

Once these selections have been made, this is what the "Chart Builder" dialog box will look like (Figure 2.34).

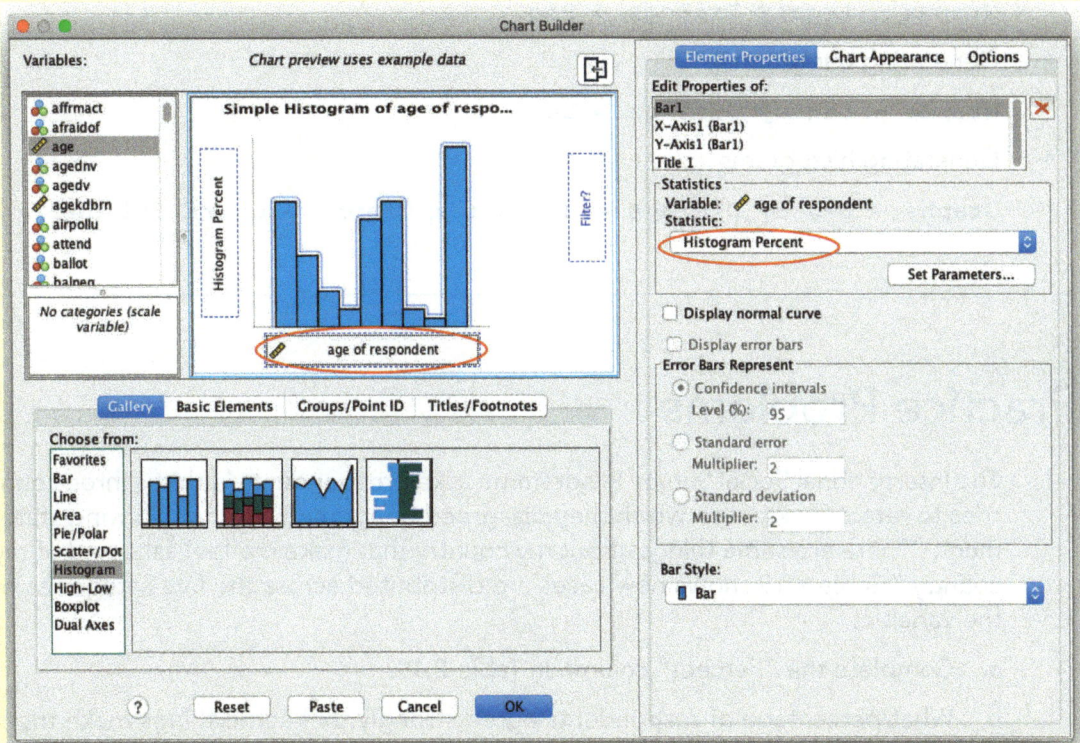

Figure 2.34

Click "OK," and SPSS will generate a histogram with a y axis that displays percentages instead of frequencies.

To save your recoded variables, save the data file before exiting SPSS. To save graphs, you can save the output file or save the graph from the "chart editor" window that pops up when you double-click the graph in the output file.

Review of SPSS Procedures

- Generate a frequency distribution

  `Analyze → Descriptive Statistics → Frequencies`

- Recode a variable

  `Transform → Recode into Different Variables`

- Attach labels to the categories of a variable

  Go to "Variable View" in "Data Editor" window and access "Value Labels" dialog box by clicking on appropriate cell in "Values" column.

- Generating pie charts

  `Graphs → Legacy Dialogs → Pie`

- Generating bar graphs

  `Graphs → Legacy Dialogs → Bar`

- Generating histograms

  `Graphs → Legacy Dialogs → Histogram Graphs → Chart Builder`

Practice Problems

1. The International Social Survey Programme asked respondents in thirty-three countries to rate the extent to which they disagreed or agreed with the following statement: "There are some things about my country that make me feel ashamed of my country." Table 2.19 shows how cases are distributed across the five categories of the variable.

    a. Complete the "Percent" column in Table 2.19.

    b. What percentage of respondents agree strongly that some things make them feel ashamed of their countries? What percentage of respondents disagree strongly?

    c. Complete the "Cumulative Percent" column in Table 2.19.

d. Use the figures in the "Cumulative Percent" column to explain whether the majority of the sample agrees or disagrees that there are things about their countries that make them feel ashamed.

Table 2.19 Agreement That Some Things about Respondents' Countries Make Them Feel Ashamed

	Frequency	Percent	Cumulative Percent
Agree strongly	8,507	19.41%	19.41%
Agree	17,730	40.46%	
Neither agree nor disagree	8,452	19.29%	
Disagree	6,324	14.43%	
Disagree Strongly	2,810	6.41%	
Total	43,823	100%	

Source: 2013 International Social Survey Programme: National Identity III.

2. The 2021 GSS asked respondents how often they felt they were treated with less respect than others. Table 2.20 shows how the responses are distributed across the six categories of the variable.

   a. Complete the "Percent" column in Table 2.20.
   b. What percentage of respondents say that they never feel they are treated with less respect than others? What percentage of respondents feel that they are treated this way every day?
   c. Complete the "Cumulative Percent" column in Table 2.20.
   d. Use the figures in the "Cumulative Percent" column to explain whether the majority of the sample feels that they are treated with less respect than others a few times a month or more.

Table 2.20 Frequency of Feeling Treated with Less Respect Than Others

	Frequency	Percent	Cumulative Percent
Almost every day	136	5.23%	
At least once a week	231	8.89%	
A few times a month	327	12.57%	
A few times a year	801	30.80%	
Less than once a year	552	21.22%	
Never	554	21.30%	
Total	2,601	100%	

Source: 2021 General Social Survey.

3. A researcher is giving a presentation on trends in national pride across the world and wants to show her audience a simplified version of the frequency distribution shown in Table 2.19.

   a. Use Table 2.19 to create a new frequency distribution with all three columns—"Frequency," "Percent," and "Cumulative Percent"—that combines the two "agree" categories into a new category for general agreement and the two "disagree" categories into a second new category for general disagreement.

   b. The researcher concludes that "the majority of respondents agree that something about their countries makes them feel ashamed," and an audience member asks how many respondents declined to answer this question. After checking her notes, the researcher tells the audience member that 1,474 respondents skipped this item. Do these 1,474 cases with missing data challenge the researcher's conclusion? Explain your answer.

4. We want to see a simplified version of Table 2.20 that gives us a general sense of how often people feel that they are treated with respect compared to others.

   a. Create a simplified version of Table 2.20 by collapsing the categories of the variable into the following three new categories: (1) "Frequently" combines at least once a week and almost every day, (2) "Sometimes" combines a few times a month and a few times a year, and (3) "Hardly Ever" combines less than once a year and never.

   b. There are 1,431 cases with missing data on this variable. If we include missing data as a category in the frequency distribution, what is the percentage of people who say they "Hardly Ever" feel disrespected?

5. Table 2.21 shows the frequency distribution for students' scores on the first exam in a statistics course at a large university.

   a. A student who scored an 85 on the exam felt that he was doing much worse in the class than most of his peers and wondered whether he should drop the course. Use the frequency distribution in Table 2.21 to determine the percentile for the student's exam score of 85. Based on the percentile, would you counsel the student to drop the course?

   b. The professor of the statistics course decides to share information about the distribution of grades for the first exam with students in the class. He prepares a histogram for the distribution of raw scores, as shown in Figure 2.35. He also prepares a bar chart, shown in Figure 2.36, using a recoded version of the scores, where ranges of scores are collapsed into five categories: A (90–100), B (80–89), C (70–79), D (60–59), and F (below 60). What level of measurement is the variable shown in the histogram? In the bar graph?

   c. Which visual representation do you think would be more useful for students to see and why?

Table 2.21 Distribution of Exam Scores in a Statistics Course

Score	Frequency	Percent	Cumulative Percent
36	1	0.69	0.69
40	1	0.69	1.39
43	1	0.69	2.08
50	1	0.69	2.78
52	1	0.69	3.47
59	1	0.69	4.17
60	1	0.69	4.86
63	1	0.69	5.56
64	1	0.69	6.25
65	2	1.39	7.64
66	1	0.69	8.33
67	1	0.69	9.03
68	1	0.69	9.72
73	4	2.78	12.50
74	3	2.08	14.58
75	8	5.56	20.14
76	6	4.17	24.31
77	4	2.78	27.08
78	1	0.69	27.78
79	1	0.69	28.47
80	12	8.33	36.81
81	6	4.17	40.97
82	8	5.56	46.53
83	1	0.69	47.22
84	7	4.86	52.08
85	18	12.50	64.58
86	10	6.94	71.53
87	12	8.33	79.86
88	15	10.42	90.28
89	5	3.47	93.75
90	4	2.78	96.53
92	2	1.39	97.92
93	1	0.69	98.61
95	1	0.69	99.31
99	1	0.69	100.00
Total	144	100	

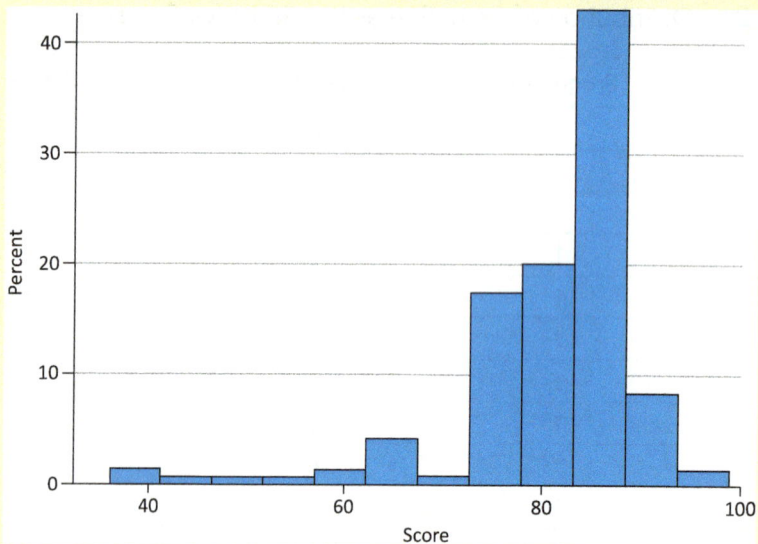

Figure 2.35

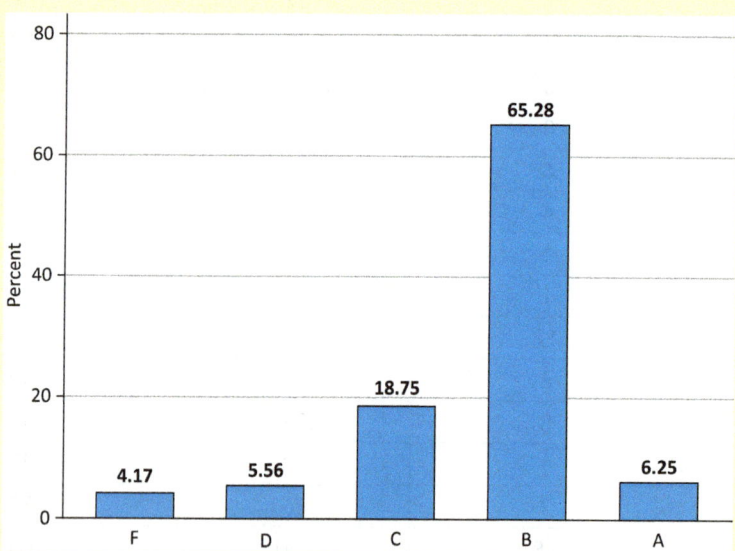

Figure 2.36

6. Table 2.22 shows the frequency distribution for prices that people paid for the most recent pair of shoes that they purchased from a retail store.

   a. A person who paid $45 for her most recent pair of shoes thinks that she is a more frugal shopper than most other people. Use the frequency distribution in Table 2.22 to determine the percentile for the $45 shoe purchase. Based on the percentile, would you agree that this person is more frugal than most when it comes to buying shoes?

   b. A histogram (Figure 2.37) and bar chart (Figure 2.38) show the distribution of prices for people's most recent shoe purchase. The histogram shows the

Table 2.22 Prices Paid for Most Recent Pair of Shoes Purchased

Price	Frequency	Percent	Cumulative Percent
16	2	2.4	2.4
17	1	1.2	3.6
21	1	1.2	4.8
22	4	4.8	9.6
25	2	2.4	12.0
27	1	1.2	13.3
30	5	6.0	19.3
31	1	1.2	20.5
43	1	1.2	21.7
45	2	2.4	24.1
50	5	6.0	30.1
55	5	6.0	36.1
60	6	7.2	43.4
65	2	2.4	45.8
80	1	1.2	47.0
85	5	6.0	53.0
90	5	6.0	59.0
95	1	1.2	60.2
100	6	7.2	67.5
115	5	6.0	73.5
125	4	4.8	78.3
130	3	3.6	81.9
135	2	2.4	84.3
150	2	2.4	86.7
155	1	1.2	88.0
175	3	3.6	91.6
200	2	2.4	94.0
225	1	1.2	95.2
250	1	1.2	96.4
275	2	2.4	98.8
300	1	1.2	100.0
Total	83	100.0	

distribution of prices, while the bar chart shows the distribution of ranges of prices. What level of measurement is the variable shown in the histogram? In the bar graph?

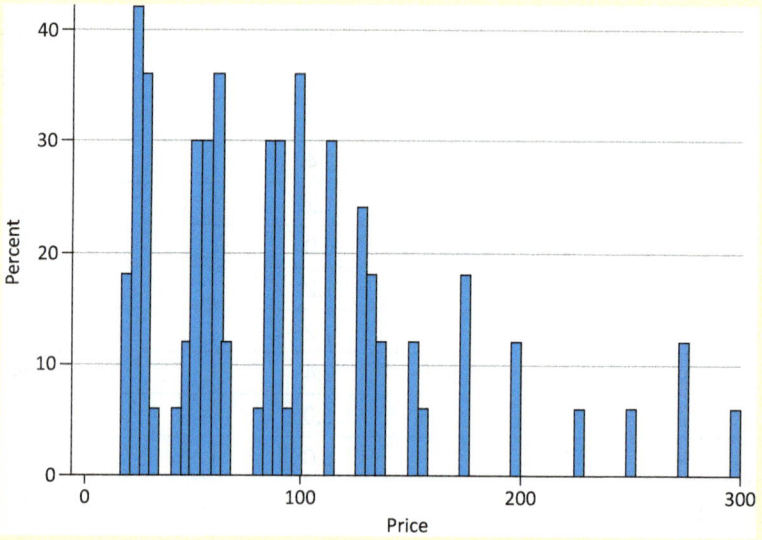

Figure 2.37

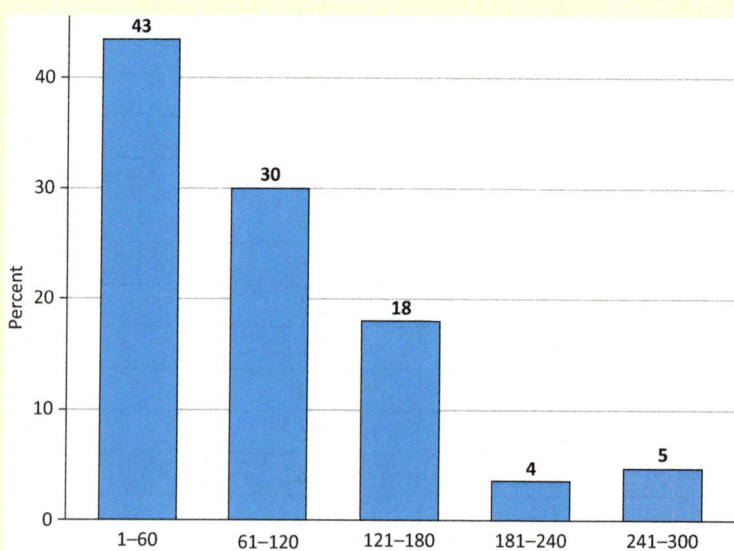

Figure 2.38

7. A local nonprofit organization employs a staff that are majority women. A small group of employees at the nonprofit have expressed concern to management that men are less likely to receive promotions. Management has hired an outside consultant to evaluate its promotion practices. The nonprofit provides the consultant with the following promotion and employment data from the last five years:

- Seventy-two women have been promoted in the last five years.
- Fifty men have been promoted in the last five years.
- The company has employed a total of three hundred women in the last five years.
- The nonprofit has employed a total of 125 men in the last five years.

a. Using this information, calculate the nonprofit's promotion rate for men. Calculate the promotion rate for women. Are these rates consistent with the employees' claim of gender inequity in promotion practices?

b. The consultant decides that she is dissatisfied with the promotion rates from part a because most promotions in this industry require at least one year of employment. However, many of the men and women employed at the nonprofit during the five-year period under examination left the nonprofit before they'd been there for a year. She finds that over five years, 150 women and 94 men stayed at the nonprofit longer than one year. Use these new numbers to calculate revised rates of promotion for men and women in the nonprofit, assuming that all employees who received promotions were at the nonprofit for more than one year. Why are these rates different from those found for part a? What can the consultant conclude about gender inequity in promotion practices at the nonprofit from these modified rates?

c. Using the information provided in parts a and b, calculate the ratio of women to men who leave the nonprofit in their first year of employment. State what this ratio means in words.

d. Over the last five years, what percentage of women have left the nonprofit within the first year? What percentage of men?

e. Given what the consultant has learned about rates of promotion, the ratio of women to men who leave the nonprofit early, and percentages of women and men who do so, how do you think the consultant should advise the nonprofit to improve gender equity?

8. A podcast that reviews movies and television had 100,000 unique listeners during the last year. One of the hosts wants to do only movie reviews in the next season, while the other cohost wants to do only TV reviews. They decide to review movies or TV based on their listeners' preferences. They find the following information about their listeners:

- 65,000 of the listeners preferred to hear movie reviews over TV reviews.
- 35,000 of the listeners preferred to hear TV reviews over movie reviews.
- 10,000 of the listeners have been listening to the podcast for only one month, with 4,300 of them preferring movie reviews and 5,700 of them preferring TV reviews.

a. Using this information, calculate the percentage of all listeners who prefer movie reviews. Calculate the percentage of all listeners who prefer TV reviews.

b. Of those listeners who have been listening to the show for one month, what percentage prefer movie reviews? What percentage prefer TV reviews?

c. Calculate the ratio of those who prefer movie reviews to those who prefer TV reviews among all listeners. State what this ratio means in words.

d. Calculate the ratio of those who prefer movie reviews to those who prefer TV reviews among new listeners. State what this ratio means in words.

9. Table 2.23 shows fifteen US cities ranked by the numeric increase in total population between 2016 and 2017. For example, San Antonio's population grew by 24,208 people between 2016 and 2017, the largest numeric increase in the table.

Table 2.23   Population Growth for Fifteen US Cities, 2016 and 2017

Rank	City	2016 Population	2017 Population	Numeric Increase	Percent Change
1	San Antonio, TX	1,487,738	1,511,946	24,208	
2	Phoenix, AZ	1,602,042	1,626,078	24,036	
3	Dallas, TX	1,322,140	1,341,075	18,935	
4	Fort Worth, TX	855,504	874,168	18,664	
5	Los Angeles, CA	3,981,116	3,999,759	18,643	
6	Seattle, WA	707,255	724,745	17,490	
7	Charlotte, NC	843,484	859,035	15,551	
8	Columbus, OH	863,741	879,170	15,429	
9	Frisco, TX	163,816	177,286	13,470	
10	Atlanta, GA	472,967	486,290	13,323	
11	San Diego, CA	1,406,682	1,419,516	12,834	
12	Austin, TX	938,200	950,715	12,515	
13	Jacksonville, FL	880,893	892,062	11,169	
14	Irvine, CA	266,385	277,453	11,068	
15	Henderson, NV	292,005	302,539	10,534	

Source: US Census Bureau Press Release CB18-78, accessed at https://www.census.gov/newsroom/press-releases/2018/estimates-cities.html.

a. Calculate the percentage change in population from 2016 to 2017 for each city in the table and enter those figures in the last column, "Percent Change."

b. Rank the cities by percentage change in descending order. Which city is at the top of the new ranking? Which is at the bottom?

c. What is the advantage of presenting rankings by percentage change instead of numeric increase?

10. According to the US Census Bureau's Annual Retail Trade Survey, annual e-commerce sales in the United States increased from $491,788,000 in 2018 to $820,843,000 in 2021.

    a. What is the numeric change in e-commerce sales in the United States from 2018 to 2021?

    b. What is the percent change in e-commerce sales in the United States from 2018 to 2021?

    c. Which figure —the numeric change or percent change—better communicates the magnitude of the change in e-commerce sales and why?

11. Figure 2.39 shows the percentage of older men and women living below the poverty line in the United States from 1966 to 2016.

    a. What kind of figure is this?

    b. How many variables are represented in the figure, and what are they?

    c. Are older men or women more likely to live below the poverty line? Offer one possible reason for this.

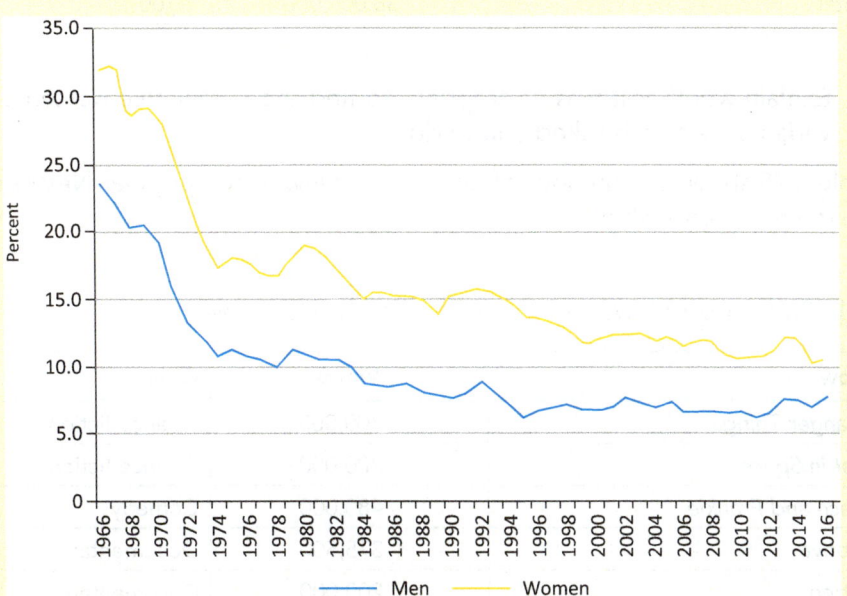

Figure 2.39 Percentage of Men and Women Sixty-Five and Older Living below the Poverty Line, 1966–2016

12. In 2017, Kendrick Lamar released *DAMN.*, a Pulitzer Prize–winning and commercially popular album. Apple Music, which has 36 million paying subscribers, reported that 2 million users downloaded *DAMN.*

    a. What percentage of Apple Music's paying subscribers downloaded *DAMN.*?

b. Kendrick Lamar wanted listeners to hear *DAMN.*'s songs in forward and backward order. To encourage this, he released a Collector's Edition of *DAMN.* with the songs in the reverse order of the original album. Table 2.24 shows the frequency distribution of paying Apple Music subscribers, by downloads of *DAMN.* and *DAMN. Collector's Edition*. Fill in each cell of the *DAMN.* and *DAMN. Collector's Edition* row.

Table 2.24 Distribution of Apple Music Subscribers' Downloads of *DAMN.* and *DAMN. Collector's Edition*

Paid Subscribers	Frequency	Percent	Cumulative Percent
Did not downloaded *DAMN.* or *DAMN. Collector's Edition*	34,000,000	94.4	94.4
Downloaded only *DAMN.* or only *DAMN. Collector's Edition*	1,750,000	4.9	99.3
Downloaded both *DAMN.* and *DAMN. Collector's Edition*			
Total	36,000,000	100	

c. Explain whether it was appropriate to find the cumulative percentage for this variable, as part b asked you to do.

13. Table 2.25 shows the number of times users viewed ten popular Netflix shows and the genre for each show.

Table 2.25 User Views and Genres for Ten Netflix Shows

Show	Views	Genre
Stranger Things	700,000	Science fiction
Lost in Space	300,000	Science fiction
Grace and Frankie	480,000	Comedy
Captive	375,500	Documentary
Rotten	200,000	Documentary
Girlboss	275,000	Comedy
Disjointed	450,000	Comedy
Mindhunter	400,000	Drama
Seven Seconds	250,000	Drama
Orange Is the New Black	750,000	Drama
Total	4,180,500	

a. Calculate the percentage of views for each genre represented in Table 2.25. Among these ten shows, which genre had the highest percentage of views? Which genre had the lowest percentage?

b. Sketch a bar graph and pie chart showing the percentage distribution of views across the four genres shown in the table. Do you think that the bar graph or pie chart more effectively shows how views are distributed across the genres represented in the table?

14. Figure 2.40 shows the percentage of New Jersey public school students who attend schools with various levels of racial segregation, by race. The categories of the segregation variable are: "white isolated" (more than 90% of the students are white), "apartheid/intensely segregated" (less than 11% of the students are white), and "integrated" (11% to 90% of the students are white).

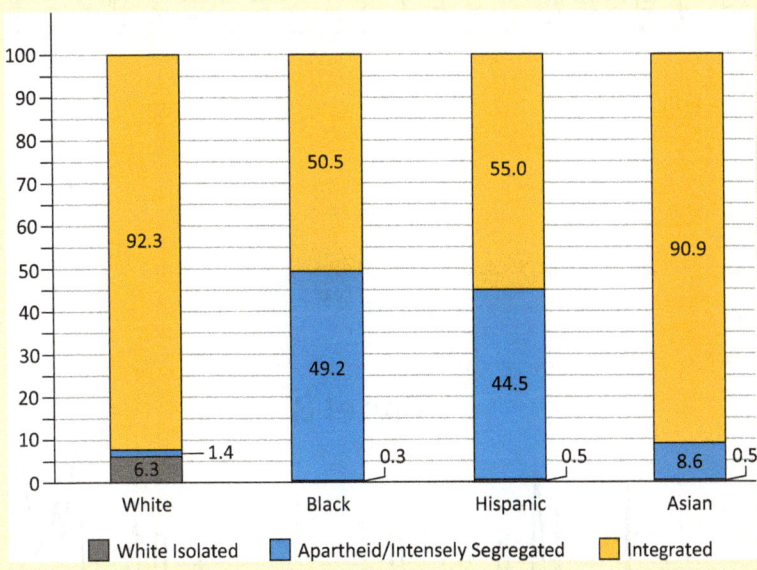

Figure 2.40 Distribution of New Jersey Students across Schools, by Level of School Segregation and Student Race

a. Which group of New Jersey students is most likely to attend schools where less than 11% of the students are white?

b. Which group of students is most likely to attend "integrated" schools?

c. Which group of students is most likely to attend schools where more than 90% of the students are white?

d. For each of the four groups of students, what percentage of students attend (1) either "apartheid or "intensely segregated" or (2) "white isolated" schools?

15. Figure 2.41 shows two charts of annual temperature changes in the forty-eight contiguous states of the United States, from 1901 to 2021 compared to the average temperature between 1901 and 2000.

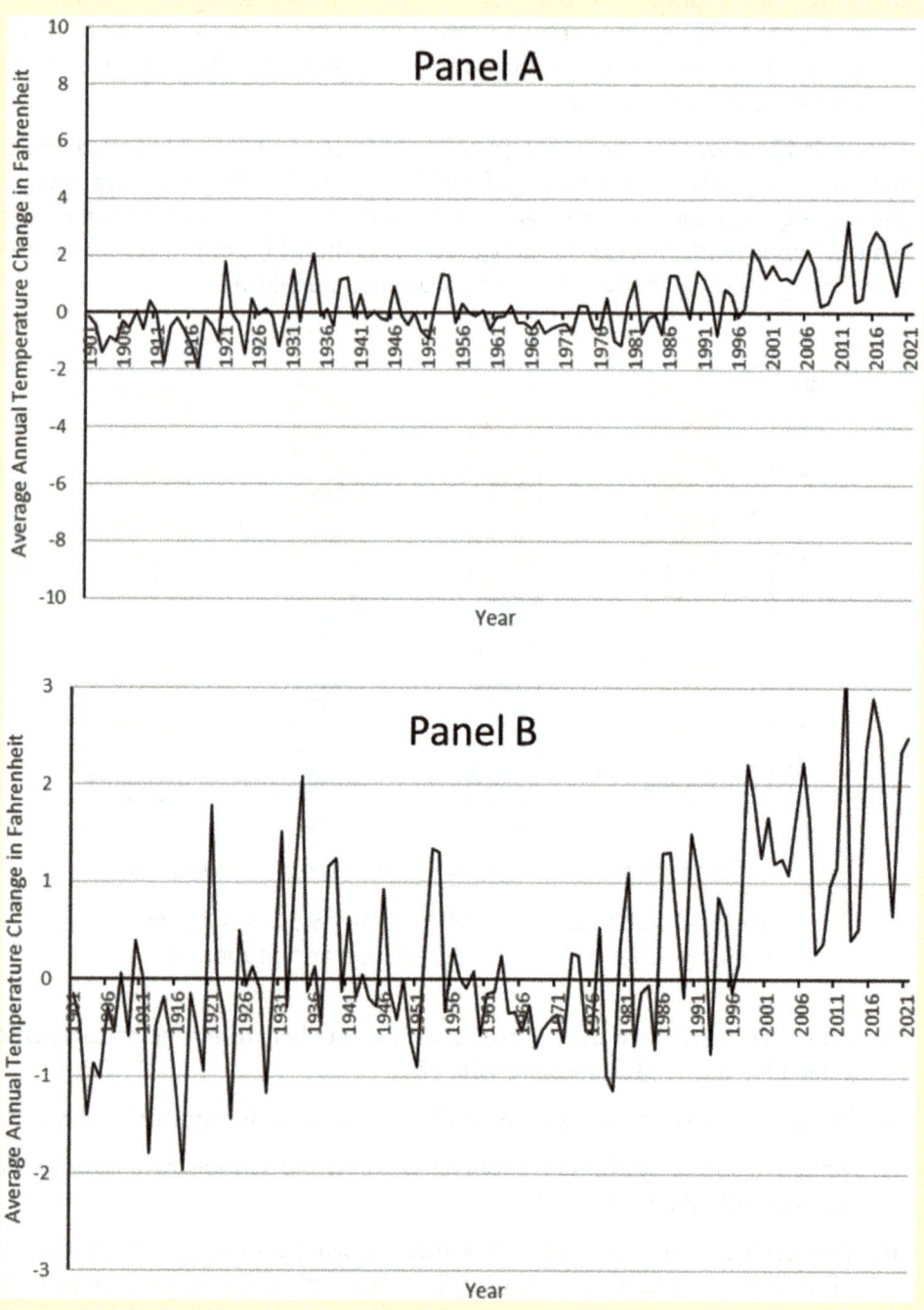

Figure 2.41   Annual Average Temperature Changes in the Contiguous Forty-Eight States, 1901–2021

a. The charts use the same data. Why do they look different?

b. Does the chart in Panel A or Panel B indicate more stability in surface temperatures compared to the 1901–2000 average? Explain why this is the case.

Stata Problems

Here, you will use data from the World Values Survey (WVS) to examine global attitudes about science and technology. You will focus on a variable called *Q163*, which measures the extent to which respondents think that science and technology are making the world better or worse.

1. Open `WVSWave7.dta`. Use the "tabulate" command to generate a frequency distribution for the variable called *Q163*. You will see that the variable is measured on a 10-point scale. Next, generate a second frequency distribution using the "tabulate" command and the "nolabel" option to view the numbers assigned to each category.

2. Use the "generate" command to create a new version of *Q163*. Call this new variable *scitech*.

3. Use the "recode" command to combine categories 1 through 10 for *scitech* into just three categories: "worse off" (1–4), "neutral" (5), and "better off" (6–10).

4. Use the "label define" and "label values" commands to create value labels for these new categories and attach them to the *scitech* variable.

5. Generate a frequency distribution for *scitech*. Explain how you can use the frequency distribution for *Q163* to confirm that the recoding proceeded properly.

6. Generate a bar graph showing the distribution of respondents across the categories of *scitech*. What does it show us about global attitudes regarding the role of science and technology in improving the world?

SPSS Problems

Here, you will use data from the World Values Survey (WVS) to examine global attitudes about science and technology. You will focus on a variable called *Q163*, which measures the extent to which respondents think that science and technology are making the world better or worse.

1. Open `WVSWave7.sav`. Use the "Frequencies" dialog box to generate a frequency distribution for the variable called *Q163*. You will see that the variable is measured on a 10-point scale. Next, use the "Variables" dialog box to view the numbers assigned to each category.

2. Use the "Recode into Different Variables" dialog box to create a new version of *Q163*. Call this new variable *scitech*. Combine categories 1 through 10 into just three categories: 1 (values 1–4), 2 (5), and 3 (values 6–10).

3. Toggle to "Data View" in the "Data Editor" window to assign the following labels to the three categories: "worse off" (1–4), "neutral" (5), and "better off" (6–10).

4. Generate a frequency distribution for *scitech*. Explain how you can use the frequency distribution for *Q163* to confirm that the recoding proceeded properly.

5. Use the "Graphs" menu to generate a bar graph showing the distribution of respondents across the categories of *scitech*. What does it show us about global attitudes regarding the role of science and technology in improving the world?

Notes

[1] Pew Research Center. "US Teens Are More Likely than Adults to Support the Black Lives Matter Movement." Retrieved from https://www.pewresearch.org/fact-tank/2022/06/15/u-s-teens-are-more-likely-than-adults-to-support-the-black-lives-matter-movement/.

[2] Pew Research Center. "Americans' Complex Views on Gender Identity and Transgender Issues." Retrieved from https://www.pewresearch.org/social-trends/2022/06/28/americans-complex-views-on-gender-identity-and-transgender-issues/.

[3] Dave Leip's *Atlas of US Presidential Elections.* Retrieved from https://uselectionatlas.org/RESULTS/national.php.

[4] J. Garfield and D. Ben-Zvi. 2007. "How Students Learn Statistics Revisited: A Current Review of Research on Teaching and Learning Statistics." *International Statistical Review* 75(3): 372–396.

[5] National Center for Health Statistics. "Births to Unmarried Women: Birth Rates for Unmarried Women by Age of Mother, 1980–2019." Retrieved from https://www.childstats.gov/americaschildren/tables/fam2a.asp.

[6] Current Population Survey. 2021. "Table 1. Educational Attainment of the Population 18 Years and Over, by Age, Sex, Race, and Hispanic Origin: 2021." Retrieved from https://www.census.gov/data/tables/2021/demo/educational-attainment/cps-detailed-tables.html.

[7] Boston Public Schools. "2019 Graduation Rate Report." Retrieved from https://www.bostonpublicschools.org/Page/694.

[8] Statista. "Marital Status of the United States Population in 2021, by Sex." Retrieved from https://www.statista.com/statistics/242030/marital-status-of-the-us-population-by-sex/.

Chapter 3
Examining Relationships between Two Variables
Cross-Tabulations

On June 24, 2022, the US Supreme Court issued a decision in a case called *Dobbs v. Jackson Women's Health Organization*. In this decision, the Court overruled two landmark Supreme Court precedents, *Roe v. Wade* (1973), which established a constitutional right to abortion, and *Planned Parenthood v. Casey* (1992), which reaffirmed the *Roe* decision. The Court declared that these long-standing precedents were wrongly decided; and that the US Constitution does not confer a right to abortion, either explicitly or implicitly. The result of this ruling was to cede to the states the full power to regulate abortion services. In the aftermath of this ruling, numerous states took actions ranging from significant restrictions on access to abortion services to an outright ban.

Six weeks before the ruling in Dobbs, the Pew Research Center reported the results of its most recent poll surveying Americans' attitudes on abortion.[1] Although there is plenty of debate surrounding the extent to which the Supreme Court should be sensitive to public opinion, the fact remains that a substantial majority of Americans have long supported abortion rights. In that poll, 61% of respondents reported that abortion should be legal in all or most cases, compared to 37% who stated that abortion should be illegal in all or most cases. That figure of 61% represented an all-time high for Pew surveys, going back to 1995. This aggregate figure of 61% represents a substantial majority, but it conceals some important group differences in support for abortion. Not surprisingly, Pew found that 80% of those who identify as Democratic believed that abortion should be legal, compared to only 38% of Republicans. Race/ethnicity was also an important dividing line in public opinion: 74% of Asian Americans and 68% of African Americans believed that abortion should be legal, compared to 59% of whites, and 60% of Latinos. Interestingly and perhaps surprisingly, the gender gap was somewhat muted: 63% of women and 58% of men believed that abortion should be legal in most or all cases, a gap of only five points. These statistics tell a story. They help us understand who is more likely to support abortion rights and how support varies depending on one's individual characteristics or group identification. But underlying these patterns are deeper explanations. Young Americans (eighteen to twenty-nine years), for example, were 20 percentage points more likely than the oldest cohort (sixty-five plus) to support abortion rights. Perhaps this has something to do with the decided Democratic leanings of younger Americans. Or perhaps this difference is rooted in the environment in which young Americans grew up. After all, the youngest cohort came of age after *Roe v. Wade* and grew up in an era in which abortion rights were legal and widely accepted. Statistics can help us see group patterns and differences in public opinion. Explaining these differences requires a deeper understanding of history and politics. Questions about whether certain social groups are more likely to support one position or another (e.g., were young people more likely to support abortion rights?) are questions about relationships between variables. In chapter 2, we examined frequency distributions as a way of describing single variables at a time. In this chapter, we examine cross-tabulations as a way of examining **bivariate relationships**, or relationships between two variables.

Cross-Tabulations and Relationships between Variables

Cross-tabulations are tables that show the relationship between two variables. Cross-tabulation tables present a frequency distribution for one variable separately for each category of another variable. They are often called "cross-tabs" for short. They are also called contingency tables, or joint frequency tables. The latter term is especially descriptive of what these tables convey. As we saw in chapter 2, we can create a frequency distribution to display the distribution of a single variable. Cross-tabulations are essentially a series of side-by-side frequency tables.

Continuing to think about what factors shaped people's views on abortion, we will examine several cross-tabulations between various factors and position on abortion. We noted above that the youngest cohort of Americans was more likely than the oldest to support the legality of abortion. But what were the differences across all ages? Table 3.1 presents the distribution of opinions based on age.

Table 3.1 Views on Abortion, by Age, 2022

	18 to 29 Years	30 to 49 Years	50 to 64 Years	65 Years Plus
Abortion should be legal	74%	62%	55%	54%
Abortion should be illegal	25%	37%	42%	44%
No answer	1%	1%	3%	2%
Total	100%	100%	100%	100%

Source: Pew Research Center.

To understand group patterns, we must examine and compare percentages. We see, for example, that 74% of respondents aged eighteen to twenty-nine years believed that abortion should be legal, compared to 62% of respondents between the ages of thirty and forty-nine. This is a fairly sizable difference in support of the view that abortion should be legal in all or most cases. On the other hand, there is a very small difference of 1 percentage point between the two oldest cohorts and suggests that their support levels for legal abortion were close to equal. As noted above, if we compare the youngest and oldest cohorts, we see larger differences: a gap of 20 percentage points between the youngest (eighteen to twenty-nine) and the oldest (sixty-five plus).ABgaps of this magnitude support the claim that age (at least young versus old) was an important influence on people's views toward the legality of abortion.

Race and ethnicity were also important shapers of opinion on abortion. Consider Table 3.2, which examines the distribution of public opinion based on race and ethnicity.

Table 3.2 Views on Abortion by Race/Ethnicity, 2022

	White	African American	Latino	Asian
Abortion should be legal	59%	68%	60%	74%
Abortion should be illegal	39%	27%	40%	26%
No answer	2%	5%	0%	0%
Total	100%	100%	100%	100%

Source: Pew Research Center.

When we compare respondents based on their race and ethnicity, we find that strong majorities of each group believe abortion should be legal. Still, there are important group differences on display in Table 3.2. Among whites and Latinos, about 60% support abortion rights. Among African Americans, 68% support legal abortion, a figure that rises to 74% among Asian Americans. So, while we may state that a majority of each group believes that abortion should be legal, there are still important differences that support the claim that there is a connection between one's race and ethnicity and one's opinion on the legality of abortion.

Independent and Dependent Variables

Each of the examples displayed in Tables 3.1 and 3.2 involves two variables. In Table 3.1, we examined the connection between age and position on abortion. Similarly, in Table 3.2, we looked at the connection between race/ethnicity and views of abortion. Underlying both of these examples is an assumed relationship between independent (or explanatory or predictor) and dependent (or outcome) variables. In the first case, we are suggesting that one's age (the independent variable) helps us explain or predict one's view on abortion (the dependent variable). And in the second case, our analysis is based on the premise that racial/ethnic identification influences one's position on abortion.

Let's focus on the example in Table 3.1, where we examine the relationship between age and attitude toward abortion. First, how do we know which is the independent variable and which is the dependent variable? Sometimes, the answer is self-evident. The independent variable is the cause, or the factor that is doing the explaining. The dependent variable is the effect, or that which we are seeking to explain. Which proposition makes more sense?

1. Age → Position on abortion
2. Position on abortion → Age

The first proposition suggests that one's age influences one's view of abortion. The second proposition holds that one's position on abortion affects one's age. As a matter of logic, the second proposition is not possible since age precedes one's attitude. So, in this case, there is only one way to model the expected relationship between age and opinion on abortion. And the evidence presented in Table 3.1, showing fairly large differences in the views of younger versus older voters, supports the proposition that age helps "explain" one's attitude toward abortion.

But consider another set of variables: political party identification and ideological identification. Which comes first? Does one's underlying partisanship influence one's subsequent ideological identification?

1. Party identification → Ideology

Or is it the other way around—with individuals choosing to identify as Republican or Democrat only after they have decided if they are conservative or liberal?

2. Ideology → Party identification

Both of these are plausible models of the relationship between partisanship and ideology. In Table 3.3, we examine the first of these models, using data from the 2020 American National Election Study (ANES).

Table 3.3 Effect of Party Identification on Ideological Identification, Column Percentages

Ideology	Democratic	Independent	Republican
Liberal	69.4%	18.6%	3.8%
Moderate	25.2%	62.2%	17.4%
Conservative	5.4%	19.2%	78.8%
Total	100%	100%	100%

Source: 2020 American National Election Study.
Total percent may not sum to 100 due to rounding.

Since party identification is the independent variable, the percentages in Table 3.3 tell us the percentage of each partisan category that identifies as liberal, moderate, or conservative. For example, the upper-left cell tells us: "69.4% of Democrats identify as liberal." Continuing down the "Democratic" column, we find that 25.2% of Democrats are moderates, and only 5.4% of Democrats call themselves conservative. The percentages in each column add up to 100%—this is because for any category of party identification, there are only three possible choices for ideology (we have excluded those who did not select liberal, moderate, or conservative); and thus, the sum of these column percentages must be 100%.

Using Table 3.3, we can easily compare the ideology of those who identify with the different parties. Whereas only 5.4% of Democrats identify as conservative, for example, 78.8% of Republicans identify as conservative, a substantial difference.

Table 3.3 presupposes that party identification is the independent variable. But what if we conceive of the relationship in the opposite direction? We can assume that one's ideological tendencies influence one's partisan identification—that is, that people select which party to identify with based on their prior ideological position. We have set up Table 3.4 to reflect this understanding of the relationship between the two variables.

Table 3.4 Effect of Ideological Identification on Party Identification, Column Percentages

Party Identification	Liberal	Moderate	Conservative
Democrat	90.4%	46.3%	6.5%
Independent	5.1%	24.0%	4.9%
Republican	4.5%	29.7%	88.6%
Total	100%	100%	100%

Source: 2020 American National Election Study.
Total percent may not sum to 100 due to rounding.

Now, if we look at the upper-left cell, we find that 90.4% of self-described liberals identify as Democrats. Compare that to what we found in the upper-left cell in Table 3.3: 69.4% of Democrats call themselves liberal. In Table 3.3, we are treating party identification as the independent variable. In Table 3.4, we are treating ideology as the independent variable.

These tables should and indeed do tell different stories. In Table 3.3, a majority of Democrats (69.4%) call themselves liberal, and a majority of Republicans (78.8%) identify as conservative. These are large majorities, but we also find that sizable minorities of each partisan category defect from the party majority's ideological view. On the other hand, when we organize the electorate by ideological category instead of by partisan identification, as we do in Table 3.4, we find a stronger association between the variables. In 2020, fully 90.4% of liberals were Democrats; only about 10% identified as Republican or Independent. Similarly, among conservatives, 88.6% identified as Republican; and about 12% of conservatives chose a different partisan identification.

How can we explain this seeming paradox—90% of liberals are Democrats, but only 69% of Democrats are liberal? The answer has to do with the composition of these groups. First, these groups are not equal in size. There are a lot more Democrats than there are liberals. And second, liberals are a fairly homogeneous group, at least when it comes to party identification. Relatively few liberals (10%) regard themselves as anything but Democrat. But there are also many moderates—46%—and even some conservatives who also regard themselves as Democrats.

Tables 3.3 and 3.4 show an important convention in constructing cross-tabulations. In each of these tables, we have placed the independent variable across the top. In other words, each column corresponds to a category of the independent variable; and each row corresponds to a category of the dependent variable. Sticking to this convention will standardize and simplify our interpretation of the percentages in these tables. We call percentages that break down the variable in the columns "**column percentages**."

In some cases, we may want to understand how two variables are related to each other without any implication that one "causes" the other. In those cases, we can put either variable in the columns. The interpretation of the column percentages remains the same.

Cross-tabulations present side-by-side frequency distributions of the dependent variable for every category of the independent variable. Returning to Table 3.4, we can view the table as conveying the frequency distributions of the dependent variable, party identification, for three different groups of people: liberals, moderates, and conservatives (the three categories of the independent variable).

How do we know if the independent variable is affecting the dependent variable, or if there is a relationship between the variables? The answer is straightforward. We simply compare the percentages across columns. When the differences between column percentages are small or nonexistent, we lack evidence to support the claim that the independent variable is influencing the dependent variable.

Look at the example in Table 3.5, using data from the 2020 ANES.

Table 3.5 Effect of Gender on Perception of Russia as a Threat to the United States, Column Percentages

	Women	Men	Total
Russia is little or no threat to the United States	18.4%	18.1%	18.2%
Russia is a moderate threat to the United States	31.6%	29.1%	30.2%
Russia is a large threat to the United States	50.0%	52.7%	51.5%
Total	100%	100%	100%

Source: 2020 American National Election Study.
Total percent may not sum to 100 due to rounding.

Does knowing a respondent's gender help us predict their perception of Russia as a threat to the United States? Not at all. Of all respondents, 51.5% believe that Russia is a large threat to the United States. Among women and men, the column percentages are very close to each other, at about 51%. We see the same pattern for the other views of the Russian threat. Knowing a respondent's gender does not help us explain their perception of Russia as a threat to the United States.

The larger the difference between column percentages, the stronger the statistical evidence that the independent variable helps "explain" the dependent variable. We have seen evidence of strong relationships between the variables in some of the prior examples in this chapter. We saw, for example, that partisanship is a good predictor of ideology (Table 3.3) and that, conversely, ideology is a very strong predictor of partisanship (Table 3.4).

Remember that we set up Tables 3.3 and 3.4 to reflect two different models of whether ideology affects partisanship or partisanship affects ideology. Nothing in the cross-tab can tell us which of these models is correct or whether the variables are simply associated, with neither being causal. In chapter 13, we will discuss how to understand causality in detail.

Calculating Column Percentages

How exactly are the percentages in the cells in a cross-tabulation calculated? The first step is assigning cases to the cells of the table.

Let's look at an example involving two variables: how often people worship (i.e., attend religious services) and how often they pray. Imagine we have ten respondents, and, for each one, we record their value on each of the variables, as shown in Table 3.6.

Table 3.6 Values for Frequency of Worship and Prayer for Ten Respondents

Respondent	How Often Does Respondent Worship?	How Often Does Respondent Pray?
1	Often	Often
2	Often	Occasionally
3	Often	Often
4	Never	Occasionally
5	Occasionally	Occasionally
6	Occasionally	Never
7	Never	Often
8	Often	Occasionally
9	Never	Never
10	Occasionally	Occasionally

Next, we construct a table, as shown in Table 3.7, that contains a cell for each possible combination of categories of the two variables. There are three categories of the worship variable and three categories of the prayer variable. Thus, there are nine (3 × 3) possible combinations of the two variables. Next, we place each respondent in the appropriate cell. For example, respondent number 1 worships often and prays often. This respondent is counted in the upper-left cell. In this example, respondent 3 is the only other individual who worships often and prays often, so this respondent is also counted in the upper-left cell. Of the ten respondents, two are counted in the upper-left cell.

We classify each respondent in the same manner and place them in the appropriate cell. This is how we populate the cells of a cross-tabulation. The number in each cell represents the raw frequency, or the number of cases that fall into the combined categories of the two variables.

Now let's use real data to examine the true relationship between attending religious services and praying and to illustrate how the percentages in the cells of cross-tabulations are calculated. Table 3.8, using the full sample from the World Values Survey (WVS), shows the relationship between how often Americans pray and how often they attend religious services.

We posit frequency of worship as the independent variable, and so we place it across the top—that is, each column represents one category of frequency of worship. Frequency of prayer is the dependent variable. To construct a cross-tabulation, we begin with the raw frequencies, displayed in Table 3.8.

Table 3.7 Raw Frequencies for Each Combination of Categories for Frequency of Prayer and Worship, N = 10

	Worships Often	Worships Occasionally	Never Worships	Total
Prays often	2	0	1	3
Prays occasionally	2	2	1	5
Never prays	0	1	1	2
Total	4	3	3	10

The interpretation of these raw frequencies is straightforward—541 respondents who worship often also pray often. But with only the raw frequencies on display, it is essentially impossible to see patterns in how the frequency of prayer differs for those who worship more or less often. This is because the total number of cases in each category of the independent variable is not the same. To see patterns across worship categories, we need to know how large each cell frequency is relative to the total number of cases in each category. Column percentages are standardized for differences in the size of each category. As our goal is to determine the extent to which the independent variable helps explain the dependent variable, we must examine and compare the column percentages.

How do we calculate the column percentages? The procedure is exactly the same as the one we saw in chapter 2, when we looked at the calculation of percentages in a frequency distribution. Only here, we employ that procedure for each column in the cross-tabulation. To calculate column percentage (col%), we divide the raw frequency of a cell (f) by the total number of cases in that column (col total).

$$col\% = \frac{f}{coltotal}(100)$$

We see, from Table 3.8, that there are 703 respondents who stated that they worship often. This is the column total (col total). Of the 703 respondents who reported worshipping often, 541 also reported praying often. That is the raw frequency of the cell (f).

$$\frac{541}{703} = 0.77$$

Table 3.8 Effect of Frequency of Worship on Prayer, Raw Frequencies

Frequency of Prayer	Worships Often	Worships Occasionally	Never Worships	Total
Prays often	541	309	122	972
Prays occasionally	160	477	167	804
Never prays	2	100	137	239
Total	703	886	426	2,015

Source: World Values Survey (Wave 7).

Of those who worship often, 77% also pray often.

Similarly, among the 426 respondents who never worship, 122 of them report that they pray often.

$$\frac{122}{426} = 0.286$$

Of those who never worship, 28.6% pray often.

Applying this procedure to every raw frequency in the cross-tabulation produces a table that contains all the column percentages we need to determine whether frequency of worship influences frequency of prayer. Take a moment to calculate the column percentages for the rest of the table. These figures are shown in Table 3.9.

Table 3.9 Effect of Frequency of Worship on Prayer, Column Percentages

Frequency of Prayer	Worships Often	Worships Occasionally	Never Worships	Total
Prays often	77.0% (541)	34.9% (309)	28.6% (122)	48.2% (972)
Prays occasionally	22.8% (160)	53.8% (477)	39.2% (167)	39.9% (804)
Never prays	0.3% (2)	11.3% (100)	32.2% (137)	11.9% (239)
Total	100% (703)	100% (886)	100% (426)	100% (2,015)

Source: World Values Survey (Wave 7).
Total percent may not sum to 100 due to rounding.

A cross-tabulation (such as Table 3.9) also provides us with a frequency distribution of the dependent variable for all respondents. This appears in the right column titled "Total" and is also known as the "marginal distribution" or "marginal percentages." Here, 48.2% of all respondents pray often, 39.9% pray occasionally, and 11.9% never pray. These are calculated by dividing the number of respondents in each category of the dependent variable by the total number of respondents for the sample. For example, 972 out of the full 2,015 respondents report they pray often (972/2,015 = 0.482, or 48.2%).

With the inclusion of the column percentages, we are in a much better position to assess the effect of the independent variable on the dependent variable. By comparing the column percentages, we see very large differences in the likelihood of praying often, depending on how regularly an individual worships. The marginal percentages tell us that about 48% of all respondents pray often. But among those who worship often, 77% pray often. As worship becomes less frequent, the likelihood of praying often also becomes lower. Similarly, those who never worship are much more likely than those who do worship to never engage in prayer. It is noteworthy that about 28% of respondents who never attend religious services still pray often. But at the opposite end of the spectrum, less than 1% of those who worship often never pray.

It is important to be clear about what the percentages in a cross-tabulation represent. Column percentages always represent the percentage of the group in the column that falls into a category of the dependent variable. In this case, that means that the column percentages are the percentage of those who worship often (for example) who pray often, occasionally, or never.

Row Percentages

We have constructed this example with worship as the independent variable and prayer as the dependent variable. But what if we want to reverse causality and examine the frequency distribution of the worship variable for each of the three categories of the prayer variable? The assumption here would be that praying often makes people inclined to go to religious services more often. Using the same raw frequencies shown in Tables 3.8 and 3.9, we can construct a table that uses "row percentages" instead of "column percentages." **Row percentages** show the percentage of the group in the row that falls into each category of the other variable. This version is contained in Table 3.10.

Table 3.10   Effect of Frequency of Prayer on Worship, Row Percentages

Frequency of Prayer	Worships Often	Worships Occasionally	Never Worships	Total
Prays often	55.7% (541)	31.8% (309)	12.6% (122)	100% (972)
Prays occasionally	19.9% (160)	59.3% (477)	20.8% (167)	100% (804)
Never prays	0.8% (2)	41.8% (100)	57.3% (137)	100% (239)
Total	34.9% (703)	44.0% (886)	21.1% (426)	100% (2,015)

Source: World Values Survey (Wave 7).
Total percent may not sum up to 100 due to rounding.

Now, the percentages in each cell add up to 100% horizontally, not vertically. Look at the upper-left cell. How is that 55.7% calculated? It is the raw number of respondents who pray often and worship often (541), divided by the number of people who pray often (972).

$$row\% = \frac{f}{row\ total}(100)$$

$$\frac{541}{972} = 0.557$$

Of those who pray often, 55.7% also worship often.

The marginal percentages in this case are shown in the bottom row. They give the percentage of the full sample who worship at each rate. In other words, 34.9% of all respondents report they worship often (703/2,015 = 0.349).

When we use row percentages, we are in a sense reversing our independent and dependent variables. We are looking at the three categories of prayer and asking, for each of these categories: What is the distribution of the worship variable? Tables 3.9 and 3.10 may look alike, but they present the data differently, and one needs to be very attentive to the difference between a column percentage and a row percentage.

Comparing Column and Row Percentages

If the independent variable is arrayed in the columns across the top of the table, and the dependent variable is presented across the rows, then the column percentage allows us to assess the effect of the independent variable on the dependent variable. For that kind of table setup, think about comparing differences between columns. This is what Table 3.9 shows.

On the other hand, if the independent variable is in the rows, this calls for the row percentage. This is what we have done in Table 3.10. For this kind of table setup, think about comparing differences between rows.

In some cases, even if the independent variable is in the columns, a researcher may want to look at the composition of the independent variable in terms of categories of the dependent variable and would use row percentages. For example, in Table 3.5 (effect of gender on perception of the Russian threat), we might want to know the gender makeup of those who believe that Russia is a large threat. Row percentages would tell us what percentage of people with that particular opinion are women versus men.

Total Percentages

There is one additional form a cross-tabulation can take. Instead of using the column or row percentage, we can use the **total percentage**, which gives the percentage of the full sample that falls into a given cell. This answers a different question. Here, we are seeking to understand the joint distribution of independent and dependent variables among all respondents.

Table 3.11 shows an example of a cross-tabulation with total percentages in each cell. Let's begin with the upper-left cell. The entry in that cell reads 25.6%. This is the raw number of respondents who pray often and worship often (541), divided by the total number of respondents (2,115).

$$tot\% = \frac{f}{total\ N}(100)$$

$$\frac{541}{2115} = 0.256$$

Table 3.11 Joint Distribution of Prayer and Worship Variables, Total Percentages

Frequency of Prayer	Worships Often	Worships Occasionally	Never Worships	Total
Prays often	25.6% (541)	14.6% (309)	5.8% (122)	46.0% (972)
Prays occasionally	7.6% (160)	22.6% (477)	7.9% (167)	38.0% (804)
Never prays	0.09% (2)	9.5% (200)	6.5% (137)	16.0% (339)
Total	33.2% (703)	46.6% (986)	20.1% (426)	100% (2,115)

Source: World Values Survey (Wave 7).
Total percent may not sum to 100 due to rounding.

Of all respondents, 25.6% are people who worship often and pray often. There are nine cells in the body of this table, and if we add up the percentages in each of these nine cells, the sum will be 100%. Notice that there are also marginal percentages (labeled "Total" because they refer to the total respondents in each row or column) on both the right side and the bottom of the table; they show the same thing they did in Tables 3.9 and 3.10: the percentage of all respondents who fall into each category of each variable.

Comparing Column, Row, and Total Percentages

Column, row, and total percentages each answer different questions. If we stick to the convention of presenting the independent variable in the columns, column percentages let us examine how the dependent variable differs among categories of the independent variable. If the independent variable is in the rows, row percentages tell us the same thing. It can be useful to ask yourself "percentage of what group?" when interpreting a percentage in a cross-tabulation. In the case of column percentages, the answer to that question is always "the percentage of the column." In the case of row percentages, the answer is always "the percentage of the row." In the case of total percentages, the answer is always "the percentage of the total sample."

BOX 3.1: APPLICATION

Practice Calculating Column, Row, and Total Percentages

We might predict that people's marital status (and thus their experience with marriage and divorce) might affect their opinion on divorce law. Table 3.12 uses 2021 GSS data to show the raw frequencies for the relationship between marital status and opinion about whether divorce should be easier, more difficult, or stay the same as it is now.

Table 3.12 Opinion on Divorce Law, by Marital Status, Raw Frequencies

Divorce Should:	Married	Widowed/Divorced/Separated	Never Married	Total
Be easier	178	132	150	460
Be more difficult	132	73	36	241
Stay the same	366	166	124	656
Total	676	371	310	1,357

Source: 2021 General Social Survey.

Using these data, practice calculating column percentages. The correct percentages are shown in Table 3.13.

Table 3.13 Opinion on Divorce Law, by Marital Status, Column Percentages

Divorce Should:	Married	Widowed/Divorced/Separated	Never Married	Total
Be easier	26.3%	35.6%	48.4%	33.9%
Be more difficult	19.5%	19.7%	11.6%	17.8%
Stay the same	54.1%	44.7%	40.0%	48.3%
Total	100%	100%	100%	100%

Source: 2021 General Social Survey.
Total percent may not sum to 100 due to rounding.

Now, practice calculating row percentages and total percentages for each cell. The correct answers are shown below in Table 3.14. (Row percentages are on top; total percentages are below in parentheses.)

Table 3.14 Opinion on Divorce Law, by Marital Status, Row, and (Total) Percentages

Divorce Should:	Married	Widowed/Divorced/Separated	Never Married	Total
Be easier	38.7% (13.1%)	28.7% (9.7%)	32.6% (11.1%)	100.0%
Be more difficult	54.8% (9.7%)	30.3% (5.4%)	14.9% (2.7%)	100.0%
Stay the same	55.8% (27.0%)	25.3% (12.2%)	18.9% (9.1%)	100%
Total	49.8%	27.3%	22.8%	100%

Source: 2021 General Social Survey.
Total percent may not sum to 100 due to rounding.

Does marital status affect opinion on divorce law? How? How can you tell?

BOX 3.2: IN DEPTH

Collapsing Categories in Cross-Tabulations

When we analyze the relationship between variables with many categories, we often must condense the categories into a manageable number. Let's say we are analyzing the effect of age (independent variable) on electoral turnout (dependent variable). Turnout is easy to use; it has only two categories (voted and didn't vote). But age runs from our youngest respondent (usually eighteen years old) to our oldest respondent (ninety or older in many surveys). We obviously need to condense age into several categories. There is no hard-and-fast rule about how many categories is too many for cross-tabulations, but too many categories will produce tables that are unwieldy and hard to interpret. When collapsing, or combining, categories, we need to make sure that each new category retains enough cases to conduct a meaningful cross-tabulation and that the new categories combine the original categories in a way that makes logical sense.

Take a look back at Table 3.3 (page 120). It examines the relationship between party identification and ideological identification. Each variable has three categories, and the resulting table, with nine cells, is easy to read. In this instance, both variables are condensed versions of the original party identification and ideological identification variables. In original form, each variable had seven categories, and, if we had not recoded them, this would have produced a 7 × 7 table, with forty-nine cells! Variables with this many categories would yield a table with so many cells that we would be unable to recognize patterns in the data. When you construct a cross-tabulation, make sure that the variables have a "reasonable" number of categories. Five categories or fewer is usually reasonable. But depending on sample size and the scale of the original variable, you may need to retain more or fewer categories.

Interpreting the Strength of Relationships

Examining differences across columns (or rows, if the table uses row percentages) shows whether a relationship exists between two variables in a cross-tabulation. If percentages differ across columns, this is evidence that the two variables are related. But how do we know whether the cross-tabulation reveals a strong relationship between the two variables? As we noted above, the larger the difference in percentages, the stronger the relationship. But how large does the difference need to be to indicate a strong relationship?

One helpful way to conceptualize the strength of a relationship is to think about strength as having three categories: weak, moderate, and strong. While there are no clear boundaries separating weak from moderate, and moderate from strong, relationships, it is helpful to use these categories as a guide to assessing the strength of the relationship. Let's look at some examples of relationships to think about how to assess their strength.

Americans have long been hostile to foreign aid. Majorities consistently overestimate the percentage of the US national budget that goes to foreign aid*; and majorities consistently express the view that the United States spends too much on foreign aid. In 2021, the

---

* A recent article puts the public's median estimate of foreign aid spending at 10% of the annual budget. In fact, it is around 1%. See https://worldpublicopinion.net/american-public-vastly-overestimates-amount-of-u-s-foreign-aid/.

General Social Survey found that 56% of Americans believed that government spends too much on foreign aid, compared to 25% who stated that the expenditure level is about right (and only 9% who believed that United States spends too little). This hostility to foreign aid among Americans may stem from two factors: a diminished interest in foreign affairs and a pervasive sense that the US government should take care of Americans before choosing to spend money abroad. However, this opposition to foreign aid may not be felt uniformly among Americans. It stands to reason that Americans who immigrated from other countries may have a greater interest in foreign affairs, and also a heightened sensitivity to some of the benefits of foreign aid. Proceeding from these assumptions, we examine attitudes toward foreign aid, broken down by nativity status (whether or not the respondent was born in the United States). Table 3.15 posits nativity status as the independent variable, and views of US government spending on foreign aid as the dependent variable.

Examining the percentage differences across the columns for "US spends about right"

Table 3.15 Effect of Nativity Status on View of Foreign Aid, Column Percentages

	Born in the United States	Not Born in the United States	Total
United States spends too little	8.5%	12.7%	8.9%
United States spends about right	32.7%	51.3%	34.6%
United States spends too much	58.8%	36.0%	56.5%
Total	100%	100%	100%

Source: 2021 General Social Survey.

and "US spends too much" indicates that there is a relationship between nativity status and views of foreign aid. A majority of those born in the United States thought that the United States spends too much, while a majority of those born abroad believed that US spending on foreign aid is about right. But how strong is this relationship? About 59% of those born in the United States said that foreign aid spending was too high, compared to only 26% of those born elsewhere, a difference of more than 20 percentage points. The difference between non-immigrants and immigrants in stating that spending is about right is similar in size, about 19 percentage points. These differences are notable, though they do not indicate diametrically different views across the two groups. Based on these differences, we might characterize the strength of the relationship between nativity status and views of foreign aid as a moderate one.

Refer back to Table 3.4 for an example of an even stronger relationship than that shown in Table 3.15. There, we saw an 84-percentage-point difference between the percentage of liberals and conservatives who identify as Democrats and also an 84-percentage-point difference between the percentage of liberals and conservatives who identify as Republicans. We could characterize the relationship between political ideology and party affiliation as a strong relationship, much stronger than the relationship between nativity and opinions about foreign aid.

Another way to think about the strength of a relationship in a cross-tabulation is to compare the size of the differences in the table to the size of the differences for the strongest (or weakest) possible relationship that could exist. Table 3.16 presents what the strongest possible relationship between these two variables would look like.

Table 3.16 Effect of Nativity Status on View of Foreign Aid, Column Percentages (Hypothetical)

	Born in the United States	Not Born in the United States	Total
United States spends too little	0%	0%	0%
United States spends about right	0%	100%	13.7%
United States spends too much	100%	0%	86.3%
Total	100%	100%	100%

The hypothetical results in Table 3.16 show that nativity status perfectly predicts opinions on foreign aid. All those born in the United States think that the government spends too much on foreign aid, while all those born outside the United States think spending is about right. There is a 100-percentage-point difference in views of foreign aid between those who were and were not born in the United States. (Because 13.7% of the US population was born in other countries, 13.7% of the overall hypothetical sample believes that spending is about right, as shown in the "Total" column.[2]) We would rarely observe such a strong relationship with real data because there are many factors in the social world that affect any given dependent variable besides the independent variable. For example, along with nativity status, partisanship probably shapes what people think of foreign aid.

Whereas the strongest possible difference between categories in a cross-tabulation is 100 percentage points, the weakest possible difference is 0 percentage points, which happens when the same percentage of cases falls into the two cells being compared. Characterizations of relationships as weak, moderate, or strong should be informed by the hypothetical strongest and weakest differences as well as comparisons to similar relationships found in past research.

In addition to characterizing the strength of a relationship between variables, we also always describe the nature of that relationship. In other words, you would not only say that there is a moderate relationship between whether respondents were born in the United States and their attitudes about foreign aid. That does not tell us which group is more supportive of foreign aid. You would say that respondents born outside the United States are moderately more supportive of foreign aid than those born in the United States.

Interpreting the Direction of Relationships

Whereas the strength of a bivariate relationship in a cross-tabulation can be assessed for relationships between variables measured at any level, the **direction of a relationship** can be assessed only between ordinal or interval-ratio variables. A **positive relationship** occurs when the values of both variables increase together, and a **negative relationship** (also called an inverse relationship) occurs when the values of one variable increase while the other variable's values decrease (and vice versa). In other words, in a positive relationship, the values of both variables move together in the same direction. In a negative or inverse relationship, the values move in opposite directions.

If the categories of either variable are not rank-ordered, as with nominal variables, then the relationship can have no direction.

Returning to the 2021 General Social Survey (GSS), we examine the relationship between two ordinal variables: respondents' social class and their general happiness, which respondents rated as "very happy," "pretty happy," or "not too happy." Social class has two categories: (1) lower class or working class and (2) middle class or upper class.* In colloquial terms, can money buy happiness? What would you hypothesize about this relationship? One might imagine that higher social class status would lead people to feel happier. This is a hypothesis about a positive relationship between the variables, in which higher class status would be associated with higher levels of happiness and (conversely) lower class status would be associated with lower levels of happiness. The values of the variables would go up or down in tandem. Table 3.17 gives the result of the cross-tabulation.

Table 3.17 Effect of Social Class on Happiness, Column Percentages

	Lower or Working Class	Middle or Upper Class	Total
Very happy	15.0%	23.3%	19.5%
Pretty happy	55.4%	59.4%	57.6%
Not too happy	29.5%	17.3%	22.9%
Total	100%	100%	100%

Source: 2021 General Social Survey.
Total percent may not sum to 100 due to rounding.

Table 3.17 shows that there is a positive relationship between class and happiness. Compare the percentages of lower-/working- and middle-/upper-class respondents who say they are very happy. Only 15% of the lower-class group, compared to 23.3% of the higher-class group, are very happy. On the other end of the scale, 29.5% of lower-/working-class respondents say they are not too happy, compared with 17.3% of middle-/upper-class respondents. (There is a much smaller gap [4 points] in the "pretty happy" category.) As class status increases, happiness increases. And as class status decreases, happiness decreases. It may seem confusing to say that both of these statements are true or that decreasing values of the two variables are characteristic of a positive relationship. Remember, the key rule is that when variables are positively related, their values move in the same direction.

We can use another example to see what a negative or inverse relationship looks like. We will examine how education is related to television viewing. It is certainly possible that those with more education are less likely to watch a lot of television. Education imparts cognitive and analytical skills that may motivate people to turn to other and more varied

---

* For the purpose of illustration, we have combined the four categories of the original GSS variable into two categories.

sources of information and entertainment. Using data, we can try to answer the question: as education increases, does television viewing decrease? In other words, is education negatively related to television viewing? The 2021 GSS asked respondents how often they watch television and also how much schooling they have had. Table 3.18 presents the cross-tabulation of these two variables.

Table 3.18 Effect of Education on TV Viewing, Column Percentages

	Less Than High School	High School	Some College	College Degree or More	Total
0 or 1 hour	17.8%	15.8%	18.0%	28.7%	22.6%
2 hours	19.1%	18.1%	20.5%	25.8%	22.4%
3 or 4 hours	31.2%	35.0%	33.6%	29.6%	31.8%
5 hours or more	31.8%	31.2%	28.0%	15.9%	23.2%
Total	100%	100%	100%	100%	100%

Source: 2021 General Social Survey.
Total percent may not sum to 100 due to rounding.

Table 3.18 shows that there is a negative relationship between education and television viewing. As education increases, television viewing decreases. Compare the 31.8% of respondents with less than a high school education who report that they watched five or more hours of television per day, to the much lower 15.9% of those with a college degree who do so. At the other end of the scale, 17.8% of those with less than a high school education watch one hour of television or less, compared to 28.7% of those with a college degree. The differences in the middle category (high school and some college) are fairly minor.

Putting the Results into Words

Producing numerical results, such as a cross-tabulation, is not the end of a statistical analysis. There is another step, which is to think carefully about the story that the data tell and how that story helps us to answer the research question that motivated the analysis. This interpretation skill is as important in statistics as the ability to conduct a statistical analysis. After all, we use statistics as a means to answer interesting questions. We need to be able to explain the answers that the results offer us.

Unlike with many of the statistical analyses that we examine in this book, there are few clear rules about how to explain the results of cross-tabulations in words. But here are some general guidelines for interpreting the results of a cross-tabulation:

- First, state the general qualities of the relationship between the two variables, the strength and direction (if appropriate), and do so in terms that a nonexpert would understand.
- Second, after stating the general qualities, give specific examples from the table to illustrate them.
- Finally, state how the relationship between the two variables answers the research question that guided the analysis.

We can think about this process of explaining results in words as telling the data's story. Let's tell the story of the cross-tabulation shown in Table 3.17:

- *State the direction of the relationship in clear language.* There is a positive relationship between social class status and self-reported happiness. As class status increases, people are more likely to say that they are very happy.
- *State the strength of the relationship in clear language.* This is a moderately strong relationship, with class status associated with a noticeable, but not extremely large, difference in self-reported happiness.
- *Give (a) specific example(s) from the data to illustrate the direction and strength.* For example, whereas 15% of lower-status respondents said they were very happy, 23% of higher-status respondents said they were very happy, a difference of 8 percentage points. Conversely, only 17% of higher-status respondents were "not too happy," compared with 30% of lower-status respondents, a 13-percentage-point difference.
- *Relate the findings back to the research question.* We expected that people of higher social class status would be happier because of their higher social status and economic security, while people of lower-class status would report lower happiness because of economic insecurity or social stigma attached to their class status. The results are consistent with this expectation, but the differences between the two groups are less dramatic than one might have expected. This suggests that factors other than class status also affect how people rate their happiness.

What story do the data in Table 3.18 tell us? While we see that there is a moderate negative relationship between education and television viewing, we must exercise caution in interpreting this bivariate association as a causal relationship. There could be other factors driving the relationship between education and television viewing. Individuals who by nature are curious and knowledge-seeking may be both more likely to go to college, and also less likely to watch a lot of television. Or individuals who are highly educated may live in neighborhoods more conducive to outdoor activities or be more able to afford to go to movies, concerts, or sporting events.

Following these general steps for explaining the results of a cross-tabulation in words will yield a description that includes all of the necessary elements and is understandable to non-specialists.

Graphical Representations of Bivariate Relationships

Cross-tabulations are one way of visually representing the relationship between two variables, but for some people graphical representations are easier to digest. In this section, we cover **clustered bar graphs** and **stacked bar graphs** as alternative modes of representing the relationship between two variables. Figures 3.1 and 3.2 show a clustered bar graph and

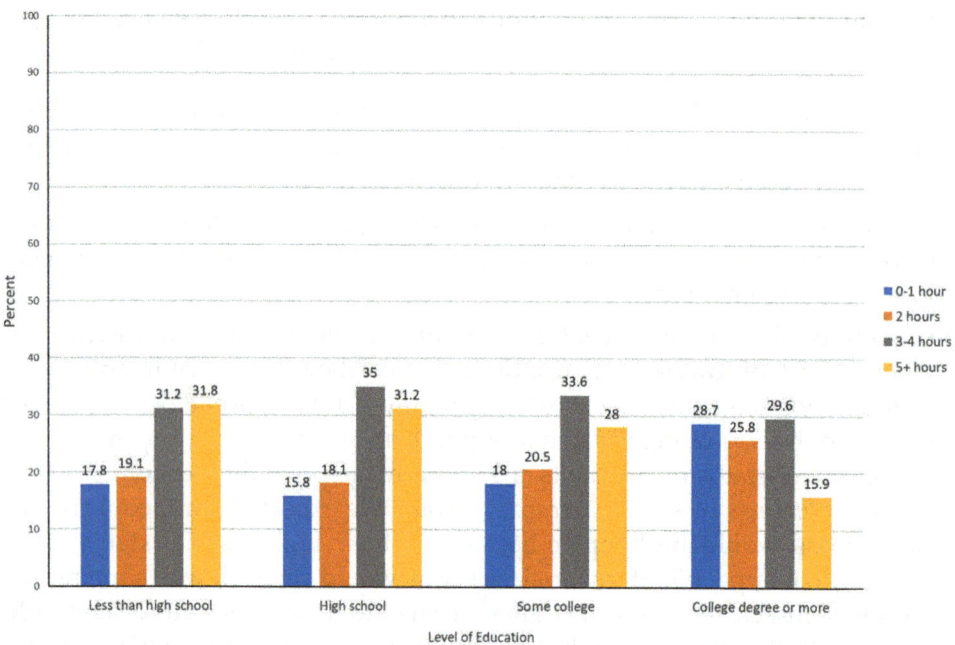

Figure 3.1 Clustered Bar Graph, Distribution of Television Viewing, by Education.
Source: 2016 General Social Survey.

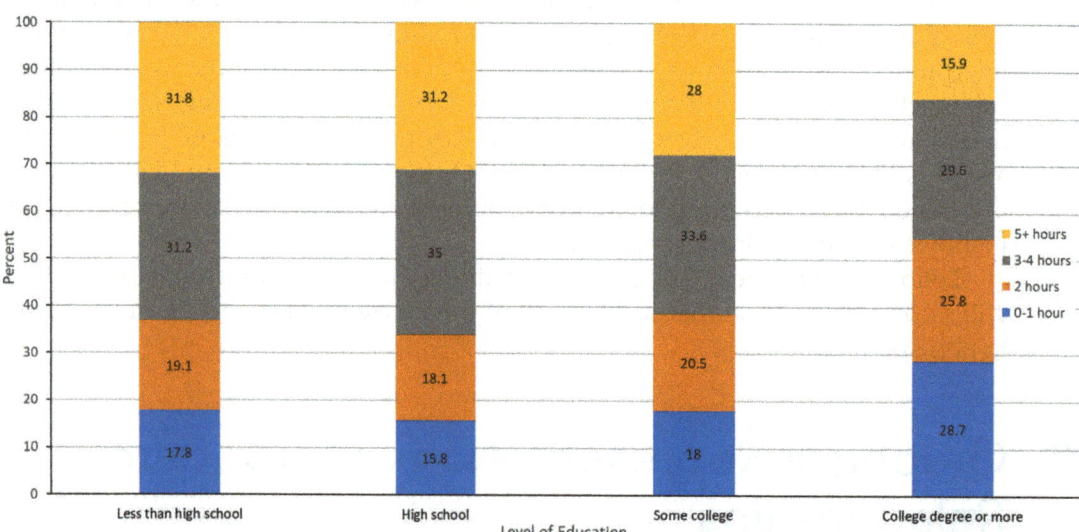

Figure 3.2 Stacked Bar Graph, Distribution of Television Viewing, by Education.
Source: 2016 General Social Survey

a stacked bar graph, respectively, representing the same relationship between education and television viewing shown in Table 3.18.

Figures 3.1 and 3.2 present the same information, but they use different methods for doing so. The difference between the clustered and stacked bar graphs is that the distribution of cases across each category of education is shown in separate bars in the clustered graph (Figure 3.1) and on the same bar in the stacked graph (Figure 3.2). In Figure 3.1, the sum of the percentages for the four bars in each cluster is 100%. In Figure 3.2, the sum of the sections of each bar is 100%. Notice that each cluster of bars in Figure 3.1, and each single bar in Figure 3.2, corresponds to a column in Table 3.18. Looking at the figures, we can see graphically the same story presented by Table 3.18: there is relatively little difference among the first three categories of education in how much television people watch. The category that stands out as markedly different is those who have a college degree; this group was somewhat less likely to watch a lot of television. There is no inherent advantage to any of these three modes of presenting bivariate distributions—cross-tabulations, clustered bar graphs, or stacked bar graphs. Supplementing a cross-tabulation with either a clustered or stacked bar graph offers a visual representation of the data that some readers find more intuitive.

Chapter Summary

This chapter introduced you to examining bivariate relationships, or relationships between two variables, by means of cross-tabulation. Below, we review the definitions of terms, steps for generating a cross-tabulation, and graphical representations of bivariate relationships covered in this chapter.

- **Cross-tabulations** are tables that present frequency distributions for the dependent variable separately for each category of the independent variable.
- The **independent variable** is the variable that we suspect causes changes in the other variable.
- The **dependent variable** is the variable that we seek to explain.
- The **strength** of the relationship between two variables in a cross-tabulation refers to the sizes of the differences in percentages across categories of the independent variable.
- The **direction** of the relationship for variables that are ordinal or interval-ratio indicates whether values of the dependent variable increase (a positive relationship) or decrease (a negative relationship) as values of the independent variable increase.

The steps for generating a cross-tabulation between two variables are:

1. Make sure your variables do not have too many categories. Collapse categories if necessary.
2. Identify which variable is the independent variable and which is the dependent variable. Decide whether you want to arrange the categories of the independent variable across the columns or down the rows of the table. The convention is to arrange them across the columns, and the rest of the steps assume this approach.
3. Count the number of cases that occur for each combination of the two variables' categories. (Multiply the number of categories of the independent variable by the number of categories of the dependent variable to determine the number of combinations.) Enter the number of cases that occur at each combination of the two variables into the cells of the table. With large samples, we rely on statistical software to perform this sorting.
4. Count the number of cases that fall into each category of the independent and dependent variables. Enter the number of cases that fall into each category of the independent variable in the bottom row. Enter the number of cases that fall into each category of the dependent variable in the last column.
5. Convert frequencies for each cell into percentages. For **column percentages**, divide the frequency in each cell by the number of cases that fall into that column. For the last column, divide the frequencies of the cells by the total sample size. For the last row, sum the percentages of each cell in the column. For **row percentages**, divide the frequency in each cell by the number of cases that fall into that row. For **total percentages**, divide the frequency in each cell by the total number of cases.

$$col\% = \frac{f}{col\ total}(100)$$

$$row\% = \frac{f}{row\ total}(100)$$

$$tot\% = \frac{f}{total\ N}(100)$$

- A **clustered bar graph** shows the distribution of the dependent variable with a clustered set of bars for each category of the independent variable. The bars within each cluster show the percentage of cases that fall within each category of the dependent variable for a given category of the independent variable.

- A **stacked bar graph** shows the distribution of the dependent variable with just one bar for each category of the independent variable. Each bar shows the percentage of cases for a given category of the independent variable that fall within each category of the dependent variable.

Using Stata

How does going to college affect students? The answer to this question often focuses on the specific skills that students learn, but research suggests that college also shapes the way people see the world. In this section, we will use cross-tabulation to examine this question, using data from Wave 7 of the World Values Survey (WVS). Specifically, we will ask how level of education is related to confidence in television as an organization. Level of education is the independent variable that we suspect shapes the views measured by the dependent variable.

We will use two variables from the WVS. The independent variable, *college*, measures whether respondents attended college, with two categories—"yes" and "no." The dependent variable, confidence in television (*conftv*), indicates the level of confidence respondents have in television as an organization, divided into two categories: confidence and little to no confidence.*

Open `WVSWave7.dta`. To generate a cross-tabulation between *college* and *conftv*, we use the same "tabulate" command that we used to generate frequency distributions for single variables in chapter 2. Type the following command into Stata:

`tabulate conftv college, column`

---

* The *college* variable is derived from the original WVS variable called Q275. The *conftv* variable is derived from the original WVS variable called Q67.

With the "tabulate" command, the first variable following the command will always be arranged down the rows of the table, and the second variable will always be arranged across the columns. Following the convention of placing the independent variable across the columns of the table, *college* is the second variable listed. The "column" option after the comma tells Stata that it should calculate column percentages. The output is shown in Figure 3.3.

```
. tabulate conftv college, column
```

Key
frequency
column percentage

conftv	college		Total
	No colleg	College	
Confidence	29,074	7,810	36,884
	45.05	36.64	42.96
No confidence	35,468	13,504	48,972
	54.95	63.36	57.04
Total	64,542	21,314	85,856
	100.00	100.00	100.00

Figure 3.3

Stata provides us with a key in the upper-left corner of the output telling us that the top value in each cell is the frequency and the bottom value is the column percentage. Is there a relationship between attending college and confidence in television as an organization? The results in the table suggest that there is a moderate negative relationship between these variables. For example, about 63% of those who attended college have little to no confidence in television, compared to 55% of those who did not attend college. These results suggest that college might condition people to be somewhat more skeptical of television than people who never attended college.

Let's say we wanted to generate another cross-tabulation between college and confidence in television, but this time with row percentages. What syntax do you think we might use? The difference is simple. Instead of specifying the "column" option, we specify "row" at the end of the command:

`tabulate conftv college, row`

The output is shown in Figure 3.4.

The data in this table tell a different story from the cross-tabulation that calculated column percentages. Here we see that 79% of the respondents who were confident in television did not attend college but also that 72% of those who did not trust television did not attend college. This reflects the fact that there are many more non–college attenders (64,542) in the sample than college attenders (21,314). Therefore, it is not surprising that the majority of those who trusted television, or did not, had not attended college.

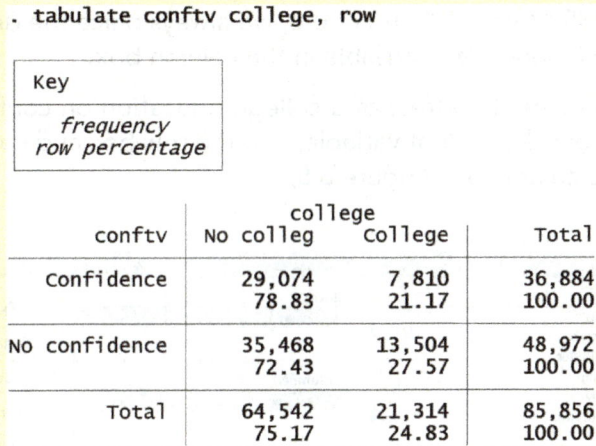

Figure 3.4

Review of Stata Commands

- Generate a cross-tabulation, specifying the option of column or row percentages

    `tabulate row variable column variable, (column)`

    `tabulate row variable column variable, (row)`

Using SPSS

How does going to college affect students? The answer to this question often focuses on the specific skills that students learn, but research suggests that college also shapes the way people see the world. In this section, we will use cross-tabulation to examine this question, using data from Wave 7 of the World Values Survey (WVS). Specifically, we will ask how level of education is related to confidence in television as an organization.

We will use two variables from the WVS. The independent variable, *college*, measures whether respondents attended college, with two categories—"yes" and "no." The dependent variable, confidence in television (*conftv*), indicates the level of confidence respondents have in television as an organization, divided into two categories: confidence and little to no confidence.*

Open `WVSWave7.sav`. To generate a cross-tabulation between *college* and *conftv*, we use the following sequence of menu commands:

`Analyze → Descriptive Statistics → Crosstabs`

---

* The *college* variable is derived from the original WVS variable called Q275. The *conftv* variable is derived from the original WVS variable called Q67.

When we use the SPSS "crosstab" procedure, we always place the dependent variable in the row box and the independent variable in the column box.

Since we want to examine the effect of a college education on confidence in television, we will place *conftv*, our dependent variable, into the row box and *college*, our independent variable, into the column box (Figure 3.5).

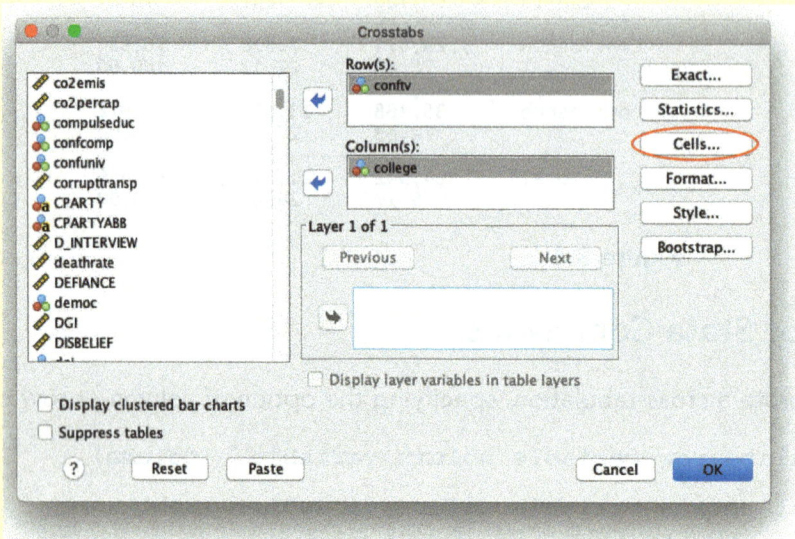

Figure 3.5

If we were to run the cross-tabulation now, SPSS would produce output containing only the raw frequencies in each cell of the table—this is the default setting for SPSS. But we want SPSS to calculate the column percentages as well. So before clicking "OK" in the "Crosstabs" dialog box, we click on "Cells" (circled above). This will open a new "Crosstabs: Cell Display" dialog box. In this box, we tell SPSS that we want column percentages by checking the small box next to "Column" (circled in Figure 3.6).

After we click "Continue," SPSS returns us to the "Crosstabs" box, where we click "OK." This will produce the desired output. Each cell will contain the raw frequency (or count) and the column percentage (Figure 3.7).

The top value in each cell is the frequency (or count), and the bottom value is the column percentage. Is there a relationship between attending college and confidence in television as an organization? The results in the table suggest that there is a moderate negative relationship between these variables. For example, about 63% of those who attended college have little to no confidence in television, compared to 55% of those who did not attend college. These results suggest that college might condition people to be somewhat more skeptical of television than people who never attended college.

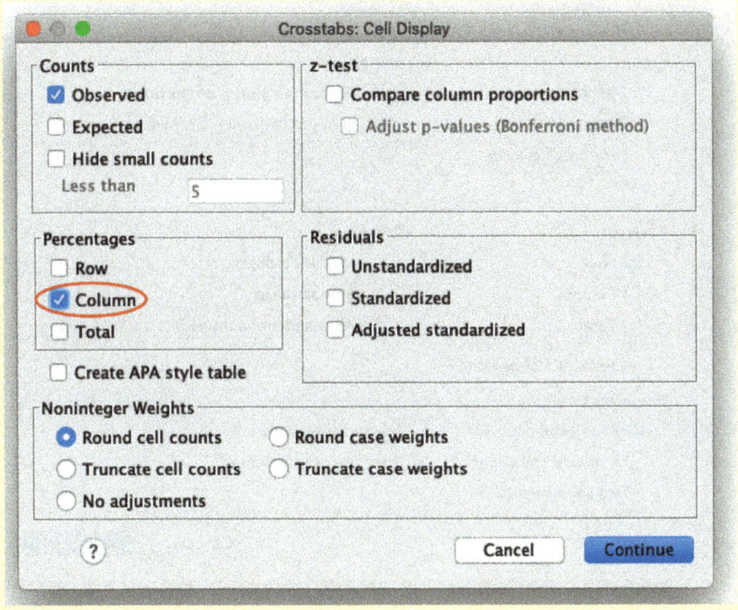

Figure 3.6

conftv * college Crosstabulation

			college		Total
			No college	College	
conftv	Confidence	Count	29074	7810	36884
		% within college	45.0%	36.6%	43.0%
	No confidence	Count	35468	13504	48972
		% within college	55.0%	63.4%	57.0%
Total		Count	64542	21314	85856
		% within college	100.0%	100.0%	100.0%

Figure 3.7

Let's say we wanted to generate another cross-tabulation between college and confidence in television, but this time with row percentages. The steps are exactly the same as we have seen so far, with one exception. In the "Cell Display" dialog box, we would select "Row" percentage instead of "Column" percentage (Figure 3.8).

The output is shown in Figure 3.9.

The data in this table tell a different story from the cross-tabulation that calculated column percentages. Here we see that 79% of the respondents who were confident in television did not attend college but also that 72% of those who did not trust television did not attend college. This reflects the fact that there are more noncollege attenders (64,542) in the sample than college attenders (21,314). Therefore, it is not surprising that the majority of those who trusted television, or did not, had not attended college.

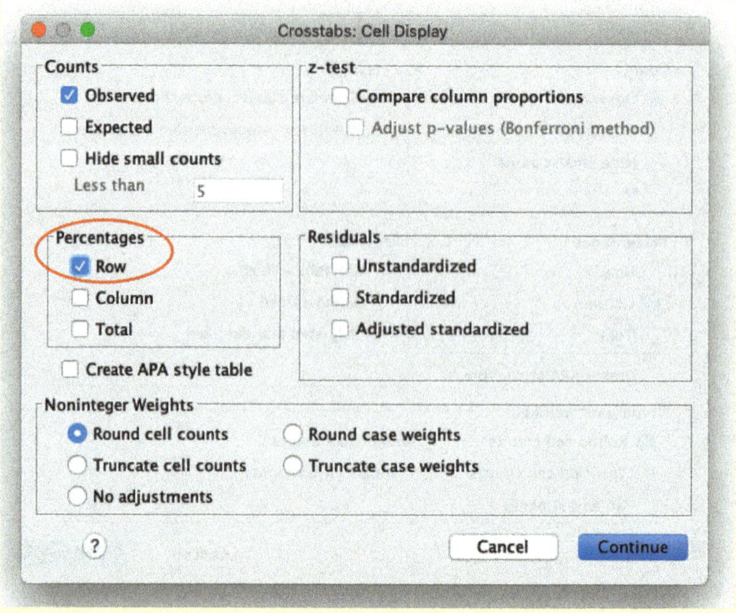

Figure 3.8

confv * college Crosstabulation

			college		Total
			No college	College	
confv	Confidence	Count	29074	7810	36884
		% within confv	78.8%	21.2%	100.0%
	No confidence	Count	35468	13504	48972
		% within confv	72.4%	27.6%	100.0%
Total		Count	64542	21314	85856
		% within confv	75.2%	24.8%	100.0%

Figure 3.9

Review of SPSS Procedures

- Generate a cross-tabulation

    `Analyze → Descriptive Statistics → Crosstabs`

- Dependent variable in "Row" box; independent variable in "Column" box
- Click on "Cells" to select column and/or row percentage

Practice Problems

1. Coachella has been a popular summer music festival in the United States since it started in 1999. When Beyoncé headlined the festival in 2018, she became the first Black woman to headline the festival as an individual performer. Table 3.19 lists the race and gender of each artist who performed as an individual headlining act (i.e., excluding bands) at Coachella from 1999 to 2022.

   Table 3.19   Coachella Individual Headliners, 1999 to 2022

Name	Race	Gender
Beck	White	Man
Björk	White	Woman
Prince	Black	Man
Roger Waters	White	Man
Jack Johnson	White	Man
Paul McCartney	White	Man
Jay-Z	Black	Man
Kanye West	Black	Man
Drake	Black	Man
Jack White	White	Man
Calvin Harris	White	Man
Lady Gaga	White	Woman
Kendrick Lamar	Black	Man
The Weeknd	Black	Man
Eminem	White	Man
Beyoncé	Black	Woman
Childish Gambino	Black	Man
Ariana Grande	White	Woman
Harry Styles	White	Man
Billie Eilish	White	Woman

   a. Use this information to create a cross-tabulation of the race and gender of Coachella headliners. Place gender on the columns and race on the rows. Include column percentages, total percentages, and frequencies in the cells. Be sure to include marginal frequencies and percentages in the last column of the table. Properly label all columns and rows and give the table an appropriate title.

   b. What does the total percentage in the cell for white men mean? What does the column percentage in that cell mean?

c. Use the table's total percentages to identify the groups that have been most often and least often represented among headliners at Coachella.

d. Has representation of Black performers relative to white performers been higher among men or women headlining the festival?

e. Can you assess the strength and direction of the relationship between the two variables shown in the cross-tabulation? Explain your answer.

f. Imagine that Coachella had only one individual headliner the following year, a Black woman. Recalculate the column percentages for each cell in the table that you made with the addition of this performer.

g. How has the addition of this performer changed the strength of the relationship between race and gender among Coachella's headliners?

h. What does the addition of this single case to the group of headliners tell us about how sample size can influence cross-tabulation?

2. A survey asked a sample of two hundred parents with at least one child living at home how they felt technology influenced their children's lives as well as their own lives. The survey yielded the following information:

- Fifty parents said that they saw technology as a dangerous influence on their children, and 150 said they saw it as a useful learning tool for their children.
- Seventy-seven parents said that technology regularly threatened their own sense of well-being, while the remaining 123 said that technology usually had either no effect or a positive effect on their lives.
- Of the fifty parents who saw technology as a dangerous influence on their children, thirty-seven of them said that technology regularly threatened their own well-being. Of the 150 parents who saw technology as a useful learning tool for their children, forty of them felt their well-being regularly threatened by technology.

a. Generate a cross-tabulation between parents' feelings about how technology affects their own well-being and how it affects their children's well-being. It should report total percentages and the appropriate frequencies in each cell. Place parents' attitudes about how technology affects them in the columns and how it affects their children in the rows.

b. Using the cross-tabulation that you created for Part A, explain what the numbers in these two cells mean: (1) parents who feel positively affected or unaffected by technology AND that technology is a dangerous influence on their children and (2) parents who feel positively affected or unaffected by technology AND that technology is a useful learning tool for their children.

3. In 2017, the Pew Research Center asked respondents about their experiences taking time off work for the birth or adoption of a child. In a sample of 1,106 people who took time off from work, a large number of people reported that their leaves had been unpaid. Here is how the sample broke down by gender and paid leave:

- Women who took leave totaled 732; 410 of them received pay during their leave, while 322 of them received no pay.
- Men who took leave totaled 374; 263 of them received pay, and 111 of them did not.

  a. Create a cross-tabulation of gender and paid leave status for the sample of parents who took time off from work. Place gender in the columns and paid leave status in the rows. Include the appropriate frequency and column percentage in each cell. Be sure to include marginal frequencies and percentages in the last column of the table. Properly label all columns and rows and give the table an appropriate title.

  b. Which is the independent variable in the table? Which is the dependent variable?

  c. Look at the cell for women who received pay while on leave from work. Explain what the frequency in that cell means. Explain what the percentage means.

  d. Use the results of the cross-tabulation to answer this question: Does gender influence paid leave status for new parents? Be sure to give specific examples from the table that help you to answer this question and state the strength of the relationship.

4. In 2021, the GSS asked Americans whether they agreed that claims about environmental threats have been exaggerated. Table 3.20 shows agreement that claims of environmental threats have been exaggerated by gender.

Table 3.20 Cross-Tabulation of Gender and Thinking Environmental Threats Are Exaggerated

	Men	Women	Total
Agree	26.5	18.1	21.8
Neither agree nor disagree	16.8	22.4	19.8
Disagree	56.8	59.6	58.3
Total	100	100	100

Source: 2021 General Social Survey.

  a. Which is the independent variable? Which is the dependent variable? How do you know?

  b. Explain in words what the results tell us about the relationship between these two variables. Be sure to use specific examples from the table.

5. A blogger wrote a blog post about gun culture in the United States that went viral. In it, he claimed that American gun culture was "steeped in sexism." He didn't show any data to support his claim. Table 3.21 shows a cross-tabulation of the

importance of gun ownership to gun owners' identities and how well they think being a "supporter of women's rights" describes them.

Table 3.21   Cross-Tabulation of Importance of Gun Ownership to Identity and Supporting Women's Rights

"Supporter of Women's Rights" Describes Respondent:	Importance of Gun Ownership to Identity				
	Not at All Important	Not too Important	Somewhat Important	Very Important	Total
Extremely well	75.36	63.98	54.48	59.79	64.15
	(N = 263)	(N = 270)	(N = 146)	(N = 171)	(N = 850)
Somewhat well	17.19	28.91	36.19	28.32	27.17
	(N = 60)	(N = 122)	(N = 97)	(N = 81)	(N = 360)
Not at all well	7.45	7.11	9.33	11.89	8.68
	(N = 26)	(N = 30)	(N = 25)	(N = 34)	(N = 115)
Total	100	100	100	100	100
	(N = 349)	(N = 422)	(N = 268)	(N = 286)	(N = 1,325)

Source: 2017 Pew Research Center American Trends Panel.

a. You decide to use the data in the cross-tabulation to evaluate the blogger's claim. Specifically, you ask the question: Is there a relationship between identifying strongly as a gun owner and identifying as a supporter of women's rights? According to the table, what is the direction of the relationship between these two variables? Would you characterize it as a strong relationship? Why or why not?

b. Do you think that these results are consistent with the blogger's claim that American gun culture is sexist? Why or why not?

6. The Pew Research Center surveyed a nationally representative sample of US residents and found generational differences in support for expanding the wall along the US border with Mexico, as shown in Figure 3.10 (see facing page). (Millennials were born between 1981 and 1996, Generation Xers between 1965 and 1980, Baby Boomers between 1946 and 1964, and members of the Silent Generation between 1928 and 1945.)

a. What variables are shown in the figure?

b. Which generation is most supportive of extending the border wall between the United States and Mexico? Which is the least supportive?

7. A local reporter notices that Figure 3.10 includes a "don't know" category for opinions about extending the border wall. She suspects that those "don't know" respondents lean toward support and decides to report the results with those respondents categorized as supporting extending the wall.

a. Redraw the figure with the respondents who said they don't know what they think included in the "favor" category.

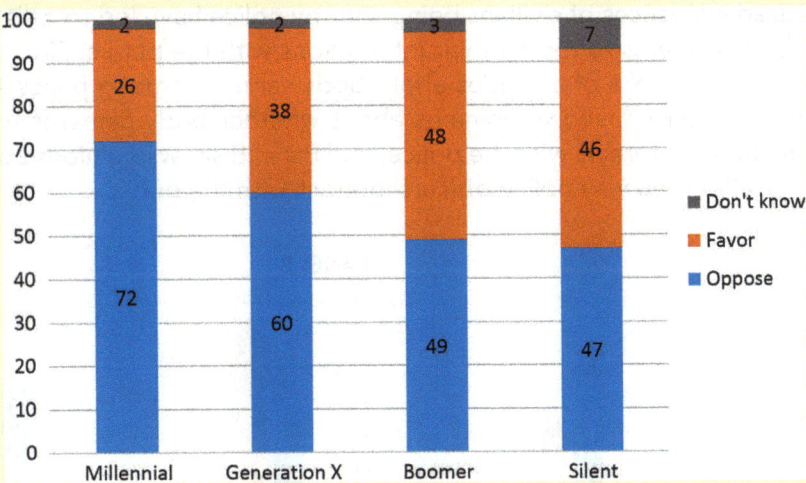

Figure 3.10

b. The reporter uses this new figure to write a story for the local newspaper with a headline that reads: "Majority of Older Americans Support Border Wall." Do you agree with her use of this headline?

8. Table 3.22 shows how family income is related to smartphone and computer access at home for US teenagers in 2022.

Table 3.22 Effect of Family Income on Smartphone and Computer Access for US Teenagers

	< $30,000	$30,000 to $74,999	$75,000 and Higher
Smartphone access:			
Yes	93%	94%	96%
No	7%	6%	4%
Total	100%	100%	100%
Computer access:			
Yes	79%	85%	94%
No	21%	15%	6%
Total	100%	100%	100%

Source: 2022 Pew Research Center, "Teens, Social Media, and Technology 2022."

a. Is there a relationship between family income and smartphone access at home? generally yes
b. Is there a relationship between family income and computer access at home? yes
c. Offer one reason why family income has different effects on smartphone access and computer access.

9. Publicized incidences of civilians being shot by police have led to calls for police to wear body cameras to record their interactions with the public. Figure 3.11 shows clustered bar graphs of attitudes about body cameras among police and the general public. Panel A shows opinions about whether body cameras will affect the way the public interacts with the police, and Panel B shows opinions about whether they will affect the way police officers interact with the public.

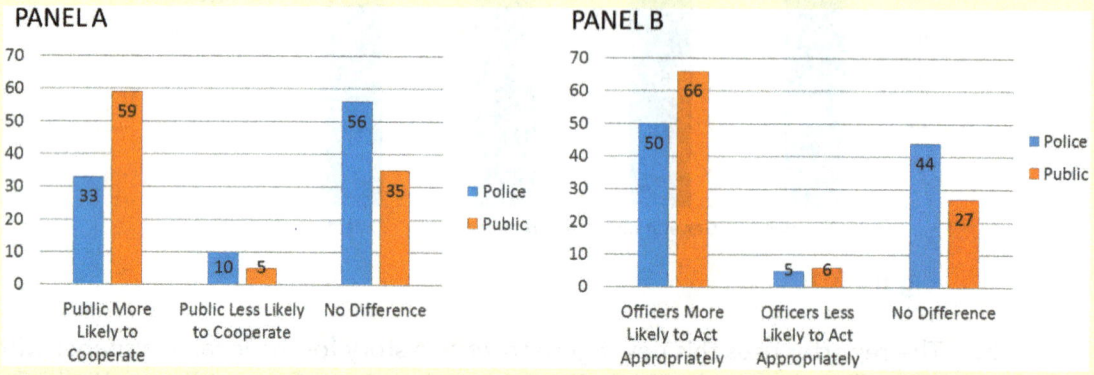

Figure 3.11

   a. What are the two variables being shown in Panel A? In Panel B?
   b. In Panel A, what percentage of police officers think that body cameras would make the public more likely to cooperate? What percentage of the public agree with that statement?
   c. In Panel B, what percentage of police officers think that body cameras would make police officers more likely to act appropriately? What percentage of the public agrees with that statement?
   d. Do we see a stronger relationship between whether someone is a member of the police and attitudes about body cameras in Panel A or B?

10. The 2021 General Social Survey asked Americans whether they thought government is responsible for raising the living standard for poor people, or whether people need to help themselves. Figure 3.12 shows the distribution of respondents' opinions by how often they reported having contact with poor people in their own lives.

   a. What are the variables shown in Figure 3.12 (see facing page)?
   b. Are these two variables related to each other, according to Figure 3.12? Explain your answer.

Stata Problems

Debates about science and religion sometimes represent the two as opposing forces. Here, you will use data from the World Values Survey (WVS) to address the question: Is religious belief incompatible with favorable views of science and technology?

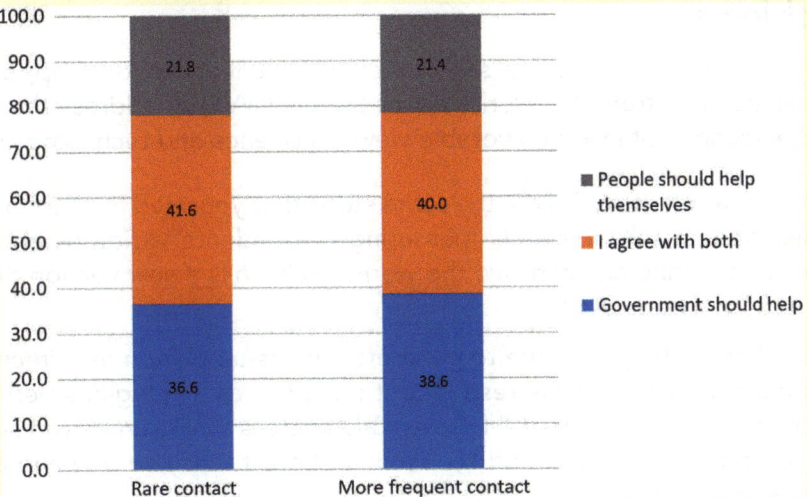

Figure 3.12

1. Open `WVSWave7.dta`. Follow the same steps that you used in chapter 2 to generate the variable called *scitech*, measuring respondents' attitudes about whether science and technology improve the world. (*Scitech* is a new version of *Q163*. See page 115 for a reminder.)

2. Use the "tabulate" command to generate a cross-tabulation of *scitech* and *Q173*, which indicates whether the respondent identifies as a "religious person," "not a religious person," or "atheist." Display *Q173* across the columns and *scitech* down the rows. Use the "row" and "column" options in your command to include row and column percentages in the cells of the table.

3. Examine the table shown in the output. Be sure to pay attention to the table's Key.

    a. Examining column percentages, which is the dependent variable, and which is the independent variable?

    b. Show by hand how the column percentage was calculated in the cell for respondents who are religious and feel that science and technology have improved the world. Explain what this percentage means.

    c. Use at least one comparison of column percentages to explain whether these results suggest that religious belief is incompatible with favorable views of science and technology.

    d. Now, examining row percentages, which is the dependent variable, and which is the independent variable?

    e. Show by hand how the row percentage was calculated in the cell for respondents who are religious and feel that science and technology have improved the world. Explain what this percentage means.

    f. Use at least one comparison of row percentages to explain whether these results suggest that religious belief is incompatible with favorable views of science and technology.

SPSS Problems

Debates about science and religion sometimes represent the two as opposing forces. Here, you will use data from the World Values Survey (WVS) to address the question: Is religious belief incompatible with favorable views of science and technology?

1. Open `WVSWave7.sav`. Follow the same steps that you used in chapter 2 to generate the variable called *scitech*, measuring respondents' attitudes about whether science and technology improve the world. (*Scitech* is a new version of *Q163*. See page 115 for a reminder.)

2. Use the "crosstab" procedure to generate a cross-tabulation of *scitech* and *Q173*, which indicates whether the respondent identifies as a "religious person," "not a religious person," or "atheist." Display *Q173* across the columns and *scitech* down the rows. Select the "row" and "column" options to include row and column percentages in the table.

3. Examine the table shown in the output.

    a. Examining column percentages, which is the dependent variable, and which is the independent variable?

    b. Show by hand how the column percentage was calculated in the cell for respondents who are religious and feel that science and technology have improved the world. Explain what this percentage means.

    c. Use at least one comparison of column percentages to explain whether these results suggest that religious belief is incompatible with favorable views of science and technology.

    d. Now, examining row percentages, which is the dependent variable, and which is the independent variable?

    e. Show by hand how the row percentage was calculated in the cell for respondents who are religious and feel that science and technology have improved the world. Explain what this percentage means.

    f. Use at least one comparison of row percentages to explain whether these results suggest that religious belief is incompatible with favorable views of science and technology.

Notes

[1] Pew Research Center. 2022. Retrieved from https://www.pewresearch.org/religion/fact-sheet/public-opinion-on-abortion/.

[2] US Census. 2019. Retrieved from https://data.census.gov/cedsci/table?q=ACSDT1Y2018.B05002&tid=ACSDT1Y2019.B05002&hidePreview=true.

Chapter 4
Typical Values in a Group
Measures of Central Tendency

Many sociologists who study higher education argue that student loan debt often prevents students from completing college degrees, tends to increase race and class inequalities, and delays or prevents decisions to have children, marry, buy homes, or even retire.[1] An estimated 43 million people, including about one-third of adults aged twenty-five to thirty-four, owe money on student loans.[2] One important step to understanding the effects of student loan debt is to understand exactly how much money people owe. What is a typical amount of student debt? A survey of borrowers (drawn from a random sample of people with credit reports at one credit reporting company) found that the mean amount—commonly called the average—owed in 2021 was $36,245.[3] President Biden's 2022 student loan forgiveness program would have forgiven $10,000 or $20,000 of debt (with the higher amount available to students who received Pell Grants, who are the lowest income students).[*] Since the average debt is markedly higher, more than $36,000, we might wonder how many borrowers would still own a lot of money after debt forgiveness.

Different ways of measuring what is a typical amount of student debt give us different understandings of this question. Instead of a mean, we might look at the median amount of debt. The median gives us the middle value, the halfway point in the distribution of student debt. The median debt in 2021, reported by the same survey, was $18,767. This means that half of debtors owe less than $18,767, and half owe more. Why is the median debt so much lower than the mean debt? Which is the best measure of typical debt? This chapter answers that question.

The mean and median are both measures of central tendency, the focus of this chapter. Each **measure of central tendency** mean, median, and mode—gives us information about typical values for a variable. The mean is the *average* value; the median is the *middle* value; and the mode is the *most frequent* value in a distribution. Whereas the descriptive statistics covered in chapter 2 focused on comparing frequencies and percentages across all

---
[*] The Supreme Court struck down Biden's debt forgiveness program in the summer of 2023.

categories of a variable, measures of central tendency condense information about a variable into a single descriptive statistic that summarizes that variable. The survey of people who owe student loan debt indicates typical debt burden, either as a mean or a median. That is what measures of central tendency do: they summarize data by indicating typical values for variables.

What Does It Mean to Describe What Is Typical?

Chapters 2 and 3 focused on frequency distributions and cross-tabulations, which we use to examine how cases are distributed across all of the categories of a variable. We can examine the distribution for one variable at a time, as with frequency distributions, or we can use cross-tabulations to examine how cases are distributed across the categories of two variables. However, as the number of categories of a variable grows, it becomes increasingly difficult to recognize patterns in frequency distributions or cross-tabulations.

Unlike frequency distributions and cross-tabulations, a measure of central tendency offers us a single number that describes a typical value for a variable. Rather than examining a frequency distribution for student loan debt, we can look at a mean and immediately get a sense of the typical student loan debt from just one number.

Knowing what is typical for a population or a group also can be very useful for understanding the value of a variable for a single case, because part of making sense of a single value involves comparing that value to the typical value for that variable. Measures of central tendency can serve as standards for assessing whether the value of a single case is high or low. One might conclude that values that are less than the typical value are "low," and values that fall above the typical value are "high."

Researchers regularly use measures of central tendency, and so do people in their daily lives. Returning to the topic of student loan debt, as students make plans for how to finance college, they may wonder whether they are taking on too much debt. But how do they know how much debt is too much? Some probably feel less concerned about their own debt load if they know that it is not too far from the typical debt load. According to the Pew Research Center, the average student loan taken out in a year increased from $2,100 in 1990–91 to $6,276 in 2019–20 (in constant 2020 dollars).[4] A student taking out a student loan of $5,000 in 1990 may have felt concerned that his loan was higher than the mean student loan, but a student in 2019 may have felt reassured that his loan of $5,000 was lower than the typical student loan. Making sense of the individual case's value for a variable ($5,000 in student debt per year) involves comparing the value to a measure of central tendency.

Consider the example of an employee who suspects that her pay is unfair. How would she begin to assemble evidence for this claim? Perhaps her first step would be to ask her coworkers how much money they make. If she can collect that information from every member of her department, she can calculate measures of central tendency for pay in her department. She can then determine whether her own pay falls above or below the typical pay in her department. If her pay is lower than what is typical, she might decide to ask for a raise.

There are different ways of thinking about—and calculating—what the typical value of a variable is. Which measure you choose depends on the nature of the variable you're studying and the purpose of your analysis. Specifically, it depends on the variable's level of measurement (nominal, ordinal, or interval-ratio) and whether the distribution of the variable is skewed, as we will explain in this chapter. In the next three sections, we take a close look at the three measures of central tendency: mean, median, and mode.

Mean

The most commonly used measure of central tendency is the **mean**, also referred to as the average. The mean is the sum of all values for a variable divided by the number of cases. Calculating a mean is two-step process. First, we sum all of the values for the variable in our sample or population. Second, we divide the sum by the number of cases in the sample or population. The equation for calculating a mean is:

$$\bar{y} = \frac{\sum y}{N}$$

The capitalized Greek letter sigma, $\sum$, indicates that we are adding together the values of the variable (y). N indicates the number of cases.

Means should be calculated only for interval-ratio variables. Why? When we add the values together, we are assuming that the actual value of the variable is numerically meaningful and that the distance between values is consistent. A value of 4 for an interval-ratio variable is 6 units lower than a value of 10, and that 6-unit difference is the same as the distance between a 10 and a 16 for that variable. We should not calculate the mean for ordinal or nominal variables because in both cases the categories are not separated by quantitatively equal intervals and the categories are not numerically meaningful. (As we discussed in chapter 1, in practice many researchers treat ordinal scales with five or more categories as if they were interval-ratio variables. For example, a researcher might treat an item that asks people to rate their agreement with a statement on a scale from 1 [strongly disagree] to 10 [strongly agree] as interval-ratio.)

BOX 4.1: IN DEPTH

Why the Mean Has No Meaning for Nominal-Level Variables

Consider a nominal variable for religious affiliation with the following eight categories: Christian, Jewish, Muslim, Buddhist, Hindu, folk religion, other religion, and unaffiliated. Recall that software packages assign numbers even to the categories of nominal variables. Imagine that your statistical software assigned values to the categories of the religious affiliation variable such that Christian = 1, Jewish = 2, Muslim = 3, Buddhist = 4, Hindu = 5, folk religion = 6, other religion = 7, and unaffiliated = 8. Now imagine that we asked the software to calculate the mean for this variable, which it would do, and it returned a mean of 2.3. Would it make any sense to say that the mean value for religious affiliation is 2.3? Or that the average religion is somewhere between Jewish (coded as 2) and Muslim (coded as 3)? No. We can see that this mean has no meaning because the categories are neither rank-ordered nor spaced at quantitatively equal intervals.

Let's turn to an example that many of us are familiar with from our own lives: using technology to track one's health practices. Consider, for instance, an app that tracks users' steps throughout the day. Research finds that taking at least eight thousand steps a day is associated with many health benefits, although most adults in the United States walk far fewer steps.[5] The app reports the daily total number of steps the user walked and calculates the mean number of steps over specified time periods so that users can track their activity levels. After their first week of using the app, one user wants to get a sense of their typical number of steps for the week. Table 4.1 shows their daily steps.

What is their mean number of daily steps for the week? We calculate the user's mean daily step number by summing all seven of their daily step totals and dividing by the number of scores, in this case seven.

$$\bar{y} = \frac{\Sigma y}{N} = \frac{4870 + 4831 + 4655 + 4711 + 5034 + 4900 + 4829}{7}$$

$$= \frac{33{,}830}{7} = 4{,}832.9$$

Table 4.1 Daily Total Steps, Week 1

Day	Daily Steps
Monday	4,870
Tuesday	4,831
Wednesday	4,655
Thursday	4,711
Friday	5,034
Saturday	4,900
Sunday	4,829

The user's mean number of daily steps for their first week is 4,833. (Note that we rounded to the nearest whole number.)

Should the user be satisfied with this typical number of daily steps of 4,833? In order to answer that question, we need to know what standard they want to use. Do they want their typical steps to reflect an unyielding commitment to a high level of daily activity, or do they prefer to achieve more of a balance between a strong commitment to a high level of activity and a total lack of commitment? The latter view would reflect Aristotle's view that virtue can be found in moderation. Known as the "golden mean rule," the idea is that we achieve virtuous character by striking a balance between deficiency and excess. Compared to the recommended 8,000 or more steps per day, a mean number of steps of 4,833 is certainly fewer, but still represents a considerable amount of activity. If the user wants their activity to be consistent with Aristotle's golden mean, then they might conclude that they have achieved this goal for the first week.

Now consider the user's second week of using the app. Table 4.2 shows their scores for Week 2.

Table 4.2 Daily Total Steps, Week 2

Day	Daily Steps
Monday	12,657
Tuesday	11,504
Wednesday	3,218
Thursday	3,100
Friday	2,024
Saturday	789
Sunday	538

What is their mean number of steps for the second week? Summing the steps and dividing the sum by the total number of days yields the mean:

$$\bar{y} = \frac{12{,}657 + 11{,}504 + 3218 + 3100 + 2024 + 789 + 538}{7}$$

$$= \frac{33{,}830}{7} = 4{,}832.9$$

The user's mean number of daily steps is 4,833 for the second week, the same as the mean for the first week. They might conclude that they have continued to abide by the golden mean, maintaining a moderate level of activity.

However, a close inspection of the daily scores in Table 4.2 shows that the user has not actually walked a moderate number of steps each day. The user began the week with a high number of steps, 12,657, on Monday—well over the recommended level of 8,000 steps. Their steps on the next day were still high, but they declined steadily over the week. By the weekend, the user abandoned their healthy activity level, walking only 789 steps on Saturday and 538 on Sunday.

Comparing the weekly steps in Tables 4.1 and 4.2 shows that a different set of values can yield exactly the same mean. In samples or populations of equal size, as long as the sums of the values are the same, the mean will be the same. Notice that the sum of daily steps for both weeks was 33,930. In the first week, most days' steps were not too far above or below the mean, but in the second week, most days' steps diverged a great deal.

Table 4.3 shows us how much each day's number of steps diverged from the mean for each week. The numbers in the table were obtained by subtracting the mean from each day's steps. For example, for Monday of week 1, 4,870 − 4,833 = 37. A positive number indicates that day's steps were above the mean, while a negative number indicates that they were below the mean.

Table 4.3 Divergence Daily Steps from Weekly Means

	Week 1 Divergence from Mean	Week 2 Divergence from Mean
Monday	37	7,824
Tuesday	−2	6,671
Wednesday	−178	−1,615
Thursday	−122	−1,733
Friday	201	−2,809
Saturday	67	−4,044
Sunday	−4	−4,295

Remember that the mean number of steps for both weeks was 4,833. We can see that the user's daily steps for Week 1 did not diverge dramatically from the mean. The largest divergence from the mean was Friday's steps, which were 201 steps above the mean. During Week 2, we see much larger divergences from the mean. Despite having the same mean for both weeks, the user did not have a moderate level of activity on each day of the second week. During Week 2, we might say that the user's activity did not follow Aristotle's golden mean rule but instead followed a saying attributed to Ralph Waldo Emerson: "Moderation in all things, especially moderation."

We see here that measures of central tendency tell us something about typical values in a population or sample, but they do not tell us about how individual values are spread out around those typical values. This idea, variability, is the topic of chapter 5.

BOX 4.2: IN DEPTH

The Mean as Redistribution: Spreading the Resources

One way of looking at the mean is to think about it as redistributing the attribute measured by the variable equally across all members of a group. When a variable is measuring an attribute such as income, height, hours spent working, or number of children, we can think of the mean as taking the resources possessed by each group member (their income, height, hours of work, or number of children) and putting them into a common pool. We can think of this pool as the collective resources of the group. The mean then divides these pooled resources equally among each member of the group. The mean tells us how much of the resource each group member would possess if the collective resources of the group were distributed equally across the members.

Take the example of trees in cities. City blocks with more trees have cooler temperatures and residents of areas with more trees report better health and well-being. Figure 4.1 offers a visual representation of four hypothetical city blocks with different numbers of trees. As you can see, the number of trees on these blocks varied, with Block 1 having the highest number of trees (6) and Block 2 having the lowest number (2).

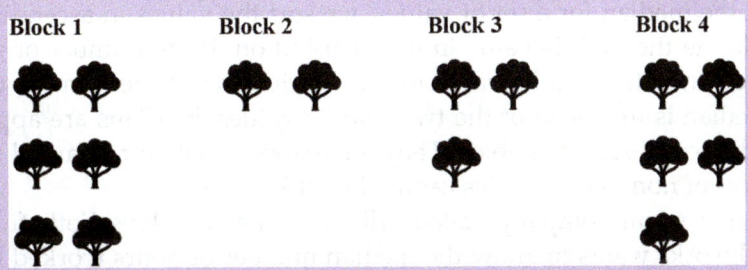

Figure 4.1

What if all the blocks shared the resource of trees equally? How many trees would each block have? Figure 4.2 shows how we use the formula for a mean to find the answer.

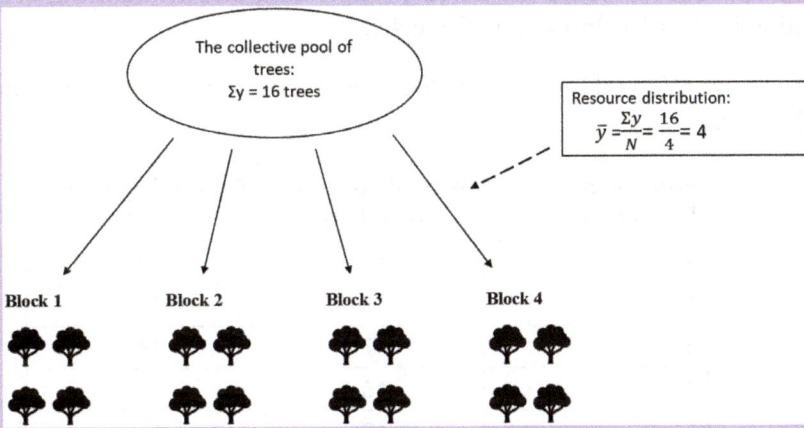

Figure 4.2

If we sum all of the trees on these four blocks, we see that they collectively possess 16 trees. To find out how much of this resource each block would possess if it were distributed equally, we divide the collective amount (16 trees) by the number of blocks (4). This yields a mean number of trees of 4. Each block would have four trees if this resource were shared equally among them.

Median

Another measure of central tendency is the **median**. Like the mean, it tells us something about the typical value for a variable in a sample or population. Unlike the mean, the median is not the result of a calculation that involves all of the values for a variable. The median is the typical value of a variable in the sense that it is the middle value. The median is the value that lies in the middle of a sorted distribution of values, such that half of the values lie below the median and half lie above it.

To find the median for a set of values, we sort the values in ascending or descending order and locate the middle value in the distribution. If the number of values, N, is odd, then the median is the middle value in the distribution. If the number of values is even, then the median is the mean of the two middle values. Medians are appropriate for both ordinal and interval-ratio variables. They are inappropriate for nominal variables because the categories of nominal variables cannot be rank-ordered.

Consider a small company called Allworks that employs both full- and part-time workers. Allworks wants to know the median number of hours worked by its twenty-one employees during the last week. The values for the twenty-one employees are as follows:

25, 20, 55, 39, 40, 17, 60, 18, 40, 29, 18, 43, 12, 48, 30, 27, 15, 45, 32, 26, 40

The first step in finding the median is to arrange the values in either ascending or descending order. The values are listed in ascending and descending order, along with their position in the distribution, in Table 4.4.

Table 4.4 Hours Worked for Twenty-One Employees, Ranked in Ascending and Descending Order

Position in Distribution	Ascending Order of Values	Descending Order of Values
1	12	60
2	15	55
3	17	48
4	18	45
5	18	43
6	20	40
7	25	40
8	26	40
9	27	39
10	29	32
11	30	30
12	32	29
13	39	27
14	40	26
15	40	25
16	40	20
17	43	18
18	45	18
19	48	17
20	55	15
21	60	12

The second step is to determine where the middle value lies. The middle value occupies the position in the distribution at which exactly half of the values lie below it and half of the values lie above it. When N is an odd number, as it is in this case (N = 21), the middle value is located by adding 1 to N and dividing by 2. The formula for finding the middle position in a list of values where N is odd is:

$$Median\ case\ number = \frac{N+1}{2}$$

In this case, the middle position in our set of hours worked is:

$$\frac{(21+1)}{2} = \frac{22}{2} = 11$$

Ten values lie below the 11th position, and ten values lie above it.

The last step in finding the median is to identify the value that occupies the middle position in the distribution. In Table 4.4, we see that 30 occupies the 11th position. Thus, 30 is the median number of hours worked last week for the twenty-one employees at this company. In other words, half of the workers at Allworks worked more than 30 hours a week, and half worked fewer than 30 hours a week. Notice that the median is the same whether we rank the list in ascending or descending order. Remember, the median is the value of the variable (30) not its position in the rank-ordered distribution (11).

When finding the median for a set of values where N is an even number, we follow the same steps, but there is a slight variation in how we identify the middle position in the distribution of values. This is because when N is even, there is no case that is in the exact middle of the distribution.

Imagine that Allworks had twenty-two employees instead of twenty-one. Unlike when there were twenty-one employees, when there are twenty-two of them, there is not a single position in the distribution for which half of the values fall above it and half fall below it. Instead, there are two middle positions in a distribution with an even number of cases. We use the same formula as above to identify the middle position in an even distribution:

$$Median\ case\ number = \frac{N+1}{2} = \frac{(22+1)}{2} = \frac{23}{2} = 11.5$$

$\frac{8+1}{2} = \frac{9}{2} = 4.5$

Of course, there is no 11.5th position in the distribution. Instead, 11.5 tells us that the two middle positions in the distribution are 11 and 12. Half of the values in the distribution lie below 11 and 12, and half of them lie above 11 and 12. When N is even, the median is equal to the mean of the values that occupy the two middle positions in the distribution.

If the extra employee at Allworks worked 62 hours during the week for which employees reported their hours, the rank-ordered distribution of hours now looks like Table 4.5.

We identified 11 and 12 as the two middle positions in the distribution. We see that the two values occupying those positions are 30 and 32 hours (regardless of whether the distribution is in ascending or descending order). The median number of hours worked for this group of twenty-two employees is:

$$\frac{30+32}{2} = 31 \text{ hours}$$

Half of the employees worked more than 31 hours a week, and half worked fewer than 31 hours a week.

Table 4.5 Hours Worked for Twenty-Two Employees, Ranked in Ascending and Descending Order

Position in Distribution	Ascending Order of Values	Descending Order of Values
1	12	62
2	15	60
3	17	55
4	18	48
5	18	45
6	20	43
7	25	40
8	26	40
9	27	40
10	29	39
11	30	32
12	32	30
13	39	29
14	40	27
15	40	26
16	40	25
17	43	20
18	45	18
19	48	18
20	55	17
21	60	15
22	62	12

Take a moment to calculate the mean number of hours worked for these twenty-two employees and compare it to the median.* Why is there a difference? The difference occurs because the mean value is influenced by all of the other values in the distribution in a way that the median is not. Unusually high or low values affect the mean, making it higher or lower, respectively. But the median is the same regardless of whether the cases on either side of the median have values very close to or much higher or lower than the median.

---

* $\bar{y} = 33.68$

In our Allworks example, the median would be 31 regardless of whether the employees who worked more than 31 hours a week all worked 70 hours a week or if they all worked 32 hours a week. We will say more about this topic later in the chapter, in the section on choosing an appropriate measure of tendency, but first we will examine a final measure of central tendency, the mode.

> **BOX 4.3: IN DEPTH**
>
> ### When Values for the Mean and Median Do Not Exist in the Population
>
> You may have observed something interesting from the previous examples for means and medians: it is possible for a mean or a median to be a value that does not actually exist in the sample or population. The user's mean number of steps of 4,833 was not an actual value for their daily score in either of the weeks that they used the app (see Tables 4.1 and 4.2). When Allworks had twenty-two employees, the median number of hours worked, 31, was not an actual number of hours worked by any of those employees in the previous week (see Table 4.5).
>
> This observation that means and medians do not have to be actual values in the sample or population reminds us that these measures of central tendency do not tell us about the typical case in a sample or population (e.g., typical people, typical countries, typical organizations). They identify the typical *value* for a variable. In fact, this is an important distinction to keep in mind for all kinds of statistics. In statistics, what we are directly studying is the distributions of variables and the relationships between them. Although variables of interest in the social sciences often measure attributes of people, we do not study people directly in statistics. We study variables that measure attributes of people. In the social sciences, we use what we learn about these variables, and relationships between them, to draw conclusions about how people operate in the social world.

Mode

The final measure of central tendency, the **mode**, is the value that occurs most frequently in a sample or population. Unlike means and medians, modes can be found for any kind of variable—nominal, ordinal, and interval-ratio. In fact, they are more commonly used for nominal than ordinal or interval-ratio variables.

Recall our example of the religious affiliation variable from Box 4.1. The eight categories of the variable are Christian, Jewish, Muslim, Buddhist, Hindu, folk religion, other religion, and unaffiliated. Since this is a nominal variable, finding the mean or median would be inappropriate, but it could be quite useful to identify the modal category for the variable. Table 4.6 presents the frequency distribution of religious affiliation for the world population, according to the Pew Research Center.[6]

What is the modal category for the religious affiliation variable? Christian is the mode because this value occurs more frequently than any other, with 31.2% of the world population claiming this affiliation. Interestingly, when we examine this same variable but limit it to residents of the United States, the modal category is still Christian, but it accounts for a

Table 4.6   Frequency Distribution for Global Religious Affiliation, 2015

Religion	Percent
Christian	31.2
Muslim	24.1
Unaffiliated	16.0
Hindu	15.1
Buddhist	6.9
Folk religion	5.7
Other religion	0.8
Jewish	0.2
Total	100

Source: Pew Research Center.

much larger share of the population of the United States (63% in 2021) than it does for the world population.[7]

In addition to indicating the category that occurs most frequently for a given variable, modes sometimes reflect what is considered to be normative in a group or society. For example, a team of researchers used a sample of almost 4,989 undergraduates at a large university in 2020 to study the sexual practices of college students.[8] They found that 1,677 of the women in the sample had ever had vaginal intercourse, while 559 had not. Among the men in the sample, 1,577 had ever had intercourse, and 603 had not. What is the variable that we are examining here, and what are its categories? What is the modal category for women? For men? What might the modal category say about normative sexual practices among college students?

The variable is having ever had vaginal intercourse, with the categories being having done so or not. The modal category for both women and men is having had intercourse. That is, it was more common for both genders to have had vaginal intercourse than not. To compare the genders, you can calculate the percentage for each: For women, the total N = 2,236. $\frac{1,677}{2,236}(100) = 75$. Of the women, 75% reported having had intercourse. For men, the total N is 2,180 and the percentage is 72. This suggests that having had sexual intercourse (or at least reporting having had that kind of sex!) is normative for both college men and women.

Similarly, we can use the mode to identify winners of elections. Imagine a race for governor with three candidates running. Candidate A wins 60% of the vote, candidate B wins 30% of the vote, and candidate C wins 10% of the vote. The modal category is the one that received the highest share of the votes—candidate A in this example. Most elections in the United States function this way, with the winner being the candidate who has won the most votes, or the "modal" vote-getter. American presidential elections are a notable exception because it is not the national popular vote that determines the winner. Rather, the Electoral College determines the winner, and this creates the possibility that the candidate who has received the most votes nationally (the "modal" vote-getter) may lose the election. In American history, the winner of the national popular vote has lost the presidential election on five occasions, most recently in 2016.

Finding the Mode, Median, and Mean in Frequency Distributions

So far, we have covered how to find the mean, median, and mode using raw data. What about when the data are presented in a frequency distribution? For example, in 2021, the General Social Survey (GSS) asked respondents how many times a week they ate beef, lamb, or products that contain them.[9] The question was asked as part of a module of questions about activities related to climate change. Their responses are presented in Table 4.7.

Table 4.7 Number of Days a Week People Eat Beef or Lamb

Number of Days	Frequency	Cumulative Frequency	Percent	Cumulative Percent
0	214	214	11.9	11.9
1	315	529	17.6	29.5
2	365	894	20.3	49.8
3	342	1,236	19.1	68.9
4	206	1,442	11.5	80.4
5	163	1,605	9.1	89.5
6	67	1,672	3.7	93.2
7	123	1,795	6.9	100.1
Total	1,795		100	

Source: 2021 General Social Survey.

Finding the Mode in a Frequency Distribution

It is easy to find the mode in a frequency distribution. It is the category with the largest number and the highest percentage. In Table 4.7, it is 2. That is, the most common number of days a week people eat beef or lamb is two; 365 people, or 20.3%, eat beef or lamb two days a week.

Finding the Median in a Frequency Distribution

To find the median in a frequency distribution, we need to find the value of the case that is in the middle of the distribution. We can use cumulative frequency with the same formula we used earlier to find that case:

$$\text{Median case number} = \frac{N+1}{2}$$

In this distribution, this is: $\frac{1795+1}{2} = 898.$

The median case number is 898. Looking at the cumulative frequencies, we see that case number 898 would fall among those who eat beef or lamb on three days a week. Case numbers 1 through 214 eat beef or lamb zero days a week; case numbers 215 through 529 eat beef or lamb one day a week; case numbers 530 through 894 eat beef or lamb two days a week; and case numbers 895 through 1,236 eat beef or lamb three days a week.

Therefore, we can say that the median is three. Half of respondents ate beef or lamb three or fewer days a week, and half ate beef or lamb three or more days a week.*

We could also use cumulative percentage to find the value of this case, because the median is the case that is at a cumulative percentage of 50. Looking at Table 4.7, we see that 49.8% of people ate beef or lamb two or fewer days a week and 68.9% ate it three or fewer days. The 50th percentile would fall within the cases that ate beef or lamb three or fewer days a week.

Finding the Mean in a Frequency Distribution

Finding the mean in a frequency distribution is a little more complicated than finding the median. We use the same principle as we did when finding the mean with raw data. We sum the values of the variable for all of the respondents and divide by N. In order to sum the values of the variable for all respondents, we have to multiply each value of the variable by the frequency for that value.

Why is this? Imagine that instead of a frequency distribution, we had a list of the individual values for all 1,795 respondents, which we would add up. Multiplying each value of the variable (y) by the number of respondents (f) for that value yields the same result. Look at the first category in Table 4.7: 214 people eat beef or lamb 0 days a week. In order to account for all 214 of these households, we multiply 214 by 0. This is faster than adding 0 + 0 + 0 and so forth until we had added 214 cases. The same is true for each value in Table 4.7.

The formula for finding the mean in a frequency distribution is:

$$\bar{y} = \frac{\Sigma fy}{N}$$

Pay attention to the order of operations in this formula. First, multiply each value of the variable (y) by the frequency for that value (f), and then sum the results. For the number of days a week people eat beef or lamb, Table 4.8 shows *fy* and *Σfy*.

$$\text{Thus: } \bar{y} = \frac{\Sigma fy}{N} = \frac{4,973}{1,795} = 2.8$$

The mean number of days a week that people ate beef or lamb was 2.8. We could also say that, on average, people ate beef or lamb 2.8 days a week. A mean calculated in this

---

* You may have noticed that, in this example, we say that half of the cases fell *at* or below the median of three, and half of the cases fell *at* or above it. But on page 159 we said that half of the cases fall below the median, and half fall above it. Why is this description slightly different? It is because there are many cases in this sample with the same value. In the present example, we used a cumulative frequency distribution to find the median. We see that only 20% of the cases lie below three, and 31% fall above three. Thus, half of the values fall *at* or below three, and half fall *at* or above three.

Table 4.8 Finding the Mean in a Frequency Distribution

Number of Days	Frequency	fy
0	214	0
1	315	315
2	365	730
3	342	1,026
4	206	824
5	163	815
6	67	402
7	123	861
Total	1,795	Σfy = 4,973

Source: 2021 General Social Survey.

way is also referred to as a "weighted mean," because we are weighting each value of the variable by the number of cases that have that value.

Remember, the mean is a characteristic of the sample as a whole, not of individual respondents. Of course, there are no fractions of days a week; no respondent ate beef or lamb exactly 2.8 days a week. Also notice that the mean and the median are very close to each other. The mean is just a little lower lower than the median of 3, suggesting that the sample is not particularly skewed. This is one issue to consider when choosing a measure of central tendency.

Choosing the Appropriate Measure of Central Tendency

The mean, median, and mode can all be useful in understanding patterns in data, but it is important to choose the best measure for the job. First, consider the *variable's level of measurement*: Is it nominal, ordinal, or interval-ratio? The mode can be used with any variable but will be most useful if the variable has a relatively small number of categories. The median can be used for ordinal or interval-ratio variables. The mean can be used only with interval-ratio variables. But remember that researchers frequently treat ordinal scales with with five or more categories as interval-ratio.

The other major factor to consider is the *shape of the frequency distribution*. If a variable is distributed symmetrically with the highest numbers in the middle, the mean, median, and mode will be identical, as in Figure 4.3.

On the other hand, if a variable has cases at one extreme of the distribution, the mean and median will differ from each other. Cases at the very high or very low end of a distribution are called "outliers." When a variable has outliers on one extreme, the distribution is skewed. The median is not affected by such scores, but the mean is pulled in their direction. In other words, if there are a small number of cases with very high values, the mean will be higher than the median; if the small number of cases have very low values, the mean will be lower than the median. Figure 4.4 shows a skewed distribution. Like Figure 4.3, Figure 4.4 contains seventeen cases. Unlike Figure 4.3, there are now three outlier cases, with values of 1 and 3.

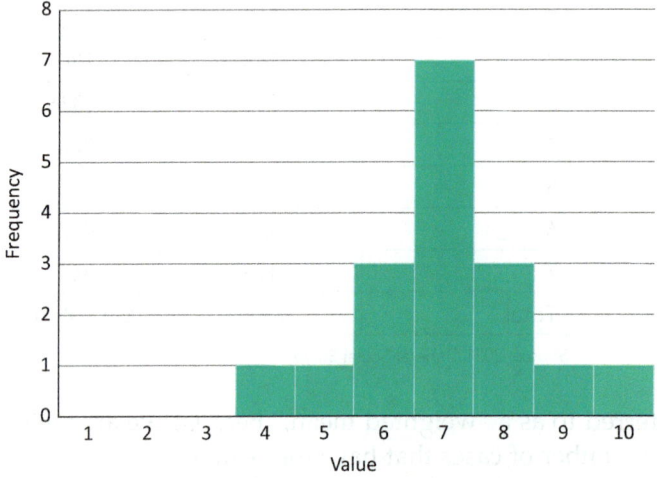

Figure 4.3 Symmetrical, Unimodal Distribution

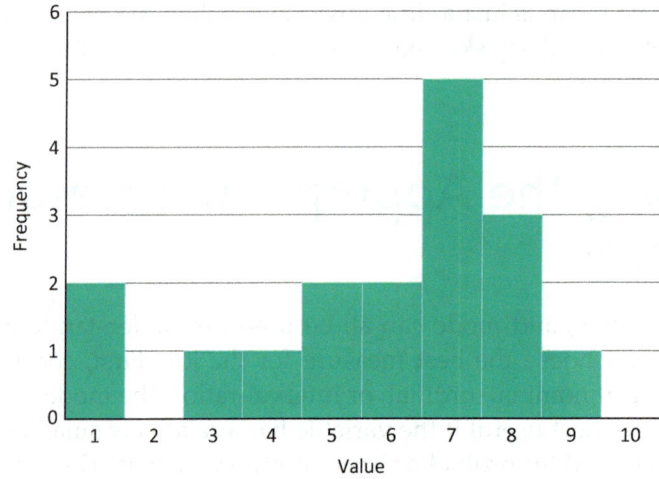

Figure 4.4 Distribution with Outliers

Notice that the median and mode for Figure 4.4's distribution are the same as in Figure 4.3. The median and mode in both distributions are 7, reflecting the fact that the median and mode are unaffected by the presence of a small number of outliers. However, the mean for Figure 4.4 (5.82) is lower than the mean for Figure 4.3 (7), because the outliers pull the mean toward them. In the case of Figure 4.4, the small number of outliers with low values pull the mean downward. It is skewed.

When deciding whether to rely on the mean or the median, it is useful to look at the shape of the distribution to see if it is skewed. The easiest way to do this is with a histogram.

For example, look at Figure 4.5, which is a histogram for the number of hours worked last week from the 2021 GSS.

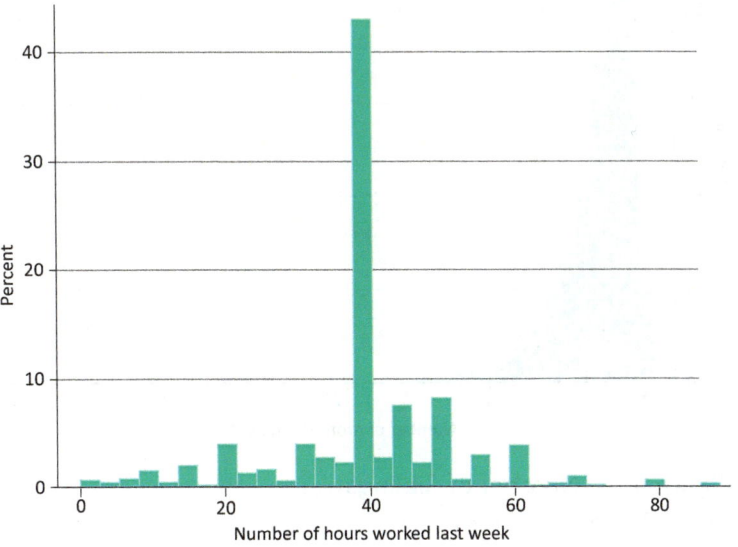

Figure 4.5 Number of Hours Worked Last Week
Source: 2021 General Social Survey.

While the distribution is not exactly symmetrical, it is relatively close. (Very few social science variables will be exactly symmetrical.) Most of the scores are in the middle, with decreasing numbers of scores on both low and high ends. There are outliers on both sides of the distribution. As expected in a fairly symmetrical distribution, the mean of 40.0 (which we rounded from 39.99) hours and the median of 40 hours are the same.

In this case, we could use either the median or the mean. Reporting our results for the mean, we would say, "On average, respondents worked 40 hours last week." For the median, we would say, "Half of the respondents worked 40 or more hours last week, and half worked 40 or fewer hours." Note that the modal category is also 40. We would report these results by saying, "The most common number of hours worked was 40." Since 40 hours a week is considered working full-time in the United States, it is not surprising that all three measures of central tendency are near 40.

In contrast, Figure 4.6 shows the histogram for the number of siblings that respondents to the 2021 GSS reported.

We can see that this distribution is quite skewed, with the majority of respondents reporting that they have one, two, or three siblings, smaller percentages reporting four to nine siblings, and very small percentages reporting more siblings; although not visible on the graph, the maximum number of siblings reported was thirty-five.

The outliers, or the "tail" of the distribution, are to the right, so we say that it is skewed to the right or that it is positively skewed. The mean is always pulled in the direction of the skew, so, in this distribution, we would expect the mean to be higher than the median. And in fact, the mean is 3.13, and the median is 2. When distributions are skewed like this, the median is the preferable measure. If you were reporting these results, you would say, "Half of respondents had two or fewer siblings, and half had

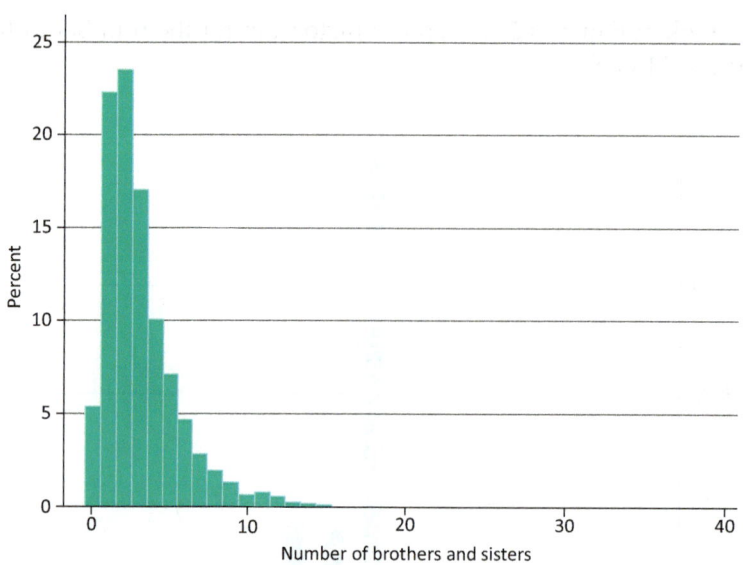

Figure 4.6 Number of Siblings
Source: 2021 General Social Survey.

two or more siblings." (You would also present a frequency distribution, as discussed in chapter 2.)

There is one more consideration in choosing a measure of central tendency, also related to the shape of the distribution: Does the distribution have a single mode, or does it have more than one value that are much more common than others? When a variable has two modes, we call the distribution bimodal. The frequencies for the most common values do not have to be identical in order to consider the frequency bimodal. For example, look at Figure 4.7, which shows the number of children reported by respondents to the GSS.

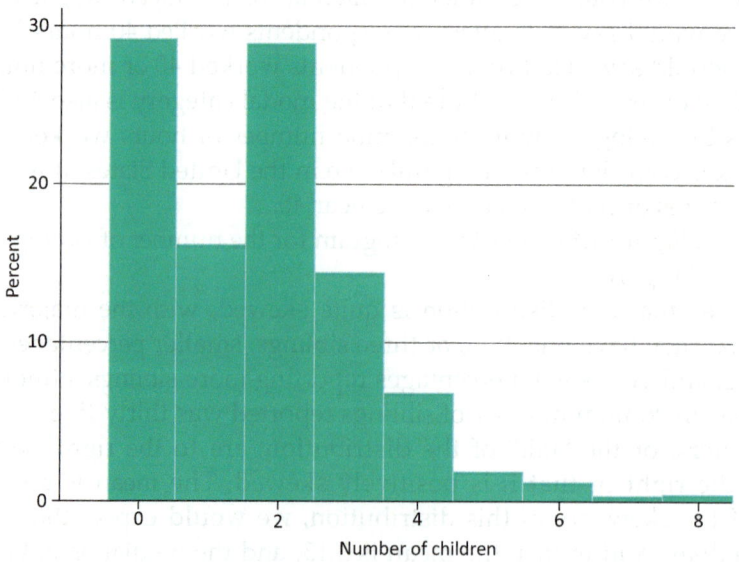

Figure 4.7 Number of Children
Source: 2021 General Social Survey.

We can easily see that there are two values that are much more common than the rest: 0 and 2. That is, respondents tended to report having either no children or two children. We could calculate the mean or the median for this distribution, but they are somewhat misleading. The mean is 1.7 children, and the median is 2. Knowing that, on average, people had 1.7 children or that half of the respondents had two children or fewer and half had two children or more gives us a different—and less accurate—impression of the distribution.

In a case like this, an analyst should report that the number of children people have is close to bimodal and then report the percentages at the most common values. In this case, 29% had no children and 29% had two children. The analyst might also report the mean or median, or might calculate the average number of children for respondents who had children (omitting those who reported having none).

Median versus Mean Income

Income is a good example of a highly skewed variable that is of great interest to social scientists. According to the US Census Bureau, the mean income of workers in the United States in 2021 was $57,143. That is the amount that each worker would receive if all pay were evenly distributed across all workers. In contrast, the median was $37,522.[10] Half of all workers earned less than $37,522, and half earned more. Why is the median so much lower than the mean? Because a few workers earn much more than others, and the distribution is therefore highly positively skewed.

The greater the difference between the mean and the median, the more skewed the variable is. Income, for example, has become more skewed over time, as income inequality has increased.

Table 4.9 shows mean and median income in the United States in 1950 and 2021 as well as the relationship between median and mean income. We look at this relationship by calculating the median as a percentage of the mean. To do this, we divide the value for median income by the value for mean income (i.e., 37,522/57,143 = .657, or 65.7%). This is a good measure of how much lower median income is than mean income, which tells us how skewed income is. When the median is less than 100% of the mean, this suggests some degree of skew in the distribution, with high values in the right tail of the distribution driving up the mean. The smaller the median is as a percentage of the mean, the more skewed the distribution is.

Table 4.9 Mean and Median Individual Income in Constant 2021 Dollars

	1950	2021
Mean	$23,408	$57,143
Median	$19,418	$37,522
Median as percentage of mean	83.0%	65.7%

Source: US Census Bureau, 2021.

As Table 4.9 shows, in 1950, income was much less skewed than it was in 2021. The 1950 median income was 83% of mean income, while by 2021 median income had dropped to approximately 66% of mean income. The income distribution became more skewed over time, meaning that income was more unequally distributed.

Which is the best measure of income, median or mean? If your purpose is to describe what is typical about income, median is the best measure. The mean is too distorted by the small number of high values.

The shape of a variable's distribution helps determine which measure of central tendency is most appropriate. When we think about a variable's distribution, it is important to remember that the values of variables have two meanings. One is as a description of a unique case, such as the employment status of your uncle or the literacy rate in California. The other meaning is as a description of a collective pattern, such as employment status or literacy rates of states in aggregate.[11] Data in the second sense are not about describing individuals but rather about describing overall patterns in the variable of interest. When we looked at frequency distributions in chapter 2, we began describing distributions in this sense. The measures of central tendency are an additional way to do so. Looking at visual pictures of distributions is another.

Chapter Summary

There are three measures of central tendency: mean, median, and mode. They all describe the "typical" value of a variable.

Mean:

- Use with interval-ratio variables only (but many researchers use ordinal scales with five or more categories as interval-ratio).
- The average.
- $\bar{y} = \dfrac{\Sigma y}{N}$
- $\bar{y} = \dfrac{\Sigma fy}{N}$ (for a frequency distribution).
- Interpretation: "The average number of pets owned was 2.5." "The mean number of pets owned was 2.5." "On average, respondents owned 2.5 pets."

Median:

- Use with interval-ratio or ordinal variables.
- The middle of the distribution or the 50th percentile; the value at which half of the cases fall below and half fall above.
- Median case number = $\dfrac{N+1}{2}$ The median is the value of that case (or the value of the mean of the values of the two cases on either side if N is an even number).
- Interpretation: "Half of respondents had more than two pets, and half had fewer."

Mode:

- Use with any level of measurement.
- The only measure that can be used with nominal variables.
- The most frequently occurring value.
- Interpretation: "Cats were the most common pet." "The most common number of pets was one."

Choosing a measure of central tendency:

- Consider level of measurement.
- For interval-ratio variables, look at the shape of the distribution with a histogram in order to choose between mean and median.

- Skewed distributions have a relatively small number of cases with values at the extreme high or low end of the distribution. Positively skewed distributions have a relatively small number of cases with values at the extreme high end—or right side—of the distribution. Negatively skewed distributions have a relatively small number of cases with values at the extreme low end—or left side—of the distribution.
- Skewed distribution: Use median.
- Not highly skewed distribution: Use mean.
- Bimodal distribution: Report the modes only.

Using Stata

In 2020, the ANES asked respondents who had heard about the presidential campaign on the internet whether they regularly visit a list of news websites. The survey asked about fifteen specific sites (with variable names in parentheses): Yahoo News (*V201634a*), CNN (*V201634b*), Huffington Post (*V201634c*), New York Times (*V201634d*), Breitbart News Network (*V201634e*), Fox News (*V201634f*), Washington Post (*V201634g*), The Guardian (*V201634h*), USA Today (*V201634i*), BBC News (*V201634j*), NPR News (*V201634k*), Daily Caller (*V201634m*), Bloomberg (*V201634n*), Buzzfeed (*V201634p*), and NBC News (*V201634q*).

For each variable, 1 indicates that the person visits the news site at least once per month, and 0 indicates that the person does not visit the site. In this section, we will explore measures of central tendency for these internet news variables to get a sense of typical internet news usage.

Generating a New Variable from a Set of Existing Variables

Before finding measures of central tendency for internet news use, we will use the fifteen news site variables to create a new interval-ratio news site variable that measures the number of news sites that the respondent regularly uses. We will combine the fifteen variables into a new variable, called *newssites*, such that the new variable counts the number of sites that a person uses regularly. Learning how to create a new variable from a set of existing variables is an important skill and one that researchers frequently use.

Without knowing what the actual values are in the data set, we can determine that this new variable could range from 0 (meaning that the respondent uses none of the sites) to 15 (meaning that the respondent uses all of the sites). How should we go about combining these variables? Notice that what we basically want to do is to sum the values for all fifteen variables for each respondent in the data set. In cases where we want to sum the values of a set of variables to create a new variable, Stata offers us a simple shortcut. Open `ANES20.dta` and type the following command into Stata to create a new variable called newssites:

```
egen newssites=rowtotal(V201634a V201634b V201634c V201634d V201634e
V201634f V201634g V201634h V201634i V201634j V201634k V201634m
V201634n V201634p V201634q), missing
```

The "egen" command is telling Stata to create a new variable called *newssites* according to a specific function that we provide after the equals sign. Here, "rowtotal" tells Stata to sum the values for the fifteen variables included in the parentheses for each case in the data set. We have specified the "missing" option after the command because we want Stata to code *newssites* as missing if the respondent has a missing value for all fifteen of the news site variables. We do this because we do not want cases with no information about their internet news use to be scored as a 0 for the new *newssites* variable, which is what Stata would do by default.

Now we have a new interval-ratio variable called *newssites*, measuring the number of news websites that respondents regularly use, and are ready to consider measures of central tendency for our new variable.

Finding the Mean

First, we will find the mean for *newssites*. We will learn two quick methods for asking Stata to find the mean for a variable. The first command asks Stata to summarize the variable:

`summarize newssites`

The output is shown in Figure 4.8.

```
. summarize newssites

    Variable |        Obs        Mean    Std. dev.       Min        Max
-------------+---------------------------------------------------------
   newssites |      5,290    2.199433    2.604685          0         15
```

Figure 4.8

As is often the case, Stata shows us more information than we needed for the task at hand. It is up to us to decide what parts of the output are relevant for answering our question. In this case, we are interested in the mean for *newssites*, but Stata also has provided us with the standard deviation and the lowest and highest values of the variable found in our sample, measures that we will discuss in the next chapter. The mean for *newssites*, 2.20, indicates that the typical respondent visits a small number of news websites on a regular basis.

We can also request the mean from Stata by using the following command:

`tabstat newssites, statistics(mean)`

The "tabstat" command allows the user to customize the descriptive statistics that Stata returns for a variable. The output, shown in Figure 4.9, is more streamlined than the output for the "summarize" command.

```
. tabstat newssites, statistics(mean)

    Variable |      Mean
-------------+----------
   newssites |  2.199433
```

Figure 4.9

The mean here matches the mean from the previous output. The difference is that with the "tabstat" command, Stata has provided us only with the specific statistic that we requested, the mean.

Finding the Median

To request the median for *newssites*, we will use the same "tabstat" command with a modified option:

`tabstat newssites, statistics(median)`

Instead of asking for the mean, we requested the median. The output is shown in Figure 4.10.

```
. tabstat newssites, statistics(median)
    Variable |       p50
   ----------+----------
    newssites|         1
```

Figure 4.10

The output looks very similar to the output from our previous "tabstat" command, with the difference being that here Stata provides us with the median rather than the mean. Do you notice something curious? Stata appears to be calling the median "p50." What does that mean? Stata uses "p50" to mean 50th percentile, which is just another way of talking about the median. The median and the 50th percentile are both the value at which half of the values in the distribution fall below it and half fall above it. The median for *newssites* is 1, showing that half of the respondents regularly visit either one or no news websites, while half of respondents regularly visit 1 or more sites.

Finding the Mode

Finding the mode in Stata is less straightforward than finding the median or mean. There is no command for requesting the mode for a variable. One might have thought that using the "tabstat" command, as we used for the mean and median, and specifying the "mode" option would return the mode. We can try it here:

`tabstat newssites, statistics(mode)`

The output is shown below:

`. tabstat newssites, statistics(mode)`

unknown statistic: mode

Instead of a table with a mode, we see an error message, denoted by red text, telling us that "mode" is an unknown statistic. See, we cannot ask Stata directly for a mode! However, the mode for a variable can be spotted relatively easily in a frequency distribution. We can request a frequency distribution for *newssites* to find the mode ourselves:

`tabulate newssites`

The output is shown in Figure 4.11.

```
. tabulate newssites

  newssites |      Freq.      Percent        Cum.
------------+-----------------------------------
          0 |      1,826        34.52       34.52
          1 |      1,007        19.04       53.55
          2 |        679        12.84       66.39
          3 |        516         9.75       76.14
          4 |        361         6.82       82.97
          5 |        266         5.03       88.00
          6 |        190         3.59       91.59
          7 |        154         2.91       94.50
          8 |        111         2.10       96.60
          9 |         72         1.36       97.96
         10 |         46         0.87       98.83
         11 |         32         0.60       99.43
         12 |         14         0.26       99.70
         13 |         12         0.23       99.92
         14 |          2         0.04       99.96
         15 |          2         0.04      100.00
------------+-----------------------------------
      Total |      5,290       100.00
```

Figure 4.11

We see that 0 is the mode for tabulate *newssites* because it has the largest number of cases. Nearly 35% of respondents do not consult any of the listed news websites on a regular basis.

Examining a frequency distribution to find the mode is not challenging, but it does introduce the possibility of human error. Every time we put ourselves in charge of making a statistical calculation, even if it is as simple as looking at a frequency distribution to determine which category for a variable has the largest number of cases, we introduce the possibility for human error. There is an interesting way to trick Stata into reporting the mode for a variable without relying on ourselves to correctly identify it. This method uses the "egen" command that we saw above. Here, we will tell Stata to create a new variable that has the same value for every case in the data set. That new value will be the mode for *newssites*. Enter the following syntax into Stata:

`egen newssitesmode=mode (newssites)`

The "egen" command tells Stata to create a new variable called *newssitesmode*. What values should this new variable have? With the "mode (newssites)" part of the command, we are telling Stata to set the variable equal to the mode for *newssites*, such that all cases in the data set will be assigned the mode for *newssites* as their value for the new *newssitesmode* variable. Note that if the variable has multiple modes this strategy will not work because Stata will return a missing value for the mode. After running the command, there is now a new variable in the data set called *newssitesmode* that has the same value for every single case in the data set—the mode for *newssites*. To get Stata to report the mode for *newssites*, we only need to ask for a frequency distribution for *newssitesmode*:

`tabulate newssitesmode`

We see the output in Figure 4.12.

```
. tabulate newssitesmode

newssitesmo
         de |      Freq.     Percent         Cum.
------------+-----------------------------------
          0 |      8,280      100.00       100.00
------------+-----------------------------------
      Total |      8,280      100.00
```

Figure 4.12

The only value for *newssitesmode*, 0, is equal to the mode for *newssites*. We have rather sneakily gotten Stata to report the mode for *newssites* directly without having to rely on correctly spotting the mode in a frequency distribution.

Using ANES data, we have seen in this section that the typical number of news websites sites that people regularly visit depends on the measure of central tendency (mean = 2.20, median = 1, and mode = 0).

Review of Stata Commands

- Creating a new variable by summing the values of existing variables and specifying that cases with missing data on all of those existing variables should be coded as missing

    `egen new variable name=rowtotal(existing variable names), missing`

- Finding the mean

    `tabstat variable name, statistics(mean) (or) summarize variable name`

- Finding the median

    `tabstat variable name, statistics(median)`

- Finding the mode

    `tabulate variable name and identify mode in frequency distribution (or)`

    `egen new variable name=mode(variable name)`

Using SPSS

In 2020, the ANES asked respondents who had heard about the presidential campaign on the internet whether they regularly visit a list of news websites. The survey asked about fifteen specific sites (with variable names in parentheses): Yahoo News (*V201634a*), CNN (*V201634b*), Huffington Post (*V201634c*), New York Times (*V201634d*), Breitbart News Network (*V201634e*), Fox News (*V201634f*), Washington Post (*V201634g*), The Guardian (*V201634h*), USA Today (*V201634i*), BBC News (*V201634j*), NPR News (*V201634k*), Daily Caller (*V201634m*), Bloomberg (*V201634n*), Buzzfeed (*V201634p*), and NBC News (*V201634q*).

For each variable, 1 indicates that the person visits the news site at least once per month, and 0 indicates that the person does not visit the site. In this section, we will explore measures of central tendency for these internet news variables to get a sense of typical internet news usage.

Calculating a New Variable

Before finding measures of central tendency for internet news use, we will use the fifteen news site variables to create a new interval-ratio news site variable that measures the number of news sites that the respondent regularly uses. We will combine the fifteen variables into a new variable, called *newssites*, such that the new variable counts the number of sites that a person uses regularly. Learning how to create a new variable from a set of existing variables is an important skill and one that researchers frequently use.

Without knowing what the actual values are in the data set, we can determine that this new variable could range from 0 (meaning that the respondent uses none of the sites) to 15 (meaning that the respondent uses all of the sites). How should we go about combining these variables?

Open `ANES20.sav`. Notice that what we basically want to do is to sum the values for all fifteen variables for each respondent in the data set. In cases where we want to sum the values of a set of variables to create a new variable, we use the following SPSS procedure:

`Transform → Compute Variable`

This opens the "Compute Variable" dialog box. We will create a new variable called newssites. We will tell SPSS to add up respondents' scores on the fifteen news sites variables. In other words, SPSS will add 1 to a respondent's score every time they register a "1" on one of the fifteen social media variables. This will create a scale that measures how many news sites the respondent uses. The scale runs from 0 (uses no news sites regularly) to 15 (uses all fifteen news sites regularly).

We put the new variable name *newssites* into the "Target Variable" box. We move all fifteen variables (*V201634a*, *V201634b*, etc.) from the variable list into the "Numeric Expression" box, each pair separated by a plus sign (as we are adding up all variables). The image in Figure 4.13 displays the "Compute Variable" dialog box with these specifications filled in.

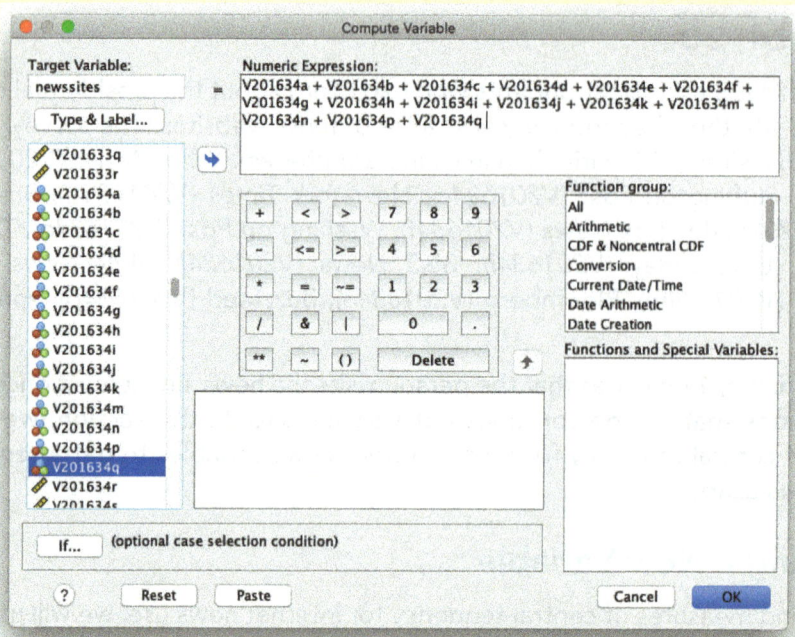

Figure 4.13

Click on "OK," and the SPSS "Output" window confirms that the variable *newssites* has been created. If we want to make sure we proceeded correctly, we can generate a frequency on the new variable. Its categories run from 0 to 15.

Now we have a new interval-ratio variable called *newssites*, measuring the number of news sites that respondents regularly visit, and are ready to consider measures of central tendency for our new variable.

Finding the Mean

First, we will find the mean for *newssites*. We will learn two quick methods for finding a variable's mean. The first command asks SPSS to produce several descriptive statistics, including the mean (Figure 4.14):

`Analyze → Descriptive Statistics → Descriptives`

This will open the "Descriptives" dialog box. Move the variable socmedia from the variable list on the left to the box labelled "Variables" on the right and then click on "OK" (Figure 4.15).

The output is shown in Figure 4.16.

As is often the case, SPSS shows us more information than we needed for the task at hand. It is up to us to decide what parts of the output are relevant for answering our question. In this case, we are interested in the mean for *newssites*, but SPSS also has provided us with the standard deviation and the minimum and maximum values for the variable, measures that we will discuss in the next chapter. The mean for *newssites*, 2.20, indicates that the typical respondent visits a small number of news websites on a regular basis.

Figure 4.14

Figure 4.15

Descriptive Statistics

	N	Minimum	Maximum	Mean	Std. Deviation
newssites	5290	.00	15.00	2.1994	2.60468
Valid N (listwise)	5290				

Figure 4.16

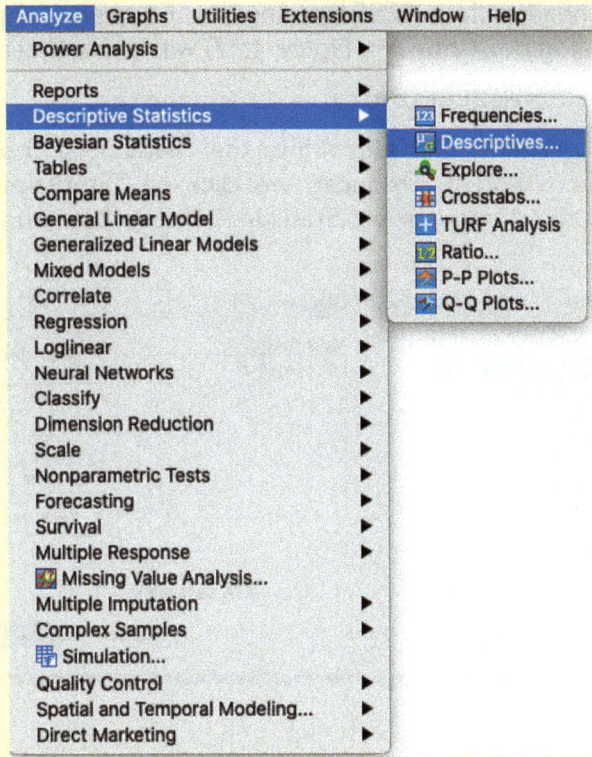

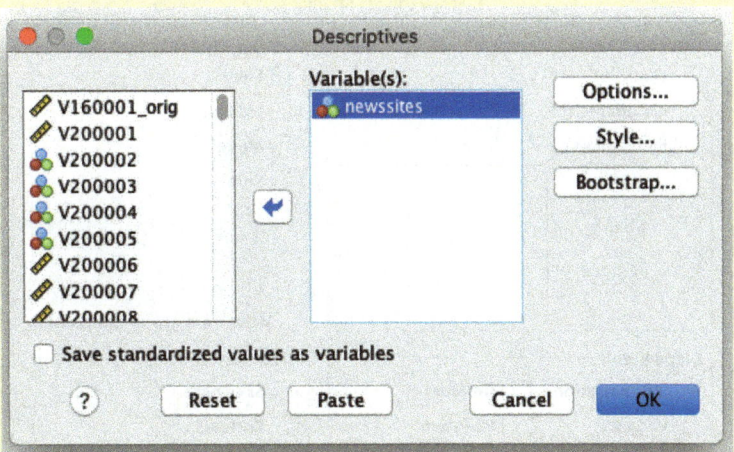

We can also request the mean from SPSS by using the "Frequencies" procedure. We open the "Frequencies" dialog box (shown in Figure 4.17) with this sequence:

`Analyze → Descriptive Statistics → Frequencies`

We place our desired variable (*newssites*) into the "Variable(s)" box. But before clicking on "OK" and submitting the request, we click on "Statistics" (circled in Figure 4.17). This will open the "Frequencies: Statistics" dialog box (shown in Figure 4.18).

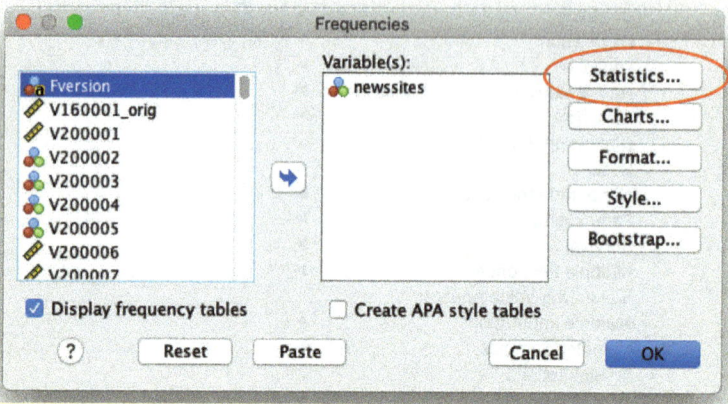

Figure 4.17

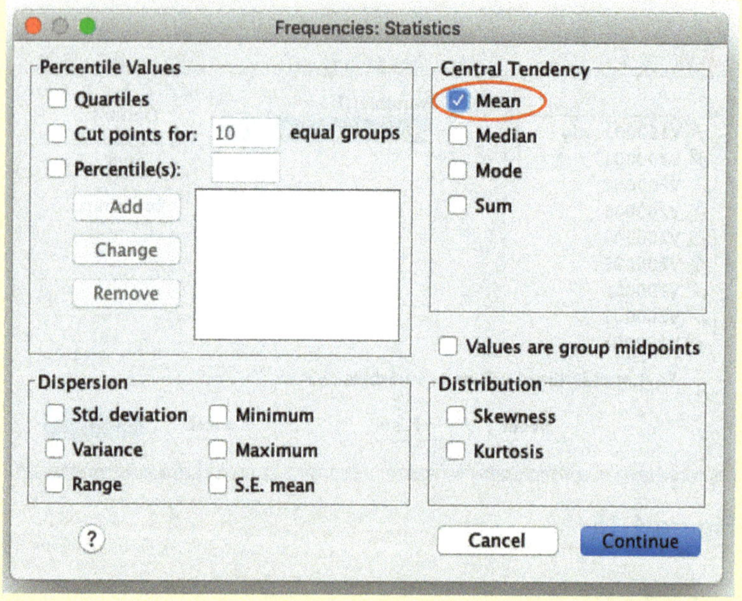

Figure 4.18

We check the small box next to "Mean" (circled), click on "Continue," and we return to the "Frequencies" box, where we click on "OK" to submit the request. Note that there are other statistics besides the mean that one can request. We will be returning to the "Frequencies: Statistics" dialog box later when we examine the median and the mode.

When we use the "Frequencies" procedure to request descriptive statistics, SPSS will generate a frequency table in addition to the requested descriptive statistics. We have not reproduced the full frequency table in Figure 4.19, only the portion of the output with the requested descriptive statistics—in this case, the mean.

Statistics

newssites

N	Valid	5290
	Missing	2990
Mean		2.1994

Figure 4.19

The mean here matches the mean from the previous output. The difference is that with the "Frequencies" command, SPSS has provided us only with the specific statistic that we requested.

Finding the Median and Mode

To request the median and mode for *newssites*, we will use the same procedure we just used to generate the mean: the "Frequencies" command.

`Analyze → Descriptive Statistics → Frequencies`

Click on "Statistics" and check the small boxes next to "Median" and "Mode." This produces the output shown in Figure 4.20.

Statistics

newssites

N	Valid	5290
	Missing	2990
Median		1.0000
Mode		.00

Figure 4.20

The median for *newssites* is 1, showing that half of the respondents in the sample visit one or no news sites and half of them visit two or more sites. We see that 0 is the mode for *newssites*.

Using ANES data, we have seen in this section that the typical number of news websites sites that people regularly visit depends on the measure of central tendency (mean = 2.20, median = 1, and mode = 0).

Review of SPSS Procedures

- Creating a new variable

    `Transform → Compute Variable`

- Generating a frequency distribution

    `Analyze → Descriptive Statistics → Frequencies`

- Finding the mean

    `Analyze → Descriptive Statistics → Descriptives`

    (or)

    `Analyze → Descriptive Statistics → Frequencies (check box for mean)`

- Finding the median

    `Analyze → Descriptive Statistics → Frequencies (check box for median)`

- Finding the mode

    `Analyze → Descriptive Statistics → Frequencies (check box for mode)`

Practice Problems

1. Eight students are comparing how far away they live from campus. Table 4.10 shows the distance for each student, in miles.

    a. Calculate the mean distance from campus. State your result in words. 1.75

    b. Find the median for distance from campus. State your result in words. 7.0

    c. What is the modal distance from campus? State your result in words. 4.0

    d. Which measure of central tendency is the best choice in this case, and why?

Table 4.10 Students' Distance from Campus

Student	Distance from Campus (Miles)
A	0.0
B	0.0
C	0.4
D	1.5
E	1.8
F	2.6
G	3.7
H	4.0

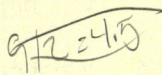

2. You want to get a third dog, and in order to convince your housemates, you investigate how many pets people have. In your apartment building, the number of pets in seven units is presented in Table 4.11.

Table 4.11 Number of Pets per Apartment

Apartment	Number of Pets
A	0
B	1
C	1
D	1
E	2
F	2
G	5

a. Calculate the mean number of pets owned by these residents. State your result in words.

b. Find the median number of pets owned by these residents. State your result in words.

c. What is the mode for number of pets owned by these residents? State your result in words.

d. Which measure of central tendency would you use to convince your housemates that it's reasonable to get a third dog? Is this the best measure of the typical number of pets that residents have? If not, what would be the best measure of central tendency to use in this case?

3. What measures of central tendency can be used with each of the following variables?

a. number of minutes spent online per day

b. favorite ice cream flavor
c. percentage of women in STEM majors
d. support for legalizing cannabis (yes, maybe, no)
e. political party affiliation
f. highest degree completed

4. What measures of central tendency can be used with each of the following variables?
    a. years of education
    b. percentage of residents in each county who rent
    c. frequency of attending religious services (often, occasionally, rarely, never)
    d. astrological sign
    e. marital status
    f. level of happiness (very happy, pretty happy, not too happy)

5. Table 4.12 shows the scores for each of the four games in the 2022 WNBA finals between the Las Vegas Aces and the Connecticut Sun.

   Table 4.12 Scores for Aces and Sun in 2022 WNBA Finals

	Las Vegas Aces	Connecticut Sun	Winner
Game 1	67	65	Aces
Game 2	85	71	Aces
Game 3	76	105	Sun
Game 4	78	71	Aces

    a. What was the mean score for the Aces in the series? For the Sun? What is the difference between the mean scores?
    b. For each game, by how many points did the winning team beat the losing team?
    c. What is the mean of the score differences between the winning and losing teams across the four games?

6. The following list shows the dollar values of assets owned by a group of twelve forty-five-year-olds: 15,000; 27,000; 0; 3,000; 100,000; 32,000; 20,500; 1,000; 37,700; 55,000; 117,000; –20,000. Positive values indicate that the person has more assets than debt, and negative numbers indicate that the person has more debt than assets.
    a. Find the median value. Be sure to show all of the steps that you followed to find the answer.

b. Find the mean.

c. Compare the median from part a to the mean from part b. What does this say about the shape of the distribution?

d. Add −200,000 to the list of values. Recalculate the median and mean.

e. How did the addition of this value change the median? The mean? Why do we see these changes?

7. In 2022, the General Social Survey found that 56% of American adults favored the death penalty for murder. Imagine a follow-up study that wanted to know how *strongly* people favored or opposed the death penalty for murder. The variable measures respondents' opinion on the death penalty, where 0 is completely disagree and 100 is completely agree. The middle of the scale, 50, suggests that the respondent does not feel strongly either way. In a sample of five hundred people, the mean score is 50.

   a. Explain in words how you would interpret the mean score of 50.

   b. You find that the median score for the sample is 59.5. Explain in words how you would interpret the median.

   c. Given that the mean is 50 and the median is 59.5, calculate the median as a percentage of the mean. What could this indicate about the shape of the distribution of scores in this sample?

   d. Which measure of central tendency would you choose to characterize the typical value in this sample—the median or the mean? Explain your answer.

8. Respondents to a hypothetical poll were asked to rate their views of the US Supreme Court on a feeling thermometer, in which 0 represents very cold or negative and 100 represents very warm or positive. They found a mean rating of 46.5.

   a. Explain in words how you would interpret the mean score of 46.5.

   b. You find that the median score for the sample is 48. Explain in words how you would interpret the median.

   c. Given that the mean is 46.5 and the median is 48, calculate the median as a percentage of the mean. What could this indicate about the shape of the distribution of scores in this sample?

9. Figure 4.21 shows the distribution of responses from a hypothetical sample who were asked whether higher education is headed in the wrong direction.

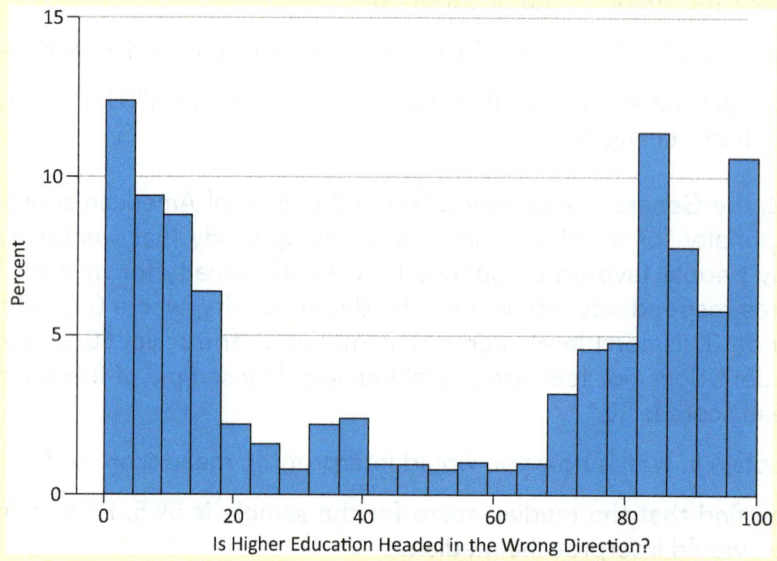

Figure 4.21 Distribution of Agreement that Higher Education Is Heading in the Wrong Direction (0 = Completely Disagree, 100 = Completely Agree)

   a. What is the shape of the distribution in Figure 4.21?

   b. Given the shape of the distribution, which measure of central tendency is most appropriate to present for this sample? Explain your answer.

10. Figures 4.22a and 4.22b show distributions of agreement that higher education is headed in the right direction separately for Republicans and Democrats, respectively.

   a. What is the shape of the distribution for Republicans? For Democrats?

   b. Discuss the differences in Republicans' and Democrats' views of the direction of higher education.

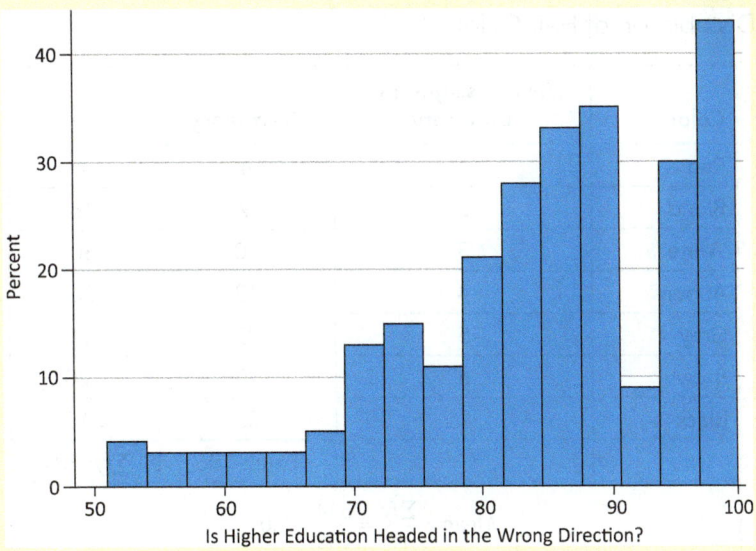

Figure 4.22a Distribution of Agreement That Higher Education Is Heading in the Wrong Direction (0 = Completely Disagree, 100 = Completely Agree), Republicans

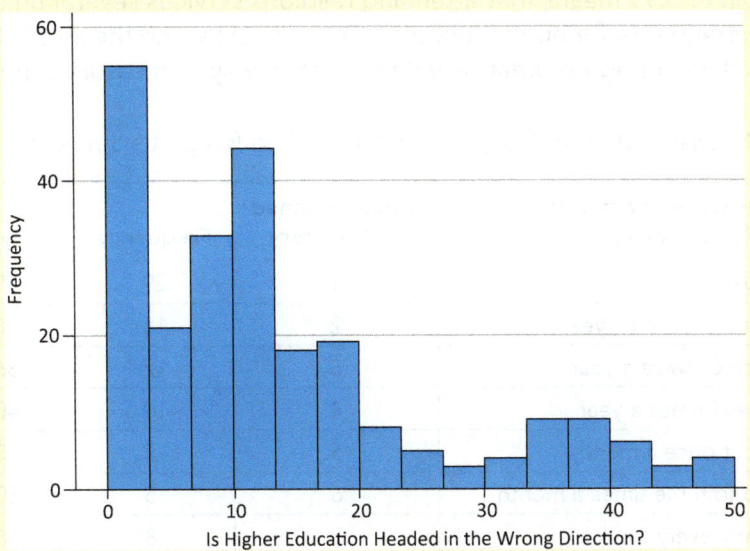

Figure 4.22b Distribution of Agreement That Higher Education Is Heading in the Wrong Direction (0 = Completely Disagree, 100 = Completely Agree), Democrats

11. A student makes Table 4.13 showing how they calculated the mean for a variable measuring respondents' hair color in their data set. The student concludes that the mean of 5.0 means that grey is the typical hair color for his respondents. Do you agree with the student? Explain your answer. If you disagree, what would be a better way to measure central tendency?

Table 4.13 Distribution of Hair Color

Color	Value Assigned to Category	Frequency	fy
Red	1	4	4
Blond	2	9	18
White	3	10	30
Auburn	4	13	52
Grey	5	20	100
Brown	6	20	120
Black	7	30	210
		Total: 106	Σfy: 534

$$\text{Mean} = \frac{\Sigma fy}{N} = \frac{534}{106} = 5.0$$

12. A student makes Table 4.14 showing how he calculated the mean for a variable measuring how often respondents attend religious services. The student concludes that the mean of 3.91 means that attending religious services several times a year is the typical attendance for his respondents. Do you agree with the student? Explain your answer. If you disagree, what would be a better way to measure central tendency?

Table 4.14 Distribution of Frequency of Attending Religious Services

How Often R Attends Religious Services	Value Assigned to Category	Frequency	fy
Never	1	29	29
Less than once a year	2	14	28
Once or twice a year	3	12	36
Several times a year	4	10	40
About once a month	5	3	15
Two to three times a month	6	5	30
Nearly every week	7	8	56
Every week	8	14	112
Several times a week	9	5	45
		Total: 100	Σfy: 391

$$\text{Mean} = \frac{\Sigma fy}{N} = \frac{391}{100} = 3.91$$

13. The *New York Times* asked its readers to suggest names for the generation after millennials, people born between roughly 1995 and 2015. Imagine that the most frequently submitted suggestions were "Generation Z" (N = 231), "Homeland Generation" (N = 20), "Post-Millennials" (N = 101), and "iGeneration" (N = 76).

Imagine that the Times decides to use the modal category among the top four suggestions as its official name for the generation after millennials. Which name should it use?

14. Your kind professor wants to bring students' favorite snack for the last class meeting. A survey of the class shows the following favorites: Doritos (N = 57), chocolate bars (N = 78), Goldfish crackers (41), and Oreo cookies (34). Which snack would the professor choose to bring? What measure of central tendency does this represent?

15. Parts a through c list variables and measurements, and Figures 4.23a through 4.23c show visual representations of how their values are distributed. For each variable, identify the measure(s) of central tendency that would be appropriate to report.

    a. Countries' gross domestic product in dollars
    b. Political party: Democrat (1), Republican (2), Independent (3), Green (4), Libertarian (5), Other (6)
    c. Strength of political engagement, ranging from 1 (no interest) to 10 (very high interest)

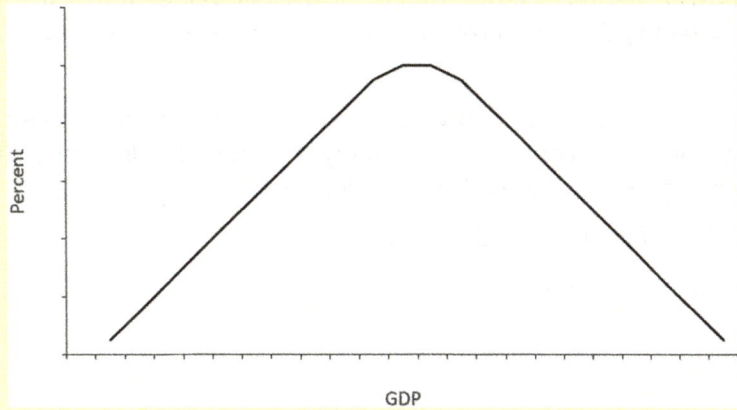

Figure 4.23a   Distribution of GDP

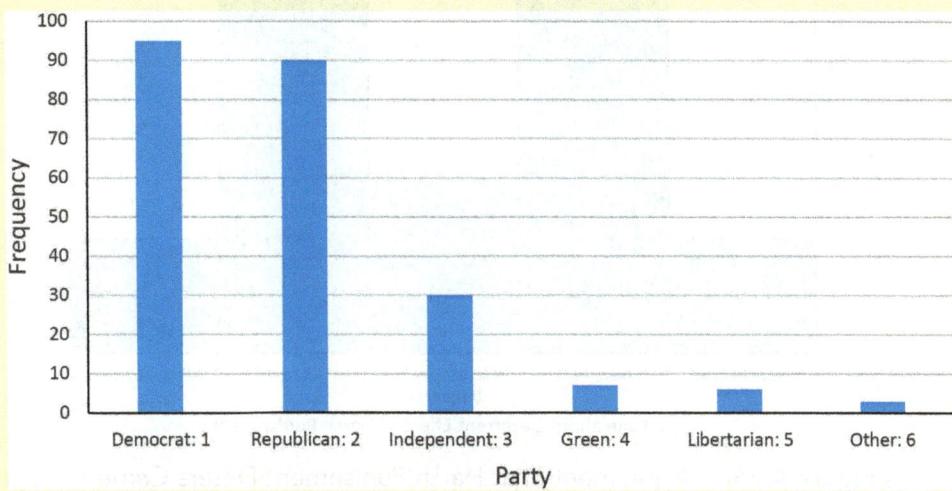

Figure 4.23b   Distribution of Party Affiliation

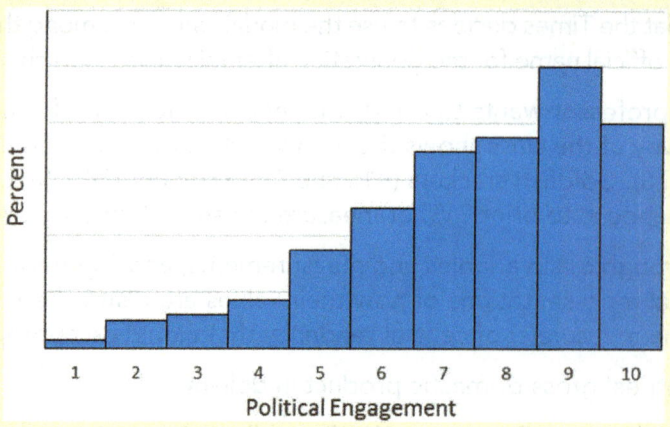

Figure 4.23c Distribution of Strength of Political Engagement

16. Parts a through c list variables and measurements, and Figures 4.24a through 4.24c show visual representations of how their values are distributed. For each variable, identify the measure(s) of central tendency that would be appropriate to report.

   a. Agreement with the following statement: "Harsh punishment deters people from committing crime" (1 = strongly disagree, 2 = disagree, 3 = neither agree nor disagree, 4 = agree, 5 = strongly agree).

   b. Number of brothers and sisters

   c. Age of respondent

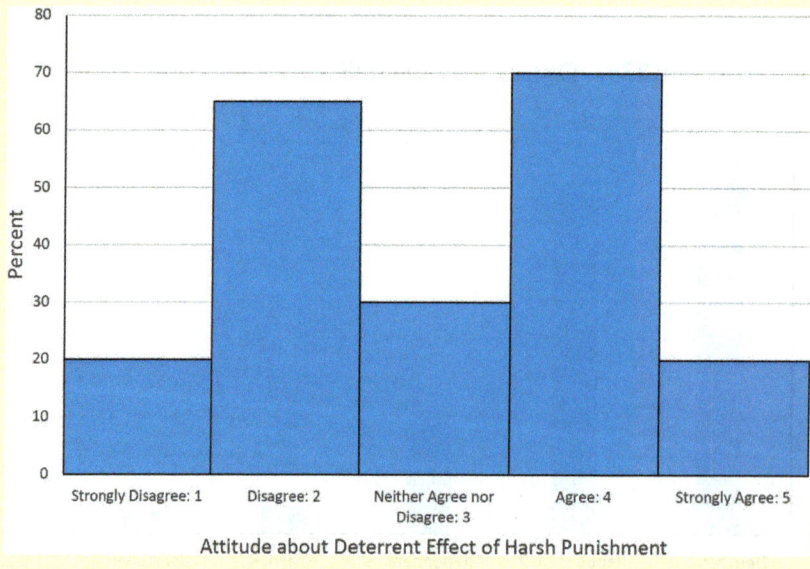

Figure 4.24a Agreement That Harsh Punishment Deters Crime

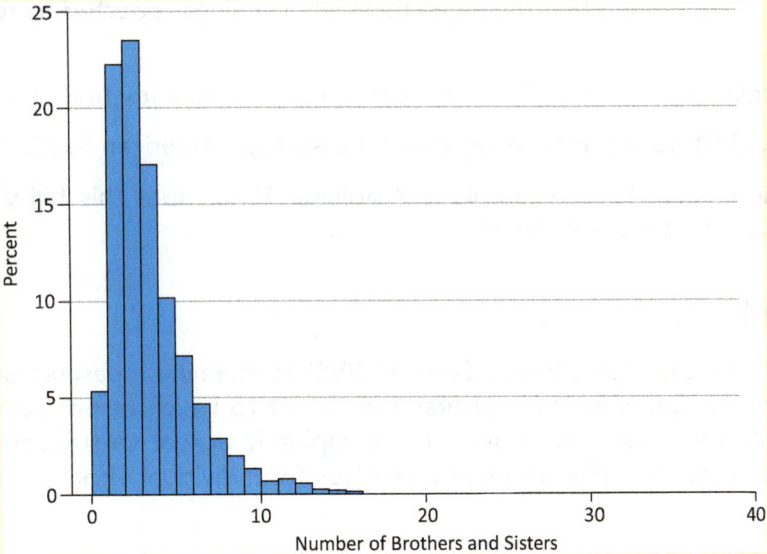

Figure 4.24b

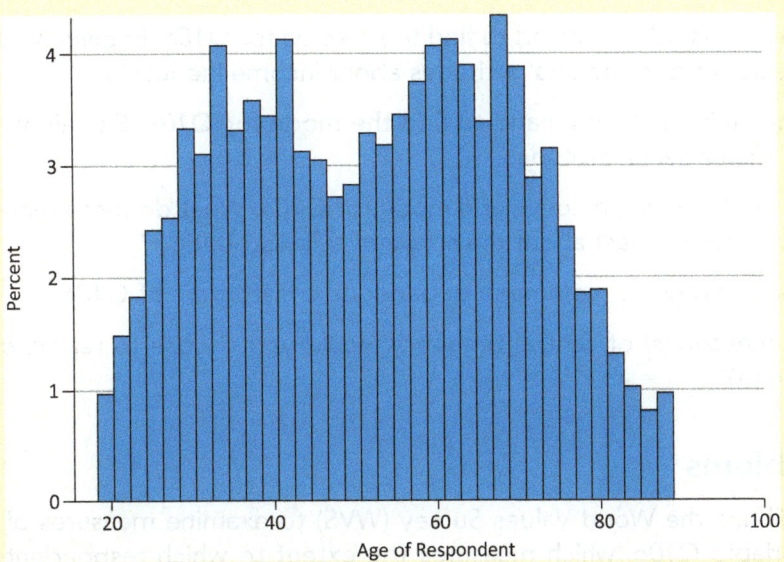

Figure 4.24c

17. A study of Americans' consumer behavior finds that the mean number of credit cards for the sample is 3.6. After reading the study, a blogger writes in a post: "The average American has 3.6 credit cards."

   a. Explain how it is possible for the mean to include a fraction of a credit card.
   b. Do you agree with the blogger's statement about "the average American"?
   c. If the median for the sample is 2 credit cards, what can you say about the shape of the distribution?

18. The 2021 GSS found that the mean number of siblings reported by respondents is 3.13.

    a. Explain how it is possible for the mean to include a fraction of a sibling.

    b. Would it be accurate to say that "The average American has 3.13 siblings?"

    c. The median for the sample is 2 siblings. What does this tell you about the shape of the distribution?

Stata Problems

Here, you will use the World Values Survey (WVS) to examine measures of central tendency for variable *Q106*, which measures the extent to which respondents agree that incomes should be made more equal (1 = completely agree) versus agree that there should be larger income differences as incentives for individual effort (10 = completely agree).

1. Open `WVSWave7.dta`. Use the "summarize" command to produce summary statistics for *Q106*. Find the mean in the output and explain what it tells us about respondents' attitudes about income inequality.

2. Use the "tabstat" command to find the median for *Q106*. Explain what the median tells us about respondents' attitudes about income inequality.

3. Use the "tabulate" command to find the mode for *Q106*. Explain what it tells us about respondents' attitudes.

4. Compare the mean, median, and mode for *Q106*. What do these measures of central tendency suggest about the shape of the distribution?

5. Use the "histogram" command to generate a histogram of *Q106*.

6. Which measure(s) of central tendency would you choose to report based on the histogram?

SPSS Problems

Here, you will use the World Values Survey (WVS) to examine measures of central tendency for variable *Q106*, which measures the extent to which respondents agree that incomes should be made more equal (1 = completely agree) versus agree that there should be larger income differences as incentives for individual effort (10 = completely agree).

1. Open `WVSWave7.sav`. Use the "Frequencies" dialog box to produce summary statistics for *Q106*.

2. Find the mean in the output and explain what it tells us about respondents' attitudes about income inequality.

3. Find the median in the output and explain what it tells us about respondents' attitudes about income inequality.

4. Find the mode in the output and explain what it tells us about respondents' attitudes.
5. Compare the mean, median, and mode for *Q106*. What do these measures of central tendency suggest about the shape of the distribution?
6. Use the "Histogram" dialog box to generate a histogram of *Q106*.
7. Which measure(s) of central tendency would you choose to report based on the histogram?

Notes

[1] Raphael Charron-Chenier, Louise Seamster, Tom Shapiro, and Laura Sullivan. 2020. "Student Debt Forgiveness Options." Roosevelt Institute Working Paper. Retrieved from https://rooseveltinstitute.org/wp-content/uploads/2020/08/RI_StudentDebtForgiveness_WorkingPaper_202008.pdf.

[2] Ben Miller, Colleen Campbell, Brent J. Cohen, and Charlotte Hancock. "Addressing the $1.5 Trillion in Federal Student Loan Debt." Center for American Progress Report, June 12, 2019. Retrieved from https://www.americanprogress.org/article/addressing-1-5-trillion-federal-student-loan-debt/.

[3] Daniel Mangrum, Joelle Scally, and Crystal Wang. 2022. "Three Key Facts from the Center for Microeconomic Data's 2022 Student Loan Update." Federal Reserve Bank of New York *Liberty Street Economics*. August 9. Retrieved from https://libertystreeteconomics.newyorkfed.org/2022/08/three-key-facts-from-the-center-for-microeconomic-datas-2022-student-loan-update.

[4] Pew Research Center. 2021. "How the Pandemic Could Affect the Rise in Student Debt." December 22. Retrieved from https://www.pewtrusts.org/en/research-and-analysis/issue-briefs/2021/12/how-the-pandemic-could-affect-the-rise-in-student-debt.

[5] National Institutes of Health Research Matters. 2020. "Number of Steps per Day More Important Than Step Intensity." March 31. Retrieved from https://www.nih.gov/news-events/nih-research-matters/number-steps-day-more-important-step-intensity.

[6] Pew Research Center. 2017. "The Changing Global Religious Landscape." April 5. Retrieved from https://www.pewresearch.org/religion/2017/04/05/the-changing-global-religious-landscape/.

[7] Pew Research Center. 2021. "About Three-in-Ten US Adults Are Now Religiously Unaffiliated." December 14. Retrieved from https://www.pewresearch.org/religion/2021/12/14/about-three-in-ten-u-s-adults-are-now-religiously-unaffiliated/.

[8] D. Herbenick et al. 2021. "Diverse Sexual Behaviors in Undergraduate Students: Findings from a Campus Probability Survey." *The Journal of Sexual Medicine* 18(6): 1024–1041.

[9] The GSS variable is EATMEAT. Beef and lamb are the forms of meat most strongly associated with climate change.

[10] US Census Bureau. Historical Income Tables: People (Table p-4). https://www.census.gov/data/tables/time-series/demo/income-poverty/historical-income-people.html.

[11] Chris Wild. 2006. "The Concept of Distribution." *Statistics Education Research Journal* 5(2): 10–26. http://www.stat.auckland.ac.nz/serj.

Chapter 5
The Diversity of Values in a Group
Measures of Variability

After reading the previous chapter, you know how social scientists determine what is "typical" about a sample. You can find and describe the mean income, the median cost of an apartment, or the modal family composition. But you also know that many individuals in a sample do not possess these typical traits. While we often think of social science or statistics as focusing on the typical or general patterns (and it does), understanding variation from these central tendencies is equally important. Measures of central tendency do not tell you how much diversity exists in a sample.

Stephen Jay Gould was a scientist, professor, and writer, whose essay on his cancer diagnosis illustrates the importance of considering variability along with central tendency. "The Median Isn't the Message," Gould's essay, recounts his discovery that the cancer he was diagnosed with in 1982 had a median survival time of only eight months.[1] Applying his understanding of statistics, however, Gould says: "When I learned about the eight-month median, my first intellectual reaction was: fine, half the people will live longer; now what are my chances of being in that half. I . . . concluded, with relief: damned good. I possessed every one of the characteristics conferring a probability of longer life: I was young, my disease had been recognized in a relatively early stage, I would receive the nation's best medical treatment, I had the world to live for, I knew how to read the data properly and not despair." Investigating further, Gould learned that the distribution of survival time was strongly skewed to the right. That is, there were a small number of cases whose survival time was many years longer than the median. Gould did, in fact, live for twenty more years before dying of an unrelated cause. Had Gould considered only the median, his outlook on his prognosis would have been much grimmer. Understanding the issue of variability in detail, including skew, allowed a more optimistic understanding of the prognosis.

One can obtain the same mean or median for samples in which the cases are tightly clustered around the mean and for samples in which there are extreme differences. For example, imagine three people of different height. In one group, they measure 64 inches, 65 inches, and 66 inches. In another group, they measure 53 inches, 65 inches, and 77 inches.

The mean height for both groups is 65 inches. Yet the groups are very different. In the first one, the highest and lowest values are only 1 inch above or below the mean, while, in the second group, they are a full foot above or below the mean.

This chapter focuses on measures of variability, which are ways of quantifying how much diversity there is in a sample or population. How are the scores distributed? How are they spread around the mean? There are several different measures of variability: range, interquartile range, variance, and standard deviation. Like the measures of central tendency discussed in the previous chapter, these are all univariate measures.

Recall the example in the previous chapter about student loan debt. Mean student loan debt in 2021 was $36,245. How much student loan debt do you have, if any? Your debt likely differs from the mean, and the answers from the other students in your statistics class would probably vary quite a bit from $36,245, too. Some values will be a little higher or lower, some much higher or lower. If we report only the mean or median student loan debt, we miss an important part of the picture. Social scientists should always present one or more measures of variability along with a measure of central tendency.

Researchers and policymakers dealing with student loan debt have been concerned with variability. A study by Terrance Martin Jr. and coauthors sought to understand how owing debt for their own education affected whether parents saved money or took out loans to help pay for their children's college.[2] The researchers found that parents' own student debt averaged $23,245. (The median was only $11,000, showing that the distribution is strongly skewed toward higher values.)

But within this group, there is substantial variation, and this is what this chapter focuses on. We know that parents' mean student loan debt was $23,345, but we also know that some of those parents must have more or less debt. The researchers found that, on average, the debt of these parents differed from the mean by $34,631.

This number—$34,631—is the standard deviation, one of several ways of measuring variability within a sample. It tells us the average amount that student loan debt varies from the sample mean. You can imagine a scenario where, on average, debt differed from the mean by a much smaller amount—say $1,000. This would indicate much less variability on parents' student loan debt. Most people would owe an amount very close to the mean. Later in this chapter we will cover standard deviation in detail, including how to calculate it. We might also ask what the highest and lowest amount of student loan debt is, which is the range. Or we might ask what range covers the middle half of the sample, bounded by the 25th and the 75th percentiles of student loan debt. This is the interquartile range.

Range

The **range** is simply the distance from the lowest value of a variable to the highest. If, for example, the lowest amount of student loan debt were $0 and the highest amount were $100,000, we would say that student loan debt ranged from $0 to $100,000 or that it had a range of $100,000 (the difference between 0 and 100,000).

The range can be used for ordinal or interval-ratio variables but never for nominal variables. When range is used for ordinal variables, you can present only the lowest and highest levels. For example, you might say that opinions about Medicare range from very satisfied to very dissatisfied or that respondents' attendance at church services ranged from never to several times a week.

With an interval-ratio variable, you also can present the distance between these values. For example, respondents to the 2021 General Social Survey (GSS) were asked how old they were when their first child was born. The youngest age reported was nine, and the oldest was fifty-seven. There is a range of forty-eight years for the age at which respondents' first child was born, from nine to fifty-seven years (57 − 9 = 38).

If you are someone who relies on formulas, you may find the formula for range useful. It is:

$$Range = highest\ value - lowest\ value$$

You should never simply present the result of this formula when presenting the range, however. You should also present the actual low and high values. This is because, as a social scientist, you are using statistics to tell you something about the actual variable you're studying.

The range has some important limitations. It is greatly affected by extreme scores, and it doesn't tell us about what is going on between those scores. If we know that the median rent for a one-bedroom apartment in San Jose, California, is $2,080,[3] knowing that the range is, say, $750 to $7,800 gives us some additional information. But it doesn't tell us whether apartments are, in general, fairly close to $2,080, with a few cheap or luxury apartments, or whether there are a lot of apartments well above or well below $2,080. Similarly, we might know the range for household income in the Seattle metropolitan area, but Bill Gates (one of the wealthiest people in the world), who lives there, is throwing off the high end of that range. Seattle is indeed an affluent city, but Bill Gates would make it look wealthier than it is, if we rely on the range.

Interquartile Range

The interquartile range (IQR) gives more information than the range. It, too, can be used with ordinal or interval-ratio variables. It relies on cumulative percentage and percentile rank, discussed in chapter 2. Recall that the 25th percentile is the value for which the cumulative percentage is 25, meaning that 25% of the cases fall below that value. The 75th percentile is the value for which the cumulative percentage is 75 and for which 75% of the cases fall below. If we divide a sample into four equal parts based on their percentile rank, they will divide at the 25th, 50th, and 75th percentiles. These parts are referred to as quartiles. The **interquartile range** gives us the values of the 25th and 75th percentiles and the distance between them. Half of the sample will have values between the 25th and 75th percentile. (The other half will have values either below the 25th or above the 75th percentile.)

Let's consider the example of scores on the critical reading component of the SAT for college-bound high school seniors in 2022.[4] As Figure 5.1 shows, the mean Evidence-Based Reading score for college-bound seniors in 2022 was 529. The score for the 25th percentile was 450, while the score for the 75th percentile was 610. This means that 25% of college-bound high school seniors (434,420 of them) received scores lower than 450 and 25%

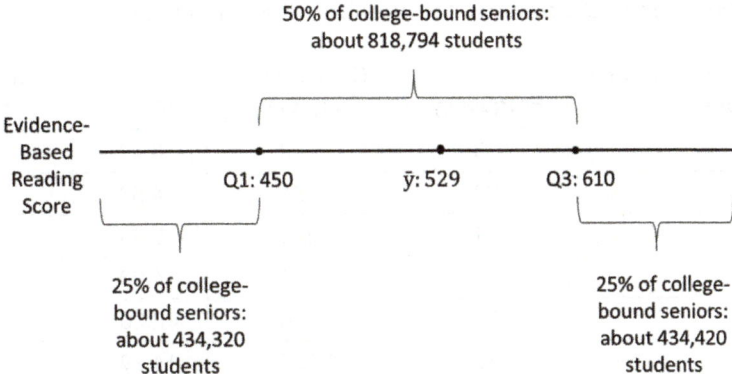

Figure 5.1 Interquartile Range for Evidence-Based Reading SAT Scores for 2022 College-Bound Seniors.
Source: College Board.

(another 434,420 of them) received scores higher than 610. The interquartile range ranges from 450 to 610. We know, therefore, that the middle half of seniors in 2022 scored between 450 and 610 on critical reading.

With interval-ratio variables, we can also calculate the size of the interquartile range, which is the difference between the two scores: 610 − 450 = 160. The interquartile range for the critical reading scores of college-bound seniors on the SAT is 160 points.

The interquartile range gives us more information than the range, and that information is less distorted by the values at the very low or very high range. It is thus the most appropriate indicator of variability for variables that are skewed. We could present the interquartile range for income in Seattle without its being affected by the presence of Bill Gates, for example. The interquartile range also has the advantage of being relatively easy to understand, making it useful to present to audiences that are not made up of statisticians.

Finding the Interquartile Range in a Frequency Distribution

Here is an example that uses a frequency distribution to find the interquartile range. Respondents to the 2021 GSS were asked how many rooms they had in their homes. Sociologically, this reflects a number of factors, including socioeconomic status and the different kinds of locations where respondents live. The median number of rooms in respondents' homes was five. (The mean was 5.34, reflecting the fact that the distribution is only mildly skewed, with a small number of respondents reporting large numbers of rooms in their homes.) Table 5.1 gives us the frequency distribution, with cumulative frequency and cumulative percentages, for their responses.

We can see that there is a wide range in the number of rooms reported, from zero to twenty. The high end of that range is the result of only one out of 1,812 respondents who reported twenty rooms. The interquartile range will tell us the range within which half of respondents fall.

How do you find the cutoff points for the 25th and 75th percentiles? Similar to finding the median (which is the 50th percentile), you must first have the data in rank order and then find the cases that are at the 25th percentile (the first quartile) and 75th percentile (the

CHAPTER 5 The Diversity of Values in a Group

Table 5.1 Frequency Distribution for Number of Rooms in Home

Number of Rooms in Home	Frequency	Cumulative Frequency	Percent	Cumulative Percent
0	3	3	0.17	0.17
1	83	86	4.58	4.75
2	189	275	10.43	15.18
3	270	545	14.90	30.08
4	221	766	12.20	42.27
5	259	1,025	14.29	56.57
6	235	1,260	12.97	69.54
7	165	1,425	9.11	78.64
8	156	1,581	8.61	87.25
9	89	1,670	4.91	92.16
10	63	1,733	3.48	95.64
11	24	1,757	1.32	95.96
12	28	1,785	1.55	98.51
13	10	1,795	0.55	99.06
14	9	1,804	0.50	99.56
15	3	1,807	0.17	99.72
16	3	1,810	0.17	99.89
17	0	1,810	0.00	99.89
18	1	1,811	0.06	99.94
19	0	1,811	0.00	99.94
20	1	1,812	0.06	100.00
Total		1,812	100	

Source: 2021 General Social Survey.
Total percent may not sum to 100 due to rounding.

beginning of the fourth quartile). The cutoff points are the values associated with those cases. By formula:

$$\text{Quartile 1 case number} = N(0.25)$$

$$\text{Quartile 3 case number} = N(0.75)$$

For this example:

$$\text{Quartile 1 case number} = 1,812(0.25) = 453$$

$$\text{Quartile 3 case number} = 1,812(0.75) = 1,359$$

To find the actual cutoff points, look at the cumulative frequency column in Table 5.1. Case number 453 will fall among the cases for 3 rooms, so 3 is the cutoff point for the first quartile. (How did we get this? The first 275 cases have 2 or fewer rooms; cases 276 through 545 have 3 rooms, and 453 falls between 276 and 545.) The cutoff point for the third quartile is the value for case number 1,359. Looking at the table, we can see that case has a value of 7 rooms. Thus, the interquartile range, within which the middle 50% of respondents fall, is from three to seven rooms. We could also say that the interquartile range is 4 (7 − 3 = 4). If you were writing a research report, you might say, "The middle half of all respondents had between three and seven rooms in their home."

If the N is not divisible by four, the formula for case number will yield a result that is not a whole number. In this case, as with when it occurs with the median, the cutoff points will be halfway between two cases. For example, a case number of 715.5 would mean that the cutoff point is between the values of case numbers 715 and 716.

We could also find the interquartile range by looking at the Cumulative Percent column of Table 5.1. We see that the case at the 25th percentile had three rooms. (15.18% of respondents had two or fewer rooms; since this is less than 25, we go to the next row.) Similarly, we see that the case at the 75th percentile had seven rooms.

Visualizing the Interquartile Range with Box Plots

One commonly used graphical display of the interquartile range of an interval-ratio variable, along with the median, is the box plot. A box plot is especially effective for comparing a variable's distribution across groups. In this example, we use data from the 2020 American National Election Study (ANES). This survey asks respondents to evaluate several groups on a 100-degree "feeling thermometer." If a respondent feels warm, or favorably, toward a group, then she or he would rate them somewhere between 50 and 100 degrees. If a respondent does not care for that group, the rating would fall somewhere between 0 and 50 degrees. A score of 50 indicates neutrality.

We will look at how Biden and Trump voters evaluated the National Rifle Association (NRA) on this 100-degree feeling thermometer. We present the information by means of a table (Table 5.2) and then by means of a box plot (Figure 5.2).

Table 5.2 Feeling Thermometer Ratings of the National Rifle Association

	Trump Voters	Biden Voters
Median	85	40
25th percentile	60	15
75th percentile	100	50
IQR	40	35

Source: 2020 American National Election Study.

Table 5.2 lists the relevant measures of position and the interquartile range. The box plots are constructed on the basis of these same measures of position but also allow for a quick visual comparison of the respective ratings of Trump and Biden supporters. The

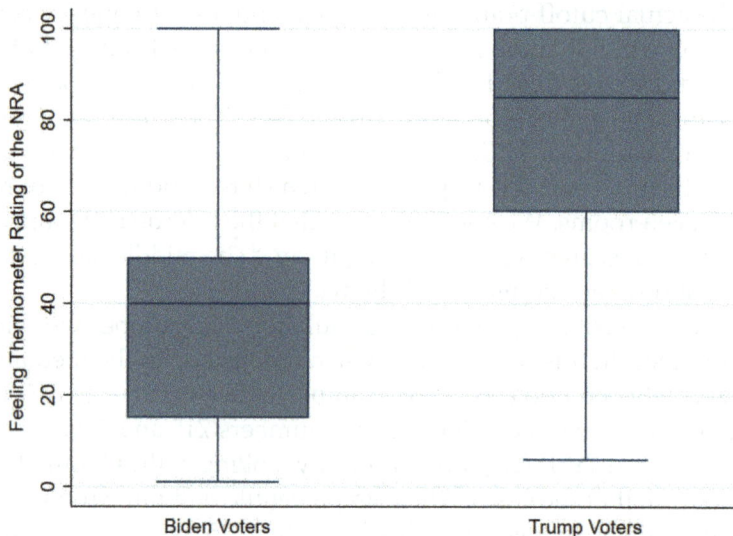

Figure 5.2 Feeling Thermometer Ratings of the National Rifle Association, Box Plots.

Source: 2020 American National Election Study

boxes in Figure 5.2 represent the interquartile range. The lower line of each box is the 25th percentile value, and the top of the box corresponds to the 75th percentile. Within each box, the solid line represents the median of the distribution. Connected to the bottom and top of each box are lines that extend 1.5 IQRs from the 25th and 75th percentiles, respectively. Horizontal lines are drawn at these points. These lines are known as "whiskers." Any points that fall below or above the whiskers are considered to be outliers.

Let's look at the box plot for Trump voters. Their median rating of the NRA is 85 degrees. The IQR runs from 60 to 100—it is equal to 40 degrees. Now compare the box plots for Trump and Biden voters. We see very large differences in their medians: the median NRA rating for Biden voters is only 40. In the case of Trump supporters, there is a high degree of compression at the higher end of the scale. The middle 50% lie between 60 and 100 degrees. The 75th percentile—the top of the box—is at 100 degrees, the warmest possible rating. In in other words, fully 25% of Trump voters rated the NRA at 100 degrees. There is somewhat more symmetry in the ratings by Biden voters with the whiskers extending in both directions of the IQR. Box plots such as this one allow us to make quick visual comparisons of how the variability of groups differs on a single variable.

⊖ Standard Deviation

The standard deviation gives us another piece of information about variability in the sample. The standard deviation is a more precise measure. It is useful both for describing the variability in a sample and for calculating other statistics that we will encounter in later chapters. As described at the beginning of the chapter, **standard deviation** is a measure of how much scores differ from the mean. It is useful to have a strong grasp on the meaning of standard deviation. Therefore, we will discuss the logic of standard deviation in some detail. Standard deviation can be calculated for interval-ratio variables only.

To begin thinking about the standard deviation, we will use a simplified example. A sample of ten people is asked how many adults live in their household. The results are as follows: 1, 1, 1, 2, 2, 2, 2, 3, 4, 6.

The mean for this sample is 2.4. (Check the calculation if you want a review on finding the mean. We say, "On average, respondents reported 2.4 adults lived in their household.") The number of adults living in the household ranged from one to six.

For each respondent, we can think about how different the number of adults in their household is from the mean. For example, respondents with one adult in their household differed from the mean by −1.4 (1 − 2.4); the negative number shows that they were below the mean. Respondents with four adults differed from the mean by 1.6 (4 − 2.4). These numbers are the deviation for each score. Deviation is the distance of a score from the mean. In order to get an overall measure of how much the scores in a sample vary from the mean, we begin with the concept of deviation.

We can imagine several ways of summarizing deviation. We could simply add together the deviations for all scores, but because some deviations are negative numbers (those for scores below the mean) and others are positive numbers (those for scores above the mean) they would cancel each other out, and we would end up with 0. We could add together the absolute values of the deviations for each score, but, in that case, the size of the sample would affect the sum. But if we add together the absolute values of the deviations for each score and divide by the sample size, N, we end up with mean deviation, which is the average of the absolute values of the deviations.

The standard deviation uses another way of getting around the problem of positive and negative numbers: squaring the deviation for each score. Table 5.3 shows the values of the variable, each value's deviation from the mean, and the squared deviation for each value.

Table 5.3 Number of Adults Who Live in Household and Deviations from the Mean

Score (y)	Deviation $(y - \bar{y})$	Squared Deviation $(y - \bar{y})^2$
1	−1.4	1.96
1	−1.4	1.96
1	−1.4	1.96
2	−.4	0.16
2	−.4	0.16
2	−.4	0.16
2	−.4	0.16
3	0.6	0.36
4	1.6	2.56
6	3.6	12.96
Total	0	22.4

If we add together the squared deviations, we get 22.4. (Adding together the unsquared deviations, as explained above, yields 0.) If we divide 22.4 by the number in the sample (10), we get 2.24. This is the average squared deviation.

Statisticians divide by N – 1 instead of N.[*] This gives us the **variance**. Dividing by N – 1 instead of N, above, we get 2.49. Variance gives us a summary of how much scores deviate from the mean that is standardized by sample size. However, it is difficult to interpret because the units are squared. In this example, we could say that, on average, respondents varied from the mean number of adults in the household by . . . 2.49 squared adults. What is a squared adult?! Fortunately, there is an easy way to get the units into comprehensible form: Take the square root. If we take the square root of 2.49, we get 1.58. We can then say that, on average, the mean number of adults in the household varied from the mean by 1.58 adults. This is the standard deviation.

Standard deviation for a sample is notated as s, and variance is notated as s². Variance is used in calculating more advanced statistics. Because standard deviation can be interpreted more logically, however, it is almost always used to report and analyze variability.

Calculating Standard Deviation

By formula, standard deviation is:

$$s = \sqrt{\frac{\Sigma(y - \bar{y})^2}{N - 1}}$$

The capital Greek letter sigma, Σ, stands for sum, so the numerator is telling us to sum the squared deviations, $(y - \bar{y})^2$, for each score. This requires careful attention to order of operations.

1) The first step is to calculate the deviation for each score, $(y - \bar{y})$. In Table 5.3, for the first household with a score of 1, we subtract the mean, 2.4 from 1: 1 – 2.4 = –1.4. That result is shown in the second column of the table. We must perform this operation individually for each case in the sample.

2) The next step is to square each of these deviations, $(y - \bar{y})^2$. Looking again at the first case with a score of 1 and deviation of –1.4, squaring –1.4 gives us 1.96, as shown in the third column of the table. Again, we must perform this operation individually for each case in the sample.

3) The next step is to sum these squared deviations, $\Sigma(y - \bar{y})^2$. Adding together all the numbers in the third column of the table gives us the total of 22.4.

4) The next step is to divide by N – 1. In this example, N = 10, so we divide by 9: 22.4/9 = 2.49.

5) The final step is to take the square root. The square root of 2.49 is 1.58, and this is the standard deviation.

---

[*] It can be mathematically proven that if we divide by N, the sample variance will be a "biased" estimator of the population variance. When we divide by N – 1, the sample variance is an "unbiased" estimator of the population variance.

Calculating the standard deviation by hand is very cumbersome with larger samples. In practice, most analysts rely on computers or scientific calculators to calculate standard deviation.

Note: If you want more practice with hand calculating the standard deviation, work "Practice Problems" 1c and 2c.

Using the Standard Deviation to Compare Distributions

The standard deviation is commonly used to assess how much diversity or variation there is in a sample or a population. Let's look at an example of how we might interpret the standard deviation in the context of some survey data from the 2022 ANES. In Table 5.4, we have presented the mean feeling thermometer ratings of several groups, along with the

Table 5.4 Feeling Thermometer Ratings of Different Groups

	Mean	Standard Deviation
Scientists	79.8	19.2
Labor unions	62.4	23.3
Black Lives Matter	65.6	27.1
Christian fundamentalists	57.9	24.5
Conservatives	53.4	23.9
Big business	50.8	19.9
Socialists	50.2	21.6

Source: 2020 American National Election Study.

standard deviations. By examining both the means and standard deviations of these feeling thermometer ratings, we can gain a sense of how high or low the public rates different groups and also how much consensus or division there is within these ratings.

One group—scientists—seems to enjoy the highest levels of public esteem, with a mean feeling thermometer rating of 79.8 degrees. This mean is a lot higher than any of the others. Moreover, the standard deviation is relatively low, at 19.2 degrees. This indicates that many respondents are within 19 degrees (above or below) of the mean score. A standard deviation of 19 degrees can be understood as the average respondent's deviation from the mean. We might conclude that Americans tend to think highly of scientists (high mean) and that there is a lot of consensus around this view (low standard deviation).

In the middle, we find three groups: labor unions, Black Lives Matter, and Christian fundamentalists. The mean ratings of these groups fall in the 60-degree or upper-50-degree range, at 62.4, 65.6, and 57.9 degrees, and the standard deviation for these groups is 23.3, 27.1, and 24.5 degrees, respectively, higher than it was for the scientists.

What does this higher standard deviation indicate? That there is less consensus in the ratings of labor unions, Black Lives Matter, and Christian fundamentalists than in those of scientists. What is particularly noteworthy here is the high standard deviation associated with ratings of Black Lives Matter (BLM). The standard deviation of 27.1 degrees suggests that there are a lot of people who think quite poorly of BLM and also that there are a lot of people who think highly of this group. In other words, with an average deviation from the mean of 27.1 degrees, there is a definite lack of consensus in the ratings of BLM, more so than in the ratings of any other group in Table 5.4. Indeed, one could argue that, on the basis of the high standard deviation, the BLM movement is more polarizing than any of the other groups that people were asked about.

Now let's look at the three groups with the lowest mean feeling thermometer ratings. Conservatives, big business, and socialists all have means closer to 50 degrees. Their standard deviations suggest more consensus in these lower ratings for big business than for socialists or conservatives.

The differences in the degree of division are also evident when we look at histograms of ratings with different standard deviations.

Figure 5.3 (next page) shows three histograms for feeling thermometer ratings of scientists (top), big business (middle), and BLM (bottom). The height of the outside bars helps us understand why the standard deviation for BLM is higher. For both big business and scientists, there is a clear modal category (the top for scientists and the middle for big business). The rating of scientists has a large drop-off in the proportion of respondents who hold much lower views (0–20, 20–40), and the rating of big business also has a large drop-off in the proportion of respondents who hold extreme views on either side of the feeling thermometer (0–20, 80–100). In other words, a lot of people are near the mean, and relatively few are far from the mean—thus we find a lower standard deviation.

When we look at the histogram for BLM, we see that the distribution takes on a different shape. We see neither a drop-off in respondents who hold much lower views of BLM (as we saw with views of scientists), nor a drop-off on both ends of the thermometer (as we saw with views of big business). Although there are a lot of respondents who rate BLM somewhere between 80 and 100 degrees, the similar height of the bars in all of the other categories tells us that there are also a lot of respondents with ratings across the board. This is why the standard deviation for BLM is so much higher.

This example demonstrates how we can use the standard deviation to compare average variation from the mean across different samples or groups. In the next section, we will see how we can use the standard deviation to compare single values from different samples or groups even when those values are measured on different interval-ratio scales.

Comparing Apples and Oranges

You're probably familiar with the old adage, "You can't compare apples and oranges." The idea is that comparing two different kinds of things will lead to faulty conclusions about potential differences between them. Apples and oranges are both round fruits, but they're also quite different. However, thanks to the standard deviation, we can "compare" the values of any two interval-ratio variables, even between two different kinds of cases, such as apples and oranges.

Let's actually compare apples and oranges. Say that the mean weight for apples is 6.5 ounces, with a standard deviation of 0.4 ounces. Say that oranges average 4.5 ounces in weight, with a standard deviation equal to 0.5 ounces. You are at the fruit store and pick one big apple and one big orange. You weigh them; the apple weighs 6.9 ounces, and the orange weighs 5.5 ounces. Which one is "bigger"?

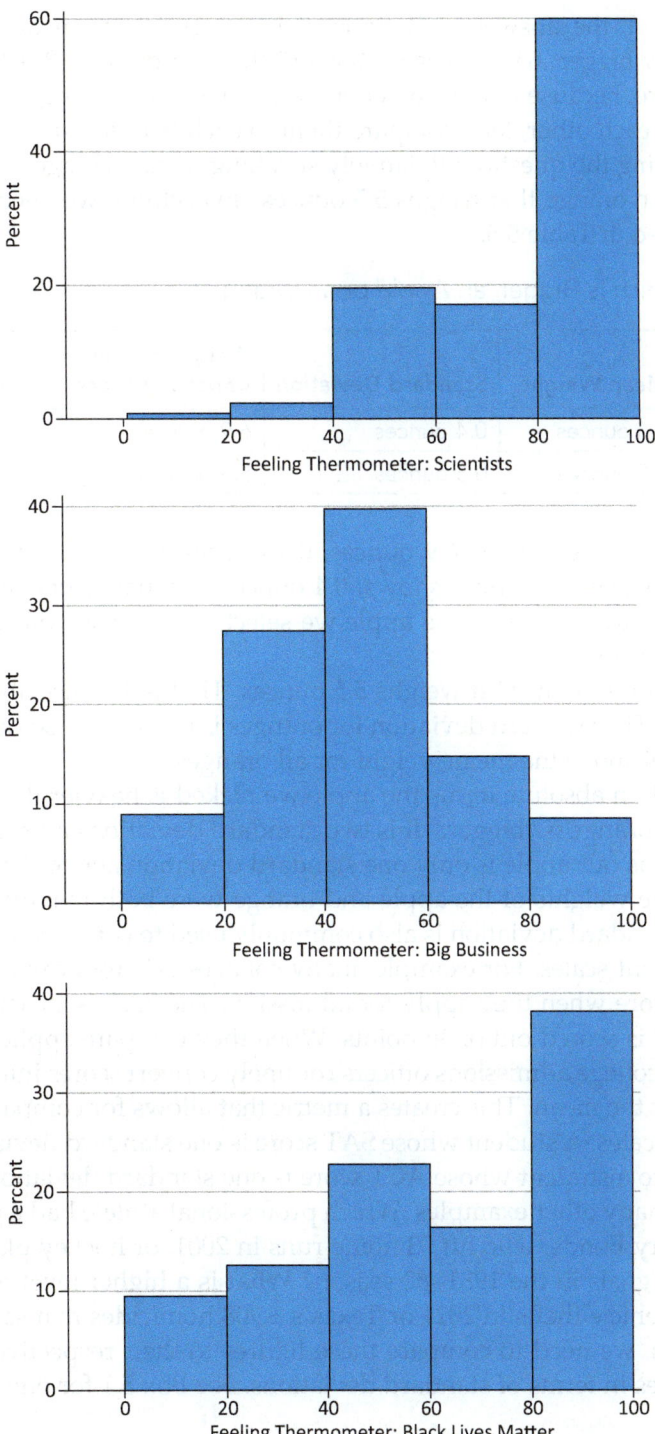

Figure 5.3 Feeling Thermometer Ratings for Scientists (top), Big Business (middle), and Black Lives Matter (bottom) (Histograms).

Source: 2020 American National Election Study.

At one level, the answer is obvious. If by "bigger" we mean heavier, the apple is unquestionably bigger: 6.9 ounces is "bigger" than 5.5 ounces. But that may not be a fair way to compare, because apples in general are bigger than oranges. Instead of comparing the weights to each other, let's compare them to each fruit's respective mean. In so doing, we are answering the question: Relatively speaking, what is bigger—an apple that weighs 6.9 ounces or an orange that weighs 5.5 ounces? Everything we need to answer this question is contained in Table 5.5.

Table 5.5 Which Is Bigger, an Apple or an Orange?

	Mean Weight	Standard Deviation	Weight of Fruit Chosen at Store	Standard Deviations above the Mean
Apples	6.5 ounces	0.4 ounces	6.9 ounces	+1 SD
Oranges	4.5 ounces	0.5 ounces	5.5 ounces	+2 SD

Our chosen apple weighs 6.9 ounces, 0.4 ounces above the mean for all apples. As the Standard Deviation column shows, 0.4 ounces is equal to one standard deviation for apples. We can say, then, that the apple we selected is one standard deviation above the mean weight for apples.

What about the orange? It weighs 5.5 ounces. That is 1 ounce above the mean weight for all oranges. The standard deviation for oranges is 0.5 ounces, so our orange is two standard deviations above the mean weight for all oranges.

So although in absolute terms the apple we picked is heavier than our orange, in relative terms, our orange is "bigger." It is two standard deviations above the mean weight for oranges, whereas our apple is only one standard deviation above the mean apple weight.

Whereas the weights of the apple and orange were both measured on the same scale (ounces), the standard deviation is also commonly used to compare variables that are measured on different scales. For example, many colleges ask applicants to report either their SAT or ACT score when they apply for admission. The SAT is measured on a 1,600-point scale. The ACT is scored out of 36 points. When they compare applicants who have taken different tests, college admissions officers routinely convert scores into standard deviations above or below the mean. This creates a metric that allows for comparison across different measurement scales. A student whose SAT score is one standard deviation above the mean is comparable to a student whose ACT score is one standard deviation above the mean.

There are many other examples. Which professional athlete had a greater season? Baseball player Barry Bonds, who hit 73 home runs in 2001, or hockey player Wayne Gretzky, who scored 92 goals in the 1981–82 season? What is a higher level of crime? California's 11,272 motor vehicle thefts in 2021 or Texas's 1,908 homicides that same year?[5] To answer these questions, we need to compare these figures to their respective means and express those differences in terms of standard deviations. See Box 5.1 for one example.

BOX 5.1: APPLICATION

Who Had a Better Season, Bonds or Gretzky?

In 1981–82, National Hockey League (NHL) player Wayne Gretzky smashed the previous record and scored 92 goals in a single season (the previous record was 76 goals). In 2001, Major League Baseball (MLB) player Barry Bonds shattered the previous record in his sport, hitting 73 home runs in a single season (the previous record was 70 home runs).

Which record is more impressive? This is a case of "apples and oranges." It's hard to argue that there is any equivalency between a goal in hockey and a home run in baseball. So instead of comparing these figures directly to each other, let's compare them to each sport's respective average for the top twenty-five goal scorers and home run hitters. Table 5.6 shows the means and standard deviations for top scorers in the NHL and MLB for the relevant seasons.

Table 5.6 Means and Standard Deviations for Top Scorers in NHL and MLB

	Top 25 Goal Scorers in 1981–82 NHL Season	Top 25 Home Run Hitters in 2001 MLB Season
Mean	48.4 goals	43.6 home runs
Standard deviation	11.3 goals	9.7 home runs

Source: The National Hockey League (www.nhl.com) and ESPN (www.espn.com)..

Now let's calculate, in standard deviations, how much Gretzky's and Bonds' record-setting performances exceeded the means among top players in their respective leagues during the relevant years.

Gretzky:

$$\frac{92 - 48.4}{11.3} = 3.86$$

Bonds:

$$\frac{73 - 43.6}{9.7} = 3.03$$

What Can We Conclude?

Compared to the other leaders in their sports, Wayne Gretzky's record season is more impressive than Barry Bonds's record season. Gretzky's 92 goals were almost four standard deviations above the mean number of goals among the top twenty-five scorers that year. Bonds's 73 home runs were closer to three standard deviations above the mean number of home runs among the top twenty-five sluggers that season. Both athletes were well above their sports' means, but, by this metric, Gretzky's season was more exceptional.

Let's try one other example. We know that younger Americans are more likely than their elders to use the Internet. In 2021, the GSS asked respondents how many hours per week they spend on the Internet. The means and standard deviations are shown in Table 5.7, and they confirm our expectations: young people spend more time on the Internet on average.

Table 5.7 Hours per Week Spent on Internet

	Mean	Standard Deviation
18 to 29 years old	24.5 hours	24.7 hours
60 years or older	10.6 hours	12.4 hours

Source: 2021 General Social Survey.

Now let's take two individuals, one young, the other older. Both report that they spend 28 hours per week on the Internet. In absolute terms, there is no denying that they spend the same amount of time on the Internet. But in relative terms, the young person is a less active Internet user than the older one. On what basis do we make this claim?

We make this claim by comparing the older respondent's score to the overall mean for older people and by comparing the young respondent's score to the mean for younger people (see Table 5.8).

Table 5.8 Weekday Hours Spent on Internet, Young versus Old

	Hours Spent on Internet	Age Group Mean	Age Group Standard Deviation	Distance from Mean	Distance Expressed in Standard Deviations
Younger respondent	28.0	24.5	24.7	3.5 hours above mean	3.5/24.7 = 0.14
Older respondent	28.0	10.6	12.4	17.4 hours above mean	17.4/12.4 = 1.40

Source: 2021 General Social Survey.

By this metric, the older person, whose mean score is 1.40 standard deviations above the mean, is a more active Internet user than the younger respondent, whose score is just barely above the mean (0.14 standard deviations above the mean) for young people. This example makes clear that, when we use the standard deviation, the reference point for evaluating whether 28 hours is a large amount of time to spend on the Internet in a day is not how much time people in general spend on the Internet but how much time one's own age group tends to spend on the Internet.

This example also underlines another important point. When we want to compare an individual score to the mean for his or her group, we follow these steps:

1. We subtract the mean from the individual's score.
2. We divide that difference by the standard deviation.

These steps allow us to express an individual score in terms of standard deviations above or below the mean. This is an important procedure that we will see again in the next chapter, when we study the normal distribution.

Skewed versus Symmetric Distributions

As we have seen, the standard deviation is sensitive to the shape of a distribution. If many of the cases are clustered around the mean and relatively few cases are on the outsides, then the standard deviation will be relatively small. On the other hand, if we have a highly skewed distribution, the standard deviation will be disproportionately influenced by cases far away from the mean.

Let's examine three hypothetical distributions of income. In the first case, we have a very egalitarian society. This is shown in Figure 5.4, Panel A. The height of the curve shows the number of people at each income level. Note that the distribution is symmetric. Many people have incomes close to the center (the mean), and very few people are extremely poor or extremely rich. Note as well that this graphic does not collapse the income variables into bins or categories. Income (measured in dollars) is a continuous variable, and continuous variables can be displayed by means of an area curve (instead of a histogram, which collapses the variable into bins). An area curve is essentially a smoothed histogram; just as in a histogram, the height of the line in an area curve represents the number or percentage of cases at that level of the variable.

The other two curves show skewed distributions. In the first (Panel B), we see a left or negative skew. In this society, there are a very small number of poor people and lots of affluent people. In Panel C, we find the opposite: lots of poor and middle-income people and a small number of wealthy people. Panel C is closest to what the income distribution in the United States looks like (as we will see in chapter 6).

The distribution in Panel A has the smallest standard deviation. This is because a large proportion of respondents are close to the mean, and so the average deviation will be relatively small. In Panel B, the mean will be pulled downward, and, as a result, the standard deviation will be larger. The same logic applies to the curve in Panel C, except in the opposite direction. Those few individuals in the high end of the curve will pull the mean upward, with the consequence that the standard deviation will rise accordingly. For these skewed distributions, the interquartile range would be a better measure of variability, since it is based only on the variable's values at the 25th and 75th percentiles.

There is one final feature of distributions that is worth noting. Two symmetric distributions may convey a similar impression, but the peakedness or flatness of a curve is an important feature to note. Compare the curves in Figure 5.5. Both are symmetric. But in the first case (solid curve), the curve is relatively tall and narrow, with few cases in the tails, whereas the second curve (dotted curve) is lower and flatter, with more cases in the tails. In the second case, the standard deviation will be higher, because more individuals have scores farther away from the mean.

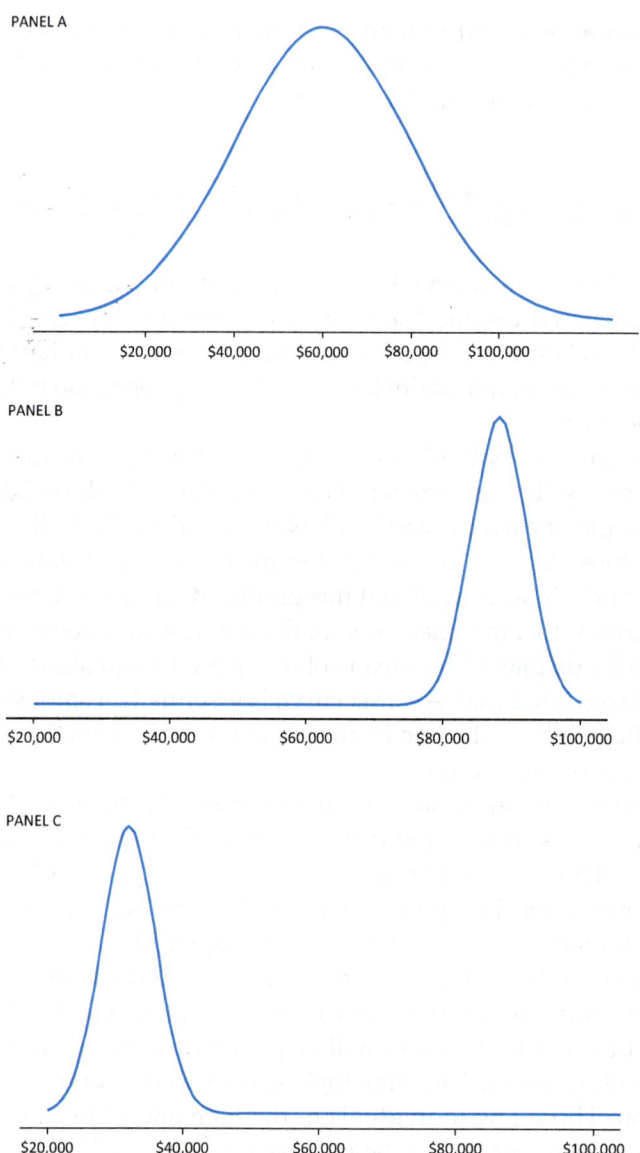

Figure 5.4 Three Hypothetical Income Distributions

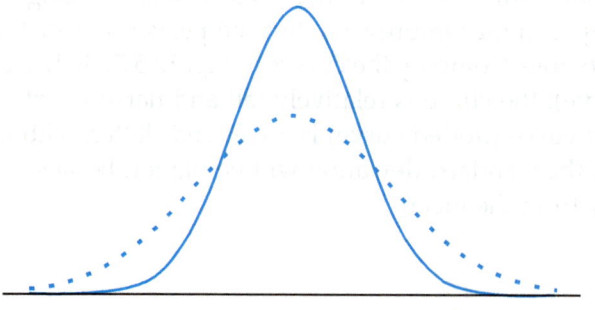

Figure 5.5 Comparing the Kurtosis of Two Curves

Chapter Summary

There are three commonly used measures of **variability**: range, interquartile range, and standard deviation. They all describe how much diversity or spread there is in some variable.

Range:

- Use with ordinal or interval-ratio variables.
- *Range = highest value − lowest value.*
- Very sensitive to extreme values.
- Example: There is a range of $40,000 in the incomes of our sample respondents, from $35,000 to $75,000.

Interquartile range:

- Use with ordinal or interval-ratio variables.
- *IQR = value at 75th percentile − value at 25th percentile.*
- Defined as the values associated with the middle 50% of the data or the distance between the values associated with the 25th and 75th percentiles.
- If we have a frequency table, we use the cumulative percentage column to locate the 25th and 75th percentiles.
- IQR is a more stable measure of variability if the variable is skewed.
- Example: The IQR of SAT critical reading scores is 410 (25th percentile) to 570 (75th percentile).

Standard deviation:

- Use with interval-ratio variables.
- By formula:

$$s = \sqrt{\frac{\Sigma(y - \bar{y})^2}{N - 1}}$$

- Standard deviation is interpreted as average deviation from the mean.
- Standard deviation squared is the variance, another measure of variability.
- By expressing individual scores as standard deviations above or below the mean, we can compare scores on variables measured on different scales.
- Standard deviation is a better measure of variability when the data are symmetric or nearly symmetric.
- Example: On average, ratings of scientists varied from the mean by 19.2 points.

Using Stata

For this section, we will use data from the World Values Survey (WVS) to explore variation in people's perceptions of how much control they have over their lives. This variable, called *Q48* in the dataset, measures how much choice and control respondents felt over their own lives, on a scale from 1 to 10, where 1 means "no choice at all" and 10 means "a great deal of choice." The Stata procedures are straightforward and rely primarily on one command: "tabstat."

Finding the Range

The variables are saved in the data set called `WVSWave7.dta`. To find the range for the variable, we use the "tabstat" command that we used for measures of central tendency in chapter 4. Type the following command into Stata:

`tabstat Q48, stat(range)`

We can request the range for the variable by listing the name of the variable after "tabstat," followed by the option "stat(range)." The output is shown in Figure 5.6.

```
. tabstat Q48, stat(range)
```

Variable	Range
Q48	9

Figure 5.6

We see that the distance between the highest and lowest scores is 9. Recall that this variable is measured on a scale from 1 (no choice at all) to 10 (a great deal of choice). The range of 9 tells us that some number of respondents chose the lowest and highest values for this variable.

Finding Percentiles

Now, we will find the values for *Q48* at the 25th and 75th percentiles. Again, we can use the "tabstat" command:

`tabstat Q48, stat(p25 p75)`

This time, we have specified "stat(p25 p75)" as an option because we want to see the values at the 25th and 75th percentiles. The output is shown in Figure 5.7.

```
. tabstat Q48, stat(p25 p75)

    variable |       p25        p75
-------------+----------------------
         Q48 |         6          9
```

Figure 5.7

The "p25" and "p75" rows show the 25th and 75th percentiles: 6 and 9, respectively.

Finding the Interquartile Range

We could find the interquartile range for our variable fairly easily by using the above Stata output showing the 25th and 75th percentiles for each variable. However, one of the major benefits of using a statistical software program like Stata is that we know the program will not make any calculation errors. We can never be as sure of ourselves as that. Use the "tabstat" command to request the IQR for the variable:

`tabstat Q48, stat(iqr)`

The output is shown in Figure 5.8.

```
. tabstat Q48, stat(iqr)

    variable |       IQR
-------------+----------
         Q48 |         3
```

Figure 5.8

Take a moment to check Stata's work by using the previous output showing the 25th and 75th percentiles for each variable. Subtracting the 25th percentile from the 75th percentile yields the same IQR displayed in the IQR output. The IQR of 3 is equal to the distance between the 25th and 75th percentiles for the variables (6 to 9). When it comes to feelings of choice and control over one's life, half of the respondents fell between values of 6 and 9, indicating that half the sample felt some degree of control over their lives.

Generating a Box Plot

Generating box plots for feelings of control over one's life provides a visual depiction of the variable's distribution. Enter the following "graph" command into Stata to generate a box plot:

`graph box Q48`

With "box," the command specifies that we want a box plot for the variable. The box plot is shown in Figure 5.9.

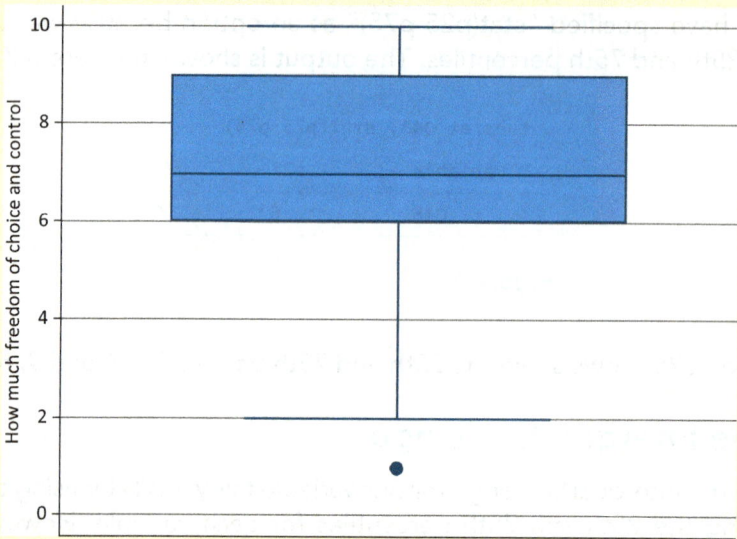

Figure 5.9

The box plot shows us visually what we already learned from the IQR. For feelings of control over life, half of respondents fall somewhere between values of 6 and 9.

Finding the Variance and Standard Deviation

We will return to the "tabstat" command to get the variance and standard deviation for the variables:

`tabstat Q48, stat(variance sd)`

The output is shown in Figure 5.10.

```
. tabstat Q48, stat(variance sd)
```

Variable	Variance	SD
Q48	5.124933	2.263832

Figure 5.10

The average deviation from the sample mean for feeling of control over one's life is 2.26 points.

Another way of requesting the standard deviation is to use the "summarize" command, which we saw in chapter 4. Type the following command into Stata:

`summarize Q48`

The output is shown in Figure 5.11.

As we saw in chapter 4, the "summarize" command returns the mean, standard deviation, and minimum and maximum values for the variables. Note that the standard deviation reported here matches the standard deviations that Stata gave us when we used the

```
. summarize Q48
    Variable |        Obs        Mean    Std. dev.       Min        Max
         Q48 |     87,022    7.202719    2.263832          1         10
```

Figure 5.11

"tabstat" command. The advantage of using the "summarize" command is that it provides us with measures of variability and central tendency at the same time. This allows us to think about how the values of the variable are distributed around a key measure of central tendency, the mean.

Review of Stata Commands

- Finding the range

    `tabstat variable name/s, stat(range)`

- Finding percentiles

    `tabstat variable name/s, stat(p25 p75)`

- Finding interquartile range

    `tabstat variable name/s, stat(iqr)`

- Finding variance and standard deviation

    `tabstat variable name/s, stat(variance sd)`

- Finding standard deviation, with mean, range, and N

    `summarize variable name/s`

- Generating a box plot

    `graph box variable name`

Using SPSS

For this section, we will use data from the World Values Survey (WVS) to explore variation in people's perceptions of how much control they have over their lives. This variable, called *Q48* in the dataset, measures how much choice and control respondents felt over their own lives, on a scale from 1 to 10, where 1 means "no choice at all" and 10 means "a great deal of choice."

Finding the Range

The variables are saved in the data set called `WVSWave7.sav`. To find the range for the variable, use the "Descriptives" procedure that we saw in the previous chapter:

`Analyze → Descriptive Statistics → Descriptives`

Move the variables *Q48* into the "Variable(s)" box and click on "Options." This will open the "Descriptives: Options" dialog box. By default, certain statistics have check marks next to them (we can uncheck them if we like). The "range" is not one of them, so we must check the small box next to "Range," which is circled in Figure 5.12.

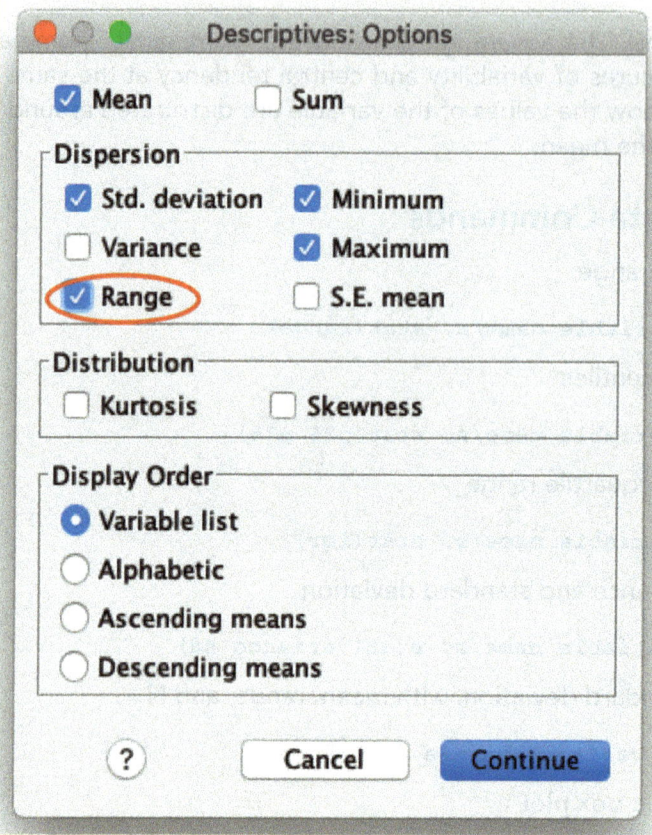

Figure 5.12

Click on "Continue," then click on "OK." This produces the desired output (shown in Figure 5.13). The range (circled) is included along with the other default statistics generated by SPSS.

We see that the distance between the highest and lowest scores is 9. Recall that this variable is measured on a scale from 1 (no choice at all) to 10 (a great deal of choice). The range of 9 tells us that some number of respondents chose the lowest and highest values for this variable.

Finding Percentiles

Now, we will find the values at the 25th and 75th percentiles (i.e., the quartiles). Here, we use the "Frequencies" procedure.

`Analyze → Descriptive Statistics → Frequencies`

Descriptive Statistics						
	N	Range	Minimum	Maximum	Mean	Std. Deviation
How much freedom of choice and control	87022	9	1	10	7.20	2.264
Valid N (listwise)	87022					

Figure 5.13

After moving Q48 into the "Variable(s)" box, we click on "Statistics." This opens the "Frequencies: Statistics" dialog box (shown in Figure 5.14). We check the small box next to "Quartiles."

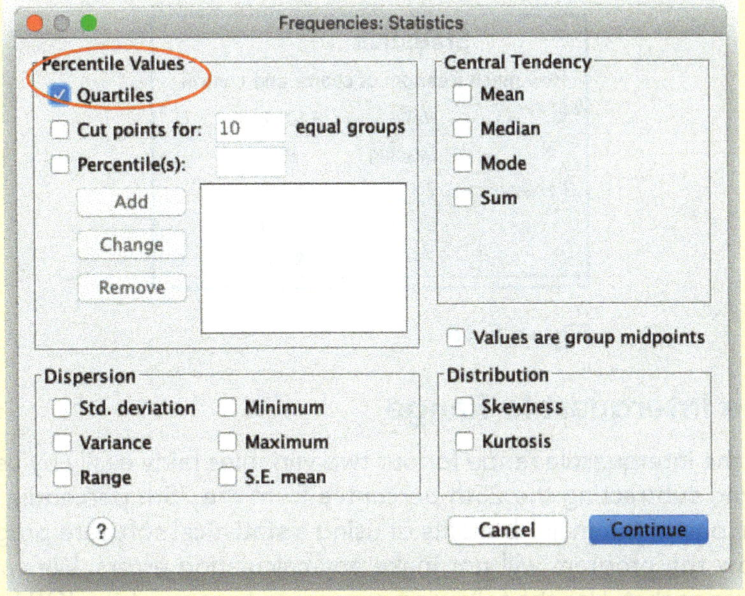

Figure 5.14

We click on "Continue." If we click on "OK," we will get the desired output, but we will also get frequency tables with a lot of categories. We can tell SPSS to suppress the tables and report only the desired statistics. In the "Frequencies" dialog box, we click on "Format." This opens the "Frequencies: Format" dialog box (shown in Figure 5.15). We place a check mark in the box next to "Suppress tables with many categories."

We click on "Continue" and then click on "OK." This produces the desired output (shown in Figure 5.16).

The table includes the 25th and 75th percentiles, in addition to the 50th percentile (the median). The 25th and 75th percentiles are 6 and 9, respectively.

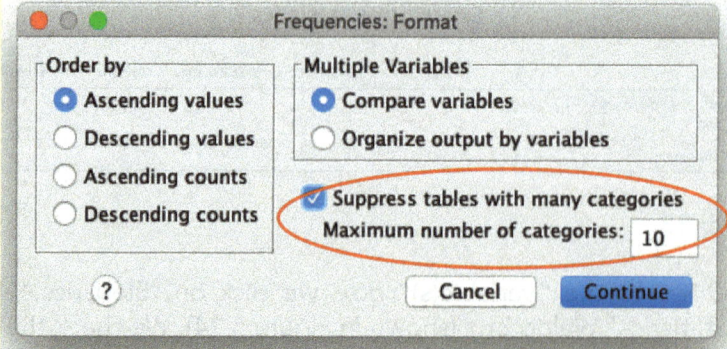

Figure 5.15

	Statistics	
How much freedom of choice and control		
N	Valid	87022
	Missing	800
Percentiles	25	6.00
	50	7.00
	75	9.00

Figure 5.16

Finding the Interquartile Range

We could find the interquartile range for our two variables fairly easily by using the above SPSS output and subtracting the 25th percentile from the 75th percentile, for each variable. However, one of the major benefits of using a statistical software program like SPSS is that we know the program will not make any calculation errors. We can never be as sure of ourselves as that. Use the following procedure to request the IQR for the variable:

`Analyze → Descriptive Statistics → Explore`

Move Q48 into the "Dependent List." Click on "OK." The output is shown in Figure 5.17.

As we can see, SPSS generates a slew of descriptive statistics, including the interquartile range. Take a moment to check SPSS's work by using the previous output showing the 25th and 75th percentiles for each variable. Subtracting the 25th percentile from the 75th percentile for the Q48 yields the same IQR displayed for that variable in the IQR output. The IQR for the variable is 3 (9 − 6). When it comes to feelings of choice and control over one's life, half of the respondents fell between values of 6 and 9, indicating that half the sample felt some degree of control over their lives.

Descriptives

			Statistic	Std. Error
How much freedom of choice and control	Mean		7.20	.008
	95% Confidence Interval for Mean	Lower Bound	7.19	
		Upper Bound	7.22	
	5% Trimmed Mean		7.35	
	Median		7.00	
	Variance		5.125	
	Std. Deviation		2.264	
	Minimum		1	
	Maximum		10	
	Range		9	
	Interquartile Range		3	
	Skewness		-.652	.008
	Kurtosis		-.045	.017

Figure 5.17

Generating a Box Plot

Generating box plots for feelings of control over one's life provides a visual depiction of the variable's distribution. We can use the exact same procedure we just used (for the IQR) to generate box plots:

`Analyze → Descriptive Statistics → Explore`

The output will include all of the descriptive statistics we saw above as well as a box plot. Figure 5.18 shows the output for the box plot for *Q48*.

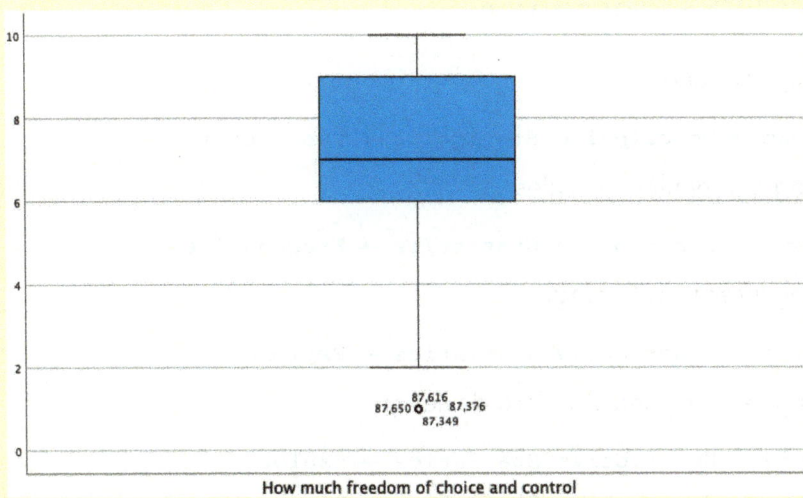

Figure 5.18

The box plot shows us visually what we already learned from the IQR. For feelings of control over life, half of respondents fall somewhere between values of 6 and 9.

Finding the Variance and Standard Deviation

We have already seen one way to ask SPSS for the variance and standard deviation for the two variables.

`Analyze → Descriptive Statistics → Explore`

As we saw in the SPSS IQR output, SPSS generated several other descriptive statistics, including both the standard deviation and the variance. Alternatively, we could use the following sequence:

`Analyze → Descriptive Statistics → Descriptives`

Under "Options," we place a check mark next to "Standard Deviation" and "Variance." The output is shown in Figure 5.19.

Descriptive Statistics

	N	Mean	Std. Deviation	Variance
How much freedom of choice and control	87022	7.20	2.264	5.125
Valid N (listwise)	87022			

Figure 5.19

The average deviation from the sample mean for feelings of control over one's life is 2.26 points.

Review of SPSS Procedures

- Finding the range

    `Analyze → Descriptive Statistics → Descriptives`

- Finding percentiles/quartiles

    `Analyze → Descriptive Statistics → Frequencies`

- Finding interquartile range

    `Analyze → Descriptive Statistics → Explore`

- Finding variance and standard deviation

    `Analyze → Descriptive Statistics → Explore`
    `Analyze → Descriptive Statistics → Descriptives`

- Generating a box plot

    `Analyze → Descriptive Statistics → Explore`

Practice Problems

1. Table 5.9 shows the rents for fourteen renters of two-bedroom apartments from two cities.

   Table 5.9 Rents for Two-Bedroom Apartments in Two Cities

New York City, NY (Mean: 5,145)	Grinnell, IA (Mean: 398)
3,350	475
2,500	500
5,375	250
2,200	325
2,199	350
2,500	450
9,000	400
12,000	300
3,300	375
4,300	600
4,000	445
7,500	300
3,800	335
10,000	475

   [handwritten note: Range for NY = 9,801; Range for IA = 350]

   a. Find the range of rents for each city. Explain what the range means in words.
   b. Find the interquartile range of rents for each city. Explain what the interquartile range means in words.
   c. Find the standard deviation of rents for each city. Explain what the standard deviation means in words.
   d. For each city, explain which measure of variability—standard deviation, range, or interquartile range—is the most appropriate measure.
   e. Use the mean, range, IQR, and standard deviation to briefly summarize the patterns of rents in the two cities.

2. Table 5.10 shows the travel time to work for ten Starbucks employees in two different cities (hypothetical data). The mean travel time for San Jose employees is 27.8 minutes and the mean travel time for Rapid City employees is 34.1 minutes.

   a. Find the range of travel time for each city. Explain what the range means in words.

Table 5.10 Travel Time to Work in Minutes

San Jose, CA (Mean: 27.8)	Rapid City, SD (Mean: 34.1)
10	5
18	8
23	19
28	34
30	40
30	40
30	45
31	45
35	52
43	53

b. Find the interquartile range of travel time for each city. Explain what the interquartile range means in words.

c. Find the standard deviation of travel time for each city. Explain what the standard deviation means in words.

d. For each city, explain which measure of variability—standard deviation, range, or interquartile range—is the most appropriate measure.

e. Use the mean, range, IQR, and standard deviation to briefly summarize the patterns of travel time in the two cities.

3. A survey asked a sample of 642 teenagers whether they thought that social media encouraged bullying, on a scale from 0 (completely disagree) to 10 (completely agree). Figure 5.20 shows the distribution of responses to this question.

   a. The mean score for this variable is 5.5, indicating that, on average, respondents slightly agree that social media encourages bullying. What does the distribution in Figure 5.20 visually indicate about how much variation there is around the mean?

   b. Which standard deviation is more likely: 1.52 or 3.27?

   c. Use Figure 5.20 to find the interquartile range for this variable. Explain what it means in words.

   d. Use the figure, the mean, the interquartile range, and standard deviation (whichever you chose in part b) to briefly describe the central tendency and variability in this distribution.

   e. Which measures of variability are appropriate for characterizing diversity of social media attitudes in the sample: range, standard deviation, or interquartile range? Explain your answer.

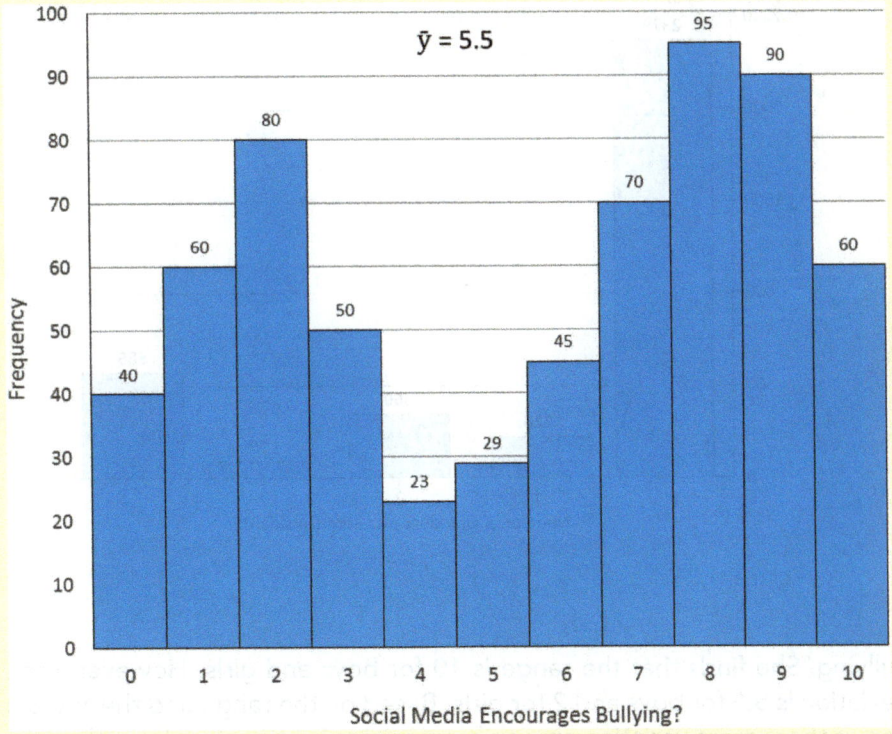

Figure 5.20 Teenagers' Attitudes about Whether Social Media Encourages Bullying

4. The GSS asked respondents how many of their grandparents were born outside the United States. Figure 5.21 shows the results from the 2021 survey.

   a. The mean score for this variable is 0.98, indicating that, on average, respondents report that about 1 grandparent was born outside of the United States. What does the distribution in Figure 5.21 visually indicate about how much variation there is around the mean?

   b. Which standard deviation is more likely: 1.52 or 3.27?

   c. Use Figure 5.21 to find the interquartile range for this variable. Explain what it means in words.

   d. Use the figure, the mean, the interquartile range, and standard deviation (whichever you chose in part b) to briefly describe the central tendency and variability in this distribution.

   e. Which measures of variability are appropriate for characterizing diversity of number of grandparents born outside the United States in the sample: range, standard deviation, or interquartile range? Explain your answer.

5. A researcher uses the data from Problem 3 to investigate whether there is more variation among boys or girls in attitudes about whether social media encourages

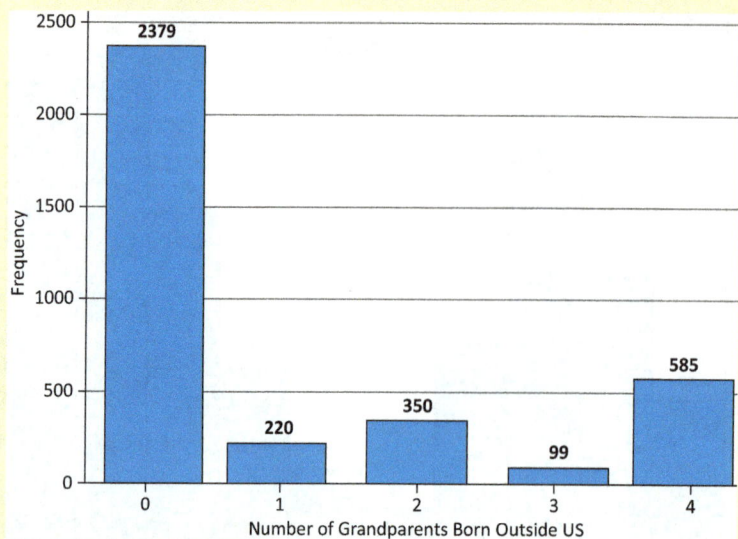

Figure 5.21

bullying. She finds that the range is 10 for boys and girls. However, the standard deviation is 5.5 for boys and 2 for girls. Based on the range and the standard deviation, is there more variation among boys or girls in attitudes about the role of social media in bullying?

6. A researcher uses the data from Problem 4 to investigate whether there is more variation among older or younger people in their number of grandparents who were born outside the United States. They find that the range is 4 for both younger and older people (those below fifty and those age fifty and older). However, the standard deviation is 1.55 for those below fifty and 1.47 for those fifty and older. Based on the range and the standard deviation, is there more variation among older or younger people in their number of grandparents who were born outside the United States?

7. The 2020 American National Election Survey (ANES) asked respondents how hopeful and how worried they feel about how things are going in the country on two separate scales from 1 (not at all) to 5 (extremely). Because these scales have at least 5 items, researchers treat them as interval-ratio level variables. For both variables, the highest value given by a respondent was 5, and the lowest was 1.

   a. What is the range for the how hopeful respondents feel about how things are going in the country? What is the range for the how worried respondents feel about how things are going in the country?

   b. Table 5.11 shows a frequency distribution for how hopeful respondents feel about how things are going in the country. Find the interquartile range and explain what it means.

   c. Use the frequency distribution in Table 5.11 to sketch a box plot of how hopeful respondents feel about how things are going in the country. Be sure to label the values for the 25th, 50th, and 75th percentiles.

Table 5.11 Frequency Distribution of Hopefulness about How Things Are Going in the Country

Score	Frequency	Percent	Cumulative Percent
1. Not at all	1,981	23.99	23.99
2. A little	2,092	25.34	49.33
3. Somewhat	2,514	30.45	79.77
4. Very	1,208	14.63	94.40
5. Extremely	462	5.60	100.00
Total	8,257	100.00	

8. The 2021 GSS asked respondents how good or bad they thought the impact of climate change will be for America and for the world, in two separate scales ranging from 0 (extremely bad) to 10 (extremely good). Because these scales have at least five items, researchers treat them as interval-ratio level variables. For both variables, the highest value given by a respondent was 10, and the lowest was 0.

   a. What is the range for what respondents think the impact of climate change will be for America? For the world?

   b. Table 5.12 shows a frequency distribution for how respondents rate the impact of climate change for America. Find the interquartile range and explain what it means.

Table 5.12 How Bad or Good Respondents Think the Impact of Climate Change Will Be for America

Impact of Climate Change for America	Frequency	Percent	Cumulative Percent
0 (extremely bad)	372	21.60	21.60
1	124	7.20	28.80
2	242	14.05	42.86
3	263	15.27	58.13
4	198	11.50	69.63
5	349	20.27	89.90
6	60	3.48	93.38
7	52	3.02	96.40
8	19	1.10	97.50
9	24	1.39	98.90
10 (extremely good)	19	1.10	100.00
Total	1,722	100.00	

c. Use the frequency distribution in Table 5.12 to sketch a box plot of how respondents rate the impact of climate change for America. Be sure to label the values for the 25th, 50th, and 75th percentiles.

9. The same 2020 ANES respondents from Problem 7 yielded the box plot shown in Figure 5.22 for how worried respondents feel about how things are going in the country.

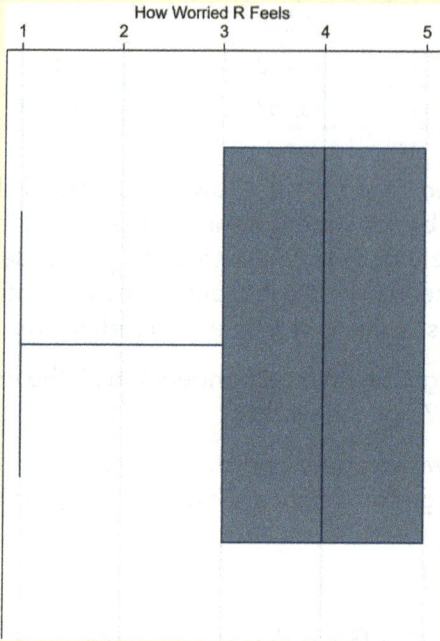

Figure 5.22

a. What is the interquartile range for how worried respondents feel about how things are going in the country?

b. Compare the interquartile ranges for the how hopeful and how worried respondents feel about how things are going in the country. What does the comparison say about variation in how hopeful and worried people feel?

10. The 2021 GSS data discussed in Problem 8 also yielded the box plot shown in Figure 5.23 for how respondents rate the impact of climate change for the world on a scale ranging from 0 (extremely bad) to 10 (extremely good).

a. What is the interquartile range for how respondents rate the impact of climate change for the world?

b. Compare the interquartile ranges for the how respondents rate the impact of climate change for America and for the world. What does the comparison say about variation in how people rate the impact for each location?

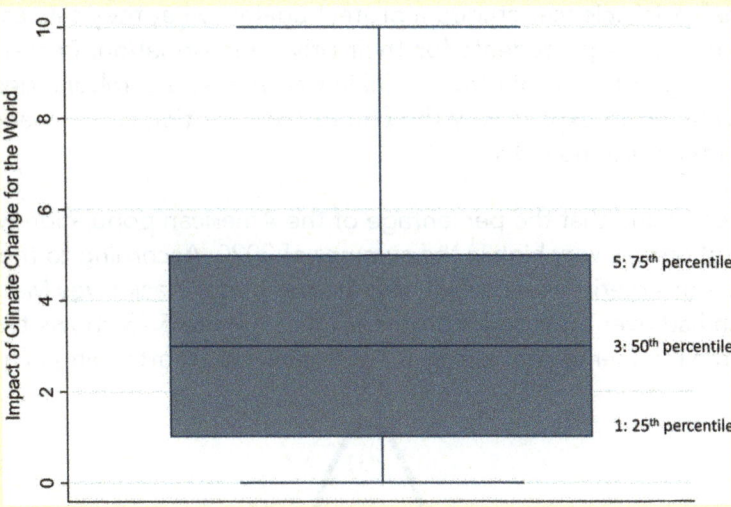

Figure 5.23

11. Shiru Café serves college students free coffee every two hours in exchange for students' names, email addresses, field of study, and professional interests. In exchange, students allow Shiru's corporate sponsors to access their information. Figure 5.24 shows two distributions: the numbers of cups of coffee consumed per day for students at Brown University the year before and the year after Shiru opened a location across the street from campus.

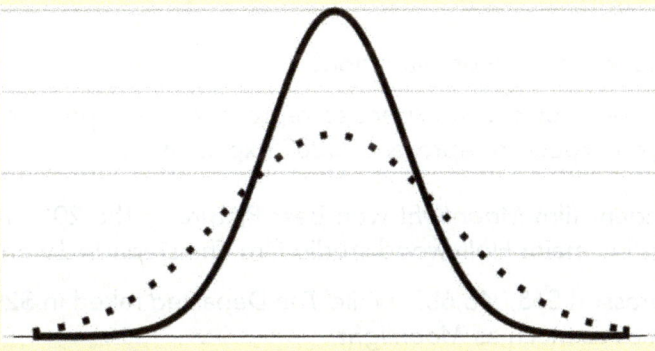

Figure 5.24

a. Health practitioners in Brown's Student Health Center found that variation in students' coffee consumption increased after Shiru opened. Which curve represents coffee consumption before Shiru, and which is the distribution for coffee consumption after Shiru?

b. It turns out that Shiru recruits its student customer base primarily from the fields of math, science, and technology. It also turns out that several departments in the social sciences staged a protest against what they saw as Shiru's exploitation of college students for their private information. Protestors were also encouraged to abstain from drinking coffee as a symbolic gesture. Use this information to explain why the coffee consumption curve changed the way it did after Shiru moved in.

12. Researchers found that the percentage of the American population who attended a demonstration was very high in the summer of 2020. According to the Pew Research Center, 6% of Americans reported having attended a Black Lives Matter protest, far more than had ever attended a protest before. Figure 5.25 shows two curves, each showing the number of protests that Americans report attending in a given year.

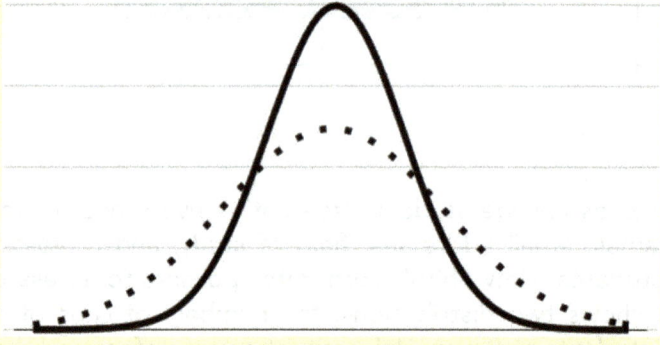

Figure 5.25

a. Which curve shows more variation?

b. Which curve would you expect to represent a year prior to 2020, and which would you expect to represent 2020? Explain why.

13. The independent film *Moonlight* won Best Picture at the 2016 Academy Awards. Ten years earlier, major Hollywood studio film *The Departed* won the award.

    Moonlight grossed $65,046,687, while *The Departed* raked in $291,465,034, more than four times as much as *Moonlight*.

    Table 5.13 Mean and Standard Deviation for Earnings of 2016 Independent Films and 2006 Hollywood Studio Films*

	Mean Earnings	Standard Deviation
Independent films, 2016	25,000,000	10,000,000
Hollywood studio films, 2006	100,000,000	45,000,000

    *Hypothetical data

    a. A film student is hesitant to declare that *The Departed* is a higher-earning film than *Moonlight*. Use the data presented in Table 5.13 to calculate how many standard deviations each film earned above or below mean earnings for independent films and Hollywood studio films in their respective release years.

b. Based on your calculations from part a, which film was the higher earner among its peers: *Moonlight* or *The Departed*?

c. Which metric do you think is a better one to use for comparing the earnings of *Moonlight* and *The Departed*: absolute earnings or relative earnings? Explain your answer.

14. Two siblings are arguing about who does better in school. Both siblings have above average GPAs. Sibling A claims they are the better student because their overall GPA is higher, 3.95, compared to Sibling B's GPA of 3.70. Sibling B claims that they are the better student because their school (B) grades more harshly than Sibling A's school (A). Table 5.14 shows the means and standard deviations for the GPAs at schools A and B. Based on the information in the table, which sibling do you agree with and why?

Table 5.14 Mean and Standard Deviation for Grades at School A and School B

	Mean GPA	Standard Deviation
School A	3.80	0.34
School B	3.63	0.14

Stata Problems

Here we will use data from the 2020 American National Election Survey (ANES) to compare variability in voters' feelings about 2020 Democratic presidential candidate Joe Biden and 2020 Republican candidate Donald Trump. Respondents rated Biden and Trump on a feeling thermometers (V202143 for Biden and V202144 for Trump). Is there more variability in voters' feelings about Biden or Trump?

1. Open the `ANES20.dta`. Use the "tabstat" command to ask Stata for the following statistics for each feeling thermometer: standard deviation; mean; interquartile range; values for the 25th, 75th, and 50th percentiles; and range.

2. Which candidate's interquartile range is larger? What is the low and high value of the interquartile range for each? What does this say about variability in voters' opinions about each of them?

3. What is the range for each candidate? What must be the highest and lowest value for each of them?

4. What is the standard deviation for each candidate? Explain what they mean in words.

5. Use the "graph box" command to generate box plots of the feeling thermometers for Biden and Trump.

SPSS Problems

Here we will use data from the 2020 American National Election Survey (ANES) to compare variability in voters' feelings about 2020 Democratic presidential candidate Joe Biden and 2020 Republican candidate Donald Trump. Respondents rated Biden and Trump on a feeling thermometer (*V202143* for Biden and *V202144* for Trump). Is there more variability in voters' feelings about Biden or Trump?

1. Open `ANES20.sav`. Use the appropriate "Descriptives" procedure to ask SPSS for the following statistics for each feeling thermometer: standard deviation; mean; interquartile range; values for the 25th, 75th, and 50th percentiles; and range.

2. Which candidate's interquartile range is larger? What is the low and high value of the interquartile range for each? What does this say about variability in voters' opinions about each of them?

3. What is the range for each candidate? What must be the highest and lowest value for each of them?

4. What is the standard deviation for each candidate? Explain what they mean in words.

5. Use the appropriate "Descriptives" procedure to generate box plots of the feeling thermometers for Biden and Trump.

Notes

[1] Stephen Jay Gould. 1985. "The Median Isn't the Message." First published in *Discover* 6 (June): 40–42. Retrieved from http://www.cancerguide.org/median_not_msg.html.

[2] Martin, Terrance, Lua Augustin, Laura Ricaldi, and Jose Nunez. 2020. "The Effect of Student Loans on Parental Views of Education Financing." *Journal of Financial Planning* 33(5): 46–55. Retrieved from https://www.financialplanningassociation.org/article/jun20-effect-student-loans-parental-views-education-financing.

[3] "Apartment List National Rent Report." October 30, 2022. Retrieved from https://www.apartmentlist.com/research/national-rent-data.

[4] College Board. 2022. "SAT Suite of Assessments Annual Report." Retrieved from https://reports.collegeboard.org/media/pdf/2022-total-group-sat-suite-of-assessments-annual-report.pdf.

[5] FBI Crime Data Explorer. Retrieved from https://crime-data-explorer.fr.cloud.gov/pages/explorer/crime/crime-trend.

Chapter 6: Probability and the Normal Distribution

In an article published in 1983, two famous psychologists who were interested in how people misinterpret probability, Amos Tversky and Daniel Kahneman, constructed the following scenario about a woman named Linda:[1]

Linda is thirty-one years old, single, outspoken and very bright. She majored in philosophy. As a student, she was deeply concerned with issues of discrimination and social justice, and also participated in anti-nuclear demonstrations. Which is most likely?

1. *Linda is active in the feminist movement. (Feminist)*
2. *Linda is a bank teller. (Teller)*
3. *Linda is a bank teller and is active in the feminist movement. (Teller and Feminist)*

They showed this description of Linda, along with the numbered descriptions of Linda's possible traits, to a sample of undergraduates and asked them to rank the statements in order of likelihood, from most likely to least likely.

Before reading on, take a minute and do the exercise yourself. Which of the three statements do you think is most likely to be true of Linda? Which is least likely to be true? Rank the three statements from most to least likely.

How did you rank these three statements? If you ranked them as follows:

Most likely: Linda is active in the feminist movement. (Feminist)
Less likely: Linda is a bank teller and is active in the feminist movement. (Teller and Feminist)
Least likely: Linda is a bank teller. (Teller)

then you agree with 85% of the authors' sample.

But this order is not mathematically possible. The probability of one outcome occurring cannot be less than the probability of that outcome and another outcome. The probability that Linda is a bank teller cannot be lower than the probability that she is a bank teller and active in the feminist movement. Using symbols, we would express this as p(Teller) ≥ p(Teller and Feminist).* P(Teller and Feminist) is a subset of p(Teller). That is, out of the universe of possible political identities for Linda-the-teller, feminist is one possibility. Later in this chapter, we will learn the basic rules of probability and demonstrate mathematically why this is so.

If you ranked the statements about Linda in the wrong order, do not feel bad. Probability is one of the most difficult things for people to estimate accurately. Kahneman and Tversky found that many people have a tendency to overestimate the probability of something happening when they consider other information that seems relevant. Because Linda had a background that seems compatible with feminism, people tend to think it's more likely that she would be a feminist and a bank teller than just a bank teller. Kahneman and Tversky also showed that people make predictable errors in estimating probability depending on how representative the information at their disposal is of what they would expect.

Probability is important to statistics in two ways. First, it is the basis for thinking about how the results of an individual sample relate to the likely results for the full population. This process of generalizing from sample to population, called **inferential statistics**, relies on a specific probability distribution, the normal curve, which is the subject of the second half of this chapter. Second, probability can be used to understand the likelihood of events or patterns in the social world.

⊖ The Rules of Probability

Many textbooks use dice or playing cards to illustrate the rules of probability. There are good reasons for using these kinds of examples—you may after all find yourself in a casino at some point in the future! But probability also can be used to understand the likelihood of more consequential events, such as the likely gender makeup of children in a family. Kahneman and Tversky conducted an experiment to see how people assessed the likelihood of different scenarios regarding the order and number of boys and girls born to families.[2] A large sample of students was given the following information and question:

All families with six children in a city were surveyed. In seventy-two families the exact order of births of boys and girls was G B G B B G, where G represents daughters and B represents sons.

What is your estimate of the number of families surveyed in which the exact order of births was B G B B B B?

In other words, are we more likely to find a family that has a *G B G B B G* birth order than a family that has a *B G B B B B* birth order?

What do you think? Do you think that the first sequence is a more likely birth order than the second? If that's what you think, then you agree with 83% of Kahneman and

---

* "p(Teller)" means the probability of being a Teller; ≥ means is greater than or equal to; p(Teller and Feminist) means the probability of being both a Teller and a Feminist.

Tversky's sample. But in fact, the probabilities of these two birth sequences occurring are approximately equal. (Later in the chapter, we will see how this is calculated.)

Most people get it wrong because the second sequence does not seem like a likely outcome—there are too many boys, and too many boys in a row—even though the probability of that happening is about the same as any other sequence of six births.

To many people probability seems counterintuitive, but, at its core, it is quite simple. Probability is a fraction, in which the numerator represents the outcome(s) of interest and the denominator represents all of the possible outcomes. Sometimes that fraction is expressed as a proportion or a percentage. But the meaning does not change.

Probability represents the likelihood of some outcome occurring, given all of the possible outcomes. The key to understanding probability, as the above examples suggest, is not to rely on your "gut sense" of what seems likely! In this chapter, we show you how to calculate probability. We also begin to show you the important application of probability to statistics through the normal distribution.

Let's begin with a simple example. You roll a fair, six-sided die. What is the probability that you will roll a 5? To answer this question, think of probability as a fraction, in which the outcome of interest (the number 5) is in the numerator and all the possible outcomes (1, 2, 3, 4, 5, 6) are in the denominator. There is one "5," so we put a "1" in the numerator, and there are six possible outcomes, so we put a "6" in the denominator. Thus, the probability of rolling a 5 is:

$$\frac{1}{6}$$

What if we want to know the probability of rolling a 5 *or* a 6? The denominator remains the same (six possible outcomes); but the numerator will change, as we now place two outcomes of interest (5 or 6) in the numerator. Thus the probability of rolling a 5 or a 6 is:

$$\frac{2}{6} = \frac{1}{3}$$

Alternatively, for a problem as simple as this one, we can construct a probability distribution of all of the possible outcomes of the roll of a single die (see Table 6.1).

Table 6.1 Probability Distribution of the Outcomes of a Single Die

Outcome	Probability of Outcome
1	1/6
2	1/6
3	1/6
4	1/6
5	1/6
6	1/6

If the die is fair, then each result is equally likely. And though it may be obvious, the sum of all of the individual probabilities is equal to 1, or 100%. In other words, if we throw a die once, we are certain that we will get one of these six outcomes.

Let's say we throw a die five times, and we do not get a 5. Are we guaranteed to get a 5 on the sixth roll? Of course not. One of the keys to understanding probability is to remember that it is best understood as an estimate of what we should expect in the long run, not what must happen on the next trial. So if our die is fair, and we have not rolled a 5 on the first five rolls, our probability of rolling a 5 on the sixth roll is unchanged: 1/6. But if we were to throw a die one hundred times, one thousand times, or more, we would *expect* to roll a 5 one-sixth of the time.

Think of a basketball player who shoots free throws and has a career success rate of 9/10, or 90%. On the next free throw, that player will either make the shot or not. That does not mean that the probability is 50%; it is still 90%. But it is hard to wrap our brains around the application of a 90% probability to a single free throw. It is more helpful to think of it over the long run. Assuming this basketball player maintains a success rate of 90%, then we expect her to make about ninety of the next one hundred free throws.

If we know the distribution of any variable, probability can be used to understand the likelihood of a given outcome in the population. For example, we can think of the proportions that we saw in chapter 2 (e.g., the proportion of respondents with two children in the 2021 General Social Survey sample) as probabilities representing the likelihood of that outcome in the general population.

In Tables 6.2a and 6.2b, we offer two frequency distributions. They contain the results from two World Values Survey (WVS) samples: one from Mexico, the other from Pakistan. In the WVS, respondents were asked how regularly they vote in national elections. The tables contain the raw frequencies and the corresponding percentages.

Table 6.2a Voting in National Elections, Mexico

	Frequency	Percent
Vote always	1,118	67.2
Vote usually	377	22.7
Vote never	169	10.2
Total	1,664	100%

Source: World Values Survey (Wave 7).

Table 6.2b Voting in National Elections, Pakistan

	Frequency	Percent
Vote always	1,101	60.9
Vote usually	364	20.1
Vote never	343	19.0
Total	1,808	100%

Source: World Values Survey (Wave 7).

We can use Tables 6.2a and 6.2b to calculate probabilities. What is the probability that a randomly selected Mexican always votes? Let's answer this with a fraction. In the numerator we put all the possible respondents who said that they vote always; there are 1,118 of them, according to Table 6.2a. And in the denominator, we place all the Mexican respondents. So, the probability that a randomly selected Mexican always votes is:

$$\frac{1{,}118}{1{,}664} = 0.6718 \text{ or } 67\%$$

Note that Table 6.2a already contains the probability associated with each response, in the Percent column, rounded to one decimal place. While we are showing one decimal place in the Tables, in our discussion below we round to the nearest whole number.

Who is more likely to be a non-voter, a Mexican or a Pakistani respondent? About 19% of Pakistani respondents reported that they never vote, compared to 10% of Mexican respondents. By comparing these probabilities, we can conclude that Pakistani respondents are almost twice as likely to abstain from voting as Mexican respondents. Respondents from Mexico are also about 3 percentage points more likely to say that they usually vote than Pakistani respondents (23% to 20%). And if you randomly select a Mexican respondent, he or she is about 6 percentage points more likely than a Pakistani respondent to report that he or she votes always (67% versus 61%).

The Addition Rule

Now let's complicate the question a little bit. If we randomly select a Mexican respondent, what is the probability that he or she will be either someone who votes always or someone who votes usually? Again, we put the two outcomes of interest in the numerator and all possible outcomes in the denominator.

$$\frac{1{,}118 + 377}{1{,}664} = 0.90 \text{ or } 90\%$$

With this example, we have now seen the first basic rule of probability, the **Addition Rule**. When we are interested in calculating the probability that one of two mutually exclusive outcomes has occurred, we add up their respective probabilities. Outcomes are mutually exclusive if they cannot both exist at the same time. For example, a respondent cannot be both someone who always votes and someone who usually votes. You will often see this rule expressed in the following form:

$$p(A \text{ OR } B) = p(A) + p(B)$$

This expression is saying that the probability (p) of either outcome A or outcome B occurring is equal to the probability of A occurring plus the probability of B occurring. Let's express this statement in words, in the context of our example:

The probability of selecting a Mexican who always votes OR a Mexican who usually votes is equal to the probability of selecting a Mexican who always votes PLUS the probability of selecting a Mexican who usually votes.

Mathematically, it is:

The probability of selecting a Mexican who always votes OR a Mexican who usually votes

$$\frac{1{,}118}{1{,}664} + \frac{377}{1{,}664} = \frac{1{,}495}{1{,}664} = 0.90 \text{ or } 90\%$$

The Complement Rule

You may have noticed that there is a quicker way to answer this question. We know that the probability associated with each of the three possible outcomes (always, usually, and never vote) must add up to 100%. Therefore, another way to calculate the probability of something occurring is to take the probability that it does *not* occur and subtract that from 1 (or from 100%). In this example, we are looking for the probability that someone votes always OR votes usually. The answer is:

One minus the probability of selecting someone who never votes.

We know from Table 6.2a that 10% of Mexican respondents said that they never vote. Therefore, the probability of selecting someone who votes always or usually is 1 − .10, or .90. This is known as the **Complement Rule**. It means that the probability of something occurring is equal to 1 minus the probability that it does not occur.

$$\text{If } p(A) + p(B) = 1, \text{ then } p(A) = 1 - p(B)$$

The Complement Rule comes in very handy when solving probability questions. It is often more straightforward to calculate the probability of something not happening than to calculate the probability that it does happen.

The examples we have examined thus far have all dealt with a single outcome or two mutually exclusive outcomes. What happens if we want to find the probability of two outcomes happening in sequence? Instead of finding the probability of a 5 or a 6 on a single roll of a die, how do we proceed when we want to calculate the probability of a 5 and a 6 on two rolls of a die? Similarly, how would we calculate the probability of randomly selecting three consecutive non-voters in our sample of Mexican respondents? This is the kind of question that the examples about birth order and the characteristics of "Linda" from the beginning of the chapter address.

> **BOX 6.1: APPLICATION**
>
> ## Finding Probability Using the Complement Rule
>
> The WVS asks respondents how often they use the Internet as a source of information. Table 6.3 presents the distribution of responses from a sample of Japanese.
>
> **Table 6.3** Frequency of Internet Use, Japan
>
> | | Frequency | Percent |
> |---|---|---|
> | Daily (A) | 804 | 60.6 |
> | Weekly (B) | 155 | 11.7 |
> | Monthly (C) | 33 | 2.5 |
> | Less than monthly (D) | 34 | 2.5 |
> | Never (E) | 301 | 22.7 |
> | Total | 1,327 | 100.0 |
>
> *Source:* World Values Survey (Wave 7).
>
> If we randomly select one individual from this sample, what is the probability that she or he ever uses the Internet as a source of information? There are two ways to answer this question.
>
> Addition Rule:
>
> $$p(A \text{ or } B \text{ or } C \text{ or } D) = p(A) + p(B) + p(C) + p(D)$$
>
> Here, we add up the column percentages for all those who say they use the Internet "less than monthly" or more:
>
> $$60.6\% + 11.7\% + 2.5\% + 2.5\% = 77.3\%$$
>
> Complement Rule:
>
> $$p(A \text{ or } B \text{ or } C \text{ or } D) = 1 - p(E)$$
>
> We subtract the percentage of those who say "Never" from 100%:
>
> $$100\% - 22.7\% = 77.3\%$$

Before answering these questions, we need to introduce the concept of **independence**. If you throw a single die, the probability that you will roll a 5 is 1/6. What about rolling the die a second time? Will the probability of a 5 change, depending on what you got on the first roll? No, it will remain 1/6 (assuming you are using a fair die). In other words, the outcome of your first roll has no effect on the outcome of your second roll. These rolls are *independent* of each other.

Scientific surveys also generally select their respondents independently of each other. When an individual is randomly selected to be part of a sample, his or her probability of being selected is not affected by, nor does it affect, the selection of other respondents. When two outcomes do not affect each other, we say that they are independent of each other. And when two outcomes are independent of each other, we can multiply their respective probabilities to calculate the probability that they will both occur. We will come back to that shortly.

Can you think of examples in which two outcomes are not independent of each other? There are lots of examples from everyday life. What is the probability that you will go out for ice cream after you have stuffed yourself at an all-you-can-eat buffet dinner? It is probably lower than it would have been had you eaten a dinner with smaller portions. How likely is it that someone has an Apple iPad if they already have an iPhone? It's probably higher than it is for an Android user. How will the probability of doing well in statistics change if you decide not to study for the next exam? There are countless examples where the probability of something changes based on a prior decision or outcome. When one outcome affects another outcome, we say that their respective probabilities are *dependent* on each other. To find the probability that both have occurred, we still can multiply their probabilities, so long as we adjust the probability of the second outcome to take into account its **dependence** on the first outcome.

The Multiplication Rule with Independence

Let's illustrate the **Multiplication Rule** with two simple examples that we have referred to already.

We want to find the probability of getting a 5 on the first roll of a die and a 6 on the second roll. We know that the probability of any outcome of the roll of a die is 1/6. We also know that what we roll the first time has no bearing on what we roll the second time. So the probability of a 5 on Roll 1 and a 6 on Roll 2 is:

$$\frac{1}{6} \times \frac{1}{6} = \frac{1}{36}$$

There is a 1/36 chance that we will obtain a 5 on the first roll and a 6 on the second roll.

For our second example, let's return to the question about voting posed to a sample of Mexican respondents (Table 6.2a). If we randomly select three individuals, what is the probability that all three are non-voters? The probability of a single non-voter, as shown in Table 6.2a, is 0.10. Assuming that each "draw" is independent, the probability of selecting three non-voters in a row is:

$$0.10 \times 0.10 \times 0.10 = 0.001$$

The probability is 0.001, or one in one thousand, that we would obtain three consecutive non-voters from this sample.*

---

* If each selection is returned to the sample before the next selection is made, the observations would be independent of each other. If we did not return the cases to the sample after each selection, we would treat them as dependent; in that case the probability of three non-voters is calculated as:

$$\frac{169}{1,664} \times \frac{168}{1,663} \times \frac{167}{1,662} = 0.00103$$

Notice that the denominator decreases by one in each successive draw because the respondent is *not* returned to the sample.

If instead we want to calculate the probability of obtaining at least one voter (always or usual) among our three selections, the Complement Rule can do this for us. If the probability of selecting three non-voters is 0.001, then the probability of selecting at least one voter is:

$$1 - 0.001 = 0.999$$

The Multiplication Rule without Independence

As we have suggested, there are many instances in the real world where two outcomes are not independent of each other. In such cases, we can still multiply their respective probabilities but only after we have adjusted the probability of the second outcome to account for its dependence on the first.

Here is an example. Five people (three women and two men) have applied to serve on a local commission. There are two slots left, and they will be filled by random selection from among these five people. What is the probability that women will be selected for both of these slots?

There are two ways to solve this problem. We can list out all of the possible combinations of candidates, or we can use the Multiplication Rule. While the first method is longer, it can be helpful in understanding the underlying logic. We will show you both methods here.

In Table 6.4, we list all of the possible outcomes of this random selection of two indi-

Table 6.4 List of Possible Combinations among Three Women and Two Men

Possible Combinations	Number of Women Selected
Woman A, Woman B	2
Woman A, Woman C	2
Woman B, Woman C	2
Woman A, Man A	1
Woman A, Man B	1
Woman B, Man A	1
Woman B, Man B	1
Woman C, Man A	1
Woman C, Man B	1
Man A, Man B	0

viduals from our group of five people.

Let's summarize the results by means of a frequency table, shown in Table 6.5.

As Table 6.5 shows, there are ten possible pairs. In exactly three of those pairs, we find two women. So, the probability of randomly selecting two women from among these five people is:

$$\frac{3}{10} = 0.30 = 30\%$$

Table 6.5 Frequency Distribution of Number of Women Selected

Number of Women Selected	Frequency	Relative Frequency
0	1	1/10
1	6	6/10
2	3	3/10
Total	10	10/10 or 100%

There is a 30% chance that two women will be randomly selected from among these five people (three women and two men) in the applicant pool.

We can also use the Multiplication Rule to solve this problem. Again, we are interested in the outcome that has two women and zero men. The probability that a woman is selected in the first slot is 3/5 (three women out of five people). We must adjust the probability of a second woman to account for the selection of a woman in the first slot. If a woman is selected in the first slot, there are four people left, and two of them are women. So the probability that a woman is selected in the second slot, if a woman has already been selected in the first slot, is 2/4. When the probability of a second outcome is dependent on the probability of a first outcome, this always means that the denominator for determining the second probability is lower than the denominator for the first probability. In this case, the denominator decreased by one.

We can multiply these two probabilities to calculate the probability that we end up with two women:

$$\frac{3}{5} \times \frac{2}{4} = \frac{6}{20} = \frac{3}{10} = 0.30$$

It is obviously much quicker to use the Multiplication Rule than it is to list all of the possible outcomes, as we have done in the above tables. Listing all of the possible outcomes is feasible only with small samples and, therefore, a limited number of outcomes.

Now that we have learned how to assess the independence or dependence of outcomes and calculate probabilities accordingly using the Multiplication Rule, we can return to the problems introduced at the beginning of the chapter.

Applying the Multiplication Rule with Independence to the "Linda" and "Birth-Order" Probability Problems

We began this chapter with two examples from the work of Tversky and Kahneman. We are now in a position to solve those two problems. The first example involved a hypothetical person named Linda. Respondents were asked to rank three statements about Linda in order of likelihood, after reading a description of Linda's background:

Statement 1: Linda is active in the feminist movement. (Feminist)
Statement 2: Linda is a bank teller. (Teller)
Statement 3: Linda is a bank teller and is active in the feminist movement. (Teller and Feminist)

We explained that it is not mathematically possible for Statement 3 to be more likely than Statement 2, although a large majority of respondents mistakenly ranked them that way. To illustrate why, let's invent a probability for each of the single outcomes (Teller, Feminist) and then use the Multiplication Rule to show why p(Teller) ≥ p(Teller and Feminist):

Probability that Linda is active in the feminist movement (Feminist) = 0.75

Probability that Linda is a bank teller (Teller) = 0.30

To calculate the probability that Linda is a bank teller and is active in the feminist movement (Teller and Feminist), we multiply the probabilities of Teller and Feminist.

$$p(Teller\ and\ Feminist) = p(Teller) \times p(Feminist) = 0.30 \times 0.75 = 0.225$$

If the probability that Linda is a bank teller is 0.30 and the probability that Linda is active in the feminist movement is 0.75, then the probability that she is a bank teller *and* active in the feminist movement is 0.225. Because we are multiplying two decimals by each other, the product will always be smaller than either of the two factors. You can assign any probability to each of these two outcomes; you will always find that the probability that Linda is a feminist *and* a teller is less than the probability that she is a teller only.

The second example involved comparing two sequences of birth orders, where G represents daughters and B represents sons:

Sequence 1: G B G B B G
Sequence 2: B G B B B B

Most subjects believed that one would find more families with the birth order shown in the first sequence than in the second sequence, and this was because the four consecutive boys in Sequence 2 appear to lack the quality of randomness. In fact, these two sequences are equally likely. If we assume equal probabilities for the birth of a boy and a girl (probability = 0.50) and we assume independence, then we apply the Multiplication Rule, as follows:

Probability of Sequence 1: $0.50 \times 0.50 \times 0.50 \times 0.50 \times 0.50 \times 0.50 = 0.016$
Probability of Sequence 2: $0.50 \times 0.50 \times 0.50 \times 0.50 \times 0.50 \times 0.50 = 0.016$

In fact, with the assumption of equal probability for the birth of a girl and a boy, any sequence of six births will have a probability equal to 0.016.*

---

* Historically there are 105 boys born to every 100 girls. See https://www.pewresearch.org/religion/2022/08/23/sidebar-sex-ratios-around-the-world/. This ratio translates into a 0.512 probability that a boy is born and a 0.488 probability

BOX 6.2: IN DEPTH

The GSS Tests Americans' Knowledge of Probability

The 2018 GSS asked respondents two questions that tested their basic knowledge of probability. Here is the first question:

A doctor tells a couple that their genetic makeup means that they've got one in four chances of having a child with an inherited illness. Does this mean that if their first child has the illness, the next three will not have the illness?

How would you respond after reading this chapter thus far? To answer, we need to think about whether the probability of inheriting an illness is independent of whether an older sibling inherited the illness. Just as the sex of a baby born to a set of parents does not depend on the sex of the previous baby born to that set of parents, neither does the chance of one baby inheriting an illness depend on whether the previous baby inherited the illness. Because these outcomes are independent, if the couple's first child inherits the illness, this will not affect the probability that their next three children will inherit the illness. The correct answer, then, is no. Do you think that most GSS respondents correctly answered this question? Yes, the vast majority of them, 88%, said that the first child's inheritance of the illness did not mean that the next three would not inherit the illness.

The second question asked GSS respondents whether each of the couple's children will have the same risk of suffering from the illness. We know that since these outcomes are independent, each child will have the same probability of inheriting the illness, .25. About three-quarters of the GSS respondents also knew that all of the couple's children would have the same risk of inheriting the illness.

Probability Distributions

Earlier in this chapter, we examined frequency distributions and used the relative frequencies (or percentage) to calculate the probability of randomly selecting an individual who falls into a particular category. We can use a histogram to find a probability in the same way.

Figure 6.1 shows the distribution of responses on a variable from the 2020 American National Election Study (ANES). The survey asks respondents to place themselves on a 7-point "health care" scale, where 1 indicates support for a government health insurance plan and 7 indicates a preference for a plan based on private insurance. This variable is ordinal. It is also "discrete" because it is measured in whole numbers and these units cannot be broken down further.

We can use the histogram shown in Figure 6.1 to calculate the probability that a randomly selected individual is in any particular category or that an individual holds a position above or below any particular value. For example, there is about a 17% chance that a randomly selected individual holds the middle position (point 4) on this ordinal scale.

---

that a girl is born. Therefore, the first sequence (*G B G B B G*) has a 0.016 probability of occurring. The second sequence (*B G B B B B*) has a 0.017 probability of occurring.

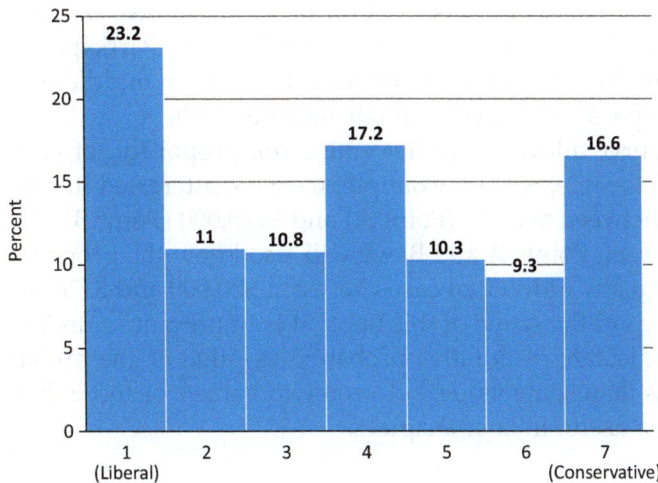

Figure 6.1 Attitudes toward Health Insurance.
Source: 2020 American National Election Study.

What if we want to find out the probability that a randomly selected individual holds a position on the "liberal" side of the scale (points 1, 2, or 3)? We use the Addition Rule and simply add up the percentages associated with each bar. In this case, we would add up 23.2% + 11% + 10.8% and conclude that there is a 45% chance that a randomly selected individual takes a liberal position on health care.

Some variables, however, are continuous, which means that the values can be continually subdivided. Income measured in dollars, schooling measured in years, and weight measured in pounds are all examples of continuous variables. And as we have already seen, when we have continuous variables, we must either collapse them into categories and display the distribution by means of a histogram or use an area curve to graphically represent the "unrecoded" distribution. As you learned in chapter 5, the height of the line of an area curve represents the number of cases that have that value of a variable.

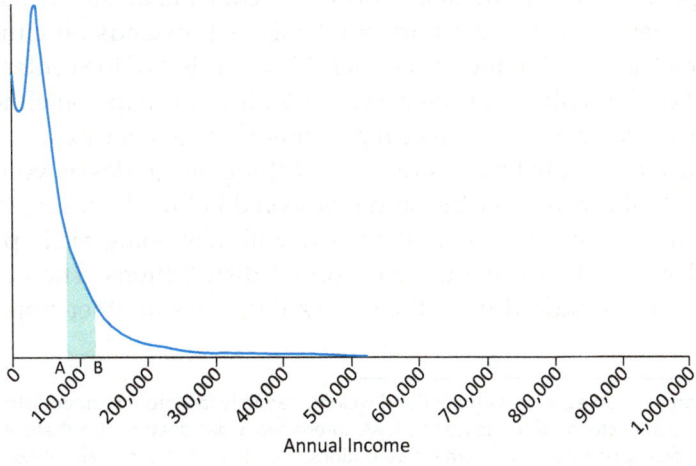

Figure 6.2 Income Distribution in the United States, 2022.
Source: 2022 Survey of Income and Program Participation.

Figure 6.2, displays the distribution of a continuous variable, income in the United States, from the 2022 Survey of Income and Program Participation (SIPP). This is a right-skewed distribution, with lots of people in the low- to middle-income category and fewer and fewer people as we move to upper-income levels.

An area curve allows us to determine the proportion of cases that fall between any two points on the line. Say, for example, we were interested in the probability that an individual earns between $80,000 (Point A) and $120,000 (Point B). If we know the area under the curve between Points A and B, we will also know the proportion of cases and thus the probability that any individual earns between $80,000 and $120,000. Although these calculations are beyond the scope of this book, it is important to understand that an area curve allows us to calculate cumulative probabilities—that is, the probability that an individual case earns less than some value, between two values, or more than some value. This basic concept will be useful in later chapters.

The Normal Distribution

There is one particular distribution that deserves special attention—the normal distribution. The **normal distribution** is an area curve. As we just saw, in area curves the height of the line represents the number of cases with that value of the variable. And the area under the curve between any two points represents the proportion of cases with values that fall between those two points.

The normal distribution is important for two reasons. First, in the real world, many variables take on the shape of the normal distribution. IQ scores, SAT scores, height, and weight are common examples of variables that are normally distributed.[*] Second, if we repeatedly draw random samples from a population, the distribution of these sample means or proportions will be normally distributed (under certain conditions). This second point is the focus of chapter 7.

Figure 6.3 shows a normal distribution. It may look familiar to you. For reasons that are obvious, it is often referred to as a bell curve.

The normal distribution has several important features. It is unimodal, symmetrical, and bell-shaped. The mean, median, and mode coincide at the center of the distribution, and the curve never touches the horizontal axis as it extends on either side. Because the curve is symmetrical, half of the area under it is to the left of the center, and half of the area is to the right of the center, and the curve is identical in shape on either side. As with any other curve, the area under the curve represents 100% of the cases.

If Figure 6.3 represented the distribution of students' grades on your last statistics exam, we would say that exactly half the students scored below the mean, half above the mean, and a lot of students scored close to the mean, with few doing really poorly or really well.

Look at Figure 6.4. It contains two normal distributions. The curves have the same mean but different standard deviations. This illustrates another important feature of the

---

[*] The values of some variables, such as height and weight, naturally fall into a normal distribution, while the values of other variables, such as standardized test scores, are intentionally designed to sort themselves into a normal distribution. For such "socially constructed" normal distributions, we should be careful about concluding that differences in values reflect innate differences.

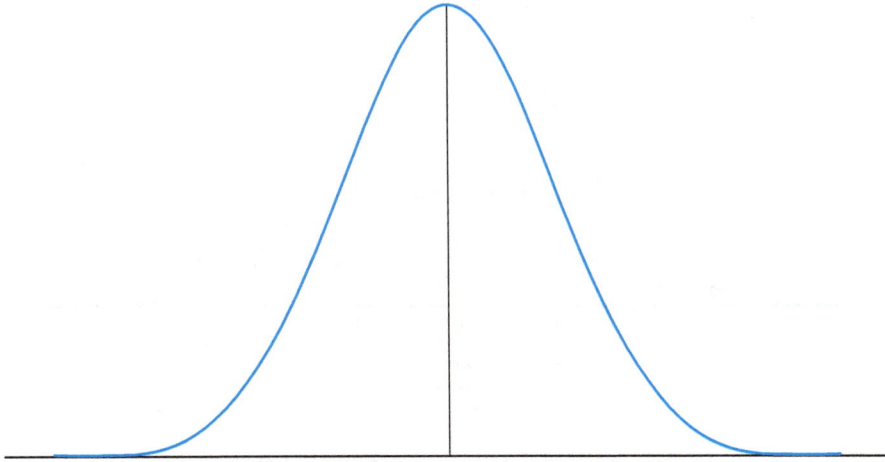

Figure 6.3 Normal Distribution

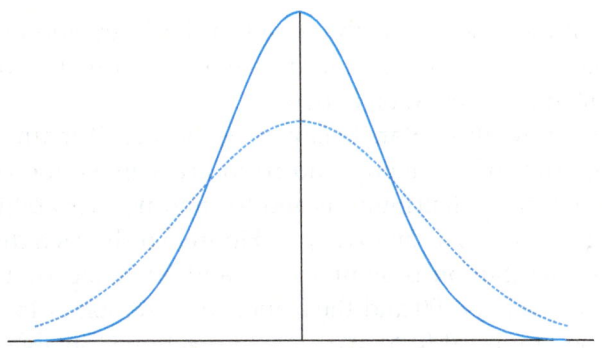

Figure 6.4 Two Normal Distributions

normal distribution. A normally distributed variable can have any combination of mean and standard deviation. In this case, the taller curve has a smaller standard deviation, whereas the larger spread on the shorter curve suggests more variability in the distribution. But they are both normally distributed.

If these curves represented exam grade distributions of two statistics classes, which one would you rather be in? The answer depends on how risk averse you are. Because the shorter curve has a larger standard deviation, there is a better chance you will do very well. By the same token, there's also a higher probability that you will get a low grade.

The Empirical Rule

One of the most important features of the normal distribution—and a feature that makes it particularly handy—is the **Empirical Rule**. This rule states that for any normally distributed variable, a fixed proportion of the cases will fall between any given standard deviations from the mean. This is illustrated in Figure 6.5.

As Figure 6.5 illustrates, for any variable that is normally distributed, about 68% of the observations will fall within one standard deviation of the mean, about 95% will fall within two standard deviations of the mean, and more than 99% will fall within three standard deviations of the mean. Because the distribution is symmetrical, we can break this down

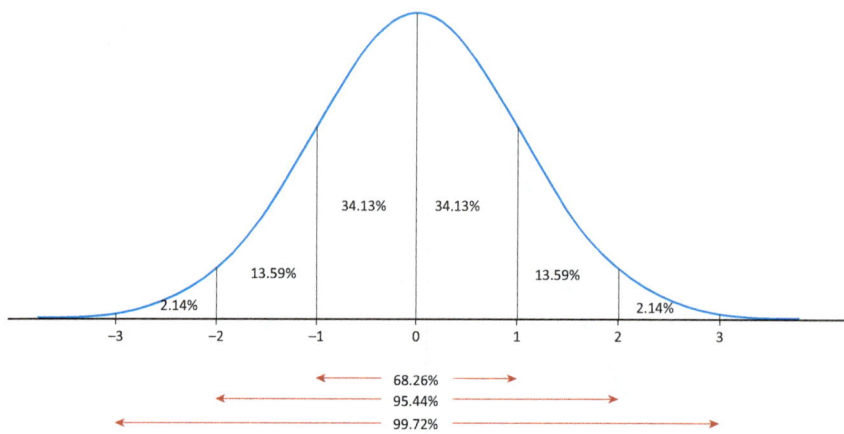

Figure 6.5 Distribution of Cases in a Normal Distribution: The Empirical Rule

further. If 68% of the cases are within one standard deviation of the mean, then 34% will fall between the mean and one standard deviation below it, and 34% will fall between the mean and one standard deviation above it.

How do we know this? Statisticians have proved it mathematically.[3] As we will see shortly, its properties of symmetry and fixed areas make the normal curve an especially handy tool for analyzing distributions and for finding probabilities.

Let's look at IQ scores as an example. Figure 6.6 shows a distribution of IQ scores. As we can see, the distribution is symmetrical and bell-shaped. The IQ test is intentionally scaled so that the mean is 100 and the standard deviation is 15. Armed with this information and with the Empirical Rule, we are in a position to analyze this distribution.

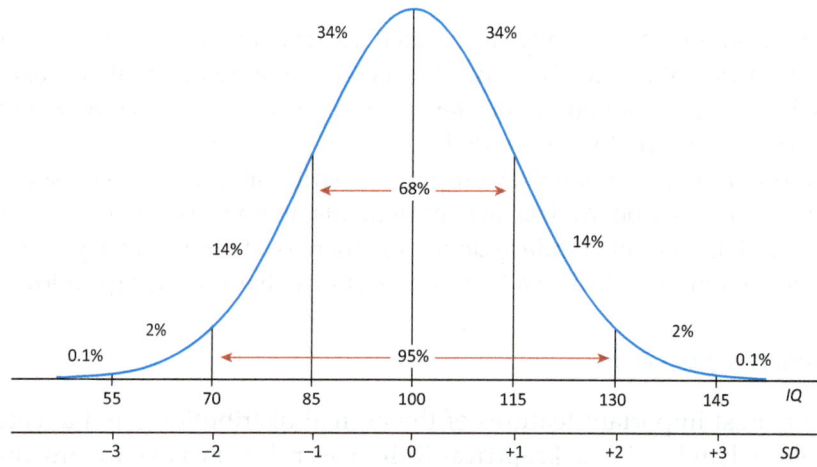

Figure 6.6 The Normal Distribution and IQ Scores

We know that 68% of individuals will have an IQ score within one standard deviation, or 15 points, of the mean. That is, 68% of individuals will have an IQ score between 85 and 115. Half of them, or 34%, will be between the mean and 15 points above it, and the other half, also

34%, will be within the mean and 15 points below it. The Empirical Rule also tells us that 95% of individuals will have IQ scores within two standard deviations of the mean; in this case, that is equal to 30 points. So we know that 95% of individuals have IQ scores between 70 and 130. If we took an individual with an IQ score of 115, we could make the following claims:

- Her score is one standard deviation above the mean.
- Her IQ score places her in the top 16% of the distribution. Why? Look at Figure 6.6. Half of all scores are above the mean, and 34% are between the mean and 115. Just subtract: 50% − 34% = 16%.
- About 84% of individuals have IQ scores below her score of 115. Why? Use the Complement Rule again (100% − 16% = 84%) or use the Addition Rule. Fifty percent are below the mean and 34% between the mean and 115: 50% + 34% = 84%.

We can also use the normal distribution to calculate probabilities. For example, say we are interested in the probability that a randomly selected individual has an IQ score between 100 and 130. We know that 100 is the mean and that 130 is two standard deviations above the mean. The Empirical Rule tells us that, in any normally distributed variable, about 47.5% (half of 95%) of the cases are found between the mean and two standard deviations above the mean. Therefore, the probability of randomly selecting an individual with a score between 100 and 130 is 47.5%.

Standardizing Variables and Calculating z-Scores

In chapter 5, we learned how to express an individual observation as a distance from the mean, measured in standard deviations. As we described it, there are two steps:

1. We subtract the mean from the individual's score. In this step, we quantify the individual score's distance from the mean.
2. We divide that difference by the standard deviation. In this step, we convert that distance into standard deviation units.

When we express a variable's scores as deviations from the mean, we are "standardizing" the variable. When we standardize a variable in this way, we call the standardized values **z-scores**. We will be using z-scores frequently here and in subsequent chapters.

Converting scores from the unit of the variable to standard deviation units is analogous to converting from inches to feet or meters to yards. We are expressing the same information in different units.* As we will see, expressing the raw values of a normally distributed variable in standard deviation units allows us to take advantage of the properties of the normal curve in assessing the probabilities attached to specific values for that variable.

---

* When we standardize a variable, we convert each observation into a z-score. One interesting property of a standardized variable is that the mean of the z-scores will always equal 0, and the standard deviation of the z-scores will always equal 1. To understand why, think about how we standardize. Since we subtract the mean from each observation, the mean of the standardized values will be *the original mean − the mean* = 0. When we standardize, we also divide by the standard deviation. Hence the standard deviation of the standardized values will be *the original standard deviation ÷ the standard deviation* = 1.

Let's return to our example with IQ scores. Take an individual who has an IQ score of 120. That score is 20 points above the mean, but how many standard deviations above the mean is it? In other words, what is that individual's z-score? By formula:

$$z = \frac{y - \mu}{\sigma}$$

where μ is the mean for the population and σ is the standard deviation for the population.*

Plugging in the values from our example:

$$z = \frac{120 - 100}{15} = \frac{20}{15} = 1.33$$

This individual's IQ score of 120 is 1.33 standard deviations above the mean. Or, we can say that this person's z-score is equal to 1.33. Instead of saying that their IQ score is 20 points above the mean, we express that difference in standardized units—that is, number of standard deviations.

What proportion of respondents have IQ scores above 120? We have just calculated that the z-score for an IQ score of 120 is 1.33. With a z-score of +1, we were able to use the Empirical Rule to figure out what percentage of cases fall below and above that point. But this time our z-score is not a whole number. With a z-score that is not a whole number, we must use either statistical software or the Normal Table in appendix A to find the proportion.

Figure 6.7 illustrates how to solve this problem. As we know, IQ scores are normally distributed, around a mean of 100. We are interested in finding out the proportion of cases above an IQ score of 120. Or, to put it differently, what proportion of cases in a normal distribution fall above a z-score of 1.33? The area we are interested in is represented by the shaded area of the curve.

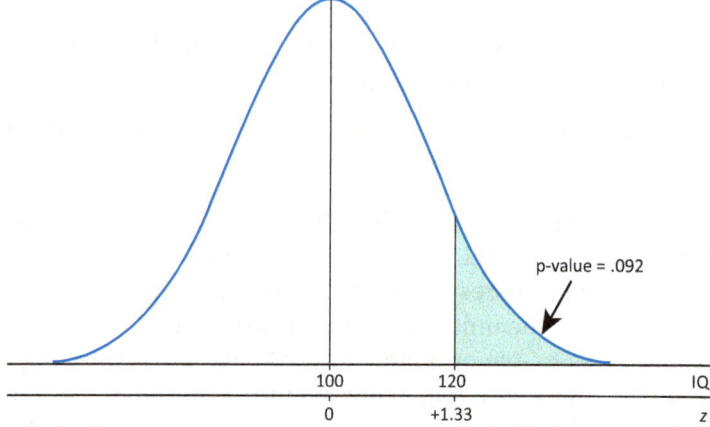

Figure 6.7 Finding Right-Tail Probability for IQ Score of 120

To find this proportion, we turn to the Normal Table in appendix A. In the left column, we locate the z-score to its first decimal point; in this case, that is the row for z = 1.3. The

---

* You may notice that we are using μ rather than $\bar{y}$ and σ rather than s; this is because statisticians use different notation for population parameters and sample statistics. We will cover this in detail in the next chapter.

second decimal point is arrayed across the columns of the table. In this case, we follow the "z = 1.3" row until it intersects with the column headed by "Second Decimal Place of z = .03." This gives us the z-score of 1.33, with a corresponding "p-value," in the body of the table, equal to 0.0918. (Box 6.3 shows this process.) The "p-value" refers to the probability that a randomly selected case would fall above the z-score. As we saw earlier, this probability is the same as the proportion of cases that fall into that range. In other words, 9.2% of the observations in this normally distributed variable will have IQ scores greater than a z-score of 1.33. Or, there is about a 9% chance that a randomly selected individual will have an IQ score greater than 120. We sometimes refer to this as the right-tail probability.

BOX 6.3: APPLICATION

Using the Normal Table

For any normally distributed variable, we can use the Normal Table to find areas to the right of a z-score, to the left of a z-score, or in between two z-scores. The Normal Table in appendix A, and excerpted in Figure 6.8, displays the area to the right of a z-score. The left column of the Normal Table shows the z-scores to the first decimal point. The top row of the table shows the second decimal point. And the area to the right of any given z is shown in the body of the table.

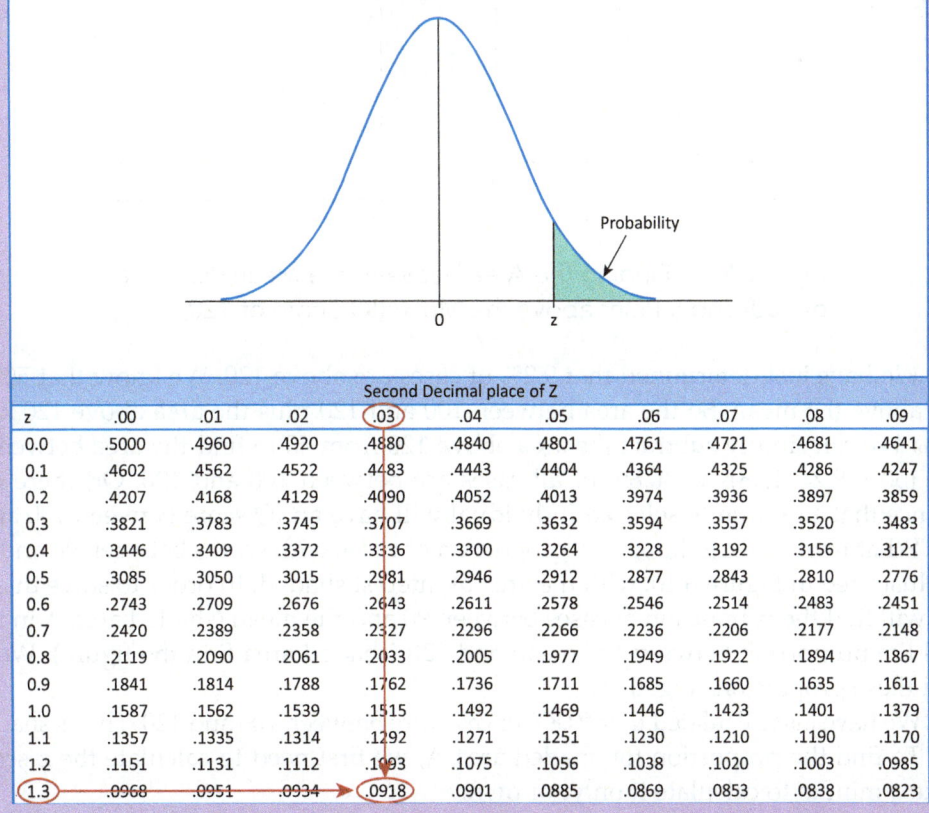

z	.00	.01	.02	.03	.04	.05	.06	.07	.08	.09
0.0	.5000	.4960	.4920	.4880	.4840	.4801	.4761	.4721	.4681	.4641
0.1	.4602	.4562	.4522	.4483	.4443	.4404	.4364	.4325	.4286	.4247
0.2	.4207	.4168	.4129	.4090	.4052	.4013	.3974	.3936	.3897	.3859
0.3	.3821	.3783	.3745	.3707	.3669	.3632	.3594	.3557	.3520	.3483
0.4	.3446	.3409	.3372	.3336	.3300	.3264	.3228	.3192	.3156	.3121
0.5	.3085	.3050	.3015	.2981	.2946	.2912	.2877	.2843	.2810	.2776
0.6	.2743	.2709	.2676	.2643	.2611	.2578	.2546	.2514	.2483	.2451
0.7	.2420	.2389	.2358	.2327	.2296	.2266	.2236	.2206	.2177	.2148
0.8	.2119	.2090	.2061	.2033	.2005	.1977	.1949	.1922	.1894	.1867
0.9	.1841	.1814	.1788	.1762	.1736	.1711	.1685	.1660	.1635	.1611
1.0	.1587	.1562	.1539	.1515	.1492	.1469	.1446	.1423	.1401	.1379
1.1	.1357	.1335	.1314	.1292	.1271	.1251	.1230	.1210	.1190	.1170
1.2	.1151	.1131	.1112	.1093	.1075	.1056	.1038	.1020	.1003	.0985
1.3	.0968	.0951	.0934	.0918	.0901	.0885	.0869	.0853	.0838	.0823

Figure 6.8 Excerpt of Normal Table, P-Value Associated with Z-Score of 1.33

In our IQ example, we found that an IQ score of 120 is 1.33 standard deviations above the mean. We run down the left column to find z = 1.3; we move across that row until we reach the column headed by "Second Decimal Place of z = .03." At the intersection of that column and that row, we find an area equal to 0.0918. That is the proportion of cases with z greater than 1.33.

If we want to know what proportion of people have IQs lower than 120, we simply use the Complement Rule. Since we know that the total area under the normal curve adds up to 100%, or a probability of 1, we can subtract 9% (the proportion that falls above 120) from 100% to find that 91% of people have IQs below 120. To put it in terms of probability, 1 − 0.09 = 0.91; there is a 0.91 probability that a randomly selected person would have an IQ below 120.

Figure 6.9 shows how to use this same information to find the proportion of cases with scores between 100 (the mean) and 120.

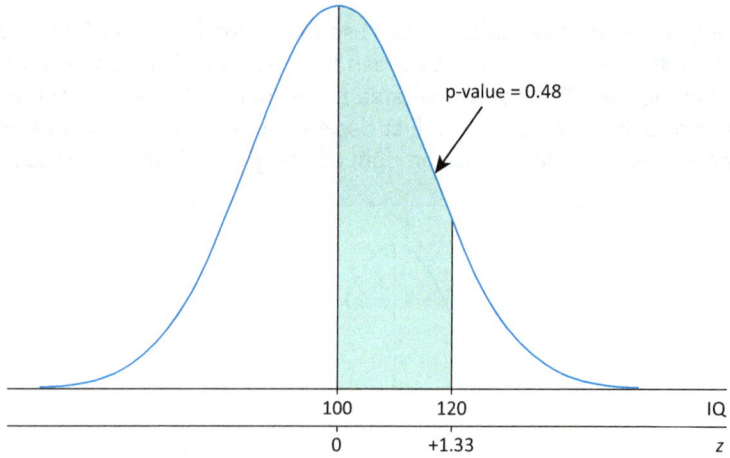

Figure 6.9 Finding the Area between the Mean (IQ score of 100) and a Point above the Mean (IQ Score of 120)

We have just determined that 9.2% of cases are above 120. We know that 50% of cases are above the mean. So the area between 100 and 120 plus the area above 120 must equal 50%. We can simply subtract the area above 120 from 50 to find the area between 100 and 120 (50 − 9.2 = 40.8), so 40.8% of all cases are between 100 and 120. Or, there is a 40.8% chance that a randomly selected individual will have an IQ score between 100 and 120.

What if we want to know the proportion of cases with scores between 90 and 120? This is illustrated in Figure 6.10, with the area of interest shaded. In order to solve this problem, we will find the proportion of cases between 90 and the mean (shaded area A in the figure) and the proportion between the mean and 120 (shaded area B in the figure). We will then add these proportions together.

We have just calculated that 40.8% of cases are between 100 and 120. This is shaded area B.

To find the proportion for shaded area A, we first need to calculate the z-score for 90. Take a minute to calculate it on your own.

$$z = \frac{y - \mu}{\sigma} = \frac{90 - 100}{15} = \frac{-10}{15} = -0.67$$

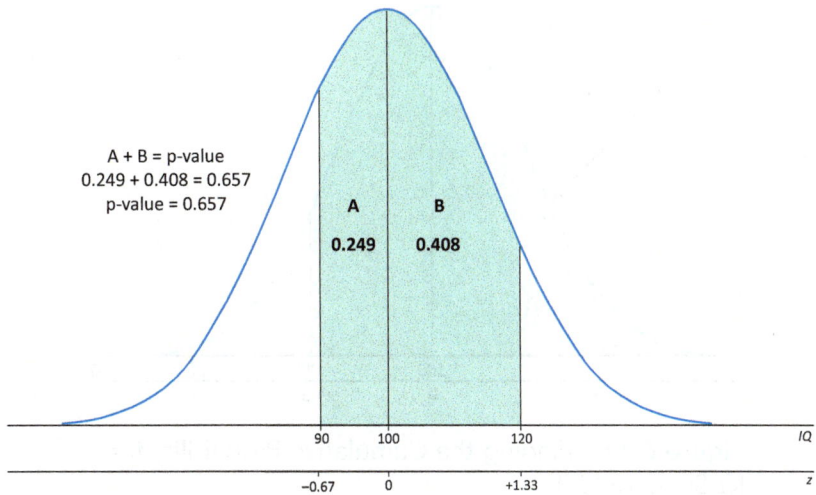

Figure 6.10 Finding the Area between Two Points (IQ Scores of 90 and 120)

The negative number reflects the fact that 90 is below the mean. We simply look up the absolute value in the Normal Table. Looking up 0.67 in the Normal Table, we find p = 0.251. What does this tell us? It tells us that 25.1% of cases are below z = – 0.67 (or below IQ = 90). But we need to know what proportion of cases fall between 90 and 100. We know that 50% of all cases fall below the mean. Simply subtract to solve: 50 – 25.1 = 24.9, so 24.9% of cases fall between 90 and 100. This is shaded area A in the figure.

We can add these two results to obtain the percentage of cases that fall between IQ scores 90 and 120:

$$24.9 + 40.8 = 65.7$$

Of all cases, 65.7% have IQ scores that fall between 90 and 120. Or, there is a 65.7% chance that a randomly selected individual would have an IQ score between 90 and 120. Note that this follows the Addition Rule of probability that we discussed earlier in the chapter.

We can use this same information to determine the proportion of cases that have IQ scores of 120 or below (see Figure 6.11). We have just determined that 40.8% of cases are between 100 and 120, and we know that 50% are below 100. Again using the Addition Rule, adding 40.8 + 50 = 90.8. About 91% of cases have IQ scores at or below 120. Or, there is a 91% probability that a randomly selected individual would have an IQ score at or below 120.

This is referred to as the **cumulative probability** for an IQ score of 120. Cumulative probability gives the total probability for scores that fall at or below (or to the left of) a given value. Cumulative probability is similar to the idea of cumulative percentage or percentile rank, discussed in chapter 2.

Sometimes we use the Normal Table to solve problems in reverse. Using this same distribution of IQ scores, we now want to find the IQ score associated with the 90th percentile. In other words, what is the raw IQ score for which 10% of cases are higher and 90% are lower? This is depicted in Figure 6.12.

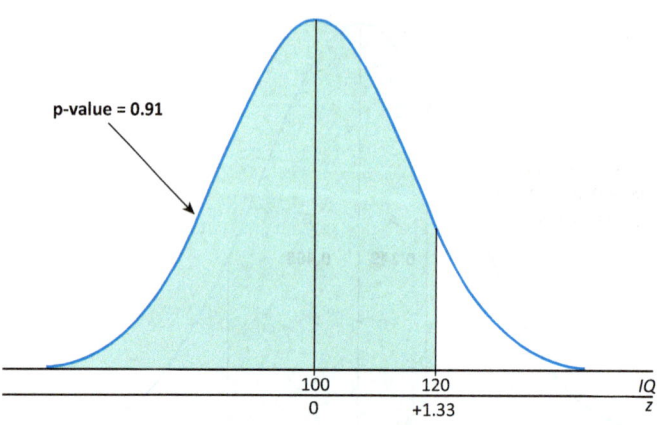

Figure 6.11 Finding the Cumulative Probability for IQ Score of 120

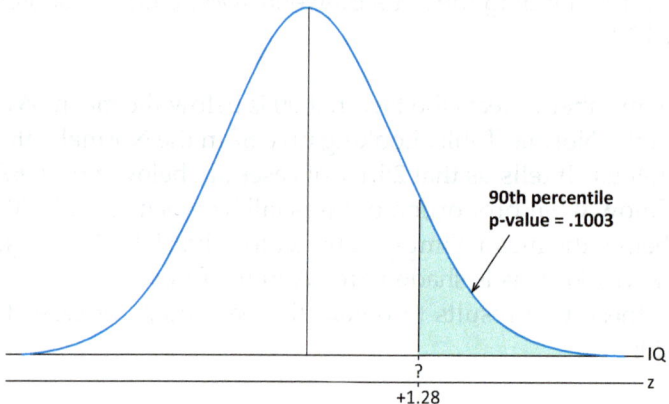

Figure 6.12 Finding the IQ Score Associated with the 90th Percentile

The shaded area represents the top 10% of the IQ distribution. But now, the IQ score associated with this area is unknown. This is what we are solving. We use the Normal Table again, only now we look for the area (or proportion) in the body of the table that comes as close as possible to 0.10 (or 10%) without exceeding it. We find that an area equal to 0.1003 is associated with a z-score of 1.28. In other words, for any normally distributed variable, the 90th percentile will always be associated with a z-score of 1.28, or the score that is 1.28 standard deviations above the mean. With a mean equal to 100 and a standard deviation equal to 15, we can use the z-score formula to find the IQ score at the 90th percentile, leaving the individual score (y) blank and solving for it.

$$z = \frac{y - \mu}{\sigma} \rightarrow y = \mu + \sigma(z)$$

$$y = 100 + (1.28 \times 15) = 119.2$$

The IQ score associated with the 90th percentile is 119.2.

BOX 6.4: APPLICATION

What IQ Score Is Required to Join Mensa?

Mensa is, in its own words, a "high IQ society." To become a member, one has to score in the top 2% (the 98th percentile) on an approved intelligence test. What IQ score is required to join Mensa?

We have seen that IQ scores are normally distributed, with a mean of 100 and a standard deviation of 15.

Using the Normal Table, we find that the 98th percentile is associated with a z-score of 2.05. In other words, the IQ score at the 98th percentile is 2.05 standard deviations above the mean.

$$z = \frac{y - \mu}{\sigma} \rightarrow y = \mu + \sigma(z)$$

$$y = 100 + 2.05(15) = 130.75$$

The required IQ score for Mensa membership is 130.75.

BOX 6.5: APPLICATION

Practicing Terminology for the Normal Curve

The normal curve is foundational to inferential statistics, the topic of the next few chapters of this book, so it is worth taking a moment to work on mastering some standard terminology associated with it. Figure 6.13 shows a visual representation of "cumulative probability" (left) and "left-tail" and "right-tail" probability (right).

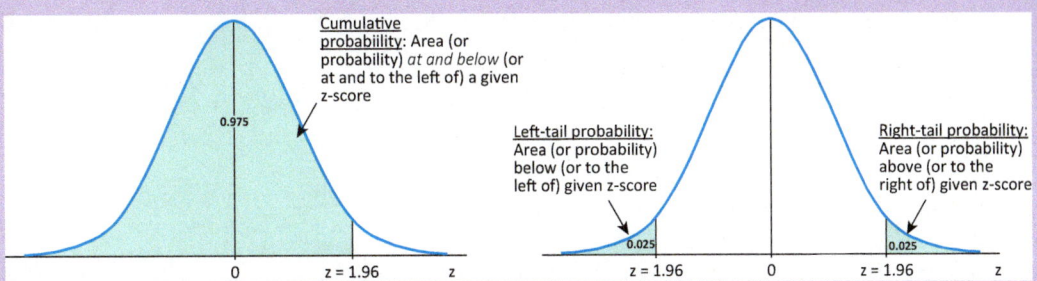

Figure 6.13 Cumulative, Left-Tail, and Right-Tail Probabilities

"Cumulative probability" (the left-hand normal curve) refers to the area of the curve that falls at and to the left of a given z-score. The left-hand curve shows a z-score of 1.96. If you consult the Normal Table in appendix A, you will see that it lists a probability (sometimes also referred to as the "area") of 0.025. Since the Normal Table gives the probability to the right of the z-score, this means that the area to the left of the z-score of 1.96 is equal to 0.975 (1 − 0.025). Note that cumulative probability is inclusive of the given z-score.

Left-tail probability and right-tail probability (the right-hand normal curve in Figure 6.12) refer to the area of the curve that falls below or above a given z-score, respectively. Note that these probabilities are *not* inclusive of the given z-score. Sometimes we refer to this as the area "to the left of" or "to the right of" a given z-score. As we discussed above, according to the Normal Table, the probability to the right of the z-score of 1.96 is 0.025, which means that the probability to the left of −1.96 is also 0.025. You will notice from this paragraph that when referring to the normal curve, "probability," "proportion," and "area" function as synonyms.

Chapter Summary

This chapter covered probability and the normal distribution.

- The four basic rules of probability are:
    - **Addition Rule**: The probability of either outcome A or outcome B is the sum of the probability of A and B. *p(A or B) = p(A) + p(B)*. (This assumes A and B are mutually exclusive.)
    - **Complement Rule**: If the probability of outcome A plus the probability of outcome B is 100%, the probability of outcome A is equal to one minus the probability of outcome B. If *p(A) + p(B) = 1*, then *p(A) = 1 − p(B)*.
    - **Multiplication Rule**, with independence: If the probabilities of outcomes A and B are independent of each other, the probability of both outcome A and outcome B occurring is found by multiplying the probability of outcome A by the probability of outcome B. *p(A and B) = p(A) × p(B)*.
    - **Multiplication Rule**, without independence: If outcomes A and B are not independent of each other, the probability of both outcome A and outcome B occurring is found by multiplying the probability of outcome A by the probability of outcome B, after adjusting the probability of outcome B for its dependence on A. *p(A and B) = p(A) × p(B adjusted for its dependence on A)*.
- The **normal distribution** can be used to find the probability, or proportion of cases, at specified values or ranges of values for variables that are normally distributed. It is also important for inferential statistics.
- The normal distribution's characteristics are:
    - It is unimodal, bell-shaped, and symmetrical.
    - The mean, median, and mode coincide at the center of the distribution.
    - A **z-score** expresses a value of a normally distributed variable in terms of how many standard deviation units it falls from the mean. Positive z-scores indicate the value falls above the mean; negative z-scores, below.
    - We convert individual values to z-scores and use the Normal Table to find the proportion of cases that are below a z-score, between two z-scores, or above a z-score.
    - The area under the curve shown in the Normal Table is the same as the proportion of cases that fall above that z-score and the probability of randomly selecting a case with a value that falls above that z-score.
- To convert an individual value to a z-score, subtract the mean from the value of the individual value and divide by the standard deviation. By formula,

$$z = \frac{y - \mu}{\sigma}$$

- To find the individual value that is associated with any given percentile, use the formula for the z-score to solve for y, using the z-score associated with the given percentile.

$$z = \frac{y - \mu}{\sigma} \rightarrow y = \mu + \sigma(z)$$

Using Stata

The "Using Stata" section of this chapter does something that we have yet to see in this book: we will use Stata with no data in use (i.e., we will not use a data set). We can do this because Stata contains all of the same information included in the Normal Table at the back of this book—the probabilities associated with every z-score on the normal curve, from −5 to 5. This allows us to provide Stata with a z-score and obtain the associated cumulative probability. We also can provide it with a cumulative probability and obtain the associated z-score. If you lose your textbook, at least you know that Stata contains a Normal Table!

In this section, we will use information about the SAT scores of college-bound high school seniors in the class of 2021. We will use Stata to accomplish two tasks: (1) identify the probability of drawing a score above and below certain cutoff scores and (2) find the score associated with a particular percentile in the distribution.

In 2021, 41% of high school seniors took the SAT. The mean scores and standard deviations for the Evidence Based Reading and Writing and Math tests are shown in Table 6.6.[4]

Table 6.6

SAT Test	Mean (μ)	SD (σ)
Evidence-Based Reading and Writing	533	108
Mathematics	528	120

Finding the Probability Associated with a z-Score

Here, we will use Stata to find the probability of drawing a score of 700 or higher from the distributions for the two SAT tests. In order to do that, we will need to provide Stata with the z-score associated with a score of 700 for each test. We can use the z-score formula to calculate the z-score associated with 700 for the two tests, as shown in Table 6.7.

Table 6.7

SAT Test	z-Score Associated with Score of 700
Evidence-Based Reading and Writing	$z = \dfrac{700-533}{108} = \dfrac{167}{108} = 1.55$
Mathematics	$z = \dfrac{700-528}{120} = \dfrac{172}{120} = 1.43$

What is the probability of getting a score above 700 on both tests? To get started, enter the following commands into Stata:

`display normal(1.55)`

`display normal(1.43)`

The "display" commands tell Stata to display the cumulative distribution or the area to the left, for each of the z-scores. In other words, Stata will show us the probability of scoring 700 or lower on each of the SAT tests. The output is shown in Figure 6.14.

```
. display normal(1.55)
.93942924

. display normal(1.43)
.92364149
```

Figure 6.14

We see that the probability of scoring 700 or lower on each exam is very high—0.94 for Reading and Writing, and 0.92 for Math. (You can check the normal distribution in the back of the book to confirm the cumulative probabilities associated with each of these z-scores.) These probabilities mean that a score of 700 occupies the 94th and 92nd percentiles for the Reading and Writing and Math tests, respectively. In other words, students who did not score above 700 have a lot of company. While the two percentiles are similar, they remind us that the relative meaning of the same score varies across SAT tests depending on the distribution of scores for each test.

However, remember that we wanted to know the probability of scoring above a 700 on each test in 2021. There is one more step in our calculation, which we can do using the Complement Rule discussed earlier in the chapter: If $p(A) + p(B) = 1$, then $p(A) = 1 - p(B)$. Since we know the probability of scoring 700 or lower plus the probability of scoring higher than 700 is equal to 1, the probability of scoring above 700 is equal to

1 − p(scoring 700 or lower). For the Reading and Writing and Math tests, then, the probability of scoring higher than 700 is 0.06 and 0.08, respectively.

There is another way of asking Stata to do this calculation that eliminates the final step that we just performed manually. We can ask Stata directly for the probability to the right of the z-score associated with 700 by invoking the Complement Rule directly in the syntax. Enter the following commands into Stata:

`display 1-normal(1.55)`

`display 1-normal(1.43)`

This time, the "display" commands tell Stata to subtract the cumulative probability for each z-score from 1, just as we did above manually with the Complement Rule. The output is shown in Figure 6.15. The right-tail probabilities for each z-score, listed beneath each command, match the results of our manual calculations.

```
. display 1-normal(1.55)
.06057076

. display 1-normal(1.43)
.07635851
```

Figure 6.15

Finding the z-Score Associated with a Specific Probability

Next, suppose a student knows that he scored in the 75th percentile for the Math test, but he lost his detailed test report from the College Board and does not remember the actual score associated with the 75th percentile. Again, we can enlist the help of Stata because it contains the normal distribution. Enter the following command into Stata:

`display invnormal(.75)`

The "display invnormal" command provides Stata with a cumulative probability, 0.75. Stata will return the z-score for which 75% of the curve's area falls at or to the left. The output is shown in Figure 6.16.

```
. display invnormal(.75)
.67448975
```

Figure 6.16

Stata returns the z-score of 0.67. (This should be approximately the same z-score that you find for a 0.75 probability in the body of the normal distribution at the back of the book.) The student now knows that his z-score for the Math test is 0.67, which means that he scored 0.67 standard deviations above the mean score. But remember that he wants to know his actual score on the test. We can solve the z-score formula for y to find the student's score:

$$z = \frac{y - \mu}{\sigma} \rightarrow y = z\sigma + \mu = 0.67(120) + 528 = 608.4$$

Now the student knows that his score on the Math test is 608.

Review of Stata Commands

- Find the cumulative probability associated with a z-score

    `display normal(z-score)`

- Find the right-tail probability associated with a z-score

    `display 1-normal(z-score)`

- Find the z-score associated with a cumulative probability

    `display invnormal(probability)`

Using SPSS

The SPSS demonstration in this chapter will focus on only one procedure: standardizing a variable. SPSS will take variables, standardize all of their observations, and save these standardized values as a separate variable. The standardized values are the z-scores for the values of the variable. To illustrate how this works, we will use data from the 2020 American National Election Study. We will focus on two variables: the feeling thermometer ratings of vice presidential candidates Kamala Harris (V201153) and Mike Pence (V201154). These variables are measured on a scale that runs from 0 (very cold) to 100 (very warm) degrees.

Open `ANES20.sav`. To standardize a variable, we use the "Descriptives" procedure we have used in prior chapters.

`Analyze → Descriptive Statistics → Descriptives`

When the "Descriptives" dialog box opens (shown in Figure 6.17), we move the desired variables into the "Variable(s)" box. Before we click on "OK," we place a check mark into the small box next to "Save standardized values as variables."

After we click on "OK," the SPSS "Output" window displays the descriptive statistics generated by this procedure. To locate the standardized values, make visible the SPSS "Data Editor" window. Switch to "Variable View," and you will see two new variables

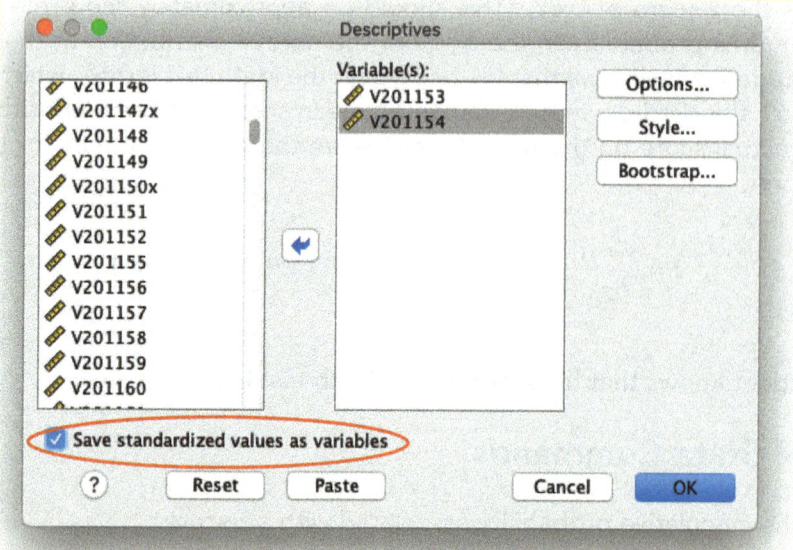

Figure 6.17

listed, each starting with the letter "Z." The new variables' names are the original variable names, with a Z placed before the name. In this example, our two new variables are *ZV201153* (standardized Harris feeling thermometer scores) and *ZV201154* (standardized Pence feeling thermometer scores). If instead we switch the "Data Editor" window to "Data View," we can see the standardized scores on both variables for each respondent in the survey.

When we switch to "Data View," we can see the standardized scores and the original values for each respondent in the survey. For example, a respondent who rated Harris at 60 degrees on the original feeling thermometer scale shows a value of 0.33 on the standardized version of the variable. This is because 60 degrees is equal to about one-third of a standard deviation above the mean of the Harris feeling thermometer variable.

Review of SPSS Procedures

- Standardize the values of a variable

    `Analyze → Descriptive Statistics → Descriptives`

    Make sure to place a check mark next to "Save standardized values as variables."

Practice Problems

1. The 2020 *American National Election Study* asked a random sample of respondents how often they pay attention to politics and elections. Table 6.8 shows the distribution of responses.

   Table 6.8 Distribution of Respondents according to Level of Interest in Politics and Elections, 2020

	Frequency
Always	1,798
Most of the time	3,007
About half the time	1,408
Some of the time	1,176
Never	63
Total	7,452

   Source: 2020 American National Election Study.

   a. What percentage of respondents fell into each category?
   b. What is the probability that we would randomly choose someone who always follows politics? Someone who follows politics most of the time? Someone who follows politics half of the time? Someone who follows politics some of the time?
   c. What is the probability that a randomly selected respondent follows politics *at least* half of the time?
   d. In the 2012 *American National Election Study*, 51% reported that they voted most of the time or always. Compare that figure to the corresponding figure in 2020. Is there evidence to support the claim that the electorate in 2020 is more active?

2. The World Values Survey (Wave 7) asked respondents to state the degree to which they agree with this statement: "Whenever science and religion conflict, religion is always right." Table 6.9 shows the distribution of responses for random samples of respondents in Canada and in the United States.

   Table 6.9 Agreement or Disagreement with Statement: "Whenever Science and Religion Conflict, Religion Is Always Right"

	Canada	United States
Strongly agree	197	325
Agree	360	475
Disagree	1,207	962
Strongly disagree	1,809	795
Total	3,573	2,557

   Source: World Values Survey (Wave 7).

a. What percentage of Canadian respondents fall into each category?

b. What percentage of American respondents fall into each category?

c. What is the probability that a randomly selected Canadian respondent strongly disagrees with the statement? How does that compare to the probability that a randomly selected American strongly disagrees with the statement?

d. What is the probability that a randomly selected American respondent agrees or strongly agrees with the statement?

e. Who is more likely to feel strongly (agree or disagree) about the statement? A randomly selected American or a randomly selected Canadian?

3. A group of American scientists is trying to understand Americans' attitudes toward science. They decide to conduct a focus group discussion with a small group of American respondents.

   a. Use the information from Problem 2 to find the probability that the first American invited to their focus group will be one person who strongly agrees, and that the second American will be someone who strongly disagrees? (Assume independence.)

   b. What rule of probability needs to be used to calculate the probability for part a?

   c. What is the probability that the first two invited will both be Americans who strongly agree? (Assume independence.)

   d. Use the Complement Rule to calculate the probability that the first two invitations will include at least one American who strongly agrees. (Assume independence.)

4. A woman is frustrated with her online dating experience, feeling that it has made her define her ideal partner too specifically. She wants someone who likes vegetables, graduated from college, and loves dogs. She has agreed to invite one person on a date from a list of ten people generated by her friends. She will choose the date at random. She will know nothing about the person before the date. Table 6.10 gives information about each of the ten potential dates.

   a. Make a table listing all of the possible combinations of the three characteristics for the potential dates. Label each combination with a number (e.g., "Combination 1").

   b. Make a frequency distribution showing how the ten potential dates are distributed across the possible combinations of the three characteristics that you identified in part a. Include a column showing the frequency of each combination and a column showing each combination's relative frequency.

   c. Use your frequency distribution from part b to find the probability that the woman will go on a date with someone who possesses all three of her desired traits. Give the probability as a fraction, a proportion, and a percentage.

Table 6.10 Characteristics of Potential Dates

	Likes Vegetables?	Graduated from College	Loves Dogs?
Person 1	Yes	Yes	No
Person 2	Yes	Yes	No
Person 3	Yes	Yes	Yes
Person 4	Yes	Yes	No
Person 5	Yes	Yes	Yes
Person 6	Yes	No	No
Person 7	No	Yes	Yes
Person 8	No	Yes	Yes
Person 9	Yes	No	No
Person 10	Yes	Yes	Yes

5. The friends of the woman in Problem 4 persuade her to go on one date with all ten people on the list from Table 6.10. She will go on dates with each of the ten people in random order.

   a. Which Multiplication Rule must be used to calculate the probability from part a: with independence or without independence? Explain why.

   b. Use the Multiplication Rule to find the probability that the five first dates will all be with someone who has graduated from college.

6. There was a mad rush when tickets for Taylor Swift's ERAS Tour went on sale in November 2022. According to one estimate, the demand for tickets was so high that fans had only a 2 percent chance of successfully buying a ticket.

   a. What is the probability that two randomly selected fans were both able to purchase a ticket for Taylor Swift's ERAS tour? (Assume independence).

   b. Among five randomly selected fans, what is the probability that at least one of them was able to secure a ticket? (Assume independence.)

7. Figure 6.18 shows the distribution of wealth for American households, according to the 2019 Survey of Consumer Finances, administered by the Federal Reserve Board.

   a. What is the probability that a value for household wealth chosen randomly from the sample will fall between $21,345 and $16,721,760?

   b. Use the Complement Rule to calculate the probability that a household randomly chosen from the sample has wealth of at least $16,721,760.

8. According to the wealth data from Problem 7, the ratio of the value at the 90th percentile to the value at the 25th percentile is 783. In other words, households

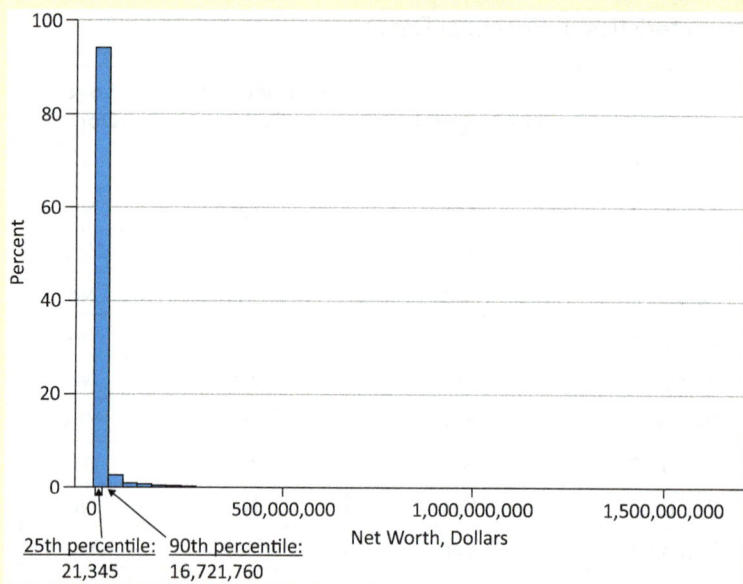

Figure 6.18 Distribution of Wealth among American Households.

Source: 2019 Survey of Consumer Finances.

that fall at the 90th percentile for wealth have about 783 times as much wealth as households at the 25th percentile. We know from Figure 6.18 that the distribution of household wealth in the United States is extremely skewed to the right. However, imagine that we asked American adults to estimate the ratio of the 90th to the 25th percentile for household wealth. Imagine that these estimates were normally distributed, as shown in Figure 6.19.

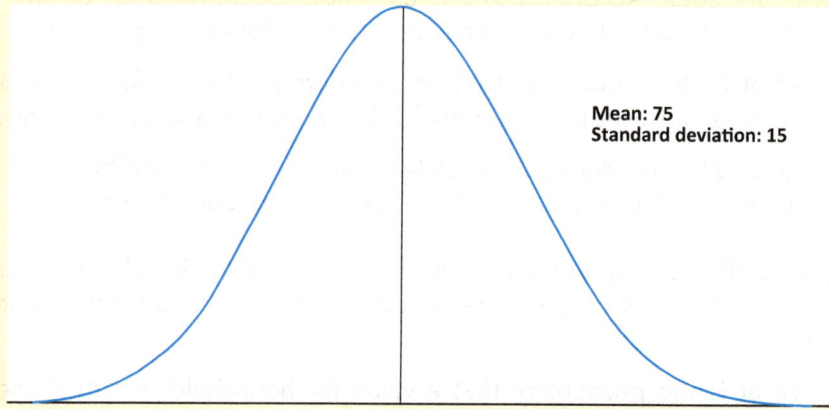

Figure 6.19 Distribution of Adults' Estimates of the 90:25 Ratio of U.S. Household Wealth

a. The mean of the distribution in Figure 6.19 is 75, and the standard deviation is 15. Draw Figure 6.19 with two x axes: 90:25 ratio and standard deviations. On each axis, label the appropriate values for the mean and one, two, and three standard deviations away from the mean.

b. What percentage of respondents estimated the 90:25 ratio to fall between 60 and 90? 45 and 105? 30 and 120? How do you know?

c. The actual 90:25 wealth ratio for American households is 783. How much higher is 783 than the mean estimate of the ratio among American adults? How much higher is 783 than the mean estimate in standard deviation units?

d. Use the Normal Table to find the probability that an estimate of the 90:25 ratio would fall above 783.

e. What is the probability that a randomly selected adult will offer an estimate of the 90:25 ratio that falls below the actual ratio of 783? Why do you think this is?

9. Say that the mean number of hours per day that teenagers around the world listen to music is normally distributed, with a mean of 5 and a standard deviation of 1.5 hours.

   a. What is the probability that a randomly selected teenager will listen to music for less than 1.5 hours per day? Explain how you found the probability.

   b. What is the probability that a randomly selected teenager will listen to between 2.25 and 6 hours of music per day? Show how you found the probability by sketching a normal curve.

   c. What is the probability that a randomly selected teenager will listen to between 5 and 8.5 hours of music per day? Explain how you found the probability.

   d. Find the score associated with the 75th percentile for the distribution.

10. Imagine that the distribution of daily hours of music listened to by teenagers from Problem 9 was still normally distributed and had the same mean, 5, but a standard deviation of 1 hour. Find the score associated with the 75th percentile for the distribution and explain why it's different from the value that you found for part d in 9.

11. This year, two sections of Metaphysics are being taught at State University during the fall semester, one by a new professor who would like to see more students master the material and the other by a veteran professor who believes that he must hold a very high standard or students will not try to grapple with the material. The professors give identical first exams, but the newer teacher allows students to use their books and notes during the exam, while the veteran teacher allows no outside materials. Table 6.11 shows the scores for the students in both sections.

   a. Originally, both professors had planned to give exam grades according to their university's grading scale using students' raw scores. The university's grading scale is as follows:

    - A: 90 to 100
    - B: 80 to 89
    - C: 70 to 79
    - D: 60 to 69
    - F: lower than 60

   Use Table 6.11 to assign the students in the two sections letter grades according to the university's grading scale, using their raw scores. What proportion of students in each section received F grades? A grades?

Table 6.11 Exam Grades for Twenty Students in Two Metaphysics Sections

Student	Veteran Professor	New Professor
1	45	85
2	37	90
3	80	87
4	25	83
5	45	98
6	50	100
7	60	90
8	65	93
9	70	75
10	41	84
11	47	89
12	48	90
13	55	99
14	54	86
15	60	84
16	65	87
17	70	88
18	68	90
19	67	93
20	18	91
Mean score	53.5	89.1

b. The professors are dismayed by the low grades in the veteran's section compared to those in the new professor's. The veteran proposes that they pool the grades for the students across the two sections, convert the raw scores into z-scores, and assign letter grades to all of the students according to the following scale:

- A: above 1.5 standard deviations above the mean
- B: between the mean and 1.5 standard deviations above the mean
- C: between the mean and −.5 standard deviations below the mean
- D: between −.5 and −1.5 standard deviations below the mean
- F: lower than −1.5 standard deviations below the mean

After pooling the students from the two sections, they find that the overall mean score is 71.3. All of the students' raw scores exceeded this mean in the new professor's section, but only one student's score in the veteran's section did. How many students in each section would receive a letter grade of either

A or B using the veteran professor's proposed grading scale? Explain how you know. Does the veteran's proposed solution address the disparity in grades across sections? Explain why.

c. The new teacher proposes that they do not pool the grades and instead convert the grades to z-scores within each section. Then they can use the scale proposed by the veteran professor in part b. Will this lessen the problem of lower grades in the veteran's section? Explain why.

12. Over the last few seasons, a college track coach has been disappointed with her long jumpers' performance. In preparation for recruiting jumpers for the incoming class, she decides that she will recruit only high school long jumpers whose longest jump would be beat by no more than 20% of college long jumpers nationally. The coach knows that the mean long jump length for college jumpers nationally is 5.5 meters, with a standard deviation of 0.25 meters.

a. The coach remembers her statistics class from college and assumes that these jump lengths are normally distributed. Based on this assumption, she decides to use the Normal Table to find the z-score for the jump length that would be beat by 20% of all college long jumpers. Use the Normal Table to find this z-score.

b. Once again assuming that jump lengths are normally distributed, the coach uses the z-score formula to find the jump length associated with the z-score that would be beat by only 20% of college long jumpers. Find this jump length.

c. The coach digs a little deeper and finds a graphic presentation of the distribution of jump lengths. That distribution is shown in Figure 6.20.

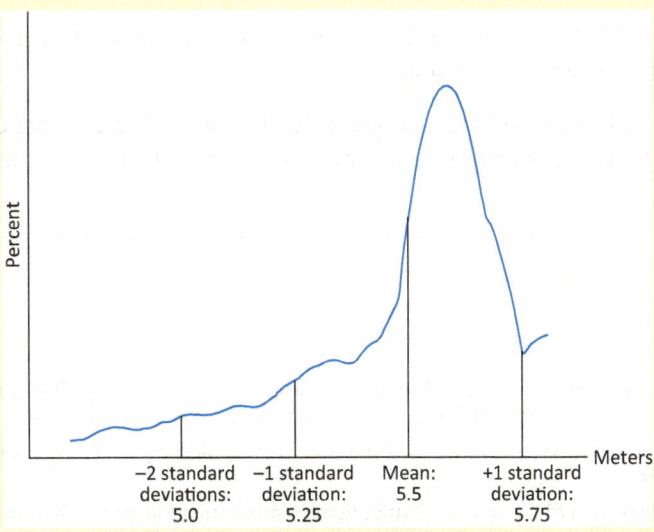

Figure 6.20 Distribution of Length of Jumps for College Long Jumpers, Nationally

The coach finds a potential recruit whose best high school jump was 5.9 meters. The coach is happy to see that this jump exceeds the cutoff length found in part b. She concludes that the jump would be bested by less than 20% of college jumpers and decides to recruit the student. Do you agree with her reasoning based on all of the information presented in this problem?

Stata Problems

Here, we will use Stata without any data stored in its memory to answer questions about registered nurses' hourly pay. According to the Bureau of Labor Statistics, in 2021 mean hourly pay for registered nurses in the United States was $37.31. Say that hourly pay for registered nurses follows a normal distribution and the standard deviation is $5.50.

1. Use the "display" command to find the probability that a randomly selected registered nurse would make more than $38.00 per hour.
2. Use the Complement Rule directly in the "display" command to find the same probability from Problem 1.
3. We want to know the median hourly wage for registered nurses associated with the 30th percentile. Use the "display" command to find the wage.

SPSS Problems

Here we will use the 2021 General Social Survey (GSS) to standardize the variable measuring the number of hours that respondents spend on the internet each week, *wwwhr*.

1. Open `GSS2021.sav` and use the "Descriptives" procedure to generate a standardized version of the *wwwhr* variable.
2. Look at the "Data View" window to find the z-score that corresponds to 0 hours spent on the internet per week.
3. Confirm that the value that you gave for Problem 2 is correct using the following information: Mean internet hours per week is 14.8, with a standard deviation of 17.4.

Notes

[1] Amos Tversky and Daniel Kahneman. 1983. "Extensional versus Intuitive Reasoning: The Conjunction Fallacy in Probability Judgment." *Psychological Review* 90(4): 293–315. We have simplified the example.

[2] Daniel Kahneman and Amos Tversky. 1972. "Subjective Probability: A Judgment of Representativeness." *Cognitive Psychology* 3: 430–454.

[3] S. Stahl. 2006. "The Evolution of the Normal Distribution." *Mathematics Magazine* 79(2): 96–113.

[4] National Center for Education Statistics. Table 226.40. SAT mean scores of high school seniors, standard deviations, and percentage of the graduating class taking the SAT, by state: Selected years, 2017 through 2021. Retrieved from https://nces.ed.gov/programs/digest/d21/tables/dt21_226.40.asp.

Chapter 7

From Sample to Population

Sampling Distributions and the Central Limit Theorem

Social science researchers all face a similar problem. They are interested in social phenomena in populations that are usually so large that it is impractical to study every member. If we want to know the opinion of residents of a state or a country, the effects of educational programs on whether incarcerated people stay out of prison, or the proportion of nurses who are women, we rarely study every resident, incarcerated person, or nurse. The amount of time and money required is too great. Researchers usually rely on data from a sample—a subset of the population. If a sample is properly collected, it will resemble the population from which it comes.

So far, you have learned a range of statistical techniques for describing the results of social and political research. All of this has been based on describing the data in a sample. The next several chapters cover various aspects of how to generalize the results from a sample to the population. This is the branch of statistics known as inferential statistics (compared with descriptive statistics, the focus of chapters 2 through 5). It addresses the question: How can we infer the characteristics of a population from more limited sample data?

We discussed the importance of random sampling in chapter 1. In a random sample, every member of a population has an equal probability of being selected for the sample. Under these conditions, the sample is likely to yield a result that is somewhat close to the result for the population.

But even when a sample is random, its results will rarely be identical to the characteristics of the population. This is why even using the same sampling techniques on the same population can yield different results. Political polls are a good example of this issue. Even when a poll is conducted perfectly, another identical poll might produce a slightly different result. That is why political polls are reported with a "margin of error." We will discuss what this means and how to calculate it in much more detail in chapter 8. In this chapter, we explain the underlying logic and the relationship between samples and populations.

Repeated Sampling, Sample Statistics, and the Population Parameter

A given statistic for a population (such as a mean or proportion) is called the **population parameter**, while the same mean or proportion for a sample is called a **sample statistic**.

> **BOX 7.1: IN DEPTH**
>
> ### Statistical Notation for Samples and Populations
>
> Here are the symbols used to denote means, proportions, and standard deviations for samples and populations:
>
> - mean: $\bar{y}$ (sample); μ (population)
> - proportion: p (sample); π (population)
> - standard deviation: s (sample); σ (population)
>
> The symbols for population parameters are lowercase Greek letters. (If you attend a university with fraternities or sororities, you may be familiar with the capital letters.) Here is how you pronounce the Greek letters for mean, proportion, and standard deviation:
>
> - μ (mu, sounds like "mew")
> - π (pi)
> - σ (sigma)

Inferential statistics take advantage of some specific characteristics of random sampling, probability, and the normal curve. In brief, statisticians have shown that if we could draw repeated random samples from a population, their sample statistics would fall into a normal curve in which the center or mean of the curve is the true population parameter; the most frequent (or probable) sample statistics are close to the population parameter; and statistics that are further from the population parameter become increasingly improbable. We can then use the procedures covered in chapter 6 to estimate the probability of observing given sample statistics.

Let's break this down step by step. First, let's examine the relationship of sample statistics to the population parameter. The seventh wave of the World Values Survey (WVS) asked respondents in different countries their opinion about whether hard work brings success. They were asked to rank their views from 1—"In the long run, hard work usually brings a better life"—to 10—"Hard work doesn't generally bring success—it's more a matter of luck and connections." For this example, we will look at respondents in one country only: Mexico. There were 1,726 respondents from Mexico who answered this question, and their mean response to this question was a 4.16. Because lower numbers mean greater agreement that hard work brings a better life, this suggests moderate support for that idea, on average. (By contrast, the mean for residents in the United States was 3.74, suggesting slightly more support for the idea.)

For purposes of illustration, we are going to treat these 1,726 respondents as if they constitute a "population." We will randomly sample 200 of them and see how close the sample mean on this WVS question is to the overall "population" mean. We use statistical software to draw a random sample of two hundred respondents and calculate the mean ranking for this variable. With our sample of two hundred, we obtain a mean of 4.10.

Why is this mean different from the mean of 4.16 for the full group? Because, by random selection, the individuals in our sample happened to have slightly lower rankings, on average. Taking a second random sample of two hundred respondents, we find a mean of 3.87. Why are the results for these two random samples different? Because, by random chance, the individuals in each one are slightly different from each other. If we repeat this process, each sample will yield a slightly different result. Table 7.1 lists the results, and Figure 7.1 shows the means from these twenty samples.

Table 7.1 Mean Opinion about Whether Hard Work Brings Success in Twenty Samples of Two Hundred Mexican Respondents

Sample	Mean
1	4.10
2	3.87
3	4.34
4	4.09
5	4.16
6	4.45
7	4.05
8	4.27
9	3.93
10	4.22
11	3.77
12	4.36
13	4.07
14	4.08
15	3.89
16	4.18
17	4.19
18	4.17
19	3.78
20	4.25

Source: World Values Survey (Wave 7).

The sample means vary from a low of 3.77 to a high of 4.45. As Figure 7.1 shows, some are below the overall "population" mean of 4.16 (drawn as a solid red line) and some are above. Many are quite close, and one is exactly 4.16.

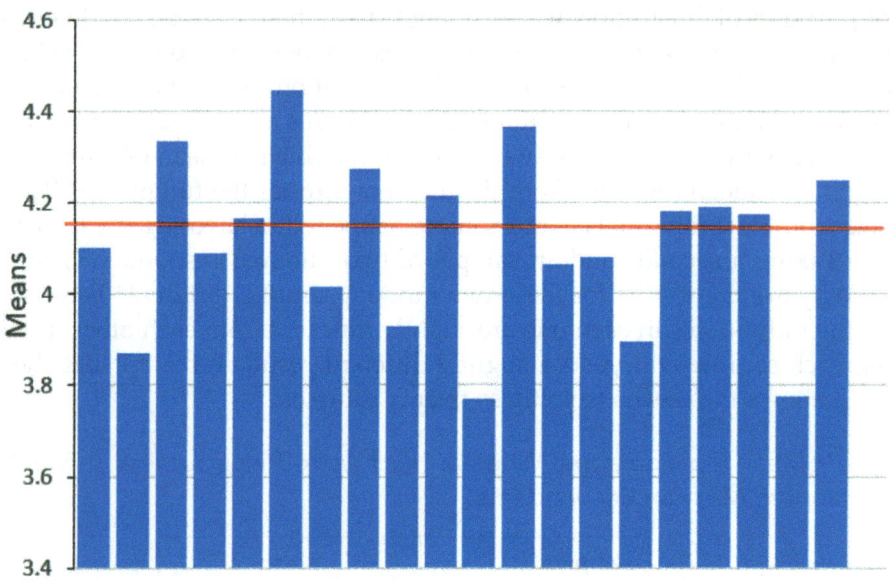

Figure 7.1 Mean Opinion about Whether Hard Work Brings Success in Twenty Samples.

Source: World Values Survey (Wave 7).

Let's get more specific. Of the 20 sample means, how many are less than 0.1 points away from the population mean of 4.16? 11. How many are between 0.1 and 0.19? 3. How many are between 0.2 and 0.29? 4. How many are between 0.3 and 0.39? 2. Going further, we can see that half (10) of the sample means are below the population mean, one sample mean is exactly 4.16, and the others are above it. Graphing this, we can see the emerging shape does not yet resemble a normal curve, as shown in Figure 7.2.

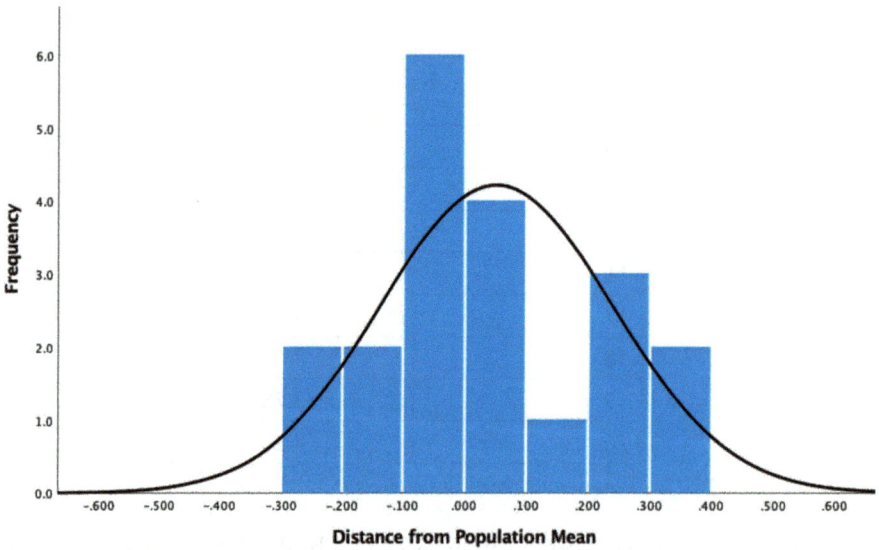

Figure 7.2 Distribution of 20 Sample Means.

Source: World Values Survey (Wave 7).

As we select more samples, our histogram will begin to take on the appearance of a normal distribution. We repeated the simulation described above, this time with two hundred samples (each with two hundred respondents) and computed the mean for each of those two hundred samples (see Figure 7.3). The histogram in Figure 7.3 shows how close the sample means are to the true population mean. Compare this histogram to the one shown in Figure 7.2. As the number of samples grows, our histogram looks increasingly "normal." If you'd like to prove this to yourself, the "Using Stata" and "Using SPSS" sections at the end of this chapter show you how to take repeated samples and graph them against the normal curve.

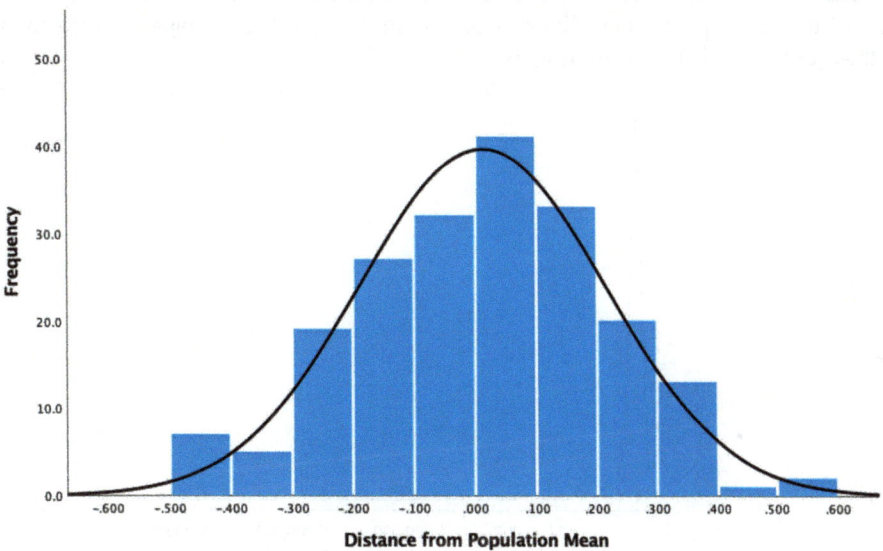

Figure 7.3 Distribution of Fifty Sample Means.

Source: World Values Survey (Wave 7).

● Sampling Distributions

If we drew enough random samples from the population, the graph of the sample means would more and more closely approximate the normal curve. If we imagine doing it an infinite number of times, we would see exactly the normal curve. The frequency distribution of the statistic (in this case mean) obtained through this theoretical repeated sampling process is called the **sampling distribution**; when the statistic is the mean, we call it the **sampling distribution of the means** (When the statistic is a proportion, we call it the **sampling distribution of the proportions**.)

The **Central Limit Theorem** tells us that, with large enough sample sizes, the sampling distribution will approximate a normal curve. *This is the case regardless of whether the actual variable is normally distributed in the population.* For example, we know that income is not normally distributed. But a sampling distribution of mean income would be normally distributed. (How do we know this is true? Statisticians proved the Central Limit Theorem, in part by actually taking repeated random samples and charting their results. It can also be proven mathematically.) If you want to know more about this, Box 7.2 gives an example.

BOX 7.2: IN DEPTH

Sampling from a Skewed Population

One of the amazing features of the Central Limit Theorem is that, with a large enough sample size, even a highly skewed population will produce a sampling distribution that is normally distributed. Figure 7.4 shows the distribution of income in the United States in 2022, from the Survey of Income and Program Participation (SIPP). Clearly, income is a highly skewed variable.

Out of this population, we drew one hundred samples, each one with 200 respondents, and we calculated the mean income for each of those samples. The histogram below shows the distribution of those sample means (Figure 7.5). As we expected, the shape is close to symmetric. This is the Central Limit Theorem at work.

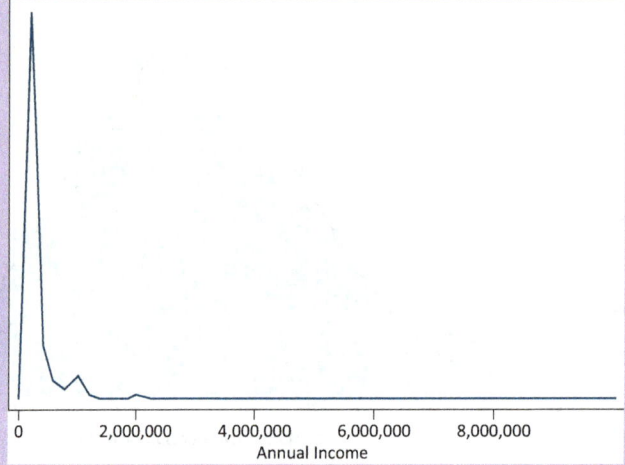

Figure 7.4 U.S. Income Distribution.

Source: 2022 Survey of Income and Program Participation.

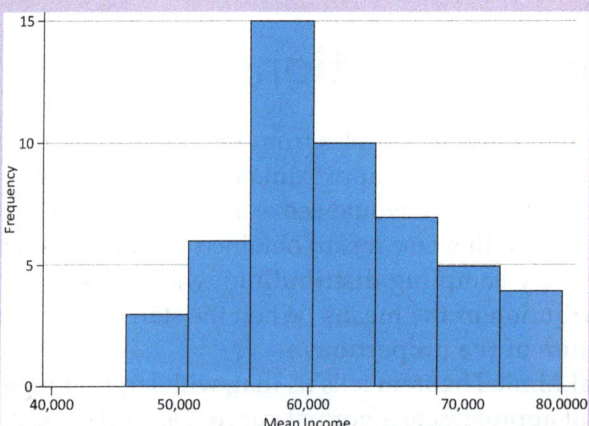

Figure 7.5 Distribution of One Hundred Sample Means, Drawn from U.S. Income Distribution.

Source: 2022 Survey of Income and Program Participation.

Chapter 6 discussed the properties of the normal curve. To review: The normal curve is a probability distribution that has known qualities. It is bell-shaped, symmetrical, and unimodal, with the number of cases on either side of the mean gradually decreasing at a specific rate and the tails extending indefinitely on either side never reaching the baseline. The area between the mean and one standard deviation (or any fraction of a standard deviation) is always the same.

What does it mean that the sampling distribution is a normal curve? Applying the properties of the normal curve to the sampling distribution, we know that the mean (or center) of this curve is the true population parameter. The probability of obtaining a sample statistic that is close to the population parameter is high, with decreasing probability of obtaining sample statistics further from the population parameter. Just like any frequency distribution, a sampling distribution has a standard deviation. We call the standard deviation of the sampling distribution the **standard error**.

Figure 7.6 shows the theoretical sampling distribution, with an infinite number of samples.

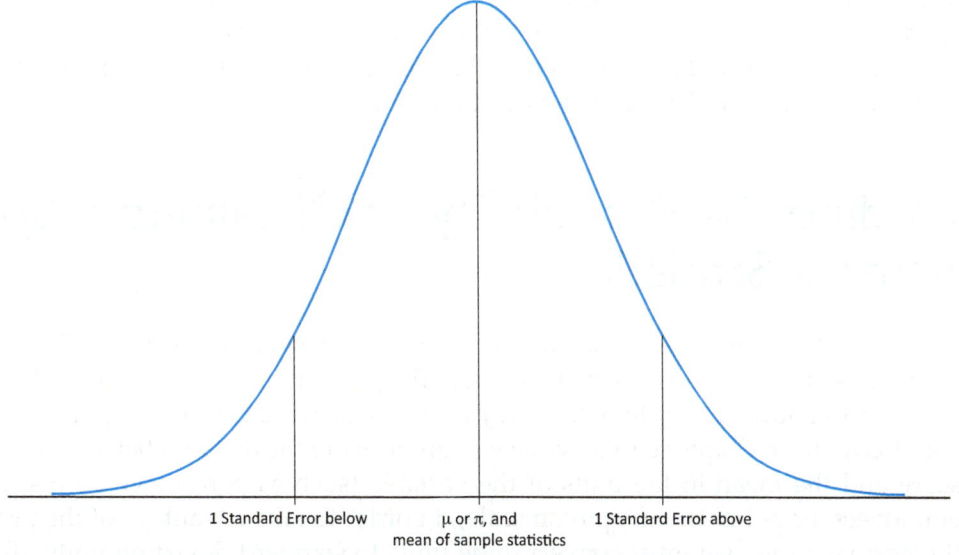

Figure 7.6 Theoretical Sampling Distribution (Infinite Number of Samples)

Take a minute to remind yourself what this curve shows. It is not a distribution of the values of the variable. It is a distribution of sample statistics. Each point on the curve is a hypothetical sample statistic. The standard error tells us how much, on average, sample statistics vary from the population parameter. For example, if we are studying the number of sick days taken by employees of a particular organization, each sample of one hundred employees would yield a mean number of sick days for the employees in that sample. If we take an infinite number of repeated samples, they would yield different means, just because of random sampling error. If we graph the means of the different samples, the Central Limit Theorem tells us that they would form a normal curve. The mean of that curve would be the mean of the means of the repeated samples, and it would also be the true population mean. The standard deviation of that curve is called the standard error, which is the average difference between a single sample score and the mean of the distribution. Because each single score is a sample mean and the mean of the distribution is the true population mean, the standard error is the average difference between a sample mean and the population mean.

Read that again: The mean of the curve is the mean of the means of the repeated samples. It is always also the population mean. The standard error (which is the standard deviation for the sampling distribution) is the average difference between a sample mean and the population mean. This is one of the most important properties of the sampling distribution and is the basis for much of inferential statistics.

If the sample statistic we are interested in is a proportion (such as the proportion of young people who favor legalizing marijuana), the curve shows the distribution of the sample proportions, if we were to take repeated samples. The center of the curve is the true population proportion. The standard error is the average difference between a sample proportion and the population proportion.

Using the properties of the normal curve, we know that half of samples will yield a statistic above the population parameter and half below. Of the samples, 68.26% will yield a result within one standard error of the population mean; 95.44% will be within two standard errors; and 99.74% within three standard errors. (Remember, the standard error is what we call the standard deviation of the sampling distribution.) If you want a refresher on this, look back at "The Normal Distribution" in chapter 6.

Finding the Probability of Obtaining a Specific Sample Statistic

Also using the properties of the normal curve, we can specify the probability of obtaining a sample statistic any specific distance from the population parameter. In order to do this, we need to know exactly how far away a sample statistic is from the population parameter. Recall from chapter 6 that, while we can measure the distance between an individual score and the mean in the units of the variable (such as percentage points, number of employees, or points on an opinion scale), in order to take advantage of the properties of the normal curve, we must convert these units to standard deviation units. (Remember, the standard error is the standard deviation for the sampling distribution curve, so when working with sampling distributions, we will be converting the distance from the mean into standard error units.) Instead of saying that a sample mean is $20 below the population mean, we want to be able to say how many standard errors that mean is below the population mean.

Just as in chapter 6, we can calculate a z-score for a sample statistic. The z-score is the distance between that sample statistic and the center of the sampling distribution, which is the population parameter. In chapter 6, you used the standard deviation to calculate a z-score and find the probability of obtaining a given score. We can do the same thing with the sampling distribution. Using standard error, we can calculate a z-score for a sample statistic and determine the probability of obtaining that sample statistic with a given population parameter.

Estimating the Standard Error from a Known Population Standard Deviation

You may have realized that we do not know the standard error. It is not the same value as the standard deviation of the sample. There is much less variation in a sampling distribution than there is in a single sample, and the standard error is consequently smaller than the standard deviation. The means of samples will always vary less than the scores of individuals, because means literally average out individual differences. You could imagine an individual respondent giving a rating of 10 on the WVS question about whether hard work brings success but not a sample in which the mean was 10.

We have several methods of estimating the standard error. If we know the standard deviation for the population, we can use that to estimate the standard error. If we do not know the standard deviation for the population (the more common situation), we use the sample standard deviation to estimate standard error. This requires some minor adjustments in other procedures, which we will discuss in chapter 8. For now, we will discuss the often-hypothetical situations in which we know the population standard deviation.

Let's look at an example for which the population mean and standard deviation are known. Many standardized tests fit this description. For example, we know that the mean IQ score for the population is 100 and the standard deviation is 15.[1]

The standard deviation for the population is not the standard error, which is the standard deviation for the sampling distribution. We use the population standard deviation to calculate the standard error for the sampling distribution of mean IQ scores. Remember, this is a hypothetical distribution that models what would happen if we drew repeated samples and charted their means.

The standard error is the population standard deviation divided by the square root of N (the number of cases in the sample). By dividing the standard deviation by the square root of the sample size, the size of the standard error comes to depend on the sample size. Specifically, the larger the sample size, the smaller the standard error. This makes sense because there is less sampling variability in large samples. We will notate the standard error as SE.

$$SE = \frac{\sigma}{\sqrt{N}}$$

In the case of IQ scores, if the sample size is 150:

$$SE = \frac{15}{\sqrt{150}} = \frac{15}{12.25} = 1.22$$

Finding and Interpreting the z-Score for Sample Means

Now that we have estimated the standard error, we can use it to mark the normal curve in both IQ score units and standard error units. In chapter 6, we converted individual scores to z-scores (that is, converting the units of the variable to standard error units). This allowed us to see how far away on the normal curve a given individual score was from the sample mean. The same logic applies to the sampling distribution. If one standard error is 1.22, that means that the sample mean that is one standard error above the population mean of 100 is 101.22 (100 + 1.22 = 101.22). The curve is shown in Figure 7.7.

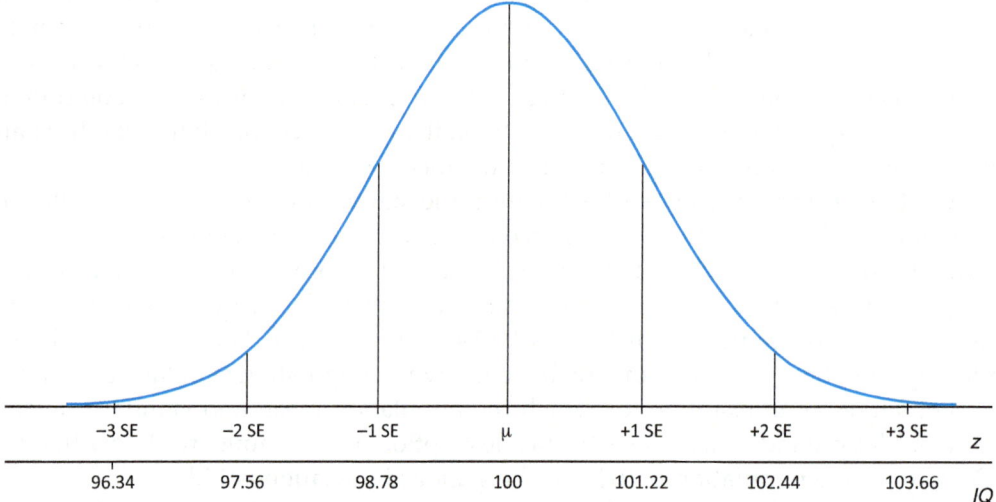

Figure 7.7 Sampling Distribution of Mean IQ Scores

Recall from chapter 6 that the proportion of scores at each distance from the mean is always the same in a normal curve. Half are above the mean (in this case, 100) and half below. Further, recall that, for example, 34.13% of the scores are between the mean and one standard deviation above the mean. The same is true for the sampling distribution. Figure 7.8 shows the sampling distribution of mean IQ scores with the proportions of samples falling within each range marked.

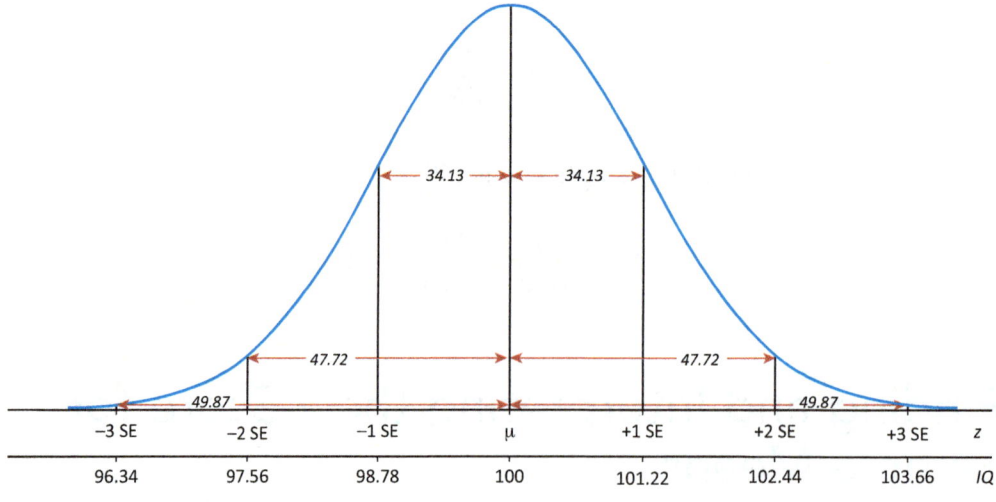

Figure 7.8 Sampling Distribution of Mean IQ Scores

We can see in Figure 7.8 that a little more than one-third of samples would yield means between 100 and 101.22. Of the samples, 68% would yield means between 98.78 and 101.22. And virtually all samples would produce means between 96.34 and 103.66. This is the case even though there are many more individuals with IQs well above or well below 100.

We can also use the properties of the sampling distribution to determine the probability of obtaining a specific sample mean. Let's return to the WVS variable that measures people's opinion about whether hard work brings success, discussed at the beginning of the chapter. We know that the mean for the full sample of people from Mexico was 4.16. Let's treat that as a population mean and see what the probability is of obtaining a result that is at least as far from that mean as one of our sample means, 4.36. In other words, how likely is it that we would randomly get a sample for which the mean is at least 4.36? Figure 7.9 presents this on the normal curve.

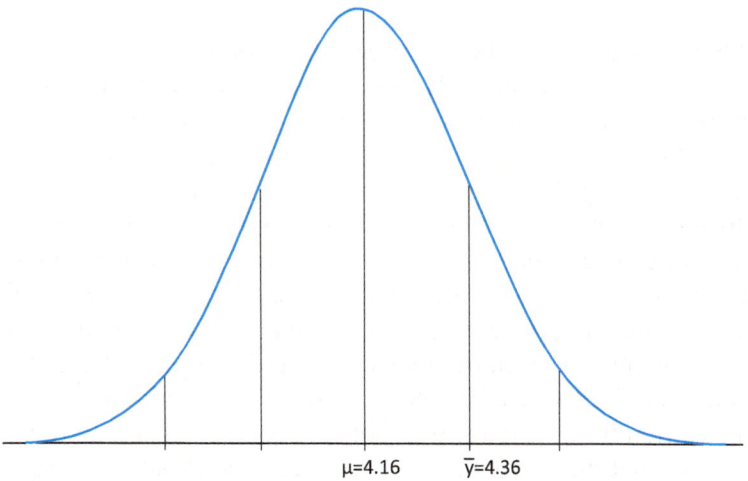

Figure 7.9 Sampling Distribution of Means, Opinion of Whether Hard Work Brings Success, Mexican Respondents to WVS. Source: World Values Survey (Wave 7).

Just as with the IQ example, we must calculate the standard error (the standard deviation for the distribution of sample means) in order to determine the proportion of sample results that fall within one, two, three, or any fraction of a standard error's distance from the population mean. The standard deviation for the full group of Mexican respondents to the WVS was 3.00. We use this information to estimate the standard error for the sampling distribution of means.

$$SE = \frac{\sigma}{\sqrt{N}}$$

Our WVS sample size (N) was 200.

$$SE = \frac{\sigma}{\sqrt{N}} = \frac{3.00}{\sqrt{200}} = \frac{3.00}{14.14} = 0.21$$

As we did in chapter 6, we can now calculate the z-score associated with the sample mean of 4.36. A z-score is always the distance between the specific score and the mean of the normal distribution. In chapter 6, we used the z-score to indicate the distance between an individual's score and the mean score for the sample. For the sampling distribution, remember, the specific score is a sample mean, and the mean of the normal distribution is the population mean. The formula looks slightly different as a result.

$$z = \frac{\bar{y} - \mu}{SE}$$

$$z = \frac{4.36 - 4.16}{0.21} = \frac{0.20}{0.21} = 0.95$$

Our sample mean is 0.95 z-scores—or 0.95 standard errors—above the population mean. Looking up the z-score in the Normal Table (appendix A), we see a probability of 0.1711. That is, with a population mean of 4.16, there is a 17% chance of getting a sample with a mean above 4.36. To put it differently, out of every ten random samples of two hundred cases drawn from this population, we would expect about one or two to have a mean above 4.36.

In the two hundred samples of size two hundred drawn earlier from the WVS, we actually observed twenty samples with a mean above 4.36. We would have predicted that about thirty-four of those samples (17% of 200 is thirty-four) would have had a mean above 4.36. Did something go wrong? No, it is important to remember that these predictions hold in the long run, over many, many repeated samples. Recall that the sampling distribution takes on the shape of the normal curve only under an infinite number of samples. So, with two hundred samples, it is not surprising that we do not observe the exact number of samples that the normal curve predicted would fall to the right of a particular z-score.

What about a sample mean that is just slightly different from the population mean of 4.16? One of our samples produced a mean of 4.10. What is the probability of obtaining that sample result or less? We can calculate z using the standard error that we calculated above:

$$z = \frac{\bar{y} - \mu}{SE} = \frac{4.10 - 4.16}{0.21} = \frac{-0.06}{0.21} = -0.29$$

Notice that the z-score is a negative number. This is because the sample mean is less than the population mean. A negative z-score always means that the sample mean is that distance below the population mean, while a positive z-score tells us that the sample mean is that distance above the population mean. We ignore the sign when looking up the probability for the z-score because it is the same regardless of whether z is above or below the mean.

Looking up 0.29 in the table of normal values, we see a probability of 0.3859. There is a 38.6% chance of obtaining a sample mean below 4.10. In other words, we would expect to find a sample mean at least that distance from the population mean quite often.

What if we obtained a sample mean of 4.68? In this case:

$$z = \frac{\bar{y} - \mu}{SE} = \frac{4.68 - 4.16}{0.21} = \frac{0.52}{0.21} = 2.48$$

Looking up 2.48 in the table, we find probability equals 0.0066. We would obtain a sample mean of at least 4.68 about 0.7% of the time.

Take a moment to think about this. We know that sampling error is inevitable, due to the random selection of cases, which are likely to vary in some ways from the overall population. Concretely, we now have a sense of what sampling error looks like. If a researcher obtained a sample mean of 4.10, would this be the wrong result? In one sense, of course: the true population mean is not 4.10. But it would not represent any error on the part of the researcher. The same is true if the researcher obtains a sample mean of 4.68 or 3.9.

The identical process of random sampling can produce all of these sample means. One is a closer approximation of the population mean than the others, and sample statistics that are closer approximations of the population parameter will be produced more often. However, if we do not know the population parameter, we can never know with certainty whether our sample statistic is, in fact, close to the population parameter.

Finding and Interpreting the z-Score for Sample Proportions

Let's take another example. Let's assume we know that 38% of the residents of Los Angeles support a particular candidate (let's call her Jane Doe) for city council. What is the probability of selecting a random sample of one hundred people in which at least 45% support Ms. Doe?

Just as with the sampling distribution of the means, sample proportions will be normally distributed around the true population proportion. We call the resulting distribution the sampling distribution of the proportions. The standard deviation of this distribution—just as with the means—is the distance, on average, that sample proportions would be from the population proportion.

In order to find the probability, we need the z-score for .45. First, we have to calculate the standard error. The formula for standard error of the proportions is different from the formula for means.* It is:

$$SE = \sqrt{\frac{\pi(1-\pi)}{N}}$$

In this case, $\pi = .38$, the population proportion.

$$SE = \sqrt{\frac{0.38(1-0.38)}{100}} = \sqrt{\frac{0.2356}{100}} = \sqrt{0.002356} = 0.049$$

We can then use the standard error to calculate z. The formula looks a little different because we are using proportions rather than means. It still has the difference between the

---

* Proportions do not have standard deviations the way means do, because they are calculated on categorical variables. That is why the formula is different.

sample statistic and population parameter in the numerator and the standard error in the denominator.

$$z = \frac{p - \pi}{SE} = \frac{0.45 - 0.38}{0.049} = \frac{0.07}{0.049} = 1.43$$

Looking this up in the table of normal values, we see that the probability is 0.0764. Approximately 7.6% of samples will yield a proportion of 0.45 or higher.

So, if you are a pollster for Jane Doe, you will want to be aware that there is a substantial likelihood that your polls will significantly overestimate the proportion of the population who supports her. Of course, they might also underestimate it.

The Impact of Sample Size on the Standard Error

The Central Limit Theorem tells us that the sampling distribution of the means (or proportions) is normally distributed even when the variable itself is not normally distributed, as long as the sample size is large enough. Even a skewed or bimodal variable will almost always produce a normally distributed sampling distribution with an N of 30 or more.[*]

Sample size does affect the standard error, however, with larger samples producing smaller standard errors. Standard error, in turn, affects z-scores.

Returning to the pollster for Jane Doe, imagine that a poll has a sample size of thirty instead of one hundred. When we plug this into the formula for standard error, we get a result that is almost twice the size that it was with N of 100:

$$SE = \sqrt{\frac{0.38(1 - 0.38)}{30}} = \sqrt{\frac{0.2356}{30}} = \sqrt{0.007853} = 0.089$$

With this smaller sample size, sample results that are further from the population proportion become much more frequent. The proportion of samples that showed a result of 45% or higher was 7.6% with a sample size of one hundred. What happens with a sample size of only thirty?

$$z = \frac{p - \pi}{SE} = \frac{0.45 - 0.38}{0.089} = \frac{0.07}{0.089} = 0.786$$

With a z-score of 0.786 (rounding to 0.79), 21.48% of samples will yield a result of 45% or higher. The smaller sample size makes the range of samples much wider in percentage points. Figure 7.10 shows the sampling distribution of proportions, with the poll results (sample proportions) associated with each z-score.

---

[*] For proportions, if the number of cases across the two categories is extremely lopsided, you will need more than thirty cases for the central limit theorem to hold.

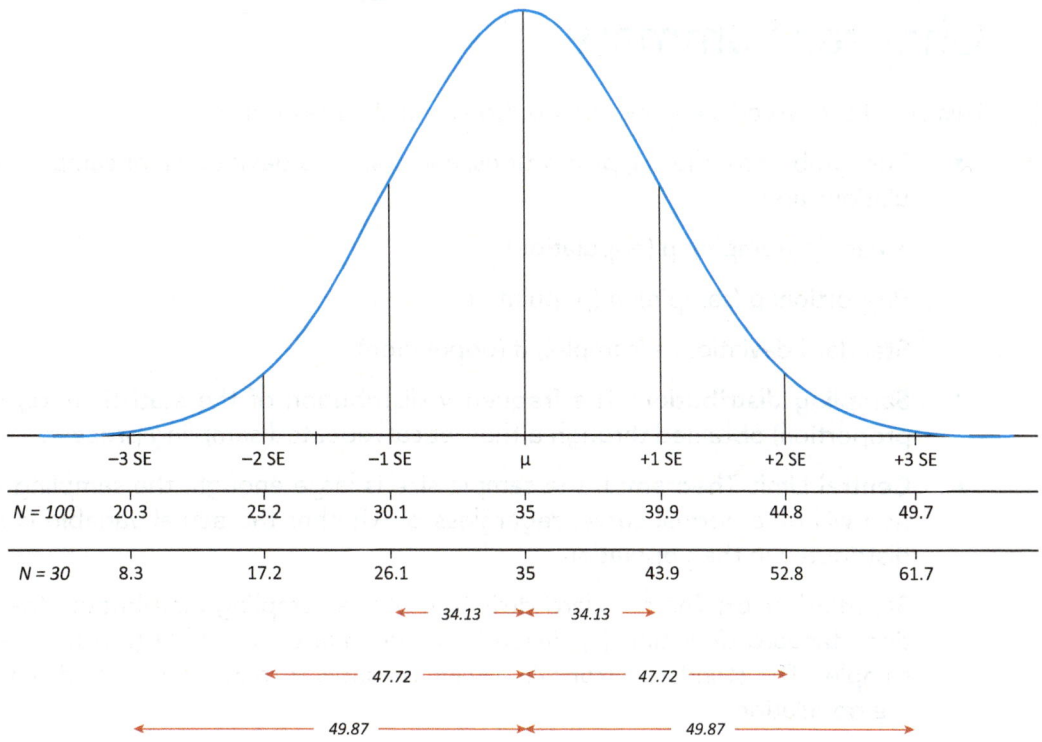

Figure 7.10 Sampling Distribution of Proportions, with N = 100 and N = 30

We can see that at N = 100, more than 95% of samples will yield proportions of support for Candidate Doe between 25.2% and 44.8%. At N = 30, more than 95% of samples will yield results between 17.2% and 52.8%. In practical terms, this means that a sample of thirty cannot reliably tell the candidate whether she is losing badly or ahead, while most samples of one hundred will produce a result that is closer to her actual standing. As the sample size increases, the level of precision increases as well.

In virtually all social science research, researchers do not have information about the population mean or standard deviation. In the next chapter, we will explain how to use the logic of the sampling distribution to infer population parameters without knowing either the population mean or proportion or the population standard deviation.

Chapter Summary

This chapter covered sampling distributions and their properties.

- The symbols for means, proportions, and standard deviations for samples and populations are:

    Mean: $\bar{y}$ (sample); μ (population)

    Proportion: p (sample); π (population)

    Standard deviation: s (sample); σ (population)

- **Sampling distribution**: The frequency distribution of the statistic (e.g., mean or proportion) obtained through a theoretical repeated sampling process.
- **Central Limit Theorem**: If the sample size is large enough, the sampling distribution will be a normal curve, regardless of whether the actual variable is normally distributed in the population.
- **Standard error**: The standard deviation of the sampling distribution; the population standard deviation (σ) divided by the square root of N (the number in the sample). The standard error will always be smaller than the standard deviation in the population.

$$\text{For means}: SE_{\bar{y}} = \frac{\sigma}{\sqrt{N}}$$

$$\text{For proportions}: SE_p = \sqrt{\frac{\pi(1-\pi)}{N}}$$

- **z-score** for a sample mean or proportion in a sampling distribution: The distance between the sample mean or proportion and the population mean or proportion, measured in standard errors:

$$\text{For means}: z = \frac{\bar{y} - \mu}{SE}$$

$$\text{For proportions}: z = \frac{p - \pi}{SE}$$

Using Stata

In this section, we will take repeated random samples in order to demonstrate the properties of the sampling distribution. We will use the same variable from the WVS that we saw at the beginning of the chapter—the extent to which people think success is the result of hard work. The variable, named Q110, measures respondents' opinions about whether hard work brings success on a scale from 1—"In the long run, hard work usually brings a better life"—to 10—"Hard work doesn't generally bring success—it's more a matter of luck and connections." In the beginning of this chapter, we examined sample means for respondents from Mexico; here, we examine means for respondents from the United States.

We will examine two methods in Stata for drawing random samples from a larger sample. The first method is to draw a single random sample, and the second draws many samples and saves requested sample statistics in a new data file.

Open `WVSWave7.dta`. Before we begin, we must drop all cases from our data from non-US respondents. To do this, enter the following "keep" command into Stata:

```
keep if B_COUNTRY==840
```

With this command, we are keeping only those cases that have a value of 840, the code for US respondents, for the variable called *B_COUNTRY*, which identifies the country of the respondent. We can confirm that Stata kept only the observations from the US respondents by requesting a frequency distribution for the country ID variable, *B_COUNTRY*:

```
tabulate B_COUNTRY
```

As we can see from the output, shown in Figure 7.11, the only value for *B_COUNTRY* is the United States, which means that we can proceed with our sampling procedure.

```
. tabulate B_COUNTRY

   ISO 3166-1
      numeric
 country code │    Freq.    Percent       Cum.
──────────────┼───────────────────────────────────
United States │    2,596     100.00     100.00
──────────────┼───────────────────────────────────
        Total │    2,596     100.00
```

Figure 7.11

Drawing a Single Random Sample

To draw a random sample from the data set, enter the following command into Stata:

```
sample 150 if Q110 ~= ., count
```

The "sample" command tells Stata to draw a random sample of 150 cases from the data set. The "count" option tells Stata that the number "150" refers to the number of cases in the sample. Without the "count" option, Stata's default approach would be to draw a sample that has x% of the cases from the original data set. For example, the command "sample 15" would draw a sample that had 15% of the total cases in the data set. (Note that "if Q110 ~=." tells Stata to omit cases with missing data on our variable of interest from its sampling procedure.) We can use the "summarize" command to confirm that Stata has drawn 150 cases:

`summarize Q110`

Recall that the "summarize" command returns the number of observations, mean, standard deviation, and range for a variable, as shown in the output in Figure 7.12.

Variable	Obs	Mean	Std. dev.	Min	Max
Q110	150	4.033333	2.75977	1	10

Figure 7.12

The output shows that *Q110* has only 150 observations, indicating that the sampling procedure proceeded correctly. We can also see that the mean is 4.0. Just as we saw with respondents from Mexico at the beginning of this chapter, if we were to repeatedly draw random samples of 150 from the full sample of US respondents, the mean for *Q110* would differ each time. Try drawing another random sample with 150 cases—what is the sample mean?

In order for the sampling distribution of sample means to take on the shape of a normal distribution, we would need to draw a large number of samples of the same size from our population. To construct the distribution of sample means, we would have to enter the "sample" command many times into Stata and keep track of the mean for *Q110* for each of those samples. This sounds like tedious, error-prone work, but there is a way that Stata can help us, which leads us to the second random sampling method.

Drawing Repeated Samples and Saving Sample Statistics in New Data File

The second method of drawing random samples tells Stata to take repeated random samples of equal size from our data set, calculate sample means for *Q110* each time, and save those sample means in a new data file. The new data file will contain a single variable that contains the sample means of *Q110*. The variable will have as many values as samples that we ask Stata to draw, or the sample statistic for each sample (in this case sample means). With this method, we do not need to run the same command repeatedly, nor do we need to keep track of the means for all of those samples. Before running our second sampling method, we must return to the original WVS data set (our previous command drawing a sample of 150 left us with that sample data instead of the full data set) and then keep only the data from the US respondents:

`keep if B_COUNTRY==840`

To conduct the sampling procedure, enter the following command into Stata:

```
bootstrap if Q110 ~= ., reps(1000) size(150) saving("WVS7Simulation
.dta", replace): mean Q110
```

Here, we use a "bootstrap" command, which tells Stata to draw multiple random samples from our data set.* The options "reps" and "size" tell Stata how many samples to draw and the size of each sample, respectively. Here, we are telling Stata to draw one thousand random samples, each with 150 cases. The "replace" option tells Stata to overwrite the original file if this command has been previously run, meaning that observations that are drawn in one sample are returned to the full sample so that they have a chance to be drawn in subsequent samples. "Mean Q110" tells Stata to calculate the mean for *Q110* for each of the one thousand samples and record the means in a new data file called `WVS7Simulation.dta`. The output is shown in Figure 7.13.

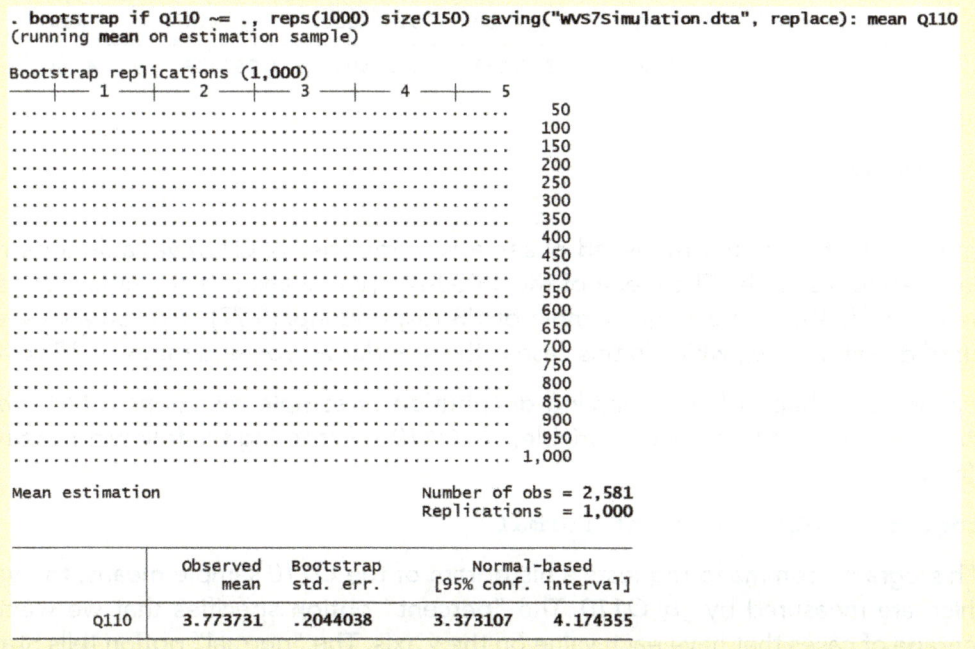

Figure 7.13

The top portion of the output shows that Stata drew one thousand samples. In the table at the bottom of the output, we see that Stata has returned the full sample mean for all US respondents, 3.77, and has produced a standard error, which is the standard deviation of the 1,000 means. The last two columns of the output provide a confidence interval for the sample mean based on the estimated standard error, a topic to which we will turn in the next chapter. Note that the statistics reported in your output will differ slightly from those shown in the output here. This is because Stata draws random samples each time.

---

* "Bootstrap" is a special term in statistics that refers to the practice of using multiple random samples from a single sample to make inferences about the underlying population.

But this output from the "bootstrap" command is not our central interest. We want to know whether the sample means for Q110 approximate a normal distribution. To answer that question, we must open the new file that contains all 1,000 of the sample means:

```
use "WVS7Simulation.dta", replace
```

We can see that this data file has only one variable, _b_Q110. It is a strange variable name, but it is just the name that Stata has assigned to the variable that measures all 1,000 sample means for Q110. We can view the mean and standard deviation of the sample means by using the "summarize" command:

```
summarize _b_Q110
```

The output is shown in Figure 7.14.

. summarize _b_Q110

Variable	Obs	Mean	Std. dev.	Min	Max
_b_Q110	1,000	3.77618	.2044038	3.153333	4.48

Figure 7.14

We see that there are one thousand observations for the _b_Q110 variable, one mean for each random sample. The mean of the 1,000-sample means is 3.78, almost the same as the mean for the entire single sample of US respondents (3.77). We can also see the standard deviation (.20), which in this case is the standard error, and range (3.15 to 4.48).

To examine the shape of the sampling distribution of sample means of Q110, we can construct a histogram for the new variable, _b_Q110, by entering the following command into Stata:

```
histogram _b_Q110, percent normal
```

The "histogram" command requests a histogram of the Q110 sample means, the values of which are measured by _b_Q110. The "percent" option specifies that we want the percentage of cases that have each value on the y axis. The "normal" option tells Stata to overlay a normal curve on the histogram, as shown in Figure 7.15.

The histogram of 1,000 sample means, or the sampling distribution, is very close to a normal distribution, as shown by the overlaid normal curve. If we asked Stata to draw 2,000 or more sample means, it would be even closer.

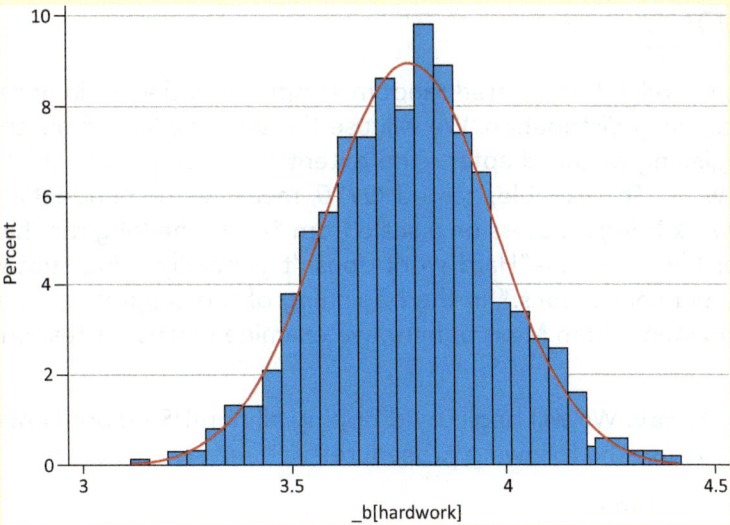

Figure 7.15

Review of Stata Commands

- Keep only observations with a specific value for a variable

  `keep if variable name==value`

- Draw a single random sample from a data file

  `sample # of cases, count` (or) `sample % of cases`

- Generate a frequency distribution for a variable

  `tabulate variable name`

- View number of cases, mean, standard deviation, and minimum and maximum values for a variable

  `summarize variable name/s`

- Draw multiple samples, with replacement, and record sample means to a new data file as values of a new variable

  `bootstrap, reps(# of samples) size(# of cases per sample) saving("location and name of new data file", replace): mean variable name`

- Produce a histogram for an interval-ratio variable

  `histogram variable name`

Using SPSS

In this section, we will take repeated random samples in order to demonstrate the properties of the sampling distribution. We will use the same variable from the WVS that we saw at the beginning of the chapter—the extent to which people think success is the result of hard work. The variable, named *Q110*, measures respondents' opinions about whether hard work brings success on a scale from 1—"In the long run, hard work usually brings a better life"—to 10—"Hard work doesn't generally bring success—it's more a matter of luck and connections." In the beginning of the chapter, we examined sample means for respondents from Mexico; here, we examine means for respondents from the United States.

Open `WVSWave7.sav`. We will begin by dropping all non-US respondents from our data set. To do this, we use the following procedure:

`Data → Select Cases`

This opens the "Select Cases" dialog box (shown in Figure 7.16). We click on the small circle next to "If condition is satisfied." We then click on "If . . . "

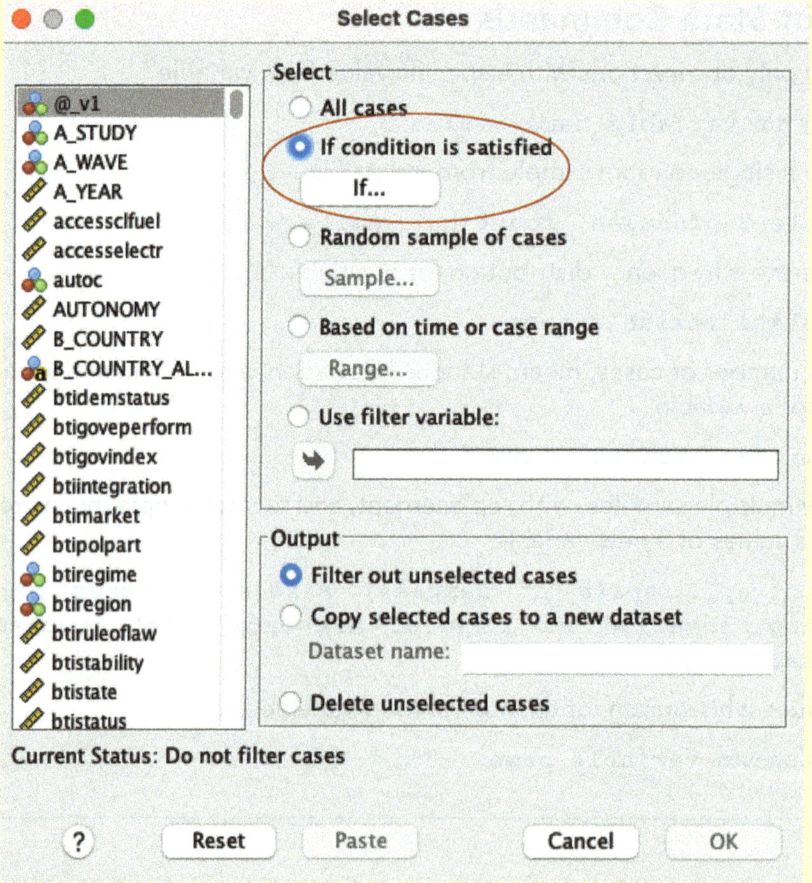

Figure 7.16

This opens the "Select Cases: If" dialog box (shown in Figure 7.17). Here we must specify that we are keeping only those cases that have a value of 840, the code for US respondents, on variable *B_COUNTRY*, which identifies the country of the respondent. We enter "B_COUNTRY = 840" into the empty box, click on "Continue," then "OK."

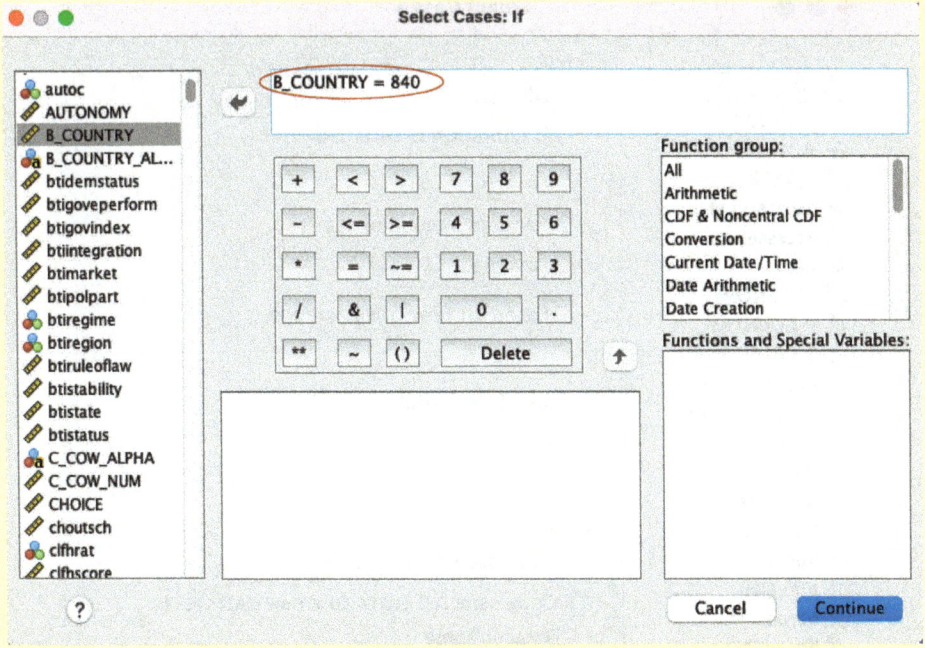

Figure 7.17

This procedure has the effect of keeping for further analysis all respondents whose country is the United States. (If, at some point, you want to restore the data set to include all respondents, go back to the "Select Cases" dialog box and check "Select All Cases.")

If we want to verify that our sample is limited to American respondents, we can ask SPSS to generate a frequency of the variable *B_COUNTRY*, country of respondent. The output is shown in Figure 7.18.

ISO 3166–1 numeric country code				
		Frequency	Percent	Valid Percent	Cumulative Percent
Valid	United States	2596	100.0	100.0	100.0

Figure 7.18

As we can see, the only value for *B_COUNTRY*, country, is the United States.

We will now proceed to take a random sample of 150 respondents, drawn from the 2,596 American respondents. To do that, it is easiest if we instruct SPSS to first create a new dataset with only the 2,596 American respondents. Toward that end, go back to

`Data → Select Cases`

and select "Copy selected cases to a new dataset." This is shown in Figure 7.19, in the output section of the "Select Cases" dialog box (circled in red). Give the dataset with American respondents a name.

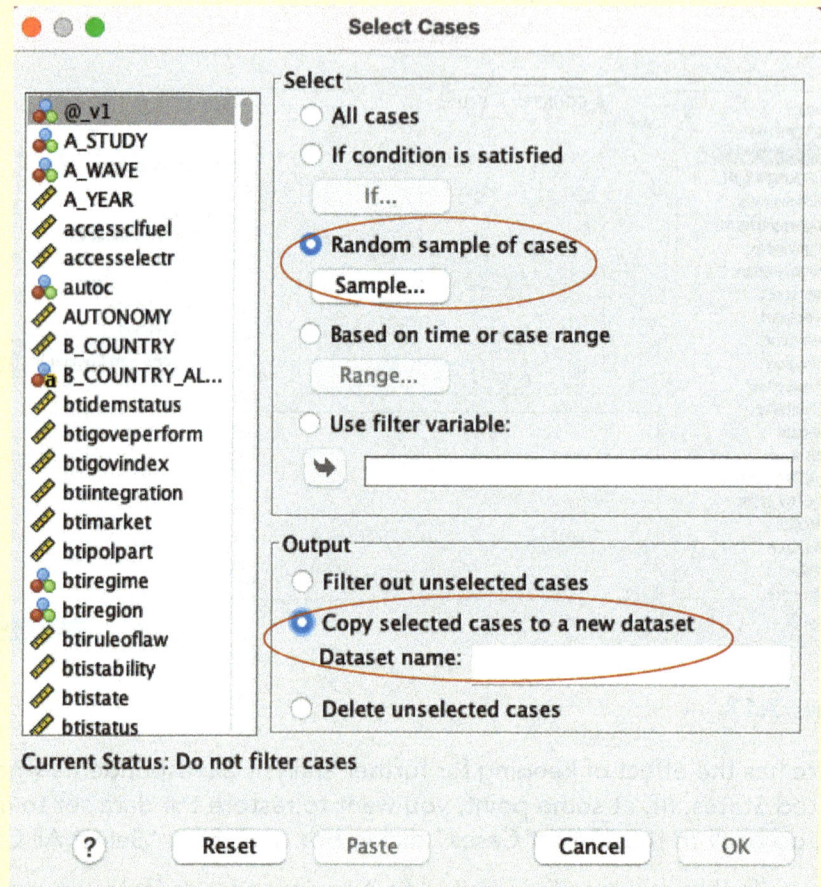

Figure 7.19

Working from this new SPSS dataset that contains only the US respondents, we can now easily draw a random sample.

To do that, go back to:

`Data → Select Cases`

Select the "Random sample of cases" option and click on "Sample" (as shown in Figure 7.19, circled in red).

This opens the "Select Cases: Random Sample" dialog box (Figure 7.20).

For our example, we will tell SPSS that we want it to randomly select 150 respondents out of the 2,596 (American) respondents that make up the full sample. We click on "Continue" and then "OK." We now have a randomly selected sample of 150 Americans.

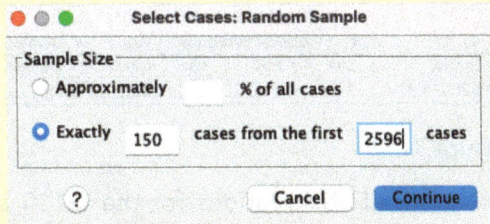

Figure 7.20

Any statistical procedures we employ now will be limited to this sample of 150, until we specify otherwise.

Let's now ask SPSS to calculate a mean score on the *Q110* variable for this sample of 150. We use the following procedure, which we have seen before:

`Analyze → Descriptive Statistics → Descriptives`

Recall that this procedure returns the number of observations, mean, standard deviation, and minimum and maximum values for this variable, as shown in the output in Figure 7.21.

Descriptive Statistics

	N	Minimum	Maximum	Mean	Std. Deviation
Success: hard work vs luck	150	1	10	3.83	2.516
Valid N (listwise)	150				

Figure 7.21

The output shows that *Q110* has only 150 observations, indicating that the sampling procedure proceeded correctly. We can also see that the mean is 3.83. Just as we saw with respondents from Mexico at the beginning of this chapter, if we were to repeatedly draw random samples of 150 from the full sample of US respondents, the mean for *Q110* would differ each time. Try drawing another random sample with 150 cases—what is the sample mean?

In order for the sampling distribution of sample means to take on the shape of a normal distribution, we need to draw a large number of samples of the same size from our population. We drew five hundred samples (manually!), each with 150 respondents, and used SPSS to calculate the mean for each of those five hundred samples. Using the "Descriptives" procedure again, SPSS calculated the mean of those five hundred sample means. The output is shown in Figure 7.22. (Note that we do not expect students to go through this whole process!)

Descriptive Statistics					
	N	Minimum	Maximum	Mean	Std. Deviation
Q110_smpl	500	3.05	4.42	3.7712	.19907
Valid N (listwise)	500				

Figure 7.22

We see that there are five hundred observations for the *Q110_smpl* variable, one mean for each random sample. The mean of the 500 sample means is 3.77, the same as the mean for the entire sample of US respondents (also 3.77).

To examine the shape of the sampling distribution of sample means of *Q110*, we can construct a histogram for the new variable, *Q110_smpl*.

Graphs → Legacy Dialogs → Histograms

We place the desired variable into the "Variable" box and click on "OK." We can also overlay a normal curve on the histogram. To do that, double-click on the histogram that SPSS generated in the output window. This puts the histogram into "Chart Editor" mode. Right-click on the histogram and choose "Show Distribution Curve." The default curve is a normal curve. The output is shown in Figure 7.23.

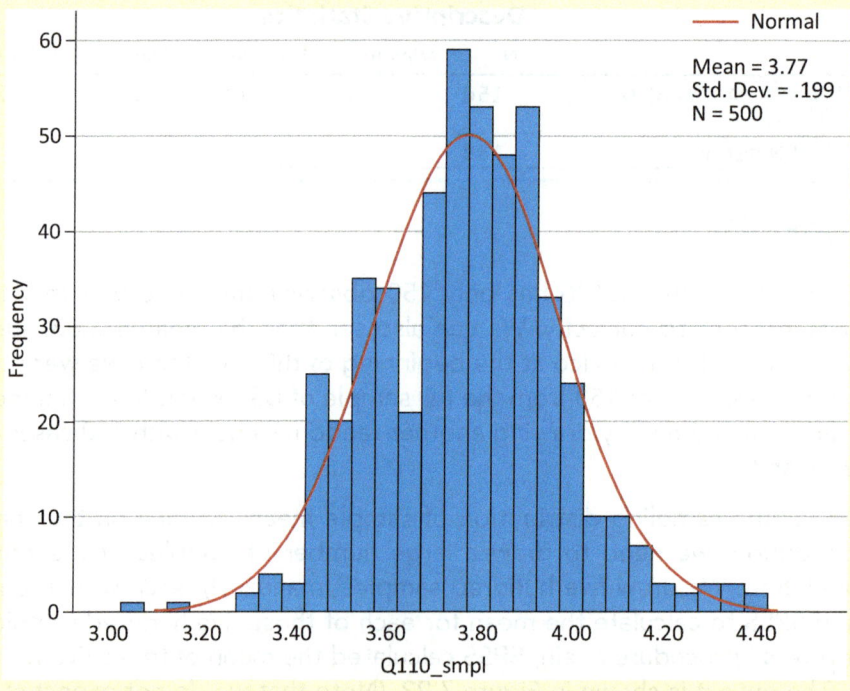

Figure 7.23

The histogram of 500 sample means, or the sampling distribution, is very close to a normal distribution, as shown by the overlaid normal curve. If we used SPSS to draw 1,000 or more sample means, it would be even closer.

Review of SPSS Procedures

- Keep only cases with a specific value for a variable. Write cases to a separate SPSS file

    `Data → Select Cases`

    Click on "If condition is satisfied"

    Click on "Copy selected cases to a new dataset"

- Draw a single random sample from a data file

    `Data → Select Cases`

    Click on "Random sample of cases"

- Generate a frequency distribution for a variable

    `Analyze → Descriptive Statistics → Frequencies`

- View number of cases, mean, standard deviation, and minimum and maximum values for a variable

    `Analyze → Descriptive Statistics → Descriptives`

- Generate a histogram for an interval-ratio variable

    `Graphs → Legacy Dialogs → Histogram`

Practice Problems

1. The average GPA of students graduating from the University of Mars is 3.12, with a standard deviation of 0.47. You draw a random sample of one hundred Martian sociology majors and find a mean GPA of 3.25.

    a. Estimate the standard error for a sampling distribution of mean GPAs.

    b. Calculate the z-score for your sample mean.

    c. What is the probability of selecting a random sample with a mean above 3.25? State your result in words.

    d. Sketch the sampling distribution, including the population mean, standard error, and sample mean. Indicate the part of the normal curve that your obtained probability applies to.

2. On average, all students assigned this textbook rank it at an 8.6 on a scale from 1 (not useful) to 10 (extremely useful), with a standard deviation of 1.2. You draw a random sample of one hundred students assigned this textbook and find that the average ranking in your sample is 8.3.

    a. Estimate the standard error for a sampling distribution of mean rankings.

b. Calculate the z-score for your sample mean.

c. What is the probability of selecting a random sample with a mean below 8.3? State your result in words.

d. Sketch the sampling distribution, including the population mean, standard error, and sample mean. Indicate the part of the normal curve that your obtained probability applies to.

3. Of students at the University of Texas, Austin, 90% are required to be in-state students, by law. You draw a random sample of one hundred students and find that 87% of them are in-state.

   a. Estimate the standard error for a sampling distribution of the percentage of in-state students.

   b. Calculate the z-score for your sample percentage.

   c. What is the probability of selecting a random sample with a percentage of in-state students below 87? State your result in words.

4. Spotify reported 456 million monthly listeners in 2022. Of these, 90,401,583, or 20%, listened to The Weeknd, the artist with the most monthly listeners.* You draw a random sample of two hundred Spotify listeners and find 22% listened to The Weeknd.

   a. Estimate the standard error for a sampling distribution of the percentage of The Weeknd listeners.

   b. Calculate the z-score for your sample percentage.

   c. What is the probability of selecting a random sample with a percentage of The Weeknd listeners above twenty-two? State your result in words.

5. Say that the proportion of American adults who are unaffiliated with any religion is 23%.

   a. What is the probability of selecting a random sample of two hundred people in which at least 30% are religiously unaffiliated?

   b. What is the probability of selecting a random sample of two hundred people in which between 23% and 30% of the sample identifies as religiously unaffiliated?

   c. Imagine that the random sample had only thirty-five cases. What is the probability of selecting a random sample of thirty-five people in which at least 30% are religiously unaffiliated?

   d. Compare the probabilities from parts a and c. Explain why the probabilities are different.

---

* Rohit Shewale. 2023. "Spotify Stats 2023: Subscribers, Revenue, and Other Insights." Demandsage, August 29. https://www.demandsage.com/spotify-stats/.

6. Say that the mean income of Michigan residents is $47,000 a year, with a standard deviation of $17,800.

   a. What is the probability of selecting a random sample of two hundred Michigan residents in which the mean income is $50,000 or less?

   b. What is the probability of selecting a random sample of two hundred Michigan residents with a mean income between $47,000 and $50,000?

   c. Imagine that the random sample had only thirty-five cases. What is the probability of selecting a random sample of thirty-five people with a mean income of at least $50,000?

   d. Compare the probabilities from parts a and c. Explain why the probabilities are different.

7. A polling firm has declared Beyoncé among the top five respected global celebrities. The distribution of her favorability rankings, measured on a scale from 0 to 100, in a random sample of one hundred respondents is shown in Figure 7.24.

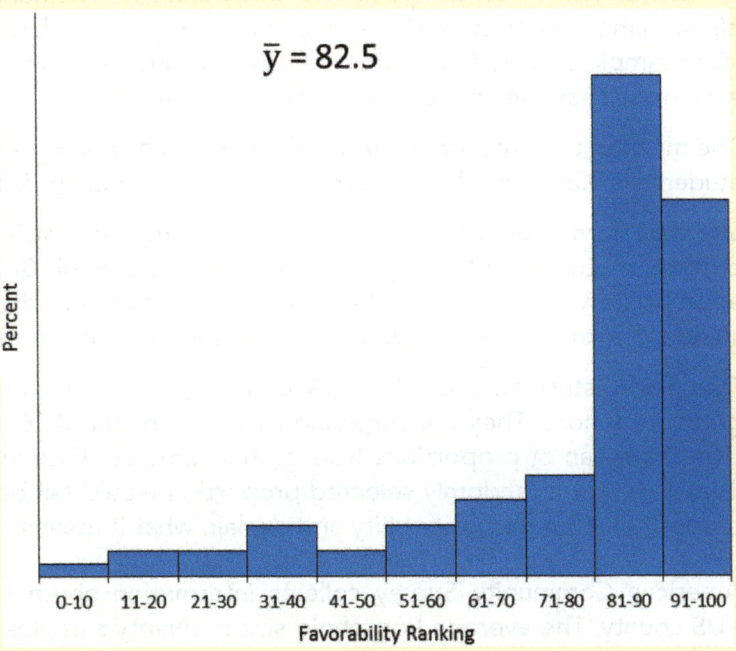

Figure 7.24 Distribution of Favorability Rankings for Beyoncé

   a. Beyoncé's mean favorability ranking in the sample is 82.5. Imagine that the polling firm drew many repeated random samples and measured Beyoncé's favorability ranking in each. They want to know if, by random chance, their original sample included too many diehard Beyoncé fans. Assume that the mean of the mean favorability rankings from repeated samples is 85, with a standard deviation of 1.8. What is the standard error of the sampling distribution? Explain what the standard error means in words.

b. Use the information from part a to sketch the sampling distribution for Beyoncé's favorability ratings if the polling firm were to take many repeated samples. Be sure to label the appropriate values at the center of the distribution and the first and second standard errors. Show how you found the values for the standard errors.

c. Describe the difference between the appearance of the distribution of values from the sample of one hundred respondents shown in Figure 7.24 and the sampling distribution that you drew for part b. Why do we observe the difference?

d. What is the probability that a mean greater than the sample mean of 82.5 would be randomly drawn from all of the random samples conducted by the polling firm?

8. The students in a large statistics course are completing an extra credit project that seeks to demonstrate how sampling distributions work. One hundred students are participating in the project, and they break up into pairs, with each pair surveying a random sample of forty fellow students. Each group must find the proportion of students in their sample who plan to go to graduate school.

a. The mean of the proportions from all fifty samples conducted by the statistics students is 0.25. Find the standard error of the sampling distribution.

b. Use the information from part a to draw the sampling distribution for the proportion of students planning to go to graduate school. Be sure to label the appropriate values at the center of the distribution and the first and second standard errors. Show how you found the values for the standard errors.

c. One pair of students finds that 16% of students in their sample plan to go to graduate school. They are surprised that their result, 0.16, is so much lower than the mean of proportions from all the samples. They want to know how likely it is that a randomly selected proportion would fall below their sample proportion. Find this probability and explain what it means.

9. The American Community Survey collects information about the population in every US county. The average household size in twenty samples of fifty randomly selected counties are shown in Table 7.2.

a. An incredulous observer doubts that these county means are correct. After all, he knows of many households that have more than five people, including his own, which has eight people. Explain to him how it is possible that his own observations of household size can be so different from the county means shown in Table 7.2.

b. If you arranged all of the sample means from Table 7.2 in a distribution, what kind of distribution would it be?

Table 7.2 Mean Household Size in Twenty Random Samples of Counties, N = 50

Sample	Sample Mean
1	2.56
2	2.68
3	2.67
4	2.48
5	2.64
6	2.77
7	2.44
8	2.49
9	2.62
10	2.63
11	2.69
12	2.71
13	2.59
14	2.63
15	2.64
16	2.48
17	2.40
18	2.55
19	2.53
20	2.39

c. Because we have data for all counties in the United States, we know that the mean household size for all counties is 2.53. What proportion of the twenty samples in Table 7.2 fall below the population mean? What proportion of sample means fall at or above the population mean?

d. Based on what you know about normal curves, what proportion of sample means would you have expected to fall above the population mean? What proportion would you have expected to fall below it? Are the proportions in part c different from what you would expect to find? Explain why or why not.

10. The parent teacher organizations (PTOs) at all ten high schools in a large school district decided to sponsor a study of students' social media use. Random samples of students were drawn from each school. Table 7.3 shows the proportion of students who have ever taken a complete break from social media for longer than one month for the ten high schools.

   a. Assume that social media behavior across these schools is actually very similar. Explain how knowledge of sampling offers an understanding of why we still

Table 7.3 Sample Proportions of Students Who Have Taken Social Media Breaks, by High School

High School	Proportion Who Have Taken a Social Media Break
1	0.10
2	0.12
3	0.13
4	0.09
5	0.08
6	0.12
7	0.10
8	0.15
9	0.08
10	0.07
	Mean = ?

observe differences across schools in the percentages of students who have taken a break from social media.

b. What is the mean of proportions for the ten high schools?

c. Is the mean that you calculated for part b equal to the proportion of all students in the district's high schools who have taken a social media break?

11. For each part below, tell whether the data are population parameters or sample statistics, specify whether the information of interest is a mean or proportion, and identify the symbol that should be used to denote it.

a. A study of seventy-five employed adults in Portland, Oregon, finds that seven of them bike to work.

b. In the 2010 US Census, 2.7% of the population chose to identify themselves with at least two racial categories.

c. In 2016, 2,229,872 people were interviewed for the American Community Survey to produce estimates of the US population. Mean household income for the group was $77,866.

d. Sociology faculty at California State University, Fullerton, wanted to know if students in their major felt that their sociology coursework had improved their understanding of the world. A survey of all sociology majors at the school found that average satisfaction with the major was 8.3 on a scale from 0 to 10.

12. For each part below, tell whether the data are population parameters or sample statistics, specify whether the information of interest is a mean or proportion, and identify the symbol that should be used to denote it.

   a. An organization advocating for prison reform sent an anonymous survey to the warden of every private prison in the United States. Among the 33% of wardens who returned the survey, 67% said that the prison system in the United States needs some kind of reform.

   b. The mean high school GPA for students admitted to the University of Wisconsin, Green Bay, is 3.16.

   c. Of respondents, 24.16% on the GSS reported they had never been married.

   d. Of students attending the University of California, Riverside, about 77% graduate within six years.

13. Figure 7.25 shows two pairs of distributions, with each pair drawn from the same population of means.

   - In Pair A, the distribution on the left has 25 sample means (N = 100 for each sample), and the one on the right has 100 sample means (N = 100 for each sample).
   - In Pair B, the distribution on the left has 200 sample means (N = 40 for each sample), and the one on the right has 200 sample means (N = 100 for each sample).

   Which set is Pair A, and which is Pair B? Explain how you know.

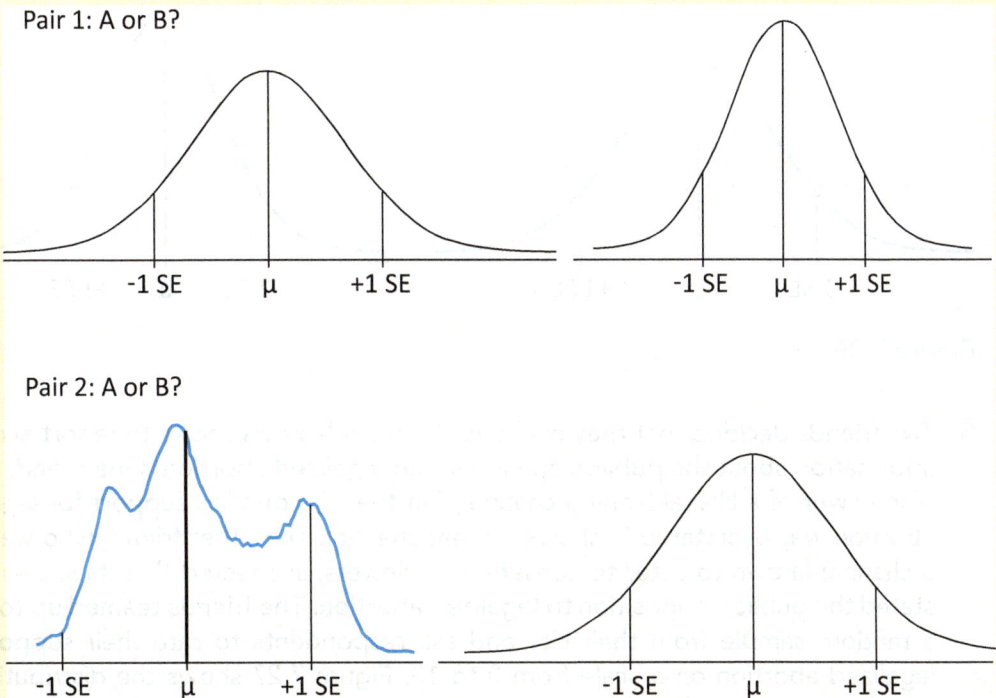

Figure 7.25 Two Pairs of Distributions

14. Figure 7.26 shows two pairs of distributions, with each pair drawn from the same population of means.
    - In Pair A, the distribution on the left has 20 sample means (N = 50), and the one on the right has 200 sample means (N = 50).
    - In Pair B, the distribution on the left has 200 sample means (N = 40), and the one on the right has 200 sample means (N = 100).

    Which set is Pair A, and which is Pair B? Explain how you know.

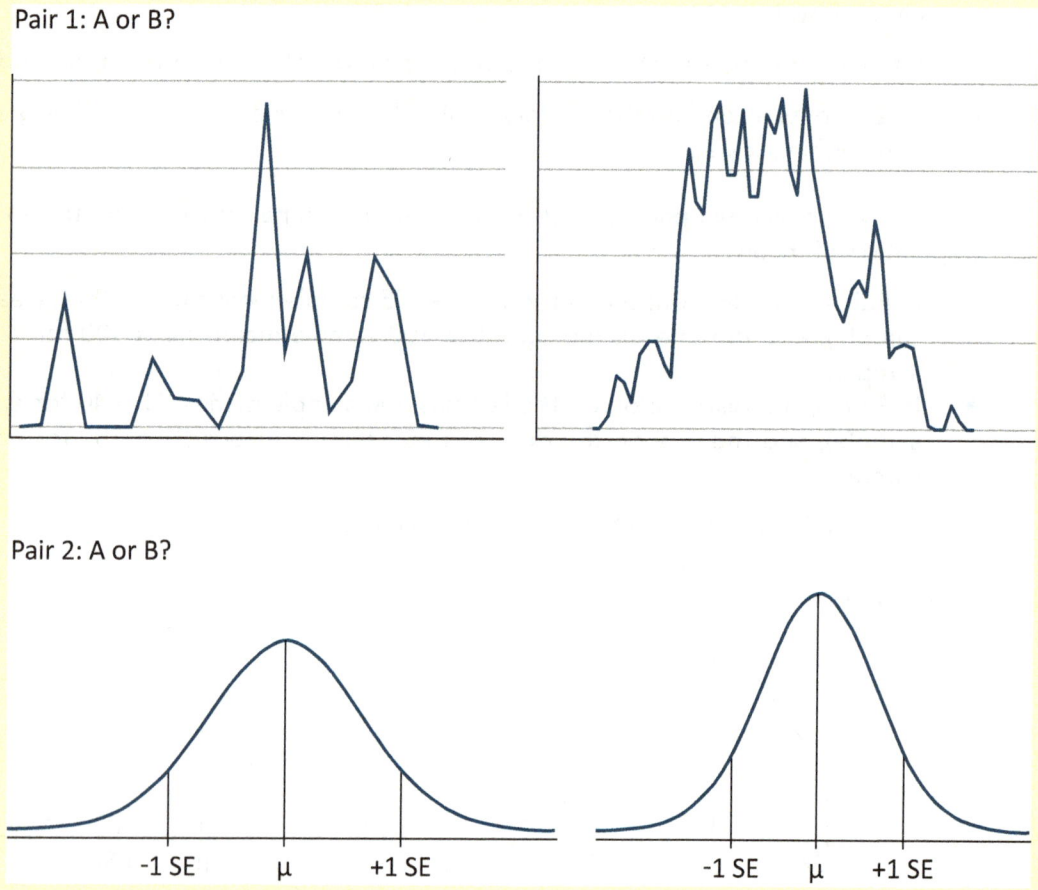

Figure 7.26

15. Two friends decided that they could not trust cable news shows to report accurate information about the public's opinion about legalized abortion. One friend, a regular viewer of a liberal-leaning channel, felt that the public's support for legalized abortion was overstated in shows on that channel. The other friend, who watched a channel known to cater to conservative viewers, suspected that its shows overstated the public's opposition to legalized abortion. The friends teamed up to draw a random sample from their city and ask respondents to rate their support for legalized abortion on a scale from 0 to 10. Figure 7.27 shows the distribution of responses.

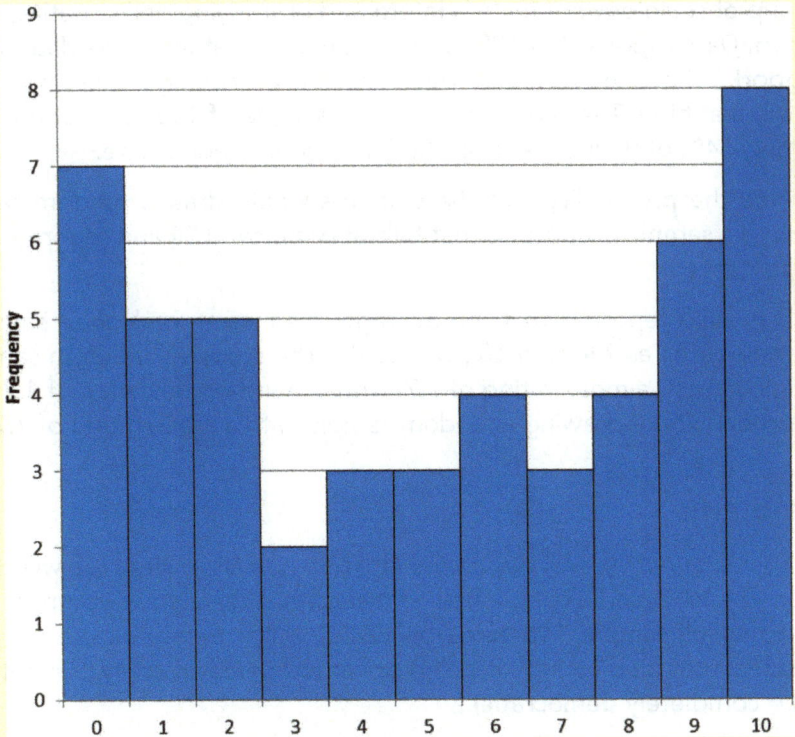

Figure 7.27 Support for Abortion, Random Sample

a. How many respondents were in the sample, according to Figure 7.27?

b. Use Figure 7.27 to find the sample mean.

c. If overall mean support for legalized abortion in the friends' city is 4.95, and the population standard deviation is 0.7, find the probability that the friends would draw another random sample of the same size that falls at or above the original sample mean that you calculated for part b.

d. What proportion of the sample collected by the friends strongly supports legalized abortion (9 or 10) *or* strongly opposes it (0 or 1)?

e. If the actual percentage of people in the friends' city who hold strong opinions about legalized abortion is 51.5%, find the probability that the friends would draw another random sample of the same size that falls above the original sample proportion that you calculated for part d.

f. Which is more unlikely: drawing a random sample with a mean larger than the original sample mean or drawing a random sample with a proportion larger than the original sample proportion? Give one reason that one is more likely than the other.

g. Do the data support the friends' suspicion that the news stations they watch overstate support for or opposition to legalized abortion?

16. A group of sociology students is interested in studying online dating. A major dating app, Datr, reports that 63% of their users have at least one date that they rank as "good." The students are skeptical that the percentage of users who have good dates is that high. They conduct a random sample of 100 users of the app and find that only 54% of their sample has had at least one good date.

    a. Find the probability that the students would draw a random sample of 100 with a sample proportion that falls at or below 0.54 and report your result in a sentence.

    b. Datr also reported that, on average, their users rank their satisfaction with their dates as 7.9 on a 10-point scale. The students' random sample yielded, in contrast, a mean rating of 7.7, with a standard deviation of 1.7. What is the probability of drawing a random sample with a mean rating of 7.7 or lower?

Stata Problems

The World Values Survey (WVS) has a very large sample size. Here we will draw random samples from the WVS and examine how statistics for the samples compare to the same statistics for the full sample. We focus on *Q251*, which measures the extent to which respondents feel that their country is being governed democratically (1 = not at all democratic to 10 = completely democratic).

1. Open the `WVSWave7.dta` and use the "summarize" command to generate summary statistics for *Q251* in the full sample.

2. What are the mean and standard deviation for *Q251*?

3. Use the "sample" command to draw a single random sample of one hundred cases from the full sample.

4. What are the mean and standard deviation for *Q251* in this smaller sample? Compare them to those statistics for the full sample.

5. Are there 100 values for *Q251* in the sample of one hundred? If no, explain why.

6. Now close this smaller sample WVS data file and reopen the original WVS data file. Use the "bootstrap" command to generate one hundred samples, each with one hundred cases. In the same command, tell Stata to record the mean of *Q251* for each of the 100 samples and save them in a new data file called "`WVSsimulationCh7.dta`." Open the new data file.

7. Use the "summarize" command to generate the mean and standard error of the 100 sample means.

8. Use the "histogram" command to generate a histogram of the means for *Q251*. Does it look like a normal distribution?

SPSS Problems

The World Values Survey (WVS) has a very large sample size. Here we will draw random samples from the WVS and examine how statistics for the samples compare to the same statistics for the full sample. We focus on *Q251*, which measures the extent to which respondents feel that their country is being governed democratically (1 = not at all democratic to 10 = completely democratic).

1. Open the `WVSWave7.sav` and use the "Descriptives" procedure to generate summary statistics for *Q251* in the full sample.
2. What are the mean and standard deviation for *Q251*?
3. Use the "Select Cases" procedure to draw a single random sample of one hundred cases from the full sample.
4. What are the mean and standard deviation for *Q251* in this smaller sample? Compare them to those statistics for the full sample.
5. Are there 100 values for *Q251* in the sample of one hundred? If no, explain why.
6. Reopen the original WVS data file and use the "Select Cases" procedure to draw a second single random sample of one hundred cases from the full sample.
7. Compare the mean for this random sample to the first random sample. Are the means different? If so, why?

Note

[1] Earl Hunt. 2011. *Human Intelligence.* Cambridge: Cambridge University Press. Technically, 100 is the median IQ.

Chapter 8
Estimating Population Parameters
Confidence Intervals

The movement to legalize marijuana in the United States continues to gain momentum, with more than half of all states having enacted legislation that either decriminalizes or legalizes the possession of small amounts of marijuana.[1] Have Americans' attitudes about the legalization of marijuana kept pace with these changes in the legal landscape? In this chapter, we introduce you to **confidence intervals**, a method for using sample results to estimate a range of values within which a population value probably falls.

A team of researchers—Jacob Felson, Amy Adamczyk, and Christopher Thomas—addressed this question in a 2019 study examining changes in public opinion about marijuana legalization.[2] Felson and his colleagues traced changes in Americans' opinions about the legalization of marijuana using representative samples of US residents from the General Social Survey (GSS) and Gallup polls between 1969 and 2016. Consistent with increased legal lenience, the proportion of respondents who support the legalization of marijuana has increased over time. In 1969, a small minority of Americans supported legalization (around 13%), but by 2016, the majority of Americans (around 61%) were in favor of it.

There is little doubt that public support for the legalization of marijuana really did increase over those four decades. But notice how the support cited for 1969 and 2016 says "around" 13% and "around" 61%. This is because these are the proportions of *sample* respondents who supported the legalization of marijuana, not the proportion of the *population* of Americans who support such legislation.

As we saw in chapter 7, a single sample will yield a statistic that, in all likelihood, is neither exactly the same as the value for the full population nor the same as the statistics that additional samples drawn from the same population would yield. To account for the uncertainty produced by sampling variation, Felson and his colleagues reported a "margin of error" for each year's proportion of support for the legalization of marijuana. Each margin of error is subtracted from and added to the proportion of the sample that supported legalization, yielding a range. In other words, the researchers estimated the rates of support for the legalization of marijuana as ranges, which account for the uncertainty produced by sampling variability. For example, they estimated the support for legalization in 2016 as the interval ranging from 59% to 63%.

With these margins of error placed around each year's rate of legalization support, we can more confidently evaluate how rates have changed over time, especially from year to year, when the magnitude of changes in support are not as large as they tend to be over larger spans of time. The proportion of the sample that supported marijuana legalization changed from year to year, sometimes increasing by a lot, as in recent years, and sometimes increasing more modestly, as in the 1990s. But when we focus on the range of support rather than the single sample proportion, we see that sometimes these estimated ranges overlap, even when the sample proportions differ.

For example, from 1998 to 2000, support for legalization increased from 30% to 33%. The sample proportions alone reflect an increase. But when we take into account the margins of error for each proportion, we see that the 1998 proportion actually falls within the margin of error for the level of support in 2000. This casts doubt on the claim that public support in the population really increased between 1998 and 2000. However, when we examine the increase in support from 2011 (48%) to 2014 (56%), we see that the proportion of respondents who supported legalization in 2011 falls outside the margin of error for 2014, suggesting that true population support for the legalization of marijuana really did increase from 2011 to 2014.

These ranges estimating support for the legalization of marijuana in the United States are known as confidence intervals, the focus of this chapter. Confidence intervals are the range of values that result from the specific value found in a sample plus and minus the margin of error. Using confidence intervals to estimate population support—rather than relying on the single value found in a sample—accounts for the sampling variation that occurs in any single sample from a population of interest.

In the rest of the chapter, we will explain the underlying logic of confidence intervals, show how to estimate them, and explore applications for confidence intervals in social sciences.

Inferential Statistics and the Estimation of Population Parameters

Chapters 2 through 5 of this book examined various ways to describe variables, and relationships among them, for data that we actually have. But, as we saw in chapter 7, many fundamental questions of the social sciences concern populations for which we lack complete data. What proportion of people have ever experienced workplace discrimination based on gender or race? What is the average level of job satisfaction in the population? We simply lack complete population data to answer those questions directly. Consider this: If you have ever held a job, has a researcher ever asked you whether you have experienced workplace discrimination or how satisfied you have been with your job? Most readers probably have never reported their responses to those questions to a researcher. For those two topics and countless more, social scientists lack population data.

As chapter 7 made clear, using a single sample as a substitute for population data introduces a particular kind of error into our estimates. The source of this error is sampling variation. In chapter 7, we saw that the mean response to whether hard work brings a better life in twenty repeated random samples of equal size from the World Values Survey yielded different sample means every time. This makes clear that, no matter how well a researcher constructs a sample, a statistic for any one sample is unlikely to be equal to the population parameter. (Remember, a population parameter is a piece of information [such as a mean or proportion] about a population of interest that is often unknown.)

When sample statistics are used to estimate population parameters, we refer to them as **point estimates**, which conveys the fact that a single sample statistic is only ever an estimate of the population parameter. The rest of this book focuses on inferential statistics, which use probability theory and theoretical sampling distributions (such as the normal curve) to assess the extent to which statistics from a single sample approximate the corresponding unknown population parameters.

The question addressed by inferential statistics is: How can we use information from a single sample to generalize to the larger population?

Figure 8.1 presents a visual representation of how we use inferential statistics to generalize what we know about a single sample to a population of interest.

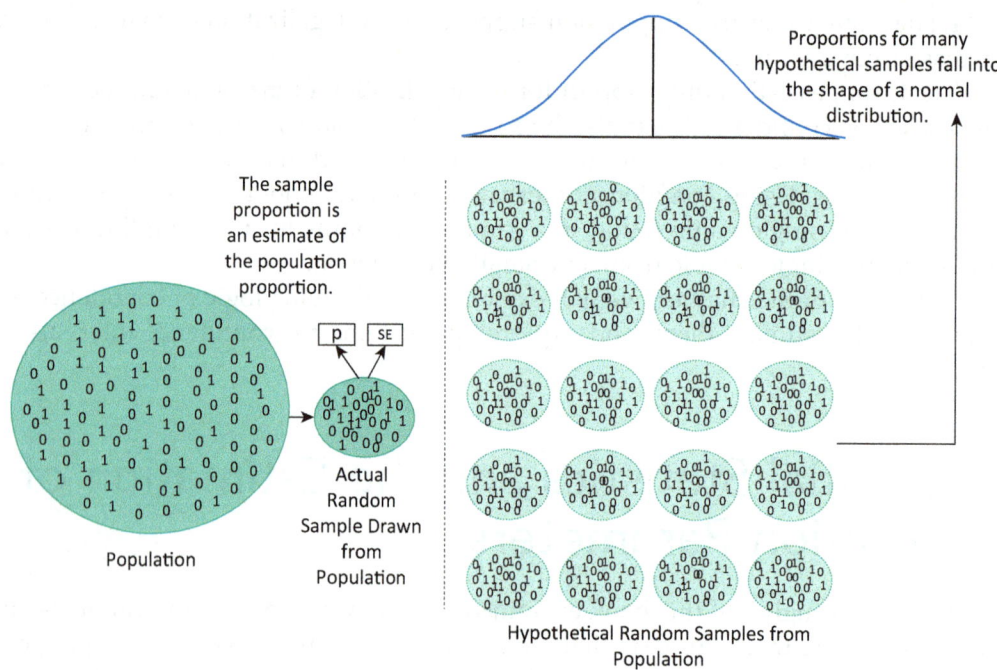

Figure 8.1 A Visual Representation of How Inferential Statistics Works: Generalizing from a Single Sample to a Population

The left-hand side of Figure 8.1 shows a population of values for a variable with two possible values, 1s and 0s, and the single random sample drawn from that population. (The two values for this variable could have been represented by any label—e.g., yes/no, high/low, etc.) The sample proportion and standard error are calculated based on the single sample. The single sample proportion is just one among many, many theoretical random samples that could be drawn from the population. The light blue samples shown on the right-hand side of the figure represent some of these theoretical samples. The proportions from these sample statistics take on the shape of a normal distribution as we draw a very large number of samples, as shown by the normal curve situated above the theoretical samples.

Figure 8.1 reminds us that our estimate of a sample proportion's proximity to the true population proportion depends on what we know about the theoretical distribution of many, many sample proportions—represented by the light blue circles. We use the standard error to estimate the shape of the sampling distribution—flat and widely dispersed or tall and narrowly dispersed—which tells us how much, on average, our sample statistic varies from the population parameter.

Confidence Intervals Manage Uncertainty through Margins of Error

Felson and his colleagues employed a statistical tool called estimation when they calculated confidence intervals for their point estimates of public support for the legalization of marijuana. Instead of the single value provided by a point estimate, the confidence interval specifies a range of values within which the population parameter may lie. Confidence intervals give the range of values within which there is a high probability (e.g., 95%) that the estimated interval includes the population parameter (e.g., a mean or proportion). The **confidence level** is the probability that the range of values includes the population parameter.

Estimation of these ranges relies on the logic of sampling distributions, which tell us the probability of obtaining results at specified distances from the population parameter. (Review chapter 7, pages 275–278, if you want a refresher.) A confidence level of 95% means that if we were to take repeated samples, 95% of those samples would yield a confidence interval where the range of values included the population parameter. That is, at a confidence level of 95%, there would be a .95 probability that the confidence interval for a given sample includes the population parameter.

We know that a point estimate is likely to vary from the population parameter, but how far away from the point estimate might the population parameter lie? Confidence intervals provide a margin of error around the point estimate of the population parameter. For example, in the legalization of marijuana study, the researchers found that 61% of respondents supported the legalization of marijuana in 2016. The margin of error for this point estimate was 2%. They calculated the confidence interval by subtracting the margin of error from, and adding it to, the point estimate of 61%. Thus, the confidence interval for supporting marijuana legalization in 2016 was 59% to 63%.

Certainty and Precision of Confidence Intervals

Even though the confidence interval estimates a population parameter as a range of values instead of a single value, we still cannot be completely sure that the range actually includes the population parameter. Therefore, a confidence interval also gives the probability that it is correct. The uncertainty that is an inherent part of sampling is thus built into estimating confidence intervals.

The confidence interval for support for the legalization of marijuana in 2016 (59% to 63%) is specifically a 95% confidence interval. The "95%" means that under repeated sampling, 95% of the confidence intervals for marijuana support would contain the true rate of support for the population. This means that there is a 95% chance that the confidence interval of 59% to 63% contains the true proportion of population support. The probability attached to a confidence interval, such as 95%, refers to the **certainty** that the estimation is correct (that is, that it contains the population parameter). The higher the certainty of a confidence interval, the greater the likelihood that the interval includes the population parameter. We can never know with total certainty whether a single confidence interval contains the population parameter, but we can estimate the probability of its accuracy. As we will see below, the researcher chooses the level of accuracy for the confidence interval, commonly 90%, 95%, or 99%.

Whereas certainty refers to the likelihood of confidence intervals including the true population parameter, the **precision** of a confidence interval refers to the range of values that it covers, or its width. The wider the interval, the less precise is the estimation. There is a trade-off between certainty and precision. A wider confidence interval is more likely to contain the population parameter. At the same time, widening the interval decreases the precision of the estimate.

Should you compromise precision of a confidence interval in order to increase the chances that it will actually include the population parameter? It depends on the goal. We will discuss this later in this chapter. In the sections that follow, we focus on the calculation of confidence intervals, which is quite straightforward. First, we cover the steps for calculating a confidence interval for a proportion. After that, we cover how to calculate a confidence interval for a mean. The logic and basic procedure are the same in both cases.

Confidence Intervals for Proportions

To calculate a confidence interval for a proportion, we need two pieces of information: the sample proportion and the sample size. Our work here relies on the characteristics of the sampling distribution: the sample proportions we would obtain if we drew repeated random samples would be normally distributed around the true population parameter. According to the Empirical Rule, there is a known probability that sample proportions fall between the population parameter and any given z-score (e.g., about 68% of all sample proportions will fall between the population parameter and $z = \pm 1$). We will find the z-score associated with the probability of our desired confidence level (e.g., 95%).

In order to calculate the margin of error, we must first calculate the standard error. We then use the z-score, the standard error, and the characteristics of the normal curve to calculate the margin of error and the confidence interval.

The calculation proceeds in the following steps:

Step 1: Calculate the standard error.

Recall from chapter 7 that the standard error is the standard deviation of the sampling distribution. The standard error is the average difference between a sample statistic (e.g., a sample proportion) and the population parameter (e.g., the true population proportion). When we construct a confidence interval for a proportion, we calculate the standard error of the sampling distribution, SE_p, on the basis of the sample statistic, p, and the sample size, N:

$$SE_p = \sqrt{\frac{p(1-p)}{N}}$$

As we saw in chapter 7, the standard error provides an estimate of how dispersed the sample statistics would be from the population parameter. While we cannot know whether our sample statistic is especially close to or distant from the population parameter, we use the standard error to establish the characteristics of the sampling distribution.

Step 2: Select a confidence level and identify the associated z-score.

As noted above, the confidence level of a confidence interval is the proportion of confidence intervals, under repeated sampling, that will contain the population parameter. The confidence level is equal to 1 − α, where the Greek letter α ("alpha") is the proportion of confidence intervals that will not contain the population parameter. Thus, alpha is equal to one minus the confidence level. For example, if the confidence level is 95%, alpha is 5%. We can think of alpha as the level of tolerable risk because it represents the area of the curve in which a population parameter that is not covered by the confidence interval will fall. The area of alpha is distributed equally across the two tails of the distribution.

The 95% confidence level is commonly used, but any level of confidence can be chosen. In choosing a confidence level, the analyst must decide what level of risk is appropriate given the topic at hand. If there are grave consequences of generating a confidence interval that misses the population parameter, then the analyst should opt for a lower level of risk.

Figure 8.2 shows the area of the confidence level (1 − α) and the risk that a confidence interval will not include the population parameter (α), for a 95% confidence interval on a normal curve. We use the normal curve because, as we saw in chapter 7, the distribution of repeated sample proportions is approximately normal with sufficient sample size, per the Central Limit Theorem. Remember: the sampling distribution is a theoretical distribution of repeated sample proportions that forms a normal curve, centered on the true population parameter, here a proportion, or the Greek letter pi (π).

To calculate the confidence interval, we need to find the z-score that corresponds with 2.5% of the curve's area falling into each tail. Figure 8.3 shows an excerpt of the Normal Table. (The table is printed in full in appendix A.) We can find the z-score for which the right-tail probability is equal to .025. That z-score is 1.96. Look at Figure 8.2. The upper and lower limits of the confidence interval are at z-scores of 1.96 and −1.96.

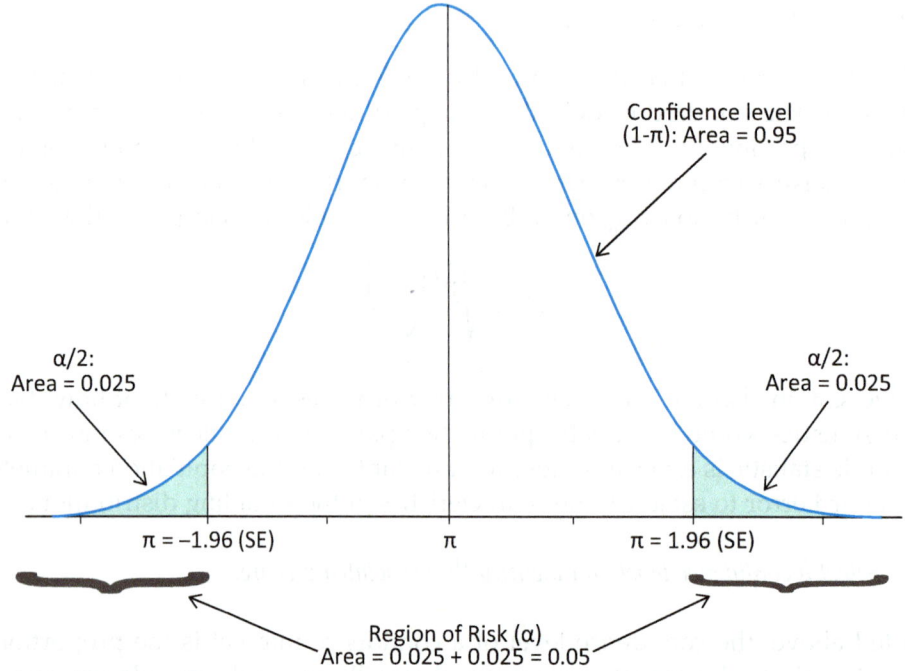

Figure 8.2 Using the Normal Curve to Set up a 95% Confidence Interval for Proportions

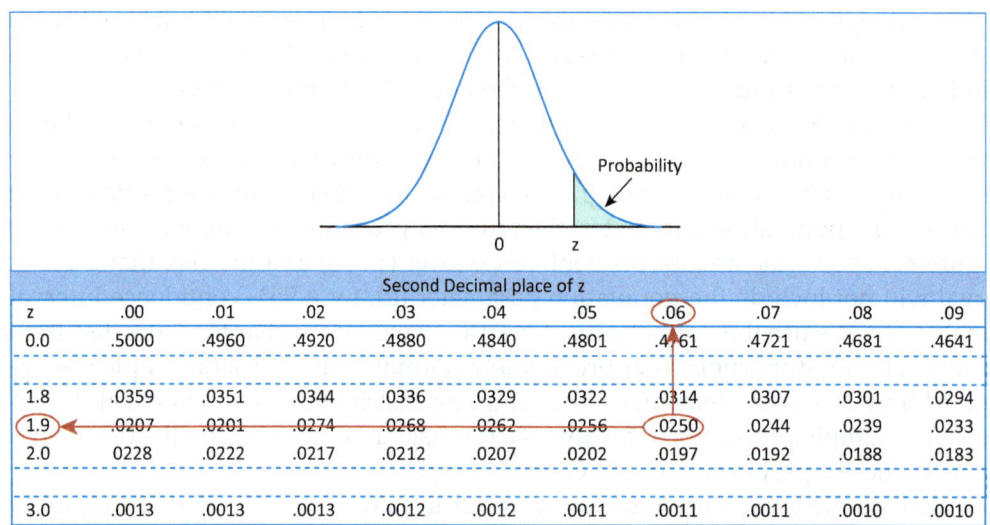

Figure 8.3 Finding the z-Score for a Right-Tail Probability of .025 in the Normal Table

Step 3: Calculate the margin of error.

The next step in calculating the confidence interval is to calculate the margin of error. This step is what allows us to create an interval estimate of the population parameter that covers a range of values. The margin of error is the product of the z-score and the standard error, expressed in the units of the variable (percentage points in the case of proportions). By formula, the margin of error is:

$$\text{Margin of error} = z(SE_p)$$

Step 4: Calculate the lower and upper bounds of the confidence interval.

In this step, we subtract the margin of error from, and add it to, the sample proportion to produce the range of values covered by the confidence interval:

$$CI = p \pm \text{margin of error}$$

Subtracting the margin of error from the sample proportion forms the lower bound of the confidence interval, and adding it forms its upper bound. We typically present the confidence interval with the lower and upper bound enclosed in parentheses, like this: (LB, UB).

Figure 8.2 shows how the product of z and standard error determine where the upper and lower boundaries of the confidence interval fall.

We use this same set of steps to construct confidence intervals for proportions at any level of confidence. The only difference is that we need to find the appropriate z-score in the Normal Table for the corresponding level of confidence. For example, to find the z-score for a 99% confidence interval, we know that 99% of the curve's area should fall between the z-scores. If we subtract 0.99 from 1, we will see how much of the curve's area falls outside of those z-scores. Since 1 − 0.99 = 0.01, we know that 1% of the curve's area lies outside of the range covered by the z-scores. Half of that "region of risk" falls in either tail of the distribution. Dividing the remaining area (0.01) by 2 tells us that the right-hand probability is 0.005. We can find 0.005 in the body of the Normal Table to identify the corresponding z-score, 2.58. Look at the Normal Table in appendix A to see how we found that z-score.

BOX 8.1: APPLICATION

Commonly Used Levels of Confidence and Their Associated z-Scores

Table 8.1 contains z-scores associated with some of the more commonly used confidence intervals. The z-score is associated with the right-tail probability of α/2, where α is equal to 1 minus the confidence level expressed as a proportion.

Table 8.1 Confidence Levels and Their Associated z-Scores

Confidence Level	z-Score
90%	1.65
95%	1.96
98%	2.33
99%	2.58
99.9%	3.29

Next, we use two examples—the extent to which people's neighborhoods have been affected by extreme weather events in the last year and the study of a hypothetical drug's effectiveness at reducing relapse rates among former opioid users—to practice calculating confidence intervals for proportions.

Constructing a Confidence Interval for Proportions: Examples

The first example uses data from the 2021 GSS. Respondents were asked to what extent in the last year an extreme weather event (such as severe storms, droughts, floods, etc.) had affected their neighborhood. Response categories range from "not at all" to "a very great extent." For this example, we will focus on the proportion of people who say their neighborhood was affected by extreme weather events to either "a great extent" or "a very great extent." Table 8.2 presents the frequency distribution for the variable.

Table 8.2 Frequency Distribution for Neighborhood Affected by Extreme Weather Event

	Frequency	Percent
To a great or very great extent	310	17.7
To a lesser extent or not at all	1,446	82.3
Total	1,756	100

Source: 2021 General Social Survey.

We can use Table 8.2 to determine the sample proportion, p, for this example. We see that 17.7% of the sample reported that their neighborhood was affected by an extreme weather event in the last year, which makes the sample proportion 0.177. We see that the number of cases is 1,756. Thus, we have the information we need to construct a confidence interval around our sample proportion of 0.177.

After finding the sample proportion, we follow the same four steps as outlined in the previous section.

Step 1: Calculate the standard error.

$$SE_p = \sqrt{\frac{p(1-p)}{N}} = SE_p = \sqrt{\frac{0.177(1-0.177)}{1756}} = 0.009$$

Step 2: Select a confidence level and identify the associated z-score.

We will calculate a 98% confidence interval. This means that we should find the z-score for which the right-tail probability is equal to $\alpha/2$, where $\alpha = 1 - 0.98$. Since $0.02/2 = 0.01$, we find the probability closest to, but not greater than, 0.01 in the body of the Normal Table and find the associated z-score. The z-score at which 1% of the normal curve's area lies to the right is 2.33.

Step 3: Calculate the margin of error.

$$\text{Margin of error} = z(SE_p) = 2.33(0.009) = 0.021$$

Step 4: Calculate the lower and upper bounds of the confidence interval.

$$CI = p \pm \text{margin of error} \rightarrow 0.177 \pm 0.021 \rightarrow (0.156, 0.198)$$

The 98% confidence interval for the proportion of the population that lives in a neighborhood affected to a great extent by an extreme weather event in the last year is (0.156, 0.198). What does this mean? The distribution of sample proportions is approximately normal, as we saw in chapter 7, so over repeated sampling, 98% of the confidence intervals would include the population parameter. This means that there is a 98% chance that the population proportion falls within the current interval: 15.6% to 19.8%.

According to our estimate, there is a very high chance that between 15.6% and 19.8% of Americans feel that their neighborhoods have been greatly affected by extreme weather events in the last year. While this variable does not directly measure the occurrence of extreme weather events, it is an indicator of the extent to which people in the United States perceive that they are affected by extreme weather. The confidence interval shows that the percentage of US residents affected in this way approaches one-fifth of the population, consistent with reports of increased incidences of extreme weather events around the globe.

In the next example, imagine that researchers are testing the effectiveness of a new drug meant to decrease the withdrawal symptoms and relapse rates for people trying to quit opioids. The researchers randomly assign ninety participants at an inpatient drug treatment program to either a treatment or placebo group—forty participants to the treatment group and fifty to the placebo group. Participants in the treatment group will receive the new drug, and those in the placebo group will receive a placebo. The researchers want to know whether the drug reduces the relapse rate in the month after patients finish the treatment program, a critical period in drug rehabilitation. At the end of the thirty-day treatment program, the researchers continue to follow the two groups for thirty more days. Table 8.3 presents a cross-tabulation of the study participants' status as members of the control or placebo groups and whether they relapsed within the month following the treatment program.

Table 8.3 Relapse Rates for Treatment and Placebo Groups, Thirty Days after Treatment

	Treatment (N = 40)	Placebo (N = 50)
Relapse	70%	76%
No relapse	30%	24%
Total	100%	100%

As Table 8.3 shows, after thirty days, 76% of patients in the placebo group had relapsed compared to 70% of patients taking the new drug. The researchers want to know whether the different relapse rates for the treatment and placebo groups are the result of sampling variability or whether they reflect a real effect of the treatment. To learn more, the researchers construct 95% confidence intervals around the point estimates of the relapse rates among the two groups. If the confidence intervals do not overlap, this provides stronger evidence that the rates of relapse in the two populations are actually different. Table 8.4 presents the steps in calculating the confidence intervals for both proportions.

Table 8.4 Calculating 95% Confidence Intervals for Rates of Relapse among Treatment and Placebo Groups

Step	Treatment	Placebo
Calculate standard error	$\sqrt{\dfrac{p(1-p)}{N}} = \sqrt{\dfrac{0.7(1-0.7)}{40}} = 0.072$	$\sqrt{\dfrac{p(1-p)}{N}} = \sqrt{\dfrac{0.76(1-0.76)}{50}} = 0.060$
Identify z-score for 95% confidence level	Find area in each tail of distribution: α/2 = 0.05/2 = 0.025; Identify corresponding z-score: 1.96	
Calculate margin of error	$z(SE_p)$ = 1.96(0.072) = 0.141	$z(SE_p)$ = 1.96(0.060) = 0.118
Calculate lower bound	p − margin of error = 0.70 − 0.141 = 0.56	p − margin of error = 0.76 − 0.118 = 0.64
Calculate upper bound	p + margin of error = 0.70 + 0.141 = 0.84	p + margin of error = 0.76 + 0.118 = 0.88

The last two rows in Table 8.4 show the lower and upper bounds for the proportion of relapses for the treatment and placebo groups. The confidence interval for the treatment group is (0.56, 0.84); for the placebo group it is (0.64, 0.88). Over repeated sampling, the true population proportion of relapses after thirty days of treatment would be contained within 95% of the confidence intervals. Thus, there is a 0.95 probability that the proportion of relapses for a population receiving treatment falls between 0.56 and 0.84 and that the proportion of relapses for a population receiving a placebo falls between 0.64 and 0.88.

As Figure 8.4 shows, these confidence intervals overlap. When taking sampling variation into account, the range of possible relapses for each group is wide enough that the relapse rates for the groups are not clearly different from each other. Based on the data from these two samples, the researchers decide they do not have enough evidence to conclude that the new drug is more effective at preventing relapse among opioid users than the placebo.

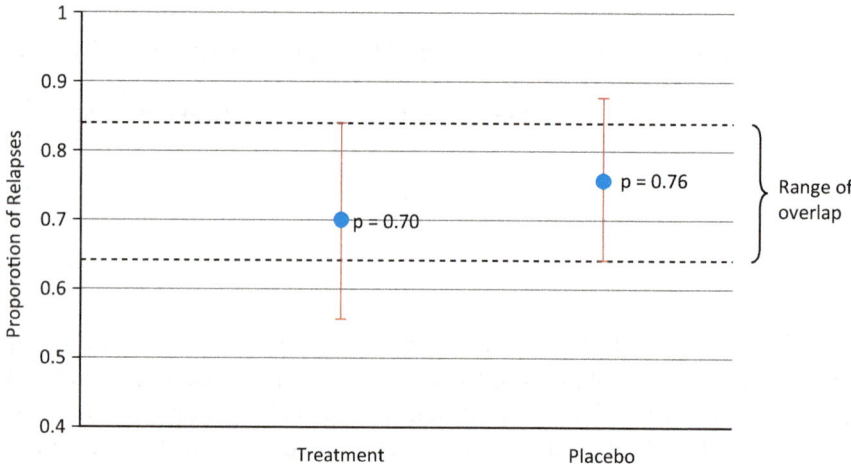

Figure 8.4 95% Confidence Intervals for Proportion of Relapses in Treatment and Placebo Groups.

BOX 8.2: IN DEPTH

Election Polling and Confidence Intervals

In election poll reporting, journalists often report a poll's margin of error along with the proportion of those polled who say they will vote for candidate A or candidate B. This is the same margin of error that we use in constructing a confidence interval: the product of the standard error and z-score, which is then subtracted from and added to the sample proportion. When a poll reports a margin of error of 3 points, this means that the confidence interval covers a range of 6 percentage points, 3 points above and 3 points below the point estimate. However, observers often focus on the "horse race" aspect of these polls by comparing the point estimates of the candidates' support without accounting for the margin of error. As some examples in this chapter demonstrate, confidence intervals can overlap even when point estimates themselves differ. A poll may report that 48% of respondents will vote for candidate A and 46% for candidate B. Candidate A, therefore, may be perceived as "leading" in the polls. But if the margin of error for both proportions is 3 percentage points, what would the confidence interval be for each candidate? (It would be 45% to 51% for candidate A and 43% to 49% for candidate B.) These intervals overlap, undermining the claim that anyone is in the lead.

It is important to keep in mind that the margin of error reported for a poll reflects only one kind of error: that which is produced by sampling variability. Point estimates of candidates' support among voters can be affected by other kinds of error, which are not accounted for by confidence intervals. In the 2020 election (as in 2016), Republican candidates appeared to outperform many polls. There are several possible reasons for this pattern, reasons that have nothing to do with sampling variability.[3] One possible explanation has to do with the difficulties in identifying likely voters. It is not enough for pollsters to ask respondents which candidate they intend to vote for. They must also identify which respondents are likely to actually vote in an election. And there is evidence that pollsters underestimated the turnout among voters who leaned toward Republican candidates. Another explanation has to do with Republicans being less willing to respond to pollsters, a phenomenon that pollsters have called "the silent Trump voter." And yet another suggests that Republicans were harder to reach than Democrats since Democrats were more likely to be at home during the pandemic.

Errors caused by sources other than sampling variation are difficult to trace. Recent elections serve as a reminder that polling predictions are only as good as the methods used to collect polling samples and measure candidate preferences.

Confidence Intervals for Means

The general steps for calculating a confidence interval for a population mean are the same as for proportions. However, there is an important difference: We use a slightly different curve called a t-distribution, instead of a normal distribution, to estimate confidence intervals for means.

The t-Distribution

In chapter 7 and the first part of this chapter, we saw how the theoretical sampling distribution of proportions and means follows the normal curve. In this section, we examine another theoretical distribution, the **t-distribution**. We almost always use the t-distribution instead of the normal distribution when calculating a confidence interval for means. This is because of how we estimate the standard error for means. In chapter 7, when we calculated the z-score for means, we used the population standard deviation (σ) to calculate the standard error. In practice, when we are dealing with sample data, we almost never know the true population standard deviation. If we did, we would probably already know the true population mean as well and have no need to estimate!

Instead, we use an *estimate* of the population standard deviation to find the standard error. Specifically, we substitute the sample standard deviation, s, for the population standard deviation, σ, in order to estimate the standard error:

$$SE_{\bar{y}} = \frac{s}{\sqrt{N}}$$

The standard error of the sample means ($SE_{\bar{y}}$) is equal to the sample standard deviation (s) divided by the square root of the sample size (N). Just as in chapter 7, this standard error is the standard deviation of the sampling distribution of the means, but it is an estimate rather than an exact measure.

Relying on an estimate increases the error (or variability) in our calculations. This means that, instead of following a normal distribution, the sampling distribution of means falls into a different family of distributions, the t-distribution. The smaller our sample size, the greater the uncertainty introduced into our estimations. To correct for the differing amounts of sampling variability at different sample sizes, the shape of the t-distribution depends on sample size.

This is an important difference between the t- and normal distributions: the shape of the normal distribution does not change with sample size, but the shape of the t-distribution depends on the sample size. The smaller the sample size, the flatter the t-distribution is relative to the normal curve. Another way to think about the shape of the t-distribution relative to the normal distribution is that the tails of the t-distribution thicken as sample size decreases, with greater areas of the curve concentrated in its tails.

We describe the influence of sample size on the shape of the t-distribution as **degrees of freedom**. Degrees of freedom (DF) is equal to the sample size minus 1 (DF = N − 1). Figure 8.5 shows the shapes of the t-distribution with different degrees of freedom. The distribution with the solid yellow line, with a sample size of infinity, is exactly the same shape as the normal distribution. As sample size decreases, as denoted by decreasing degrees of freedom, the t-distributions flatten, with the areas in their tails increasing compared to the curve with the solid line. As degrees of freedom increase, the t-distribution approaches the normal distribution. Like the normal distribution, the t-distribution is symmetric and unimodal, regardless of degrees of freedom.

This is an important point, so it's worth restating. When we are estimating confidence intervals for means, if we do not know the population standard deviation (σ) and instead

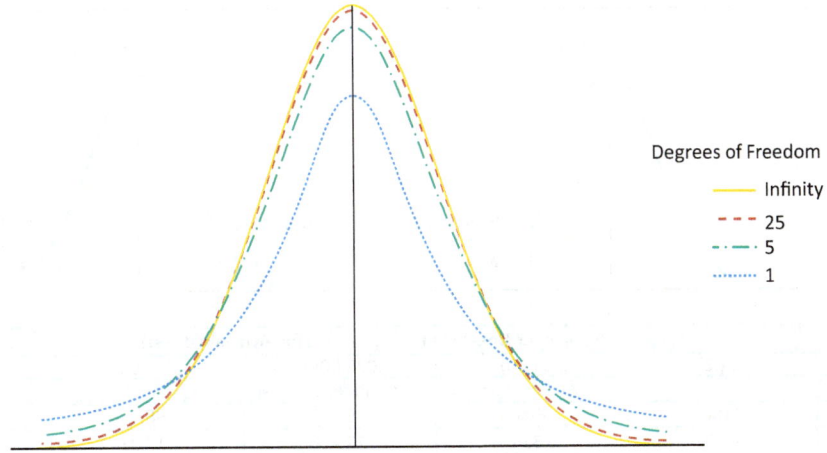

Figure 8.5 t-Distributions for Varying Degrees of Freedom

use the sample standard deviation (s) to estimate standard error, the sampling distribution of means follows the t-distribution; and the exact shape of the distribution changes depending on the size of the sample, as reflected by degrees of freedom.

In practice, we use t similarly to z. When calculating a confidence interval for a mean, instead of finding the z-score associated with the designated level of confidence, we find the **t-value**. The t-value will be larger than the z-score for the same level of confidence, with that difference increasing as sample size decreases.

Imagine that you are finding two 95% confidence intervals: one for a proportion and one for a mean. For the proportion confidence interval, we need to find the appropriate z-score. To find it, we turn to the Normal Table (in appendix A) and find the z-score that corresponds to 2.5% of the curve's area in the right tail of the distribution. The z-score is 1.96.

To find the corresponding t-value for a 95% confidence interval for a mean, we need one additional piece of information. Remember, the shape of the t-distribution depends on sample size, so we need to know the sample size. Say the sample size is 50. Before turning to the t-table, in appendix B, we need to calculate the degrees of freedom, which is simply $N - 1$. In this case $50 - 1 = 49$. To find the t-value, first locate the row in the table for 49 degrees of freedom.

Look at the excerpt from the t-table in Figure 8.6.

You will notice that there is no row for 49 degrees of freedom. In that case, we use the next smallest degrees of freedom available, which in this case is 40. We choose the closest value that is lower than our degrees of freedom because that provides us with a more conservative, cautious estimate.

Next, locate the column for 95% confidence level. Now find the t-value that lies at the intersection of the 40 degrees of freedom row and the 95% confidence level column: 2.021. This is the t-value that we will use to calculate our 95% confidence interval. Notice that this t-value is slightly higher than the z-score (1.96) at the same confidence level. As degrees of freedom decrease, the difference between t-values and z-scores widens. Larger t-values, as compared with the corresponding z-scores, produce wider confidence intervals, because the margin of error is the product of the standard error and the t-value. Using the t-distribution yields wider confidence intervals in order to adjust for the additional error introduced by estimating the standard error.

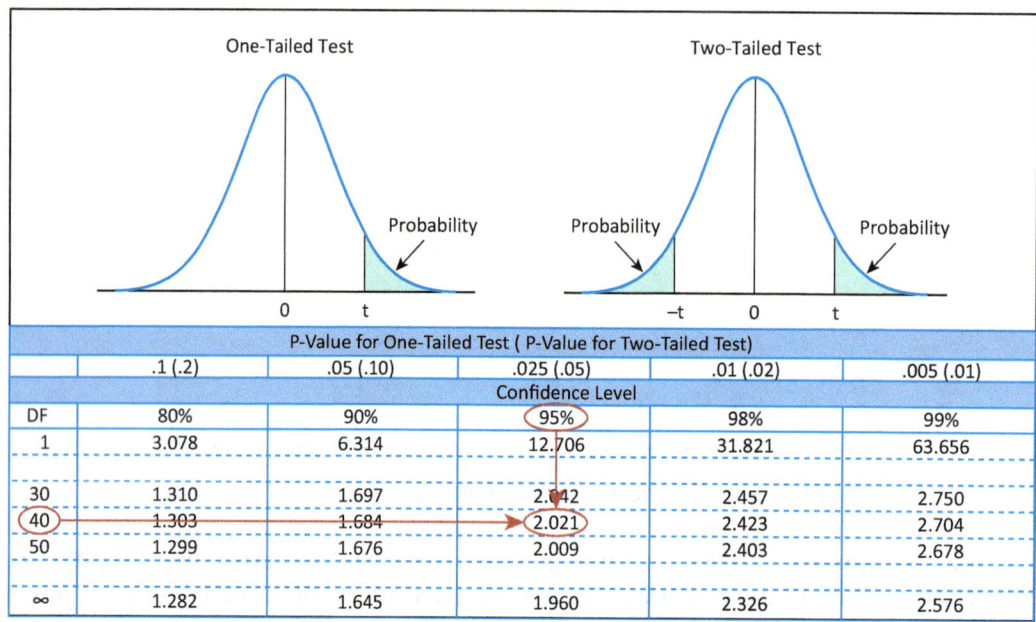

Figure 8.6 Using the t-Table to Find the t-Value for a 95% Confidence Interval with 49 Degrees of Freedom

In the next section, we demonstrate two examples of calculating confidence intervals for means, the first for mean social adjustment scores among a sample of formerly incarcerated people and the second for the mean net worth of recent immigrants to the United States.

Calculating Confidence Intervals for Means: Examples

In the first example, we will calculate a confidence interval of the mean score on a social adjustment assessment for a sample of formerly incarcerated people. Despite modest declines in recent years, the United States still has both the highest rate of incarceration in the world and the largest number of people incarcerated.[4] Given that most people who are incarcerated will be released into the community at some point, many argue that it is in the public's interest to facilitate successful societal reentry as much as possible. Recent years have seen growing interest in programs meant to rebuild and strengthen former prisoners' relationships with family, with numerous studies suggesting that this effort pays off in lower recidivism rates.[5]

Imagine that a private foundation is willing to fund the expansion of one of these reentry programs called Family Strong. Before funding this expansion, the foundation funds a small pilot study to test the effectiveness of the program, with a team of researchers assessing its effectiveness by measuring participants' scores on a social integration scale. The scale ranges from 0 to 100, with higher scores indicating closer relationships with family and higher levels of social support. The foundation has set the benchmark for success at a mean score of 40. That is, if the mean social integration score for Family Strong participants is at least 40, the foundation will fund the program's expansion. Instead of relying on just the sample mean, the researchers will construct a 90% confidence interval, and the program will be considered successful if the confidence interval includes the benchmark score of 40.

Here, we will calculate the 90% confidence interval of the participants' mean social integration score to determine whether Family Strong has met the benchmark outcome. The mean score from the sample is 38, and the standard deviation of the scores is 10. The sample size, or the number of participants who have been randomly chosen to participate in the program, is 150.

Step 1: Estimate the standard error.

$$SE_{\bar{y}} = \frac{s}{\sqrt{N}} = \frac{10}{\sqrt{150}} = 0.816$$

Step 2: Set confidence level. Find degrees of freedom and the corresponding t-value.

We have already established that our confidence level is 90%.

$$DF = N - 1 = 150 - 1 = 149$$

With large samples, the shape of the t-distribution is the same as the shape of the normal distribution. As a general rule, when DF is greater than 120, we can use the infinity row in the t-distribution (in which the t-values are the same as z-scores). DF is 149, so we use the row in the table for infinite (∞) degrees of freedom. Looking at the table under the 90% confidence level column:

$$t = 1.645$$

Step 3: Find the margin of error.

$$\text{Margin of error} = t(SE_{\bar{y}}) = 1.645(0.816) = 1.34$$

Step 4: Calculate lower and upper bounds of confidence interval.

$$\bar{y} \pm \text{margin of error} = 38 \pm 1.34 = (36.66, 39.34)$$

In repeated sampling, 90% of confidence intervals would include the true population mean social integration score for Family Strong participants. Thus, there is a 90% chance that participants' true mean score falls between 36.66 and 39.34. Has the program met the predetermined benchmark that would indicate the program's success? No, unfortunately the program has not met the goal, because the benchmark mean score of 40 falls outside the 90% confidence interval.

The next example uses data from a study examining wealth inequality among recent immigrants to the United States, in which the researchers predicted that immigrants belonging to racial and ethnic minority groups would have less wealth than immigrants who identified as white.[6] The researchers used data from the New Immigrant Study (NIS), with a sample size of 8,573 immigrants. The study provides descriptive statistics for the net worth of recent immigrants, by racial and ethnic background, as shown in Table 8.5.

Table 8.5 Mean and Standard Deviation for Net Worth for Immigrants, by Race

	Asian	Black	Latino	White
Mean	83,500	34,318	40,073	92,965
Standard deviation	388,797	171,041	213,566	442,564
N	2,486	1,029	3,343	1,715

Source: M. A. Painter and Z. Qian. 2016. "Wealth Inequality among New Immigrants." *Sociological Perspectives* 59(2): 368–394.

Are the data in Table 8.5 consistent with the prediction that white immigrants have greater net worth than other groups? If we base our answer on comparisons of the point estimates for the population wealth of each group, we would say that the data are consistent with the prediction. The sample mean for white immigrants ($92,965) is higher than the sample means for the other three groups. However, we know that due to sampling variation, the means presented in the first row of the table likely vary from the true population means for each group.

Using the data in Table 8.5, we can calculate confidence intervals around each group's point estimate of net worth. If the confidence intervals for any of the groups overlap, this casts doubt on the claim that the true population wealth differs for those groups. Here, we will calculate 95% intervals. Table 8.6 summarizes the steps for determining the confidence intervals for the net worth of Latino and white immigrants.

Table 8.6 Calculating 95% Confidence Intervals for Net Worth for Immigrants, by Race (Latino and White Immigrants)

Step	Latino	White
Calculate standard error	$SE = \frac{s}{\sqrt{N}} = \frac{213,566}{\sqrt{3,343}} = 3,693.72$	$SE = \frac{s}{\sqrt{N}} = \frac{442,564}{\sqrt{1,715}} = 10,686.71$
Find degrees of freedom and t-value for 95% confidence level	DF = N − 1 = 3,342 t-value: 1.96	DF = N − 1 = 1,714 t-value: 1.96
Calculate margin of error	$t(SE_{\bar{y}}) = 1.96(3,693.72) =$ 7,239.69	$t(SE_{\bar{y}}) = 1.96(10,686.71) =$ 20,945.95
Calculate lower bound	$\bar{y}$ − margin of error = 40,073 − 7,239.69 = 32,833.31	$\bar{y}$ − margin of error = 92,965 − 20,945.95 = 72,019.05
Calculate upper bound	$\bar{y}$ + margin of error = 40,073 + 7,239.69 = 47,312.69	$\bar{y}$ + margin of error = 92,965 + 20,945.95 = 113,910.95

Source: M. A. Painter and Z. Qian. 2016. "Wealth Inequality among New Immigrants." *Sociological Perspectives* 59(2): 368–394.

Step 1: Estimate the standard error.

Since we do not know the population standard deviation, we will estimate the standard errors using the sample standard deviations for each group.

$$SE_{\bar{y}Asian} = \frac{s}{\sqrt{N}} = \frac{388{,}797}{\sqrt{2{,}486}} = 7{,}797.80$$

$$SE_{\bar{y}Black} = \frac{s}{\sqrt{N}} = \frac{171{,}041}{\sqrt{1{,}029}} = 5{,}332.03$$

$$SE_{\bar{y}Latino} = \frac{s}{\sqrt{N}} = \frac{213{,}566}{\sqrt{3{,}343}} = 3{,}693.72$$

$$SE_{\bar{y}White} = \frac{s}{\sqrt{N}} = \frac{442{,}564}{\sqrt{1{,}715}} = 10{,}686.71$$

Step 2: Set confidence level. Find degrees of freedom and the corresponding t-value.

Our confidence level is 95%.
 Each sample has well over 120 cases, so we use the row in the t-table in appendix B for infinite (∞) degrees of freedom. For a 95% confidence interval with ∞ degrees of freedom, the t-value is 1.96.

Step 3: Calculate the margin of error.

The margin of error is the product of the standard error and the t-value:

$$\text{Margin of error}_{Asian} = t(SE\bar{y}) = 1.96(7{,}797.80) = 15{,}283.69$$

$$\text{Margin of error}_{Black} = t(SE\bar{y}) = 1.96(5{,}332.03) = 10{,}450.78$$

$$\text{Margin of error}_{Latino} = t(SE\bar{y}) = 1.96(3{,}693.72) = 7{,}239.69$$

$$\text{Margin of error}_{White} = t(SE\bar{y}) = 1.96(10{,}686.71) = 20{,}945.95$$

Step 4: Calculate the lower and upper bounds of the confidence interval by subtracting the margin of error from, and adding it to, the sample mean.

$$95\% \ CI = \bar{y} \pm \text{margin of error}$$

$$95\% \ CI_{Asian} = 83{,}500 \pm 15{,}283.69 = (68{,}216.31, 98{,}783.69)$$

$$95\% \ CI_{Black} = 34,318 \pm 10,450.78 = (23,867.22, 44,768.78)$$

$$95\% \ CI_{Latino} = 40,073 \pm 7,239.69 = (32,833.31, 47,312.69)$$

$$95\% \ CI_{White} = 92,965 \pm 20,945.95 = (72,019.05, 113,910.95)$$

Figure 8.7 plots each confidence interval on a graph with the sample means designated by the dots in the center of each interval.

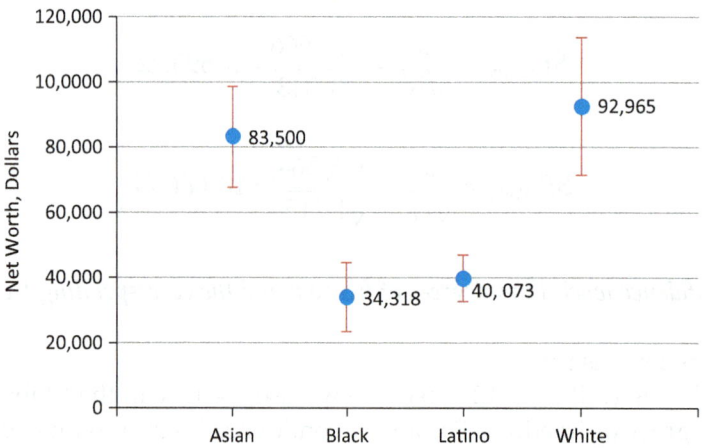

Figure 8.7 95% Confidence Intervals for Immigrants' Net Worth in NIS Sample, by Race

We see that two sets of confidence intervals overlap: those for Asian and white immigrants and those for Black and Latino immigrants. Estimating the population parameters for these four groups as intervals rather than point estimates suggests that we cannot conclude that whites are, in fact, wealthier than all other groups of new immigrants. Specifically, the overlapping confidence intervals between whites and Asians give us reason to doubt that their population means are actually different. However, the confidence intervals for Black and Latino immigrants do not overlap with the other two groups (although they do overlap with each other), suggesting that the true mean net worth for Black and Latino immigrants is lower than that of whites or Asians.

The Relationship between Sample Size and Confidence Interval Range

Sample size has a substantial effect on the range for confidence intervals for both proportions and means. Larger samples yield narrower—more precise—ranges. We will conduct two simulations using 2021 childhood poverty rates at the county level to demonstrate how sample size affects the range of confidence intervals. The United States experienced a reduction in childhood poverty after the expansion of anti-poverty programs during COVID. In 2021, the national childhood poverty rate was 15.3% among youth under the age of eighteen in 2021.[7] Here, we will focus on whether counties exceeded the national childhood poverty rate.

In both simulations, we draw fifty random samples of counties from the entire population of 3,194 counties in the United States. The only difference between the simulations is that, whereas each sample in Simulation 1 will include twenty-five counties, each sample in Simulation 2 will include fifty counties. We categorize each county as either exceeding the national childhood poverty rate of 15.3%, or falling at or below it. For each sample, we will calculate 90% confidence intervals for the proportion of counties in which childhood poverty rates are above the national average.

Table 8.7 presents the frequency distribution for this variable, showing that over half of all counties (62%) exceed the national poverty rate for youth under the age of eighteen. Notice here that the population proportion, π, is known: 0.62. This will allow us to evaluate the accuracy of the confidence intervals in our simulations.

Table 8.7 Frequency Distribution for Exceeding National Childhood Poverty Rate, County-Level Data

	Frequency	Percent
At or below national rate	1,988	62.2
Above national rate	1,206	37.8
Total	3,194	100

Source: 2021 US Census Bureau.

In Simulation 1, we draw fifty random samples from the 2021 US Census data, each with twenty-five counties. In Simulation 2, we draw fifty random samples, each with fifty counties. In both simulations we construct fifty 90% confidence intervals, one for each sample proportion. Table 8.8 shows the following information for the fifty samples in Simulations 1 and 2, respectively:

- sample proportions (p)
- lower and upper bound (LB and UB) of the confidence intervals
- range of the confidence intervals

The circles in the table indicate the confidence intervals that do not include the known population parameter, 0.62. For example, the confidence interval for sample 2 in Simulation 1 (0.625, 0.918) excludes the population parameter of 0.62.

Comparing the "CI Range" columns for Simulations 1 and 2 in Table 8.8 shows that the confidence intervals are narrower for the larger samples in Simulation 2. The mean range for all fifty confidence intervals in Simulation 1 (with sample sizes of twenty-five) is 0.337, compared to only 0.239 for Simulation 2 (with sample sizes of fifty). Why is this? Sample size is in the denominator in the formula for standard error. Thus, larger samples yield smaller standard errors, all else equal. Since the margin of error is the product of the

Table 8.8 90% Confidence Intervals for Proportion of Counties above National Childhood Poverty Rate, Simulations 1 (Sample Size = Twenty-Five Counties) and 2 (Sample Size = Fifty Counties)

	Simulation 1: N = 25				Simulation 2: N = 50			
Sample	p	LB	UB	CI Range	Sample	p	LB	UB	CI Range
1	.59	.417	.764	.348	1	.56	.434	.680	.246
2	.77	.625	.918	.293	2	.69	.576	.805	.229
3	.55	.379	.730	.352	3	.56	.434	.680	.246
4	.59	.417	.764	.348	4	.65	.535	.770	.236
5	.63	.456	.798	.342	5	.48	.357	.605	.247
6	.74	.580	.890	.309	6	.61	.494	.735	.241
7	.63	.456	.798	.342	7	.73	.619	.839	.220
8	.66	.496	.830	.333	8	.63	.514	.753	.239
9	.63	.456	.798	.342	9	.73	.619	.839	.220
10	.66	.496	.830	.333	10	.48	.357	.605	.247
11	.8	.665	.935	.270	11	.60	.474	.717	.243
12	.63	.456	.798	.342	12	.65	.535	.770	.236
13	.7	.538	.861	.323	13	.60	.474	.717	.243
14	.63	.456	.798	.342	14	.54	.415	.662	.247
15	.48	.305	.659	.353	15	.56	.434	.680	.246
16	.55	.379	.730	.352	16	.65	.535	.770	.236
17	.55	.379	.730	.352	17	.56	.434	.680	.246
18	.55	.379	.730	.352	18	.65	.535	.770	.236
19	.48	.305	.659	.353	19	.63	.514	.753	.239
20	.63	.456	.798	.342	20	.44	.320	.566	.246
21	.55	.379	.730	.352	21	.58	.454	.699	.245
22	.59	.417	.764	.348	22	.58	.454	.699	.245
23	.55	.379	.730	.352	23	.58	.454	.699	.245
24	.74	.580	.890	.309	24	.58	.454	.699	.245
25	.7	.538	.861	.323	25	.69	.576	.805	.229
26	.52	.341	.695	.353	26	.71	.597	.822	.225
27	.63	.456	.798	.342	27	.67	.555	.788	.233
28	.66	.496	.830	.333	28	.67	.555	.788	.233
29	.48	.305	.659	.353	29	.63	.514	.753	.239
30	.59	.417	.764	.348	30	.71	.597	.822	.225
31	.81	.670	.943	.273	31	.58	.454	.699	.245
32	.63	.456	.798	.342	32	.63	.514	.753	.239

(continued)

Table 8.8 90% Confidence Intervals for Proportion of Counties above National Childhood Poverty Rate, Simulations 1 (Sample Size = Twenty-Five Counties) and 2 (Sample Size = Fifty Counties) **(Continued)**

	Simulation 1: N = 25				Simulation 2: N = 50				
Sample	p	LB	UB	CI Range	Sample	p	LB	UB	CI Range
33	.48	.305	.659	.353	33	.52	.395	.643	.247
34	.52	.341	.695	.353	34	.60	.474	.717	.243
35	.63	.456	.798	.342	(35)	.44	.320	.566	.246
36	.55	.379	.730	.352	36	.61	.494	.735	.241
37	.63	.456	.798	.342	37	.67	.555	.788	.233
38	.7	.538	.861	.323	38	.60	.474	.717	.243
39	.55	.379	.730	.352	39	.67	.555	.788	.233
40	.66	.496	.830	.333	40	.67	.555	.788	.233
41	.59	.417	.764	.348	41	.58	.454	.699	.245
42	.63	.456	.798	.342	(42)	.77	.662	.871	.209
43	.63	.456	.798	.342	43	.69	.576	.805	.229
44	.52	.341	.695	.353	44	.60	.474	.717	.243
45	.7	.538	.861	.323	45	.61	.494	.735	.241
46	.55	.379	.730	.352	46	.50	.376	.624	.248
47	.59	.417	.764	.348	47	.56	.434	.680	.246
48	.63	.456	.798	.342	48	.65	.535	.770	.236
(49)	.81	.670	.943	.273	49	.5	.38	.624	.247
50	.55	.379	.730	.352	50	.505	.38	.63	.251

z-score (or t-value) and the standard error, smaller standard errors mean smaller margins of error. In turn, smaller margins of error mean smaller confidence intervals and, therefore, increased precision.

There is one more reason why larger sample sizes yield narrower, more precise confidence intervals. When we use the t-distribution to estimate confidence intervals for means with unknown population standard deviations, larger sample sizes mean more degrees of freedom. Recall that the shape of the t-distribution narrows with more degrees of freedom. t-values with more degrees of freedom are slightly smaller than t-values with fewer DF. When we estimate the margin of error for larger samples, then, we are multiplying a smaller t-value by a smaller standard error.

Samples circled in Table 8.8 indicate confidence intervals that exclude the population parameter, 0.62. Since these are 90% confidence intervals, we would expect 10% of the confidence intervals, in this case five, to exclude the population parameter. The simulations performed roughly as expected in terms of accuracy, with four confidence intervals in Simulation 1 excluding the population parameter and five in Simulation 2. This shows that, while the accuracy of confidence intervals does not depend on sample size, the precision of the intervals, as indicated by the interval's range, increases with larger samples.

The Relationship between Confidence Level and Confidence Interval Range

In the simulations above, we saw that the range of a confidence interval decreases as sample size increases, making the estimate more precise. What happens to the range when we increase the confidence level—for example, from 95% to 99%?

Here, we will return to the example of the confidence interval for social integration scores for participants in Family Strong, the reintegration program for people being released from prison. Recall that a foundation would fund the program's expansion if the mean score for a sample of the program's participants met or exceeded the benchmark score of 40 on a social integration scale. The researchers used a 90% confidence interval to estimate the true social integration score of all participants in the program across the country. Since the 90% confidence interval fell below the benchmark of 40, the foundation did not fund the program's expansion. In the example, the sample included 150 program participants, with a sample mean of 38 and a standard deviation of 10. This yielded a 90% confidence interval that ranged from 37 to 39.

What happens if we estimate a 99% confidence interval for the sample's true population mean social integration score? The calculation would proceed as before with one change: The t-value will reflect a 99% level of confidence. With the updated t-value, we can recalculate a confidence interval for Family Strong participants at 99%:

Step 1: Calculate the standard error.

The standard error (0.816) stays the same as in the previous calculation because the sample standard deviation and sample size have remained the same.

Step 2: Find the t-value for ∞ DF (because N > 120) and 99% confidence.

Look at the t-table (appendix B). The t-value for ∞ DF and a 99% confidence level is 2.576. (Note that the t-value for the 90% confidence interval was lower, at 1.645.)

Step 3: Calculate the margin of error.

$$\text{Margin of error} = t(SE_{\bar{y}}) = 2.576(0.816) = 2.10$$

Step 4: Calculate lower and upper bounds of confidence interval.

$$\bar{y} \pm \text{margin of error} = 38 \pm 2.10 = (35.90, 40.10)$$

Recall that the 90% confidence interval (36.66, 39.34) did not include the benchmark score of 40, resulting in the foundation's decision not to fund the program's expansion. However, a 99% level of confidence yields a wider confidence interval: (35.90, 40.10). Unlike the 90% confidence interval, the 99% confidence interval includes the benchmark score of 40. With the 99% confidence interval, the foundation would choose to fund the expansion of the program. This example illustrates how increasing the level of certainty for a confidence interval entails a sacrifice in precision, or a widening of the confidence interval. This happens because the margin of error is a product of the t-value (or z-score) and the standard error, and higher levels of confidence mean larger t-values (or z-scores).

With the 90% confidence interval, Family Strong does not receive funding to expand its use, but with the 99% confidence interval the program does receive funding. How do we know which confidence interval is the right one to use? Neither confidence interval is wrong. They both indicate the chance that the given interval will include the true population mean. The answer about which confidence interval one should choose to assess the program's success depends on the level of risk involved in the confidence interval not including the true population mean. The foundation has a limited amount of money to use for its ultimate goal of reducing recidivism rates in the United States, and from that point of view, it might be very important to produce a narrower, or more precise, confidence interval. From a societal perspective, using the wider 99% confidence interval might be worth the risk of the program not being as successful as the foundation would like it to be, because even small reductions in recidivism are beneficial for society.

Researchers should consider these issues in advance and select a confidence level before actually conducting their statistical analyses. Adjusting the confidence level in order to produce a desired result is an example of an unethical manipulation of data.

Interpreting Confidence Intervals

Confidence intervals can be tricky to interpret. In an article called "Robust Misinterpretation of Confidence Intervals," a group of researchers reports that students and researchers alike frequently fall victim to misinterpreting the meaning of confidence intervals.[8] There are two common misinterpretations of confidence intervals. The first is a clear misinterpretation, and the second could be construed as one because it omits important information.

The first common misinterpretation is that the confidence interval offers an estimate of the *sample* statistic. Here, one might say that there is a 95% chance that the sample mean lies within the confidence interval. This misinterpretation is understandable because we subtract the margin of error from, and add it to, the sample statistic to calculate the confidence interval. However, a confidence interval is never an estimate of a sample statistic—we know the exact value of the sample statistic! The confidence interval is always an estimate of a *population* parameter.

Another issue is not technically incorrect but can be misleading. When we choose a confidence level for a confidence interval, we are deciding what proportion of a large number of confidence intervals would miss the population parameter. The confidence level (for example, 90%) does not refer to a single confidence interval but to a large group of confidence intervals under repeated sampling.

We can return to the example of childhood poverty rates for counties in the United States to look closely at how this works. Figure 8.8 plots the fifty confidence intervals generated for the samples drawn in Simulation 1 of the proportion of counties with childhood poverty rates higher than the national rate.

As shown in Figure 8.8, four out of the fifty confidence intervals in Simulation 1, or 8% of them (circled in red), excluded the population parameter of 0.62. When interpreting

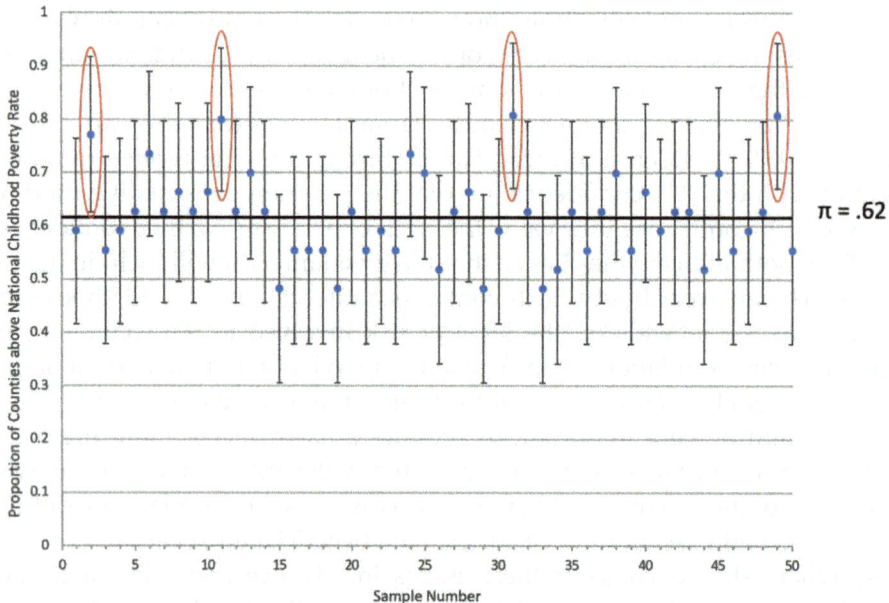

Figure 8.8 Fifty 90% Confidence Intervals for the Proportion of Counties with Childhood Poverty Rates above the National Childhood Poverty Rate, for Sample Sizes of Twenty-Five

a confidence interval, envision something like Figure 8.8—a large number of confidence intervals plotted around their point estimates.

Over repeated sampling, we expect 90% of the confidence intervals to include the population parameter. Therefore, we know that, if 90% of confidence intervals contain the population parameter, then there is a 0.9 probability that any single confidence interval includes the population parameter. Technically, then, we can say that there is a 0.90 probability that a specific confidence interval includes the population parameter. We would report the results of one confidence interval by saying, "There is a 90% chance that the proportion of counties with childhood poverty rates that exceed the national rate is between 0.555 and 0.788." However, we should always remember that a different random sample would likely yield different values for the lower and upper bounds.*

● How Big a Sample?

With a single random sample, we can construct an interval that, with high probability, contains the true population parameter. But how big a sample is required? This is one of the first questions a researcher must answer.

There are non-mathematical considerations in determining the required sample size. Although the Internet is growing in popularity as a method of conducting a survey,

---

* There is debate among statisticians about the specifics of how to report confidence intervals, given this point. See Andrew Gelman. 2016. "Abraham Lincoln and Confidence Intervals." November 23. Retrieved from http://andrewgel-man.com/2016/11/23/abraham-lincoln-confidence-intervals/.

telephone interviewing is still a common form of polling in many fields. Indeed, many pollsters now use multiple methods when they conduct a poll.[9] Regardless of the polling method, it takes time and money to administer any poll. If you've ever watched a presidential debate, you have probably seen the networks broadcast the results of their post-debate "Who Won?" survey within an hour or so of the end of the debate. Such polls are typically done by telephone; they are administered quickly, and although they randomly select respondents, they can interview only a limited number of people. For example, after the first presidential debate of the 2020 campaign, CNN drew a random sample of 568 respondents and interviewed them all within minutes of the close of the debate.[10] Some surveys are even conducted in person, and you can imagine that it is quite expensive and time-consuming for polling organizations to send their interviewers into the field. By necessity, this limits sample sizes as well.

But there are also important mathematical considerations. Most important is the margin of error. How large an error can the researcher tolerate? If the goal is to simply describe the distribution of public opinion on some issue, you can probably get away with a smaller sample and a larger margin of error. On the other hand, if you were polling for a candidate in a close race, you would probably want a small margin of error so that you could tell your candidate, with some precision and with a high degree of confidence, whether she is likely to win or lose the race. In order to obtain a smaller margin of error, you would need a larger sample size.

When working with proportions, there is a simple formula for calculating the required sample size. To use it, we must first answer two questions: How much error are we willing to tolerate? What is our desired level of confidence?

Let's try an example. It is a few days before the election, and you are conducting a pre-election poll for your candidate. The race seems close, and your candidate demands a high level of certainty and a high degree of precision. You point out that both of these demands will translate into a larger sample size. Your candidate gives you the green light; after all, she has raised a lot of money from her Super PAC, so she can afford it. She needs to know whether she is going to win.

To meet her demands, you decide on a 99% degree of certainty and a margin of error equal to 2 percentage points. You plug the appropriate values into the following formula:[*]

$$\text{Required sample size}(N) = 0.25 \left(\frac{z}{B}\right)^2$$

Where

$$z = z\text{-score associated with the desired confidence level}$$

$$B = \text{margin of error}$$

In this case, because we want a 99% confidence level, z will equal 2.58. And since we have decided on a 2-point margin of error, B will be equal to 0.02. Pay attention to the order

---

[*] In the formula, 0.25 is a constant. It assumes the largest possible population division—that is, that $\pi = 0.5$.

of operations in the formula. (You need to square the result of z/B before multiplying by 0.25.)

$$\text{Required sample size}(N) = 0.25\left(\frac{2.58}{0.02}\right)^2 \approx 4{,}160$$

In this case, we would need 4,160 respondents. This is a very large sample size. Most polls rarely exceed one thousand respondents. But in this example, our candidate's constraints made it impossible to keep the sample size down.

When we are instead working with a confidence interval for a mean, the approach is similar, though a bit trickier. In order to estimate the required sample size when we are dealing with means instead of proportions, we must know (or make an educated guess of) the population standard deviation. In fact, we are unlikely to know the population standard deviation. If we did, we would likely know the population mean as well, and we would not need to sample. But there are cases where we have a good idea (perhaps from prior research) of the population standard deviation, and so it makes sense to see how we estimate the required sample size for means.

Here is the formula for estimating required sample size when we are working with means:

$$\text{Required sample size}(N) = \sigma^2 \left(\frac{z}{B}\right)^2$$

where

$$\sigma = \text{population standard deviation}$$

$$z = z\text{-score associated with the desired confidence level}$$

$$B = \text{margin of error (in units of the variable)}$$

Here is an example. A researcher needs to estimate the mean income of the population in a city in order to apply for a state grant to support the construction of a new library.* Since the amount of the grant depends on residents' income, she needs a fairly small margin of error. How large a sample size is required to estimate the mean income of a population with a margin of error equal to $1,000 and with a 95% level of confidence? We can assume from prior research that the standard deviation of income is around $20,000.

Plugging the appropriate values into the formula:

$$\text{Required sample size}(N) = 20{,}000^2 \left(\frac{1.96}{1{,}000}\right)^2 \approx 1{,}537$$

---

* Note that, because income is skewed, researchers often use a median instead of a mean. It is possible to calculate a confidence interval for a median, but that is beyond the scope of this book, and many researchers will choose to calculate a confidence interval for a mean even with skewed data. Similarly, many of the statistical techniques you will learn in the next few chapters require use of means instead of medians. Nevertheless, researchers should keep the degree of skew in mind as they interpret their final results.

We will need a sample of 1,537 individuals to meet the standards specified in the problem.

You may be wondering why we used z instead of t in this problem. After all, we have seen that, when we work with a sampling distribution of means, we must use the t-distribution and adjust for sample size. The problem facing us, however, is that we must know the sample size so that we can estimate the appropriate degrees of freedom in the t-distribution; but we cannot know the sample size, since this is precisely what the problem is asking us for. Because of this dilemma, we use the normal distribution and the appropriate z-score when we need to estimate sample size for an interval-ratio variable.

Unless the sample size is small, such as below two hundred or so, the consequences of using z instead of t are not all that grave. With larger samples, the differences between a z-score and a t-value at the same confidence level are minimal. You can verify this by looking at the t table in appendix B. The z-score for a 95% confidence interval is 1.96, which you will find in the t table in the bottom row (at DF = ∞). The next smaller DF value, 120, yields a t-value of 1.98.

Assumptions for Confidence Intervals

When estimating confidence intervals, certain assumptions about the data must hold. It is very important to think carefully about whether your data conform to the relevant assumptions before you estimate a confidence interval. If an assumption does not hold, you must decide whether to continue with the analysis.

When we work with confidence intervals, we make the following assumptions:

- The sample was collected randomly from a population.
- When we construct a confidence interval around a sample mean, the variable must be measured at the interval-ratio level. Recall from chapter 1 that survey researchers often treat certain kinds of ordinal variables, such as Likert scale items, as if they are interval-ratio. Typically, ordinal variables should have at least five categories in order to be treated as interval-ratio. The means for such variables can be used to calculate a confidence interval. With other types of ordinal or nominal variables, we select one category and construct the confidence interval around the proportion of cases in that category.
- When population distributions are approximately normal, samples of any size are appropriate for estimation of a confidence interval around a sample mean. However, as we have seen from the Central Limit Theorem, once sample sizes are large enough, the sampling distribution of means or proportions is approximately normal regardless of the population distribution. For means, a sample size of thirty is sufficient. For proportions, a general rule is that there should be at least fifteen cases in the category of interest (the category for which we estimate the confidence interval) and at least fifteen cases outside the category of interest.[11]

Chapter Summary

This chapter focused on how to estimate confidence intervals for point estimates using the logic of inferential statistics.

- Inferential statistics use probability theory and the properties of theoretical sampling distributions (such as the normal curve and the t-curves) to assess the extent to which statistics from a single sample approximate the corresponding unknown population parameters.

Key terms from the chapter include:

- **confidence interval**: an interval estimate of a population parameter (proportion or mean) that covers a range of values.
- **point estimate**: a single value estimate (e.g., mean, proportion) of a population parameter.
- **lower bound**: the lowest value in a confidence interval.
- **upper bound**: the highest value in a confidence interval.
- **confidence level**: under repeated sampling, the proportion of confidence intervals that will contain the population parameter (90%, 95%, and 99% are commonly chosen confidence levels).
- **precision**: the width, or range, of the confidence interval; the smaller the range, the higher the precision.
- **confidence or certainty**: the probability that our confidence interval contains the population parameter.
- **margin of error**: the amount of error above and below the point estimate of the population parameter caused by sampling variability (the product of the standard error and z or t).
- **t-distribution**: a probability distribution used to find the t-value for means when the population standard deviation, σ, is unknown; corrects for the increased uncertainty introduced by estimating the standard error in smaller samples.
- Steps in constructing a confidence interval for a proportion:
  1. Calculate standard error:

  $$SE_p = \sqrt{\frac{p(1-p)}{N}}$$

  where p is the sample proportion and N is the sample size.
  2. Set confidence level (e.g., 90%, 95%, 99%, etc.) and find associated z-score in the Normal Table. Find z-score associated with right-tail probability, α/2.
  3. Calculate margin of error: $z(SE_p)$.

4. Calculate lower and upper bounds of confidence interval: *CI = p ± margin of error*.

- Steps in constructing a confidence interval for a mean:
  1. Calculate standard error:

  $$SE_{\bar{y}} = \frac{s}{\sqrt{N}}$$

  where s is the sample standard deviation and N is the sample size.
  2. Set confidence level (e.g., 90%, 95%, 99%, etc.), calculate degrees of freedom (DF), and find the associated t-value.

  $$DF = N - 1$$

  Find the t-value at the intersection of the degrees of freedom and confidence level.
  3. Calculate margin of error: $t(SE_{\bar{y}})$.
  4. Calculate lower and upper bounds of confidence interval: *CI = ȳ ± margin of error*.

Table 8.9 summarizes all the formulas for confidence intervals of proportions or means.

Table 8.9 Summary of Formulas for Confidence Intervals

	Proportions	Means
Standard error	$SE_p = \sqrt{\dfrac{p(1-p)}{N}}$	$SE_{\bar{y}} = \dfrac{s}{\sqrt{N}}$
z or t?	z (associated with right-tail probability, α/2)	t (with DF = N − 1)
Margin of error	$z(SE_p)$	$t(SE_{\bar{y}})$
CI	CI = p ± margin of error	CI = ȳ ± margin of error
Necessary sample size Where B = acceptable margin of error and σ = population standard deviation	$N = 0.25\left(\dfrac{z}{B}\right)^2$	$N = \sigma^2\left(\dfrac{z}{B}\right)^2$

- To calculate the necessary sample size for a given confidence level and margin of error:
  1. For proportions:

  $$N = 0.25\left(\frac{z}{B}\right)^2$$

2. For means:

$$N = \sigma^2 \left(\frac{z}{B}\right)^2$$

where N is sample size, z is the z-score associated with the desired level of confidence, B is the acceptable margin of error (in percentage points for proportions; in units of variable for means), and σ is the population standard deviation (for means only).

Using Stata

In this section, we will learn how to ask Stata to calculate confidence intervals for a mean and for a proportion. We will construct confidence intervals for two versions of the same variable from the 2020 ANES data set. The variable *V202187* asks respondents to rate their feelings about the Centers for Disease Control and Prevention (CDC) on a feeling thermometer from 0 (very cold or unfavorable) to 100 (very warm or favorable). First, we will calculate a confidence interval for the mean CDC thermometer rating. Next, we will use an alternative version of the variable, called *V202187b*, to calculate a confidence interval for the proportion of people who feel warmly about the CDC (gave a rating over 50 degrees). *V202187b* is a recoded version of the original variable, coded 0 (rated the CDC 50 degrees or lower) and 1 (rated the CDC higher than 50 degrees).

Calculating Confidence Intervals of Means

We will first use Stata to calculate the 95% confidence interval of the mean temperature rating of the CDC. Open `ANES20.dta` and enter the following command into Stata:

`ci means V202187, level(95)`

The "ci" command asks for a confidence interval for *V202187*. After "ci" we enter "means" to specify that we want a confidence interval for the mean of the listed variable. At the end of the command, we specify the option "level(95)" to tell Stata the level of confidence for the confidence interval. The output is shown in Figure 8.9.

```
ci means v202187, level(95)
```

Variable	Obs	Mean	Std. err.	[95% conf. interval]	
v202187	7,273	70.19634	.2769825	69.65338	70.73931

Figure 8.9

Before examining the confidence interval reported in the output, note that the output reports most of the information we would need to calculate the confidence interval manually, along with the confidence interval itself. First, we see the sample mean for the variable: 70.2, indicating that the sample felt on average somewhat warmly toward the CDC. Second, the output reports the standard error and sample size for the variable. The only piece of information that we would still need to calculate the confidence interval by hand is the t-value. We can find the confidence interval in the last two columns of the output: (69.7, 70.7). The confidence interval is quite precise, as indicated by its narrow range.

Calculating Confidence Intervals of Proportions

What proportion of respondents feel warmly about the CDC? Here, we will calculate the 95% confidence interval for the alternative version of the variable, *V202187b*. Enter the following command into Stata:

`ci proportions V202187b, level(95)`

The "ci" command here differs from the command we used for means in only one way: we have specified "proportions" rather than "means."*

The output is shown in Figure 8.10.

```
ci proportions V202187b, level(95)

                                                      Binomial exact
   variable  |     obs   Proportion   Std. err.    [95% conf. interval]
   V202187b  |   7,273    .7516843     .005066      .7415868    .7615798
```

Figure 8.10

The output for the confidence interval of the proportion shows the same information as the output for the confidence intervals of means, as shown above. We see the sample statistic: the proportion of respondents who feel warmly about the CDC (.75). The 95% confidence interval, shown in the last two columns of the output, is (0.74, 0.76).**

The confidence interval suggests a high probability that the true proportion of people who feel warmly about the CDC falls between 74% and 76%. This confidence interval is consistent with the confidence interval for the mean rating of the CDC (69.7, 70.7), which also falls on the warm side of the feeling thermometer. Since the outbreak of COVID-19, the CDC has been a consistent object of public attention. At least according to this measure, the public feels generally positively toward the agency.

---

* With the "ci proportions" command, Stata will allow variables with only two categories, coded 1 and 0.
** In the Stata output above the columns reporting the confidence intervals, it says "binomial exact." This refers to the fact that Stata uses the binomial distribution instead of the normal distribution to calculate confidence intervals of proportions. In large samples, these two distributions are nearly identical.

Review of Stata Commands

- Find the confidence interval for a mean

    `ci means variable name/s, level(specify confidence level)`

- Find the confidence interval for a proportion

    `ci proportions variable name/s, level(specify confidence level)`

Using SPSS

In this section, we will learn how to ask SPSS to calculate confidence intervals for a mean and for a proportion. We will construct confidence intervals for two versions of the same variable from the 2020 American National Election Study (ANES) data set. The variable *V202187* asks respondents to rate their feelings about the Centers for Disease Control and Prevention (CDC) on a feeling thermometer from 0 (very cold or unfavorable) to 100 (very warm or favorable). First, we will calculate a confidence interval for the mean CDC thermometer rating. Next, we will use an alternative version of the variable, called *V202187b*, to calculate a confidence interval for the proportion of people who feel warmly about the CDC (gave a rating over 50 degrees). *V202187b* is a recoded version of the original variable, coded 0 (rated the CDC 50 degrees or lower) and 1 (rated the CDC higher than 50 degrees).

Calculating Confidence Intervals of Means

We will first use SPSS to calculate the 95% confidence interval of the mean temperature rating of the CDC. `Open ANES20.sav`. To generate a confidence interval for a mean, we use the "Explore" procedure that we have seen previously.

`Analyze → Descriptive Statistics → Explore`

This opens the "Explore" dialog box. We move the variables of interest, *V202187*, into the "Dependent List" box and click on "OK." The output is shown in Figure 8.11.

Before examining the confidence interval reported in the output, note that SPSS generates many other descriptive statistics with the "Explore" procedure. We see the sample mean for the variable: 70.2, indicating that the sample felt on average somewhat warmly toward the CDC. The confidence interval is circled in the output (69.7, 70.7) The confidence interval is quite precise, as indicated by its narrow range.

Calculating Confidence Intervals of Proportions

What proportion of respondents feel warmly about the CDC? Here, we will calculate the 95% confidence interval for the alternative version of the variable, *V202187b*.

Descriptives

			Statistic	Std. Error
POST: Feeling thermometer: Center for Disease Control (CDC)	Mean		70.20	.277
	95% Confidence Interval for Mean	Lower Bound	69.65	
		Upper Bound	70.74	
	5% Trimmed Mean		71.96	
	Median		70.00	
	Variance		557.980	
	Std. Deviation		23.622	
	Minimum		0	
	Maximum		100	
	Range		100	
	Interquartile Range		30	
	Skewness		-.820	.029
	Kurtosis		.444	.057

Figure 8.11

SPSS does not directly calculate confidence intervals for proportions, the way it does with means. But we can "trick" the software into providing us with a confidence interval when we have a variable with two values coded as 0 and 1. That is because, with a 0/1 variable, the proportion of cases that score a 1 is equal to the mean of the variable. So, if we construct a confidence interval around the mean of a 0/1 variable, we are in effect calculating the upper and lower limits of a confidence interval around a proportion.

We can use the same "Explore" procedure we used when we calculated a confidence interval on a mean. We place the variable *V202187b* into the "Dependent List" box and click on "OK." The relevant portion of the output is shown in Figure 8.12.

Descriptives

			Statistic	Std. Error
V202187b	Mean		.7517	.00507
	95% Confidence Interval for Mean	Lower Bound	.7418	
		Upper Bound	.7616	
	5% Trimmed Mean		.7796	
	Median		1.0000	
	Variance		.187	
	Std. Deviation		.43207	
	Minimum		.00	
	Maximum		1.00	
	Range		1.00	
	Interquartile Range		.00	
	Skewness		-1.165	.029
	Kurtosis		-.642	.057

Figure 8.12

We see the sample statistic: the proportion of respondents (or the mean of a 0/1 variable) who gave the CDC a rating of higher than 50 degrees (0.75). The 95% confidence interval, shown in the circled section of the output, is (0.74, 0.76). The confidence interval suggests a high probability that the true proportion of people who feel warmly about the CDC falls between 74% and 76%. This confidence interval is consistent with the confidence interval for the mean rating of the CDC (69.7, 70.7), which also falls on the warm side of the feeling thermometer. Since the outbreak of COVID-19, the CDC has been a consistent object of public attention. At least according to this feeling thermometer, the public feels generally positively toward the agency.

Review of SPSS Procedures

- Find the confidence interval for a mean

    `Analyze → Descriptive Statistics → Explore`

- Find the confidence interval for a proportion (0/1 variable only)

    `Analyze → Descriptive Statistics → Explore`

Practice Problems

1. In 2021, 2,650 respondents told the GSS the extent to which they have confidence in organized religion. Of the respondents who answered the question, 879 people reported having "hardly any" confidence in organized religion.

    a. What proportion of the sample has "hardly any" confidence in organized religion?

    b. Calculate a 95% confidence interval for the proportion of all US residents who have "hardly any" trust in organized religion. Show all of your steps.

    c. State what this confidence interval means in words.

2. In 2021, the GSS asked respondents how good or bad they thought the impacts of climate change would be for the world as a whole. Out of 1,737 respondents, 428 said that the impact of climate change on the world would be "extremely bad."

    a. What proportion of respondents said that the impact of climate change on the world would be something other than "extremely bad?"

    b. Calculate a 90% confidence interval for the proportion of all US residents who think that the impact of climate change on the world, as a whole, will be something other than "extremely bad." Show all of your steps.

    c. State what this confidence interval means in words.

3. In 2020, the ANES asked a sample of 8,152 respondents to rate their feeling about the Democratic Party on a feeling thermometer ranging from 0 (least favorable) to 100 (most favorable). The mean feeling thermometer score for the sample is 45.3, and the sample standard deviation is 32.3.

   a. Calculate a 95% confidence interval for the mean thermometer rating of the Democratic Party among all US residents. Show all of your steps.

   b. State what this confidence interval means in words.

4. The Gini coefficient is a measure of inequality, ranging from 0 to 1, with 0 indicating perfect equality, with everyone earning the same income, and 1 indicating total inequality, which means one person earns all of the income. The higher the Gini coefficient, the greater the inequality. Sometimes the coefficient is expressed as a percentage ranging from 0% to 100%. Figure 8.13 shows 90% confidence intervals for the Gini coefficients for income in the United States between 1967 and 2015.

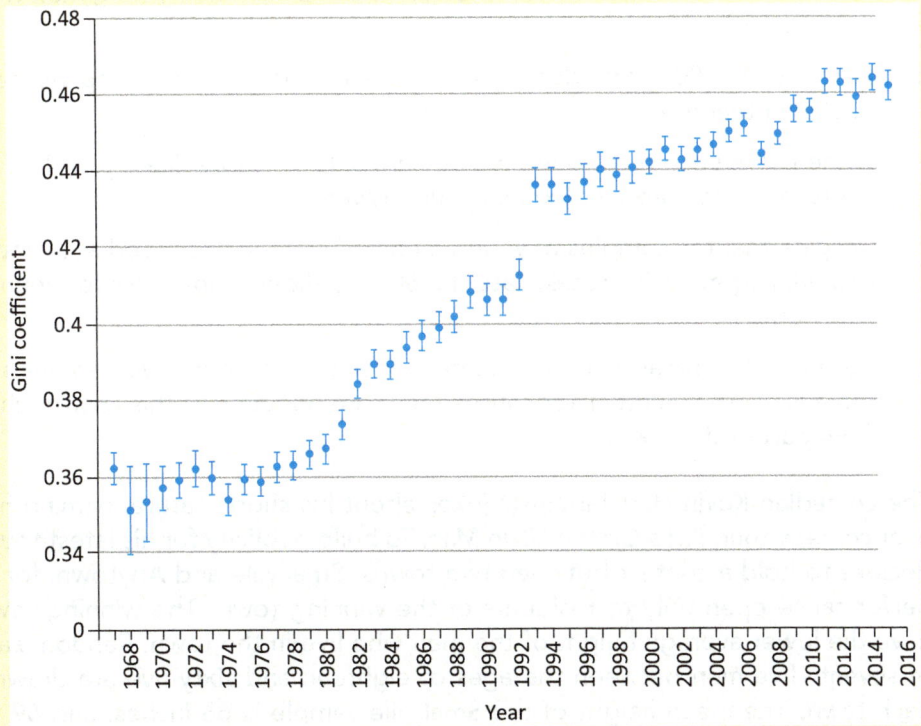

Figure 8.13   90% Confidence Intervals for Gini Coefficients, 1967–2015.
Source: U.S. Census, Current Population Survey.

   a. In 2016 the average of the Gini coefficients from random samples of US residents in all fifty states was 0.464. If we want to estimate a confidence interval around this point estimate of the Gini coefficient, should we estimate a confidence interval for a proportion or a mean?

b. If the standard deviation for the 2016 Gini coefficient is 0.32 and the total N for all fifty states is fifty thousand, estimate the appropriate 90% confidence interval. Show all of the steps.

c. Figure 8.13 does not include the 2016 confidence interval. Based on your calculation of this confidence interval for part b, do the 2015 and 2016 confidence intervals overlap?

d. Based on your answer for part c, can we say that income inequality in the United States, as indicated by the Gini coefficient, increased from 2015 to 2016?

5. In a probability sample of 2,023 people aged 15 to 24, the Public Religion Research Institute (PRRI) found that 37% of young women and 30% of young men had volunteered for a group or cause that they care about. Say that PRRI surveyed 1,072 women and 951 men.

a. Calculate the 80% confidence interval for the proportion young men who had volunteered.

b. Calculate the 80% confidence interval for the proportion young women who had volunteered.

c. Is there evidence that young women are actually more likely to volunteer for causes they care about? Explain your answer.

d. Imagine that the samples of young men and women described above were 70 and 38, respectively. Recalculate the 80% confidence intervals for young men and women.

e. Based on this recalculation, is there evidence that young women are actually more likely to volunteer for causes they care about? Has this answer changed from part c? If so, why?

6. The comedian Kevin Hart famously jokes about his short stature, naming his first ever comedy tour, *I'm a Grown Little Man*. To build publicity for his latest tour, Hart decides to hold a contest between two towns, Smallville and Anytown, for a free performance open only to residents of the winning town. The winning town will have the lower average height of the men who live in the town. Random samples of seventy-five men between the ages of eighteen and sixty-five are drawn from each town. The mean height of the Smallville sample is 68 inches, and 69 inches for Anytown. The standard deviations are 2 inches for Smallville and 3.75 inches for Anytown.

a. Find the 80% confidence intervals for the mean height of men in both towns.

b. Is there a winner of the contest, according to these confidence intervals?

c. The 99% confidence intervals for the mean height of men in Smallville and Anytown are 67.4 to 68.6 inches for Smallville and 67.8 to 70.2 inches for Anytown. Compare the range of each town's 99% confidence interval to the ranges for the 80% confidence intervals from part a. Which confidence intervals are wider, the 80% or 99%? Explain why.

7. A researcher is analyzing data from a random sample of cigarette smokers over the age of sixteen. She uses the sample data to estimate the average age of all cigarette smokers. Imagine that the researcher's confidence interval is the one shown in blue, in Figure 8.14.

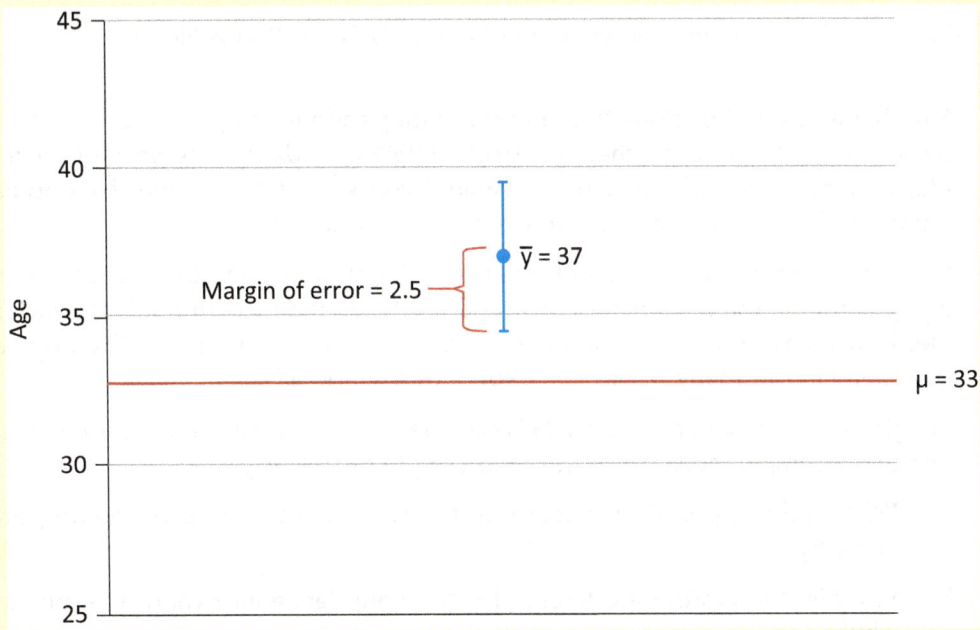

Figure 8.14 Confidence Interval for Mean Age of Smokers

   a. According to Figure 8.14, what is the mean age of smokers in general?
   b. What is the mean age of the smokers in the researcher's sample?
   c. What is the value for the bottom of the confidence interval? The top? Does the confidence interval from the researcher's study include the population mean? How do you know?

8. A random sample of American teenagers spends an average of 8 hours per day on smartphones. Using these sample data, an analyst estimates the average daily smartphone use, in hours, for all American teenagers. Imagine that the analyst produces the confidence interval shown in Figure 8.15.

   a. What is the mean hourly use of smartphones by teenagers in the sample?
   b. What is the value for the bottom of the confidence interval? The top?

c. Say that, among all American teenagers, mean daily smartphone time is 7.2 hours. Does the confidence interval from the researcher's study include the population mean?

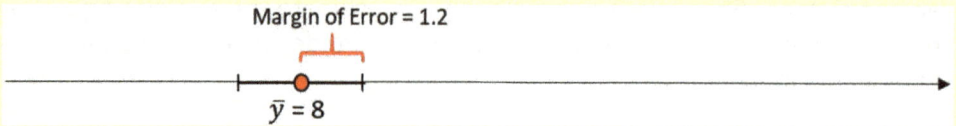

Figure 8.15 Confidence Interval for Mean Daily Smart Phone Hours

9. Say that a researcher drew two hundred independent samples of equal size from a population. If the researcher constructed 95% confidence intervals for each sample, how many confidence intervals would you expect to include the population parameter? How many would you expect to exclude it?

10. Say that a researcher drew one thousand independent samples of equal size from a population. If the researcher constructed 90% confidence intervals for each sample, how many confidence intervals would you expect to include the population parameter? How many would you expect to exclude it?

11. Imagine that you are calculating 99% confidence intervals around a sample proportion and a sample mean from the same sample of size fifty.

    a. What is the appropriate z-score for the confidence interval of the proportion, and why?

    b. What is the appropriate t-value for the confidence interval of the mean, and why?

    c. Which is larger, the z-score or the t-value? Explain why.

12. Imagine that you are calculating 95% confidence intervals around a sample proportion and a sample mean from the same sample of size twenty-five.

    a. What is the appropriate z-score for the confidence interval of the proportion, and why?

    b. What is the appropriate t-value for the confidence interval of the mean, and why?

    c. Which is larger, the z-score or the t-value? Explain why.

13. A polling firm wants a margin of error of 3 percentage points for a 90% confidence interval estimating the percentage of people who plan to vote for a statewide ballot measure. Calculate the size of the sample that would have to be collected to meet these conditions.

14. A large nonprofit organization wants a margin of error of 2.5 and a 95% confidence interval for mean support for LGBTQ causes across the state, on a 0 to 100 feeling

thermometer. The population standard deviation is estimated to be 15. Calculate the size of the sample that would have to be collected to meet these conditions.

15. The Pew Research Center posted an article on their website called "When Writing about Survey Data, 51% Might not Mean a Majority." Use the logic of confidence intervals to explain why this title is true.

16. According to the *Merriam-Webster* dictionary, one definition of the word "confidence" is "the quality or state of being certain." A statistician looks at this definition and suggests that "confidence intervals" should be renamed "probability intervals." Explain why she has made this suggestion.

Stata Problems

Here we will use the 2021 GSS to examine how Americans perceive their social standing relative to others. The variable *rank* measures where respondents would place themselves on a social scale relative to others, with 10 being at the very bottom and 1 being at the very top. We will use this interval-ratio version of the variable and a second version, called *lowrank*. *Lowrank* has two categories: 0 for respondents who placed themselves on the higher part of the social scale (from 1 to 5 for *rank*) and 1 for respondents who placed themselves on the lower part of the social scale (from 6 to 10 for *rank*).

1. Open the GSS data file (`GSS2021.dta`). Use the appropriate "ci" command to generate a 90% confidence interval for *rank*.
2. According to the output, how many respondents answered this question, and what is the mean response? Does the mean suggest, generally, that people see themselves as ranking higher or lower than others?
3. What is the standard error of the sample mean, according to the output?
4. What is the 90% confidence interval for the mean, according to the output? Explain what this particular confidence interval means in words.
5. Use the appropriate "ci" command to generate a 90% confidence interval for *lowrank*.
6. According to the output, what percentage of respondents said that they rank in the bottom half of the scale?
7. According to the output, what is the 90% confidence interval for the percentage of people who said that they rank on the bottom half of the scale? Explain what the confidence interval means in words.

SPSS Problems

Here we will use the 2021 GSS to examine how Americans perceive their social standing relative to others. The variable *rank* measures where respondents would place themselves on a social scale relative to others, with 10 being at the very bottom and 1 being at the very top. We will use this interval-ratio version of the variable and a second version, called *lowrank*. *Lowrank* has two categories: 0 for respondents who placed themselves on the higher part of the social scale (from 1 to 5 for *rank*) and 1 for respondents who placed themselves on the lower part of the social scale (from 6 to 10 for *rank*).

1. Open the GSS data file (`GSS2021.sav`). Use the "Explore" procedure to generate a 90% confidence interval for *rank*.

2. According to the output, how many respondents answered this question, and what is the mean response? Does the mean suggest, generally, that people see themselves as ranking higher or lower than others?

3. What is the standard error of the sample mean, according to the output?

4. What is the 90% confidence interval for the mean, according to the output? Explain what this particular confidence interval means in words.

5. Use the appropriate "Explore" procedure to generate a 90% confidence interval for *lowrank*.

6. According to the output, what percentage of respondents said that they rank in the bottom half of the scale? Is this percentage surprising, given the sample mean of *rank*? Explain your answer.

7. According to the output, what is the 90% confidence interval for the percentage of people who said that they rank on the bottom half of the scale? Explain what the confidence interval means in words.

Notes

[1] Michael Hartman. 2022. "Cannabis Overview." National Conference of State Legislatures. May 31. Retrieved from https://www.ncsl.org/civil-and-criminal-justice/cannabis-overview.

[2] J. Felson, A. Adamczyk, and C. Thomas. 2019. "How and Why Have Attitudes about Cannabis Legalization Changed So Much?" *Social Science Research* 78: 12–27.

[3] See Emily Ekins. 2021. "Why Did Republicatns Outperform the Polls Again? Two Theories." *FiveThirtyEight*. March 2. Retrieved from https://fivethirtyeight.com/features/why-did-republicans-outperform-the-polls-again-two-theories/. See also Judy Woodruff. 2020. "What Went Wrong with Polling in 2020?" PBS. November 19. Retrieved from https://www.pbs.org/newshour/show/what-went-wrong-with-polling-in-2020.

[4] John Gramlich. 2021. "America's Incarceration Rate Falls to Lowest Level since 1995." Pew Research Center. August 20. Retrieved from https://www.pewresearch.org/fact-tank/2021/08/16/americas-incarceration-rate-lowest-since-1995/.

[5] Christy A. Visher and Shannon M. E. Courtney. 2007."One Year Out Experiences of Prisoners Returning to Cleveland." *Returning Home: Understanding the Challenges of Prisoner Reentry*. Urban Institute, Justice Policy Center. April 2007. 3–8. Retrieved from https://www.urban.org/sites/default/files/publication/43021/311445-One-Year-Out-Experiences-of-Prisoners-Returning-to-Cleveland.PDF.

[6] M. A. Painter and Z. Qian. 2016. "Wealth Inequality among New Immigrants." *Sociological Perspectives* 59(2): 368–394.

[7] Kalee Burns, Liana Fox, and Danielle Wilson. 2022. "Expansions to Child Tax Credit Contributed to 46% Decline in Child Poverty since 2020." *US Census Bureau*. September 13. Retrieved from https://www.census.gov/library/stories/2022/09/record-drop-in-child-poverty.html#:~:text=Child%20poverty%2C%20calculated%20by%20the,Census%20Bureau%20data%20released%20today.

8. R. Hoekstra, R. D. Morey, J. N. Rouder, and E. J. Wagenmakers. 2014. "Robust Misinterpretation of Confidence Intervals." *Psychonomic Bulletin & Review* 21(5): 1157–1164.
9. Courtney Kennedy, Dana Popky, and Scott Keeter. 2023. "How Public Polling Has Changed in the 21st Century." Pew Research Center. April 19. Retrieved from https://www.pewresearch.org/methods/2023/04/19/how-public-polling-has-changed-in-the-21st-century/.
10. Jennifer Agiesta. 2020. "Post-Debate CNN Poll: Six in 10 Say Biden Won the Debate." *CNN Politics*. September 30. Retrieved from https://www.cnn.com/2020/09/29/politics/donald-trump-joe-biden-debate-poll/index.html.
11. Alan Agresti. 2018. *Statistical Methods for the Social Sciences*, fifth edition. New York: Pearson, 112.

Chapter 9

Differences between Samples and Populations
One-Sample Hypothesis Tests

In chapter 8, we discussed how to use sample data to make interval estimates, called confidence intervals, of population proportions and means. Along with using confidence intervals, many researchers also use a tool called hypothesis testing to assess the likelihood that a sample statistic reflects a population parameter. For instance, we could ask whether a particular segment of college students (such as students at your college) is similar to college students in general in the mean number of years it takes to complete their degree. Hypothesis tests use the logic of probability and sampling distributions. Tests of the relationship between a sample and a larger population are called **one-sample tests**. Researchers use the same procedures to test hypotheses about the difference between two samples, in two-sample tests, which we discuss in chapter 10.

The results of hypothesis tests are sometimes used to assess the effectiveness of new vaccines or medications. In 2020, after COVID-19 began to spread worldwide, pharmaceutical companies accelerated their research efforts to develop a vaccine that would protect people against the new virus.

In November 2020, several months into this effort to develop a vaccine, the pharmaceutical company Pfizer reported the results of a large clinical trial. The results indicated that their new COVID-19 vaccine was highly effective. On November 20, 2020, the *New York Times* ran a story with this headline: "2 Companies Say Their Vaccines Are 95% Effective. What Does That Mean?"[1] We can use the information contained in this article to illustrate how a hypothesis test might work.

During this clinical trial, Pfizer randomly assigned 43,661 volunteers into one of two groups. The groups were about the same size: one half (about 21,830 participants) received the new vaccine, and the other half (also about 21,830 participants) received a placebo. The researchers then waited for participants to experience symptoms of COVID-19 and to test positive. If the vaccine was ineffective, we would expect roughly the same number of positive cases in the vaccine group and in the placebo group. But out of the 170 people who came down with COVID during the trial period, 162 of them came from the placebo group, and only eight participants came from the vaccine group. This appears to be very strong evidence that the vaccine was effective.

How effective? Here is another way to look at it, and a way that would allow us to conduct a one-sample hypothesis test. Consider the information in Table 9.1.

Table 9.1   COVID-19 Vaccination Trials

Group	Number of Participants	Number of Positive Cases	Rate of Infection
Vaccine	21,830	8	0.04%
Placebo	21,830	162	0.74%

The fourth column, with the heading "Rate of Infection," gives us the percentage of cases in each group that contracted the COVID-19 virus. Within the random sample of vaccine recipients, 0.04% came down with COVID. This is calculated as follows:

$$\frac{8}{21,380} = 0.00037 \cong 0.0004$$

To compute this percentage, we multiply 0.0004 by 100, and end up with 0.04%, the infection rate among vaccinated participants. Following these same steps, we found the infection rate within placebo group to be 0.74%. Let's now use this information to illustrate how a one-sample hypothesis test might work.

We will use 0.04% as our sample proportion in a hypothesis test. The essence of the hypothesis test is a comparison between the sample proportion we actually found, and some hypothetical claim about the population from which that sample is drawn. In this example, we want to know if the vaccine is effective in preventing COVID. If the vaccine is not effective, then we would expect the infection rate within a sample of vaccinated people to be the same as or close to the infection rate for unvaccinated people.

We will begin with an assumption: that the vaccine is ineffective. In that case, the incidence of COVID in the full population of vaccinated people would be the same as it is among unvaccinated people. We know from the Pfizer study that 0.74% of the placebo sample came down with COVID. Let's use that sample proportion of 0.74% to generate a hypothesis (that may or may not be correct): that in the full population of vaccinated people, the COVID infection rate is 0.74% (the same as unvaccinated people). In other words, our hypothesis is that the vaccine is not effective. This starting hypothesis—that there is no difference between the vaccine group and the overall population—is called the null hypothesis. (Null refers to zero or no difference.)

The logic of the hypothesis test relies on the Central Limit Theorem, which tells us that repeated random samples will yield sample statistics normally distributed around the population parameter. We will compare our obtained sample proportion (0.04%) to a hypothetical claim about the population (0.74%). To put it differently, we will ask this question: If the COVID infection rate among all vaccinated people in the United States is 0.74%, how likely is it that we would have gotten a vaccinated sample proportion of only 0.04%? In this chapter, we will learn how to ask and answer questions such as this one.

But without doing the calculations right here, it turns out that in this example, it would have been almost impossible to obtain a vaccinated sample infection rate of only 0.04% if the true infection rate among all vaccinated people was 0.74%. This hypothesis test would provide us with evidence to support the claim that the true infection rate among all vaccinated people is in all likelihood less than 0.74%; in other words, we would have evidence that the vaccine is effective.

A hypothesis test such as this one has real-world ramifications. The sample results reported in Table 9.1 show the infection rates for both groups—vaccinated and placebo—to be quite low, 0.04% and 0.74% respectively. But we use samples to make claims about the population. The difference in infection rates of 0.7% (0.74 − 0.04) represents 0.7% of the population of the United States. If the US population is 332 million people, and we multiply that by 0.7%, the product is about 2.3 million. That represents the potential reduction in the number of COVID cases in the full population if everyone had been vaccinated.

The Logic of Hypothesis Testing

Hypothesis testing relies on a counterfactual logic. Instead of trying to prove one thing, we try to disprove the opposite. Specifically, instead of trying to prove that our sample is different from a population of interest, we see whether we can *disprove* that our sample is the same as that population. We begin by generating a null hypothesis. The **null hypothesis** states that an unknown population parameter is equal to a specified value. **Hypothesis testing** determines how likely it is that we would obtain a sample statistic at least as far from the population parameter as our sample statistic, if that null hypothesis were correct. We often think about tests as providing us with decisive answers to questions. But hypothesis tests depart from our common understanding of how tests work because they do not give us definitive answers about whether our hypotheses are correct.

That may sound convoluted, but this kind of reasoning is something many of us use in daily life. For example, imagine that a sister and her brother are allowed to share one bag of M&Ms after dinner. Their parents leave the brother in charge of dividing them fairly. The sister has a strong preference for blue M&Ms and expects to get many in her half. After her brother divides the M&Ms and delivers his sister her half, she sees that she has just one blue candy. Remembering the many blue M&Ms that she has seen in the past when she has had her own bag, she suspects her brother of hoarding the blue M&Ms for himself. She therefore believes that her "sample" is not representative of the "population" of M&Ms that made up the entire bag. The sister complains to her parents. Her parents, however, tell her not to assume that her brother has been unfair. Unlike their daughter, who wants to prove that her brother has treated her unfairly, her parents want to assume that their son has treated his sister fairly until they are convinced otherwise. Their null hypothesis is that the proportion of blue M&Ms in the population—that is, the full bag—is the same as the proportion in the sample—that is, the sister's share. Casting doubt on the null hypothesis would require the parents to consider whether the proportion of blue M&Ms in their daughter's half is sufficiently different from the proportion of M&Ms in the full bag to suggest that the difference is unlikely to have occurred due to random chance. Our conclusion at the end of a more precise statistical process would state whether there is a statistically significant difference between the proportion of blue M&Ms in the sample (i.e., the sister's share) and the proportion of blue M&Ms in the population (i.e., the entire bag).

A hypothesis always makes a statement about a population with an unknown parameter. Hypothesis tests use the idea of a sampling distribution to imagine what would happen if we repeatedly collected samples. We know from chapter 7 that repeated samples would yield results that follow a normal distribution, with the true population value at the center of the distribution. One-sample hypothesis tests place the value specified in the null hypothesis at the center of that normal distribution. In other words, hypothesis tests utilize sampling distributions to tell us how likely it is that we would have observed a statistic as extreme as the one we saw in our sample (e.g., the COVID infection rate among vaccinated participants) if the null hypothesis were true.

Null Hypotheses (H_0) and Alternative Hypotheses (H_a)

The first step in hypothesis testing is to generate a null hypothesis, which is the hypothesis being tested. The null hypothesis, denoted by H_0, makes a statement about the value of a population parameter. We never hypothesize about a sample because we know the sample statistic. The null hypothesis will always state that the unknown parameter for the population represented by the sample is equal to a specific value. Where do researchers get the specific value for the null hypothesis? Sometimes researchers have data on a full population, such as an election result (which includes all voters), data from the US Census (which includes all residents), or government crime reports (which include all reported crimes), and they want to know whether their sample resembles that known population. Sometimes researchers use an expected finding as the specific value for the population parameter. This expected finding might come from prior research or from a prediction or claim. For example, prior research might show that 40% of all college students engage in binge drinking; the corresponding null hypothesis would use 40% as the population parameter. A study of whether there is a gender difference in declared majors in science might use 58% women as the population parameter for the null hypothesis because 58% of all undergraduates are women.[2] A researcher might wonder whether Americans eat the recommended five servings of vegetables a day; in this case, the null hypothesis would use five servings as the population parameter. In all cases, the null hypothesis is a statement of no difference between population and sample, thus explaining why this statement is called the "null" hypothesis.

The next step is to formulate the **alternative hypothesis** (H_a). It, too, is a hypothesis about the population. The alternative hypothesis states that the true population value for the group being studied (i.e., the group from which we drew the sample) is different from the value established for the population parameter in the null hypothesis. The null and alternative hypotheses make competing and mutually exclusive claims about the value of a population parameter; they cannot both be true. The alternative hypothesis is sometimes referred to as the research hypothesis because it often refers to what the researcher expects to find, a difference between the hypothesized population parameter and the statistic from the sample being studied. Unlike the null hypothesis, the alternative hypothesis always states that the population parameter is equal to a range of values. For example, a researcher studying the gender composition of science majors knows that 58% of all undergraduate students are women and sets up the null hypothesis that the percentage of science majors who are women is equal to 58%. The researcher's alternative hypothesis is that the percentage of women in science majors is not equal to 58%. Using the properties of the sampling distribution, the hypothesis test will tell us the probability of obtaining a result as extreme as our sample result if the null hypothesis were true.

One-Tailed and Two-Tailed Tests

There are two versions of hypothesis tests: one-tailed and two-tailed tests (sometimes called one-sided and two-sided). In a **two-tailed test**, the alternative hypothesis states that the population parameter is not equal to the value stated in the null hypothesis—it may be higher or lower than that value. Here, the researcher does not have a specific expectation of the direction of difference for the alternative hypothesis. In a **one-tailed test**, however, the researcher asks more specifically whether the population parameter is less than or greater than a specified value. The alternative hypothesis in a one-tailed test specifies that the population parameter is either higher or lower than the value specified by the null hypothesis. For example, the researcher studying gender and science majors might hypothesize that the proportion of women majoring in science is lower than 0.58, the proportion of women undergraduates, rather than hypothesizing that it is simply different. In contrast, in a two-tailed test, the alternative hypothesis would be that the proportion of women in science majors is either greater or less than 0.58.

The decision of whether to employ a one-tailed or a two-tailed test is usually dictated by the question at hand. If the researcher is looking for the probability of a sample result either only above or only below the null hypothesis, then a one-tailed test is called for. A researcher would do so if she had a strong reason, such as findings from previous research, to expect a result that was either above or below the population parameter. On the other hand, a two-tailed test is required if the problem directs the researcher to look for an effect both below and above the null hypothesis—that is, a sample result that differs from the population in either direction. Alternatively, researchers may choose to employ a two-tailed test because it doubles the p-value and thus adds a higher degree of certainty if the null hypothesis is ultimately rejected. We will show you examples of both one- and two-tailed hypothesis tests.

Hypothesis Tests for Proportions

Next, we will conduct a hypothesis test for proportions. Let's begin with an example. Imagine that your state is holding a referendum on the legalization of recreational marijuana. A poll of two hundred randomly selected young citizens, aged eighteen to thirty, interviewed right before the election shows that 59% of them support the legalization initiative. On Election Day, 54% of residents statewide voted to support the legalization initiative. Are young citizens really more supportive of the legalization of marijuana than voters in general?

Why do we even ask this question? At one level, it seems fairly obvious that young citizens are indeed more supportive of legalization. After all, if we just compare the percentages, it is clear that 59% of the young group support legalization, compared to 54% of all voters. That's a difference of 5 percentage points.

But we cannot hastily jump to this conclusion. Fifty-nine percent is a sample statistic. We have seen in chapter 8 that we should not rely on a point estimate because there is a range of possible population values attached to any sample statistic. If 59% of our sample of young people supports legalization, we know that there is a very high probability that the true proportion of young people in the population who support legalization will be 59% plus or minus a margin of error. That interval may include 54%, in which case we could not assert with confidence that young people are more likely than the general public to support legalization.

Hypothesis testing is a systematic way of addressing the question: Are young people more supportive of legalization than voters in general? We have two pieces of information at our disposal:

- In a random sample of two hundred young people, 59% support legalization.
- In the full population, 54% of voters support legalization.

Let's begin by assuming that—in reality—young people in the population support legalization at the same rate as all voters, which is 54%. Remember, this is an assumption, or a hypothesis. Where does it come from? We know that 54% of all voters support legalization. So, if young people support legalization at the same rate as everyone else, then 54% of young people in the population will support legalization.

This is our null hypothesis:

H_0: *The proportion of young people in the population who support legalization is 54%*

In statistics, we use Greek letters to represent population parameters and Roman letters to represent sample statistics—in this case, π represents the proportion for the population (see Box 7.1 for a refresher on notation). Thus, the null hypothesis in statistical notation is:

$$H_0: \pi = 0.54$$

Next, we formulate the alternative hypothesis (H_a). In this case, we are interested in the question of whether young people in the population are more supportive of legalization than the rest of the electorate. So our alternative hypothesis is:

H_a: *The proportion of young people in the population who support legalization is greater than 54%*

In statistical notation:

$$H_a: \pi > 0.54$$

Unlike the null hypothesis, the alternative hypothesis always states that the population parameter is equal to a range of values. In this case, we are suggesting that the proportion of young people who support legalization is greater than 54%. This is a one-tailed test.

Let's review what we have done. Using information about the full population (54% of them support legalization), we are testing the claim that young people in the population support legalization at the same rate as everyone else. Alternatively, young people may be more supportive of legalization.

The hypothesis test always proceeds under the assumption that the null hypothesis is true. We test the null hypothesis by seeing whether our sample data provide strong evidence against it. The evidence we use is the probability that we could obtain our sample statistic if the null hypothesis were true. If the evidence against the null hypothesis is strong enough, this leads us to support its opposite, the alternative hypothesis. In the context of this example, this means that we are going to answer the following question:

> If 54% of all young people support the legalization of marijuana, what is the probability that, out of a randomly selected sample of two hundred young people, at least 59% of them would support legalization?

Figure 9.1 graphically displays the question we are asking. The normal curve in Figure 9.1 is a theoretical distribution of sample proportions, centered on the hypothesized population parameter ($\pi = 0.54$).

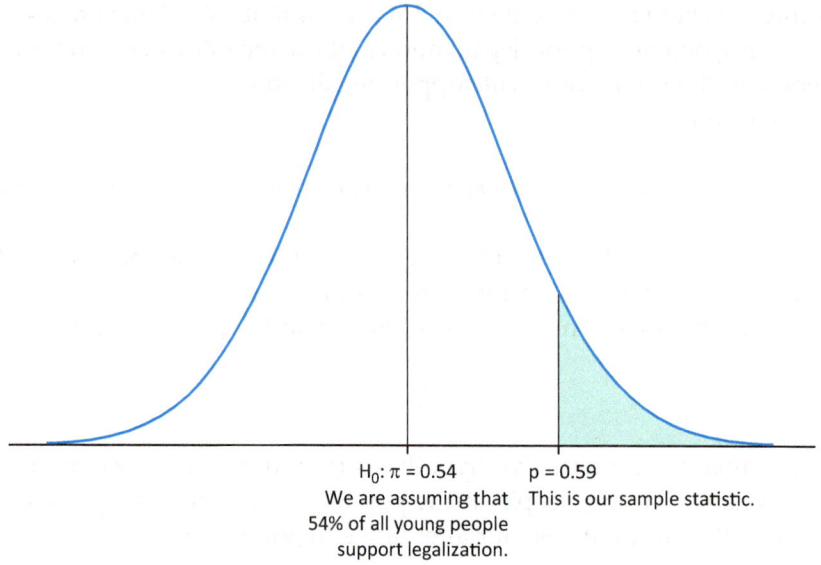

Figure 9.1 The Null Hypothesis and Sample Statistic on the Normal Curve, Young People's Support for Legalization of Marijuana

Using what we know from our discussion of the Central Limit Theorem and sampling distributions, we can calculate the probability of obtaining a sample proportion of 0.59 or higher under the assumption that the true population proportion is 0.54. Toward this end, we must calculate the z-score for $p = 0.59$ and use the Normal Table (see appendix A) to find the corresponding p-value, which is the probability associated with that z-score. In order to calculate z, we must calculate the standard error. You learned how to calculate standard errors and z-scores in chapter 8, in order to calculate confidence intervals. But there is one important difference between the construction of the standard error for a confidence interval and for a hypothesis test. Recall that when we construct a confidence interval, we calculate the standard error on the basis of the sample statistic:

$$SE_p = \sqrt{\frac{p(1-p)}{N}}$$

When we run a hypothesis test for the difference between a sample proportion and hypothesized population proportion, we use the population parameter (π) instead of the sample statistic (p).

$$SE_p = \sqrt{\frac{\pi(1-\pi)}{N}}$$

We use the population parameter because, in a hypothesis test, we proceed from the assumption that our guess about the population parameter (the H_0) is true, and so we construct a distribution of sample proportions around that hypothesized value. In this example, we are operating from the assumption that the true population proportion for young people is 0.54.

The standard error is:

$$SE_p = \sqrt{\frac{\pi(1-\pi)}{N}} = \sqrt{\frac{0.54(1-0.54)}{200}} = \sqrt{\frac{0.54(0.46)}{200}} = 0.035$$

Next, we need to calculate a z-score for a sample proportion of at least 0.59. Or, in other words, how far away is our obtained sample proportion from the hypothesized population parameter, in standard errors? We calculate z using the same formula we used in chapter 8:

$$z = \frac{p-\pi}{SE_p} = \frac{0.59-0.54}{0.035} = 1.43$$

Figure 9.2 shows the normal curve, with the null hypothesis of $\pi = 0.54$ at the center. The sample proportion, p, is marked on the right, and we have added its z-score of 1.43 and the p-value associated with that z-score, 0.076. Remember: "p" refers to the sample proportion (0.59), and "p-value" refers to the probability of obtaining a given z-score or higher.

How do you find the p-value marked on Figure 9.2? By looking up your z-score in the Normal Table in appendix A. The Normal Table shows that the p-value corresponding to a z-score of 1.43, represented by the shaded area, is 0.076. In other words, if the null hypothesis is true—that is, if 54% of all young people support legalization—then there is about a 7.6% chance that we would have obtained a sample statistic equal to 59% or higher.

Many analysts misinterpret the p-value. They may describe it as the probability that the null hypothesis is true. Or they may characterize it as the probability of obtaining the specific sample percentage they got. These interpretations are incorrect. There is only one way to correctly interpret the p-value: It is the probability of obtaining a sample statistic

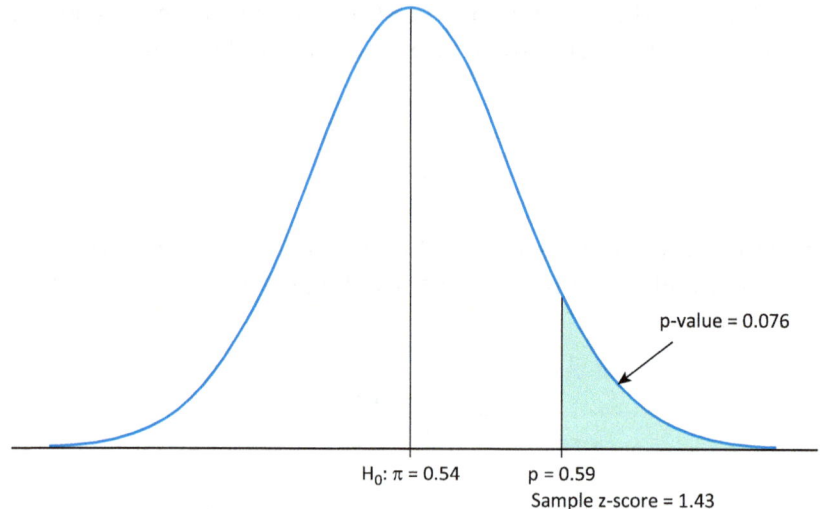

Figure 9.2 The Null Hypothesis and Sample Statistic on the Normal Curve, with Corresponding z-Score and p-Value, Young People's Support for Legalization of Marijuana

that is at least as far away from the population parameter as our sample statistic, if the null hypothesis were true. In the case of a sample statistic that is lower than the null hypothesis value, the p-value is the probability of obtaining the sample statistic or lower if the null hypothesis were true. Even some scientists who have spent their careers thinking carefully about p-values have difficulty translating the technical statement about what p-values tell us into an intuitive interpretation.[3]

Some students find the interpretation of the p-value to be counterintuitive. This is because the lower the p-value, the more evidence the researcher has against the null hypothesis. In other words, lower p-values make it more likely that the researcher will reject the null hypothesis and accept the alternative hypothesis. A very low p-value means that, if our null hypothesis about the population were true, then it would have been very unlikely to obtain the sample statistic that we got. But since we did obtain that sample statistic, we conclude that our null hypothesis about the population is probably incorrect. So, with a very low p-value, we will reject the null hypothesis. If we reject the null hypothesis, we can accept the alternative hypothesis.

How low is a low enough p-value? That is a subjective decision that the investigator makes before actually conducting the hypothesis test. Many researchers use a threshold of 5% or 1%, meaning that the obtained p-value must be lower than one or the other of these thresholds in order to reject the null hypothesis. Statisticians call this threshold **alpha**, which is the Greek letter α. In social science, most researchers choose an alpha level of 0.10, 0.05, or 0.01. Any p-value that is less than alpha allows the researcher to reject the null hypothesis and to declare that his or her finding is "statistically significant." That phrase simply means that the results of our hypothesis test have allowed us to reject the null hypothesis. If the p-value is above alpha, we cannot reject the null hypothesis and therefore must conclude that any difference between our sample statistic and the hypothesized population parameter could be due to the normal variation in results from different samples.

In this case, we would say that the difference between our sample and the population is not statistically significant. Again, researchers generally set alpha before beginning to calculate sample z or its associated p-value.

Let's return to our example. We found that if 54% of all young people supported legalization, then there's a 0.076 chance that we would get a sample percentage of at least 59%. Should we reject our null hypothesis? The answer to that question depends on what alpha level we established before conducting the hypothesis test. Table 9.2 summarizes three scenarios with different levels of alpha.

Table 9.2 Rejecting or Failing to Reject the Null Hypothesis for Different Levels of Alpha, Young People's Support for Legalization of Marijuana

	Obtained p-Value	Alpha	Decision
Scenario 1	0.076	0.10	Reject null hypothesis
Scenario 2	0.076	0.05	Do not reject null hypothesis
Scenario 3	0.076	0.01	Do not reject null hypothesis

Only in Scenario 1, with an alpha level of 0.10, would the researcher reject the null hypothesis and conclude that young people are indeed more supportive of legalization. Scenarios 2 and 3 have lower alpha levels (0.05 and 0.01, respectively), and therefore, using either of these thresholds, the p-value of 0.076 exceeds alpha. In Scenarios 2 and 3, the researcher could not conclude that young people support the legalization of marijuana at a higher rate than the rest of the population.

The Steps of the Hypothesis Test

We have discussed the process of conducting a hypothesis test for the difference between a sample proportion and a hypothesized population parameter in detail here. It can be condensed into five steps. First, set up null and alternative hypotheses. Second, set an alpha level. Third, calculate the standard error. Fourth, calculate the z-score associated with the sample proportion. Fifth, find the p-value associated with the z-score in the Normal Table and determine whether to reject the null hypothesis by comparing the p-value with the chosen alpha level. We can then draw a conclusion about whether the sample statistic is significantly different from the hypothesized population parameter. We will walk you through these steps in other examples and show you how they apply if you are looking at means rather than proportions. But first, we will look more closely at the difference between one- and two-tailed hypothesis tests and show how a two-tailed test works.

BOX 9.1: IN DEPTH

Publication Bias toward Statistically Significant Results

One of the goals of science is to share research results so that we can use the findings from many studies to build a body of knowledge about important topics. A key forum for sharing results is academic journals. However, some have expressed concern about a bias in academic publishing, in which studies with "negative" results—that is, results that are not statistically significant—are less likely to be published.

Many researchers have criticized this tendency to privilege statistically significant results because it can distort our collective knowledge about important social issues. This publication bias means that important information about key topics remains buried in file drawers, discarded in trash bins, or lost in the cloud.

Publication bias has been a concern across the sciences. For example, the World Health Organization released a statement in 2015 that raised the concern of publication bias in clinical trials that test the effectiveness of medical interventions.[4] The statement calls on researchers to make all results of clinical trials publicly available, including those that do not reach an alpha-level that would meet the threshold of statistical significance.

Scientists must continue to challenge the perception that a study has failed to yield useful information if the results are not statistically significant. In fact, it is also important to know when a hypothesized result does not occur, or a subgroup is *similar to* rather than different from the population.

One-Tailed and Two-Tailed Tests in Practice

Let's return to our legalization of marijuana example and examine the difference between one and two-tailed tests in the context of this example. Here is the original, one-tailed, set of hypotheses:

H_0: 54% of all young people support legalization.

H_a: More than 54% of all young people support legalization.

When we originally looked at this problem (see Figure 9.2), we found that the probability of obtaining our sample proportion, if the null hypothesis were true, is 0.076. Let's be more precise now about what this means. The p-value means that if 54% of all young people support legalization, there is a 0.076 chance of getting a sample proportion of at least 59%. But what if we offer a different alternative hypothesis? Consider this set of hypotheses:

H_0: 54% of all young people support legalization.

H_a: The percentage of all young people who support legalization is not equal to 54%.

The alternative hypothesis is now two-tailed. In Figure 9.3, we depict what this new set of hypotheses is suggesting.

Notice that we have doubled our p-value. It is now 0.152 (0.076 in each tail). This means that if the null hypothesis were true—that is, if 54% of all young people support

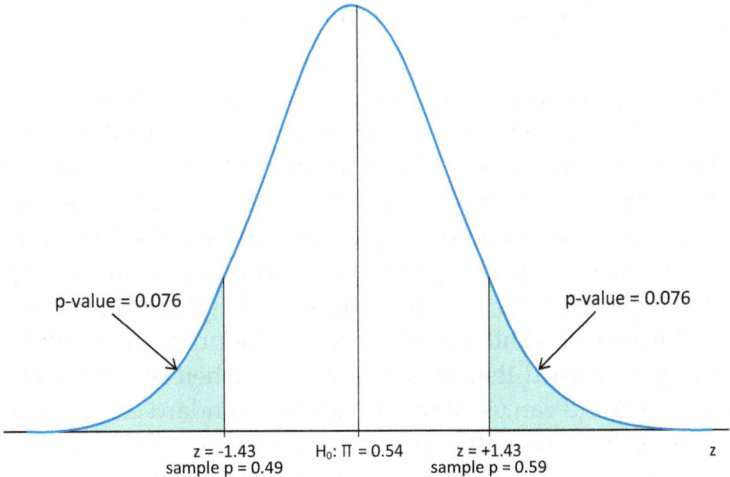

Figure 9.3 The Null Hypothesis and Sample Statistic on a Normal Curve, Two-Tailed Hypothesis Test, Young People's Support of Legalization of Marijuana

legalization—there is a 0.152 chance of getting a sample proportion as far in either direction from the null hypothesis value as the one we actually obtained. In other words, assuming that the null hypothesis is true, there would be about a 15% chance of obtaining a sample percentage of 59% or higher or 49% or lower. (49% is the same distance from the hypothesized population value—5 percentage points—as our sample proportion. With a two-tailed hypothesis, we are testing for a sample result either above or below the population value.) In our example, regardless of whether we had chosen an alpha level of 0.01, 0.05, or 0.10, we would not be able to reject the null hypothesis. Our p-value exceeds all of those alpha levels. We would therefore conclude that there is no significant difference in support for marijuana legalization between younger people and the overall population.

The mechanics of conducting a two-tailed test are identical to those of a one-tailed test, with one exception. For a two-tailed test, we double the p-value associated with the sample statistic. Naturally, this makes it harder to reject the null hypothesis. Another way to look at this is that a two-tailed test divides the alpha level between two tails, making the threshold for rejecting the null hypothesis more stringent.

BOX 9.2: IN DEPTH

Equivalence of Two-Tailed Hypothesis Tests and Confidence Intervals

Two-tailed hypothesis tests and confidence intervals are equivalent. A confidence level is equal to (1 − alpha). For example, a 95% confidence interval corresponds to a hypothesis test with an alpha-level of 5% (1 − 0.05). In other words, if a two-tailed hypothesis test rejects the null hypothesis at an alpha-level of 5%, then a 95% confidence interval around the sample proportion will fail to include the null hypothesis value.

Hypothesis Tests for Means

What if the research question you are interested in is about a mean rather than a proportion or percentage? For example, you may want to know whether the mean GPA of sociology majors is different from the mean GPA of all students at a university. Or you may want to know whether the mean income for graduates of your college is higher than the mean income for the general population. Many questions that social scientists are interested in deal with whether the mean for a specific group differs from the mean for the general population. Just as we saw with proportions, we use a one-sample hypothesis test to answer these questions. The procedure is very similar to what we saw for proportions, with one key difference.

In chapter 8, you learned that we use a z-score when we either know the population standard deviation (σ) and can use it to estimate the standard error or can estimate the standard error without using σ (as in the case of proportions). You learned that we use a t-value when we do not know the population standard deviation and must estimate standard error using the sample standard deviation. In almost all cases, when testing the difference between a sample mean and a population mean, we do not know σ and therefore must use t.

Let's look at income. Say we know that the mean income for young adults (age twenty-five to thirty-four) who are college graduates and are employed full-time is $50,000.[5] Imagine that we draw a random sample of two hundred graduates of your college who are employed full-time and ask them their income. We then calculate that the mean income for this sample is $51,750, and the standard deviation for the sample is $9,000. At first glance, it seems that graduates of your college earn more, on average, than college graduates overall. But we know that it is possible that this difference is due to sampling variation, rather than a real difference. As you know, whenever we draw a sample, our results are likely to vary somewhat from the true population mean, but most of the time the difference from the population will be small. (Remember, if you flip a coin twenty times, you will not always get ten heads and ten tails, but it would be unusual for you to get nineteen heads and one tail.)

Just as we did with proportions, we can determine how likely it is that we could get a sample mean as high as $51,750 if the mean income of graduates of your college is actually the same as that of college graduates overall. This is what a one-sample t-test does. The steps are similar to those for the hypothesis test for proportions. The main difference is that, instead of using the z-score associated with the sample statistic, we use the t-value. In both cases, we are using t or z to determine the probability of obtaining the sample result if the null hypothesis were true.

The steps for conducting the hypothesis test are shown in Table 9.3. Don't worry if you don't understand everything in Table 9.3 right now! We will go through each step in detail as we work through the example.

Step 1: Set up null and research hypotheses.

To conduct the hypothesis test, first, we set up hypotheses. In this case, our null hypothesis is that the mean income of all graduates of your college is the same as that of college graduates overall. In other words:

$$H_0 : \mu > 50,000$$

Table 9.3 Steps in Conducting One-Tailed Hypothesis Test for Mean Income of College Graduates

Step 1: Set up hypotheses.	$H_0: \mu = 50{,}000$; $H_a: \mu > 50{,}000$
Step 2: Set α, calculate DF, and find decision t.	$\alpha = 0.05$; DF = N − 1 = 200 − 1 = 199; decision t = 1.645
Step 3: estimate standard error.	$SE_{\bar{y}} = \dfrac{s}{\sqrt{N}} = \dfrac{9{,}000}{\sqrt{200}} = \dfrac{9{,}000}{\sqrt{14.14}} = 636.4$
Step 4: Calculate sample t.	$t = \dfrac{\bar{y} - \mu}{SE_{\bar{y}}} = \dfrac{51{,}750 - 50{,}000}{636.4} = \dfrac{1{,}750}{636.4} = 2.75$
Step 5: Compare sample t and decision t; draw conclusion.	2.75 > 1.645; therefore, reject H_0

Note: Because the Normal Table and t-table are set up differently (see appendices A and B), the steps in testing hypotheses about means and proportions differ slightly.

(Remember, we use Greek letters to represent the population parameters and Roman letters to represent sample statistics—in this case, μ and ȳ represent the means for the population and the sample, respectively.)

The alternative hypothesis is that the mean income of all graduates of your college is greater than that of college graduates overall. In other words:

$$H_a : \mu > 50{,}000$$

Drawing the t-curve to represent our null hypothesis, we place the known population mean, 50,000, at the location of the mean (Figure 9.4). We place the sample mean further out on the curve. (Until we calculate the t-value for the sample mean, we do not know exactly where on the curve the sample mean is located.) Recall from chapter 8 that a t-curve has the same general properties as the normal (z) curve and also represents a sampling distribution.

As with the hypothesis test for a proportion, we will estimate the probability of obtaining a sample mean of 51,750 or higher if the true population mean is 50,000. However, as we will see, while we can look up the exact p-value associated with our sample statistic when that sample statistic is z, we must proceed a little differently when using t.

Step 2: Set the alpha level, find degrees of freedom (DF), and find decision t.

We will set our alpha at 0.05. That means that we will reject our null hypothesis if the probability of obtaining our sample statistic or higher (if the null hypothesis were true) is below 0.05. Just like z, the sample t-value has a specific p-value associated with it, which is the exact probability of obtaining that score or higher if the null hypothesis were true. Unlike the Normal Table, the t-table does not tell us the specific p-value. It just gives us a cutoff value associated with our alpha level. If the t-value for our sample is beyond this cutoff point, that tells us that the probability of obtaining our sample statistic or higher is lower than 0.05 (our alpha level). We call this cutoff point **decision t** because it is the criterion

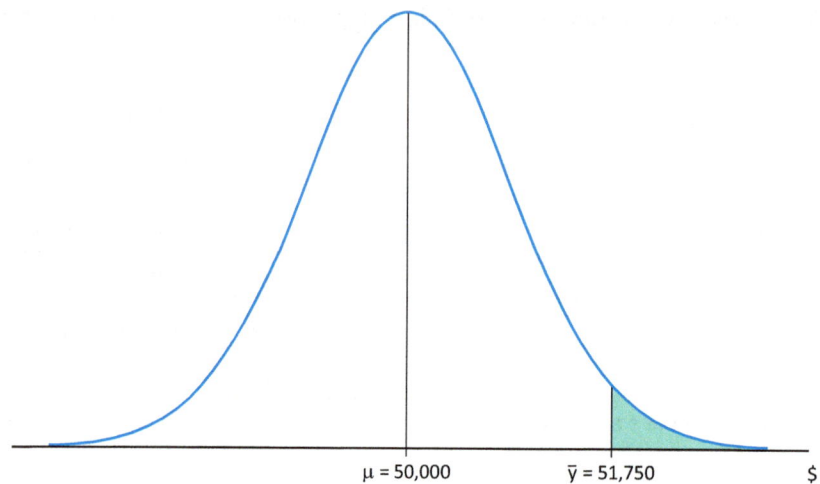

Figure 9.4 The Null Hypothesis and Sample Statistic on a Normal Curve, College Graduates' Mean Income

that we use for making a decision about our null hypothesis. (You may also see it called "critical t.") Statistical software, in contrast, will report the exact p-value for your sample t.

You learned how to find the t-value for a given alpha level in chapter 8. To recap, you must first calculate degrees of freedom, because there are many t-curves, and you use the t-curve associated with the degrees of freedom (DF) in your sample.

$$DF = N - 1$$

For our example:

$$DF = 200 - 1 = 199$$

Looking in the t-table in appendix B, we see that the highest DF is 120. Because our DF is greater than 120, we use the last line of the table, labeled with the infinity symbol, ∞. We see that a t-value of 1.96 is required for a two-tailed test at alpha = 0.05 and a t of 1.645 for a one-tailed test. Because our research hypothesis is that earnings for graduates of your college are greater than those of college graduates overall, we are conducting a one-tailed test. Our decision t is therefore 1.645.

Steps 3 and 4: Estimate the standard error and calculate the sample t.

Just as with the hypothesis test for a proportion, we calculate a standardized score for the sample value. Whereas for proportions we calculated a z-score, in this case, we are calculating a t-value. Remember from chapter 8 that both z-scores and t-values convert the distance between the sample score and the mean from the units of the variable (here, dollars of income) into standard error units. Instead of saying that the mean income for graduates of your college is $1,750 higher than the mean income for all college students, the t-value will tell us how many standard errors higher your college's mean income is than the overall mean.

The formula for t, introduced in chapter 8, is:

$$t = \frac{\bar{y} - \mu}{SE_{\bar{y}}}$$

In other words, we are calculating the difference between our sample mean and the population mean ($\bar{y} - \mu$) and then standardizing this in terms of the standard error ($SE_{\bar{y}}$).

In order to calculate the t-value, we must first estimate the standard error ($SE_{\bar{y}}$). This is the third step of the hypothesis test. We do so by using the standard deviation for our sample as a stand-in for the population standard deviation.

$$SE_{\bar{y}} = \frac{s}{\sqrt{N}}$$

$$SE_{\bar{y}} = \frac{9{,}000}{\sqrt{200}} = 636.4$$

Now we can calculate t, Step 4 in the hypothesis test:

$$t = \frac{\bar{y} - \mu}{SE_{\bar{y}}} = \frac{51{,}750 - 50{,}000}{636.4} = \frac{1{,}750}{636.4} = 2.75$$

The t-value associated with our sample mean of 51,750 is 2.75.

Step 5: Compare decision t to sample t. If sample t > decision t, reject the null hypothesis. State your conclusion.

How do we know whether the probability of obtaining a sample with this t-value, if the true population mean is fifty thousand (in other words, if the null hypothesis is correct), is below our alpha of 0.05? We know that a t-value of 1.645 (our decision t) is the cutoff point for an alpha of 0.05. If our sample t is greater than decision t, the probability of obtaining our sample result if the null hypothesis were true is less than 0.05.

Figure 9.5 shows the sample mean, sample t-value, and decision t on the curve. The sample t-value tells us that the sample mean is further out into the tail of the curve than our cutoff value, established by alpha.

We can now compare our sample t of 2.75 with the decision t-value. In this case, 2.75 is greater than 1.645. We therefore reject the null hypothesis that earnings for graduates of your college are the same as those of college graduates overall. We conclude that earnings for graduates of your college are statistically significantly greater than those of college graduates overall.

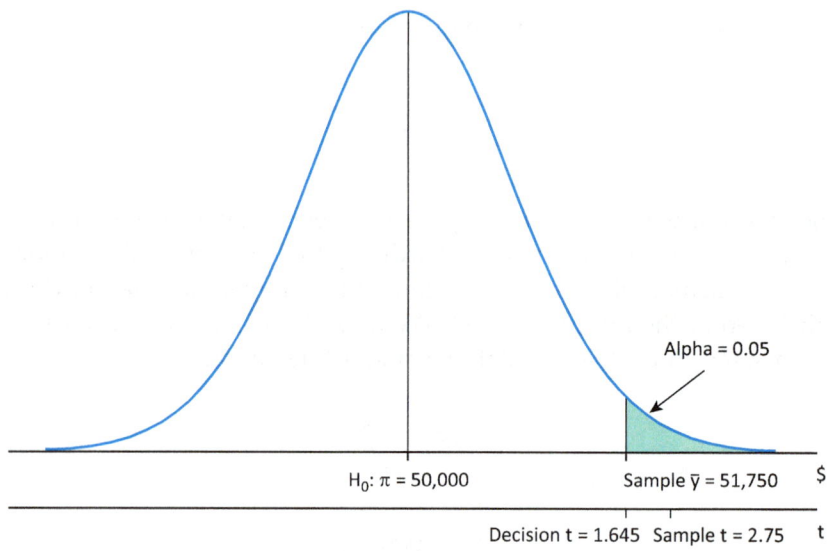

Figure 9.5 The Null Hypothesis, Sample t-Value, and Decision t-Value (College Graduates' Mean Income)

Optional follow-up: Calculate a confidence interval to estimate the range within which the population mean is likely to fall.

If we want to estimate the actual mean earnings of graduates of your college, we would conduct a confidence interval for that mean. This is exactly what we did in chapter 8. If you need a refresher, see the summary on pages 336–337 in chapter 8.

For our example:

$$CI = \bar{y} \pm t(SE_{\bar{y}}), \text{ where } SE_{\bar{y}} = \frac{s}{\sqrt{N}}$$

and t at DF = 199 and 90% confidence = 1.645.
Notice that the t-value for a 90% confidence interval is the same as the t-value associated with an alpha of 0.05 for a one-tailed test. (Remember, this is not your sample t.) We calculated SE = 636.4 above. Plugging in the numbers:

$$CI = 51{,}750 \pm 1.645(636.4)$$

$$= 51{,}750 \pm 1{,}047 \rightarrow (50{,}703.12,\ 52{,}796.99)$$

We conclude that there is a 90% chance that the mean income for graduates of your college falls between around $50,703 and $52,797. Notice that this confidence interval does not overlap with $50,000, the mean income for all college graduates.

BOX 9.3: IN DEPTH

Assumptions with Hypothesis Tests

Like confidence intervals, hypothesis tests assume that certain conditions hold. These assumptions are:

- Random sampling from a population.
- When we hypothesize about means, we use interval-ratio variables. When we construct hypotheses about proportions, we use nominal variables. Ordinal scales with at least 5 categories can be treated as interval level; otherwise, ordinal variables are treated as nominal in hypothesis testing.
- If a population distribution is approximately normal, samples of any size are appropriate for hypothesis tests of means. Since in practice we do not observe the population distribution and thus cannot know if it is normal, a rule of thumb is that the sample size must be at least thirty. For proportions, the required sample size depends on the hypothesized proportion in the population. The further the hypothesized proportion in the population is from 0.5, the larger the sample size needs to be.*

Example: Testing a Claim about a Population Mean

Let's look at one more example. Imagine that a marketer claims that consumers give a new app, on average, 9 out of 10 stars. But when you sample consumers, you find that your sample rates the app, on average, 7 stars. (Let's say the sample is eighty-five people and the standard deviation is 3.8.) You might be skeptical, rightly, of the marketer's claim. However, as a statistics student, you know that the results from a sample can vary from the population. A one-sample t-test allows you to determine the probability of getting a mean rating of 7 stars if the population really gave the app an average of 9 stars. Instead of comparing your sample to actual data about the population, you are comparing it to a claim about the population. You do this in exactly the same way as in the previous example, since you are still comparing an observed sample mean to a (claimed) population mean.

In order to conduct the hypothesis test, you need four pieces of information: the hypothesized (claimed) population mean (μ), the sample mean ($\bar{y}$), the sample standard deviation (s), and the size of the sample (N). In this case, $\mu = 9$, $\bar{y} = 7$, s = 3.8, and N = 85. The steps for the hypothesis test are as follows:

Step 1: Set up null and research hypotheses.

Note that we are setting this up as a two-tailed test.

$H_0 : \mu = 9$. In other words, the mean rating for the entire population of consumers who have used the product is 9.

---

*A rule of thumb for required sample size for proportions is that the product of the sample size (N) and the null hypothesis proportion (π) and the product of N and (1 − π) must both be greater than or equal to 10. In practice, the value of π in the null hypothesis is rarely close to 0 or 1, and thus an N of 30 is sufficient. With very small samples, a special probability distribution called the binomial distribution must be used instead of the normal distribution; this is beyond the scope of this book.

$H_a: \mu \neq 9$. In other words, the mean rating for the entire population of consumers who have used the product is different from 9.

Our sample mean of 7 is two stars away from the hypothesized population mean of 9 stars. Remember, the hypothesis test is testing the *likelihood of getting a sample result at least that far away from 9*. You are setting up the t-curve *as if the population mean is 9* in order to find the probability of obtaining a sample mean at least as far from the population mean as 7. If you find that the probability of getting your sample mean is low enough (i.e., less than your alpha level), you will conclude that the claimed population mean is probably false.

Figure 9.6 shows this setup. Because it is a two-tailed test, we have drawn in the sample mean of 7, along with another value, 11, which is the same distance above the hypothesized population mean as 7 is below it.

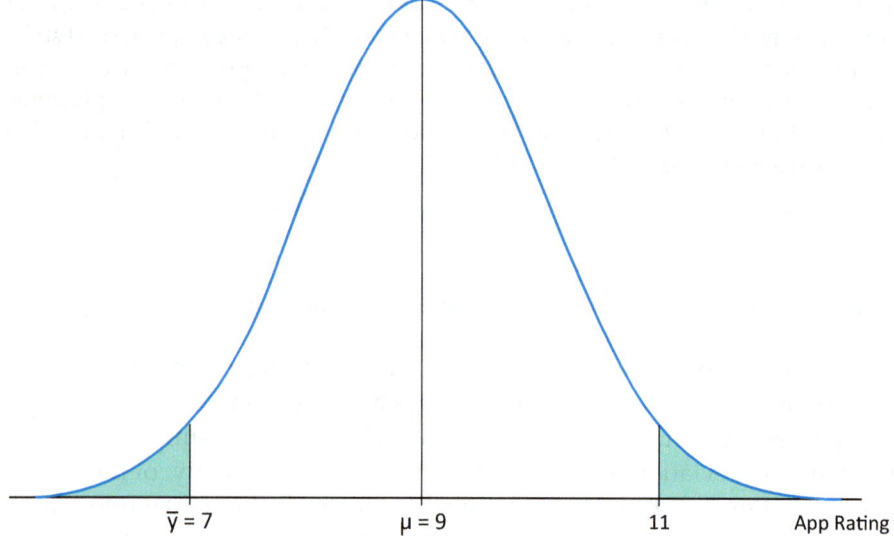

Figure 9.6 The Null Hypothesis and Sample Statistic on a t-Distribution with DF = 84, Two-Tailed Hypothesis Test (App Rating)

Step 2: Set alpha level, calculate degrees of freedom, and find your decision t in the t-table.

$$DF = N - 1 = 85 - 1 = 84$$

Looking in the t-table, we see that 84 is not listed, so we use the next lowest degrees of freedom, 80. For an alpha of 0.01 and a two-tailed test, decision t = 2.639.

Step 3: Estimate the standard error.

$$SE_{\bar{y}} = \frac{s}{\sqrt{N}}$$

$$SE_{\bar{y}} = \frac{3.8}{\sqrt{85}} = \frac{3.8}{9.22} = 0.41$$

Step 4: Calculate sample t.

$$t = \frac{\bar{y} - \mu}{SE_{\bar{y}}} = \frac{7-9}{0.41} = \frac{-2}{0.41} = -4.88$$

We see that t is a negative number because it corresponds to a score that is below the hypothesized population mean. A t-value of positive 4.88 would correspond to the score that is the same distance above the mean, in this case 11.

Step 5: Compare your sample t and your decision t and draw your conclusion.

If sample t is greater than decision t, reject the null hypothesis. (If you are conducting a t-test using statistical software, the software will compute the exact p-value for your sample t-value. In that case, you reject the null hypothesis if p < alpha.)

In this example, the absolute value of sample t is 4.88, well beyond decision t of 2.639. (Remember, you compare the absolute value of your sample t to the decision t.) We reject the null hypothesis. We have shown that there is less than a 1% chance of obtaining a sample mean 2 or more points away from the claimed mean rating of 9.

In other words, it is highly improbable that the marketer is telling the truth about customer ratings of the app. Customer ratings are statistically significantly lower than the claimed rating of 9 stars.

Optional follow-up: Calculate a confidence interval for the sample mean.

If we want to estimate the actual customer ratings for the app, we would construct a confidence interval, as we did in chapter 8. Here we choose to construct a 99% confidence interval. In this case

$$CI = \bar{y} \pm t(SE_{\bar{y}}), \text{ where } SE_{\bar{y}} = \frac{s}{\sqrt{N}} \text{ and t at DF} = 84 \text{ and } 99\% \text{ confidence} = 2.639$$

We calculated $SE_{\bar{y}} = .41$ above. Note that t is the value found in the t-table for your DF and confidence level, not your sample t. Plugging in the numbers, we find:

$$CI = 7 \pm 2.639(0.41) = 7 \pm 1.08 \rightarrow (5.92, 8.08)$$

There is a 99% probability that the mean rating for the app is between 5.92 and 8.08 stars. Notice that the confidence interval does not contain the marketer's claimed rating of 9 stars.

Error and Limitations: How Do We Know We Are Correct?

Hypothesis tests are widely used in the social sciences; their simplicity is appealing. They allow the investigator to reduce a finding to either significant (reject H_0) or not significant (do not reject H_0). But one should always keep in mind that a hypothesis test is built around a single sample statistic, and that sample statistic is not especially stable. It is highly likely that if we were to draw another random sample, we would get a different sample statistic. It would probably be fairly close to the first one we obtained but not exactly the same.

Recall the example from earlier in the chapter. A sample of two hundred young people found that 59% supported legalization of marijuana. We obtained a z-score of 1.43 and a p-value of 0.076 for that sample. Imagine that we took a second random sample of two hundred young people, and this time we find that 56% of them support legalization. This is depicted in Figure 9.7.

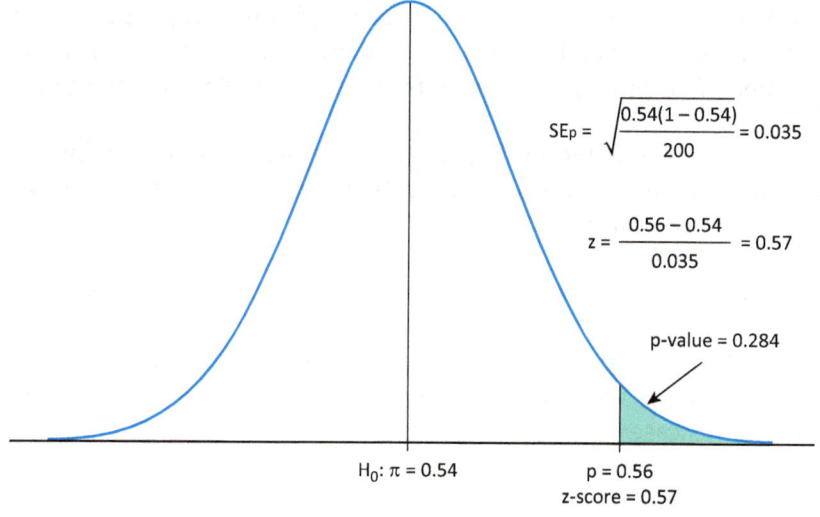

Figure 9.7 The Null Hypothesis and Sample Statistic on a Normal Curve, Young People's Support for Legalization of Marijuana (Sample Proportion = 0.56)

The z-score for a sample proportion of 0.56 is 0.57, and the corresponding p-value is 0.284. What is our conclusion now? Whereas for our earlier example the p-value of 0.076 allowed us to reject the null hypothesis at an alpha of 0.10, our p-value of 0.284 is well above even an alpha of 0.10.

In this example, we would not reject the null hypothesis no matter what alpha level we had chosen. Although in practice we draw only one sample and conduct the hypothesis test on the basis of that single sample statistic, we must always be aware that a different random sample—which would likely yield a slightly different sample statistic—might well produce a different decision about the null hypothesis.

There is an important caveat to keep in mind. In some cases, as we saw in Table 9.2, we would reject the null hypothesis with a larger alpha (Scenario 1) but not with a smaller alpha (Scenarios 2 and 3). However, the determination of alpha must always precede the execution of the hypothesis test. The researcher must choose a proper rejection threshold, and should make this decision on its merits, before calculating the p-value associated with the sample proportion. This decision depends on an understanding of the possible errors we may make.

Type I and Type II Errors

The null hypothesis cannot be directly observed. With it, we are making an educated guess about the value of a population parameter. That guess is either correct or incorrect. We make our determination by looking directly at the evidence we have—the sample statistic and sample standard deviation. Is the evidence consistent with our guess about the population parameter? If so, we do not reject the null hypothesis. If the observed evidence is inconsistent with our guess about the population parameter, then we reject the null hypothesis.

Once we have made a decision, we have to be aware of the possibility that we may have made the wrong decision. Why might this be? Mistaken decisions are the consequence of sampling variability. Any time we draw a random sample from a population, it is possible that the sample is not representative of that population. This is not likely, but it is possible. There are two types of error, "false positives," in which we wrongly reject the null hypothesis and support the research hypothesis, and "false negatives," in which we wrongly fail to reject the null hypothesis, wrongly concluding that the research hypothesis is false.

Let's return to the legalization of marijuana example. Imagine that we (somehow) are omniscient, and we know for a fact that 54% of all young people support legalization. We draw a random sample of two hundred young people. As we saw when we studied sampling distributions, we know that most of those samples will produce percentages close to 54%. But a small number of them will be well below or well above 54%. We can be precise. Of those sample proportions, 95% will be within 1.96 standard errors of 54%, and 5% will be more than 1.96 standard errors away from the true population parameter. If we were unlucky enough to get one of those samples far away from the center, we would mistakenly conclude that the null hypothesis was incorrect. We summarize the possible outcomes in Table 9.4.

Table 9.4 Outcomes of Hypothesis Testing

	Reject H_0	Do Not Reject H_0
H_0 is true	(A) Type I error	(B) Correct decision
H_0 is false	(C) Correct decision	(D) Type II error

Let's examine Scenarios A, B, C, and D, one at a time, using our legalization of marijuana example. Recall our hypotheses:

H_0: 54% of all young people support legalization.
H_a: More than 54% of all young people support legalization.

Scenario A: The null hypothesis is true, but we have rejected it. This is called a **Type I error** (also known as a false positive). Why would we have rejected it? Because, although in reality 54% of all young people support legalization, we had the bad luck of drawing a random sample in which the percentage supporting legalization was well below or well above 54%. This can happen because of sampling variability.

Scenario B: Here, we have made the correct call. The null hypothesis is true—of all young people, 54% do support legalization, and we obtained a sample proportion that was close enough to 54%. Our observed evidence—the sample proportion—is close enough to the null hypothesis proportion. So we do not reject the null hypothesis.

Scenario C: Here again, we have made the correct decision. In reality, the null hypothesis is incorrect. In the context of our example, this means that the true percentage of all young people in support of legalization is not equal to 54%. And because we obtained a sample statistic far enough away from 54%, we reject the null hypothesis.

Scenario D: Here, the null hypothesis is false, but we fail to reject it. In this case, we have committed a **Type II error** (also known as a false negative). What does this mean in context? It means that, in the population, the true percentage of young people who support legalization is not 54%. Yet we ended up with an unrepresentative sample—one that indicated that the sample proportion was indeed close to 54%.

Any time we conduct a hypothesis test, we are potentially committing a Type I or a Type II error. This is because of two factors: (1) we do not ever directly observe the population parameter; we are drawing a sample to assess the accuracy of our hypothesis; and (2) sampling variability means that there is always a chance that we end up with a sample that is not representative of our population. As a result, we should be careful in how we word our conclusions. We cannot definitively "prove" that the null hypothesis is true or false. We can only conclude, within a range of probability, that our sample findings likely reflect a true difference from the population parameter.

We mentioned earlier that, when conducting a hypothesis test, the researcher should choose an alpha level based on the merits of the case. One of the most important factors that contributes to the decision about alpha level is the gravity associated with committing a Type I or Type II error, in the context of the particular situation. When the consequences of a Type I error—a false positive—are more serious than the consequences of a Type II error—a false negative—the researcher should choose a low alpha level. In fact, the probability of committing a Type I error is equal to the alpha level. If alpha is set at 1%, the researcher must obtain a p-value less than 1% in order to reject the null hypothesis. This also means that the probability of committing a Type I error is only 1%.

The probability of committing a Type I error is inversely related to the probability of committing a Type II error. Calculating the probability of a Type II error is more complicated and depends upon the distance between the true population parameter and the hypothesized one. These calculations are beyond the scope of this book. Even so, because the probabilities of Type I and Type II errors are inversely related, we can minimize the probability of a Type II error by raising alpha. When the null hypothesis is false, as it is in Scenarios C and D, we want a high probability of rejecting it. The probability of rejecting a false null hypothesis is known as the **power** of a test. The more "powerful" a test, the less likely we are to commit a Type II error.

In our legalization of marijuana example, as we discussed above, a Type I error means that 54% of all young people support legalization, but we wrongly conclude that the true percentage is higher. A Type II error means that the percentage of young people who support legalization is higher than 54%, but we wrongly conclude that it is 54%. In the first case, we are overestimating support for legalization. In the second case, we believe that the population percentage for young people is 54%, but it isn't. It's not clear which error is a more serious one. Indeed, when we are simply measuring public opinion, the consequences of either a Type I or Type II error are usually not all that grave.

On the other hand, there are many examples where the trade-off between the two kinds of errors is very important. Our entire system of criminal justice is based upon the idea of minimizing a Type I error. Consider this set of hypotheses:

H_0: Defendant is not guilty.
H_a: Defendant is guilty.

In this context, a Type I error sends a defendant to prison even though that defendant is innocent. A Type II error frees a guilty defendant. Most people would agree that it is far worse to convict an innocent person than to free a guilty one. This is why, to convict a defendant, a jury must determine guilt "beyond a reasonable doubt." This is the threshold our legal system uses to minimize Type I errors.

There are many cases in science and social science research where a Type I or Type II error can have substantial implications. Consider the COVID-19 vaccination example from the beginning of this chapter. That clinical trial tested the effectiveness of a new vaccine, and researchers of course hoped that the vaccine would be effective in protecting people against COVID. Here is the set of hypotheses:

H_0: The incidence of COVID will be the same among vaccinated and unvaccinated populations.
H_a: The incidence of COVID will be the lower among the vaccinated population than among the unvaccinated population.

As we described, the researchers randomly assigned half of the participants to receive the vaccine, while the other half received a placebo. Given the large difference between the obtained sample proportion and the null hypothesis value, researchers in all likelihood would reject the null hypothesis and conclude that the vaccine is effective. This scenario, along with all the other possibilities, are played out in Table 9.5.

If we reject the null hypothesis, we are concluding that the vaccine is indeed effective. If the vaccine really is effective, then we have made the right call (illustrated in Scenario C). The vaccine will be developed and distributed to the public. In fact, this is exactly

Table 9.5 Outcomes of Hypothesis Testing in COVID Vaccination Example

	Reject H_0	Do Not Reject H_0
If H_0 is true	(A) **Type I error**: Vaccine is ineffective but we mistakenly conclude that it is effective.	(B) **Correct decision**: Vaccine is ineffective, and we conclude that it is ineffective.
If H_0 is false	(C) **Correct decision**: Vaccine is effective, and we conclude that it is effective.	(D) **Type II error**: Vaccine is effective, but we mistakenly conclude that it is ineffective.

what happened with the vaccine. But there is a probability, albeit a very low one, that we have made a Type I error. In this case (Scenario A), the vaccine in fact is ineffective, but by rejecting the null hypothesis, we are concluding that it is effective. This is a serious error: an ineffective vaccine would be manufactured and distributed to the general public, with grave consequences for public health. To reduce the possibility of this happening, researchers would set a very low alpha level, to minimize the probability of committing a Type I error.

The problem here is that a Type II error would also be very serious. In this scenario (Scenario D in Table 9.5), the vaccine is effective, but researchers conclude that is ineffective. With this outcome, we would have lost a chance to improve public health by means of an effective vaccine. Although the public does not hold the false belief that an ineffective vaccine will protect them against COVID (as would be the case with a Type I error), a Type II error would lead to deaths that could have been prevented by administering an effective vaccine. As this example illustrates, it is essential to evaluate the gravity of Type I and II errors in context. Deciding how to navigate the consequences of Type I and II errors is not always self-evident, and there could be disagreement about which type of error would be worse (e.g., administering an ineffective vaccine versus withholding an effective vaccine). It is important not to select an alpha level based on what you want the *answer* to be, but rather to select it before conducting your statistical tests and to base it on an even-handed consideration of the consequences of Type I versus Type II error.

What Does Statistical Significance Really Tell Us? Statistical and Practical Significance

Testing hypotheses about the relationship of sample findings to population parameters can be very useful. However, they tell us only one thing. They tell us the probability that we would obtain our sample result if the null hypothesis were true. While this allows us to support or reject the research hypothesis, as we have discussed there is always a known probability that we may be making a Type I or Type II error. In addition, our hypothesis test only shows whether our findings are statistically significant. That phrase simply means that our test has rejected the null hypothesis at our chosen alpha level. In statistical terms only, there are significant differences between the sample and the population (or the hypothesized population parameter). Such differences are not always important on a *practical or substantive* level. For example, if graduates of your college make, on average, $150 more per year than college graduates overall, is that enough of a difference to be a consideration in enrollment choices or to be touted in your school's advertising? If an educational program in prison reduces recidivism by 1%, is that a meaningful reduction? What if it reduces recidivism by 10%? The question of practical or substantive significance is one that the researcher—and those who read research results—must determine for themselves.

Moreover, the determination of **statistical significance** is extremely sensitive to sample size. It is important to understand that, if a sample size is large enough, even small differences can be statistically significant. Small samples are not necessarily bad samples, but they make it harder to reject a null hypothesis, all other factors being equal. Many students of introductory statistics will use their education to be better informed consumers of information. After reading this chapter, you know that, when research is reported, you should consider *both* whether the results are statistically significant and whether they are meaningful in other ways.

In fact, many statisticians argue that researchers should not rely on hypothesis testing and p-values. Their view is that even if Type I errors are relatively rare, they still occur often enough that many research findings claiming statistical significance are in fact false positives. Further, hypothesis testing gives us only one of two answers: either the null hypothesis is rejected, or it is not. In order to know more about the actual variable we are interested in—such as college graduates' earnings—we need to estimate those earnings using a confidence interval. This is a useful addition to hypothesis testing. Finally, statisticians remind us of the inherent uncertainty and variability that occur in sampling. Although we know in theory that repeated samples will provide different results, in practice we are usually working with one sample. There is no solution to this problem, except to be cautious in our conclusions and, where possible, to seek to replicate research findings that have serious potential consequences.[6]

Chapter Summary

One-sample hypothesis tests compare sample statistics to population parameters or to claims about a population.

- The steps for testing a one-sample hypothesis about the difference between an obtained sample statistic and a given population parameter are similar for proportions and for means. They are:
  1. Set up null and alternative hypotheses.
  2. Set the alpha level. If you are doing a hypothesis test for means (using t), calculate DF for your sample and look in the t-table to determine the decision t.
  3. Estimate the standard error.
     a. For proportions: $SEp = \sqrt{\dfrac{\pi(1-\pi)}{N}}$
     b. For means: $SE_{\bar{y}} = \dfrac{s}{\sqrt{N}}$
  4. Calculate the sample statistic, either z (for proportions) or t (for means).
     a. For proportions: $z = \dfrac{p - \pi}{SEp}$
     b. For means: $t = \dfrac{\bar{y} - \mu}{SE_{\bar{y}}}$
  5. For proportions, look in the Normal Table to find the p-value associated with your sample. For means, compare the sample t with the decision t. Determine whether you can reject the null hypothesis. If the p-value is lower than alpha (for proportions) or if sample t is higher than decision t (for means), you can reject the null hypothesis. State your conclusion.

- To estimate the range within which your population proportion or mean is likely to fall, follow the hypothesis test by calculating a **confidence interval** using the steps summarized in chapter 8.

- In a **one-tailed test**, the alternative hypothesis states that the population parameter is above or below a specified value but not both; in a **two-tailed test**, the alternative hypothesis is that the population parameter is either above or below a specified value.

- The researcher sets the **alpha level** (α), which is the threshold used to determine whether to reject the null hypothesis. If the p-value is lower than α, the null hypothesis is rejected.

- The difference between a sample statistic and a hypothesized population parameter in a one-sample test is **statistically significant** if the p-value associated with the sample statistic (z or t) is less than alpha, allowing the researcher to reject the null hypothesis.

- **Statistical significance** means that the difference between the sample statistic and the hypothesized population parameter is unlikely to be due to sampling error; it is more likely reflective of a genuine difference between the two.
- A **Type I error** occurs when the null hypothesis is true, but it is rejected. The probability of a Type I error is equal to alpha.
- A **Type II error** occurs when the null hypothesis is false, but it is not rejected.
- The **power** of a hypothesis test is the probability of rejecting a false null hypothesis.

Using Stata

In this section, you will learn how to conduct one-sample hypothesis tests for means and proportions in Stata. We will use data from the 2020 wave of the Police Public Contact Survey (PPCS), conducted by the Bureau of Justice Statistics, which interviews a nationally representative sample of US residents older than the age of sixteen. Participants were asked whether they had contact with the police in the last year and, if so, questions about the nature of that contact. We will use the following variables from the PPCS:

- *Contact* measures whether respondents who did have contact with the police had face-to-face contact with them. There are two possible values for *contact*: 1 indicates that the respondent had face-to-face contact with the police, and 0 means that the respondent had police contact that was not face-to-face.
- *Time* measures the duration (in minutes) of respondents' most recent contact with the police.

In recent years, the Black Lives Matter movement has drawn national attention to the issue of police mistreatment of African Americans. The questions in this section will focus on whether the rate of face-to-face contact with police and mean duration of police contact for Black people are equal to estimates of the respective parameters in the overall population of US residents. For this section, we limit the PPCS sample to only Black respondents. We will set an alpha level of 0.05 for all three hypothesis tests in this section.

Hypothesis Test for a Proportion

We begin by conducting a hypothesis test for a proportion. The first question uses the *contact* variable to investigate the likelihood that the proportion of Black people in the population who experienced face-to-face police contact is equal to the face-to-face police contact rate in the overall population, 53.9% (or, as a proportion, 0.539).* Our null hypothesis can be stated as:

H_0: The population rate of face-to-face police contact for Black people is equal to 0.539. (H_0: $\pi = 0.539$).

---

* We use the face-to-face contact rate in the full PPCS sample as an estimate of the population parameter.

We have three options for alternative hypotheses: a two-tailed test, a right-tailed test, and a left-tailed test. Since the dependent variable could include police contact for any reason, not punitive contact specifically, we do not have a strong reason to expect that Black people would have specifically more or less face-to-face contact with police than occurs in the general population. For example, it may be that rates of certain kinds of face-to-face contact with police (e.g., that which results from calling for help) are higher in the general population than among Blacks. Thus, we will conduct a two-tailed test, with the following alternative hypothesis:

H_a: The population rate of face-to-face police contact for Black people is not equal to 0.539. (H_a: π ≠ 0.539).

Open `PPCS2020Ch9.dta`. To run the test, type the following command into Stata:

`prtest contact = .539`

The "prtest" command tells Stata to run the test for a proportion.* Setting contact equal to a specific value tells Stata that this is a one-sample test. The "prtest" command produces the output shown in Figure 9.8.

```
. prtest contact = .539

One-sample test of proportion              Number of obs   =    351

    Variable |      Mean    Std. err.                [95% conf. interval]
    contact  |   .5982906   .0261673                  .5470037   .6495775

    p = proportion(contact)                                   z =   2.2284
H0: p = 0.539

    Ha: p < 0.539              Ha: p != 0.539              Ha: p > 0.539
 Pr(Z < z) = 0.9871        Pr(|Z| > |z|) = 0.0259       Pr(Z > z) = 0.0129
```

Figure 9.8

The output yields all of the information we need to make a decision about our hypothesis test. We see that the sample mean for face-to-face contact among Black people is 0.598. For variables that are coded 0/1, like *contact*, the mean is equal to the proportion of cases with a value of 1. Thus, 59.8% of Black people in the sample who had contact with the police experienced face-to-face contact. This is higher than the percentage in the overall sample, 53.9%. The output also provides us with the test statistic z-score: 2.23.

Stata provides us with the p-values for all three possible hypothesis tests. At the bottom of the output, the results for these hypotheses are, from left to right, for the one-tailed test that the proportion is less than 0.539; for the two-tailed test that the proportion is not equal to 0.539; and for the one-tailed test that the proportion is greater than 0.539. It is your job to pay attention to the results for the alternative hypothesis that you constructed prior to the test. In this case, our alternative hypothesis stated that the population proportion

---

* Note that Stata will conduct hypothesis tests for proportions only for variables that have two categories, which must be coded as 0 and 1.

would not be equal to 0.539, corresponding to the middle test, circled in the output. The p-value associated with the z-score for the two-tailed test is 0.0259, lower than our alpha level of 0.05. Thus, we reject the null hypothesis that the rate of face-to-face contact with the police for Black people is equal to 53.9%, the rate observed in the sample overall.

Since the sample statistic (0.598), or the rate of face-to-face police contact with police for Black people, is greater than the proposed population parameter, we can also examine the result for the right-tailed alternative hypothesis. The result on the right is for the one-tailed test with the alternative hypothesis that the population proportion for Black people is higher than 0.539. For this right-tailed hypothesis test, circled in the output, the p-value is below 0.05; the evidence suggests that the parameter for Black people is likely higher than 0.539. In other words, there is a high probability that the rate of face-to-face police contact among Blacks is higher than for the population in general.

Hypothesis Test for a Mean

Next, we will conduct a one-sample hypothesis test for a mean, using the *time* variable, which measures how many minutes respondents' most recent contact with police lasted. We will test whether the mean duration of contact for Black people in the population is the same as the mean duration of contact for the overall population, 12.34 minutes.* We use that value to generate our null hypothesis:

H_0: *The mean number of minutes of police contact for Black people in the population is equal to 12.34 (H_0: μ_y = 12.34).*

The syntax for a test for means is similar to that for the test for proportions. Instead of asking for a "prtest," we ask for a "ttest." We specify in the syntax that we will be using the t-distribution (rather than the z-distribution) because we do not know the population standard deviation:

```
ttest time = 12.34
```

The output from the test is shown in Figure 9.9. We learn from the output that the mean for the Black sample is 14.36 minutes, higher than the population mean from the null hypothesis, 12.34. But are we convinced by the results of our hypothesis test that the population mean for Blacks is significantly higher than 12.34? Our alternative hypothesis is that the population mean for Black people is higher than 12.34, and the results are circled in the bottom-right corner of the output. The sample t, 4.42, is associated with a very low p-value (p < 0.0000), smaller than our alpha level of 0.05. Although Stata will report a p-value that appears to be 0, we report that as p < 0.001 to reflect that the probability is never truly 0. This indicates that we can reject the null hypothesis in favor of the alternative hypothesis. We can conclude that the mean duration of police contact for Black people in the population is significantly higher than it is in the general population.

The results of these hypothesis tests suggest that both the rate of face-to-face police contact and duration of police contact are statistically significantly higher in the population of Black people than in the general population. In the next chapter, we will learn how

---

* We use the mean duration of police contact for the full PPCS sample as an estimate of the population parameter.

```
. ttest time = 12.34

One-sample t test

Variable  |   Obs  |   Mean   | Std. err. | Std. dev. | [95% conf. interval]
   time   |  1,624 | 14.3593  | .4567172  | 18.40519  | 13.46348   15.25512

    mean = mean(time)                                        t =   4.4213
H0: mean = 12.34                              Degrees of freedom =   1623

Ha: mean < 12.34            Ha: mean != 12.34            Ha: mean > 12.34
 Pr(T < t) = 1.0000       Pr(|T| > |t|) = 0.0000        Pr(T > t) = 0.0000
```

Figure 9.9

to make explicit comparisons between the means and proportions for two groups by using two-sample hypothesis tests.

Review of Stata Commands

- Run a one-sample test for a proportion (0/1 variable only)

  `prtest variable name=value`

- Run a one-sample test for a mean, standard deviation unknown

  `ttest variable name=value`

Using SPSS

In this section, you will learn how to conduct one-sample hypothesis tests for means and proportions in Stata. We will use data from the 2020 wave of the Police Public Contact Survey (PPCS), conducted by the Bureau of Justice Statistics, which interviews a nationally representative sample of US residents older than the age of sixteen. Participants were asked whether they had contact with the police in the last year and, if so, questions about the nature of that contact. We will use the following variables from the PPCS:

- *Contact* measures whether respondents who did have contact with the police had face-to-face contact with them. There are two possible values for *contact*: 1 indicates that the respondent had face-to-face contact with the police, and 0 means that the respondent had police contact that was not face-to-face.
- *Time* measures the duration (in minutes) of respondents' most recent contact with the police.

In recent years, the Black Lives Matter movement has drawn national attention to the issue of police mistreatment of Black people. The questions in this section will focus on whether the rate of face-to-face contact with police and mean duration of police contact for Black people are equal to estimates of the respective parameters in the overall population of US residents. For this section, we limit the PPCS sample to only Black respondents. We will set an alpha level of 0.05 for all three hypothesis tests in this section.

Hypothesis Test for a Proportion

We begin by conducting a hypothesis test for a proportion. The first question uses the *contact* variable to investigate the likelihood that the proportion of Black people in the population who experienced face-to-face police contact is equal to the face-to-face police contact rate in the overall population, 53.9% (or, as a proportion, 0.539).[*] Our null hypothesis can be stated as:

H_0: *The population rate of face-to-face police contact for Black people is equal to 0.539.* (H_0: $\pi = 0.539$).

We have three options for alternative hypotheses: a two-tailed test, a right-tailed test, and a left-tailed test. Since the dependent variable could include police contact for any reason, not punitive contact specifically, we do not have a strong reason to expect that Black people would have specifically more or less face-to-face contact with police than occurs in the general population. For example, it may be that rates of certain kinds of face-to-face contact with police (e.g., that which results from calling for help) are higher in the general population than among Blacks. Thus, we will conduct a two-tailed test, with the following alternative hypothesis:

H_a: *The population rate of face-to-face police contact for Black people is not equal to 0.539.* (H_a: $\pi \neq 0.539$).

SPSS includes hypothesis tests only for means, not proportions. But since we are using a 0/1 variable, the proportion of cases scoring a 1 is equal to the mean of the variable. So, we can use the SPSS procedure that generates a hypothesis test on a mean.

Open `PPCS2020Ch9.sav`. To run the hypothesis test, we use this SPSS procedure:

`Analyze → Compare Means → One Sample T Test`

This opens the "One Sample T Test" dialog box. We move the *contact* variable into the "Test Variable(s)" box, and in the space next to "Test Value" we type in the null hypothesis value; in this case, 0.539 (circled), as shown in Figure 9.10.

When we click on "OK," SPSS produces the output shown in Figure 9.11.

The output yields all of the information we need to make a decision about our hypothesis test. We see that the sample mean for face-to-face contact among Black people is 0.60 (circled in red). For variables that are coded 0/1, like *contact*, the mean is equal to the proportion of cases with a value of 1. Thus, 60% of Blacks in the sample who had contact with the police experienced face-to-face contact. This is higher than the percentage in the overall sample, 53.9%. The output also provides us with the test statistic t: 2.263.[**]

SPSS provides us with the p-value for both the one-tailed and the two-tailed hypothesis tests (circled in red). As we can see from the output, the two-sided p-value is 0.024. In other words, with an alpha of 0.05, we can reject the null hypothesis. We are highly confident that the proportion of Black people who had face-to-face contact with the police is not equal to 0.539 (our null hypothesis). If we wish to conduct a one-tailed hypothesis

---

[*] We use the face-to-face contact rate in the full PPCS sample as an estimate of the population parameter.

[**] SPSS automatically uses t for all hypothesis tests.

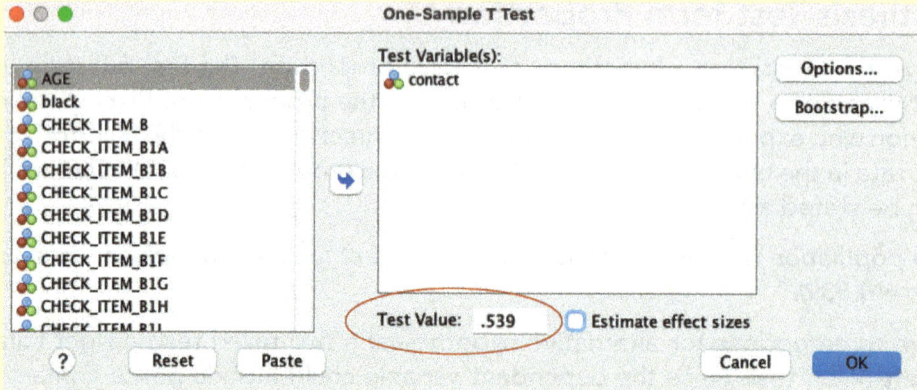

Figure 9.10

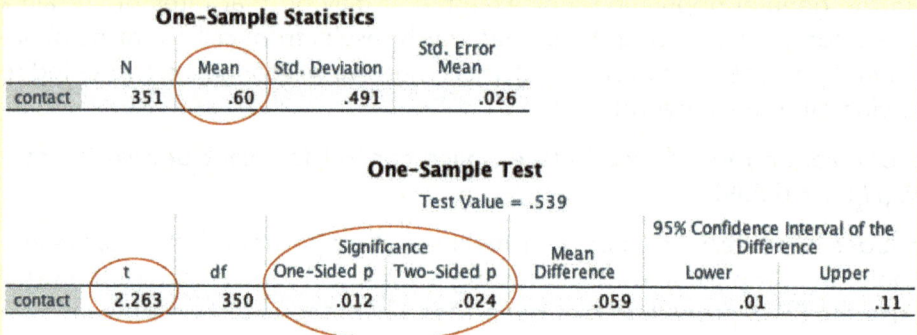

Figure 9.11

test to determine whether the contact rate for Black people is higher than it is for the full population (H_a: $\pi > 0.539$), the p-value is 0.012. Once again, we reject the null hypothesis and conclude that the contact rate for Black people is higher than it is for the general population.

Hypothesis Test for a Mean

Next, we will conduct a one-sample hypothesis test for a mean, using the *time* variable, which measures how many minutes respondents' most recent contact with police lasted. We will test whether the mean duration of contact for Black people in the population is the same as the mean duration of contact for the overall population, 12.34 minutes.* We use that value to generate our null hypothesis:

H_0: The mean number of minutes of police contact for Black people in the population is equal to 12.34 (H_0: $\mu_y = 12.34$).

The procedure we use for a test for means is the same as the one we used for the test of proportions.

```
Analyze → Compare Means → One Sample T Test
```

---

* We use the mean duration of police contact for the full PPCS sample as an estimate of the population parameter.

We indicate the variable we are using as well as the null hypothesis test value (12.34), as shown in Figure 9.12.

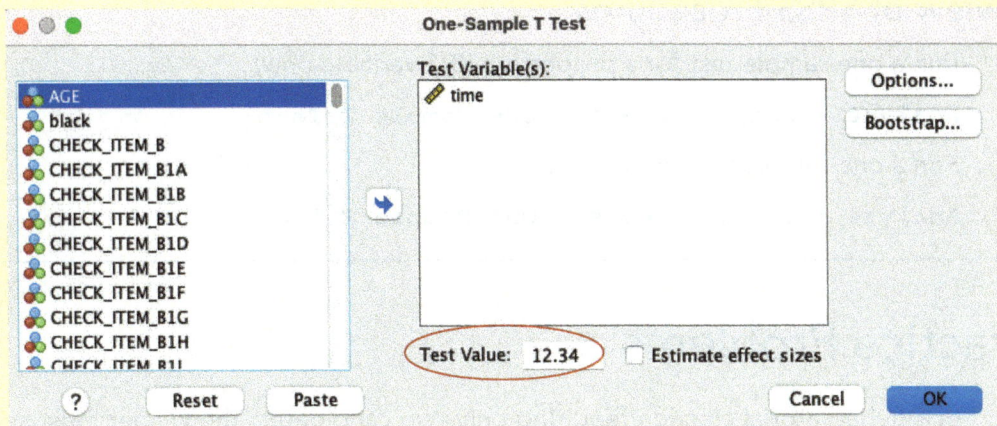

Figure 9.12

The output is shown in Figure 9.13.

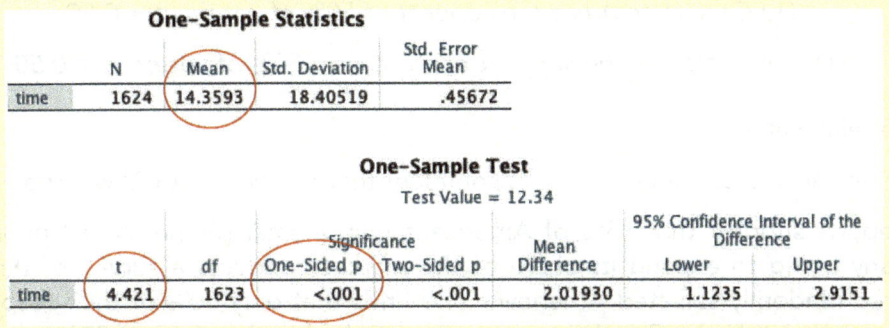

Figure 9.13

We learn from the output that the mean for the Black sample is 14.36 minutes, higher than the population mean from the null hypothesis, 12.34. But are we convinced by the results of our hypothesis test that the population mean for Black people is significantly higher than 12.34? Our alternative hypothesis is that the population mean for Black people is higher than 12.34, and the one-tailed p-value is contained in the output. The sample t, 4.42, is associated with a very low one-tailed p-value, less than 0.001. This indicates that we can reject the null hypothesis in favor of the alternative hypothesis. We can conclude that the mean duration of police contact for Black people in the population is significantly higher than it is in the general population.

The results of these hypothesis tests suggest that both the rate of face-to-face police contact and duration of police contact are statistically significantly higher in the population of Black people than in the general population. In the next chapter, we will learn how to make explicit comparisons between the means and proportions for two groups by using two-sample hypothesis tests.

Review of SPSS Procedures

- Run a one-sample test for a proportion (0/1 variable only)

    `Analyze → Compare Means → One Sample T Test`

- Run a one-sample test for a mean

    `Analyze → Compare Means → One Sample T Test`

Practice Problems

1. In a gubernatorial election featuring only two candidates, the winner must secure just over 50% of the total votes. Imagine a gubernatorial election with candidate A running against candidate B. In a poll of seven hundred randomly selected voters taken right before the election, 53% of voters state that they will support candidate A. Conduct a one-tailed hypothesis test with the following hypotheses:

    H_0: Candidate A has the support of 50% of voters; $\pi = 0.50$

    H_a: Candidate has the support of more than 50% of voters; $\pi > 0.50$

    Set alpha at 1%.

    What can you conclude? Are you confident that candidate A will win the election?

2. Reports suggest that 85% of Americans own a smartphone. A cell phone company trying to expand its share of the market conducts a survey of three hundred randomly selected Americans and finds that only 80% of their respondents own a smartphone. Can the company conclude that the claim of 85% smartphone ownership is too high? Conduct a one-tailed hypothesis test with the following hypotheses:

    H_0: 85% of all Americans own a smartphone; $\pi = 0.85$

    H_a: Fewer than 85% of all Americans own a smartphone; $\pi < 0.85$

    Set alpha at 5%.

    What is your conclusion?

3. A candidate running for re-election wants a better understanding of how her constituents feel about her. She commissions a poll and instructs the surveyors to measure how the public feels about her on a 100-degree feeling thermometer, where any score above 50 degrees represents a warm or positive feeling; a score below 50 degrees suggests a negative feeling; and a score of 50 is neutral. The

pollster selects a random sample of eighty-five constituents and finds that the candidate's mean feeling-thermometer score is 54 degrees, with a standard deviation of 25 degrees. Conduct a one-tailed hypothesis test with the following hypotheses:

H_0: The public's mean rating of the candidate is neutral; $\mu = 50$ degrees

H_a: The public's mean rating of the candidate is positive; $\mu > 50$ degrees

Set alpha at 5%.

Can the candidate conclude that her constituents feel positively toward her? Explain your answer.

4. A private high school makes the claim that its "average student" is Mensa (high IQ society) material. To qualify for membership in Mensa, one must have an IQ score of at least 132. The high school has gathered IQ scores of a sample of 55 students, and finds the mean to be 125, with a standard deviation of 15. Conduct a one-tailed hypothesis test with the following hypotheses:

H_0: The mean IQ score of students at this high school is 132, high enough to qualify for membership in Mensa; $\mu = 132$

H_a: The mean IQ score of students at this high school is less than 132, not high enough to qualify for membership in Mensa; $\mu < 132$

Set alpha at 0.05.

Can the high school legitimately claim that its "average student" is Mensa material? Explain your answer.

5. Prior to 2000, the US Census did not allow people to choose more than one racial identity. Since 2000, people have been able to mark more than one box on the Census question about race. Since then, demographers have used Census data to study the multiracial population, as measured by people who mark more than one box on the Census race question. The 2010 Census showed that 2.9% of the entire US population identified as multiracial. In a random sample of 350 people drawn from the Western region of the United States, 4.5% identified as multiracial.

   a. State the null and alternative hypotheses to test whether the multiracial population in the West is greater than the overall rate in the United States.

   b. Explain whether the test for part a would be a one- or two-tailed test.

6. As the size of the gig economy (i.e., the growth of work arrangements that depart from traditional job arrangements and rely more on short-term contracts) increases, observers have debated whether this represents a turn toward more freedom and autonomy for workers, or a dangerous loss of job security. A recent study estimates that 60% of adults see the growth of the gig economy as more negative than positive. Three colleagues want to know if gig economy workers themselves are as likely as everyone else to see the gig economy as generally negative. They intend to conduct a one-sample hypothesis test to determine the answer to this question.

a. The null hypothesis for the test is that the proportion of gig economy workers who feel negatively about the gig economy is equal to the proportion of people overall who feel that way. But, the three colleagues disagree on what the alternative hypothesis should be. They find that their differences are irreconcilable. They go their own ways and conduct their own tests. The first colleague sets up the alternative hypothesis like this: $H_a: \pi \neq 0.6$. The second opts for $H_a: \pi > 0.6$. The third colleague prefers $H_a: \pi < 0.6$. Explain in words what each of these alternative hypotheses means.

b. Which of the three alternative hypotheses do you think makes the most sense? Explain your answer.

c. Use the Normal Table to find the z-score associated with an alpha-value of 0.05 for each of the three alternative hypotheses. Explain why the z-scores differ between the one-tailed tests. Explain why the z-score is different for the one-tailed tests compared to the two-tailed test.

7. In Problem 5, it was reported that that 2.9% of the entire US population identified as multiracial. In a random sample of 350 people drawn from the Western region of the United States, 4.5% identified as multiracial. Conduct a hypothesis test at the 0.05 alpha level for proportions to assess whether the percent of people identifying as multiracial in the West is greater than the overall population rate.

8. A researcher in California is investigating how gender is related to school discipline. She is studying a random sample of five hundred students from a large high school in California with more than five thousand students. In her sample, 83% of the students who have been suspended from school are boys. In the state of California as a whole, 79% of students who have been suspended are boys, according to the Department of Education's Civil Rights Data Collection. After learning this, the principal of the school wants to know: Does this school really have a larger gender imbalance in its suspension rates than California as a whole? The principal strongly believes that there is no reason to suspect that her high school is any harsher on boys than any other school in the state.

   a. Conduct a two-tailed hypothesis test at the 0.05 alpha level. Show all the steps. Explain whether you can reject the null hypothesis.

   b. How will you explain the p-value from the hypothesis test to the principal, who is unfamiliar with statistical analysis?

   c. The principal reports to her staff that "there is only a 2.6% chance that the gender composition of the suspended population at their school is the same as it is across the state." Is this correct? Explain your answer.

9. Researchers find that a preventative health care program implemented in Anytown reduces emergency room visits to a mean of 0.50 per household per year. You sample 105 low-income households in Anytown and find that their mean number of visits to an emergency room in the past year was 0.65, with a standard deviation

of 0.41. You want to assess whether the preventative health care program is working as well for low-income residents as it is for residents overall.

   a. Conduct each of the steps of a hypothesis test (at an alpha level of 0.05) for this question.

   b. Discuss the practical significance of the findings.

   c. Estimate a 95% confidence interval for the annual number of emergency room visits for low-income households. Explain what this interval means.

10. A state passes a law prohibiting advertising targeting children (e.g., the use of cartoon characters in television ads to sell cereal). After the law's first year of implementation, a group of researchers wants to know if children have adopted less materialistic attitudes. They cannot survey all of the state's children, so they draw a probability sample of one thousand children and measure their materialistic attitudes on a scale ranging from 0 to 100. The sample mean is 68, with a standard deviation of 10. A meta-analysis of previous studies examining children's materialistic attitudes suggests that American children overall have a mean score of 70 on the scale.

    a. Conduct a hypothesis test at the 0.01 alpha level to assess whether children from the state that has banned children's advertising has the same mean score for materialistic attitudes as the nationwide mean.

    b. Draw a t distribution. Label the decision and sample t values on the x axis. Shade the appropriate area on the curve and label it.

    c. Why is the sample t a negative value? Why is the decision t a negative value?

    d. What would you conclude about the success of the elimination of advertising to children at reducing their materialistic values? Consider the statistical and practical significance of the result in your response.

    e. In this example, which type of error—Type I or Type II—could we have made? Explain how you know, and what the error would mean. Assess the consequences of making that particular error.

11. Virtually all politicians now use social media as a way of communicating with voters, and increasing numbers of people use social media as a way to engage in politics. A random sample of three hundred social media users score a mean of 6 on a political engagement scale that ranges from not at all politically active, 0, to extremely politically active, 10. Research suggests that the overall population score for political engagement is 4. Does this difference of 2 points on the political engagement scale indicate a real difference between social media users and the general population? A one-tailed t-test yields a t statistic that is larger than the decision t at an alpha of 0.01. Therefore, we reject the null hypothesis, which states that the mean score for the population of social media users is equal to 4.

    a. Which kind of error, Type I or Type II, is it possible that we have made in this example? Explain why.

b. What is the probability that we committed the kind of error that you identified in (a).

12. Consider the following two hypotheses:

    H_0: Casual users of e-cigarettes have no more health problems than the general population.

    H_a: Casual users of e-cigarettes have more health problems than the general population.

    a. Explain the consequences of making a Type I error in this case? A Type II error?
    b. Which type of error do you think is a more serious one to make in this case? Explain your answer.

13. A marketing firm believes that their advertising campaign has increased the favorability of their client's product among the coveted eighteen to thirty-four demographic. Using a probability sample of eighteen- to thirty-four-year-olds, the marketing firm found that 75% of them said that they liked the product. This compares to the less promising 50% favorability rating from this population in extensive previous market research. The firm conducts a one-sample test for proportions, with the null hypothesis that the proportion of all eighteen- to thirty-four-year-olds who like the product is equal to 0.5. The test yields a p-value of 0.001. The firm declares victory! Based on the p-value, they report to their client that they are 99.9% sure that 75% of all eighteen- to thirty-four-year-olds like the product after the ad campaign.

    a. The firm has bungled its interpretation of the p-value. Explain why its interpretation is incorrect. State in words what we can say about the p-value, given the information provided above.
    b. We learn later that the firm set up a two-tailed alternative hypothesis in this analysis. Do you agree with that decision? Explain your answer.
    c. If the firm had set up a one-tailed alternative hypothesis, which should it have chosen: left- or right-tailed?
    d. Would the choice of a one-tailed hypothesis test over a two-tailed test have changed the general findings in this case? Explain your answer.

14. Chemists use litmus paper to determine whether solutions are basic or acidic. Blue litmus paper turns red if the solution is acidic, and red litmus paper turns blue if the solution is basic. Thus, the phrase "litmus test" generally means that the result of the test adjudicates between two possible outcomes that are mutually exclusive.

    a. A litmus test decides between two mutually exclusive answers. Does a hypothesis test do the same thing? Explain your answer.
    b. A litmus test provides a definitive answer: either the first or second possible answer is certainly correct. Does a hypothesis test do the same thing? Explain your answer.

15. During his presidential campaign, one candidate gave hundreds of speeches in which he mentioned the unemployment rate, promising to fix the "broken" economy if elected. One reporter noticed that the rates the candidate cited tended to be higher than the official unemployment rate released by the Federal Reserve, 4%. The reporter didn't have the resources to analyze the hundreds of speeches that the candidate had delivered on the campaign trail, so she collaborated with a political science professor at a nearby university on analyzing a sample of the candidate's speeches. After drawing a random sample of thirty speeches, they found that the unemployment rate stated by the candidate in the sample was, on average, 12%, with a standard deviation of 2%. They set up a hypothesis test to calculate the chance that sampling variability explained why this figure was so much higher than the Federal Reserve's.

   a. Which type of hypothesis test should they conduct: a test for a mean or proportion? Explain why.

   b. Conduct the appropriate test, with an alpha level of .05. Is the result of the test statistically significant? Explain your answer.

   c. Do you find the difference between the candidate's mean cited unemployment rate, 12%, and the one given by the Federal Reserve, 4%, to have practical significance? Explain your answer.

16. A researcher is investigating whether the cost of rental housing in San Francisco is the same as the cost of rental housing in the San Francisco Bay Area overall. The mean monthly rent for a representative sample of two hundred three-bedroom rental units in San Francisco is $5,100, with a standard deviation of $1,000. Say that monthly rent for three-bedroom units in the whole Bay Area averages to $4,975. The null hypothesis is that the population mean for San Francisco's monthly rents is $4,975, the same as in the Bay Area overall. The researcher conducts a one-tailed hypothesis test and fails to reject the null hypothesis. What alpha level is the researcher using? Use the sample t-value to explain your answer.

Stata Problems

Open the World Values Survey (`WVSWave7.dta`) data file. We will work with the *Q163* variable that we saw in the practice problems for chapters 2 and 3 to ask whether people who live in China have the same feelings about science and technology as do people across the globe. *Q163* ranges from 1 (the world is a lot worse off because of science and technology) to 10 (the world is a lot better off because of science and technology). You will use the *Q163* mean for all respondents, 7.08, as an estimate of the total population mean. For the following problems, you will need to limit the sample to respondents who live in China. The variable *B_COUNTRY* indicates the country of residence for respondents, and the value for respondents who live in China is 156.

   1. Use the "ttest" command to run the hypothesis test examining whether attitudes about science and technology for all people in China are the same as attitudes for

people around the world. Be sure to limit the hypothesis test to people living in China.

2. Find the sample mean for Chinese respondents in the output.
3. What is the null hypothesis? What is the two-tailed alternative hypothesis?
4. What is the sample t-value?
5. Interpret the output for the two-tailed test. Does the evidence suggest that residents of China feel the same as the rest of the world about science and technology?

SPSS Problems

Open the World Values Survey (`WVSWave7.sav`) data file. We will work with the *Q163* variable that we saw in the practice problems for chapters 2 and 3 to ask whether people who live in China have the same feelings about science and technology as do people across the globe. *Q163* ranges from 1 (the world is a lot worse off because of science and technology) to 10 (the world is a lot better off because of science and technology). You will use the *Q163* mean for all respondents, 7.08, as an estimate of the total population mean. For the following problems, you will need to limit the sample to respondents who live in China. The variable *B_COUNTRY* indicates the country of residence for respondents, and the value for respondents who live in China is 156.

1. Use the "One Sample T Test" dialog box to run the hypothesis test examining whether attitudes about science and technology for all people in China are the same as attitudes for people around the world. Be sure to limit the hypothesis test to people living in China.
2. Find the sample mean for Chinese respondents in the output.
3. What is the null hypothesis? What is the two-tailed alternative hypothesis?
4. What is the sample t-value?
5. Interpret the output for the two-tailed test. Does the evidence suggest that residents of China feel the same as the rest of the world about science and technology?

Notes

[1] Carl Zimmer. 2020. "2 Companies Say Their Vaccines Are 95% Effective. What Does That Mean?" *New York Times*. November 20. Retrieved from https://www.nytimes.com/2020/11/20/health/covid-vaccine-95-effective.html. See also Dashiell Young-Saver. 2020. "What Does 95% Effective Mean? Teaching the Math of Vaccine Efficacy." *New York Times*. December 13. Retrieved from https://www.nytimes.com/2020/12/13/learning/what-does-95-effective-mean-teaching-the-math-of-vaccine-efficacy.html#:~:text=The%20number%20you%20got%2C%2095,of%20infection%20by%2095%20percent.

[2] National Center for Education Statistics. 2022. "Digest of Education Statistics [Table 303.70]." Retrieved from https://nces.ed.gov/programs/digest/d22/tables/dt22_303.70.asp.

[3] C. Aschwanden. 2015. "Not Even Scientists Can Easily Explain p-Values." *FiveThirtyEight*. November 4. Retrieved from http://fivethirtyeight.com/features/not-even-scientists-can-easily-explain-p-values/.

4. World Health Organization. 2015. "WHO Statement on Public Disclosure of Clinical Trial Results." April 9. Retrieved from http://www.who.int/ictrp/results/reporting/en/.
5. National Center for Education Statistics. April 2017. "Annual Earnings of Young Adults." Retrieved from https://nces.ed.gov/programs/coe/indicator_cba.asp.
6. For more information, see Ronald L. Wasserstein and Nicole A. Lazar. 2016. "The ASA's Statement on p-Values: Context, Process, and Purpose." *The American Statistician* 70(2): 129–133. DOI: 10.1080/00031305.2016.1154108.

Chapter 10

Comparing Groups
Two-Sample Hypothesis Tests

Many questions in the social sciences ask about differences between groups. To name just a few examples that we have examined in this book, social scientists are interested in differences in political attitudes between older and younger people, health outcomes across control and treatment groups, and wealth differences between racial and ethnic groups. Since COVID-19 spread across the globe in 2020, researchers have investigated differences in the pandemic's effects on different groups of people. For example, a team of psychologists focused on the mental health of people in the United Kingdom during the 2020 lockdown that prohibited people from leaving home except under very limited circumstances.[1] The researchers asked whether men and women had different levels of anxiety, depression, insomnia, and stress during this restrictive period. An online survey asked a representative sample of 1,006 people from across the United Kingdom to report information about their mental health and the quality of their sleep. Selected results are presented in Table 10.1.

The first two columns of Table 10.1 show mean scores for anxiety, stress, and insomnia for men and women in the sample. The last column shows gender differences between these mean scores. All three differences are positive, indicating that women had higher average scores on these three measures than men. For instance, on average, women scored 2.72 points higher on the anxiety scale than men. While the analysis cannot determine whether the pandemic lockdown (or the pandemic itself) *caused* these gender differences in mental health and insomnia, the means indicate that women in the sample reported higher levels of anxiety, stress, and insomnia than did men during an acutely stressful period of the pandemic. But how do we know whether the sample differences between women's and men's well-being reflect actual differences in the population and not just sampling variability?

Whereas the hypothesis tests described in chapter 9 tested whether a single sample mean or proportion differed from a mean or proportion for a population, the hypothesis tests in this study tested the *difference* between two sample means. The asterisks in the last column of Table 10.1 indicate that each difference is statistically significant at the 0.001 alpha level. As

Table 10.1 Average Anxiety, Stress, and Insomnia Severity for Women and Men in the United Kingdom during the 2020 COVID-19 Lockdown

	Women	Men	Difference
Anxiety Scale	9.28	6.56	2.72***
Stress Scale	19.45	15.66	3.79***
Insomnia Severity	11.08	9.66	1.42***

we saw in chapter 9, this means that the probability of obtaining a sample statistic (in this case the *difference* in means between women and men) as extreme as the one we observed in the sample is less than 0.001, if the null hypothesis were true. Here, the null hypothesis for each test is that there is no difference between men and women on these indicators of well-being. In each case, the null hypothesis of no gender difference is rejected.

In previous chapters, we have seen several ways to compare the values of a variable across groups (e.g., comparing the mean or proportion from one group to those of another group, examining whether the confidence intervals around the means or proportions for two groups overlap, or visually comparing histograms for a variable across groups). All of these methods involve comparing one statistic, or graph, with another. The study of gender differences in well-being during the pandemic takes a different approach by examining the differences themselves.

In this chapter, we turn our attention to comparisons of this type, which are called two-sample hypothesis tests. You are already familiar with the logic of hypothesis tests from the one-sample tests seen in chapter 9. The major difference between those tests and two-sample tests is that **two-sample tests** focus on the *difference* between two means or proportions.

⊖ Two-Sample Hypothesis Tests

When we conduct a two-sample hypothesis test, we are testing for a relationship between two variables. We can think of the variable that is used to divide the cases into separate groups as the independent variable. The dependent variable is the one for which the difference in values across the groups will be compared. In the preceding example, the independent variable is gender, and the dependent variables are anxiety, stress, and insomnia scores. If there is a difference between the groups in the means or proportions for the dependent variable, this suggests that the independent variable may cause changes in the dependent variable, though we should be cautious about making causal claims, as we will see in chapter 13.

The Logic of the Null and Alternative Hypotheses in Two-Sample Tests

We saw in chapter 9 that a hypothesis test assesses the likelihood that, over repeated sampling, we would observe a value at least as far from the hypothesized population parameter as our sample statistic if the null hypothesis were true. In a one-sample test, the null hypothesis states that the population parameter—a mean or proportion—is equal to a specific value. For a two-sample test, the null hypothesis also states that the population parameter is equal to a specific value, but the population parameter in question is the difference between the proportions or means for two groups. The null hypothesis typically states that the difference between the two groups is equal to 0, which would indicate no relationship between the independent, or grouping, variable and the dependent variable.[*]

As an example, consider research comparing the mental health of the general population to those who are incarcerated. Comparing samples of prisoners and non-prisoners, a report found that 14.5% of prisoners and 4.6% of non-prisoners reported symptoms of "serious psychological distress."[2] The difference in proportions between the two samples is 0.099 (0.145 − 0.046 = 0.099). In statistical notation, with p denoting the sample proportions and the subscripts indicating prisoners (p) and non-prisoners (np), we write the difference as:

$$p_p - p_{np} = 0.099$$

If we were to conduct a hypothesis test of the difference between the proportions of prisoners and non-prisoners who experience serious psychological distress, how would we set up the null hypothesis? Incarceration status is the independent variable, and serious psychological distress is the dependent variable. The null hypothesis is that there is no relationship between these variables. Specifically, it states that there is no difference between the proportions of prisoners and non-prisoners experiencing serious psychological distress. In statistical notation, the null hypothesis looks like this:

$$H_0: \pi_p - \pi_{np} = 0$$

Some researchers prefer a different, though equivalent, statement of the null hypothesis:

$$H_0: \pi_p = \pi_{np}$$

As we saw in chapter 7, we use the Greek letter pi, π, to denote that we are hypothesizing about population proportions in the null hypothesis.

Just as with one-sample tests, the alternative hypothesis can state either that there is a difference in proportions between the two groups or, more specifically, that the proportion for one group is higher or lower than for the other. As we saw in chapter 9, the former option corresponds to a two-tailed hypothesis test and the latter to a one-tailed test. In statistical notation, the options for the alternative hypotheses are stated like this:

$$H_a: \pi_p - \pi_{np} \neq 0 \text{ (two-tailed test)}$$

$$H_a: \pi_p - \pi_{np} > 0 \text{ (one-tailed test, right-tail)}$$

$$H_a: \pi_p - \pi_{np} < 0 \text{ (one-tailed test, left-tail)}$$

---

[*] While the null hypothesis could state that the difference in means or proportions between two groups is equal to a value other than 0, there should be a strong reason, informed by prior research or theory, to hypothesize a non-0 value.

When we examine the difference in rates of serious psychological distress between prisoners and non-prisoners, the right-tail hypothesis makes more sense. Given the disproportionate arrest rate among those who struggle with mental illness and the stress caused by incarceration, the most reasonable alternative hypothesis would be that the rate of serious psychological distress is higher in the population of prisoners than in the population of non-prisoners.[3]

Notation for Two-Sample Tests

Looking carefully at the null and alternative hypotheses for the difference in the proportion of those experiencing serious psychological distress between prisoners and non-prisoners, we see that notation for two-sample tests is more complicated than for one-sample tests. In two-sample tests, we must keep track of information from two samples instead of just one. For example, in a one-sample test of a mean, we distinguish the population mean, μ, from the sample mean, $\bar{y}$, and the population standard deviation, σ, from the sample standard deviation, s.

In addition to maintaining the distinction between population parameters and sample statistics, in a two-sample test we must distinguish the two groups from each other. The most straightforward way to manage the extra information involved in two-sample tests is to assign separate subscripts to each group. The choice of subscripts is arbitrary—common approaches are to use "1" and "2" for each group or particular letters associated with the groups—but consistent application of the subscripts to the correct groups throughout all of the steps in the hypothesis test is essential for avoiding confusion. In the prisoner/non-prisoner example, we used the subscript "p" to denote the sample of prisoners and "np" to denote the sample of non-prisoners. When we are working with specific examples in the chapter, we present formulas with letter subscripts to denote the specific groups from the examples; when we present formulas without reference to specific examples, we use the subscripts "1" and "2" to denote two separate groups. (It doesn't matter which group you label "1" or "2," as long as you are consistent throughout your calculations for the hypothesis test.) The generic versions of each formula are shown in the "Chapter Summary."

The Sampling Distribution for Two-Sample Tests

Conducting a two-sample test follows the same basic process as we saw for one-sample tests in chapter 9. The difference is that, whereas the hypothesized population parameter for a one-sample test is a single sample mean or proportion, the population parameter for a two-sample test is the difference between two means or proportions, each drawn from two separate samples.[*] With one-sample tests, we used the sampling distribution of means or proportions. With two-sample tests, we use the sampling distribution of the differences between means or proportions. Figure 10.1 shows the sampling distribution of the differences between means.

As with the sampling distribution of means or proportions, the sampling distribution of the difference between means or proportions centers on the population parameter, which

---

[*] Note that these "separate samples" are often subgroups within a larger sample. As long as cases were selected independently of each other during sampling, we can treat subgroups within a sample as separate samples. For example, we can conduct hypothesis tests on differences between men and women in the General Social Survey (GSS) because individuals were selected independently of each other for inclusion in the GSS.

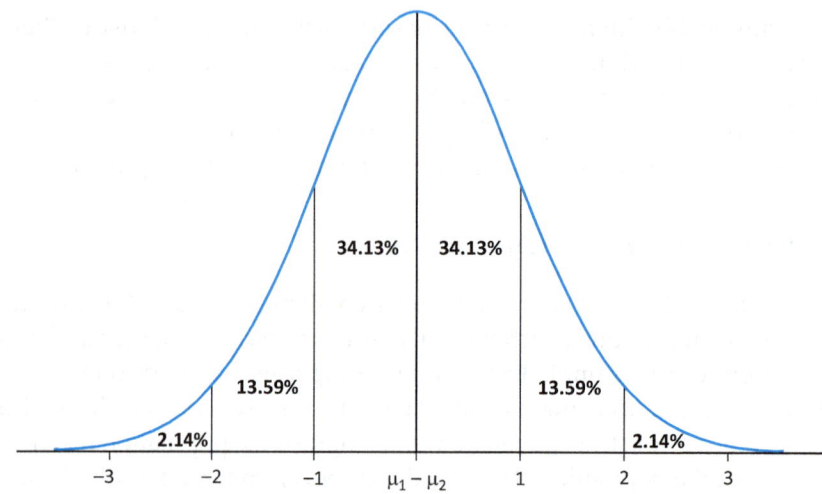

Figure 10.1 Sampling Distribution of Differences in Means

for two-sample tests is the population difference (here, $\mu_1 - \mu_2$). (For the sampling distribution of differences between proportions, the population parameter is $\pi_1 - \pi_2$.) Figure 10.1 is set up to reflect the null hypothesis, in which the true difference between the population means is 0. According to the Central Limit Theorem, this distribution of differences will be approximately normal with a large enough sample size. As we know from the Empirical Rule, about 68% of the sample differences will fall within one standard error of the mean, 95% within two standard errors, and almost 100% within three standard errors. Just as we saw in chapter 9, we will use the normal distribution for two-sample tests of proportion differences and the t-distribution with the appropriate degrees of freedom for two-sample tests of mean differences (because standard deviations of population means, σ, are almost always unknown).

The assumptions for two-sample hypothesis tests are the same as those for one-sample tests, as described in Box 9.3. The assumptions must apply for both samples.

Hypothesis Tests for Differences between Means

In this example, we focus on income differences between rural and urban counties in the United States. Using county-level data from the US Department of Commerce, we will examine whether mean incomes differ between counties located in metropolitan areas compared to rural counties that fall outside of metropolitan areas.[4] For this example, we will use random samples of forty counties from the population of all counties in the United States. We categorize each county as "metropolitan" or "non-metropolitan," according to the Census Bureau's designation. Note that the unit of analysis is the county—that is, each case in our data is one county. For each county, we know the median income of its residents; we can then average these values to find the overall mean of the median incomes of people living in metropolitan and non-metropolitan counties.

Table 10.2 County-Level Mean Household Income, Samples of Forty Counties in Metropolitan Areas and Non-Metropolitan Areas

	Mean Household Income*	SD
Metropolitan (m)	64,864	14,475
Non-metropolitan (nm)	53,543	10,335

*This is the mean of the median incomes of households in all forty counties in each category.
Source: 2020, Census Bureau, Small Area Income and Poverty Estimates (SAIPE) Program.

The independent variable in this example is metropolitan area, with each county categorized as falling inside or outside of a metropolitan area, because this is the variable by which we are grouping cases. The dependent variable is the mean of the median household incomes. Table 10.2 presents the mean and standard deviation for the dependent variable for the samples of forty counties in metropolitan and non-metropolitan areas. We see that mean median household income is $11,321 higher in the sample of metropolitan counties than in the sample of non-metropolitan counties ($64,864 − $53,543 = $11,321).

We will conduct a two-sample t-test to assess the likelihood that we would observe at least a $11,321 difference in the mean median household income for the samples of metropolitan and non-metropolitan counties even if there is no actual difference in mean median household income for the populations from which they were drawn. Table 10.3 shows the step-by-step procedure for conducting the test. The formulas for estimating standard error and calculating t are somewhat different from those for one-sample tests that we saw in chapter 9, but the steps and logic are the same. We will go through the steps one by one.

Table 10.3 Steps in Conducting Hypothesis Tests for County-Level Income Differences between Metropolitan and Non-Metropolitan Areas

Step 1: Set up hypotheses.	$H_0: \mu_m - \mu_{nm} = 0$; $H_a: \mu_m - \mu_{nm} \neq 0$
Step 2: Set α, calculate DF, and find decision t.	$\alpha = 0.05$; $DF = (N_m + N_{nm}) - 2 = 78$; decision $t = 2$
Step 3: Estimate standard error.	$SE_{\bar{y}_m} = \dfrac{s_m}{\sqrt{N_m}} = \dfrac{14475}{\sqrt{40}} = 2288.7$; $SE_{\bar{y}_{nm}} = \dfrac{s_{nm}}{\sqrt{N_{nm}}} = \dfrac{10335}{\sqrt{40}} = 1634.11$; $SE_{\bar{y}_m - \bar{y}_{nm}} = \sqrt{SE_{\bar{y}_m}^2 + SE_{\bar{y}_{nm}}^2}$ $= \sqrt{(2288.7)^2 + (1634.11)^2}$ $= 2812.2$
Step 4: Calculate sample t.	$t = \dfrac{(\bar{y}_m - \bar{y}_{nm})}{SE_{\bar{y}_m - \bar{y}_{nm}}} = \dfrac{(64,864 - 53,543)}{2812.2} = \dfrac{11,321}{2812.2} \approx 4$
Step 5: Compare sample t and decision t; draw conclusion.	4.03 > 2.0; therefore, reject H_0

Step 1: Set up hypotheses.

The first row in Table 10.3 shows the null and alternative hypotheses. Notice that we have assigned the subscript "m" to denote the sample of counties from metropolitan areas and "nm" to denote the sample of counties from non-metropolitan areas. The null hypothesis is that there is no difference, on average, in the mean of median household incomes between the populations of metropolitan and non-metropolitan counties. The alternative hypothesis states that there is a difference, on average, in the mean of median household incomes between the two populations, but not that one type of county has a higher or lower mean income than the other. This means that we are conducting a two-tailed test.

Step 2: Set alpha level, calculate degrees of freedom, and find decision t.

Step 2 in Table 10.3 shows that we have chosen an alpha level of 0.05, meaning that we are risking rejecting the null hypothesis in 5% of repeated samples when the null hypothesis is actually true. (Recall from chapter 9 that rejecting the null hypothesis when it is actually true is called a Type I error.) Figure 10.2 shows the t-distribution for our hypothesis test.

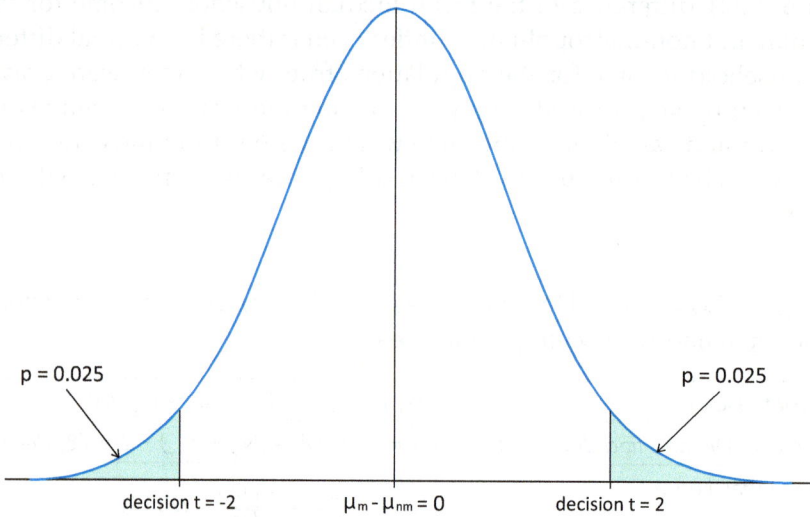

Figure 10.2 t-Distribution for Two-Sample Test with 78 Degrees of Freedom and Alpha = 0.05

We see from Figure 10.2 that the mean of the sampling distribution is 0, the hypothesized difference between the population means in the null hypothesis. Recall that, in a two-tailed test, we distribute the alpha value equally across the left and right tails of the distribution. In this case, since alpha is equal to 0.05, this means that 2.5% of the curve's area falls into each tail. We also see that the decision t at the 0.05 alpha level and 78 degrees of freedom is 2. (We will discuss how we determined the degrees of freedom below.)

How do we know the value of decision t? Look at the t-table in appendix B. It has no row for 78 degrees of freedom, so we move down to 60 in order to be more conservative in our estimate. The table shows that, for a two-tailed hypothesis test, where alpha = 0.05

and degrees of freedom is 60, decision t is 2. At this point, we do not yet know where our difference between sample means of $11,321 falls on the curve. We will know its location on the curve only after we have converted that raw difference into a t-value, which occurs in Step 3 of the hypothesis test.

But first, how did we calculate the degrees of freedom for this test? Recall that, in a one-sample test, the degrees of freedom are equal to the sample size minus one because we are testing the value of a single unknown population parameter. In the second row of Table 10.3, we have provided the following formula for calculating degrees of freedom for two-sample tests:

$$DF = (N_m + N_{nm}) - 2$$

This approach seems reasonable because there are two unknown parameters in two-sample tests instead of one.

However, we must pause to acknowledge that estimating the degrees of freedom for a two-sample test is slightly more complicated than what we have shown. When we cannot assume that the variances of the two populations are equal, an assumption that we usually cannot make, degrees of freedom are estimated by the following formula:

$$DF = \frac{\left(\frac{s_1^2}{N_1} + \frac{s_2^2}{N_2}\right)^2}{\frac{1}{N_1 - 1}\left(\frac{s_1^2}{N_1}\right)^2 + \frac{1}{N_2 - 1}\left(\frac{s_2^2}{N_2}\right)^2}$$

This formula is cumbersome. Luckily, you do not have to use it when calculating by hand. Using this formula to calculate the degrees of freedom for the difference in metropolitan and non-metropolitan county incomes yields 70.6, rounding to 71. This is lower than the 78 degrees of freedom yielded by the simpler formula. In practice, though, the decision t for a two-tailed test with alpha of 0.05 at 78 and 71 degrees of freedom is the same—2.0—due to the fact that we have to round both down to 60 degrees of freedom to use the t-table. In fact, this is what happens most of the time: The formula results in a very small or no difference in the decision t. Fortunately, statistical software programs will estimate the degrees of freedom for us using the complicated formula provided above. We recommend using the abbreviated formula ($DF = N_1 + N_2 - 2$) when calculating by hand and allowing the statistical program to use the more cumbersome formula.

Step 3: Estimate the standard error of the difference.

The third row of Table 10.3 shows the calculation of the standard error. The standard error of the difference between means is the square root of the sum of the variances of each mean (i.e., the square root of the sum of the squared standard errors of each mean):*

---

* You can calculate the standard error for each group first and then plug those into the above formula. Or you can combine the calculations: If you put the formulas for the two standard errors into the formula for the standard error of the difference, you get the following formula:

$$SE_{\bar{y}_1 - \bar{y}_2} = \sqrt{\left(\frac{s_1}{\sqrt{N_1}}\right)^2 + \left(\frac{s_2}{\sqrt{N_2}}\right)^2}$$

$$SE_{\bar{y}_m - \bar{y}_{nm}} = \sqrt{SE_{\bar{y}_m}^2 + SE_{\bar{y}_{nm}}^2}$$

The standard error for each group is calculated with the formula:

$$SE_{\bar{y}} = \frac{s}{\sqrt{N}}$$

Plugging in the numbers from our example, we get:

$$SE_{\bar{y}_m - \bar{y}_{nm}} = \sqrt{SE_{\bar{y}_m}^2 + SE_{\bar{y}_{nm}}^2} = \sqrt{(2288.7)^2 + (1634.11)^2} = 2812.2$$

Step 4: Calculate sample t.

Once we find the standard error of the difference, 2812.2, we place it in the denominator of the formula for sample t (see row four of Table 10.3):

$$t = \frac{(\bar{y}_m - \bar{y}_{nm})}{SE_{\bar{y}_m - \bar{y}_{nm}}} = \frac{(64{,}864 - 53{,}543)}{2812.2} = \frac{11{,}321}{2812.2} \approx 4.$$

Before we move forward, it is important to explain what the formula for sample t is doing. We presented a simplified formula for t above. The full formula is:

$$t = \frac{(\bar{y}_m - \bar{y}_{nm}) - (\mu_m - \mu_{nm})}{SE_{\bar{y}_m - \bar{y}_{nm}}}$$

That is, the numerator calculates the difference between the actual difference in our sample means and the hypothesized difference between population means. Because the population difference always equals 0 in our null hypothesis, we can drop $\mu_m - \mu_{nm}$ from the formula.

Step 5: Compare the absolute value of the sample t and decision t. If sample t > decision t, reject the null hypothesis. State your conclusion.

The value for sample t is higher than the decision t of 2.0, as shown in Figure 10.3. This means that we can reject our null hypothesis of no difference between the incomes in metropolitan and non-metropolitan counties.

If there were no population difference in incomes between metropolitan and non-metropolitan areas, we would have been very unlikely to have observed a sample

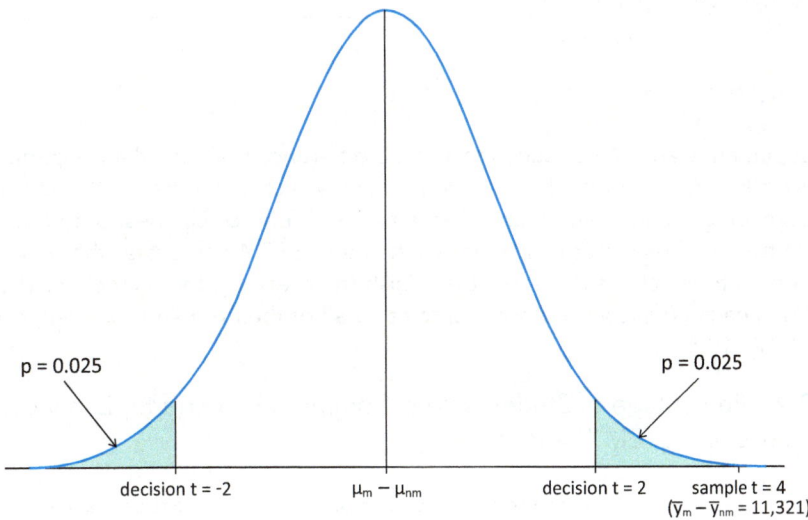

Figure 10.3

difference of at least $11,321. We can conclude that there is an actual population difference between metropolitan and non-metropolitan incomes. We would report this conclusion by saying that counties in metropolitan areas have statistically significantly higher household incomes than do counties in non-metropolitan areas or that household incomes in metropolitan counties are significantly higher than in non-metropolitan counties.

Why can we conclude that metropolitan incomes are greater than in non-metropolitan counties, when our two-tailed alternative hypothesis was simply that the incomes in metropolitan and non-metropolitan counties were different? Imagine that we had set up the test as a one-tailed test with the same alpha level of 0.05, with the alternative hypothesis that incomes in metropolitan counties were greater than in non-metropolitan counties. Our decision t would have been even lower than it was for this two-tailed test, and we would have rejected the null hypothesis, just as we did with the two-tailed test. This is always the case. Thus, we can safely conclude that the mean income for metropolitan areas is greater, not just that it is different.

We have data for every county in the country, allowing us to check whether the result of our hypothesis test (based on two random samples of forty counties) is consistent with real population differences. The mean of median incomes for the population of metropolitan counties is $65,669 compared to $52,611 for the population of non-metropolitan counties, yielding a difference of $13,058. As we would expect, due to sampling variability, the difference in our sample, $11,321, is not exactly the same as the difference in the population. Still, the standard error for our sample difference was small enough to yield a large enough sample t-value that led to the accurate conclusion of a real population difference.

BOX 10.1: APPLICATION

Is the Sample Statistic a Difference in Proportion or Mean?

A statistics student wants to conduct a test of the difference between the percentage of students at her university who complete their reading for every class in courses in the social sciences and science, technology, engineering, and mathematics (STEM) disciplines. She draws two random samples of twenty classes from all social science and STEM courses at the university and asks each student in those classes whether they finish their reading for each class. The percentages of students in each course who report completing all of their reading for every class period are shown in Table 10.4.

Table 10.4 Percentage of Students Who Complete Reading for Every Class in Twenty Social Science and STEM Courses

Course	% of Students Who Complete Reading for Every Class, Social Sciences	% of Students Who Complete Reading for Every Class, STEM
1	75	70
2	60	65
3	63	50
4	41	45
5	90	80
6	28	53
7	53	25
8	50	45
9	65	70
10	71	72
11	25	50
12	48	33
13	65	70
14	61	65
15	85	75
16	62	60
17	70	73
18	75	70
19	57	50
20	60	43

Should the student conduct a hypothesis test for a mean or proportion difference? The student's first instinct is to say that it should be a test for the difference in proportion. After all, she is measuring the *proportion* of students in each course who regularly complete their reading for each class. However, when the student gives this question a bit more consideration, she realizes that she should actually conduct a test for a difference between means. Even though each class's value is a proportion (e.g., 75% of students in the first social science course report completing their reading for every class), the student will calculate the *mean* percentage across the twenty classes for the STEM and social science samples. (The mean percentage of students who complete the reading for every class in the social sciences is 60.2, compared to 58.2 in the STEM courses.) The sample statistic, then, is not a difference in proportions but the difference in the *means* of percentages. Here, the sample statistic is 2 (60.2 − 58.2).

The question of whether one should conduct a hypothesis test for differences between means or proportions can always be answered by examining how the dependent variable is measured. We can test for differences between proportions when we are comparing only two categories of the dependent variable. (The dependent variable may initially have more than two categories, such as party identification, but for the purpose of the hypothesis test you might distinguish between one category, such as independents, and all others.) That is, the dependent variable is nominal or ordinal. If the variable is at the interval-ratio level, then we must conduct a test for a difference between means. In this example, the variable, percentage of students who complete all reading, is interval-ratio, ranging from 0 to 100.

The student could use the same data to measure a different dependent variable: the proportion of courses in the social sciences and STEM disciplines in which more than half of the students complete their reading. In that case, the student would categorize each of the twenty courses in the social sciences and STEM as "yes" or "no," indicating whether the course met the 50% threshold. In this example, the student could conduct a hypothesis test of the difference in the proportion of social science and STEM courses in which at least 50% of the students complete their readings for every class. In the sample, at least 50% of the students completed their readings for every class in 80%, or 0.8, of social science courses, versus 75%, or 0.75, of STEM courses. Thus, the sample statistic is 0.05 (0.8 − 0.75).

Confidence Intervals for Differences between Means

Estimating a confidence interval for the difference between means follows the process covered in chapter 8 for single means. Instead of estimating an interval range of a population mean, a confidence interval for a difference between means offers an interval range of the difference between means. If the range includes the value of 0, this indicates that the population parameter could be 0, meaning that there could be no mean difference between the two groups on the dependent variable.

We will calculate a confidence interval for the difference between the metropolitan and non-metropolitan counties. The results of the hypothesis test indicated that we should reject the null hypothesis of no difference at the 0.05 level. What would this mean for the 95% confidence interval of the difference? It should mean that the interval does not include 0.* The procedure for calculating the confidence interval is the same as we saw in chapter 8. If you need a refresher, see the "Chapter Summary" on pages 336–337.

---

* We saw in chapter 9 (Box 9.2) that two-tailed hypothesis tests and confidence intervals at the same alpha level are equivalent.

The formula for the confidence interval of the difference looks a little different because our point estimate is the difference between the sample means rather than a single sample mean, and the standard error, similarly, is the standard error of the difference.

$$CI = (\bar{y}_m - \bar{y}_{nm}) \pm t(SE_{\bar{y}_m - \bar{y}_{nm}})$$

When we conducted the hypothesis test, we calculated the standard error of the difference in Step 3. It is 2812. Similarly, the degrees of freedom and the t-value for a 95% confidence interval are the same values that we identified in Step 2 of the hypothesis test. Looking at the t-table in appendix B, we can see that a 95% confidence interval corresponds to an alpha level of 0.05 for a two-tailed test. At 78 DF, the t-value is 2.0. Plugging these numbers into the formula:

$$CI = (\bar{y}_m - \bar{y}_{nm}) \pm t(SE_{\bar{y}_m - \bar{y}_{nm}})$$
$$= (64,864 - 53,543) \pm 2.0(2812)$$
$$= 11,321 \pm 5624 \to (\$5,697, \$16,945)$$

The 95% confidence interval of the county-level income difference between metropolitan and non-metropolitan counties ranges from $5,697 to $16,945. Over repeated sampling, 95% of confidence intervals would contain the true population parameter. We can say, "There is a 0.95 probability that the true population difference in mean median household income between metropolitan and non-metropolitan counties lies between $5,697 and $16,945." Because the confidence interval falls above 0, we can be 95% confident that metropolitan counties have a higher median household income, on average, than do non-metropolitan counties. In fact, as we saw above, the true population difference in mean median incomes between all metropolitan and non-metropolitan counties is $13,058, which falls within the 95% confidence interval constructed here.

Hypothesis Tests for Differences between Proportions

Here, we return to the example in chapter 8 of the hypothetical study of the effectiveness of a new drug designed to treat opioid addiction (see page 317). Recall that the researchers wanted to know whether the relapse rate for a group of participants assigned to take the drug was lower than for participants assigned to a placebo group. In the first thirty days after the conclusion of a rehabilitation program, 70% of the treatment group (N = 40) and 76% of the placebo group (N = 50) had relapsed. After calculating 95% CIs for both groups in chapter 8, we saw that they overlapped to a great extent, casting doubt on the effectiveness of the new drug at reducing relapse rates for former opioid users.

Now we will conduct a hypothesis test for the difference between relapse rates in the treatment and placebo groups of this study. Conducting tests of differences between proportions follows the same basic logic as for all hypothesis tests. As we saw in chapter 9, we calculate the standard error differently for proportions, and that difference is reflected in the two-sample tests as well, as we will see below in the example.

Step 1: Set up hypotheses.

The first step is to set up our null and alternative hypotheses. The null hypothesis is that there is no difference in relapse rates between the placebo and treatment groups. In statistical notation, with the Greek letter pi, π, denoting population proportions:

$$H_0 : \pi_p - \pi_t = 0$$

Note that the subscripts "p" and "t" denote the placebo group and treatment groups, respectively. In the previous example of a two-sample test of a difference in means, we conducted a two-tailed test, in which the alternative hypothesis stated that the difference between mean incomes for metropolitan and non-metropolitan counties was not equal to 0. In this test, we will specify a direction for the difference, reflecting the hypothesis that the drug effectively lowers relapse rates among former opioid users. Specifically, the alternative hypothesis is that the difference in relapse proportions between the placebo and treatment groups is greater than 0, shown in statistical notation as:

$$H_a : \pi_p - \pi_t > 0$$

In other words, we want to test the probability that the true population relapse rate for those taking the new drug would be lower than the true population relapse rate of those who are taking a placebo. (We would specify that the difference between the placebo and treatment groups, $\pi_p - \pi_t$, would be less than 0 if we hypothesized that the relapse rate would be higher for those taking the new drug.)

Step 2: Set alpha level.

The second step is to set the alpha level. To stay consistent with the example from chapter 8, where we estimated 95% confidence intervals, we will choose an alpha level of 0.05. How is this consistent with a 95% confidence interval? Recall from chapter 8 that the confidence level for an interval is equal to one minus alpha. For a 95% CI, then, alpha is 0.05 (1 − 0.95 = 0.05).

Notice that, for the same alpha level, a one-tailed hypothesis test and a confidence interval have different corresponding z-scores. For the 95% confidence interval (and two-tailed hypothesis tests with an alpha of 0.05), the alpha value is distributed equally across the left and right tails of the distribution, such that the area of the curve that lies beyond the cutoff points (the z-scores) is in the left and right 2.5% of the tails (0.05/2 = 0.025). For a one-tailed test, however, we do not distribute the area across both tails—we choose just one tail. Here, we have specified in the alternative hypothesis that the difference is greater than 0, indicating that we want all of the probability to be in the right tail of the distribution. (It would be in the left tail if the alternative hypothesis specified a difference less than 0.) Figure 10.4 shows the locations of the probabilities for a 95% confidence interval and two-tailed test

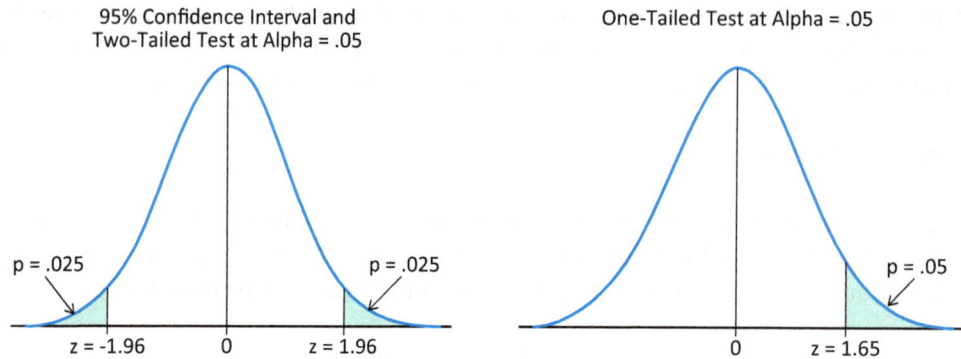

Figure 10.4 Comparing Probabilities and z-Scores for Confidence Intervals, Two-Tailed Tests, and One-Tailed Tests

with an alpha of 0.05 (left side), compared to a one-tailed test with an alpha of 0.05 (right side). In addition to the p-values, Figure 10.4 shows the different z-scores for a 95% CI/two-tailed test at alpha of 0.05 (1.96) and a one-tailed test at alpha of 0.05 (1.65).

Step 3: Estimate the pooled standard error.

The standard error for a hypothesis test of the difference between proportions works slightly differently from the standard error for a hypothesis test of a mean difference. For the difference between means test, recall that the standard error of the difference is the square root of the sum of the squared standard errors for each mean.

For the standard error of the difference between proportions, we combine the cases from the two samples to find the "pooled" proportion across the two samples. The formula for the pooled standard error of the difference between proportions is:

$$SE_{p_1-p_2} = \sqrt{\frac{\pi(1-\pi)}{N_1} + \frac{\pi(1-\pi)}{N_2}}$$

There are no subscripts attached to the proportions in this formula. This indicates that, unlike the standard error of the difference between means, the standard error of the difference between proportions for a hypothesis test does not distinguish between proportions for the first and second groups. In hypothesis testing, because we assume that there is no difference in proportions between the two groups, we combine the number of cases that fall into the category of interest (e.g., individuals who relapsed) across the two groups. We assume in the hypothesis test that the pooled sample proportion reflects the population proportion; thus, we use the Greek letter pi, π, to represent the population proportion, rather than the letter "p" to represent the sample proportion.

Before we can calculate the pooled standard error of the difference in proportion of relapses between the treatment and placebo groups, we need to "pool" the proportion of relapses across both groups. We know that twenty-eight people in the treatment group (i.e., 70% of 40) and thirty-eight people in the placebo group (i.e., 76% of 50) relapsed. The total number of relapses is sixty-six (28 + 38 = 66). We find the pooled relapse rate by dividing the total number of relapses by the total number of cases: 66/90 = .73. By formula, where f_t is the frequency of relapses in group 1 and f_p is the frequency of relapses in group 2, this would be:

$$\text{Pooled proportion} = \frac{f_t + f_p}{N_{total}} = \frac{28+38}{90} = 0.73$$

Now we can plug this information into the formula for the pooled standard error of the difference in proportions:

$$SE_{p_p - p_t} = \sqrt{\frac{0.73(1-0.73)}{50} + \frac{0.73(1-0.73)}{40}}$$

$$= \sqrt{0.0039 + 0.0049} = \sqrt{0.0088} = 0.094$$

The standard error of the difference is 0.094.

To recap, first calculate the pooled proportion. Then plug that number into the formula for the standard error.

Step 4: Calculate the sample z.

The z-score is equal to the difference in sample proportions, divided by the standard error of the difference:

$$z = \frac{(p_p - p_t)}{SE_{p_p - p_t}} = \frac{(0.76 - 0.70)}{0.094} = \frac{0.06}{0.094} = 0.64$$

As with hypothesis tests for the difference between means, the formula for z in hypothesis tests for differences between proportions is actually:

$$z = \frac{(p_p - p_t) - (\pi_p - \pi_t)}{SE_{p_p - p_t}}$$

Since the hypothesized difference between population parameters ($\pi_p - \pi_t$) is 0, we can omit that part of the equation.

Step 5: Find the p-value associated with sample z. If the p-value is less than alpha, reject the null hypothesis. State your conclusion.

As shown in the Normal Table (appendix A), the sample z of 0.64 is associated with a p-value of 0.2611. This is much higher than our alpha level of 0.05. This probability indicates that we would observe this difference of at least 6 percentage points between the relapse rates for the placebo and treatment groups about 26% of the time under repeated sampling if the null hypothesis were true. Thus, we cannot reject the null hypothesis that there is no difference in relapse rates between the treatment and placebo groups. This finding of no difference is consistent with the overlapping confidence intervals for the two groups found in chapter 8.

Next, we will use the results of this study to calculate a 95% confidence interval for the difference in proportion of relapses between the treatment and placebo groups.

Confidence Intervals for Differences between Proportions

We know that 70% of the participants in the treatment group relapsed in the first thirty days following completion of a rehabilitation program, compared to 76% of participants in the placebo group. To calculate the 95% confidence interval for the unknown population difference, the steps are the same as for other confidence intervals, but the formulas are slightly different. The formula is:

$$CI = (p_1 - p_2) \pm z(SE_{p_1 - p_2})$$

We calculate the standard error for the difference between proportions differently than we did for the hypothesis test. For the confidence interval of a difference in proportions, we do not use the pooled standard error that we used for the hypothesis test, because, with confidence intervals, we make no assumptions about differences in the population. For a confidence interval for differences between proportions, we calculate the standard error in the same way as we did for the hypothesis test for means:

$$SE_{p_p - p_t} = \sqrt{SE_{p_p}^2 + SE_{p_t}^2}$$

Just as we saw with the standard error of a mean difference, the standard error of a proportion difference for a confidence interval is the square root of the sum of the squared standard errors for the two groups. Recall that the standard error of a proportion is given by:

$$SE_p = \sqrt{\frac{p(1-p)}{N}}, \text{ where } p \text{ is the sample proportion.}$$

Thus, the standard errors for the treatment and placebo groups are:

$$SE_{p_t} = \sqrt{\frac{p_t(1-p_t)}{N_t}} = \sqrt{\frac{0.7(1-0.7)}{40}} = 0.072 \text{ and}$$

$$SE_{p_p} = \sqrt{\frac{p_p(1-p_p)}{N_p}} = \sqrt{\frac{0.76(1-0.76)}{50}} = 0.060$$

We can insert the standard errors for each group into the formula for the standard error of the difference:

$$SEp_p - SEp_t = \sqrt{SEp_p^2 + SEp_t^2} = \sqrt{0.060^2 + 0.072^2}$$

$$= \sqrt{0.0036 + 0.0052} = \sqrt{0.00878} = 0.094$$

The standard error of the difference in relapse rates between the placebo and treatment groups is 0.094.

The z-score associated with a 95% confidence interval is 1.96, as shown in Figure 10.4. Remember, the z-score for a confidence interval is higher than the z-score for a one-tailed hypothesis test at the same alpha level. Thus, the z-score that we will use here, 1.96, is higher than the z-score of 1.65 associated with the one-tailed test at the 0.05 alpha level in the previous hypothesis test.

Plugging these values into the formula, we find:

$$CI = (p_p - p_t) \pm z(SE_{p_p - p_t}) = (0.76 - 0.70) \pm 1.96(0.094)$$

$$= 0.06 \pm 0.184 \rightarrow (-0.12, 0.24)$$

The confidence interval for the difference in proportions between the placebo and treatment groups ranges from −.12 to .24, or −12 percentage points to 24 percentage points. Figure 10.5 presents a visual representation of the confidence interval.

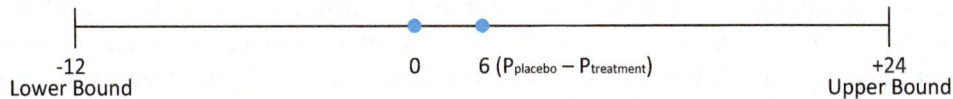

Figure 10.5 95% Confidence Interval of Difference in Proportions of Relapses for Placebo and Treatment Groups

As Figure 10.5 shows, the midpoint of the confidence interval is 6 percentage points, the sample difference between relapse rates in the placebo and treatment groups. We see not only that the confidence interval includes 0 but also that it includes negative values, which would indicate higher relapse rates in the treatment group. We know that over repeated sampling, 95% of the confidence intervals will include the population parameter. Since our sample confidence interval includes 0 and ranges into negative numbers, there is a 95% chance that the true difference in relapse rates between the two groups falls in a range that includes 0 or favors the placebo group. Thus, we cannot conclude that the 6-percentage-point difference between the treatment and placebo groups reflects anything other than sampling variability.

We now have three pieces of evidence from the study conducted on the new drug for the treatment of opioid addiction that indicate its failure to reduce relapse rates at statistically significant levels: (1) the overlapping confidence intervals for relapse rates of placebo and treatment groups found in chapter 8, (2) the failure to reject the null hypothesis of no difference in relapse rates between the placebo and treatment groups, and (3) the inclusion of 0 and negative numbers in the confidence interval of the difference in relapse rates between the two groups. Should the drug's designers return to the drawing board to design a new drug? Even though the results of these analyses are not promising, one might argue that we do not yet know if the drug is effective. The single hypothetical study conducted here had small sample sizes, forty for the treatment group and fifty for the placebo group. The researchers might consider setting up a study with larger groups to reduce the impact of sampling variability on the estimates of the drug's effectiveness.

Statistical and Practical Significance in Two-Sample Tests

Remember the asterisks that we saw in Table 10.1 of this chapter, denoting statistically significant differences between men and women on measures of anxiety, stress, and insomnia during the 2020 COVID lockdown? Researchers tend to get very excited when they see these asterisks, and, conversely, it is easy to feel disappointed when a result is not statistically significant. However, as we discussed in chapter 9, statistical significance does not necessarily signal substantive importance. It is especially important to keep this in mind when working with large samples. When sample sizes are very large, even substantively negligible results can be statistically significant. As sample size increases, the standard error decreases, and the smaller the standard error, the more likely a result is to be statistically significant.

Returning to the study described at the beginning of the chapter, we can consider whether it is substantively meaningful that the mean insomnia score for a sample of women was 1.42 points higher, on average, than for a sample of men. This mean difference was statistically significant, but that says nothing about the *substantive* meaning of the difference. How do we know whether a difference of 1.42 points is a meaningful difference? Does it indicate that women suffered to a much greater degree with insomnia during the 2020 COVID lockdown than did men? One way to assess the substantive meaning of the difference, apart from the statistical significance, is to compare the size of the difference to the range of the variable. In this case, the insomnia scale ranges from 0 to 28 points. A difference of 1.42 points covers 5% of the range of the variable (1.4/28 = 0.05). Thus, we could say that women scored 5% higher than men on the insomnia scale, given the range of the variable. We could interpret the size of the difference to be modest, yet statistically significant.

As we noted in chapter 9, the use of confidence intervals to estimate an interval range of population parameters is one way of compensating for the limited information offered by tests of statistical significance. The key point is that we should never expect the result of a single hypothesis test to provide the complete answer to a research question.

Chapter Summary

Two-sample hypothesis tests are tests of the difference between two means or two proportions.

- The steps for conducting a two-sample hypothesis test are:
    1. Set up research and null hypotheses.
    2. Set the alpha level. If you are doing a hypothesis test for the difference between means (using t), calculate degrees of freedom for your sample and look in the t-table in appendix B to determine the decision t. When calculating degrees of freedom by hand, use this formula, which technically assumes that the variances of the two means are equal:

    $$DF = (N_1 + N_2) - 2$$

    3. Estimate the standard error.
        a. For proportions: $SE_{p_1 - p_2} = \sqrt{\dfrac{\pi(1-\pi)}{N_1} + \dfrac{\pi(1-\pi)}{N_2}}$, where π is the pooled proportion of the outcome of interest for both groups.
        b. For means: $SE_{\bar{y}_1 - \bar{y}_2} = \sqrt{SE_{\bar{y}_1}^2 + SE_{\bar{y}_2}^2}$, where the standard error for each group is $\dfrac{s}{\sqrt{N}}$
    4. Calculate the sample statistic, either z (for proportions) or t (for means).
        a. For proportions: $z = \dfrac{(p_1 - p_2)}{SE_{p_1 - p_2}}$
        b. For means: $t = \dfrac{(\bar{y}_1 - \bar{y}_2)}{SE_{\bar{y}_1 - \bar{y}_2}}$
    5. For proportions, look in the z-table to find the p-value associated with your sample z. For means, compare the absolute value of the sample t with the decision t. Determine whether you can reject the null hypothesis. If the p-value is lower than alpha (for proportions) or if sample t is higher than decision t (for means), you can reject the null hypothesis.
    6. State your conclusion.

- To estimate confidence intervals of differences between proportions and means, the steps are:
    1. Estimate the standard error:
        a. For proportions: $SE_{p_1 - p_2} = \sqrt{SE_{p_1}^2 + SE_{p_2}^2}$, where the standard error for each sample proportion is $\sqrt{\dfrac{p(1-p)}{N_p}}$

b. For means: $SE_{\bar{y}_1 - \bar{y}_2} = \sqrt{SE_{\bar{y}_1}^2 + SE_{\bar{y}_2}^2}$, where the standard error for each sample mean is $\dfrac{s}{\sqrt{N}}$

2. Set confidence level (e.g., 90%, 95%, 99%, etc.) and find associated z-score or t-value.
3. Calculate margin of error: *z(SE)* or *t(SE)*.
4. Calculate lower and upper bounds of CI: *p* ± margin of error or $\bar{y}$ ± margin of error.

Using Stata

In this section, we revisit the 2020 Police Public Contact Survey (PPCS), which we used in the "Using Stata" section in chapter 9. In that section, we conducted one-sample tests for the proportion of the population who experienced face-to-face contact with police and the mean duration of people's most recent police contact. In both cases, we tested the likelihood that rates of face-to-face contact and duration of contact for Black people were equal to the population parameters for all Americans. The results suggested very small chances that we would have observed the sample rates and means that we did if the true population parameters for Black people were equal to those for all Americans. This finding suggests that, as a group, Black people are more likely to experience face-to-face contact with police and longer-lasting interactions with the police than at least some other racial groups.

We will use these two same variables from the PPCS to test whether the differences between Black and white people, in particular, are statistically significant:

- *Contact* measures whether respondents who did have contact with the police had face-to-face contact with them. There are two possible values for *contact*: 1 indicates that the respondent had face-to-face contact with the police, and 0 means that the respondent had police contact that was not face-to-face.
- *Time* measures the duration (in minutes) of respondents' most recent contact with the police.

We will conduct two hypothesis tests: (1) a two-sample test of the difference between proportions of Black and white people who experience face-to-face contact with police and (2) a two-sample test of the mean difference in duration of most recent police contact between Black and white people. Stata will also estimate confidence intervals for both differences.

Contact and *time* will serve as the dependent variables for our two-sample tests of proportions and means, respectively. The independent, or grouping, variable for both tests measures whether the respondent identified as Black (0) or white (1).

Two-Sample Test for Difference between Proportions

We begin with a two-sample test for the difference between Blacks and whites in the proportion of respondents who reported having had face-to-face contact with police. We know from the data that 59.8% of Black respondents and 52.7% of white respondents reported having had face-to-face contact with police. But is the difference statistically significant? We will test the following proportion difference in our hypothesis test:

$$p_b - p_w = 0.071$$

The null hypothesis is that there is no difference in the proportion of people who had face-to-face contact with police between Blacks and whites:

$$H_0: \pi_b - \pi_w = 0$$

Our alternative hypothesis is that the proportion of Black people who have face-to-face contact with police is higher than the proportion of whites:

$$H_a: \pi_b - \pi_w > 0$$

Thus, we are setting up a one-tailed test on the right side of the distribution. Open `PPCS2020Ch10.dta` and enter the following "prtest" command into Stata:

`prtest contact, by (white) level (95)`

The "prtest" command tells Stata to conduct a hypothesis test for proportions using the dependent variable *contact* and the independent, or grouping, variable *white*. With the "level(95)" option, we asked Stata to conduct the test at the 0.05 alpha level (1 − 0.95 = 0.05). The output, which provides a wealth of information, is shown in Figure 10.6.

The first thing to note about the output is that, in addition to the results of the hypothesis test of the difference between proportions, it presents a confidence interval for the difference. This is a nice feature that reminds us of the utility of using these two statistical tools together.

First, we will focus on the parts of the output that address the hypothesis test. The "Mean" column of the table shows the proportions of whites (0.527) and Blacks (0.598) who experienced face-to-face police contact as well as the difference in proportions between the two groups (0.0717698). As we saw above, this is the difference for which Stata is conducting the hypothesis test.

Do you notice that Stata is reporting two separate standard errors for the difference? This is because, as we saw earlier in the chapter, the standard error for a hypothesis test of the difference between proportions is calculated differently than the standard error for the confidence interval of a difference between proportions. The standard error for the hypothesis test, which pools counts across the two groups, is 0.028 (marked "under H_0").

```
. prtest contact, by (white) level (95)

Two-sample test of proportions                    0: Number of obs =      351
                                                  1: Number of obs =     3337

      Group |       Mean   Std. err.      z    P>|z|    [95% conf. interval]
------------+---------------------------------------------------------------
          0 |   .5982906   .0261673                      .5470037   .6495775
          1 |   .5265208   .0086433                      .5095802   .5434614
------------+---------------------------------------------------------------
       diff |   .0717698   .0275578                      .0177574   .1257821
  under H0: |              .027994    2.56    0.010                z =  2.5638

       diff = prop(0) - prop(1)                                   z =   2.5638
   H0: diff = 0

   Ha: diff < 0               Ha: diff != 0              Ha: diff > 0
 Pr(Z < z) = 0.9948       Pr(|Z| > |z|) = 0.0104      Pr(Z > z) = 0.0052
```

Figure 10.6

Recall from chapter 9 that, when we request a hypothesis test from Stata, it reports the results of the one-tailed test to the left, the two-tailed test, and the one-tailed test to the right, in that order, at the bottom of the output. We specified with our alternative hypothesis that the difference in proportions would be greater than 0, which corresponds to the result at the bottom-right corner. The p-value for this test is 0.0052. This is far smaller than our alpha level of 0.05 and indicates a very small probability that we would have observed a sample difference of at least 7.2 percentage points if the population difference were 0. We conclude that a significantly higher proportion of Black than white people experienced face-to-face contact with police.

We turn now to the confidence interval for the difference reported by Stata, located in the "95% Conf. Interval" column. Note that the non-pooled standard error, .0275578, was used to calculate the margin of error for the confidence interval. The confidence interval ranges from 0.018 to 0.126, excluding 0, which is consistent with the results of the hypothesis test.

Two-Sample Test for Difference between Means

Turning to our hypothesis test for a difference in means, we will examine whether the mean duration of most recent police contact is longer for Blacks than whites. The null hypothesis is that there is no difference in the mean duration of the most recent contact (in minutes) between these two groups:

$$H_0: \mu_b - \mu_w = 0$$

Our alternative hypothesis is that the mean duration of the most recent police contact for Black people is higher than for whites:

$$H_a: \mu_b - \mu_w > 0$$

Again, we are setting up a one-tailed test on the right side of the distribution. Enter the following "ttest" command into Stata:

```
ttest time, by(white) unequal level(95)
```

The "ttest" command tells Stata to conduct a hypothesis test for the difference in the means for time between Black and white people, as indicated with the "by(white)" option. The option "unequal" indicates that we are not making the assumption that the variances of time are equal across the two groups.* As with the test of proportions, the "level(95)" option specifies the alpha level of 0.05. The output, which follows the same format as the output for the two-sample test of proportions, is shown in Figure 10.7.

```
. ttest time, by(white) unequal level(95)

Two-sample t test with unequal variances

   Group  |    obs       Mean     Std. err.   Std. dev.   [95% conf. interval]
       0  |   1,624    14.3593    .4567172    18.40519    13.46348    15.25512
       1  |  14,317   12.11703    .1437089    17.19529    11.83534    12.39872
Combined  |  15,941   12.34546    .1372993    17.33509    12.07634    12.61458
    diff  |           2.242269    .4787931                1.303272    3.181267

    diff = mean(0) - mean(1)                                    t =   4.6832
H0: diff = 0                       Satterthwaite's degrees of freedom = 1958.12

    Ha: diff < 0              Ha: diff != 0                 Ha: diff > 0
 Pr(T < t) = 1.0000       Pr(|T| > |t|) = 0.0000         Pr(T > t) = 0.0000
```

Figure 10.7

The output provides the same information as the output produced by the two-sample test of proportions using the PPCS data. Again, we see the result of the hypothesis test and the 95% confidence interval. The "Mean" column shows the mean number of minutes of the most recent police contact for Blacks (14.4) and whites (12.1) in addition to the difference between them (2.24), which is the sample difference. The standard error of the difference, 0.48, is shown next to the difference. The sample t, 4.68, which as we know is equal to the sample difference divided by the standard error, is reported just under the table on the right side of the output. The degrees of freedom, 1,958, which Stata calculates using the longer formula shown earlier in the chapter, is shown just under the t-value. The decision t is not provided in the table, but the final result of the hypothesis test shown in the lower-right corner of the output shows us the p-value associated with the sample t (.0000), which indicates that we should reject the null hypothesis of no difference in mean duration of police contact between Blacks and whites. (Remember that, although Stata will present a p-value that appears to be 0, we report that as $p < 0.001$ to reflect that the probability is never truly 0.) We would conclude that police contact for Black people is significantly longer in duration than for white people.

The 95% confidence interval for the difference is the last entry in the "95% Conf. Interval" column: (1.30, 3.18). As with the confidence interval of the difference between the

---

*Earlier in the chapter, we noted that degrees of freedom can be determined with a longer or shorter formula. The longer formula, shown on page 399, does not assume equal variances across groups, and specifying the "unequal" option tells Stata to use this formula. This produces an accurate statistic for groups with and without equal variances.

proportions of Black people and white people who experience face-to-face contact with police, the confidence interval of the difference between the mean duration of the most recent contact for Blacks and whites does not include 0.

Taken together these results support the claim that the police treat Black people and white people differently. Comparing means and proportions does not give us any information about the explanations for these differences, nor can we conclude from these results that individuals' racial identities "cause" this difference in treatment. In chapter 13, we address the issue of establishing causality and consider causal mechanisms, or explanations for causal relationships.

Review of Stata Commands

- Two-sample test of a difference between proportions

    ```
    prtest dependent variable, by (independent variable) level((1-alpha) x 100)
    ```

- Two-sample test of a difference between means

    ```
    ttest dependent variable, by (independent variable) unequal level ((1-alpha) x 100)
    ```

Using SPSS

In this section, we revisit the 2020 Police Public Contact Survey (PPCS), which we used in the "Using SPSS" section in chapter 9. In that section, we conducted one-sample tests for the proportion of the population who experienced face-face contact with police and the mean duration of people's most recent police contact. In both cases, we tested the likelihood that rates of face-to-face contact and duration of contact for Black people were equal to the population parameters for all Americans. The results suggested very small chances that we would have observed the sample rates and means that we did if the true population parameters for Black people were equal to those for all Americans. This finding suggests that, as a group, Black people are more likely to experience face-to-face contact with police and longer-lasting interactions with the police than at least some other racial groups.

We will use these two same variables from the PPCS to test whether the differences between Black and white people, in particular, are statistically significant:

- *Contact* measures whether respondents who did have contact with the police had face-to-face contact with them. There are two possible values for *contact*: 1 indicates that the respondent had face-to-face contact with the police, and 0 means that the respondent had police contact that was not face-to-face.

- *Time* measures the duration (in minutes) of respondents' most recent contact with the police.

We will conduct two hypothesis tests: (1) a two-sample test of the difference between proportions of Black and white people who experience face-to-face contact with police and (2) a two-sample test of the mean difference in duration of most recent police contact between Black and white people. SPSS will also estimate confidence intervals for both differences.

Contact and *time* will serve as the dependent variables for our two-sample tests of proportions and means, respectively. The independent, or grouping, variable for both tests measures whether the respondent identified as Black (0) or white (1).

Two-Sample Test for Difference between Proportions

We begin with a two-sample test for the difference between Black and white respondents who reported face-to-face contact with police. As we saw in previous chapters, because *contact* has two values, coded as 0/1, we can use the SPSS means procedures to test hypotheses about proportions. We know from the data that 59.8% of Black respondents and 52.7% of white respondents reported face-to-face contact. Black respondents reported a higher rate of face-to-face police contact than did whites. But is the difference statistically significant? We will test the following proportion difference in our hypothesis test:

$$p_b - p_w = 0.071$$

The null hypothesis is that there is no difference in the proportion of people who had face-to-face contact police between Blacks and whites:

$$H_0: \pi_b - \pi_w = 0$$

Our alternative hypothesis is that the proportion of Black people who have face-to-face contact with police is higher than the proportion of whites:

$$H_a: \pi_b - \pi_w > 0$$

Thus, we are setting up a one-tailed test on the right side of the distribution.

Open `PPCS2020Ch10.sav`. To compare two sample proportions (using 0/1 variables), we use this SPSS procedure:

`Analyze → Compare Means → Independent-Samples T Test`

This will open the "Independent-Samples T Test" dialog box, shown in Figure 10.8. We move the variable *contact* into the "Test Variable(s)" box. But before we submit the procedure, we must specify that we want SPSS to run a test on the difference between the proportion of Black and white people who have face-to-face contact with police. We move the variable *white* into the "Grouping Variable" space and click on "Define Groups."

This opens the "Define Groups" dialog box, shown in Figure 10.9.

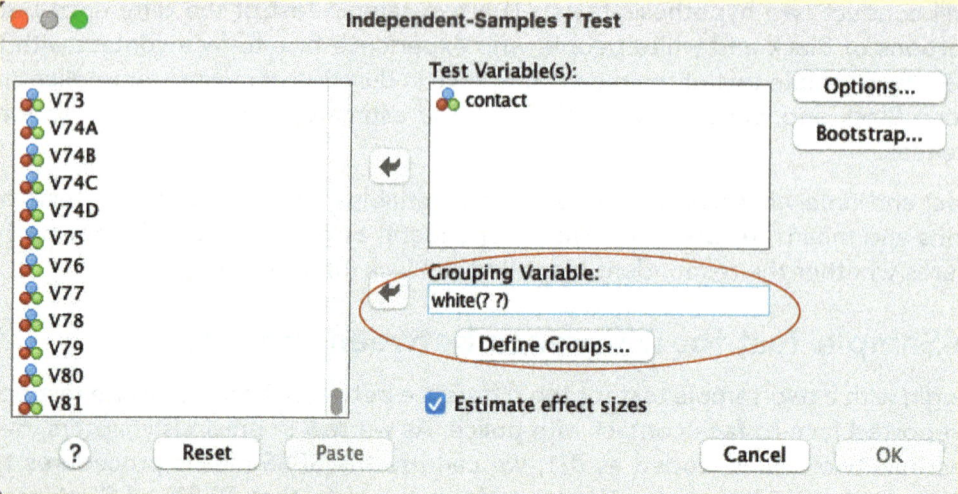

Figure 10.8

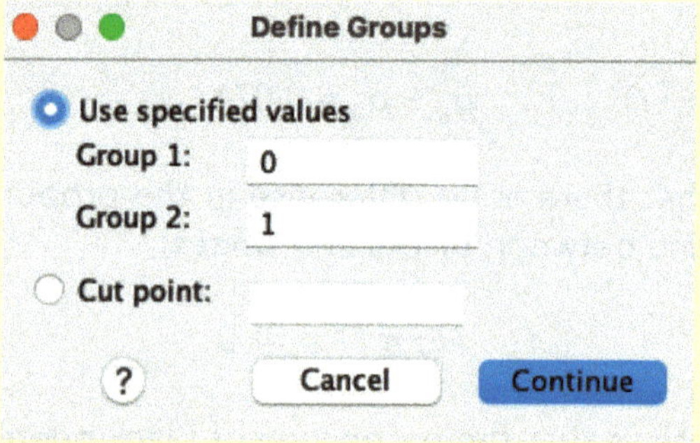

Figure 10.9

We specify value 0 (Black) for Group 1 and value 1 (white) for Group 2. Click on "Continue" and then "OK." The output, which provides a wealth of information, is shown in Figure 10.10.

The first thing to note about the output is that, in addition to the results of the hypothesis test of the difference between proportions, it presents a confidence interval for the difference. This is a nice feature that reminds us of the utility of using these two statistical tools together.

Do you notice that SPSS is reporting two values for t and two standard errors for the difference and confidence intervals of the difference (in the bottom table)? This is because there are two different ways of calculating the standard error of the difference, depending on whether the variance (or standard deviation) for the two groups is relatively equal. SPSS calculates it both ways and gives us guidance on which value to use through Levene's test. (This is under the heading "Levene's Test for Equality of Variances.") If the

Group Statistics					
	white	N	Mean	Std. Deviation	Std. Error Mean
contact	0	351	.60	.491	.026
	1	3337	.53	.499	.009

Independent Samples Test

		Levene's Test for Equality of Variances		t-test for Equality of Means							
						Significance		Mean Difference	Std. Error Difference	95% Confidence Interval of the Difference	
		F	Sig.	t	df	One-Sided p	Two-Sided p			Lower	Upper
contact	Equal variances assumed	67.090	<.001	2.565	3686	.005	.010	.072	.028	.017	.127
	Equal variances not assumed			2.601	429.790	.005	.010	.072	.028	.018	.126

Figure 10.10

significance ("Sig." in Figure 10.10) for Levene's test is below an alpha of 0.05, we reject the null hypothesis that the variances are equal, and we must then conclude that the variances are unequal. We therefore use the values of t and of the confidence interval of the difference that correspond to unequal variances; those are in the bottom line marked "Equal variances not assumed".

If you find all of this confusing, here is a simple rule: If "Sig." for Levene's test is 0.05 or lower, look at the bottom line of the table; if it is above 0.05, use the top line.

Now we will focus on the parts of the output that address the hypothesis test. The "Mean" column of the table shows the proportions of whites (0.53) and Blacks (0.60) who had face-to-face police contact, and the bottom table shows us the difference in proportions between the two groups (0.072). This is the difference for which SPSS is conducting the hypothesis test.

SPSS reports the results of both one-tailed and two-tailed tests. We specified with our alternative hypothesis that the difference in proportions would be greater than 0, which means we are using the one-tailed test. The one-tailed p-value, as shown in Figure 10.10, is 0.005. This is far smaller than our alpha level of .05 and indicates that there is virtually no chance that we would have observed a sample difference of at least 7.2 percentage points if the population difference were 0. We would conclude that a significantly higher proportion of Blacks than whites report having had face-to-face contact with police.

We turn now to the confidence interval for the difference reported by SPSS, located in the "95% Conf. Interval" columns. We use the bottom row—equal variances not assumed—as we discussed above. The confidence interval ranges from 0.018 to 0.126, excluding 0, which is consistent with the results of the hypothesis test.

Two-Sample Test for Difference between Means

Turning to our hypothesis test for a difference in means, we will examine whether the mean duration of most recent police contact is longer for Black people than white people. The null hypothesis is that there is no difference in the mean duration of the most recent police contact (in minutes) between Blacks and whites:

$$H_0: \mu_b - \mu_w = 0$$

Our alternative hypothesis is that the mean duration of the most recent police contact for Black people is higher than for whites:

$$H_a: \mu_b - \mu_w > 0$$

Again, we are setting up a one-tailed test on the right side of the distribution. We use the exact same SPSS procedure we just used:

Analyze → Compare Means → Independent-Samples T Test

We will compare the means of white (1) and Black (0) respondents on the *time* variable.

The output, which follows the same format as the output for the two-sample test of proportions, is shown in Figure 10.11.

Group Statistics

time	white	N	Mean	Std. Deviation	Std. Error Mean
	0	1624	14.3593	18.40519	.45672
	1	14317	12.1170	17.19529	.14371

Independent Samples Test

		Levene's Test for Equality of Variances		t-test for Equality of Means						
						Significance		Mean	Std. Error	95% Confidence Interval of the Difference
		F	Sig.	t	df	One-Sided p	Two-Sided p	Difference	Difference	Lower	Upper
time	Equal variances assumed	15.277	<.001	4.944	15939	<.001	<.001	2.24227	.45357	1.35322	3.13132
	Equal variances not assumed			4.683	1958.115	<.001	<.001	2.24227	.47879	1.30327	3.18127

Figure 10.11

We see the result of the hypothesis test and the 95% confidence interval. The "Mean" column shows the mean number of minutes of the most recent police contact for Blacks (14.36) and whites (12.12). The bottom table displays the difference between them (2.24), which is the sample difference, and the standard error of the difference, 0.48. The sample t, 4.68, which as we know is equal to the sample difference divided by the standard error, is reported along with the p-value, which is less than 0.001. This indicates that we should reject the null hypothesis of no difference in mean duration of most recent police contact between Black and white people. We would conclude that contact with police lasts significantly longer for Black people than for white.

The 95% confidence interval runs from 1.30 to 3.18. As with the confidence interval of the difference between the proportion of Black and white people who experience face-to-face contact with police, the confidence interval of the difference between the mean duration of most recent police contact for Blacks and whites does not include 0.

Taken together these results support the claim that the police treat Black people and white people differently. Comparing means and proportions does not give us any information about the explanations for these differences, nor can we conclude from these results that individuals' racial identities "cause" this difference in treatment. In chapter 13, we address the issue of establishing causality and consider causal mechanisms, or explanations for causal relationships.

Review of SPSS Procedures

- Two-sample test of a difference between proportions (0/1 variable only)

  `Analyze → Compare Means → Independent-Samples T Test`

- Two-sample test of a difference between means

  `Analyze → Compare Means → Independent-Samples T Test`

Practice Problems

1. Data from the 2021 American Community Survey showed that in 18% of opposite-sex couples and 30% of same-sex couples, both partners had earned at least a bachelor's degree. Say that sample sizes are 450 for opposite-sex couples and 200 for same-sex couples. Are same-sex couples really more highly educated than opposite-sex couples in the population?

   a. State the null and alternative hypotheses that would test this question.

   b. Estimate the standard error for the difference between these proportions.

   c. Calculate the sample z-score.

   d. Assuming an alpha level of 0.025, should the null hypothesis be rejected? Explain how you know.

2. Data from the 2021 American Community Survey showed that a larger share of same-sex (31.6%) than opposite-sex (18.4%) married couples were interracial. Say that sample sizes are 450 for opposite-sex couples and 200 for same-sex couples. Are same-sex couples really more likely to be interracial pairings than opposite-sex couples in the population?

   a. State the null and alternative hypotheses that would test this question.

   b. Estimate the standard error for the difference between these proportions.

   c. Calculate the sample z-score.

   d. Assuming an alpha level of 0.001, should the null hypothesis be rejected? Explain how you know.

3. Assume that the sample sizes for the two groups from Problem 1 are: $N_{opposite-sex} = 100$; $N_{same-sex} = 75$.

   a. Recalculate the standard error and sample z-score with these new sample sizes.

   b. Should the null hypothesis be rejected given these new sample sizes? (Use the same alpha level, 0.025, as in Problem 1.)

c. Do we a draw different conclusion for these smaller sample sizes than in Problem 1? Explain why or why not.

4. Assume that the sample sizes for the two groups from Problem 2 are: $N_{opposite\text{-}sex}$ = 100; $N_{same\text{-}sex}$ = 75.

   a. Recalculate the standard error and sample z-score with these new sample sizes.

   b. Should the null hypothesis be rejected given these new sample sizes? (Use the same alpha level, 0.025, as in Problem 2.)

   c. Do we a draw different conclusion for these smaller sample sizes than in Problem 2? Explain why or why not.

5. A professor is investigating dominant themes in children's picture books. She wants to know if themes differ depending on the gender of the main character. After obtaining lists of all top-selling children's picture books with a girl or boy protagonist over the last thirty years, she draws two random samples of fifty books, one sample with boy protagonists and the other with girl protagonists. Her undergraduate research assistants rate all one hundred books according to several themes, including "emphasis on appearance," which is measured on a scale ranging from 0 (no mention of physical appearance) to 50 (very strong emphasis on physical appearance). The professor asks her research team to use a two-sample test to compare emphasis on physical appearance by protagonists' gender. The mean emphasis on appearance for books with girl protagonists is 30 (standard deviation = 8); for books with boy protagonists, it is 20 (standard deviation = 2).

   a. What is the dependent variable? What is the independent variable?

   b. The team decides to set up a two-tailed hypothesis test. State the null and alternative hypotheses, in both words and statistical notation.

   c. Assuming that the alpha level is set at 0.01, sketch a t-distribution and label the following components: (1) the middle of the sampling distribution, (2) decision t, and (3) p-value.

   d. Estimate the standard error of the mean difference. Calculate sample t.

   e. Decide whether to reject the null hypothesis and draw a conclusion.

6. The 2021 GSS asked respondents how important the issue of climate change was to them, ranging from 1 (not at all important) to 5 (extremely important). The mean value of the importance of climate change to respondents who had health insurance was 3.36 (s = 1.35, N = 6,609). The mean value for respondents who did not have health insurance was 3.17 (s = 1.36, N = 704). Does having health insurance affect how important climate change is to people?

   a. What is the dependent variable? What is the independent variable?

   b. State the null and alternative hypotheses for a one-tailed hypothesis test, in both words and statistical notation.

c. Assuming that the alpha level is set at 0.05, sketch a t-distribution and label the following components: (1) the middle of the sampling distribution, (2) decision t, and (3) p-value.

d. Estimate the standard error of the mean difference. Calculate sample t.

e. Decide whether to reject the null hypothesis and draw a conclusion.

7. A study conducted on behalf of the World Cup examined the cost of participating in organized youth soccer around the globe. In two random samples of families with children participating in organized soccer in the United States and Brazil, the mean annual cost of participation per child was $1,500 in the United States and $1,000 in Brazil. (Brazilian reals were converted to US dollars for comparability.) The standard deviation for the US sample was $300 (N = 103) and $250 for the Brazil sample (N = 175).

   a. Calculate a 95% confidence interval for the mean difference between the two samples.

   b. A reporter uses the differences in sample means to write an article about the lower cost of participating in organized soccer for youth in Brazil compared to the United States. Use the confidence interval that you calculated for part a to explain whether you agree with the reporter's claim.

8. Cardi B is proud of having grown up in the Bronx and believes that residents of the Bronx are her most loyal fans. Hot 97, a popular radio station in New York City, is sponsoring a contest to see if residents of the Bronx are more familiar with the lyrics of Cardi B's hit song "Bodak Yellow" than residents of the other boroughs of New York City. If the Bronx wins, Cardi B will hold a free concert in the Bronx. Hot 97 partnered with a team of statistics students from Bronx Community College to draw two random samples, one from the Bronx (N = 50) and one from the city's other four boroughs (N = 150). Respondents were asked to rap along to the song, and the statistics students measured the percentage of the lyrics that each respondent got right. The sample from the Bronx knew, on average, 57% of the "Bodak Yellow" lyrics (standard deviation = 7), while the sample from the other four boroughs knew 55% of the lyrics, on average (standard deviation = 3).

   a. Calculate a 90% confidence interval for the difference between the two samples.

   b. Hot 97 wants to declare the Bronx the winner based on the outcome of the contest and start arranging the free Cardi B concert. Do you agree with this decision, given the confidence interval from part a?

9. Polling firm Buendía & Laredo collected random samples of all registered voters in Mexico in 2017 and 2018. Each sample included one thousand respondents. In the 2018 sample, 31% said that they had "very unfavorable" views of the United States, compared to 22% of those surveyed in 2017. A 2018 report issued by the US House Foreign Affairs Committee said it did not believe that Mexican sentiment toward the United States worsened between 2017 and 2018 because it expected short-term polling fluctuations.

a. Calculate the 95% confidence interval and explain whether it challenges the Foreign Affairs Committee's belief that Mexican sentiment toward the United States has not declined.

b. Do you think that the percentage difference between 2017 and 2018 is a substantively important difference? Explain your answer.

10. A group of nongovernmental organizations (NGOs) are collaborating to learn about whether men and women are affected differently by armed conflict in their countries. The group focuses their efforts on the effects of the civil war in Yemen. The researchers collected samples of five hundred men and women. In these samples, 90% of women experienced malnutrition at some point since the start of the conflict in Yemen, compared to 89% of men.

a. Calculate the 99% confidence interval for the difference between men and women who experienced malnutrition.

b. What does the confidence interval from part a suggest about differences in malnutrition rates among Yemeni men and women?

11. Many observers have claimed that millennials are more interested in astrology than members of Generation X. Imagine you collect random samples of one hundred millennials and one hundred Gen -ers in your town. You find that 48% of millennials and 37% of Gen Xers say they are "very interested" in astrology. Use your sample data to conduct a one-tailed test of the claim that millennials and Gen Xers have similar views about astrology. (Set the alpha level at 0.01.) State what the outcome of the hypothesis test means about the relationship between the variables.

12. The 2020 American National Election Study (ANES) asked respondents whether they agreed that limits placed on public activity due to COVID-19 were not nearly strict enough. Among the 3,829 respondents who rated their own health as excellent, 12.5% agreed that restrictions were not nearly strict enough. Among the 4,332 respondents who rated their own health as worse than excellent, 15.9% agreed that restrictions were not nearly strict enough. Use these sample data to conduct a two-tailed hypothesis test. (Set the alpha level at 0.05.) State what the outcome of the hypothesis test means about the relationship between the variables.

13. Calculate an 80% confidence interval of the difference in millennials' and Gen Xers' opinions about astrology, using the data described in Problem 11. State what the confidence interval means.

14. Calculate a 99% confidence interval for the difference in opinions about the strictness of restrictions on public activity during COVID between those in excellent health and those who rate their health as worse than excellent, as described in Problem 12. State what the confidence interval means.

15. The 2020 ANES asked voters to rate their feelings about Vice President Kamala Harris, from 0 (very cold) to 100 (very warm). The mean rating of Vice President Harris for white respondents was 47.0 (s = 38.2, N = 5,350). The mean rating of Vice President Harris for Black respondents was 80.2 (s = 25.4, N = 638). Use these data to conduct a two-tailed hypothesis test at the 0.01 alpha level. Show all of your steps and state your conclusion.

16. The 2020 ANES asked voters to rate their feelings about the Democratic Party, from 0 (very cold) to 100 (very warm). The mean rating of Democrats for white respondents was 40.5 (s = 31.8, N = 5,889). The mean rating of Democrats for Black respondents was 71.4 (s = 23.9, N = 713). Use these data to conduct a two-sample hypothesis test at the 0.05 alpha level. Show all of your steps and state your conclusion.

17. Use the data about voters' views of Vice President Kamala Harris from Problem 15 to calculate a 99% confidence interval of the difference between white and Black voters' feelings about the vice president.

18. Use the data from Problem 16 about voters' views of the Democratic Party to calculate a 99% confidence interval of the difference between white and Black voters' feelings about this party.

Stata Problems

Here, we will use the 2021 GSS to examine how fear of economic security is related to voting. Are people more likely to vote when they are worried about their economic security or when they are not? We will use two variables called *voted16* (1 = voted, 0 = did not vote), which measures whether the respondent voted in the 2016 presidential election, and *endsmeet2* (1 = hard to make ends meet; 0 = not hard to make ends meet) which measures whether respondents find it difficult to make ends meet in their households.*

1. Open `GSS2021.dta`. Decide whether to use the "prtest" or "ttest" command to run a two-sample test with *endsmeet2* as the independent variable and *voted16* as the dependent variable. Specify that you want Stata to provide a 95% confidence interval for the difference.

2. Use the output to answer the following questions:

    a. What is the null hypothesis?

    b. What proportion of people who are worried about making ends meet said they voted in the 2016 election? What proportion of those who are not worried said they voted?

    c. What is the difference in proportions? What is the z-score associated with the difference? What is the p-value associated with that z-score?

---

* *Voted16* was generated from the original GSS variable called *vote16*, and *endsmeet2* was generated from the original variable called *endsmeet*.

d. Assume that you were conducting a two-tailed test at an alpha level of 0.05. Would you conclude that the difference is "statistically significant"? Explain which p-value on the output allows you to answer this question.

e. Assume that you were conducting a one-tailed test at an alpha level of 0.05. Would you conclude that the difference is "statistically significant"? Explain which p-value on the output allows you to answer this question.

f. What is the 95% confidence interval for the difference in proportions between those who are worried and not worried about losing their main source of economic support?

SPSS Problems

Here, we will use the 2021 GSS to examine how fear of economic security is related to voting. Are people more likely to vote when they are worried about their economic security or when they are not? We will use two variables called *voted16* (1 = voted, 0 = did not vote), which measures whether the respondent voted in the 2016 presidential election, and *endsmeet2* (1 = hard to make ends meet; 0 = not hard to make ends meet) which measures whether respondents find it difficult to make ends meet in their households.*

1. Open GSS2021.sav. Use the "Independent-Samples T Test" procedure to run a two-sample test with *endsmeet2* as the independent variable and *voted16* as the dependent variable.

2. Use the output to answer the following questions:

    a. What is the null hypothesis?

    b. What proportion of people who are worried about making ends meet said they voted in the 2016 election? What proportion of those who are not worried said they voted?

    c. What is the difference in proportions? SPSS offers two t-values associated with this difference. Which should you use? What is the p-value associated with that t-value?

    d. Assume that you were conducting a two-tailed test at an alpha level of 0.05. Would you conclude that the difference is "statistically significant"? Explain how the appropriate p-value on the output allows you to answer this question.

    e. Assume that you were conducting a one-tailed test at an alpha level of 0.05. Would you conclude that the difference is "statistically significant"? Explain how you can use the p-value for the two-tailed test to answer this question.

---

* *Voted16* was generated from the original GSS variable called *vote16*, and *endsmeet2* was generated from the original variable called *endsmeet*.

f. What is the 95% confidence interval for the difference in proportions between those who are worried and not worried about making ends meet?

Notes

1. Christoph Pieh, MD; Sanja Budimir, PhD; Jaime Delgadillo, PhD; Michael Barkham, PhD; Johnny R. J. Fontaine, PhD; and Thomas Probst, PhD. 2021. "Mental Health during COVID-19 Lockdown in the United Kingdom." *Psychosomatic Medicine* 83(4): 328–337.

2. Jennifer Bronson and Marcus Berzofsky. 2017. "Indicators of Mental Health Problems Reported by Prisoners and Jail Inmates, 2011–2012." Bureau of Justice Statistics. June. Retrieved from https://www.bjs.gov//index.cfm?ty=pbdetail&iid=5946.

3. F. E. Markowitz. 2006. "Psychiatric Hospital Capacity, Homelessness, and Crime and Arrest Rates." *Criminology* 44(1): 45–72. M. Massoglia. 2008. "Incarceration as Exposure: The Prison, Infectious Disease, and Other Stress-Related Illnesses." *Journal of Health and Social Behavior* 49(1): 56–71.

4. Data available from the Economic Research Service. Retrieved from https://www.ers.usda.gov/data-products/county-level-data-sets/county-level-data-sets-download-data.

Chapter 11

Testing Mean Differences among Multiple Groups
Analysis of Variance

You now know how to conduct a hypothesis test to see if there is a statistically significant difference between two groups. But what if you want to compare more than two groups? For example, you might want to know whether there are significant differences in income among multiple racial/ethnic groups. Or you might want to know whether religious groups vary in their support for a political candidate or the number of children they have. These are examples in which the independent variable you are interested in has more than two categories and the dependent variable is interval-ratio (or an ordinal scale that we can treat as interval-ratio).

Analysis of variance, nicknamed **ANOVA**, allows us to see whether there are significant differences for independent variables with more than two categories.

ANOVA is commonly used in the fields of psychology and social psychology, where researchers frequently employ experimental methods in which research participants can be assigned to categories that are exposed to different conditions.[1] For example, a team of researchers wondered how people feel toward robots and artificial intelligence.[2] This is an important question, as we interact more and more often with these nonhuman actors. The researchers conducted an experiment to measure how much people trust robots, artificial intelligence, and humans. They recruited participants to play a game in which players receive a hypothetical $1,000 and can choose whether to keep it all or send some to their opponent. The opponent would receive triple the amount the participant sent and could, in turn, decide whether to return money to the participant. The dependent variable for the study was how much money the participant sent to their opponent. Many studies use some variation of this game to measure trust.

The researchers randomly assigned participants to one of six groups, each of which played against a different hypothetical opponent: an AI nicknamed Michael, an AI nicknamed jdrx894, a robot nicknamed Michael, a robot nicknamed jdrx894, and controls described simply with the nicknames Michael or jdrx894. They found that participants in these six groups gave different amounts of money, on average, depending on how their opponents were described. They gave the most money, on average, to the robot nicknamed jdrx894 and the least to the opponent simply described as Michael. The robot jdrx894 received an average of $492.78 while Michael received an average of $299.23. This suggested that participants trusted the opponent described in the most human way the least!

The researchers used ANOVA to test whether there were statistically significant differences across the six groups in the mean amount of money given. The ANOVA test showed that there were statistically significant differences in mean money given across the six groups. According to the researchers, this finding suggests that participants may consider humans more selfish and thus less trustworthy than robots or AI.

ANOVA is sometimes referred to as an omnibus test because it tests whether there are overall mean differences in the dependent variable across categories of the independent variable, but it does not indicate which categories have significant differences from each other. As we will see later in this chapter, analysts must conduct a second test, called a **post hoc test**, to investigate which differences are significant when they have a significant ANOVA result. In the study discussed here, the researchers found that the only significant difference in money given was between the groups with the opponent "Michael" and the robot jdrx894. In other words, the mean amounts given to the other groups were not significantly different from each other. (Further testing showed that there were no significant differences between robot, AI, and control opponents, but that opponents named Michael were given significantly less money, on average, than those named jdrx894.)

Comparing Variation within and between Groups

Let's backtrack to talk about how exactly analysis of variance works. We will take the age at which people have their first child as the dependent variable. There are many factors that shape the age at which people have their first child. One such factor is education. In general, the higher people's education, the older they tend to be when they have their first child. It would be possible to divide people into two groups—say, high and low education—and use a t-test to see whether there is a significant difference in the age at which they first have a child. But dividing the education variable into more categories can get at finer nuances in how childbearing age varies. Let's divide education into three categories: less than a high school degree, high school graduate and some college, and a bachelor's degree or higher.

Can you imagine the perfect relationship? Romantic walks on the beach, a partner who does the housecleaning . . .

No, not that kind of perfect relationship! Can you imagine a perfect relationship between the two variables? What if education completely determined age of childbearing? If you knew someone's education level, you would automatically know how old they were when they first had a child. For example, imagine a hypothetical sample of three hundred people older than eighteen who had at least one child, with one hundred people in each of three different categories of education level. Table 11.1 shows a hypothetical cross-tabulation of age at which people had their first child and education level.

Table 11.1 Age of First Childbearing, by Education (Frequencies), All Variation between Groups

Age of First Childbearing	Less than High School $\bar{y} = 19, s = 0$	High School Degree or Some College $\bar{y} = 21, s = 0$	Bachelor's Degree or Higher $\bar{y} = 25, s = 0$
19	100	0	0
21	0	100	0
25	0	0	100

Table 11.1 shows that all one hundred people with less than a high school education had their first child at nineteen, all one hundred of those with a high school degree or some college had their first child at twenty-one, and all one hundred of those with a BA or higher had their first child at twenty-five. In this hypothetical sample, there is no variation in age at which people became parents within levels of education. In other words, within each level of education, the standard deviation is 0. The only variation in the age of first childbearing exists between groups with different educational levels, indicated by the fact that the mean age of becoming a parent differs across the three groups.

In the hypothetical "perfect relationship" shown in Table 11.1, all of the variation in the age at which people have their first child is between educational groups. There is no variation within a given educational group—everyone in a group had their first child at exactly the same age. In practice, of course, there is always variation within groups even when the independent variable has a strong effect on the dependent variable.

Of course, reality does not look like this. Individuals have their children at different ages, and social scientists want to try to understand what shapes these differences. Actual data would show that there is considerable variation within each group. Individuals with the same level of education differ from each other in the age at which they first have a child. People with a high school degree, for example, have children at all different ages. Analysis of variance works by comparing the amount of variation in the dependent variable within groups to the amount of variation between groups. The closer the relationship is to the perfect relationship—that is, the more variation there is between groups compared to within groups—the stronger the relationship is between the variables.

The central statistic for analysis of variance is the **F-statistic**. F is the ratio of between groups variation to within groups variation. Although the underlying procedures for ANOVA and F are mathematically and conceptually different from those for t or z, you can think of ANOVA as analogous to a t-test, except that ANOVA can be used where the independent variable has more than two categories. A t-test compares the mean values of the dependent variable for *two* categories of an independent variable, ANOVA compares the mean values of the dependent variable for *two or more* categories of an independent variable. Just as with t, once you find the F associated with your analysis of variance, you can look up its value in a table of F-values to determine whether the relationship between the variables is statistically significant (or your statistical software can calculate the p-value for your F). In the rest of this chapter, we show you how to do this and give examples of analysis of variance in the social sciences.

Hypothesis Testing Using ANOVA

Hypothesis testing using ANOVA is based on the same logic of sampling variation explained in chapter 7 and applied in chapters 8, 9, and 10. This means that any one sample will likely differ from the population, but, with (hypothetical) repeated samples, the sampling distribution of F-values would follow a known curve, with the most frequent F-values corresponding to the population parameter or being close to it. The population parameter is the F-value that indicates the actual ratio of between groups variation to within groups variation in the population.

The null hypothesis with ANOVA is generally that there is no relationship between the two variables. If there were no relationship between the two variables being studied, the means for all of the groups would be the same. All of the variation would be within groups, and the F would be approximately 1.[3] Our hypothesis test examines the probability of obtaining our sample results if the null hypothesis were true. In other words, how probable is it that we would obtain our sample F if the means for the different groups were in fact equal to each other in the population?

Just as we did in previous chapters, we use the properties of the sampling distribution to identify the probability of obtaining specific values for the sample statistic (in standardized units, in this case F) if there were no differences among the groups being compared. As with t, there is a family of F-curves for different degrees of freedom, as shown in Figure 11.1. (Unlike t or z, the F-curve is right-skewed, not bell-shaped.)

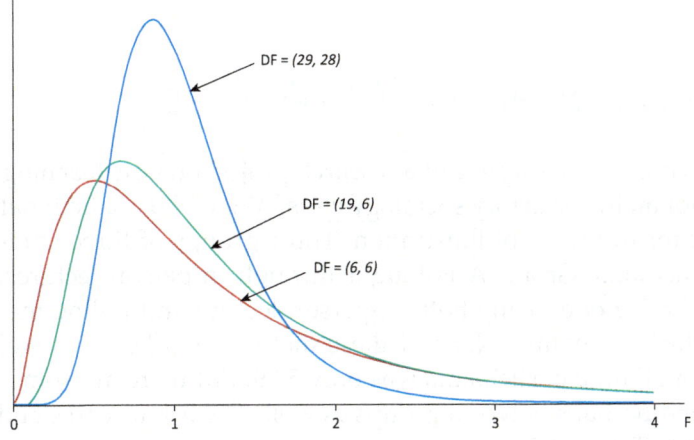

Figure 11.1   F-Curves at Different Degrees of Freedom

We set up an ANOVA hypothesis test using the same steps as we used in chapters 9 and 10:

1. Set up null and alternative hypotheses.
2. Set α, calculate degrees of freedom, and find decision F.
3. Calculate the sample F-statistic.

4. Compare sample F to decision F and draw your conclusion. If sample F is greater than decision F, reject the null hypothesis. Or compare p-value for sample F to alpha. If p-value is less than alpha, reject the null hypothesis.

Analysis of Variance Assumptions

ANOVA requires either that the dependent variable is normally distributed in the population or that the sample size is large enough that the Central Limit Theorem applies. One rough rule is that a sample size of seven in each group is the minimum, if the groups are of equal size.[4] ANOVA assumes that the data come from a random sample and that the standard deviation for the dependent variable is approximately the same for each group. ANOVA also assumes roughly equal variances across the groups; you will learn how to assess this in the Stata or SPSS section.

There are a number of different applications of the ANOVA procedure. The method of ANOVA we will present here assumes that each group is an independent random sample. When the groups are not independent—for example, as in measures of a variable in the same group at different times—somewhat different statistical calculations apply, and we call the procedure "repeated measures ANOVA."

ANOVA is used with categorical independent variables with two or more values and interval-ratio dependent variables (or ordinal scales that we treat as interval-ratio). Although ANOVA can be used for independent variables with only two values, a t-test is more commonly used in that case.

The Steps of an ANOVA Test

Let's take a hypothetical example of a research project on how teaching method affects final exam scores for an introductory sociology class.[5] We will use a fictional data set with a very small sample for purposes of illustration. Three groups of three students each are taught in different methods. Group A is taught through in-person lectures; Group B is taught through a hybrid model using both in-person lecture and online methods; and Group C is taught exclusively online. None of these methods worked that well. The overall mean exam score for all nine participants was only 56.9. But there are some differences between individuals and perhaps between groups as well. The exam scores and each group's mean are presented in Table 11.2.

We can see from Table 11.2 that there is variation within each group. That is, the three students taught through in-person lecture earned different scores from each other; the same is true for the students in the classes using the other two teaching methods. We can see that there is also variation between the groups: The mean scores for the three groups differ. You know by now that the differences in these means could be due to sampling error. That is, just through the normal variation inherent in random sampling, students in each group could differ for reasons that have nothing to do with the teaching method. Maybe the student receiving a 62 in the hybrid method was having an especially good day or had a more relevant background than others, for example. The analysis of variance is

Table 11.2 Exam Scores for Nine Students in Three Statistics Classes

In-Person Lecture	Hybrid	Online
55	56	50
57	60	52
60	62	60
$\bar{y} = 57.33$	$\bar{y} = 59.33$	$\bar{y} = 54$

how we test to see whether the differences between the groups are large enough—when taking everything else into account—to be statistically significant.

Step 1: Formulate hypotheses.

The null hypothesis is always that there is no difference between the groups in the population, or that the independent variable has no effect on the dependent variable. In other words, the mean test scores would be the same for students taught in each of the three methods.

$$H_0 : \mu_1 = \mu_2 = \mu_3$$

The alternative hypothesis is that at least one of the groups has a different population mean than at least one of the others. Note that it is not necessary for all of the groups to differ from all of the others.

$$H_a : \mu_1 \neq \mu_2 \neq \mu_3$$

Step 2: Set alpha level, calculate degrees of freedom, and find decision F.

We first determine what risk of Type I error is acceptable and set our alpha level. In this case, we will select an alpha of .05. To find decision F, we look in the table of F-values (appendix C), which gives the F-values for the .05 alpha level at various degrees of freedom. The F-table requires two values for degrees of freedom, DF 1 and DF 2. For ANOVA, DF 1 is $DF_{between}$ and DF 2 is DF_{within}, where

$$DF_{between} = number\ of\ groups - 1$$

$$DF_{within} = N_{total} - number\ of\ groups$$

"k" is used to stand for number of groups. In that case, the formulas are:

$$DF_{between} = k - 1;\ DF_{within} = N_{total} - k$$

So, $DF_{between} = 3 - 1 = 2$ and $DF_{within} = 9 - 3 = 6$

Figure 11.2 shows an excerpt of the F-table. Use the table excerpt to find $DF_{between}$ of 2 across the top and DF_{within} of 6 down the side. For these two values, decision F is 5.14.

		\multicolumn{10}{c}{$DF_1 (DF_{between})$}									
		1	2	3	4	5	6	7	8	9	10	20
	1	161.45	199.50	215.71	224.58	230.58	230.16	233.99	236.77	238.88	241.88	248.01
$DF_2 (DF_{within})$	5	6.61	5.79	5.41	5.19	5.19	5.05	4.95	4.88	4.82	4.74	4.56
	6	5.99	5.14	4.76	4.53	4.53	4.39	4.28	4.21	4.15	4.06	3.87
	7	5.59	4.74	4.35	4.12	4.12	3.97	3.87	3.79	3.73	3.64	3.44
	∞	3.84	2.99	2.60	2.37	2.37	2.21	2.09	2.01	1.94	1.83	1.57

Figure 11.2 Finding F-Value at $DF_1 = 2$ and $DF_2 = 6$, Alpha = 0.05

The decision F is the F-value for which fewer than 5% of samples would yield results greater than F due to sampling variation. Recall that this is the same way we used decision t in chapter 10.

Step 3: Calculate sample F-statistic.

You will rarely calculate an analysis of variance by hand. We will see in the software sections of this chapter how to generate the results using data analysis software. But it is useful to see an example of how F is calculated in order to understand where the statistic comes from.

Analysis of variance is based on comparing the differences within each group to the difference between the groups. We must quantify each of these. Within groups variation is quantified by comparing individual scores to the group mean. Between groups variation is quantified by comparing the group means to the overall mean. Specifically, we do this by calculating *sums of squares*. For the **within groups sum of squares**, we subtract the group mean from each individual score, square each result, and add them together. This is similar to the calculation for standard deviation (see chapter 5). In both cases, we are quantifying the overall amount of variation in a sample by summing the squared differences between each individual score and its group mean. Sums of squares are also called sums of squared deviations. For the **between groups sum of squares**, we subtract the overall mean from each group mean, square each result, multiply it by the number of cases in that group, and add them together. This is a measure of how much variation there is between groups.

By formula:

$$SS_{within} = \Sigma(y - \bar{y}_{group})^2$$

$$SS_{between} = \Sigma N_{group}(\bar{y}_{group} - \bar{y}_{total})^2$$

Table 11.3 shows the calculation for the within groups sum of squares. We first sum the squares for each group. For the first case, we subtract their exam score, 55, from the group mean exam score for those taught in person, 57.33. We get –2.33, which we square to get 5.43. We do this for the other students in that group and sum those results to get 12.67. We then do the same thing for the other two groups. The results are shown in the bottom row. You can see that this would be very time-consuming with a larger sample.

Table 11.3 Within Groups Sums of Squares, Different Teaching Methods and Exam Scores

In-Person ($\bar{y}_I = 57.33$)			Hybrid ($\bar{y}_H = 59.33$)			Online ($\bar{y}_O = 54$)		
y	$y - \bar{y}$	$(y - \bar{y})^2$	y	$y - \bar{y}$	$(y - \bar{y})^2$	y	$y - \bar{y}$	$(y - \bar{y})^2$
55	−2.33	5.43	56	−3.33	11.09	50	−4	16
57	−.33	0.11	60	0.67	0.45	52	−2	4
60	2.67	7.13	62	2.67	7.13	60	6	36
$\Sigma(y_I - \bar{y}_I)^2 = 12.67$			$\Sigma(y_H - \bar{y}_H)^2 = 18.67$			$\Sigma(y_O - \bar{y}_O)^2 = 56$		

We then add together the sum of squares for each group: 12.67 + 18.67 + 56 = 87.34.

$$SS_{within} = 87.34$$

Next, shown in Table 11.4, we find the between groups sum of squares by subtracting the overall mean score for all students from each group mean, squaring each difference, and multiplying each squared difference by the number of cases in its group. The overall mean is 56.89. For example, the first row of Table 11.4 shows how we subtract the overall mean of 56.89 from the mean score of 57.33 for the in-person group, to get 0.44. We square that to get 0.1936 and multiply that by the number of students in that group, 3, to get 0.5808. We do this for the other two groups and add the results together.

Table 11.4 Between Groups Sums of Squares, Different Teaching Methods and Exam Scores

	$(\bar{y}_{group} - \bar{y}_{total})$	$(\bar{y}_{group} - \bar{y}_{total})^2$	$N_{group}(\bar{y}_{group} - \bar{y}_{total})^2$
In-Person	57.33 − 56.89 = 0.44	0.194	3(0.194) = 0.582
Hybrid	59.33 − 56.89 = 2.44	5.954	3(5.954) = 17.862
Online	54 − 56.89 = −2.89	8.352	3(8.352) = 25.056
			$\Sigma N_{group}(\bar{y}_{group} - \bar{y}_{total})^2 = 43.500$

As shown in the bottom-right cell of Table 11.4, $SS_{between} = 43.50$ (rounding from 43.4979). (Note that we have rounded our calculations along the way in order to show them in the table. If you do not do so, you will get a slightly different answer.)

The sum of squares is larger if there is more variation (for either between or within groups). But the sum of squares is also larger if there is a larger N. Think about how within groups sum of squares is calculated: a term is added for each case in the sample. This is also true for between groups sum of squares, where a new term is added for each category of the independent variable (i.e., each group) and each term is multiplied by the group N.

To adjust for the number of cases and groups, we calculate *mean squares* for each form of variation, by dividing each sum of squares by the corresponding degrees of freedom. We divide the sum of squares for within groups by the degrees of freedom for within groups and we divide the sum of squares for between groups by the degrees of freedom for between groups. The degrees of freedom are the same ones we used to find the decision F in Step 2. As a reminder:

$$DF_{within} = N_{total} - k \text{ (where k is the number of groups)} = 6$$

$$DF_{between} = k - 1 = 2$$

We can now find mean squares.

$$MS_{within} = \frac{SS_{within}}{DF_{within}}$$

$$MS_{between} = \frac{SS_{between}}{DF_{between}}$$

For our example:

$$MS_{within} = \frac{87.34}{6} = 14.56$$

$$MS_{between} = \frac{43.50}{2} = 21.75$$

F is the ratio of $MS_{between}$ to MS_{within}.

$$F = \frac{MS_{between}}{MS_{within}}$$

$$F = \frac{21.75}{14.56} = 1.49$$

What does an F of 1.49 mean? Essentially, F tells us how much variation there is between groups, compared to the variation within groups.

Step 4: Compare sample F to decision F and draw your conclusion. If sample F > decision F, reject the null hypothesis. Or compare p-value for sample F to alpha. If p-value < alpha, reject the null hypothesis.

We compare our obtained, or sample, F with the decision F to determine statistical significance. We found decision F in Step 2, above. It is 5.14. Since our F, 1.49, is smaller than the decision F, we are unable to reject the null hypothesis. It is possible that the differences in exam scores that we found between our sample groups are simply due to sampling variation, rather than instructional method. In other words, there is not a statistically significant relationship between instructional method and exam score. This is the same result found in the actual study that this example is based on. The author found no significant difference in learning between students taught online, through in-class lecture, or through a hybrid of the two.

Let's look at another example. This time, instead of calculating ANOVA by hand, we will just look at how to interpret it. This example comes from the 2021 General Social Survey (GSS) and looks at the relationship between age when people have their first child (the dependent variable) and whether they have been geographically mobile (the independent variable). Specifically, the independent variable asks where respondents live at the time they were surveyed compared to when they were 16: in the same state and same city, same state and different city, or different state. We might expect to see that people who are more geographically mobile have children at a later age. This is the case because geographic mobility is associated with higher socioeconomic class and higher levels of education, both of which lead to later childbearing. In fact, this is what we find, as shown in Table 11.5.

Table 11.5 Age When First Child Was Born by Geographical Mobility

Geographic Mobility since Age 16	Mean	SD	Frequency
Same state, same city	23.93	6.03	733
Same state, different city	25.41	5.92	912
Different state	26.43	6.36	862
Total	25.33	6.19	2,507

Source: 2021 General Social Survey.

We see from Table 11.5 that there are differences in the average age at which people had their first child, with those who live in the same state and city as when they were sixteen having their first child at the youngest age (23.93) and those who live in a different state having their first child at the oldest age (26.43). The second column in Table 11.5 also shows that all of the geographic mobility groups have standard deviations greater than 0, indicating that there is variation in the age at which they had their first child. In other words, not all people who have moved to different states have their first child at the same age. Are the differences between geographic mobility groups large enough, relative to the variation within each geographic mobility group, to be statistically significant? Going through the steps of a hypothesis test:

Step 1: Formulate hypotheses.

H_0: Mean age at which first child is born is the same for different geographic mobility groups. Or,
$H_0: \mu_1 = \mu_2 = \mu_3$

H_a: Mean age at which first child is born differs for at least one of the different geographic mobility groups. Or H_a: $\mu_1 \neq \mu_2 \neq \mu_3$

Step 2: Set alpha level, calculate degrees of freedom, and find decision F.

We will set alpha at .05.

$$DF_{within} = N_{total} - k = 2{,}507 - 3 = 2{,}504$$

(Remember, k is the number of groups—that is, the number of categories of the independent variable. N_{total} is the total number of cases, which is shown in the bottom row of Table 11.5.)

$$DF_{between} = k - 1 = 3 - 1 = 2$$

Looking in the table of F-values in appendix C, find the decision F. Locate $DF_{between}$ of 2 across the top; because our DF_{within} of 2,504 is much larger than the highest value included in the table, use the infinity line. The value of decision F is 2.99.

Step 3: Calculate sample F-statistic.

With an N of 2,507, calculating F by hand would be extremely time-consuming. We will calculate the sample F-statistic using statistical software. All software packages will give the sums of squares, degrees of freedom, and mean squares for between and within groups variation, along with F and the p-value associated with F. Table 11.6 presents this information for the ANOVA test of the relationship between geographic mobility and age at which first child was born.

Table 11.6 ANOVA for Age When First Child Was Born by Geographic Mobility

Source of Variation	SS	DF	MS	F	p
Between groups	2479.90	2	1239.95	33.24	0.00
Within groups	93393.65	2504	37.30		
Total	95873.55	2506	38.26		

Source: 2021 General Social Survey.

Table 11.6 shows the sums of squares (SS), degrees of freedom (DF), and mean squares (MS) for between and within groups variation. The MS for within groups variation is 37.30, and the MS for between groups variation is 1239.95. The ratio of these, which produces F, is 33.24.

Step 4: Compare sample F to decision F and draw your conclusion. If sample F > decision F, reject the null hypothesis. Or compare p-value for sample F to alpha. If p-value < alpha, reject the null hypothesis.

We can compare our sample F-statistic with the decision F of 2.99 that we obtained in Step 2. Our sample F is well above that decision F, telling us that we can reject the null hypothesis at an alpha of .05. When we are using statistical software, however, the output tells us directly that the p-value for our sample F is 0.00. Our obtained p-value of 0.00 is less than our alpha of .05, indicating that differences of this size are exceedingly unlikely to arise simply by chance. We would therefore reject the null hypothesis of no differences between geographic mobility groups in the average age when people have their first child. We can then conclude that there are statistically significant differences by geographic mobility in the average age when people have their first child. Or, to put it differently, there is a statistically significant relationship between geographic mobility and age at which people have their first child.

We know from the results of the ANOVA test that there are significant differences in the average age at which people have their first child among the three geographic mobility groups. But ANOVA cannot tell us which particular geographic mobility groups are significantly different from each other. In the next section, we discuss a post hoc test that allows analysts to investigate which particular means differ from each other.

Determining Which Means Are Different: Post Hoc Tests

ANOVA tells us if at least one mean is significantly different from at least one other mean, but it does not tell us which means are significantly different from each other. We need to employ a **post hoc test** to indicate which means are significantly different from each other. ("Post hoc" simply means "after this" or "after the fact" and in statistics generally refers to a follow-up test that is done after obtaining results from another test.) There are various post hoc tests for ANOVA that yield similar results. We will show the application of **Tukey's test**. We will use statistical software to run Tukey's post hoc test.

In the example of age of first childbearing by geographic mobility, there are three geographic mobility groups, and three pairs of mobility groups to be compared. When we use statistical software to run Tukey's test, we find the three comparisons listed. Table 11.7 presents these results.

Table 11.7 Truncated Results of Tukey's Post Hoc Comparison of Mean Differences in Age When First Child Was Born by Geographic Mobility

Mobility Comparison	Difference	t	p
Same state, different city versus same state, same city	1.48	4.89	0.00
Different state versus same state, same city	2.50	8.14	0.00
Different state versus same state, different city	1.02	3.51	0.00

Source: 2021 General Social Survey.

The table shows the difference between the age of first childbearing for each pair of mobility groups. For example, the first row shows that those who lived in the same state, different city had first children 1.48 years later than those who lived in the same state, same city. It also shows the t-value for the difference, and the probability for the differences between the groups, p.[*] Those t-values for which the p is below alpha indicate pairs that have a statistically significant difference in their means. Looking at Table 11.7, we see that each of the three pairs has a p below 0.05. Thus, each geographic mobility group is significantly different from each other (at alpha = 0.05).

In other words, the ANOVA told us that geographic mobility has a significant effect on age at which people have their first children. The post hoc test tells us that, specifically, the mean age of first childbearing for those who live in the same state, same city is significantly lower than for those who have moved to a different city in same state; and that age of first childbearing is significantly lower for those who live in the same state, different city than for those who have moved to a different state.

Note that, when you interpret the post hoc test, it is not enough to say that some groups are significantly different from each other. You should also say what the difference is. In our example, we explain which groups are significantly higher or lower in their age of first childbearing.

ANOVA Compared to Repeated t-Tests

You may have wondered why we conduct ANOVA instead of simply using several t-tests to compare each pair of groups. The answer is that each time we use a t-test, we run a risk of Type I error—that is, wrongly concluding that there is a significant difference between groups. The more t-tests, the greater the likelihood that at least one of them will lead us to conclude there is a statistically significant difference between groups when there is not. For example, if we have three groups, there are three pairs, which would require three t-tests, each with a Type I error of 0.05. ANOVA, in contrast, sets an alpha level for the overall test, allowing us to set an overall risk of Type I error of 0.05 (or whatever our alpha level is). Post hoc tests can increase the risk of Type I error in the same way as multiple t-tests; Tukey's test adjusts for this.

---

[*] t for Tukey's or other post hoc tests is calculated somewhat differently from the calculation for t covered in chapters 8 through 10. The details are beyond the scope of this book.

Chapter Summary

Analysis of variance (ANOVA) is used to compare means among groups and is most commonly used when there are three or more groups.

- The steps of the hypothesis test using ANOVA are:
  - *Step 1*: Formulate hypotheses.
  - *Step 2*: Set alpha level, calculate degrees of freedom, and find decision F.
  - *Step 3*: Calculate sample F-statistic.
  - *Step 4*: Compare sample F to decision F and draw your conclusion. If sample F > decision F, reject the null hypothesis. Or, compare p-value for sample F to alpha. If p-value < alpha, reject the null hypothesis.

- Degrees of freedom:
  - $DF_{within} = N_{total} - k$ (where k is the number of groups)
  - $DF_{between} = k - 1$

- The **F-statistic** is the ratio of between groups variation to within groups variation. To calculate F, calculate **between groups** and **within groups sums of squares (SS)**, calculate *mean squares* (MS), and calculate *F*:

$$SS_{within} = \Sigma\left(y - \bar{y}_{group}\right)^2$$

$$SS_{between} = \Sigma N_{group}\left(\bar{y}_{group} - \bar{y}_{total}\right)^2$$

$$MS_{within} = \frac{SS_{within}}{DF_{within}}$$

$$MS_{between} = \frac{SS_{between}}{DF_{between}}$$

$$F = \frac{MS_{between}}{MS_{within}}$$

- Follow ANOVA with a **post hoc** test to indicate which group means are statistically significantly different from each other. **Tukey's test** is one such test.

Using Stata

In this example, we use the 2021 GSS to ask whether social position affects the extent to which people trust the government. Are people who see themselves as having a relatively high social position less likely to have confidence in the government? Are those who see themselves lower on the social hierarchy more likely to have confidence in the government? Can you think of reasons why we might expect to find a relationship between these variables?

The GSS asked respondents to place themselves on a vertical scale indicating their social position in society, with ten categories from top to bottom.* We created a variable called *rankcat* that divided respondents into four categories: highest, higher, lower, and lowest. *Rankcat* serves as the independent variable. The dependent variable, confidence in the government (*confgov*), is the sum of respondents' answers to three questions about confidence in each of the three branches of the US government: the executive branch, Congress, and the Supreme Court.** *Confgov* ranges from 3 (least trust) to 9 (most trust). In this example we will conduct an ANOVA to test whether there is a relationship between social rank and confidence in the government. In other words, we are testing whether the mean score on the confidence in government variable differs across the four categories of the social ranking variable.

Open `GSS2021.dta` and enter the following command into Stata:

`oneway confgov rankcat, tabulate`

The "oneway" command asks Stata to perform a one-way ANOVA, with "one-way" referring to the fact that we have only one dependent variable. After "oneway," we list the dependent (*confgov*) and independent (*rankcat*) variables, in that order. We have specified the "tabulate" option, which requests the means and standard deviations for the dependent variable across the categories of the independent variable. The output is shown in Figure 11.3.

The top section of the output provides the information requested by the "tabulate" option. We see that the mean confidence rating of respondents in the "lowest" category of social rank (7.2) is higher than the mean for any other group. Mean confidence in the government is lower among those in the "higher" and "highest" social ranks (6.4 and 6.5, respectively).

The second part of the output shows the results of the ANOVA test. The first row of the table presents information for between groups variation and the second row for within groups variation. We see that the mean squares for within groups variation (2.0) is much lower than for between groups variation (22.0), yielding the F-statistic of 11 (22.0/2.0). The p-value is lower than 0.0001, indicating a very small probability of observing this ratio of between to within groups variation by chance, if there actually were no relationship between these variables in the population.

---

* The original GSS variable used to generate *rankcat* is called *rank*.
** The three original GSS variables used to construct the dependent variable are *confed*, *conlegis*, and *conjudge*.

```
. oneway confgov rankcat, tabulate

                    Summary of confgov
      rankcat       Mean       Std. dev.      Freq.

      highest     6.5088757    1.3977803        169
       higher     6.4433962    1.3386574        106
        lower     7.0833333    1.3976328        228
       lowest     7.1901408    1.4436611        142

        Total    6.8511628    1.4315173        645

                    Analysis of variance
      Source           SS          df       MS              F        Prob > F

Between groups      66.0317003     3    22.0105668        11.25       0.0000
 Within groups      1253.67993    641    1.95581892

       Total        1319.71163    644    2.04924166

Bartlett's equal-variances test: chi2(3) =    0.6795      Prob>chi2 = 0.878
```

Figure 11.3

The last line of the output shows the results of a test for equal variances across the four groups (called "Bartlett's test for equal variances"). The high p-value (.878) for the test for equal variances tells us that differences in variances in governmental trust across social rank groups likely reflects sampling variation. In other words, we can assume that variances among groups are equal. This is an assumption of ANOVA. In order to be statistically reliable, ANOVA requires that the variances among groups be equal. If Bartlett's test for equal variances shows a p-value of 0.05 or less, we must be cautious in interpreting the ANOVA results.

The ANOVA test indicates that ratings of confidence in government do depend on how people assess their social position (relatively high or relatively low), with those who see themselves as toward the higher end of the social hierarchy reporting less trust in the government and those at the lower end reporting more trust.

To run Tukey's post hoc test, we use the generic command:

`pwmean dependent variable, over(independent variable) mcompare (tukey) effects`

The output, not shown here, looks much like Table 11.7. Stata labels the column reporting the difference between the means of each pair of social positions "contrast." In addition to reporting the standard error of each difference, with its associated t and p-value, Stata also reports the confidence interval for the difference. You can interpret these results the same way as those presented in Table 11.7.

For our example, the command is:

`pwmean confgov, over(rankcat) mcompare (tukey) effects`

If we look at the P > |t| column, we see that all pairs have significantly different ratings of their confidence in government *except* the highest versus higher and lowest versus lower pairs. The contrast column shows that the difference in mean rating between the higher and highest social position groups was only –0.06, for example, while the difference in mean rating between the lowest versus higher groups was 0.75.

Review of Stata Commands

- ANOVA test

    `oneway dependent variable independent variable, tabulate`

- Tukey's post hoc text

    `pwmean dependent variable, over(independent variable) mcompare (tukey) effects`

Using SPSS

In this example, we use the 2021 GSS to ask whether social position affects the extent to which people trust the government. Are people who see themselves as having a relatively high social position less likely to have confidence in the government? Are those who see themselves lower on the social hierarchy more likely to have confidence in the government? Can you think of reasons why we might expect to find a relationship between these variables?

The GSS asked respondents to place themselves on a vertical scale indicating their social position in society, with ten categories from top to bottom.* We created a variable called *rankcat* that divided respondents into four categories: highest, higher, lower, and lowest. *Rankcat* serves as the independent variable. The dependent variable, confidence in the government (*confgov*), is the sum of respondents' answers to three questions about confidence in each of the three branches of the US government, the executive branch, Congress, and the Supreme Court.** *Confgov* ranges from 3 (least trust) to 9 (most trust). In this example we will conduct an ANOVA to test whether there is a relationship between social rank and confidence in the government. In other words, we are testing whether the mean score on the confidence in government variable differs across the four categories of the social ranking variable.

We will use SPSS to run a one-way analysis of variance, with "one-way" referring to the fact that we have only one dependent variable. Open `GSS2021.sav`. We use the following procedure:

`Analyze → Compare Means → One-Way ANOVA`

---

* The original GSS variable used to generate *rankcat* is called *rank*.
** The three original GSS variables used to construct the dependent variable are *confed*, *conlegis*, and *conjudge*.

This opens the "One-Way ANOVA" dialog box. We move the dependent variable *confgov* into the "Dependent List" box and move the independent variable, *rankcat*, into the "Factor" box (shown in Figure 11.4).

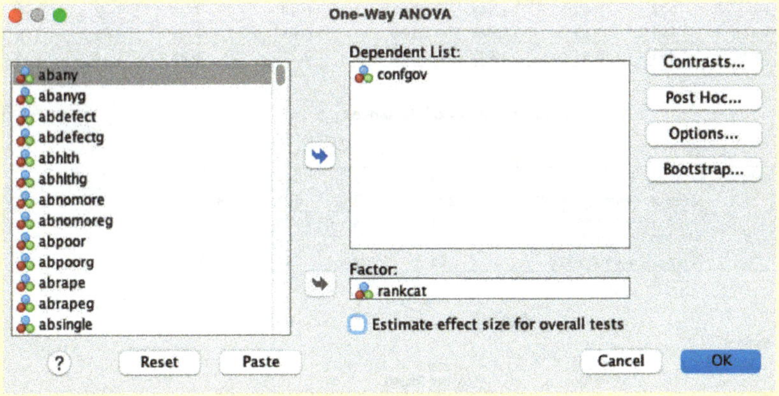

Figure 11.4

In addition to running an ANOVA test, we want SPSS to calculate some descriptive statistics and to test the assumption of equal variances across groups. Before we click on "OK," we click on "Options." This opens the "Options" dialog box (shown in Figure 11.5).

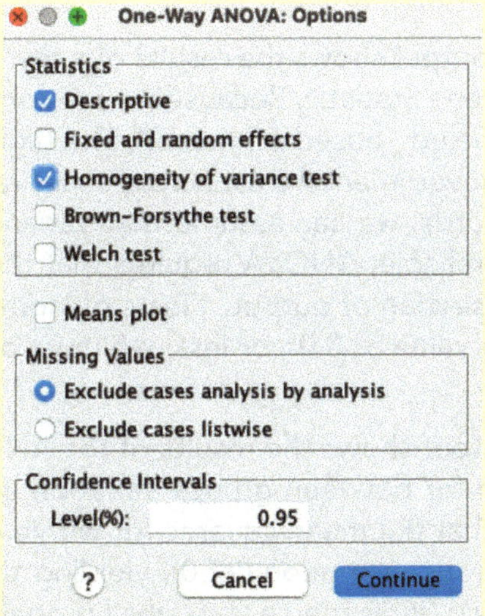

Figure 11.5

We place check marks in the small boxes next to "Descriptive" and "Homogeneity of Variance Test," click on "Continue," and then click on "OK." The output is shown in Figure 11.6.

The top section of the output provides the information requested when we selected "Descriptive." We see that the mean confidence rating of respondents in the "lowest"

Descriptives

confgov

	N	Mean	Std. Deviation	Std. Error	95% Confidence Interval for Mean		Minimum	Maximum
					Lower Bound	Upper Bound		
highest	169	6.51	1.398	.108	6.30	6.72	3	9
higher	106	6.44	1.339	.130	6.19	6.70	3	9
lower	228	7.08	1.398	.093	6.90	7.27	3	9
lowest	142	7.19	1.444	.121	6.95	7.43	3	9
Total	645	6.85	1.432	.056	6.74	6.96	3	9

Tests of Homogeneity of Variances

		Levene Statistic	df1	df2	Sig.
confgov	Based on Mean	.184	3	641	.908
	Based on Median	.271	3	641	.846
	Based on Median and with adjusted df	.271	3	635.188	.846
	Based on trimmed mean	.185	3	641	.906

ANOVA

confgov

	Sum of Squares	df	Mean Square	F	Sig.
Between Groups	66.032	3	22.011	11.254	<.001
Within Groups	1253.680	641	1.956		
Total	1319.712	644			

Figure 11.6

category of social rank (7.2) is higher than the mean for any other group. Mean confidence in the government is lower among those in the "higher" and "highest" social ranks (6.4 and 6.5, respectively).

The middle portion of the output shows the results of a test for equal variances across the four groups, called the Levene Statistic. Because we are comparing means, we look at the p-value in the first row, labeled "based on mean." The high p-value (.908) suggests that differences in variance in governmental trust across social rank groups likely reflects sampling variation. In other words, we can assume that variances among groups are equal. In order to be statistically reliable, ANOVA requires that the variances among groups be equal. Always look at this section of output, "Tests of Homogeneity of Variances." If the Levene Statistic shows a p-value of 0.05 or less, we must be cautious in interpreting the ANOVA results.

The bottom part of the output shows the results of the ANOVA test. The first row of the table presents information for between groups variation and the second row for within groups variation. We see that the mean squares for within groups variation (2.0) is much lower than for between groups variation (22.0), yielding the F-statistic of 11 (22.0/2.0). The p-value is less than 0.001, indicating a very small probability of observing this ratio of between to within groups variation by chance.

The ANOVA test indicates that ratings of confidence in government do depend on how people assess their social position (relatively high or relatively low), with those who see themselves at the top end of the social hierarchy reporting less trust in the government and those at the lower end reporting more trust.

To run Tukey's post hoc test, we click on "Post Hoc" in the "One-Way ANOVA" dialog box (Figure 11.4) and place a check mark in the small box labeled "Tukey." This produces output (not shown here) similar to Table 11.7. In addition to the difference between the means for each pair of social positions, SPSS reports the standard error of that difference, its p-value (labeled "Sig."), and the confidence interval for the difference. You can interpret these results the same way as those presented in Table 11.7. If you run the Tukey post hoc test for this example, look at the "Sig." column. You will see that all pairs have significantly different ratings of their confidence in government *except* the highest vs. higher and lowest vs. lower pairs. The "Mean Difference" column shows that the difference in mean rating between the higher and highest social position groups was only –0.06, for example, while the difference in mean rating between the lowest vs. higher groups was 0.75.

Review of SPSS Procedures

- ANOVA test

    Analyze → Compare Means → One-Way ANOVA

    In the "Options" dialog box, select "Descriptive" to generate descriptive statistics; select "Homogeneity of variance test" to test the assumption of equal variance across groups.

    In the "Post Hoc" dialog box, select "Tukey" to generate the Tukey post hoc test.

Practice Problems

1. A United Nations staffer wants to understand which groups support the UN more strongly. Using the 2020 American National Election Survey (ANES), they conduct analysis of variance looking at political party registration and scores on a feeling thermometer for the UN (where 0 represents the "coldest" and 100 the "warmest" feeling). Table 11.8 shows the mean ratings and standard deviations for each category.

    Table 11.8 Feeling Thermometer Ratings for the United Nations by Political Party Registration

Political Party	Mean	Standard Deviation	Frequency
Democratic	64.2	30.4	1,804
Republican	41.7	30.3	1,294
Independent	52.5	30.8	1,000
Total	54.2	30.8	4,098

    Source: 2020 American National Election Survey.

a. Which category reports the highest rating for the United Nations, on average?
b. Is there variation in ratings within each category? How can you tell?
c. If there were no variation within any category, which cells of the table would have to change, and how?
d. Table 11.9 shows the calculations for one of the two kinds of sums of squares used to calculate the sample F-statistic. Which sums of squares is the table demonstrating? Explain how you know.
e. Complete Table 11.9 to find the missing sum of squares.

Table 11.9 _?_ Groups Sums of Squares, Feeling Thermometer Ratings of United Nations and Political Party Registration

Political Party (N)	_?_ Groups Sum of Squares		
	$\bar{y}_{group} - \bar{y}_{total}$	$(\bar{y}_{group} - \bar{y}_{total})^2$	$N_{group}(\bar{y}_{group} - \bar{y}_{total})^2$
Democratic (1,804)	64.2 − 54.2 =	?	
Republican (1,294)	41.7 − 54.2 =	?	
Independent (1,000)	52.5 − 54.2 =	?	
			$\Sigma N(\bar{y}_{group} - \bar{y}_{total})^2 = $ _?_

Source: 2020 American National Election Survey.

2. A researcher studying family structure wondered whether people with different marital statuses have different numbers of children. Using the 2021 GSS, they conduct analysis of variance looking at mean number of children by marital status.

Table 11.10 Mean Number of Children by Marital Status

Marital Status (N)	Mean	SD	Frequency
Married	2.00	1.37	1,979
Widowed	2.41	1.56	298
Divorced	1.96	1.55	652
Separated	2.44	1.77	95
Never married	0.60	1.21	952
Total	1.70	1.53	3,976

Source: 2021 General Social Survey.

Table 11.10 shows the mean ratings and standard deviations for each category.

a. Which category reports the highest number of children, on average?
b. Is there variation in ratings within each category? How can you tell?

c. If there were no variation within any category, which cells of the table would have to change, and how?

d. Table 11.11 shows the calculations for one of the two kinds of sums of squares used to calculate the sample F-statistic. Which sums of squares is the table demonstrating? Explain how you know.

e. Complete Table 11.11 to find the missing sum of squares.

Table 11.11 _?_ Groups Sums of Squares, Number of Children and Marital Status

Marital Status (N)	$\bar{y}_{group} - \bar{y}_{total}$	$(\bar{y}_{group} - \bar{y}_{total})^2$	$N_{group}(\bar{y}_{group} - \bar{y}_{total})^2$
	_?_ Groups Sum of Squares		
Married (1,979)	2.00 – 1.70	?	
Widowed (298)	2.41 – 1.70		
Divorced (652)	1.96 – 1.70		
Separated (95)	2.44 – 1.70		
Never married (952)	0.60 – 1.70		
			$\Sigma N(\bar{y}_{group} - \bar{y}_{total})^2 = $ _?_

Source: 2021 General Social Survey.

3. The data in Problem 1 showed ratings of the UN, based on political party registration. The within groups sum of squares is 3,892,974.6. Use this information along with relevant information from Problem 1 to complete the steps of the hypothesis test for each part below.

   a. What is the null hypothesis for an ANOVA test in this case? What is the alternative hypothesis? Give both hypotheses in words and statistical notation.

   b. Find the decision F for an alpha level of 0.05.

   c. Calculate the sample F-statistic.

   d. Decide whether to reject the null hypothesis.

   e. What does your finding from part d indicate about differences in ratings of the UN depending on political party registration?

4. The data in Problem 2 showed mean number of children by marital status. The within groups sum of squares is 7678.7. Use this information along with relevant information from Problem 2 to complete the steps of the hypothesis test for each part below.

   a. What is the null hypothesis for an ANOVA test in this case? What is the alternative hypothesis? Give both hypotheses in words and statistical notation.

   b. Find the decision F for an alpha level of 0.05.

c. Calculate the sample F-statistic.

d. Decide whether to reject the null hypothesis.

e. What does your finding from part d indicate about differences in number of children depending on marital status?

5. Problems 1 and 3 found a statistically significant relationship between political party registration and feeling thermometer ratings of the United Nations.

   a. What would a Tukey's post hoc test tell you about this relationship?

   b. Tukey's post hoc test showed a p = 0.000 for each of three pairs of party registration categories. What are the three pairs? What does the p-value indicate?

   c. State the results of the ANOVA and post hoc tests in words.

6. Problems 2 and 4 found a statistically significant relationship between marital status and number of children.

   a. What would a Tukey's post hoc test tell you about this relationship?

   b. There are ten possible pairs of marital status categories. List the ten pairs.

   c. Tukey's post hoc test showed a p-value of 1.00 for separated versus widowed, 0.936 for divorced versus married, 0.024 for separated versus married, 0.013 for separated versus divorced, and 0.000 for the remaining five pairs. Which of the ten pairs have statistically significant differences?

   d. State the results of the ANOVA and post hoc tests in words.

7. The "word gap" refers to the finding that children raised in households with more resources tend to hear more words at home than do children in households with fewer resources. For example, one study found mean differences by social class in the number of unique words heard per hour by three-year-olds at home. Table 11.12 gives the number of unique words per hour that five three-year-olds heard at home in the three social class groups.

   Table 11.12 Within Groups Sums of Squares, Social Class, and Unique Words per Hour

Poverty			Working Class			Professional		
y	$y - \bar{y}$	$(y - \bar{y})^2$	y	$y - \bar{y}$	$(y - \bar{y})^2$	y	$y - \bar{y}$	$(y - \bar{y})^2$
150			250			400		
100			300			375		
200			240			310		
180			195			450		
205			270			375		
$\Sigma(y - \bar{y})^2 = ?$			$\Sigma(y - \bar{y})^2 = ?$			$\Sigma(y - \bar{y})^2 = ?$		

a. Complete the table and find the within groups sum of squares for each social class category.

b. What is the overall within groups sum of squares for the three social classes shown in Table 11.12?

c. The overall mean for the full sample is 267.7. Find the between groups sum of squares.

d. Find DF within and between and find the corresponding decision F in appendix C.

e. Find the within and between mean squares.

f. Calculate F and interpret your result in words.

g. Can you tell which social class groups have statistically significant differences in words heard per hour?

8. A random sample of five people from each of three generational groups (Gen Z, millennials, and Gen X) were asked to record how many minutes they spent on YouTube in a week. Table 11.13 gives the number of minutes for each generation.

Table 11.13 Within Groups Sums of Squares, Generation and Minutes Spent on YouTube

Gen Z			Millennial			Gen X		
y	$y - \bar{y}$	$(y - \bar{y})^2$	y	$y - \bar{y}$	$(y - \bar{y})^2$	y	$y - \bar{y}$	$(y - \bar{y})^2$
420			0			120		
100			50			350		
600			580			20		
30			260			0		
0			620			600		
$\Sigma(y - \bar{y})^2 = ?$			$\Sigma(y - \bar{y})^2 = ?$			$\Sigma(y - \bar{y})^2 = ?$		

a. Complete the table and find the within groups sum of squares for each generation.

b. What is the overall within groups sum of squares for the three generations shown in Table 11.13?

c. The overall mean for the full sample is 253.0. Find the between groups sum of squares.

d. Find DF within and between and find the corresponding decision F in appendix C.

e. Find the within and between mean squares.

f. Calculate F and interpret your result in words.

g. Can you tell which generations spend significant different numbers of minutes on YouTube?

9. Six evangelical Protestants, six mainline Protestants, and six people who are unaffiliated with any religion rate their agreement (from 0 = completely disagree to 100 = completely agree) that lesbians and gays should be able to adopt children. Figure 11.7 (next page) shows three versions (Panels A, B, and C) of the distributions of responses, by religious affiliation.

   a. Which panel shows the set of values most likely to yield a statistically significant sample F in an ANOVA? Explain your answer.

   b. Which panel shows the set of values least likely to yield a statistically significant sample F in an ANOVA? Explain your answer.

10. Four dog owners, four cat owners, four people with both cats and dogs, and four people with no pets wore fitness trackers to track their daily step count. Figure 11.8 (p. 454) shows three versions (Panels A, B, and C) of the distribution of daily steps by pet ownership.

    a. Which panel shows the set of values most likely to yield a statistically significant sample F in an ANOVA? Explain your answer.

    b. Which panel shows the set of values least likely to yield a statistically significant sample F in an ANOVA? Explain your answer.

11. A political scientist is interested in whether experience of travel or living abroad shapes American college students' support for increasing immigration. The researcher and their team compared support for increasing immigration on a scale from 1 (strong opposition) to 10 (strong support) for students who had lived outside the United States for at least one semester, those who had traveled outside the United States but not lived abroad, and those who had done neither. Table 11.14 (p. 455) shows the results.

    a. Do the mean differences shown in Table 11.14 suggest that living or traveling outside the United States is related to support for immigration? What does the effect of living or traveling outside the United States appear to be on support for immigration?

    b. Table 11.15 shows the results from an ANOVA test of differences in support for immigration by experience living or traveling outside the United States. What is the null hypothesis being tested here?

    c. What do the ANOVA results indicate about mean differences in support for immigration by experience living or traveling outside the United States?

    d. Dissatisfied with the results of the omnibus ANOVA test, a student generates a post hoc test to examine precisely which specific groups differ from each other, on average, in support for increasing immigration (continued on p. 454).

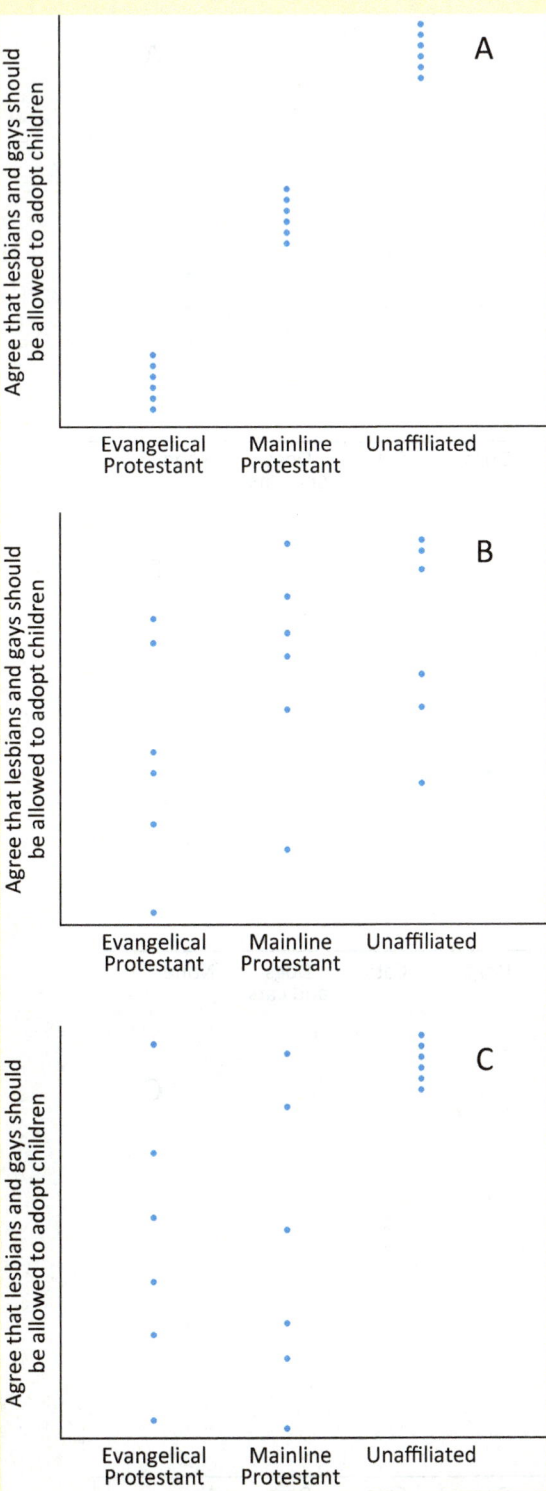

Figure 11.7 Agreement That Lesbians and Gays Should Be Allowed to Adopt Children, by Religious Affiliation, Three Sets of Values

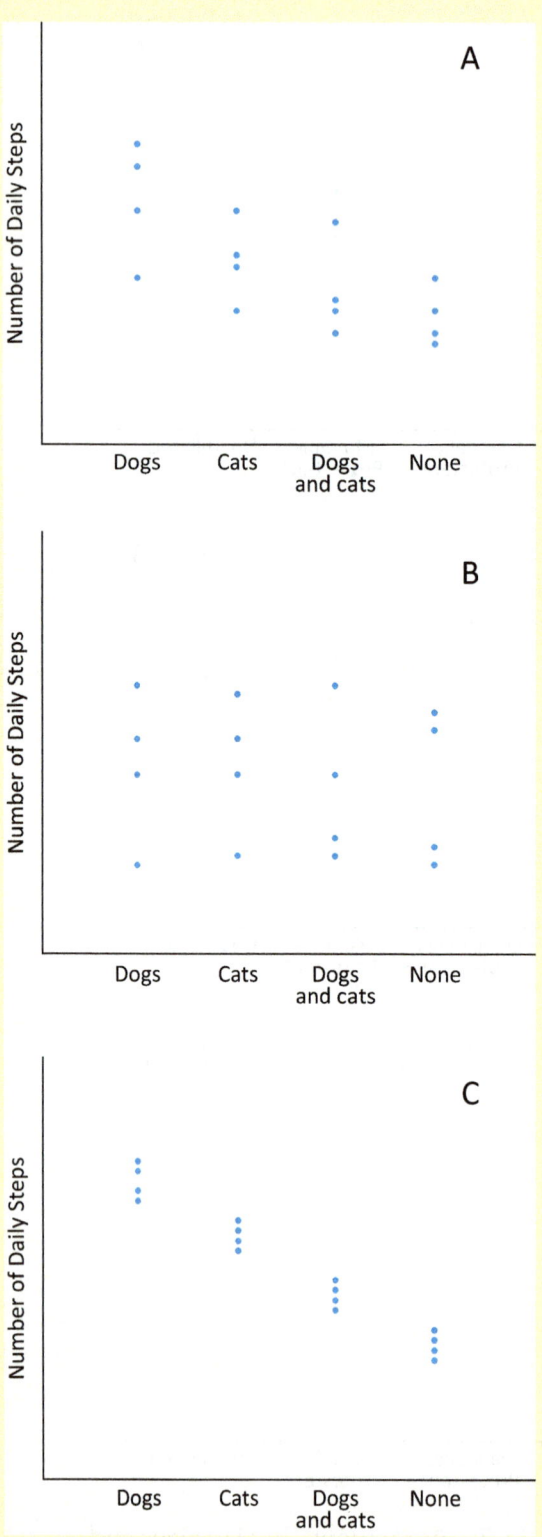

Figure 11.8

Table 11.14 Mean Support for Increasing Immigration, by Experience Abroad

Experience Abroad	Mean
1. Lived outside the United States for at least one semester	5.9
2. Traveled outside the United States but never lived abroad	4.5
3. Neither traveled nor lived outside the United States	4.0

Table 11.15 ANOVA for Support for Increasing Immigration by Experience outside the United States

Source of Variation	SS	DF	MS	F	p
Between groups	37	2	18.5	15.4	0.000
Within groups	4802	3875	1.2		

Table 11.16 Results of Tukey's Post Hoc Comparison of Mean Differences in Support for Increasing Immigration by Experience Living outside the United States

Experience Comparison	Difference
Lived—Traveled	1.4*
Lived—Neither	1.9***
Traveled—Neither	0.5

*** $p < 0.001$; ** $p < 0.01$; * $p < 0.05$

11d. (continued) Table 11.16 shows the results of the Tukey's test. According to Table 11.16, which particular mean differences are statistically significant? How can you tell?

12. At the end of a long day spent online, a researcher wondered what affects the time people spend on work-related online tasks such as email. Using the General Social Survey (GSS), she looked at the relationship between education level (respondents' highest degree) and time spent on email per week.* Table 11.17 shows the results.

Table 11.17 Mean Number of Hours Spent on Email per Week, by Highest Degree

Highest Degree	Mean
Less than high school	2.7
High school	3.8
Associate's degree	5.9
Bachelor's degree	5.8
Graduate degree	6.4

* The email variable was calculated from the GSS variables *emailhrs* and *emailmin*.

Table 11.18 ANOVA for Hours Spent on Email per Week by Highest Degree

Source of Variation	SS	DF	MS	F	p
Between groups	1341.2	4	335.3	5.83	0.0001
Within groups	50,987.6	886	57.6		

Table 11.19 Results of Tukey's Post Hoc Comparison of Mean Numbers of Hours Spent on Email per Week by Highest Degree

Experience Comparison	Difference
High school versus less than high school	1.1
Associate's versus less than high school	3.2
Bachelor's versus less than high school	3.1*
Graduate versus less than high school	3.7*
Associate's versus high school	2.1
Bachelor's versus high school	2.0*
Graduate versus high school	2.6**
Bachelor's versus Assoc	−0.1
Graduate versus Assoc	0.5
Graduate versus Bach.	0.6

*** $p < 0.001$; ** $p < 0.01$; * $p < 0.05$

a. Do the mean differences shown in Table 11.17 suggest that education level is related to time spent on email? What does the effect of education appear to be on time spent on email?

b. Table 11.18 shows the findings from an ANOVA test of differences in time spent on email by highest degree. What is the null hypothesis being tested here?

c. What do the ANOVA results indicate about mean differences in time spent on email by highest degree?

d. Dissatisfied with the results of the omnibus ANOVA test,w the researcher generates a post hoc test to examine precisely which specifc groups differ from each other, on average, in time spent on email. Table 11.19 shows the results of the Tukey's test. According to Table 11.19, which particular mean differences are statistically significant? How can you tell?

Stata Problems

Here, we will use the American National Election Survey (ANES) to conduct an ANOVA test of the relationship between political party, *pid3*, and feeling thermometer ratings of rural Americans, *V202184*. Political party is grouped into three categories: (1) Democrat;

(2) Republican; and (3) Independent or no party. The feeling thermometer ranges from 0 (cold) to 100 (warm). Is political affiliation related to feelings about rural Americans?

1. Open `ANES20.dta` and use the "oneway" command to generate an ANOVA test examining mean differences in feelings about rural Americans across political affiliations.
2. According to the output, what are the sample means for Republicans, Democrats, and Independents?
3. What is the null hypothesis for this ANOVA? The alternative hypothesis?
4. What is the sample F? Use information from the output to show how the sample F was calculated.
5. Use information from the output to find the decision F for this ANOVA test. Decide whether to reject the null hypothesis by comparing decision F to sample F.
6. What is the actual p-value associated with the F-statistic reported on the output?
7. What does the result of Bartlett's test for equal variances mean?
8. Use the "pwmean" command to run Tukey's post hoc test.
9. Which of the particular group means are different from each other at the 0.05 alpha-level?

SPSS Problems

Here, we will use the ANES to conduct an ANOVA test of the relationship between political party, *pid3*, and feeling thermometer ratings of rural Americans, *V202184*. Political party is grouped into three categories: (1) strong, not very strong, and independent-leaning Democrats; (2) Independents; and (3) strong, not very strong, and independent-leaning Republicans. The feeling thermometer ranges from 0 (cold) to 100 (warm). Is political affiliation related to feelings about rural Americans as a group?

1. Open `ANES20.sav` and use the "One-Way ANOVA" procedure to generate an ANOVA test examining mean differences in feelings about rural Americans (*V202184*) across political affiliations. Use the appropriate dialog boxes to generate descriptive statistics, a test of equal variances, and Tukey's post hoc test.
2. According to the output, what are the sample means for Republicans, Democrats, and Independents?
3. What is the null hypothesis for this ANOVA? The alternative hypothesis?
4. What is the sample F? Use information from the output to show how the sample F was calculated.
5. Use information from the output to find the decision F for this ANOVA test. Decide whether to reject the null hypothesis by comparing decision F to sample F.
6. What is the actual p-value associated with the F-statistic reported on the output?

7. According to the output, can we assume that the variances of the feeling thermometer are the same across groups?

8. Which of the particular group means are different from each other at the 0.05 alpha-level?

Notes

[1.] For complex reasons, it is simpler to set up experimental research designs that will not violate the assumptions of ANOVA. See Andrew Gelman. 2015. "Analysis of Variance, Why It Is More Important Than Ever." *The Annals of Statistics* 33(1): 1–53; Tue Tjur. 2015. "Discussion." *The Annals of Statistics* 33(1): 1–53.

[2.] A. Oksanen, N. Savela, R. Latikka, and A. Koivula. 2020. "Trust toward Robots and Artificial Intelligence: An Experimental Approach to Human-Technology Interactions Online." *Frontiers in Psychology* 11: 568256.

[3.] There is considerable debate over the details of the expected F-value in the case of no relationship between the variables. See M. C. Voelkle, P. L. Ackerman, and W. W. Wittmann. 2007. "Effect Sizes and F Ratios < 1.0." *Methodology: European Journal of Research Methods for the Behavioral and Social Sciences* 3(1): 35–46.

[4.] More skewed data may require a larger sample size, while more normal data may permit a smaller one. See C. R. Wilson and B. L. Morgan. 2007. "Understanding Power and Rules of Thumb for Determining Sample Sizes." *Tutorials in Quantitative Methods for Psychology* 3(2): 43–50; and M. J. Blanca, R. Alarcon, J. Arnau, R. Bono, and R. Bendayan. 2017. "Non-Normal Data: Is ANOVA Still a Valid Option?" *Psicothema* 29(4): 552–557.

[5.] This example is based on a study conducted by two professors at Coastal Carolina University, in which they compared the performance of students taking three versions of an Introduction to Sociology course: a traditional lecture course, an online course, and a hybrid course employing both face-to-face and online learning. See S. Brallier and M. McIlreavy. 2016. "Does Method of Course Delivery Matter: A Comparison of Student Performance in Hybrid, Online, and Lecture-Based Introductory Sociology Courses." *Infonomics Society* 7: 2413–2417.

Chapter 12

Testing the Statistical Significance of Relationships in Cross-Tabulations

The Chi-Square Test

The gender gap is a well-known phenomenon in American politics. Women are less likely to endorse policies and actions that involve violence or aggression, such as supporting capital punishment or taking a hard line on border security. Women tend to be more supportive of "compassionate" policies, such as a government role in health insurance or increased spending on assistance to poor people. And since 1980, there has been a notable gender gap on voting. Women have tended to support Democratic candidates, while men have disproportionately supported Republican candidates.[1]

How do analysts ascertain whether there is a gender gap on a given issue or on vote choice? We began to answer that question in chapter 3, when we looked at cross-tabulations. Recall that we examined how groups differ from each other in the opinions they express. We saw, for example, that young Americans were more likely to be pro-choice than older Americans. We also saw that foreign-born Americans were more likely to support foreign aid than respondents born in the United States. On the other hand, when we looked at perceptions of a Russian threat broken down by gender, we found the differences between women and men to be very small. In short, to ascertain whether group differences exist or not, we simply compare column percentages.

Let's look at another example. In the 2020 US presidential election, women were more likely than men to support Democratic presidential candidate Joe Biden. According to exit polls, 57% of women voted for Biden, compared to 45% of men. This gap—12 points—is a large one. But how large is large enough? How big does a gap have to be before we can claim that this is a "statistically significant" difference?

We ask this question now because, in chapters 7 through 11, we have become very well acquainted with the notion of statistical significance. We know, for example, that, if 57% of women in a randomly selected sample supported Biden, the percentage of Biden supporters among women in the population is very likely close to 57% but maybe a bit below it or a bit above it. In other words, because of random sampling variation, any particular sample statistic or any particular difference between two samples may or may not appear in the population.

In this chapter, we are going to examine the concept of statistical significance as it applies to cross-tabulations. Let's continue with a discussion of the gender gap and look at three cross-tabulations from the 2020 American National Election Study (Tables 12.1, 12.2, and 12.3).

Table 12.1   Effect of Gender on Support for Death Penalty

	Men	Women	Total
Favor strongly	44.4% (1,480)	36.8% (1,447)	40.3% (2,927)
Favor not strongly	19.9% (663)	20.7% (812)	20.3% (1,475)
Oppose not strongly	18.8% (625)	22.3% (875)	20.7% (1,500)
Oppose strongly	16.9% (564)	20.3% (796)	18.7% (1,360)
Total	100% (3,332)	100% (3,930)	100% (7,262)

(χ^2 = 47.75, p = 0.00)
Source: 2020 American National Election Study.

Table 12.2   Effect of Gender on Support for Government Spending on Health Care

	Men	Women	Total
Increase spending	53.3% (1,769)	56.1% (2,212)	54.7% (3,981)
No change	33.3% (1,111)	32.4% (1,279)	32.8% (2,390)
Decrease spending	13.7% (456)	11.5% (453)	12.5% (909)
Total	100% (3,336)	100% (3,944)	100% (7,280)

(χ^2 = 10.41, p = 0.005)
Source: 2020 American National Election Study.

Let's begin with Table 12.1, which examines gender differences on the death penalty. Are women more opposed to the death penalty, as we would expect? If we compare the column percentages, the answer is yes. Men are about 8 percentage points more likely than women to strongly favor it (44.4% to 36.8%). Women are more likely than men to oppose and to strongly oppose. Are these differences large enough to overcome random sampling variation? In other words, how certain are we that the observed sample differences exist in the population? The answer to that question can be found in the parentheses below the table. They contain the χ^2 (chi-square, pronounced as "kye"-square) statistic and

Table 12.3 Effect of Gender on Level of Satisfaction with Democracy in the United States

	Men	Women	Total
Very satisfied	8.9% (296)	8.2% (326)	8.5% (622)
Fairly satisfied	53.6% (1,788)	53.1% (2,102)	53.3% (3,890)
Not very satisfied	30.4% (1,013)	31.6% (1,249)	31.0% (2,262)
Not at all satisfied	7.2% (239)	7.1% (279)	7.1% (518)
Total	100% (3,336)	100% (3,956)	100% (7,292)

(χ^2 = 1.80, p = 0.62)
Source: 2020 American National Election Study.

its corresponding p-value. Later in this chapter, we will examine how these are found, but, for now, let's focus on the p-value. This p-value is similar to the p-values we have examined in the preceding chapters. It tells us the likelihood of obtaining a sample result under the assumption that a null hypothesis of no relationship is true. In Table 12.1, the p-value is equal to 0. In the context of this cross-tabulation, this means that, if there were no relationship in the population between gender and support for the death penalty, or if the distribution of opinion on the death penalty were identical for women and men (the null hypothesis), then there would be virtually no chance (p = 0) of obtaining the sample differences that we obtained. We would therefore conclude that there is a "statistically significant" relationship between gender and position on the death penalty. How strong is that relationship? The answer to that question is the same answer we provided in chapter 3. To gauge the strength of the relationship, we examine the magnitude of the differences between column percentages. In this case, women are about 8 points less likely than men to strongly favor the death penalty—a moderately strong relationship.

Table 12.2 looks at the effect of gender on a different kind of issue, a "compassion" issue. The question here is: Should the federal government spend more on health care? The gaps between the column percentages for women and men are not all that large, especially for those who say that the government should decrease spending or that there should be no change. But women are about 3 percentage points more likely to say that the government should increase spending, whereas men are a little more likely to say that the government should decrease spending. Although these differences are not very large, they are large enough to achieve statistical significance. We can make this claim on the basis of the χ^2 p-value at the bottom of the table, which is 0.005. This means that, if there were no relationship in the population between gender and support for government spending on health care (the null hypothesis), there would be a very low probability (0.005) of obtaining the sample differences we got. We conclude that there is a statistically significant relationship between these variables. It is not a very strong one (compare the column percentages), but we are confident that there are differences between women and men in the population.

Our third example (see Table 12.3) uses a different kind of question and will lead us to a different conclusion. Here, we examine the effect of gender on respondents' level of satisfaction with American democracy. Compare the column percentages. They are very close to each other. For example, 8.9% of men are very satisfied with democracy in the United States, while 8.2% of women hold this view. For all the responses, the gaps between women and men range from .1 to 1.2 percentage points. These are very small gaps, and they are reflected in the χ^2 p-value, which is equal to 0.62 (see parentheses under Table 12.3). This means that, if there were no relationship in the population between gender and satisfaction with American democracy, there would be a 62% chance of obtaining our observed sample results. This is a very high p-value. Using any conventional alpha (0.01, 0.05, or 0.10), we would come to the same conclusion. Since our obtained p-value is well above alpha, we would conclude that there is no significant relationship between gender and level of satisfaction with American democracy.

These three examples illustrate an important point. Although the column percentages alone give us a good sense of the relationship between independent and dependent variables, on their own they cannot tell us if a relationship is statistically significant. Toward this end, we must use the chi-square statistic.

The Logic of Hypothesis Testing with Chi-Square

In each of these examples, we have used a particular terminology to explain how we interpret the p-value. Based on the logic of the null hypothesis discussed in previous chapters, we have stated that the p-value can be interpreted as the probability of obtaining our observed sample differences "if there were no relationship in the population between the independent variable and dependent variable." What do we mean when we state that there is "no relationship"? When using a t-test to compare the means of two groups, as in earlier chapters, "no relationship" means that the two groups have the same mean level of the dependent variable. With chi-square, it is a little different. "No relationship" means that the categories of the independent variable are identical in their distribution across the dependent variable.

In Table 12.4, we present some hypothetical data that illustrate the condition of "no relationship." Our independent variable is gender, and our dependent variable is whether the respondent would vote for a third-party candidate in the upcoming election.

Table 12.4 Effect of Gender on Support for Third-Party Candidate, Hypothetical Population Data

	Men	Women	Total
Would support third-party candidate	25%	25%	25%
Not sure	15%	15%	15%
Would not support third-party candidate	60%	60%	60%
Total	100%	100%	100%

In this example, there is no relationship between gender and support for a third-party candidate. Women and men are identical in their opinions. For example, among both men and women, 25% respondents would support the candidate. The differences between men and women for the other two categories of the dependent variable are also equal to 0. Knowing a respondent's gender does not provide us with any ability to predict how an individual will vote. In other words, in this example, there is no relationship between the independent and dependent variables. Or one might say that these variables are statistically independent of each other. **Statistical independence** simply means that the variables are not related to each other. This is why the chi-square test is sometimes called the **chi-square of independence**.*

Let us further assume that the cross-tabulation shown in Table 12.4 is based on a population, not on a sample of individuals. (In reality, we are not likely to ever know the true population breakdown, and it is also highly unlikely that, in any sample or population, the column percentages would be identical.) Out of this population, we draw a random sample, and we survey them on their support for third-party candidates. Here are two different scenarios, shown in Tables 12.5 and 12.6.

Table 12.5 Effect of Gender on Support for Third-Party Candidate, Hypothetical Sample Data, Scenario I

	Men	Women	Total
Would support third-party candidate	24%	26%	25%
Not sure	17%	13%	15%
Would not support third-party candidate	59%	61%	60%
Total	100%	100%	100%

Table 12.6 Effect of Gender on Support for Third-Party Candidate, Hypothetical Sample Data, Scenario II

	Men	Women	Total
Would support third-party candidate	10%	40%	25%
Not sure	20%	10%	15%
Would not support third-party candidate	70%	50%	60%
Total	100%	100%	100%

By comparing Tables 12.4 and 12.5, we come to one obvious conclusion. The percentages in these tables are very close to each other. In our hypothetical population, 25% of men stated that they will vote for a third-party candidate. Because of sampling variation, it should be no surprise to obtain a sample in which 24% of men say they support a

---

* Chi-square can also be used to test whether sample results conform to expected frequencies based on a model rather than expected frequencies based on no relationship. In that case, it is called a chi-square test for goodness-of-fit rather than a chi-square test of independence. We discuss that test in Box 12.3.

third-party candidate (Table 12.5). We can say the same thing about any other percentage in these tables. If 60% of women in the population say they would not support a third-party candidate, then how surprised would you be if your sample of women found that 61% would not support the third-party candidate? Not surprised at all. Again, because of sampling variation, we would expect our sample proportions to be close to our population proportions but not necessarily equal.

Let's examine the second scenario. Out of this same population, we draw a second random sample and present the results in Table 12.6.

Compare the percentages in Tables 12.4 and 12.6. Now, they are far apart. In our population, shown in Table 12.4, 25% of men say they would vote for the third-party candidate. But, in our sample drawn from that population, the corresponding percentage was only 10%. We will find similar discrepancies for any other percentage. In our population, 60% of women would not support the third-party candidate. In our sample, that figure was 50%. What should we make of these discrepancies?

There are two possibilities. The first is that, by random bad luck, we drew a sample that is not representative of the population. This can happen any time we draw a sample out of a population. It is not likely to happen, but it is possible. The second possibility is that there actually is a difference between men and women in the population.

Thinking about these two possibilities is key to understanding the logic of the chi-square test of independence. Remember, in this example, we provided hypothetical data about a population (Table 12.4) and then drew a sample with results that were strongly at odds with that population (Table 12.6). In reality, we never know the true population percentages. In hypothesis testing, we make an assumption (reflected in the null hypothesis) that there is no relationship between variables in the population, and we see whether our sample results are inconsistent with that assumption. That assumption is at the core of the chi-square test. The assumption is that there is no relationship in the population between independent and dependent variables ("statistical independence"). In other words, we set up the null hypothesis to assume that, in the population, the column percentages are identical to each other. That is the assumption reflected in Table 12.4. We then pose the question: If there were no relationship between independent and dependent variables in the population, how likely is it that we would have obtained the sample differences that we did obtain?

If indeed there were no relationship in the population (as in Table 12.4), it is highly probable that we would have obtained sample percentages "close" to that distribution. This is what we display in Table 12.5. In this case we end up with a "high" p-value. We would therefore *not* reject the claim that there is no relationship in the population. On the other hand, if we obtain sample results that are "far" from the percentages we would have obtained if there were no relationship in the population (as in Table 12.6), our p-value will be "low," and we would reject the claim that there is no relationship in the population.

As with the z-test, t-test, and analysis of variance, the chi-square test relies on the notion of a sampling distribution. Recall from the past several chapters that, with repeated sampling, a sample statistic will be distributed in a predictable curve, most frequently close to the population parameter, less frequently a moderate distance from that parameter, and more infrequently further from the parameter.

The chi-square test of independence allows us to determine, with a high degree of certainty, whether there is a relationship between two variables in the population. It is one of the most commonly used statistical tests. Again, it is not a measure of the strength of a relationship. There are other statistics that measure strength of association, although we will not cover them in this book.* As we have already indicated, the best way to measure the strength of association between two nominal or ordinal variables is to compare the column percentages.

The Steps of a Chi-Square Test

In conducting the chi-square test of independence, we follow several steps. To illustrate each of these steps, we will use an example from the seventh wave of the World Values Survey. In this example, we will examine the effect of gender (independent variable) on how religious people claim to be (dependent variable), among American respondents. Table 12.7 presents a cross-tabulation with the results from the sample. Each cell contains the column percentage as well as the number of respondents in that category (the observed frequency). For example, 718 men, which was 52% of men, said they are religious.

Table 12.7 Effect of Gender on Religiousness

	Men	Women	Total
Religious person	52.0% (718)	59.2% (707)	55.3% (1,425)
Not a religious person	36.8% (508)	33.3% (398)	35.2% (906)
Atheist	11.2% (155)	7.5% (90)	9.5% (245)
Total	100% (1,381)	100% (1,195)	100% (2,576)

Source: World Values Survey (Wave 7).

Step 1: State the null and alternative hypotheses.

When we conduct a chi-square test of independence, we begin by stating the null and alternative hypotheses. Recall that we always hypothesize about the population. There is no need to hypothesize about the sample, since we can directly observe the sample statistics. The null hypothesis, H_0, will always state that, in the population, there is no relationship between the independent and dependent variables. In the context of chi-square, if there

---

* There are several measures of association, although they are less frequently reported by social scientists. "Lambda" and "Cramér's V," for example, measure the association between two nominal variables. "Gamma" and "Kendall's Tau-b" measure the association between pairs of ordinal variables. These measures of association typically run from 0 to 1, where 0 indicates no association whatsoever between the variables and 1 is the strongest possible association. Although there are differences in the way these measures are calculated, they are all based on the concept of "proportionate reduction of error" (PRE). A PRE-measure of association is a summary of our ability to predict values of one variable using the values of the other variable. Most statistical software programs allow the user to display these measures of association with their cross-tabulation procedure.

is no relationship between the variables, the observed frequency in each cell of the cross-tabulation would equal the frequency that would be expected if there were no relationship, as illustrated in Table 12.4. The alternative hypothesis, H_a, states the inverse of that: in the population, there is a relationship between independent and dependent variables. In the context of our example, here are the hypotheses:

H_0: *In the population, there is no relationship between gender and degree of religiousness. Or, gender and religiousness are "statistically independent" of each other.*

H_a: *In the population, there is a relationship between gender and religiousness. Or, gender and religiousness are "statistically dependent."*

Step 2: Set alpha, calculate the degrees of freedom for the cross-tabulation, and find decision chi-square.

We will set the alpha level at 0.05. As with the t- and F-statistics, the chi-square statistic falls into a family of known distributions that vary depending on degrees of freedom. Recall that with the t-distribution, degrees of freedom depend on the sample size, and, with F, degrees of freedom depend on sample size and the number of groups being compared. The shape of the chi-square distribution depends on the number of categories in each of your variables. The formula for calculating degrees of freedom for the chi-square statistic is:

$$DF = (r-1)(c-1)$$

where r = number of rows in the cross-tabulation (or the number of categories of the dependent variable); and c = number of columns in the cross-tabulation (or the number of categories of the independent variable).

In this example, the religiousness variable has three categories ($r = 3$), and gender has two categories ($c = 2$). We plug these into the formula:

$$DF = (r-1)(c-1) = (3-1)(2-1) = 2$$

You will use alpha and degrees of freedom to look up the value of your decision chi-square in a chi-square table, as we discuss below.

Step 3: Calculate the expected frequencies for each cell of the table.

The **expected frequencies** are the frequencies we would get in our sample if the null hypothesis were true. Or, to put it differently, they are the number of respondents that would be in each cell of the table if the column percentages were identical. There is a simple formula to calculate the expected frequency (F_e) for any cell of a cross-tabulation.

$$F_e = \left(\frac{row\ frequency}{total\ frequency}\right)(column\ frequency)$$

Essentially, what you are doing is calculating what percentage of the total sample falls into the row (in this case, what percentage of the total sample prays at a given level). You are then taking that percentage of the total number of cases in each column to determine how many cases would fall into a given cell. Using the data in Table 12.7, let's calculate the expected frequency for the number of men who say they are religious.

$$F_e = \left(\frac{1,425}{2,576}\right)(1,381) = 763.9$$

Note that this formula tells us, first, that 55.3% $\left(\frac{1,425}{2,576}\right)$ of the full sample say that they are religious and, next, that 55.3% of the 1,381 men in the sample who pray once a day or more are 763.9 men (the expected frequency).

If the null hypothesis were true, or if there were no relationship between gender and frequency of prayer in the population, we would expect that 763.9 of the 1,381 men in our sample would say that they are religious. Of course, we cannot have a fraction of a man or a woman, but the expected frequencies are not real or observed. They represent the number of men or women we would have obtained if the column percentages were the same for men and women, or if these variables were statistically independent.

Let's try another example. What is the expected frequency for women who are atheist? Use the formula and see if you can figure it out.

$$F_e = \left(\frac{245}{2,576}\right)(1,195) = 113.7$$

If there were no relationship in the population between gender and religiousness, our cross-tabulation would have shown 113.7 women who are atheist. From our sample, we actually observed a frequency of 90—that is, 90 women who are atheist. We call this the **observed frequency**.

We use this same method to calculate the expected frequencies for the four other cells in the cross-tabulation. These expected frequencies are shown in Table 12.8 (column 2). The table also displays the observed frequency for each cell (column 1) along with other calculations required to conduct the chi-square test of independence, which we will describe below.

How close are these expected frequencies to our observed frequencies? That question is at the essence of the chi-square test. For each cell, the test compares the observed frequency to the expected frequency. When, in the aggregate, the expected and observed frequencies are "close" to each other, we conclude that there is probably no relationship in the population between independent and dependent variables. When, on the other hand, there are large discrepancies between observed and expected frequencies, we are more likely to reject the null hypothesis and conclude that there is an association between the two variables in the population.

Table 12.8 Calculation of the Chi-Square Statistic, Effect of Gender on Religiousness

	1	2	3	4	5
	F_o	F_e	$F_o - F_e$	$(F_o - F_e)^2$	$\dfrac{(F_o - F_e)^2}{F_e}$
Men who are religious	718	763.9	−45.9	2,106.81	2.76
Men who are not religious	508	485.7	22.3	497.29	1.02
Men who are atheist	155	131.3	23.7	561.69	4.28
Women who are religious	707	661.1	45.9	2,106.81	3.19
Women who are not religious	398	420.3	−22.3	497.29	1.18
Women who are atheist	90	113.7	−23.7	561.69	4.94
$\chi^2 = \Sigma \dfrac{(F_o - F_e)^2}{F_e}$					17.37

Step 4: Calculate the chi-square statistic. This has three additional steps.

Calculate the difference between expected and observed frequencies and square each difference.

For each cell, subtract the expected frequency from the observed frequency. In statistical notation, this is $(F_o - F_e)$; these differences are displayed in column 3 of Table 12.8. We then square each difference $(F_o - F_e)^2$; these squared differences appear in column 4 of Table 12.8. For example, among men who are religious, the difference between observed and expected frequency is -45.9 (718 − 763.9). When we square that difference, we get 2,106.81.

Divide each squared difference by the expected frequency.

For each cell, take the squared difference that you calculated in Step 4 and divide it by the expected frequency. The resulting quotient for each cell is known as the chi-square component. These are displayed in column 5 of Table 12.8.
Among men who are religious, the chi-square component is 2.76, found by: ($\dfrac{2,106.81}{763.9}$).

Calculate the chi-square statistic by summing up the chi-square components.

The next step is to sum up the chi-square components. The final sum is the chi-square statistic. Here is the formula:

$$\chi^2 = \Sigma \dfrac{(F_o - F_e)^2}{F_e}$$

In Table 12.8, the chi-square statistic is displayed at the bottom of column 5. In this example, the sum of the six chi-square components is 17.37. This is the chi-square statistic. It will always be a positive number.

If observed and expected frequencies were equal to each other—the condition specified in our null hypothesis—the chi-square statistic would be 0.

Step 5: Compare your decision chi-square to your sample chi-square. If sample chi-square > decision chi-square, reject the null hypothesis. OR, if you are using statistical software that generates a p-value, compare the p-value associated with your chi-square to alpha. If p < α, reject the null hypothesis.

How do we interpret the chi-square statistic that we have just calculated and make a decision about our hypotheses? To answer that question, we return to Step 2 of the hypothesis test to find the decision chi-square at our alpha level and degrees of freedom. As we have seen, the chi-square statistic is built from the difference between observed and expected frequencies in our sample, and it allows us to determine whether these observed and expected frequencies are sufficiently far apart so that we can claim a significant relationship in the population.

The distribution of the chi-square statistic follows a known curve. It is similar to the F-distribution (chapter 11) in that it is always a positive number, positively skewed, one-sided, and takes on a shape that varies depending on the degrees of freedom.* As degrees of freedom increases, the chi-square distribution becomes less skewed. Low chi-square values will fall near the null hypothesis. A chi-square statistic equal to 0 ($\chi^2 = 0$) would mean that the column percentages were identical across categories of the independent variable, and you would conclude that the variables are statistically independent of each other in the population. As the chi-square statistic moves further away from 0, the p-value decreases because it is progressively less likely that a chi-square of that value could be obtained if there were no relationship between the variables in the population and the likelihood of rejecting the null hypothesis increases. Figure 12.1 displays the shape of the chi-square distribution, at various degrees of freedom.

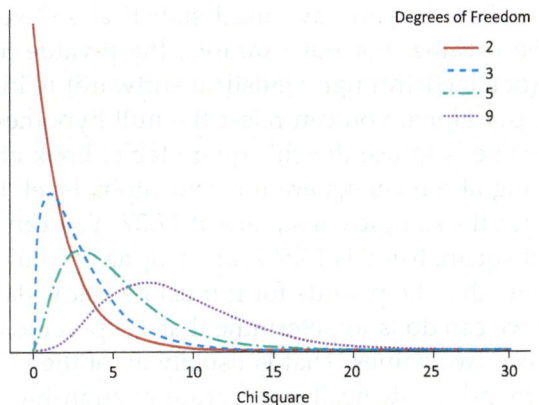

Figure 12.1 Shape of Chi-Square Distribution at Various Degrees of Freedom

---

* The chi-square test is always a one-tailed test. The null hypothesis is always that $\chi^2 = 0$ (in other words, that the column percentages in the population are equal to each other.) A χ^2 equal to 0 can be found at the left end of the curve. Deviations from that null hypothesis will always be positive and move the chi-square statistic away from 0 in a positive direction.

Our example has two degrees of freedom (Look back at Step 2, page 466, to see this.)

$$DF = (r - 1)(c - 1) = (3 - 1)(2 - 1) = 2$$

The chi-square table in appendix D is similar to the t-table. The left column contains the degrees of freedom. The top row displays the p-values. And the chi-square statistics are shown in the body of the table. We have excerpted a portion of the chi-square table in Figure 12.2.

DF	\multicolumn{5}{c}{P-Value}				
	0.1	.05	.025	.01	.005
1	2.706	3.841	5.024	6.635	7.879
2	4.605	5.991	7.378	9.210	10.597
3	6.251	7.815	9.348	11.345	12.838
50	63.167	67.505	71.420	76.154	79.490

Figure 12.2 Using the Chi-Square Table to Find Chi-Square at Alpha = 0.05 and Degrees of Freedom = 2

Looking in the chi-square table for DF = 2 and alpha = 0.05, we see the decision chi-square is 5.991.

In our example, we obtained a chi-square statistic equal to 17.37, with 2 degrees of freedom. That is greater than our decision chi-square, so we can reject the null hypothesis at this alpha level (0.05).

Let's approach that decision from another angle. Instead of comparing a sample statistic to a decision statistic, you have seen in previous chapters that you can also compare the p-value for your sample statistic to your alpha level. But how do you know the p-value? There are two main ways. First, if you have used statistical software for your analysis, the output will report the p-value. For our example, the p-value for the religiousness/gender cross-tabulation (obtained through statistical software) is infinitesimal (less than 0.001). You know that, if p < alpha, you can reject the null hypothesis. The other way to at least estimate your p-value is to use the chi-square table. Look at the table again, but, this time, instead of looking at the chi-square for your alpha level, look through the line for 2 degrees of freedom for the sample chi-square of 17.37. You can see that, at 2 degrees of freedom, the largest chi-square listed is 10.597, at an alpha of 0.005. Since our chi-square is larger than that, we know that the p-value for it must be below 0.005. When we use the chi-square table, the best we can do is to determine that the p-value is greater than or less than some value or between two values. That is usually all of the information we need to decide whether we have found a statistically significant relationship.

In order to visualize what this means in our sample, look at the chi-square distribution for 2 degrees of freedom, shown in Figure 12.3.

As we can see, our sample chi-square statistic (17.37) is in the right tail of the curve. As we know, the corresponding p-value is essentially 0. With a p-value well below our alpha threshold of 0.05, we reject the null hypothesis. We can reach the same conclusion by noting that our sample chi-square statistic, 17.37, is larger than the decision chi-square value,

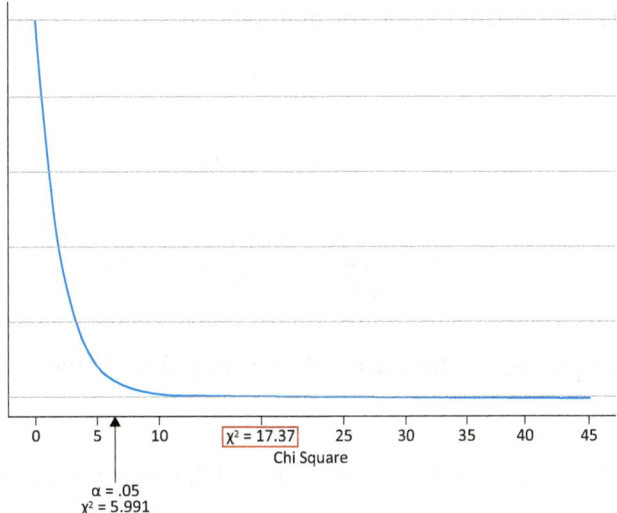

Figure 12.3 Shape of Chi-Square Distribution at 2 Degrees of Freedom

which was 5.991. We can conclude that there is a significant relationship between these two variables in the population. With a p-value of essentially 0, it would have been virtually impossible to obtain the sample differences that we found if there were no relationship in the population between gender and degree of religiousness. We conclude that gender and religiousness are significantly related.

Size and Direction of Effects: Analysis of Residuals

It is important to note that the chi-square statistic and its corresponding p-value address only the question of whether the two variables are related. Once a chi-square statistic indicates that two variables are related, we can take the analysis one step further and calculate the **standardized residuals** for each cell. These residuals will tell us where we find the largest deviations between what we expected and what we observed and also the direction of that deviation. By doing so, the residuals show the relative contributions of each cell to the overall chi-square statistic. Here is the formula for computing the standardized residuals:

$$R = \frac{F_o - F_e}{\sqrt{F_e}}$$

where:

R = standardized residual value
F_o = observed frequency
F_e = expected frequency

Let's compute the residual value for men who are religious (see Table 12.8 for values):

$$F_o = 718$$

$$F_e = 763.9$$

$$R = \frac{F_o - F_e}{\sqrt{F_e}} = \frac{718 - 763.9}{\sqrt{763.9}} = -1.7$$

Repeating this procedure for each cell, we populate Table 12.9 with these standardized residuals.

Table 12.9 Standardized Residuals, Effect of Gender on Religiousness

	Men	Women
Religious	R = −1.7	R = +1.8
Not religious	R = +1.0	R = −1.1
Atheist	R = +2.1	R = −2.2

Source: World Values Survey (Wave 7).

Some of the residuals in Table 12.9 are positive, and some are negative. The positive ones indicate that the observed frequencies were above the expected frequencies. Negative residuals tell us that the observed frequencies were below the expected frequencies. The absolute value of the residuals denotes the relative contribution that the cell makes to the results of the chi-square test. In our example, we have found the largest residuals in the categories of "atheist" and "religious." What does this tell us? It suggests that we have found a statistically significant relationship largely because it is in these particular categories that our observed and expected frequencies show the largest deviations. Or, in context, it is in these categories that men and women differ most sharply. The smaller residuals in the middle category—among those who say they are not religious—reflect column percentages for men and women that were fairly close to each other (about 3 points apart). As a general rule, when the absolute value of a chi-square standardized residual equals or exceeds 2.0, one can conclude that this particular category is an important contributor to the finding of a statistically significant relationship.

Looking at the chi-square statistic and the residuals, we then complete our interpretation by explaining how men and women differ in their degree of religiousness. We can see from the column percentages that women are more likely than men to say they are religious and men are more likely than women to say they are atheists. The residuals tend to support this finding, though the absolute value of the residuals for religious respondents is just below 2.0.

Example: Gender and Perceptions of Health

Let's put all of this together by looking at one other example. We will examine the effect of gender on Americans' perceptions of their own health (from the World Values Survey, Wave 7). The data are shown in Table 12.10.

Table 12.10 Effect of Gender on Perceptions of Health

	Men	Women	Total
Very good health	24.1% (333)	22.1% (266)	23.2% (599)
Good health	49.4% (684)	49.7% (598)	49.6% (1,282)
Fair health	21.5% (297)	22.9% (276)	22.1% (573)
Poor or very poor health	5.1% (70)	5.2% (63)	5.1% (133)
Total	100% (1,384)	100% (1,203)	100% (2,587)

(χ^2 = 1.75, DF = 3, p-value = 0.63)
Source: World Values Survey (Wave 7).

Men are a little more likely than women to say they are in very good health; women are slightly more likely to say that they are in fair health; but the column percentages are very close to each other. The null hypothesis, as usual, is that there is no relationship between gender and perceptions of health. Set an alpha of 0.05. Because Table 12.10 provides the chi-square and its p-value, there is no need to find a decision chi-square. We can just compare the p-value to alpha directly. The chi-square information is reported in the parentheses below the table. Most important to note is the high p-value (0.63). This means that, if the null hypothesis were true—that is, if gender and perceptions of health were independent of each other in the population—there would be a 63% chance of obtaining the sample differences we obtained. Figure 12.4 illustrates our finding.

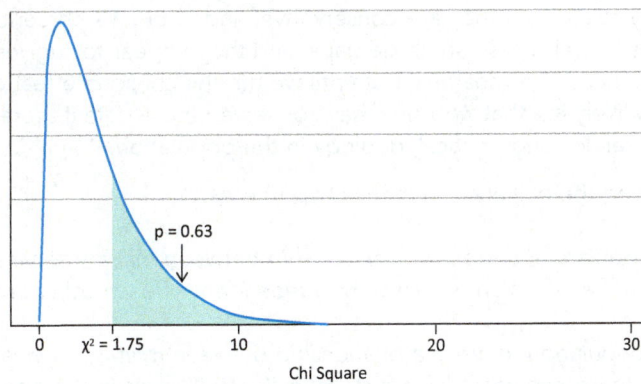

Figure 12.4 Chi-Square Distribution for 3 Degrees of Freedom

By any measure, this is a very high p-value, and it precludes us from rejecting the null hypothesis. Our conclusion is that, in spite of minor differences found in the sample, there is just not enough evidence to assert that gender influences perceptions of one's own health in the population. We could also say that there is no statistically significant difference between men and women in their perceptions of their own health.

BOX 12.1: APPLICATION

Using a Chi-Square Test to Investigate the Relationship between Military Service and Political Ideology

The 2020 American National Election Study asked respondents to place themselves on a liberal–conservative scale. It also asked respondents if they have ever served in the military. Table 12.11 presents the results, in a cross-tabulation. Military service is the independent variable, and ideology is the dependent variable. Each cell contains the column percentage and the observed frequency.

Table 12.11 Effect of Military Service on Political Ideology

	Served in Military	Never Served in Military	Total
Liberal	22.1% $F_o = 164$	37.7% $F_o = 2,121$	35.8% $F_o = 2,285$
Moderate	24.1% $F_o = 179$	25.7% $F_o = 1,448$	25.5% $F_o = 1,627$
Conservative	53.8% $F_o = 400$	36.6% $F_o = 2,062$	38.6% $F_o = 2,462$
Total	100% 743	100% 5,631	100% 6,374

Source: 2020 American National Election Study.

A quick look at Table 12.11 suggests that there are differences between respondents who have served in the military and those who have not. Military respondents are about 17 percentage points more likely to say that they are conservative, and about 15 percentage points less likely to call themselves liberal. These are large gaps, and they appear to suggest that there is a relationship between these two variables. But until we run the chi-square test of independence, we cannot know how likely it is that we would have observed these data if there were no relationship between military service and political ideology in the population.

We begin by stating the null and alternative hypotheses:

> H_0: In the population, there is no relationship between military service and ideology. Or, these two variables are statistically independent of each other in the population.

> H_a: In the population, there is a relationship between military service and ideology. Or, these two variables are statistically dependent in the population.

We then calculate the expected frequency for each cell, along with the other elements required to conduct the chi-square test of independence. These are all shown in Table 12.12.

The chi-square statistic is 95.6. Our dependent (row) variable has three categories, and our independent (column) variable has two categories. So our degrees of freedom is (3 − 1) × (2 − 1), or 2.

Table 12.12 Calculation of the Chi-Square Statistic, Effect of Military Service on Ideology

	F_o	F_e	$F_o - F_e$	$(F_o - F_e)^2$	$\dfrac{(F_o - F_e)^2}{F_e}$
Military liberals	164	266.4	−102.4	10,485.76	39.36
Military moderates	179	189.7	−10.7	114.49	0.60
Military conservatives	400	287	113	12,769	44.5
Non-military liberals	2,121	2018.6	102.4	10,485.76	5.19
Non-military moderates	1,448	1,437.3	10.7	114.49	0.08
Non-military conservatives	2062	2,175	−113	12,769	5.87
$\chi^2 = \sum \dfrac{(F_o - F_e)^2}{F_e}$					95.6

We can consult the chi-square table at the back of the book to find the p-value. With a chi-square statistic equal to 95.6 and with degrees of freedom equal to 2, we find that our p-value is less than 0.005. We can conclude, as we suspected, that military service and ideology are indeed related in the population.

Our final step would be to calculate the standardized residuals for this chi-square analysis, following the formula on page 472. This will allow us to isolate which factors in particular account for our finding of statistical significance. Table 12.13 presents the standardized residuals.

Table 12.13 Standardized Residuals, Effect of Military Service on Political Ideology

	Served in Military	Never Served in Military
Liberal	$R = -6.3$	$R = +2.3$
Moderate	$R = -0.8$	$R = +0.3$
Conservative	$R = +6.7$	$R = -2.4$

Recall our rule of thumb: that a standardized residual of 2 or more (absolute value) is a major contributor to a finding of statistical significance. In this example, there are a lot more military conservatives and a lot fewer military liberals than we expected. It is above all because of these differences that we find military service to be a strong predictor of ideology.

BOX 12.2: IN DEPTH

The Relationship between Chi-Square and Two-Sample z-Tests for 2 × 2 Tables

There is an important connection between the normal distribution and the chi-square distribution with 1 degree of freedom (i.e., in a 2 × 2 cross-tabulation). For a 2 × 2 table, a chi-square statistic will give you the exact same p-value as a two-tailed z-test of the difference between two sample proportions. Think about it: in both cases, you are testing whether the proportion for a dependent variable differs between two categories. While the calculations are different, the end result is the same.

Here is an example. In 2021, the General Social Survey asked respondents if they think a book written by a communist should be allowed in a library. Table 12.14 shows the distribution of responses, broken down by gender.

Table 12.14 Effect of Gender on Allowing a Communist's Book in the Library

	Men	Women	Total
Remove book	21.4% (126)	22.6% (163)	22.0% (289)
Do not remove book	78.6% (464)	77.4% (558)	78.0% (1,022)
Total	100% (590)	100% (721)	100% (1,311)

(χ^2 = 0.30, DF = 1, p = 0.60)
Source: 2021 General Social Survey.

Although women are more likely to support removing a communist's book, the difference is very slight (0.226 − 0.214 = 0.012), and the low chi-square statistic (χ^2 = 0.30) and the corresponding high p-value (0.60) indicate that this is not a statistically significant relationship in the population.

Now let's examine the same problem by means of a two-sample hypothesis test (covered in chapter 10).* Our null hypothesis is that, in the population, there is no difference in the proportion of men and women who support removing a communist's book, or $H_0: \pi_{men} - \pi_{women} = 0$. Our alternative hypothesis is that, in the population, there is a difference in support for book removal between men and women. Or,

$$H_a: \pi_{men} - \pi_{women} \neq 0.$$

$$SE_{\text{difference between p(women) and p(men)}} = \sqrt{\frac{0.222(0.778)}{590} + \frac{0.222(0.778)}{721}} = 0.23$$

---

* Note that we use the pooled proportion to conduct this hypothesis test on the difference between two proportions, as explained in chapter 10.

$$z = \frac{(0.226 - 0.214) - 0}{0.023} = \frac{0.012 - 0}{0.023} = 0.52$$

We now turn to the normal table and find that a z-score of 0.52 is associated with a one-tailed p-value of 0.30. Since our alternative hypothesis is two-sided, we must double our p-value. Our final result is 2 × 0.30 = 0.60. This is the same p-value we obtained using the chi-square distribution, with 1 degree of freedom.

It is always the case that the two-tailed p-value associated with any z-score is the same as the p-value associated with a chi-square statistic that is the square of that z-score.

So when you have a 2 × 2 table, you have a choice between conducting a two-sample hypothesis test with the normal distribution or conducting a chi-square test of independence. The results are equivalent, so long as the alternative hypothesis is two-tailed.

Assumptions of Chi-Square

There are four basic assumptions that must be met in order to use the chi-square test of independence:

- The sample must be randomly selected. Cases must be independent of each other.
- Both variables should be nominal or ordinal.[*]
- We must have observed frequencies. (We cannot compute chi-square statistics on the basis of percentages.)
- The sample size must be large enough. The rule we employ here is that the expected frequencies must be greater than 5 in every cell of the cross-tabulation. If the expected frequency is lower than 5 in even one cell, the chi-square statistic is not accurate. If you encounter this problem, you may be able to recode a variable to combine categories in order to obtain larger expected frequencies in small cells. Most software programs will alert you if this assumption is violated.

Statistical Significance and Sample Size

In some of our earlier discussions in this book, we have made a point of distinguishing between statistical and practical significance. In chapter 10, for example, we argued that statistical significance does not automatically equate with practical significance. With a large sample, one might reject a null hypothesis even though the difference between the two groups is fairly small. Large samples may reveal statistically significant results, but

---

[*] It is technically possible to conduct a chi-square test with interval-ratio-level variables, but variables with numerous categories produce cross-tabulations with many cells, making them very difficult to read; the p-value would then be the only meaningful result of the analysis. In addition, having numerous categories increases the likelihood that some cells will have expected frequencies below 5.

these results may not be of practical significance. This is an important point to consider in connection with the chi-square statistic, which is extremely sensitive to sample size. With the chi-square statistic, we may find that we cannot reject a null hypothesis even though we have found fairly large differences between column percentages, and this is mainly because of smaller sample sizes.

Consider this example, which involves hypothetical data. In Scenario I, we present a cross-tabulation based on a sample size of five hundred respondents (Table 12.15a). In Scenario II, we present a cross-tabulation with the same percentages but with a sample of 1,500 respondents (Table 12.15b). In Tables 12.15a and 12.15b, gender is the independent variable, and party identification is the dependent variable.

Table 12.15a Effect of Gender on Party Identification, Hypothetical Data, Scenario I, N = 500

	Men	Women	Total
Democrat	47% (118)	52% (130)	49.5% (248)
Independent	14% (35)	10% (25)	12% (60)
Republican	39% (97)	38% (95)	38.5% (192)
Total	100% (250)	100% (250)	100% (500)

(χ^2 = 2.27, DF = 2, p = 0.32)

Table 12.15b Effect of Gender on Party Identification, Hypothetical Data, Scenario II, N = 1,500

	Men	Women	Total
Democrat	47% (353)	52% (390)	49.5% (743)
Independent	14% (105)	10% (75)	12% (180)
Republican	39% (292)	38% (285)	38.5% (577)
Total	100% (750)	100% (750)	100% (1,500)

(χ^2 = 6.93, DF = 2, p = 0.03)

The column percentages across tables are identical. In Scenario I (Table 12.15a), for example, 52% of women are Democrats, 10% are Independent, and 38% are Republican. In Scenario II (Table 12.15b), the distribution of partisanship for women is exactly the same. In both samples, women are 5 points more likely than men to identify as Democrat, and men are 4 points more likely than women to call themselves Independent. The only differences between these two tables are the sample size and, consequently, the chi-square statistic and p-value. In the first scenario, with a sample size of five hundred, we found that column percentage differences of 4 and 5 percentage points were not sufficiently large to obtain statistical significance. Our chi-square p-value is 0.32; by any standard this is very high, and it leads us to conclude that, in the population, there is no gender gap on party identification. Yet if we got the exact same results with a sample of 1,500 respondents, we would likely draw the opposite conclusion. Our p-value is 0.03. Assuming an alpha of 0.05, our p-value is sufficiently low to reject the null hypothesis, and, as a result, we would conclude that, in the population, women and men differ with respect to their party identification.

BOX 12.3: IN DEPTH

The Chi-Square Goodness-of-Fit Test

The chi-square statistic can also be used to test whether sample results conform to expected frequencies based on a model rather than expected frequencies based on no relationship between the two variables. In that case, it is called a **chi-square test for goodness-of-fit** rather than a chi-square test of independence.

We follow the same steps that we have seen with the chi-square test of independence, with a few modifications. We will illustrate these steps by means of an example.

Our legal system relies on the jury system, and there is an expectation that juries be reflective of the population from which they are drawn. A 2007 study examined the racial representativeness of jury pools in Manhattan.[2] The study compared the racial composition of a sample of 14,229 Manhattan residents that made up criminal and civil jury pools in 2006 to the racial composition of all Manhattan residents based on the 2000 US Census.

1. *State the null and alternative hypotheses.*

   When we conduct a chi-square goodness-of-fit test, our null hypothesis states that the population distribution conforms to a particular model. In this case, our null hypothesis is that the racial composition of the population from which the jury pool is drawn is the same as the racial composition of the population of Manhattan. Table 12.16 shows the racial breakdown of Manhattan residents according to the 2000 Census.

Table 12.16 Racial Composition of Manhattan Borough

Race	Percent
White	54.4%
Black/African American	17.4%
Asian	9.5%
Other	18.7%

Source: 2000 US Census. Several racial categories have been combined to simplify the example. Respondents who identify as Hispanic or Latino can be classified as any racial group in the 2000 US Census.

Here are our hypotheses:

H_0: In the population from which the jury pool is drawn, the racial distribution is the same as it is for the population of Manhattan.

H_a: In the population from which the jury pool is drawn, the racial distribution is not the same as it is for the population of Manhattan.

2. *Set alpha and calculate the degrees of freedom for the cross-tabulation.*

   We will set the alpha-level at 0.05. For a goodness-of-fit test, the degrees of freedom is equal to the number of rows (or categories) in the variable minus one. $DF = (r - 1)$. Since we have four categories of race, degrees of freedom is equal to 3.

3. *Calculate the expected frequencies for each cell of the table.*

   The expected frequencies are the frequencies we would get in our sample if the null hypothesis were true. Or, to put it differently, they are the number of respondents that would be in each cell of the table if the racial composition of the population from which the jury pool is drawn were the same as the racial composition of the population of Manhattan. It is this difference in how expected frequencies are conceptualized and calculated that distinguishes the chi-square goodness-of-fit test from the chi-square test for independence.

   How do we calculate the expected frequency for each cell? We know, for example, that whites constitute 54.4% of the Manhattan population (see Table 12.16). If the null hypothesis is true, whites should make up 54.4% of the sample of 14,229 individuals. Thus, the expected frequency for whites is: 7,740.6.

   We use the same method to find the expected frequencies for the other racial categories. These expected frequencies, along with the other elements necessary to calculate the chi-square statistic, are shown in Table 12.17.

Table 12.17 Calculation of the Chi-Square Statistic, Racial Composition of Manhattan Jury Pools

	F_o	F_e	$F_o - F_e$	$(F_o - F_e)^2$	$\dfrac{(F_o - F_e)^2}{F_e}$
White	11,055	7,740.58	3314.42	10,985,379.94	1,419.19
Black/African American	1,430	2,475.85	−1,045.85	1,093,802.22	441.79
Asian	929	1,351.76	−422.76	178,726.02	132.22
Other	815	2,660.82	−1845.82	3,407,051.47	1,280.45
$\chi^2 = \sum \dfrac{(F_o - F_e)^2}{F_e}$					3,273.65

If we compare the observed frequencies to the expected frequencies, one conclusion stands out. There are a lot more whites in the jury pool than we expected and a lot fewer people from the other racial groups. But do these disparities reflect random sampling variation? Or are they statistically significant?

4. *Calculate the chi-square statistic.*

   To answer these questions, we calculate the chi-square statistic, following the same steps and using the same formula we used with the test of independence.

   $$\chi^2 = \Sigma \frac{(F_o - F_e)^2}{F_e}$$

   All of the calculations are displayed in Table 12.17.

5. *Interpret the chi-square statistic.*

   We have obtained $\chi^2 = 3{,}273.65$ with 3 degrees of freedom. This translates into a p-value of essentially 0. We reject the null hypothesis. We conclude that, in the population from which the jury pool is drawn, the racial distribution is not the same as it is for the population of Manhattan. There appears to be evidence of racial bias in the composition of jury pools.

Chapter Summary

In this chapter, we learned how to conduct and interpret the chi-square test of independence. Below, we review the steps involved in conducting this test and other concepts covered in the chapter.

- A **chi-square test of independence** is used to assess the statistical significance of a relationship presented in a cross-tabulation. It examines whether there is a relationship between two variables by assessing the differences between observed frequencies and the frequencies that would be expected if there were no relationship between the variables.

- The **expected frequency** for each cell is the number of respondents that would be in each cell if there were no relationship between the two variables (i.e., if the column percentages were identical).

- The **observed frequency** is the number of cases in a cell.

- The chi-square test is very sensitive to sample size and cannot be conducted if expected frequency in any cell is less than 5.

- The chi-square test of independence follows these steps:

  1. State the null and alternative hypotheses. Our null hypothesis asserts that, in the population, there is "statistical independence" between two variables, or that there is no relationship between the two variables. The alternative hypothesis states that, in the population, the two variables are associated with each other.

  2. Set the alpha level and calculate the degrees of freedom. If using, find decision chi-square in the chi-square table.

  $$DF = (rows - 1) \times (columns - 1)$$

  3. Calculate the expected frequencies for each cell.

  $$F_e = \left(\frac{row\ frequency}{total\ frequency}\right)(column\ frequency)$$

  4. Calculate chi-square. By formula:

  $$\chi^2 = \sum \frac{(F_o - F_e)^2}{F_e}$$

  Step by step, to calculate chi-square:

  - For each cell, subtract the expected frequency from the observed frequency and square each difference. $(F_o - F_e)^2$

- Divide each squared difference by the expected frequency. The result is the chi-square component, one for each cell: $\dfrac{(F_o - F_e)^2}{F_e}$
- Add up the chi-square components. The sum is the chi-square statistic.
- Compare your decision chi-square to your sample chi-square. Alternatively, compare the p-value associated with your chi-square to alpha. (Use statistical software or a chi-square table to obtain the p-value.) If sample chi-square > decision chi-square OR if p < α, reject the null hypothesis.
- **Standardized residuals** show the size and direction of the difference between the expected and observed frequency for each cell, indicating the relative contribution of each cell to the overall chi-square statistic. By formula:

$$R = \frac{F_o - F_e}{\sqrt{F_e}}$$

where F_o = observed frequency and F_e = expected frequency.

- The **chi-square test for goodness-of-fit** assesses whether sample results conform to expected frequencies based on a model rather than expected frequencies based on no relationship between the two variables. It is calculated the same way as the chi-square test for independence, using the expected frequencies for the model instead of the expected frequencies for no relationship.

Using Stata

In this section, we use data from the 2021 General Social Survey (GSS) to assess the relationship between people's level of education and their political views. We created a variable called *edlevel* that divides years of formal education into four categories: high school or less, some college, college degree, and graduate degree.* We have also collapsed a seven-category measure of political views into a variable, called *libcons*, with three categories: liberal, moderate, and conservative.**

In this chi-square analysis, we treat *edlevel* as the independent variable and *libcons* as the dependent variable. The null and alternative hypotheses for this test are:

H_0: In the population, there is no relationship between level of education and political views.

H_a: In the population, there is a relationship between level of education and political views.

---

* *Edlevel* is derived from the original variable called *educ* in the GSS.
** *Libcons* is derived from the original variable called *polviews* in the GSS.

Open `GSS2021.dta` and enter the following command into Stata to run the chi-square test:

`tabulate libcons edlevel, column chi2 expected cchi2`

Just as we saw in chapter 3, we use the "tabulate" command to request a cross-tabulation of our two variables, listing the dependent variable (*libcons*) before the independent variable (*edlevel*). Just as before, we use the "column" option to tell Stata to calculate column percentages. The next option, "chi2," tells Stata that we want it to perform a chi-square test on this contingency table. The next two options, "expected" and "cchi2," request detailed information from the chi-square test. "expected" tells Stata to include the expected frequencies in each cell, or the frequencies that we would observe if there were no relationship between level of education and political views. "cchi2" tells Stata to include the contribution that each cell makes to the chi-square statistic.[*]

The output for the chi-square test is shown in Figure 12.5.

```
. tabulate libcons edlevel, column chi2 expected cchi2
```

Key

| frequency |
| expected frequency |
| chi2 contribution |
| column percentage |

	edlevel				
libcons	HS or les	some coll	college d	graduate	Total
Liberal	226	247	385	462	1,320
	344.0	339.0	311.4	325.7	1,320.0
	40.5	25.0	17.4	57.1	139.9
	21.88	24.26	41.18	47.24	33.30
Moderate	440	416	268	253	1,377
	358.8	353.6	324.8	339.7	1,377.0
	18.4	11.0	9.9	22.1	61.4
	42.59	40.86	28.66	25.87	34.74
Conservative	367	355	282	263	1,267
	330.2	325.4	298.9	312.6	1,267.0
	4.1	2.7	1.0	7.9	15.6
	35.53	34.87	30.16	26.89	31.96
Total	1,033	1,018	935	978	3,964
	1,033.0	1,018.0	935.0	978.0	3,964.0
	62.9	38.7	28.3	87.1	217.0
	100.00	100.00	100.00	100.00	100.00

Pearson chi2(6) = 216.9758   Pr = 0.000

Figure 12.5

The output shows the cross-tabulation of *edlevel* and *libcons*. Because we specified a number of options in the command, each cell of the table contains quite a bit of information. Thus, it is important to consult the key in the upper-left corner of the output (circled), which tells us what each cell entry means, as we examine the table. For each cell, the top entry is the number of

---

[*] Note that we have not asked Stata to produce the residuals for each cell. Stata does not offer this as an option for chi-square tests.

respondents that fall at the intersection of those two categories (or the observed frequency), the second entry is the frequency that we would have observed if the null hypothesis were true, the third entry is the cell's contribution to the overall chi-square statistic, and the final value is the column percentage. For example, if we examine the upper-left cell in the body of the table (circled), we can see that 226 respondents completed high school or less and identify as liberal; we would have expected 344 respondents in this cell if the null hypothesis were true. The contribution of the cell to the chi-square is 40.5, and 21.9% of those who completed high school or less identify as liberal. (If you wanted to generate a streamlined version of this table, you could omit the "expected" and "cchi2" options from the command.)

Examining the column percentages (the last entry in each cell) across levels of education shows some large differences, especially among liberals and conservatives. For example, almost half of those with a graduate degree identify as liberal compared to about one-quarter of those who completed some college. But do these differences yield a statistically significant chi-square statistic?

Below the table (circled), we see the chi-square statistic, 217, at 6 degrees of freedom and its associated p-value = 0.000. If there were no relationship between level of education and political views, we would have been very unlikely to observe the data that we see in the cross-tabulation.

Review of Stata Commands

- Conduct a chi-square test, with options to view expected values and contributions to chi-square statistic in each cell:

    ```
    tabulate dependent variable independent variable, column chi2 expected cchi2
    ```

Using SPSS

In this section, we use data from the 2021 General Social Survey (GSS) to assess the relationship between people's level of education and their political views. We created a variable called *edlevel* that divides years of formal education into four categories: high school or less, some college, college degree, and graduate degree.[*] We have also collapsed a seven-category measure of political views into a variable, called *libcons*, with three categories: liberal, moderate, and conservative.[**]

In this chi-square analysis, we treat *edlevel* as the independent variable and *libcons* as the dependent variable. The null and alternative hypotheses for this test are:

H_0: In the population, there is no relationship between level of education and political views.

---

[*] *Edlevel* is derived from the original variable called *educ* in the GSS.
[**] *Libcons* is derived from the original variable called *polviews* in the GSS.

H_a: *In the population, there is a relationship between level of education and political views.*

In chapter 3, we learned how to generate a cross-tabulation in SPSS. Here, we will generate another cross-tabulation, using the same method. But, this time, we will tell SPSS to generate the relevant chi-square statistics.

Open `GSS2021.sav`, To generate a cross-tabulation in SPSS, we use the following procedure:

`Analyze → Descriptive Statistics → Crosstabs`

Remember to put the dependent variable (*libcons*) into the "Row(s)" box and the independent variable (*edlevel*) into the "Column(s)" box (shown in Figure 12.6). We click on "Statistics."

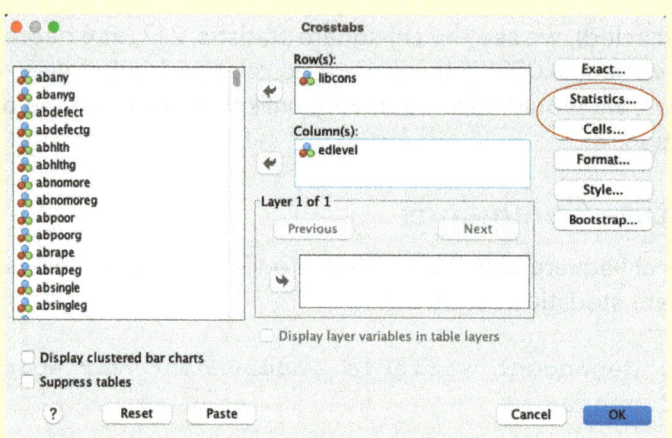

Figure 12.6

In the "Crosstabs: Statistics" dialog box that opens (shown in Figure 12.7), we check the small box next to "Chi-Square."

Click on "Continue"; this brings us back to the "Crosstabs" dialog box (see Figure 12.6). We now click on "Cells" so that we can tell SPSS exactly what it should include in each cell of the cross-tabulation. Recall that, by default, SPSS includes the observed frequencies (or counts). We will tell SPSS to also generate the expected frequencies, the column percentage, and the standardized residuals. To do that, we place a check mark next to each of these statistics, shown in Figure 12.8.

We click on "Continue" and then "OK." This generates the output shown in Figure 12.9.

The top portion of the output shows the cross-tabulation of *edlevel* and *libcons*. Because we specified a number of statistics in the "Cells" box, each cell of the table contains quite a bit of information. For each cell, the top entry is the number of respondents that fall at the intersection of those two categories (or the observed frequency), the second entry is the frequency that we would have observed if the null hypothesis were true, the

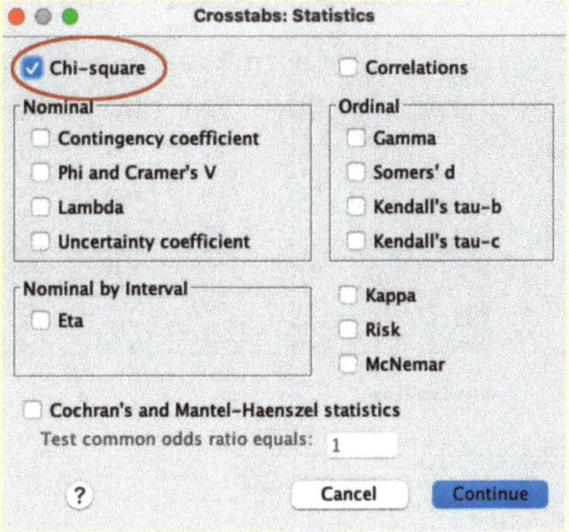

Figure 12.7

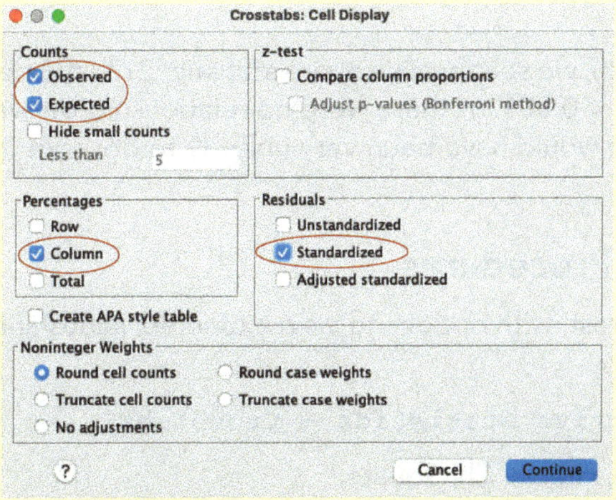

Figure 12.8

third entry is the column percentage, and the final value is the standardized residual. For example, if we examine the upper-left cell in the body of the table (circled), we can see that 226 respondents completed high school or less and identify as liberal; we would have expected 344 respondents in this cell if the null hypothesis were true. Of respondents who completed high school or less, 21.9% identify as liberal, and the standardized residual for this cell is –6.4.

Examining the column percentages (the third entry in each cell) across levels of education shows some large differences, especially among liberals and conservatives. For example, almost half of those with a graduate degree identify as liberal compared to about one-quarter of those who completed some college. But do these differences yield a statistically significant chi-square statistic?

libcons * edlevel Crosstabulation

			HS or less	some college	college degree	graduate degree	Total
libcons	Liberal	Count	226	247	385	462	1320
		Expected Count	344.0	339.0	311.4	325.7	1320.0
		% within edlevel	21.9%	24.3%	41.2%	47.2%	33.3%
		Standardized Residual	-6.4	-5.0	4.2	7.6	
	Moderate	Count	440	416	268	253	1377
		Expected Count	358.8	353.6	324.2	339.7	1377.0
		% within edlevel	42.6%	40.9%	28.7%	25.9%	34.7%
		Standardized Residual	4.3	3.3	-3.2	-4.7	
	Conservative	Count	367	355	282	263	1267
		Expected Count	330.2	325.4	298.9	312.6	1267.0
		% within edlevel	35.5%	34.9%	30.2%	26.9%	32.0%
		Standardized Residual	2.0	1.6	-1.0	-2.8	
Total		Count	1033	1018	935	978	3964
		Expected Count	1033.0	1018.0	935.0	978.0	3964.0
		% within edlevel	100.0%	100.0%	100.0%	100.0%	100.0%

Chi-Square Tests

	Value	df	Asymptotic Significance (2-sided)
Pearson Chi-Square	216.976[a]	6	<.001
Likelihood Ratio	218.591	6	<.001
Linear-by-Linear Association	116.514	1	<.001
N of Valid Cases	3964		

a. 0 cells (0.0%) have expected count less than 5. The minimum expected count is 298.85.

Figure 12.9

Below the table (circled), we see the chi-square statistic, 217, at 6 degrees of freedom and its associated p-value, < 0.001.[*] If there were no relationship between level of education and political views, we would have been very unlikely to observe the data that we see in the cross-tabulation.

Review of SPSS Procedures

Conduct a chi-square test, with options to view expected values and standardized residuals in each cell:

`Analyze → Descriptive Statistics → Crosstabs`

In the "Statistics" box, check "Chi-Square."

In the "Cells" box, check "Expected Frequencies," "Column Percentages," and "Standardized Residuals."

Practice Problems

1. In 2022, the Pew Research Center found that 5,697 people in a sample of 12,147 US adults got at least some of their news from social media. Imagine that a researcher wanted to use these data to test whether there was a relationship between age and getting news from social media. She divided the sample into four age categories:

---

[*] By default, SPSS generates several other chi-square test statistics. The chi-square test we cover in this chapter is the Pearson chi-square test. This is in the top row of the chi-square output.

people ages eighteen to twenty-nine, thirty to forty-nine, fifty to sixty-four, and sixty-five and older. Table 12.18 shows the number of people in each of the age groups.

Table 12.18   Expected Frequencies for Cross-Tabulation of Age and Getting News from Social Media

	18 to 29	30 to 49	50 to 64	65+	Total
Got news from social media	_____	_____	_____	_____	5,697
Did not get news from social media	_____	_____	_____	_____	6,450
Total	1,477	4,157	3,377	3,136	12,147

a. The researcher conducts a chi-square test of independence. Identify the independent and dependent variables.
b. What is the null hypothesis for this test?
c. Find the decision chi-square at a 0.05 alpha level.
d. Enter the expected frequencies for of each cell shown in Table 12.18.
e. Table 12.19 shows a cross-tabulation of age and getting news from social media adapted from Pew Research data, with observed frequencies and percentages.[3] Follow each of the three steps to calculate the sample chi-square statistic. Show your work for each step.
f. Should the null hypothesis be rejected? Explain how you know.
g. State your findings in words.

Table 12.19   Cross-Tabulation of Age and Getting News from Social Media

	18 to 29	30 to 49	50 to 64	65+	Total
Got news from social media	76% 1,122	58% 2,411	39% 1,317	27% 847	5,697
Did not get news from social media	24% 355	42% 1,746	61% 2,060	73% 2,289	6,450
Total	1,477	4,157	3,377	3,136	12,147

Source: Adapted from Pew Research Center, 2022.

2. Racial segregation in housing is widespread in the United States and a basis for many social science studies and interventions. The 2021 GSS asked white respondents if any Black people lived in their neighborhood and asked Black respondents if any white people lived in their neighborhood. Those who answered "no" to this question indicate an extreme level of racial segregation. Table 12.20 shows the number of respondents who answered "yes" or "no" to the racial segregation question and the number in each of the two racial categories.

Table 12.20 Expected Frequencies for Cross-Tabulation of Race and Whether Any People of the Other Race Live in One's Neighborhood, for Blacks and Whites

Any of Other Race Live in Neighborhood	White	Black	Total
Yes			2,635 82.7
No			553 17.3
Total	2,740 100.00	448 100.00	3,188 100.00

a. Which variable is the dependent variable?
b. For a chi-square test of independence, what is the null hypothesis?
c. Find the decision chi-square at a 0.05 alpha level.
d. Enter the expected frequencies for of each cell shown in Table 12.20
e. Table 12.21 shows the results of the cross-tabulation of these two variables. Follow each of the three steps to calculate the sample chi-square statistic. Show your work for each step.
f. Should the null hypothesis be rejected? Explain how you know.
g. State your findings in words.

Table 12.21 Cross-Tabulation of Race and Whether Any People of the Other Race Live in One's Neighborhood, for Blacks and Whites

Any of Other Race Live in Neighborhood	White	Black	Total
Yes	2,221 81.1	414 92.4	2,635 82.7
No	519 18.9	34 7.6	553 17.3
Total	2,740 100.00	448 100.00	3,188 100.00

3. A phenomenon known as "summer melt" occurs when students who were admitted and scheduled to start college directly after high school do not show up for their first semester of college. A researcher for the University of Indiana draws a random sample of 2,800 recent high school graduates that is representative of the entire state of Indiana. Figure 12.10 shows the relationship between family income and summer melt in this statewide sample.

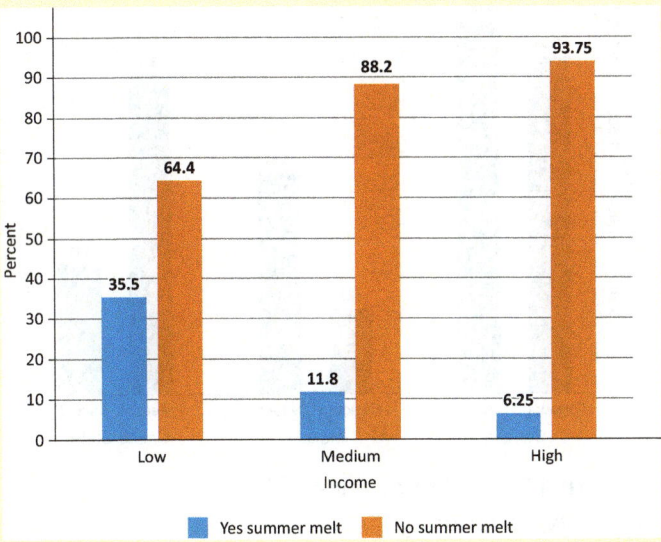

Figure 12.10 Summer Melt for Recent High School Graduates, by Family Income, Percentages (State-Wide Sample)

    a. Which group is most likely to experience summer melt in this statewide sample, according to Figure 12.10?

    b. This sample includes 900 students from low-income families, 1,100 from middle-income families, and 800 from high-income families. Determine the observed frequency for each of the six groups.

    c. The sample chi-square statistic for the relationship between family income and summer melt shown in Figure 12.10 is 276.7. Does this mean that the observed frequencies that you found for part b are different enough from the expected frequencies to conclude that there is a statistically significant relationship between the variables? Explain how you know.

    d. A college that draws most of its students from within the state wants to address the summer melt issue. Based on the findings from the statewide data, would their resources be better spent targeting all students who have accepted offers of admission or focusing specifically on students from specific parts of the income distribution? Explain your answer.

4. The World Values Survey asked respondents which is more important, freedom or security. Figure 12.11 shows the frequencies for the countries of North America—the United States, Mexico, and Canada.

    a. For each country, which value is more important, freedom or security?

    b. Which variable is the independent variable? Sketch a crosstab table with the independent variable in the columns and fill in the frequencies.

    c. Fill in the column percentages for the table.

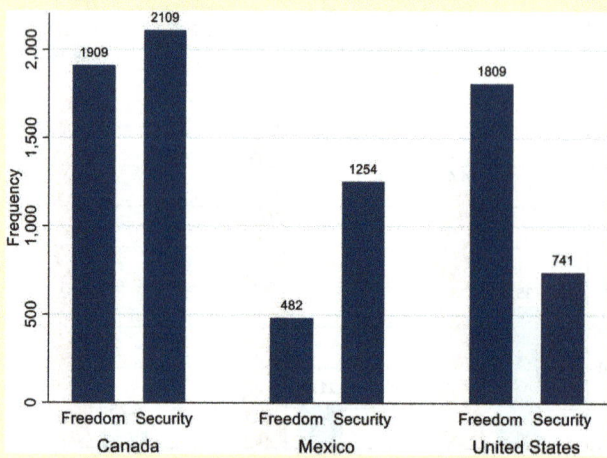

Figure 12.11 Importance of Freedom vs. Security, by Country

d. The sample chi-square statistic for the relationship between country and the importance of freedom versus security is 799.56. What does this indicate about the relationship between the two variables?

e. A researcher concludes that residents of the United States value freedom statistically significantly more than residents of Canada and Mexico. Is this interpretation of the chi-square test accurate? Explain your answer.

5. Table 12.22 shows the observed (top) and expected (bottom) frequencies for each cell of the cross-tabulation from Problem 3.

   a. Use the information in Table 12.22 to calculate the standardized residual for each cell in the body of the table.

   b. Explain why some of the residuals are positive and some are negative.

   c. What do the standardized residuals tell us about which groups contribute most strongly to the chi-square statistic?

Table 12.22 Observed (Top) and Expected (Bottom) Frequencies for Summer Melt for Recent High School Graduates, by Family Income

Summer Melt?	Income			Total
	Low	Medium	High	
Yes	320 160.7	130 196.4	50 142.9	500
No	580 739.3	970 903.6	750 616.1	2,300
Total	900	1,100	800	2,800

6. Table 12.23 shows the observed (top) and expected (bottom) frequencies for each cell of the cross-tabulation from Problem 4.

   a. Use the information in Table 12.23 to calculate the standardized residual for each cell in the body of the table.

   b. Explain why some of the residuals are positive and some are negative.

   c. Do the standardized residuals lend support to the claim that residents of the United States value freedom statistically significantly more than residents of Canada and Mexico?

   Table 12.23 Observed (Top) and Expected (Bottom) Frequencies for Importance of Freedom versus Security, by Country

Which Is More Important:	Canada	Mexico	United States	Total
Freedom	1,909	482	1,809	4,200
	2,033	878	1,289	
Security	2,109	1,254	741	4,104
	1,985	858	1,261	
Total	4,018	1,736	2,550	8,304

7. A group of social psychologists studying the effects of power and wealth on people's behavior examined the relationship between the status, or prestige, of vehicles approaching four-way stops and the likelihood that the approaching vehicle would cut off other vehicles that were already stopped at the intersection. They grouped approaching vehicles into five status categories (1 = lowest to 5 = highest), according to the vehicle's make, age, and physical appearance. They also recorded whether each approaching vehicle cut off other vehicles. In the sample of 282 approaching vehicles, 41 were in category 1, 46 in category 2, 110 in category 3, 50 in category 4, and 35 in category 5.

   a. Imagine that one of the members of the research team wants to test the idea that there is no relationship between a vehicle's status and the likelihood that it will cut off other vehicles. What kind of chi-square test should the researcher set up: a test of independence or goodness-of-fit? Give the null and alternative hypotheses for the appropriate test.

   b. Say that another member of the research team wants to test the idea that approaching vehicles will cut off other cars at the same rates that people from different income quintiles are known to cut off others in conversation from past research. For example, since past research indicates that people in the highest income quintile cut off others in conversation 37% of the time, the prediction would be that the highest status vehicles (category 5) would cut off other vehicles 37% of the time. On average, people from other income quintiles have been found to cut off others in conversation as follows: Quintile

1: 8%, Quintile 2: 15%, Quintile 3: 20%, and Quintile 4: 28%. What kind of chi-square test should the researcher set up: a test of independence or goodness-of-fit? Give the null and alternative hypotheses for the test.

c. Figure 12.12 shows the expected frequencies for each group from the chi-square test for either part a or b. With which chi-square test do these expected frequencies correspond? How can you tell?

d. The sample chi-square from the test for part c is found to be statistically significant at the 0.05 alpha level. The researcher concludes that there is no relationship between vehicle status and cutting off other vehicles. Is this conclusion accurate? If yes, explain why. If no, give an accurate conclusion.

e. Based on the expected frequencies shown in Figure 12.12, would you recommend that the analysis proceed? Explain your recommendation.

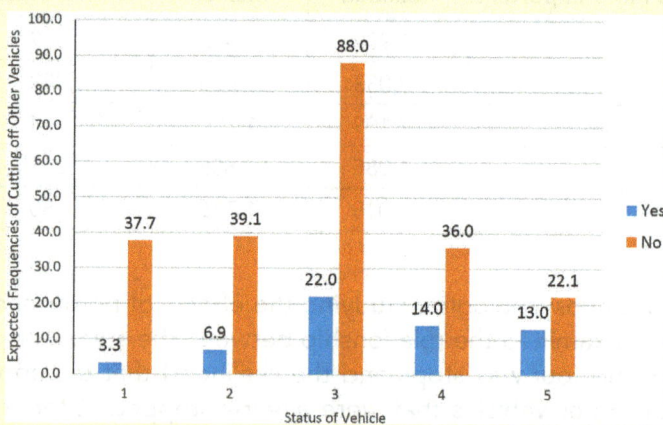

Figure 12.12 Expected Frequencies of Cutting off Other Vehicles, by Status of Approaching Vehicle

8. A statistics student is conducting research on whether expensive chocolate tastes better. They label two identical chocolate bars with either a high or a low price and ask a random sample of one hundred students to choose their favorite chocolate bar. Based on prior research, the student guesses that more status-conscious students will prefer the high-priced bar, while less status-conscious students will prefer the low-priced bar. They measure status-consciousness by whether respondents are part of a fraternity or sorority.

a. Is the measurement of status-consciousness valid? Explain your answer.

b. What kind of chi-square test should the student perform, a test of independence or goodness-of-fit?

c. Tables 12.24a and 12.24b show the expected frequencies for the correct and incorrect chi-square tests. Which table, a or b, is the correct one?

d. The sample chi-square for the correct test is found to have a p-value of 0.28. What should the researcher conclude?

Table 12.24a

	Fraternity/Sorority Member	Not a Fraternity/Sorority Member
Preferred high cost	50	0
Preferred low cost	0	50
Total		

Table 12.24b

	Fraternity/Sorority Member	Not a Fraternity/Sorority Member
Preferred high cost	25	25
Preferred low cost	25	25
Total		

9. Imagine that the cross-tabulation of vehicle status and cutting off other vehicles, as described in Problem 7, actually looked like the distribution shown in Table 12.25.

   a. Examine Table 12.25. Does there appear to be a relationship between status of vehicle and whether the vehicle cuts off other vehicles? Explain your answer.

   b. Imagine that the sample chi-square statistic in a test of independence for the cross-tabulation in Table 12.25 is significant at the 0.05 alpha level. The *New York Times* ran an article about the research called "The Rich Drive Differently, a Study Suggests." Is the result of the chi-square test consistent with the title?

Table 12.25 Approaching Vehicles That Cut Off Other Vehicles at Four-Way Stop, by Status of Approaching Vehicle (Percentages)

	Status of Approaching Vehicle					
Cut off other Vehicles?	1	2	3	4	5	Total
Yes	7%	5%	15%	10%	30%	12.4%
No	93%	95%	85%	90%	70%	87.6%
Total	100%	100%	100%	100%	100%	100%

Source: Adapted from Paul K. Piff, Daniel M. Stancato, Stéphane Côté, Rodolfo Mendoza-Denton, and Dacher Keltner. 2012. "Higher Social Class Predicts Increased Unethical Behavior." *Proceedings of the National Academy of Sciences* 109(11): 4086–4091.

10. Imagine the cross-tabulation of fraternity/sorority membership and preference for a chocolate bar labeled as high or low cost looked like Table 12.26.

    a. Examine Table 12.26. Does there appear to be a relationship between the variables? Explain.

b. The chi-square statistic for a test of independence for the cross-tabulation shown in Table 12.26 is 0.1623. Which p-value could be correct for this chi-square, 0.02 or 0.69?

c. What should a researcher concluded based on this result?

Table 12.26 Preference for Chocolate Bar Labeled as High or Low Cost, by Fraternity/Sorority Membership

	Fraternity/Sorority Member	Not a Fraternity/Sorority Member	Total
Preferred high cost	27	29	57
Preferred low cost	23	21	43
Total	50	50	100

11. Table 12.27 shows mean agreement with the statement that there are no longer significant barriers to women's professional advancement, by gender. Agreement is measured on a 10-point scale, where 1 means completely disagree and 10 means completely agree. Is it possible to conduct a chi-square test of the relationship between gender and agreement that barriers to women's professional advancement are gone, using these data? Explain your answer. If chi-square is not appropriate, what test for statistical significance would you use?

Table 12.27 Agreement That Barriers to Women's Professional Advancement Are Gone, by Gender

	Women	Men
Mean agreement	2.75	5.5

12. A human resources manager wonders if parents or non-parents take a higher number of sick days at their company. Their data shows that parents take an average of nine days a year and non-parents take an average of eight days a year. Would a chi-square test be the correct test for statistical significance? Explain your answer. If chi-square is not appropriate, what test for statistical significance would you use?

13. Horror movies are not usually critically acclaimed, but two famous horror movies were almost universally praised by critics: 2017's *Get Out* and 1960's *Psycho*. *Get Out* focuses on themes of racial inequality and violence, while *Psycho* examines sexuality, gender, and violence. A movie critic wants to know if people's racial identities affect their preferences for *Psycho* versus *Get Out*. She invites a random sample of seventy-five adults who have seen neither film to view both films. Afterward, she asks which of the two films each viewer prefers. Forty-one people in the sample identified exclusively as white, and thirty-four claimed racial identities other than white.

a. The data showed that 78% of white viewers preferred *Get Out* compared to 88% of people of color who viewed the films. What is the null hypothesis for a two-sample test for a difference in proportions? Assuming that this is a two-tailed test, state the alternative hypothesis.

b. Use the information in this problem to construct a cross-tabulation of film preference and racial identity. In each cell, include the appropriate frequency and percentage. Be sure to include a row and column for total frequencies and percentages.

c. State the null hypothesis for a chi-square test of independence. State the alternative hypothesis.

d. The p-value for the two-tailed test described in part a is 0.12. Is it correct to conclude that the sample chi-square statistic from part c would not be significant at the 0.05 alpha level? Explain your answer.

e. According to the results described in this problem, in the general population, do whites and people of color differ significantly in their preference for *Get Out* or *Psycho*?

14. Psychologist Alison Gopnik discusses two prominent modes of parenting in modern America: the gardener and the carpenter. Whereas carpenter parents try to actively shape their children into particular kinds of people, gardeners concentrate on providing environments and resources that will support the direction of their children's development regardless of the particular direction. A researcher develops a scale that measures parents' carpenter and gardening tendencies, where 1 means an exclusive carpenter orientation, 2 means a mix of carpenter and gardener, and 3 means an exclusive gardener orientation. He wants to know whether this parenting approach is related to children's anxiety. Table 12.28 shows the cross-tabulation of parenting approaches and children's anxiety in a sample of two hundred children.

Table 12.28 Children's Anxiety by Parenting Style (Percentages)

Anxiety:	Carpenter	Mix	Gardener	Total
Low	44.8	50.0	66.7	50.0
Medium	37.3	42.9	23.8	32.5
High	17.9	7.1	9.5	17.5
Total	100.0	100.0	100.0	100.0

a. What are the null and alternative hypotheses for a chi-square test of independence? What does the null hypothesis say about the distributions of anxiety levels for each parenting style compared to the others in the population?

b. What are the degrees of freedom and decision chi-square for this test, assuming a 0.025 alpha level?

c. The sample chi-square is 14.7. Should the null hypothesis be rejected?

d. Imagine that the same study described were conducted on a sample size of ninety, but the distribution of children's anxiety across parenting styles stayed exactly the same in terms of percentages, as shown in Table 12.28. The decision chi-square stays the same at the 0.025 alpha level as in the previous analysis, but the sample chi-square is now 6.6. In this smaller sample, should the null hypothesis be rejected?

e. Explain why the sample chi-squares from the larger sample (14.7) and the smaller sample (6.6) are different.

f. What can you conclude about the differences between observed and expected values across the two tests with different sample sizes?

Stata Problems

In this section, we use data from the 2021 General Social Survey (GSS) to assess the relationship between people's political views and the belief that to understand the world, we must free ourselves from old traditions. We use the *libcons* variable that measures respondents' political views with three categories: Liberal, Moderate, and Conservative. The other variable, *freemind*, indicates to what extent people agree that understanding the world requires freeing one's mind from old traditions and beliefs. Are political views related to feelings about breaking with tradition to understand the world?

1. Open `GSS2021.dta`. Use the "tabulate" command to generate a cross-tabulation between *libcons* and *freemind*, with *libcons* as the independent variable, and run a chi-square analysis of the relationship. Use the appropriate options to specify that you want Stata to include in each cell: column percentages, expected frequencies, and contributions to chi-square.

2. How many respondents are included in this chi-square analysis?

3. What percent of conservatives strongly agree that understanding the world requires freeing one's mind from old traditions? What percent of liberals share this opinion?

4. Is there a relationship between political views and attitudes about breaking from tradition in this GSS sample? How do you know?

5. According to the output, is there a relationship between political views and attitudes about breaking from tradition in the wider population? Explain how you know.

SPSS Problems

In this section, we use data from the 2021 General Social Survey (GSS) to assess the relationship between people's political views and the belief that to understand the world, we must free ourselves from old traditions. We use the *libcons* variable that measures respondents' political views with three categories: Liberal, Moderate, and Conservative. The other variable, *freemind*, indicates to what extent people agree that understanding the world requires freeing one's mind from old traditions and beliefs. Are political views related to feelings about breaking with tradition to understand the world?

1. Open GSS2021.sav. Use the "Crosstabs" procedure to generate a cross-tabulation between *libcons* and *freemind*, with *libcons* as the independent variable. Click on "Cells" and tell SPSS to include the observed frequencies, expected frequencies, column percentage, and standardized residuals in each cell. Use the appropriate dialog box to run a chi-square analysis of the relationship.

2. How many respondents are included in this chi-square analysis?

3. What percent of conservatives strongly agree that understanding the world requires freeing one's mind from old traditions? What percent of liberals share this opinion?

4. Is there a relationship between political views and attitudes about breaking from tradition in this GSS sample? How do you know?

5. According to the output, is there a relationship between political views and attitudes about breaking from tradition in the wider population? Explain how you know.

Notes

[1] Robert S. Erikson and Kent L. Tedin. 2015. *American Public Opinion*, ninth edition. New York: Pearson, 213–216.

[2] Bob Cohen and Janet Rosales. "Racial and Ethnic Disparity in Manhattan Jury Pools." Available at http://www.law.cuny.edu/academics/social-justice/clore/reports/Citizen-Action-Jury-Pool-Study.pdf.

[3] The cross-tab is adapted from separate data on the age distribution in a Pew American Trends panel and the distribution of getting news from social media.

Chapter 13

Ruling Out Competing Explanations for Relationships between Variables

Control Variables and the Logic of Causality

There are certain obligations that come with driving a car. No matter where you live, you need a driver's license. In almost every state, you need to buy auto insurance. In many localities, you have to pay property tax on the appraised value of your vehicle. And, in most states, you are required to take your car for a safety inspection. In fact, there are forty states where some form of safety inspection is required.[1] Twelve of these states require annual inspections; the rest require inspections every other year or less frequently. It would be reasonable to expect that states with mandatory inspections have lower rates of traffic deaths. After all, when an inspection finds a defect, such as a headlight or turn signal that is burned out or an axle that is cracked, you are required to make the repair, or your car will be deemed unsafe to drive. It follows that, with fewer defective cars on the road, there will be fewer accidents and, by extension, lower rates of traffic fatality in states that make safety inspections mandatory.

Political scientist and statistician Edward Tufte investigated this question.[2] He sought to determine if states with mandatory auto inspections had fewer traffic deaths and, if so, whether the difference could be explained by the presence or absence of mandatory auto inspections. He found that, in states without mandatory auto inspections, there were on average 31.9 motor-vehicle deaths per 100,000 people, whereas in states that required drivers to inspect their cars, the corresponding figure was only 26.1 deaths per 100,000 people. On the surface, this appears to support the claim that mandatory auto inspections reduce the number of traffic deaths.

Let's think of this example in terms of independent and dependent variables. We are suggesting that the presence or absence of auto inspections influences the number of traffic deaths in a state. Our independent or explanatory variable (X) is nominal (whether a state requires auto inspections), and our dependent variable (Y) is interval-ratio (traffic deaths). Using an arrow, we can model the relationship as shown in Figure 13.1. In this and in other figures in this chapter, the direction of the arrow implies the path of causality.

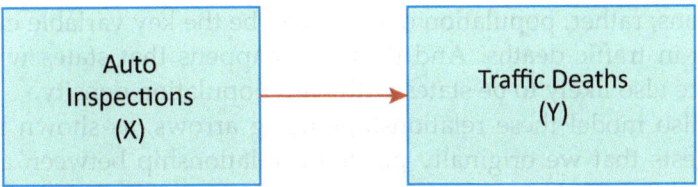

Figure 13.1 Modeling the Relationship between Independent Variable (Auto Inspections) and Dependent Variable (Traffic Deaths)

The difference in means between states with inspections and those without provides evidence of an association between auto inspections and traffic deaths. But association is not causation. For causation to be proved, the relationship must make sense logically and temporally, and we must rule out alternative explanations. (We will discuss these criteria in more detail in a few pages.) Although we are suggesting that the presence or absence of auto inspections is associated with the level of traffic fatalities, and although this makes intuitive sense, we cannot claim that mandatory auto inspections cause lower rates of traffic deaths.

In his analysis of this question, Tufte suggests that there are other variables that are also associated with differences in traffic deaths, and we must rule out alternative explanations before claiming we have found evidence of a causal relationship. For example, he finds that states with extremely low traffic death rates are not only more likely to have inspections, they are also more likely to have high population density and to have more than seven letters in their names.

How do we tease out these competing explanations? To begin, we apply some common sense to these possible independent variables. We have already argued that a relationship between auto inspections and traffic fatalities makes intuitive sense—inspections lead to fewer defective cars on the roads. What about the length of state names? Unless we are prepared to argue that having a lot of letters in a state's name somehow keeps roads safer, we must conclude that this association is coincidental; there is no inherent logic to this relationship. So we can dismiss it.

The other possible explanatory variable we suggested was the state's population density. Is there a plausible explanation for the association between population density and traffic fatalities? Consider two states that are similar in geographic area: Vermont and Massachusetts. According to the US Census Bureau, in 2020, the population density for Vermont was 69.8 people per square mile; for Massachusetts, it was 901.2 people per square mile.[3] How is this connected to traffic fatalities? In a sparsely populated state, such as Vermont, residents have to drive longer distances to see friends, go shopping, or get to work. When population density is high, more destinations are within a short drive. And it is reasonable to expect that, when people drive longer distances, there will be more accidents and, consequently, more traffic fatalities.

In his analysis of this question, Tufte suggests that there may not only be an association between population density and traffic deaths; there may also be an association between population density and auto inspections. As Tufte puts it: "The denser states tend to be the urbanized, industrialized, politically competitive states with activist state governments—governments that would be more likely to inaugurate an inspection program." In other words, we must examine the possibility that traffic fatalities are not associated with auto inspections; rather, population density may be the key variable explaining variations among states in traffic deaths. And it just so happens that states with mandatory auto inspections are also likely to be states with high population density.

We can also model these relationships using arrows, as shown in Figure 13.2. This schema suggests that we originally posited a relationship between auto inspections (X) and traffic fatalities (Y), but, when we introduced population density (Z), the original relationship between X and Y weakened or disappeared (hence the dotted line between X and Y). In this scenario, the relationship between X and Y is called a **spurious relationship**. A relationship is said to be spurious if it disappears or substantially weakens upon introduction of a third variable.

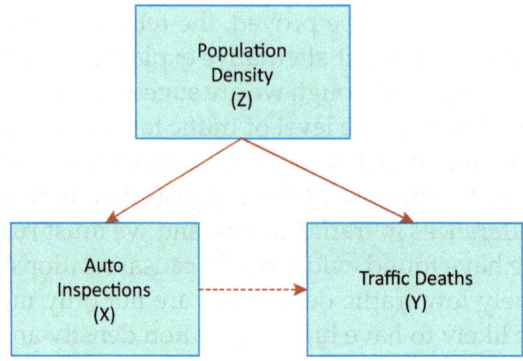

Figure 13.2 Modeling a Spurious Relationship between Independent Variable (Auto Inspections) and Dependent Variable (Traffic Deaths), Controlling for a Third Variable (Population Density)

The third variable is known as a **control variable**. When we introduce a control variable, we hold that third variable constant while we reexamine the relationship between the independent and dependent variables. When Tufte examines these relationships, he assigns three categories to the population density variable: Thin, Medium, and Thick. Table 13.1 summarizes Tufte's findings.

The bottom row of Table 13.1 displays the difference between states with and without inspections in the aggregate—that is, with no control variable. We see that there are 5.8 more deaths on average in states without inspections. Yet, when we control for population density, the gap between states with and without inspections is smaller. For thin density states the difference is 3.6; for medium density states, it is 3.1; these are large reductions in traffic fatalities when compared to the aggregate. For thick density states, the difference in traffic fatality rate is 5.3, slightly lower than the aggregate rate. But, in every instance, the impact of auto inspections is diminished, compared to the aggregate. Why is this the case? Because there is an association between population density and the presence or absence of auto inspections.

Can we determine which explanatory variable, inspections or density, is more strongly associated with traffic deaths? We have seen that, when we control for density, there are

Table 13.1 Average Traffic Death Rates by Inspections, Controlling for Population Density

	States without Inspections	States with Inspections	How Many More Deaths in States without Inspections?
Thin population density	38.5	34.9	3.6
Medium population density	31.5	28.4	3.1
Thick population density	23.6	18.3	5.3
All states	31.9	26.1	5.8

Note: Death rates are reported in deaths per 100,000 people.
Source: Edward R. Tufte. 1974. *Data Analysis for Politics and Policy*. Englewood Cliffs, NJ: Prentice-Hall, 22.

smaller differences between states with and without auto inspections, especially for thin and medium density states. But let's reverse the logic and control for auto inspections. To do that, compare death rates within columns of the table instead of across columns so that we are holding constant whether states require inspections. Look at the column with the information about states without inspections. Thin population density states have a death rate that is 14.9 (38.5 − 23.6) higher than that for thick density states. We see the same pattern in the other column, states with auto inspections. Thin density states have on average 16.6 (34.9 − 18.3) more deaths than do thick density states. These are very large differences and point to two important conclusions: that there is a stronger association between population density and death rates than between inspections and death rates; and that the aggregate differences we observed between states with and without auto inspections partly reflect the strong association between inspections and population density. Both variables—density and inspections—have an effect, but population density is a more powerful explanatory variable.

In this example, the effect of the independent variable substantially weakened when we introduced the control variable. In a perfectly spurious relationship, there would be no effect whatsoever of the independent variable after we introduced the control variable. Considering our example above, this would mean that differences in traffic death rates between states that require auto inspections and those that don't would fall to zero. In the real world, we rarely see totally spurious relationships. Here is a scenario with hypothetical data to illustrate what a perfectly spurious relationship might look like.

Dr. Strangelove is conducting a study of factors that lead to heart disease. His study takes place in a remote area of the country that is not wired for cable television; residents who want to expand their channel offerings must use satellite dishes. The doctor has detected an interesting pattern. People with satellite dishes have a much higher incidence of heart disease than people who do not own dishes. He wonders if there is something about the satellite waves that clogs arteries and leads to coronary disease. Both independent variable (whether an individual owns a satellite dish) and dependent variable (whether an individual has heart disease) are measured at the nominal level. Table 13.2 presents a summary of his findings.

The pattern here is clear. Half of the satellite dish owners have heart disease, compared to only 25% among people who don't own dishes. The doctor is tempted to conclude that satellite dishes make it more likely that an individual develops coronary disease, as modelled in Figure 13.3.

Table 13.2 Effect of Satellite Dishes on Heart Disease, Hypothetical Data

	Satellite Dish	No Satellite Dish
Heart disease	50% (10)	25% (5)
No heart disease	50% (10)	75% (15)
Total	100% (20)	100% (20)

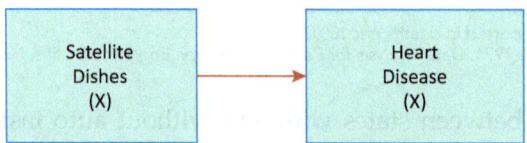

Figure 13.3 Predicted Relationship between Satellite Dish Ownership and Heart Disease

But then it occurs to the doctor that this might be a spurious relationship. Perhaps there is some other factor that simultaneously influences both heart disease and dish ownership. One such factor might be lifestyle—the degree to which one exercises and moves around. So Dr. Strangelove collects this additional information from the forty subjects in his study—and he categorizes people into two groups: those who lead an active lifestyle and those who are sedentary. He then reexamines the relationship between satellite dishes and heart disease, separately for active and sedentary people. The results are shown in Table 13.3.

Table 13.3 Effect of Satellite Dishes on Heart Disease, Controlling for Lifestyle, Hypothetical Data

		Satellite Dish	No Satellite Dish
Among people who lead *active* lifestyles	Heart disease	0% (0)	0% (0)
	No heart disease	100% (10)	100% (15)
	Total	100% (10)	100% (15)
Among people who lead *sedentary* lifestyles	Heart disease	100% (10)	100% (5)
	No heart disease	0% (0)	0% (0)
	Total	100% (10)	100% (5)

There are two panels in Table 13.3, one for each category of the control variable. Let's begin with the top panel—these are the twenty-five subjects who lead an active lifestyle. We find that, among these people, the presence or absence of a satellite dish has no effect whatsoever on the incidence of heart disease. Among those who lead active lifestyles, 100% of dish owners and 100% of those who don't own a dish are free of coronary disease.

Now look at the bottom panel. We find the same pattern, only in reverse. These fifteen sedentary people—whether they are dish owners or not—have heart disease.

Finally, notice that there is a strong association between dish ownership and lifestyle. Among the twenty-five active people, fifteen of them, or 60%, did not own a dish. Among the fifteen sedentary people, 67% (10/15) were dish owners. When Dr. Strangelove puts all of this together, he correctly concludes that satellite dishes are perfectly safe; that lifestyle is strongly associated with heart disease; and that the aggregate association he found between dishes and heart disease reflects the fact that sedentary people are far more likely than active people to own satellite dishes.

Figure 13.4 shows the schema that represents the true relationship among these three variables. This is an example of perfect spuriousness, because the relationship between independent variable (satellite dish) and the dependent variable (heart disease) totally disappears after we control for lifestyle. As we noted earlier, it is extremely rare to find perfectly spurious relationships in the real world.

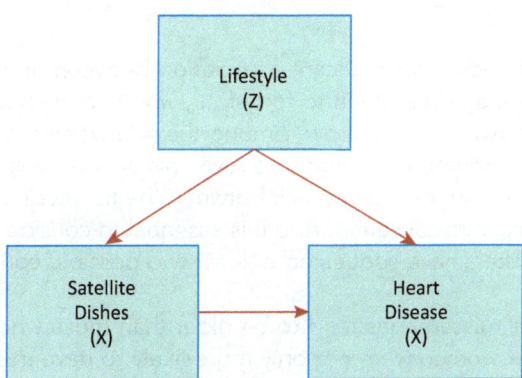

Figure 13.4 Modeling the Spurious Relationship between Satellite Dish Ownership and Heart Disease

Criteria for Causal Relationships

When we are making the case for a causal relationship between independent variable (X) and dependent variable (Y), we must make sure that three conditions are satisfied.

1. There has to be an association between the independent and dependent variables. Changes or differences in the independent variable must be associated with changes or differences in the dependent variable. If we are hypothesizing that social class influences party identification, then there must be some association between the

two variables. If they are not associated, then X cannot be influencing Y. We saw in previous chapters what it means to say that there is no association between two variables—the column percentages or the means are identical. Table 13.4 shows another example (with hypothetical data) of two variables that are not associated. Social class cannot "explain" party identification since the partisan identification of working- and middle-class people is the same.

Table 13.4 Cross-Tabulation Showing No Relationship between Social Class and Party Identification, Hypothetical Data

	Working Class	Middle Class
Democrat	50%	50%
Independent	10%	10%
Republican	40%	40%

> **BOX 13.1: APPLICATION**
>
> ### The Spurious Relationship between Flu Vaccines and Miscarriage
>
> A 2017 study reported finding a significant association between flu vaccines and miscarriage. Of 485 pregnant women, ages eighteen to forty-four, who had miscarriages, 3.5% (17/485) had received the flu vaccine two years in a row. Among the 485 women who had normal deliveries, 0.8% (4/485) had been vaccinated two years in a row. This difference is statistically significant and provides support for the claim of a causal link between the flu vaccine and miscarriage.
> But association is not equal to causation. And it is essential to consider the possibility of a spurious correlation. Researchers have suggested at least two possible confounding factors.
>
> 1. Age. Those who had miscarriages tend to be older than those who had normal deliveries. It is possible that older women were not only more likely to have miscarriages, they were also more likely to get the flu vaccine.
> 2. Bias in the makeup of the two groups. Not all women seek medical attention for a miscarriage. It is possible that those who sought medical care for their miscarriage were also more likely to get the flu vaccine. The authors also suggest that, because miscarriages often occur early in pregnancy, some women do not realize they were pregnant when they miscarry. These women may be wrongly classified into the "no miscarriage" group, and many of them would not have received the flu vaccine. Had they been correctly classified into the "miscarriage" group, the percentage having received the flu vaccine would have been lower.
>
> *Sources:* Lena H. Sun. 2017. "What to Know about a Study of Flu Vaccine and Miscarriage." *Washington Post.* September 13. Retrieved from https://www.washingtonpost.com/news/to-your-health/wp/2017/09/13/researchers-find-hint-of-a-link-between-flu-vaccine-and-miscarriage; Jon Cohen. 2017. "New Study Finds Link between Flu Vaccine and Miscarriage. But Is It Real?" *Science.* September 13. Retrieved from http://www.sciencemag.org/news/2017/09/new-study-finds-link-between-flu-vaccine-and-miscarriage-it-real.

2. If X is a cause of Y, then X must precede Y logically and temporally. Gender can be only a cause, not an effect, of party identification. Similarly, if we argue that party identification influences ideology, then one's party identification must be established before one's ideology.
3. We must rule out alternative explanations for Y. This is where we introduce control variables. If we believe that the presence of auto inspections reduces traffic fatalities, then we must be certain that there are no other "lurking" variables that are influencing X and Y and causing them to co-vary. In the example we examined at the start of this chapter, that lurking variable was population density. When we are working with observational studies, such as surveys, it is impossible to rule out all alternative explanations. The best we can do is control for several variables and present evidence in support of a causal relationship.

Modeling Spurious Relationships

As we have discussed, a spurious relationship can occur when the apparent association between independent and dependent variables is driven by a third (control) variable, which is influencing both independent and dependent variables. In such a case, the control variable must be temporally prior to the other two variables. Consider another example of a spurious relationship. Research has found a strong association between sustainable behaviors (such as driving hybrid cars, commuting shorter distances, emitting less CO_2) and vote choice in the 2008 presidential election.[4] Voters who engaged in sustainable behaviors were more likely to support the Democratic candidate. We can model the relationship as shown in Figure 13.5.

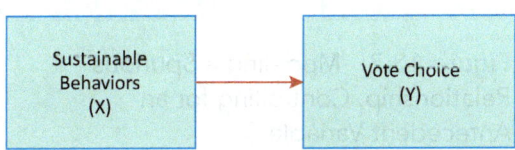

Figure 13.5 Proposed Relationship between Sustainable Behaviors and Vote Choice

But when we introduce a control variable and the original relationship breaks down, we have evidence of a spurious relationship. In this case, one obvious candidate for a control variable would be the urban/rural divide. Voters who live in urban areas are more likely to engage in sustainable behaviors and also to vote for the Democrat. Rural voters are less likely to own a hybrid vehicle or to commute short distances, and they were also more likely to vote for the Republican candidate We can model the relationship as shown in Figure 13.6.

When the control variable is temporally prior to both the independent and dependent variables and influences both of them, we call the control variable an **antecedent variable**, as shown in Figure 13.7.

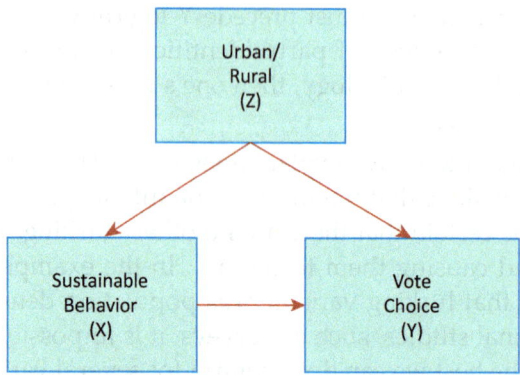

Figure 13.6 Modeling a Spurious Relationship between Sustainable Behaviors and Vote Choice

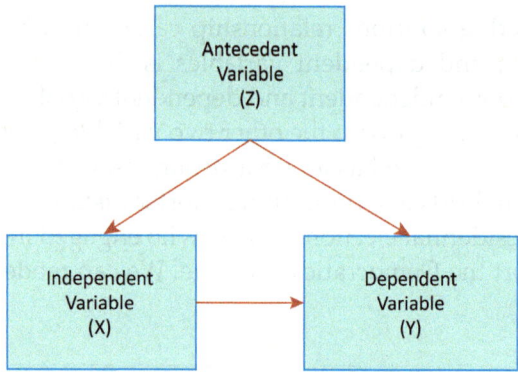

Figure 13.7 Modeling a Spurious Relationship, Controlling for an Antecedent Variable

Sometimes, the control variable affects the relationship between X and Y in a different way. Whereas the antecedent variable precedes X and Y, there are instances where the control variable intervenes between X and Y, and it is called an **intervening variable** or a **mediating variable**. In the case of an intervening variable, we would model the relationship as shown in Figure 13.8.

Consider this example. In chapter 12, when we discussed cross-tabulations and statistical significance, we described the gender gap on three specific issues. The gender gap also extends to ideological self-identification. In 2020, the American National Election Study

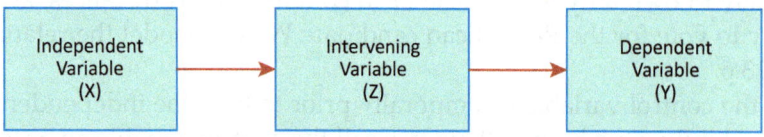

Figure 13.8 Modeling a Spurious Relationship, Controlling for an Intervening Variable

(ANES) found that women were more likely than men to identify as liberal; men were more likely to call themselves conservative. And these gaps were statistically significant (based on the chi-square test). Table 13.5 summarizes these findings.

Table 13.5   Ideology by Gender

	Men	Women	Total
Liberal	32.1%	39.4%	36.0% (2,283)
Moderate	24.3%	26.4%	25.4% (1,613)
Conservative	43.6%	34.2%	38.6% (2,450)
Total	100% (2,967)	100% (3,379)	100% (6,346)

($\chi^2 = 62.6$, $p < 0.001$)
Source: 2020 American National Election Study.

Before we assert that gender "explains" ideology, we need to consider the possibility that the relationship between these variables is spurious. Perhaps there are other variables associated with both gender and ideology. One obvious candidate is party identification. It is reasonable to expect that party identification is associated with both gender (our independent variable) and ideology (our dependent variable). Let's examine this possibility by controlling for party identification. Remember that, when we introduce a control variable, we are reexamining the relationship between independent and dependent variables, separately for each category of the control variable. Table 13.6 does this.

There are three panels in Table 13.6, one for each category of party identification (the control variable). When you examine these panels, keep in mind the aggregate breakdown of ideology by gender (Table 13.5); women were about 7 to 9 percentage points more likely to be liberal and less likely to be conservative. Now look at the first panel of Table 13.6. There is almost no difference between the proportion of women and men who call themselves liberal or conservative. The chi-square p-value at the bottom of the Democrats panel supports this finding of a non-relationship. The p-value is equal to 0.09. That is fairly high. This tells us that, if there were no relationship in the population of Democrats between gender and ideology, there would be a 9% chance of getting this sample breakdown. We conclude that, among Democrats, gender does not predict ideology. Among Republicans, our conclusion is similar. Notice the small differences between the ideological preferences of women and men and the fairly high p-value at the bottom of the table (0.08). Republicans, regardless of their gender, are overwhelmingly conservative (far more conservative than Democrats are liberal). Even among Independents there is little evidence of an ideological gender gap.

To summarize, we have found that there is an ideological gender gap in the aggregate; but, when we control for party identification, that gender gap disappears for all three categories of partisan identification. In other words, using a conventional alpha (0.05), there is no statistically significant effect of gender on ideology, after we control for party identification. Why then did we find evidence of a gender gap in the aggregate? The answer has to do with the association between gender (the independent variable) and party identification

Table 13.6 Ideology by Gender, Controlling for Party Identification

		Men	Women	Total
Democrats	Liberal	68.5%	70.3%	69.5% (2,063)
	Moderate	25.0%	25.1%	25.1% (744)
	Conservative	6.5%	4.7%	5.4% (161)
	Total	100% (1,230)	100% (1,738)	100% (2,968)
	(χ^2 = 4.85, p = 0.09)			
Independents	Liberal	19.0%	18.2%	18.6% (114)
	Moderate	60.3%	63.7%	62.1% (381)
	Conservative	20.7%	18.2%	19.4% (119)
	Total	100% (300)	100% (314)	100% (614)
	(χ^2 = 0.84, p = 0.66)			
Republicans	Liberal	3.6%	3.9%	3.8% (104)
	Moderate	16.0%	19.1%	17.5% (482)
	Conservative	80.4%	76.9%	78.7% (2,170)
	Total	100% (1,433)	100% (1,323)	100% (2,756)
	(χ^2 = 5.09, p = 0.08)			

Source: 2020 American National Election Study.

(the control variable). We found a strong aggregate association between gender and ideology because men are more likely to be Republicans and women more likely to be Democrats and because party identification is strongly associated with ideology. We can demonstrate that by way of two additional cross-tabulations.

Table 13.7 shows the relationship between party identification and ideology.

A majority of Democrats (69.4%) call themselves liberal, an overwhelming majority of Republicans (78.8%) identify as conservative, and the chi-square has a p-value of less than 0.00. This is strong evidence of an association between our control variable (party identification) and our dependent variable (ideology). What about the link between our control variable and our independent variable (gender)? That relationship is shown in Table 13.8.

As we expected, there is a fairly strong association between these two variables. A strong majority of Democrats are women (59.6%), a majority of Republicans are men (51%), and the chi-square has a p-value of less than 0.00. This is evidence of an association between our control variable (party identification) and our independent variable (gender).

Table 13.7 Ideology by Party Identification

	Democrats	Independents	Republicans	Total
Liberal	69.4%	18.6%	3.8%	35.9% (2,287)
Moderate	25.2%	62.2%	17.4%	25.5% (1,624)
Conservative	5.4%	19.2%	78.8%	38.6% (2,463)
Total	100% (2,980)	100% (625)	100% (2,769)	100% (6,374)

(χ^2 = 4,250.4, p < 0.00)
Source: 2020 American National Election Study.

Table 13.8 Party Identification by Gender

	Democrat	Independent	Republican	Total
Men	40.4%	47.4%	51.0%	45.6% (3,369)
Women	59.6%	52.6%	49.0%	54.4% (4,015)
Total	100% (3,452)	100% (859)	100% (3,073)	100% (7,384)

(χ^2 = 75.4, p < 0.00)
Source: 2020 American National Election Study.

We have one task left. We must model the relationship among the three variables in our analysis. We have established that the aggregate relationship between gender and ideology weakens substantially once we control for party identification. We also know that party identification is associated with both gender and ideology. There are two possible models to depict a spurious relationship among three variables: the antecedent model (Figure 13.9) and the intervening model (Figure 13.10).

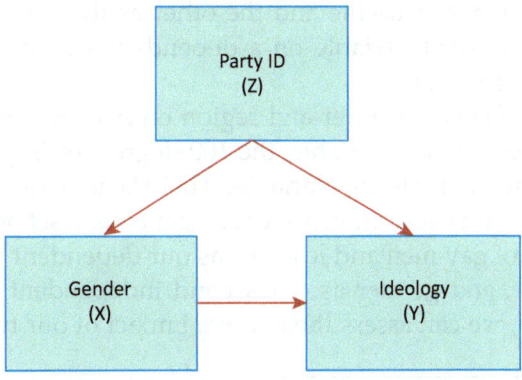

Figure 13.9 Spurious Relationship between Gender and Ideology, Antecedent Model

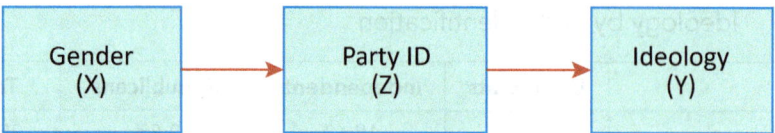

Figure 13.10 Spurious Relationship between Gender and Ideology, Intervening Model

We can rule out the antecedent model. It fails to meet one of the key criteria for causal relationships. In general, party identification (Z) will not be prior to gender (X). Although party identification is associated with ideology and with gender, the causal arrow between party identification and gender runs in the wrong direction. Thus, the only possibility is the intervening model. It makes sense, both logically and empirically. It posits that gender influences party identification, which in turn influences ideology. And, as we have seen in Tables 13.7 and 13.8, the data support this interpretation. Recall that a variable that explains all or part of the relationship between two other variables is sometimes called a mediating variable because it mediates the relationship between the other variables. In our example, party identification mediates the relationship between gender and ideology.

When we introduce a control variable and the original relationship between X and Y disappears or is substantially weakened, we must then decide if the control variable is acting as an antecedent variable or an intervening one. This decision is based on logic and on prior research. If Z precedes both X and Y, then the relationship follows the antecedent model. If X influences Z and Z in turn influences Y, we have an intervening variable model.

● Modeling Non-Spurious Relationships

In the social sciences, explanations are usually complex and require the inclusion of more than one explanatory variable. A **non-spurious relationship** between an independent and a dependent variable is one that persists even after the introduction of another explanatory variable. When two explanatory variables are not associated with each other, we can model the relationships as shown in Figure 13.11.

To measure the **independent** or **net effect** of each independent variable, we treat one of them as the independent variable and the other as the control variable. A net effect is the effect of an independent variable on a dependent variable when holding additional explanatory variables constant.

Let's look at the effects of gender and region on perceptions of gay men and lesbians. In this example, our dependent variable, the 100-degree feeling thermometer rating of gay men and lesbians, is an interval-ratio variable.* There is no reason to believe that gender and region are associated with each other. So we can generate a set of tables, treating the feeling thermometer ratings of gay men and lesbians as our dependent variable, region as our first independent variable, and gender as our second independent (and control) variable. By setting it up this way, we can assess the relative impact of our two explanatory variables.

---

* The American National Election Study asks respondents to rate various public figures and groups on a 100-degree feeling thermometer, where 0 represents "very cold or unfavorable feeling," 50 represents "no feeling at all," and 100 represents "very warm or favorable feeling." One of the groups that respondents are asked to rate is "gay men and lesbians."

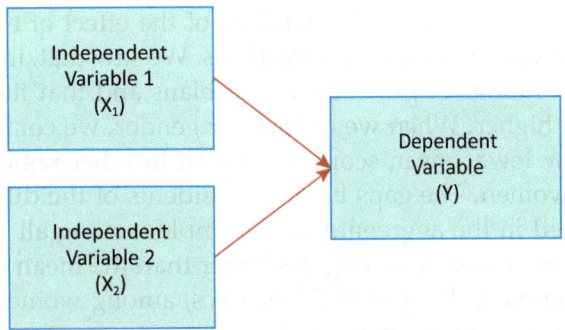

Figure 13.11 Modeling the Relationship between Two Explanatory Variables and a Dependent Variable

When conducting this kind of analysis, it is a good idea to begin by assessing the effect of the first independent variable, without controlling for the second independent variable. Then we introduce the second independent variable as a control, and we can gauge the extent to which the original relationship between independent and dependent variables changes for the various categories of the control variable.

Table 13.9 examines the relationship between our first independent variable (region) and our dependent variable (feeling thermometer scores). It shows that southerners hold the lowest mean ratings of gay men and lesbians and that Northeasterners are most supportive. The national mean is about 65 degrees.

Table 13.9 Mean Feeling Thermometer Rating of Gay Men and Lesbians by Region, in Degrees

	Northeast	Midwest	South	West	All Respondents
Mean rating	69.9	64.4	61.8	69.2	65.4

Source: 2020 American National Election Study.

Now we introduce gender as a control variable and examine the effect of region on the ratings of gay men and lesbians, separately for men and women. Table 13.10 displays our results with the control variable.

Table 13.10 Mean Feeling Thermometer Rating of Gay Men and Lesbians by Region, Controlling for Gender, in Degrees

	Northeast	Midwest	South	West	Total
Men	65.7	59.7	58.4	64.2	61.2
Women	73.6	68.7	64.7	73.4	69.1

Source: 2020 American National Election Study.

Our goal now is to gain an understanding of the effect of each variable—region and gender—on perceptions of gay men and lesbians. We saw that, in the aggregate, Southerners have the lowest ratings of gay men and lesbians and that the mean ratings of Northeasterners are a lot higher. When we control for gender, we come to the same conclusion. Southern men show lower mean scores than men in other regions; and the same can be said for Southern women. The gaps between residents of the different regions are similar to what we observed in the aggregate. For example, among all respondents, Southerners showed a mean score that was 8.1 degrees lower than the mean score for Northeasterners (Table 13.9). Among men, that gap is 7.3 degrees; among women, it is 8.9 degrees (Table 13.10). We can conclude that there is an effect of region on views of gay men and lesbians and that this effect persists among both men and women.

There is also evidence in Table 13.10 that gender impacts views of gay men and lesbians. Consider Table 13.11, using the same data from Table 13.10, but presented differently so as to highlight the effect of gender.

Table 13.11 Mean Feeling Thermometer Rating of Gay Men and Lesbians by Gender and Region, in Degrees

	Women	Men	Gender Difference
Northeast	73.6	65.7	7.9
Midwest	68.7	59.7	9.0
South	64.7	58.4	6.3
West	73.4	64.2	9.2

Source: 2020 American National Election Study.

In every region, women have higher mean ratings of gay men and lesbians than do men, by as little as 6.3 degrees (in the South) and as much as 9.2 degrees (in the West). The level of favorability is different among regions, as we saw above. When we present the data this way, we can gauge the impact of the control variable (gender) on the dependent variable (feeling thermometer means). And there is no question that gender has an effect on views of gay men and lesbians. The effect of gender varies a little across regions but it is still quite large within all four regions.

In this example, the effect of region (X) on favorability toward gay men and lesbians (Y) was similar for the different categories of gender (Z). But there are many times when the effect of an independent variable on a dependent variable changes significantly for the different categories of the control variable. In other words, the effect of the independent variable on the dependent variable depends on the control variable. This is called a **statistical interaction**. The interaction between X and Z is crucial to understanding the actual impact of X on Y. We can model this as shown in Figure 13.12.

Notice that, in the model of an interaction, the arrow from the control variable leads not to another variable but to the arrow from independent to dependent variable. This indicates that the control variable is affecting the relationship between independent and dependent variable. This kind of control variable is also called a **moderating variable**.

It is well established that age and education influence turnout in elections.[5] We know that older people are much more likely to vote than younger people and that people who have gone to college are also much more likely to turn out in an election than people who have not finished high school. Is the effect of age on turnout equal for different educational levels? Or is the effect of education so strong that, among the well-educated, age doesn't matter?

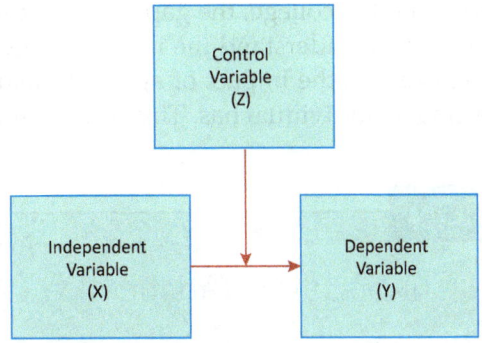

Figure 13.12 Modeling a Statistical Interaction

To answer these questions, look at Table 13.12. The figure in each cell represents the percentage of that group that reported voting in the 2020 election. For example, the figure in the upper-left cell is 42.1%. This should be understood as: "42.1% of eighteen- to twenty-four-year-olds who did not finish high school voted in the election."

The pattern in this table provides a good illustration of a statistical interaction between the two explanatory variables. Among the least educated (those who have not completed

Table 13.12 2020 Election, Percentage Reporting That They Voted by Age and Education

	18 to 24	25 to 34	35 to 54	55 to 64	65+	Difference between Oldest and Youngest
Less than high school	42.1%	26.8%	54.0%	72.6%	71.4%	29.3 points
High school diploma	53.3%	56.1%	67.3%	83.0%	90.1%	36.8 points
Some college	75.8%	69.0%	82.3%	89.0%	94.1%	18.3 points
College graduate	82.9%	89.0%	92.5%	94.0%	97.8%	14.9 points

Source: 2020 American National Election Study.

high school or those who have only completed high school), the effect of age on turnout is very strong. The percentage that reported voting tends to increase as individuals get older. For example, among those who have completed high school, the gap between the youngest and oldest is a whopping 36.8 percentage points (90.1 – 53.3).

As we move up the education ladder, that pattern changes. Age still has an effect on reported turnout, but the magnitude of that effect begins to diminish and is substantially weaker among the most educated. Look at those who have finished college (the bottom row). About 83% to 89% of the two youngest groups voted, and 93% to 98% of the three oldest groups of college graduates voted. The gaps between age groups are much smaller for well-educated people. If we examine the final column of Table 13.12, we can compare the voting rates of the youngest (eighteen to twenty-four years old) and oldest (sixty-five years or older) voters, broken down by level of education. Among those who have not studied beyond high school, the turnout gap between the youngest and oldest cohorts is about 29 to 37 percentage points. As education increases, we see a weakening effect of age.

Among those who have attended college, the gap between the youngest and oldest is only 15 to 18 points. In sum, to fully understand the impact of age on turnout, we must control for education and recognize that the impact of age is dramatically different and depends on the amount of schooling an individual has. This is the essence of a statistical interaction.

BOX 13.2: APPLICATION

Interpreting the Chi-Square Statistic in a Controlled Cross-Tabulation

When you generate a cross-tabulation with a control variable, pay close attention to the interpretation of the chi-square statistic. The key is to correctly specify the null and alternative hypotheses. Let's look at the chi-square statistic in the context of a set of controlled cross-tabulations from the 2020 American National Election Study. In this example, we examine the effect of party identification (independent variable) on respondents' frequency of worship (dependent variable), controlling for respondents' race. (To simplify, we have limited the analysis to white and African American respondents only.)

Remember that null and alternative hypotheses always make claims about a population. Since we are controlling for race in this example, our first set of hypotheses will be about whites in the population, and our second set will be about African Americans in the population. Table 13.13 presents the data.

The top panel of Table 13.13 looks at the relationship between party identification and frequency of worship among whites in the sample. The differences between the column percentages are fairly large, supporting the notion that, among whites, party identification helps us predict how often people attend religious services. The chi-square statistic and p-value support this assertion. In the case of the top panel, our null hypothesis states that, *among whites in the population*, there is no relationship between party identification and frequency of religious worship. The large chi-square statistic and the corresponding low p-value lead us to reject this null hypothesis. Our conclusion is that, *among whites in the population*, there is an association between party identification and how often people attend religious services. White Democrats are less likely than white Republicans to worship frequently.

Table 13.13 Frequency of Religious Worship by Party Identification, Controlling for Race

		Democrats	Independents	Republicans	Total
Whites	Worships every week or almost every week	48.3%	48.9%	61.3%	56.55 (1,428)
	Worships once or twice per month	17.4%	17.4%	15.0%	15.9% (401)
	Worships a few times per year or never	34.4%	33.8%	23.8%	27.7% (699)
	Total	100% (719)	100% (219)	100% (1,590)	100% (2,528)
	(χ^2 = 42.7; p = 0.00)				

Table 13.13 *Continued*

		Democrats	Independents	Republicans	Total
African Americans	Worships every week or almost every week	53.7%	59.5%	58.1%	54.5% (228)
	Worships once or twice per month	19.4%	16.2%	9.7%	18.4% (77)
	Worships a few times per year or never	26.9%	24.3%	32.3%	27.0% (113)
	Total	100% (350)	100% (37)	100% (31)	100% (418)
	($\chi^2 = 2.27$; $p = 0.69$)				

Source: 2020 American National Election Study. Total percent may not sum to 100 due to rounding.

Now look at the bottom panel. Can you figure out what the null hypothesis is here? It would state that, *among African Americans in the population*, there is no relationship between party identification and frequency of religious worship. In the bottom panel, the gaps between column percentages are much smaller, and, with a smaller sample size, these gaps are not sufficiently large to achieve statistical significance—note the chi-square p-value (0.69). If we use an alpha of 0.01 or 0.05 or even 0.10, we would not reject the null hypothesis. We would not reject the claim that, *among African Americans in the population*, there is no relationship between party identification and frequency of worship. In other words, we are not confident that party identification is a predictor of frequency of worship in the population of African Americans.

Note that these data provide an example of an interaction between party identification and race. The effects of party identification on frequency of worship are very different for white and African American respondents. Among whites, there was a fairly strong and statistically significant relationship between party identification and how often people worship. Among African Americans, however, the effect was much weaker and not statistically significant. Thus, we could represent the relationships among these variables by the model shown in Figure 13.13.

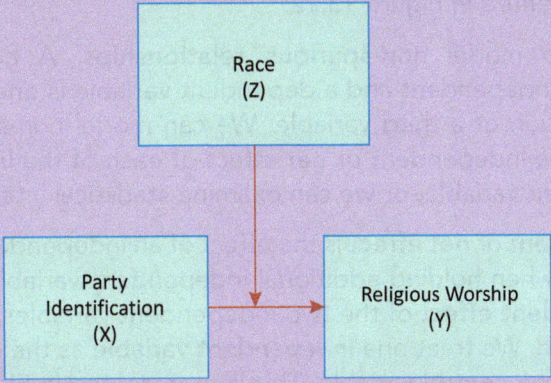

Figure 13.13 Modeling the Interaction Effect between Party Identification and Race on Frequency of Worship

Chapter Summary

This chapter explored the causal connections among variables and how to model them.

- Three conditions must be fulfilled before we can make the case for a causal connection between any two variables, X and Y.
    1. There must be an association between the two variables.
    2. If X is a cause of Y, then X must precede Y logically and temporally.
    3. We must rule out alternative explanations for Y.

- When we are analyzing the relationship among three variables, we employ one dependent variable and two independent variables, one of which we designate as a control variable.

- The dependent variable can be nominal, ordinal, or interval-ratio.

- A **control variable** is a variable that we hold constant when we examine the relationship between the independent and dependent variables.

- When we introduce a control variable and the relationship between independent and dependent variables disappears or weakens substantially, we have a **spurious relationship**.

- An **antecedent variable** is a control variable that is temporally prior to both the independent and dependent variables and influences both. It is modeled in Figure 13.7.

- An **intervening variable** is a control variable that intervenes between X and Y. It is modeled in Figure 13.8.

- A **mediating variable** is another term for an intervening variable.

- A **moderating variable** is a control variable that affects the relationship between independent and dependent variable. This is also known as a statistical interaction and is represented in Figure 13.12.

- We can also model **non-spurious relationships**. A non-spurious relationship between an independent and a dependent variable is one that persists even after the introduction of a third variable. We can model non-spurious relationships by examining the independent or net effect of each of the independent variables on the dependent variable, or we can examine statistical interaction.

- An **independent** or **net effect** is the effect of an independent variable on a dependent variable when holding additional independent variables constant. To examine the **independent effect** of the two independent variables, they must be weakly or not correlated. We treat one independent variable as the explanatory variable and the other as the control variable. This is represented in Figure 13.11.

Using Stata

In this section, we will use 2021 General Social Survey (GSS) data to learn how to run a cross-tabulation between two variables for each category of a control variable and 2020 American National Election Survey (ANES) data to examine the effect of a variable on the mean of an interval-ratio variable while controlling for a third variable.

Controlling for a Third Variable in Cross-Tabulation

For the cross-tabulation example, we will work with three variables: whether respondents think the government should spend more or less money on military and defense (*natarms2*); respondents' age, older or younger than fifty (*age2*); and whether the respondent ever served in the military (*veteran*).*

The initial research question is whether age (the independent variable) affects views on military spending (the dependent variable). The second question is whether the relationship between age and views on military spending depends on one's status as a veteran. To answer these questions, we will run a cross-tabulation between age and views on military spending, followed by separate cross-tabulations between age and views on military spending by veteran status.

Open GSS2021.dta and enter the following command into Stata to generate a cross-tabulation of age and views on military spending:

```
tabulate natarms2 age2, column chi2
```

As we saw in chapter 12, we use the "tabulate" command to request a cross-tabulation. We list the dependent variable, *natarms2*, first, followed by the independent variable, *age2*. The "column" option tells Stata to use column totals, and "chi2" requests the chi-square statistic. The output is shown in Figure 13.14.

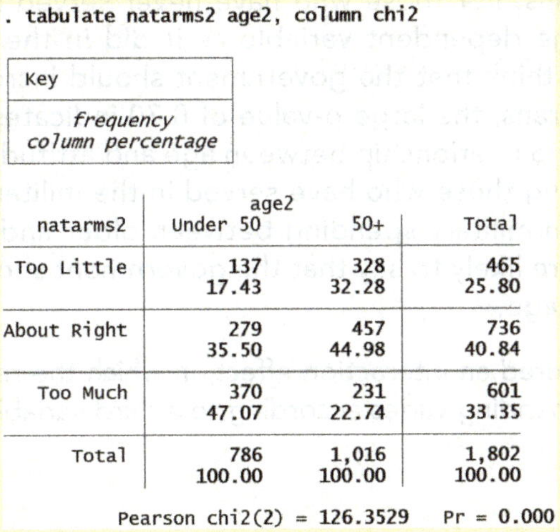

```
. tabulate natarms2 age2, column chi2
```

Key
frequency
column percentage

natarms2	age2 Under 50	50+	Total
Too Little	137 17.43	328 32.28	465 25.80
About Right	279 35.50	457 44.98	736 40.84
Too Much	370 47.07	231 22.74	601 33.35
Total	786 100.00	1,016 100.00	1,802 100.00

Pearson chi2(2) = 126.3529   Pr = 0.000

Figure 13.14

---

* *Natarms2*, *age2*, and *veteran* derive from the GSS variables *natarms*, *age*, and *vetyears*, respectively.

Looking across the columns, we see large differences in the percentages of younger and older respondents who think that the federal government allocates too little money to the military and national defense (17.4% of those younger than fifty compared to 32.3% of those fifty and older) and those who think the government spends too much on military and national defense (47.0% of those younger than fifty compared to 22.7% of those fifty and older). The chi-square statistic, 126, is associated with a p-value less than 0.001, indicating that we should reject the null hypothesis that there is no relationship between age and views on military spending. Younger people are significantly less likely than older people to support increased military spending.

Do we see the same relationship between age and views on military spending among veterans and non-veterans? To address this question, we control for veteran status by running the same cross-tabulation separately for veterans and non-veterans. Enter the following command into Stata:

```
bysort veteran: tabulate natarms2 age2, column chi2
```

Here, the "tabulate" command is preceded by the prefix "bysort veteran:". "bysort" tells Stata to sort all cases by the values of the veteran variable and run the command separately for each category of veteran. Since veteran has two categories, Stata produces two cross-tabulations, shown in Figure 13.15.

The top cross-tabulation is for respondents who have never served in the military, and the bottom cross-tabulation is for those who are serving or have served in the military. A quick glance at the p-values for the chi-square tests suggests that the relationship between age and views on military spending varies depending on whether one has served in the military. Among non-veterans the chi-square statistic is large enough to yield a p-value of less than 0.001, suggesting that we should reject the null hypothesis of no relationship between age and views on military spending among non-veterans. For those who have never served in the military, age has the same effect on the dependent variable as it did in the aggregate, with older people more likely to think that the government should increase military spending. However, among veterans, the large p-value of 0.70 indicates that we cannot reject the null hypothesis of no relationship between age and attitudes on military spending among veterans. Among those who have served in the military, there are hardly any differences in views on military spending between older and younger respondents. Veterans are much more likely to say that the government should spend more on the military, regardless of age.

This analysis has uncovered an interaction effect, in which the relationship between age and views on military spending varies according to a third variable, veteran status.

```
. bysort veteran: tabulate natarms2 age2, column chi2

-> veteran = not veteran

+------------------+
|       Key        |
|------------------|
|    frequency     |
| column percentage|
+------------------+

          |      age2       |
 natarms2 | Under 50    50+ |   Total
----------+-----------------+--------
Too Little|    118      269 |    387
          |  16.19    31.17 |  24.31
----------+-----------------+--------
About Right|   256      395 |    651
          |  35.12    45.77 |  40.89
----------+-----------------+--------
Too Much  |    355      199 |    554
          |  48.70    23.06 |  34.80
----------+-----------------+--------
    Total |    729      863 |  1,592
          | 100.00   100.00 | 100.00

         Pearson chi2(2) = 122.1103   Pr = 0.000

-> veteran = veteran

+------------------+
|       Key        |
|------------------|
|    frequency     |
| column percentage|
+------------------+

          |      age2       |
 natarms2 | Under 50    50+ |   Total
----------+-----------------+--------
Too Little|     18       58 |     76
          |  33.96    38.16 |  37.07
----------+-----------------+--------
About Right|    21       62 |     83
          |  39.62    40.79 |  40.49
----------+-----------------+--------
Too Much  |     14       32 |     46
          |  26.42    21.05 |  22.44
----------+-----------------+--------
    Total |     53      152 |    205
          | 100.00   100.00 | 100.00

         Pearson chi2(2) =   0.7034   Pr = 0.703
```

Figure 13.15

The Effect of an Independent Variable on Means of an Interval-Ratio Dependent Variable, controlling for a Third Variable

We can also use Stata to control for a third variable when we are examining the effect of an independent variable on an interval-ratio-level dependent variable. In this example, we will use data from the 2020 ANES to examine the effects of gender and party identification on the mean feeling thermometer rating of the Centers for Disease Control and Prevention (CDC), a variable measured at the interval-ratio level. We will employ the

CDC feeling thermometer rating as our dependent variable (*V202187*), party identification (*party*) as our independent variable, and gender (*gender*) as our control variable.

Open `ANES2020.dta`. To examine the effect of a party identification on feeling thermometer ratings of the CDC, type the following command into Stata:

`bysort party: tabstat V202187, statistics(mean)`

In Stata, commands preceded by "bysort" will execute the command separately for each category of the variable listed after "bysort." Stata generates the output shown in Figure 13.16, showing mean CDC ratings for the two categories of the party variable, Democrat and Republican. We see a higher mean rating of the CDC among Democrats (78.5) than Republicans (62.3).

```
. bysort party: tabstat V202187, statistics(mean)

-> party = Democrat

    Variable  |      Mean
    ----------+----------
    V202187   |  78.50815


-> party = Republican

    Variable  |      Mean
    ----------+----------
    V202187   |  62.31046
```

Figure 13.16

To control for gender, we write separate "bysort" commands for each category of the control variable. In this case, our control variable, *gender*, has two categories, 1 (man) and 2 (woman), which means that we will need two separate commands. Enter the following commands into Stata:

`bysort party: tabstat V202187 if gender==1, statistics(mean)`

`bysort party: tabstat V202187 if gender==2, statistics(mean)`

The first command tells Stata to report the means for each category of the independent variable, party identification, only for cases that have a value of 1 for gender, the control variable. The second command tells Stata to do the same, except only for cases that have a value of 2 for the control variable. Stata produces the output for men, shown in Figure 13.17, and the output for women, shown in Figure 13.18.

The tables tell us that holding gender constant does not change the relationship between political party and feelings about the CDC. Among men and women, the party difference in feelings about the CDC is almost exactly the same: about 16 points higher for

```
. bysort party: tabstat v202187 if gender == 1, statistics(mean)
```

-> party = Democrat

Variable	Mean
v202187	78.2675

-> party = Republican

Variable	Mean
v202187	61.96028

Figure 13.17

```
. bysort party: tabstat v202187 if gender == 2, statistics(mean)
```

-> party = Democrat

Variable	Mean
v202187	78.68123

-> party = Republican

Variable	Mean
v202187	62.83821

Figure 13.18

Democrats than Republicans. We see that controlling for gender does not change the overall relationship between party identification and feelings about the CDC.

Review of Stata commands

- Run a cross-tabulation between two variables, requesting chi-square statistic

    `tabulate dependent variable independent variable, column chi2`

- Run a cross-tabulation between two variables, controlling for a third variable, requesting chi-square statistic

    `bysort control variable: tabulate dependent variable independent variable, column chi2`

- Find the means of the dependent variable for each category of the independent variable

```
bysort independent variable: tabstat dependent variable,
statistics(mean)
```

- Find the means of the dependent variable for each category of the independent variable, separately for each category of a control variable (here, the control variable has two categories, 1 and 2)

```
bysort independent variable: tabstat dependent variable if control variable==1, statistics(mean)
```

```
bysort independent variable: tabstat dependent variable if control variable==2, statistics(mean)
```

Using SPSS

In this section, we will use 2021 General Social Survey (GSS) data to learn how to run a cross-tabulation between two variables for each category of a control variable and 2020 American National Election Survey (ANES) data to examine the effect of a variable on the mean of an interval-ratio variable while controlling for a third variable.

Controlling for a Third Variable in Cross-Tabulation

For the cross-tabulation example, we will work with three variables: whether respondents think the government should spend more or less money on military and defense (*natarms2*); respondents' age, older or younger than fifty (*age2*); and whether the respondent ever served in the military (*veteran*).[*]

The initial research question is whether age (the independent variable) affects views on military spending (the dependent variable). The second question is whether the relationship between age and views on military spending depends on one's status as a veteran. To answer these questions, we will run a cross-tabulation between age and views on military spending, followed by separate cross-tabulations between age and views on military spending by veteran status.

Open `GSS2021.sav`. As we have seen previously, we generate a cross-tabulation with the following sequence:

```
Analyze → Descriptive Statistics → Crosstabs
```

The variable *natarms2* is the dependent (row) variable, and *age2* is the independent (column) variable. We will indicate that SPSS should include column percentages and the chi-square statistic and p-value. The output is shown in Figure 13.19.

Looking across the columns, we see large differences in the percentages of younger and older respondents who think that the federal government allocates too little money to the military and national defense (17.4% of those younger than fifty compared to 32.3%

---

[*] *Natarms2*, *age2*, and *veteran* derive from the GSS variables *natarms*, *age*, and *vetyears*, respectively.

natarms2 * age2 Crosstabulation

			age2		Total
			Under 50	50+	
natarms2	Too Little	Count	137	328	465
		% within age2	17.4%	32.3%	25.8%
	About Right	Count	279	457	736
		% within age2	35.5%	45.0%	40.8%
	Too Much	Count	370	231	601
		% within age2	47.1%	22.7%	33.4%
Total		Count	786	1016	1802
		% within age2	100.0%	100.0%	100.0%

Chi-Square Tests

	Value	df	Asymptotic Significance (2-sided)
Pearson Chi-Square	126.353[a]	2	<.001
Likelihood Ratio	127.307	2	<.001
Linear-by-Linear Association	116.116	1	<.001
N of Valid Cases	1802		

a. 0 cells (0.0%) have expected count less than 5. The minimum expected count is 202.82.

Figure 13.19

of those fifty and older) and those who think the government spends too much on military and national defense (47.0% of those younger than fifty compared to 22.7% of those fifty or older). The chi-square statistic, 126, is associated with a p-value less than 0.001, indicating that we should reject the null hypothesis that there is no relationship between age and views on military spending. Younger people are significantly less likely than older people to support increased military spending.

Do we see the same relationship between age and views on military spending among veterans and non-veterans? To address this question, we control for veteran status by running the same cross-tabulation separately for veterans and non-veterans. We use the SPSS cross-tabulation procedure as we have done previously, but, this time, we must indicate the presence of a control variable.

`Analyze → Descriptive Statistics → Crosstabs`

This opens the "Crosstabs" dialog box (shown in Figure 13.20). As we have seen, we place the dependent variable (*natarms2*) in the "Row(s)" box, and we place the independent variable (*age2*) in the "Column(s)" box. We place the control variable (*veteran*) into the third box labeled "Layers," which is what SPSS calls control variables (circled).

The edited output is shown in Figure 13.21. Since veteran has two categories, SPSS produces two separate panels in the cross-tabulation.

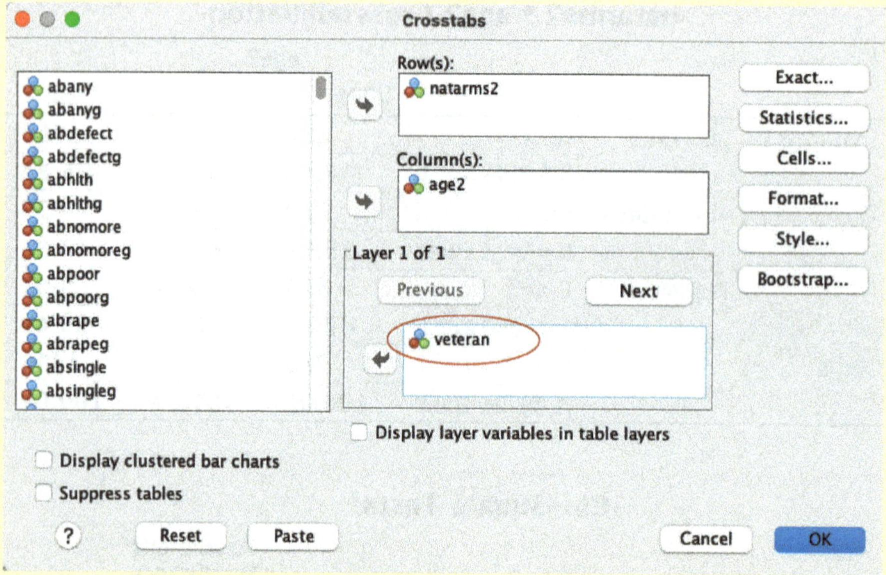

Figure 13.20

The top panel of the cross-tabulation is for respondents who have never served in the military, and the bottom panel is for those who are serving or have served in the military. A quick glance at the p-values for the chi-square tests suggests that the relationship between age and views on military spending varies depending on whether one has served in the military. Among non-veterans the chi-square statistic is large enough to yield a p-value less than 0.001, suggesting that we should reject the null hypothesis of no relationship between age and views on military spending among non-veterans. For those who have never served in the military, age has the same effect on the dependent variable as it did in the aggregate, with older people more likely to think that the government should increase military spending. However, among veterans, the large p-value of 0.70 indicates that we cannot reject the null hypothesis of no relationship between age and attitudes on military spending among veterans. Among those who have served in the military, there are hardly any differences in views on military spending between older and younger respondents. Veterans are much more likely to say that the government should spend more on the military, regardless of age.

This analysis has uncovered an interaction effect, in which the relationship between age and views on military spending varies according to a third variable, veteran status.

The Effect of an Independent Variable on Means of an Interval-Ratio Dependent Variable, Controlling for a Third Variable

We can also use SPSS to control for a third variable when we are examining the effect of an independent variable on an interval-ratio-level dependent variable. In this example, we will use data from the 2020 ANES to examine the effects of gender and party identification on the mean feeling thermometer rating of the Centers for Disease Control

natarms2 * age2 * veteran Crosstabulation

veteran				age2 Under 50	age2 50+	Total
not veteran	natarms2	Too Little	Count	118	269	387
			% within age2	16.2%	31.2%	24.3%
		About Right	Count	256	395	651
			% within age2	35.1%	45.8%	40.9%
		Too Much	Count	355	199	554
			% within age2	48.7%	23.1%	34.8%
	Total		Count	729	863	1592
			% within age2	100.0%	100.0%	100.0%
veteran	natarms2	Too Little	Count	18	58	76
			% within age2	34.0%	38.2%	37.1%
		About Right	Count	21	62	83
			% within age2	39.6%	40.8%	40.5%
		Too Much	Count	14	32	46
			% within age2	26.4%	21.1%	22.4%
	Total		Count	53	152	205
			% within age2	100.0%	100.0%	100.0%
Total	natarms2	Too Little	Count	136	327	463
			% within age2	17.4%	32.2%	25.8%
		About Right	Count	277	457	734
			% within age2	35.4%	45.0%	40.8%
		Too Much	Count	369	231	600
			% within age2	47.2%	22.8%	33.4%
	Total		Count	782	1015	1797
			% within age2	100.0%	100.0%	100.0%

Chi-Square Tests

veteran		Value	df	Asymptotic Significance (2-sided)
not veteran	Pearson Chi-Square	122.110[b]	2	<.001
	N of Valid Cases	1592		
veteran	Pearson Chi-Square	.703[c]	2	.703
	N of Valid Cases	205		
Total	Pearson Chi-Square	126.592[a]	2	<.001
	N of Valid Cases	1797		

Figure 13.21

and Prevention (CDC), a variable measured at the interval-ratio level. We will employ the CDC feeling thermometer rating as our dependent variable (*V202187*), party identification (*party*) as our independent variable, and gender (*gender*) as our control variable.

To examine the simultaneous effect of an independent and control variable on an interval-ratio dependent variable, we will follow the same logic we employed in the chapter. We will first examine the effect of the independent variable (*party*) on the dependent variable (*V202187*). Then we will re-examine this relationship, controlling for gender.

Open `ANES2020.sav`. To obtain the mean of the dependent variable broken down by the categories of the independent variable, we use an SPSS sequence that we have seen in previous chapters:

`Analyze → Descriptive Statistics → Explore …`

This opens the "Explore" dialog box. We move our dependent variable, *V202187*, into the "Dependent List" box. We move our independent variable, *party*, into the "Factor List" box. We click on "OK," and SPSS places the statistics in the output window. The "Explore" command produces a slew of descriptive statistics. In Figure 13.22, we have edited the output to show just the means. We see that the mean rating of the CDC is much higher among Democrats than Republicans.

	party		Statistic
POST: Feeling thermometer: Center for Disease Control (CDC)	Democrat	Mean	78.51
	Republican	Mean	62.31

Figure 13.22

To control for gender, we will use the same procedure we just used. But, before we do that, we must tell SPSS to execute the procedure separately for women and men (the categories of our control variable). Toward that end, we use the following SPSS sequence:

`Data → Split File …`

This opens the "Split File" dialog box (shown in Figure 13.23). We select "Compare Groups" and move our control variable, *gender*, into the box labeled "Groups Based on:."

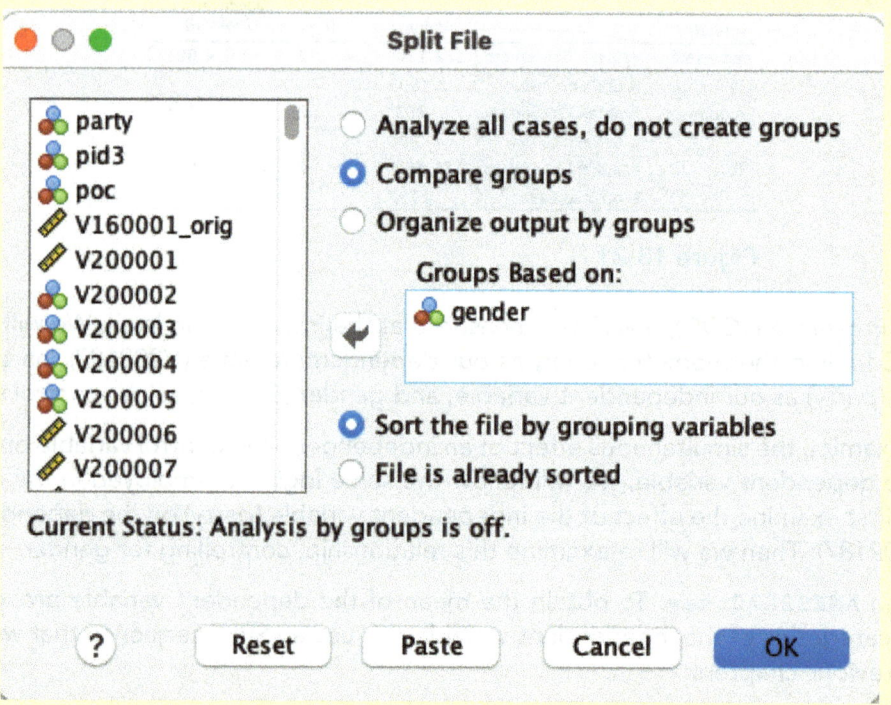

Figure 13.23

We click on "OK," and any SPSS procedure we now run will produce separate results for females and males, the two categories of our control variable. Using the same sequence we used above, our final step is to tell SPSS to once again calculate the mean feeling thermometer rating of the CDC broken down by party identification, our independent variable.

`Analyze → Descriptive Statistics → Explore …`

Recall that SPSS will now also control for gender, since our "Split File" command is still active. The edited output is shown in Figure 13.24.

Feeling Thermometer – CDC

gender	party	Statistic
Man	Democrat	79.53
	Republican	63.06
Woman	Democrat	80.49
	Republican	64.10

Figure 13.24

The tables tell us that holding gender constant does not change the relationship between political party and feelings about the CDC. Among men and women, the party difference in feelings about the CDC is almost exactly the same: about 16 points higher for Democrats than Republicans. We see that controlling for gender does not change the overall relationship between party identification and feelings about the CDC.

Remember that the "Split File" command is still active, and any SPSS procedure you employ subsequently will be conducted separately for women and men. To analyze all cases together, return to the "Split File" dialog box (above) and select "Analyze all cases, do not create groups." Click on "OK," and you can resume analyzing the full sample.

Review of SPSS procedures

- Run a cross-tabulation between two variables, requesting chi-square statistic

    `Analyze → Descriptive Statistics → Crosstabs`

    Put the dependent variable in the "Row" box and the independent variable in the "Column" box.

    Click on "Statistics" to request the chi-square.

- Run a cross-tabulation between two variables, controlling for a third variable, requesting chi-square statistic.

    `Analyze → Descriptive Statistics → Crosstabs`

    Put the dependent variable in the "Row" box, the independent variable in the "Column" box, and the control variable in the box labeled "Layers."

    Click on "Statistics" to request the chi-square.

- Analyze the joint effects of independent and control variables on an interval-ratio dependent variable.

To separate the file according to the categories of the control variable:

`Data → Split File` and select "Compare Groups."

To generate the mean of the dependent variable broken down by the categories of the independent variable:

`Analyze → Descriptive Statistics → Explore`

Practice Problems

1. In a cross-tabulation in chapter 3, we looked at the relationship between social class and happiness. That relationship is shown in Table 3.17 (p. 133). There is a strong relationship between the two, with higher class status associated with higher levels of happiness. The chi-square for this relationship is statistically significant at $p<0.001$.

   A statistics student wondered whether the positive effects of higher class on happiness applies to all groups of people. Perhaps, she thought, groups that anticipate upward mobility would not tie their happiness so strongly to social class. Because immigrants as a group tend to anticipate upward mobility, the student decided to hold constant whether respondents were born in the United States or not. Using the GSS data, they produced the cross-tabulations shown in Table 13.14.

   Table 13.14 Cross-tabulation of Happiness by Class, Controlling for Being Born inside or outside the United States (Column Percentages)

	Happiness	Class — Lower or Working Class	Class — Middle or Upper Class	Total
Born in the US	Very happy	14.2	22.8	18.9
	Pretty happy	55.2	59.0	57.3
	Not too happy	30.5	18.2	23.8
	Total	100	100	100
Born outside the US	Very happy	22.3	28.3	25.4
	Pretty happy	52.1	60.9	56.7
	Not too happy	25.6	10.9	17.8
	Total	100	100	100

   Source: 2021 GSS.

a. What is the research question addressed by the original cross-tabulation in Table 3.17 (p. 133)?

b. What is the research question addressed by the student's analysis of the data, shown in Table 13.14?

c. Identify the independent variable, dependent variable, and control variable in Table 13.14.

d. Describe the relationship between class and happiness among those born inside the United States. Describe the relationship among those born outside the United States.

e. Compare the results of the original cross-tabulation (page 133) to the student's follow-up cross-tabulations. Is there evidence of a spurious or moderating relationship between class and happiness? Explain your answer.

2. A researcher studying hate crimes notices that more hate crimes tend to happen in states with larger populations. Table 13.15 shows a cross-tabulation of states' population and prevalence of hate crimes.

Table 13.15 Cross-Tabulation of State Population by Prevalence of Hate Crimes*

Prevalence of Hate Crimes:	Population		
	Small	Medium	Large
Low	75%	15%	5%
Medium	20%	60%	25%
High	5%	25%	70%

*Hypothetical data

a. What is the relationship between population and prevalence of hate crimes, as shown in Table 13.15?

b. Tables 13.16 and 13.17 show two hypothetical cross-tabulations of state population by hate crime prevalence, controlling for whether states have hate crime laws that cover sexual orientation and gender identity. Answer the following question for both tables: Is there a relationship between state population and prevalence of hate crimes once the protection of sexual orientation and gender identity is held constant?

c. Answer this question for Table 13.16 and 13.17: Is the relationship between state population and hate crime prevalence spurious? Explain your answer.

d. Real data suggest that hate crime laws may not deter hate crimes. Is this finding contradicted by Table 13.16? by Table 13.17?

Table 13.16 Cross-Tabulation of State Population by Prevalence of Hate Crimes, Controlling for Sexual Orientation and Gender Included in Hate Crime Law*

		Population		
	Hate Crime Prevalence	Small	Medium	Large
Sexual orientation and gender identity NOT covered by hate crime law	Low	10%	11%	10%
	Medium	30%	31%	35%
	High	60%	58%	55%
	Total	100%	100%	100%
Sexual orientation and gender identity covered by hate crime law	Low	75%	80%	82%
	Medium	15%	15%	13%
	High	10%	5%	5%
	Total	100%	100%	100%

*Hypothetical data

Table 13.17 Cross-Tabulation of State Population by Prevalence of Hate Crimes, Controlling for Sexual Orientation and Gender Included in Hate Crime Law*

		Population		
	Hate Crime Prevalence	Small	Medium	Large
Sexual orientation and gender identity NOT covered by hate crime law	Low	75%	15%	9%
	Medium	20%	60%	21%
	High	5%	25%	70%
	Total	100%	100%	100%
Sexual orientation and gender identity covered by hate crime law	Low	73%	17%	10%
	Medium	22%	60%	19%
	High	5%	23%	71%
	Total	100%	100%	100%

*Hypothetical data

3. In Problem 1, we revisited the original relationship between social class and happiness. Figure 13.25 represents the relationship.

   a. Draw a figure that shows what the student's follow-up analysis was testing in the problem. Explain the figure. Was the relationship that was tested empirically supported?

   b. Consider the three criteria for determining causality discussed in this chapter: association, temporal and logical order, and ruling out alternative explanations. Discuss whether these conditions have been satisfied in order to determine a causal relationship between class and happiness, as depicted in Figure 13.25.

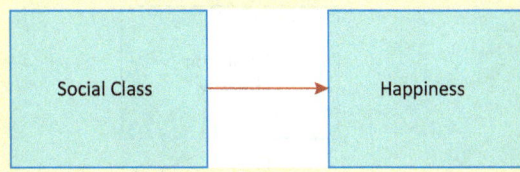

Figure 13.25 Relationship between Social Class and Happiness

4. Problem 2 showed two hypothetical cross-tabulations of state population by hate crime prevalence, controlling for whether or not states have hate crime laws that cover sexual orientation and gender identity.

   a. Draw two figures that represent the relationships depicted in Tables 13.16 and 13.17.

   b. Consider the three criteria for determining causality discussed in this chapter: association, temporal and logical order, and ruling out alternative explanations. Discuss whether these conditions have been satisfied in order to determine a causal relationship between state population and hate crime prevalence, as depicted in Table 13.16 and Table 13.17.

5. In the wake of a devastating hurricane, data collected by a relief organization showed differences in the mean amount of time that people went without power, by age, as shown in Table 13.18.

   Table 13.18 Mean Days without Power, by Age

Age	Mean Days without Power
Under 18	17
19 to 50	30
Over 50	45

   a. Describe the relationship between age and days without power shown in Table 13.18.

   b. Should we conclude that age caused the length of time that people went without power after the hurricane? Explain your answer.

   c. The organization mounted special outreach efforts to the elderly in response to the data. As relief workers traveled to the residences of people who lacked power, they noticed that many of the older people they visited lived in remote, rural areas with few people. One of the organization's workers proposed entering a third variable, population density, to test whether the relationship between age and days without power still held. The reanalysis of the data indicated that the relationship worked as shown in Figure 13.26. Explain in words what Figure 13.26 tells us about the relationship between age and days without electricity.

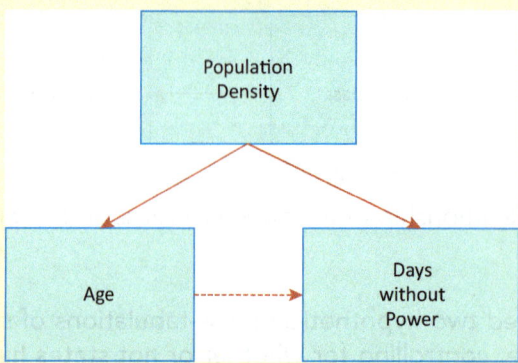

Figure 13.26 Relationship between Age and Days without Electricity, Controlling for Population Density

6. Analysis of recent data from a sample of men and women showed that younger people were far more likely to support the #MeToo movement than older people. Table 13.19 shows the relationship between age and support for #MeToo, controlling for gender. Support for #MeToo is measured on a 100-point scale, with 0 indicating no support and 100 indicating total support.

    a. A two-sample test for mean differences shows a statistically significant difference between the younger and older groups among men, but not among women. What kind of variable is gender?

    b. Draw a figure that represents the relationship among the three variables shown in Table 13.19.

Table 13.19 Mean Support for the #MeToo Movement, by Age and Gender*

	Age	Mean Support
Women	18 to 30	90
	Over 30	88
Men	18 to 30	71
	Over 30	45

*Hypothetical data

7. A researcher found that millennials who read more young adult (YA) romance books tended to have more liberal attitudes toward sexuality than those who read fewer. But does reading young adult romance books cause people to develop more liberal attitudes? Imagine that we analyzed the relationship between reading young adult romance books and attitudes toward sexuality, with the addition of parents' political ideology as a control variable, as shown in Table 13.20.

Table 13.20 Relationship between Frequency of Reading Young Adult Romance Books and Attitudes toward Sexuality, Controlling for Parents' Political Ideology

		Frequency of Reading YA Romance Novels:		
		No Books	1 to 5 Books	6+ Books
Conservative parents	Conservative	80%	82%	79%
	Moderate	15%	12%	16%
	Liberal	5%	6%	5%
	Total	100%	100%	100%
	$\chi^2 = 1.75, p > 0.20$			
Liberal parents	Conservative	7%	4%	1%
	Moderate	11%	13%	11%
	Liberal	82%	83%	88%
	Total	100%	100%	100%
	$\chi^2 = 2.0, p > 0.20$			

a. Is there still a relationship between reading YA romance and people's political ideology, after controlling for parents' political ideology? Explain your answer.

b. Is parents' political ideology an antecedent or intervening variable? Explain your answer.

c. Given the data in Table 13.20, what can you say about the relationship between parents' political ideology and how many YA romance books their children read?

8. A researcher knows that, in the past, American men were more likely than women to travel for work or to relocate for work. They wonder whether this is still the case. Using the GSS, they generated the following cross-tabulation between gender and whether the respondent lives in the same or different location compared to where they lived at age sixteen. Table 13.21 shows the results.

a. Which is the independent and dependent variable? What does Table 13.21 show about the relationship between them?

Table 13.21 Geographic Mobility since Age Sixteen, by Gender (Percentages)

Geographic Mobility since Age 16	Male	Female	Total
Same state, same city	26.2	25.5	25.8
Same state, different city	33.4	41.4	37.8
Different state	40.4	33.1	36.4
Total	100	100	100

Chi-square 8.989, p = 0.011

b. The researcher is surprised by the results and decides to control for age. Table 13.22 shows the results controlling for whether the respondent is younger than fifty or fifty or older. What does Table 13.22 show about whether and how the relationship between gender and geographical mobility is affected by age?

c. Is age an antecedent, intervening, or moderating variable?

Table 13.22 Geographic Mobility since Age Sixteen by Sex, Controlling for Age (Percentages)

Under 50	Geographic Mobility since Age 16	Male	Female	Total
	Same state, same city	33.9	37.6	36.1
	Same state, different city	34.2	32.5	33.2
	Different state	31.8	29.9	30.7
	Total	100	100	100

Chi-square = 2.18, p = 0.336

50 and older	Geographic Mobility since Age 16	Male	Female	Total
	Same state, same city	26.2	25.5	25.8
	Same state, different city	33.4	41.4	37.8
	Different state	40.4	33.1	36.4
	Total	100	100	100

Chi-square = 14.29, p = 0.001

9. An article in the *New York Times*, titled "How You Felt about Gym Class May Impact Your Exercise Habits Today," described a study that found a positive association between two variables: positive experiences in gym class and level of physical activity as an adult.

a. Do you think that the title of the article accurately represents this finding?

b. Members of a hypothetical task force charged with increasing children's physical activity were excited to read about the study. In their view, it provided evidence that getting children to enjoy physical activity (and enjoy gym class) could lead to a healthier adult population. One task force member commissioned a follow-up study to investigate factors that might explain why gym class experiences are associated with physical activity as an adult. Two of the factors considered in the follow-up study were appreciation of teamwork and self-esteem, as shown in Figure 13.27. According to the figure, what kind of variables are appreciation of teamwork and self-esteem?

c. Explain in words what each model in Figure 13.27 proposes about how the three variables are related.

d. Tables 13.23 and 13.24 show a sample of adults' physical activity by appreciation of teamwork and self-esteem, respectively. Based on the data shown in these tables, identify which of the models in Figure 13.27 cannot accurately represent the relationship among the three variables. Explain why.

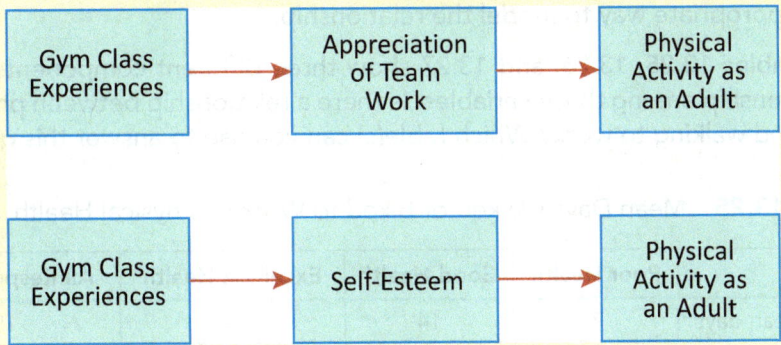

Figure 13.27 Two Models of Why Gym Class Experiences Affect Physical Activity as an Adult

Table 13.23 Adults' Physical Activity, by Appreciation of Teamwork

Physical Activity	Appreciation for Teamwork			
	Low	Medium	High	Total
Low	52%	49%	51%	50%
Medium	39%	40%	41%	40%
High	9%	11%	8%	10%
Total	100%	100%	100%	100%

Table 13.24 Adults' Physical Activity, by Self-Esteem

Physical Activity	Self-Esteem			
	Low	Medium	High	Total
Low	63%	47%	13%	50%
Medium	33%	43%	50%	40%
High	4%	10%	37%	10%
Total	100%	100%	100%	100%

10. A city planner wants to learn about the factors that affect whether people walk or bike to work. She collects data from a random sample of five hundred employed people who reside in her city. She wants to examine the effects of overall health and proximity of residence to work on the number of days that people bike or walk to work per year.

a. Assuming that people's decisions about where to live in relation to their work are unaffected by their physical health, draw the appropriate visual representation of the relationship among these three variables. Explain why this is the appropriate way to model the relationship.

b. Tables 13.25, 13.26, and 13.27 show three different components of the relationship among these variables. Is there a relationship between physical health and walking to work? Which table(s) can you use to answer this question?

Table 13.25 Mean Days Walked or Biked to Work by Physical Health

	Poor Health	Good Health	Excellent Health	All Respondents
Mean days	7	14	27	15

Table 13.26 Mean Days Walked or Biked to Work by Physical Health, Controlling for Proximity of Residence to Work

	Poor Health	Good Health	Excellent Health	All Respondents
Lives within 1 mile	14	30	40	28
Lives more than 1 mile away	2	10	22	9

Table 13.27 Mean Days Walked or Biked to Work by Proximity of Residence to Work and Physical Health

	Lives within 1 Mile	Lives More Than 1 Mile Away	Difference
Poor health	14	2	12
Good health	30	10	20
Excellent health	40	22	18

c. Is there a relationship between proximity of residence to work and walking/biking to work? Which table(s) can you use to answer this question?

d. Describe the relative effects of physical health and proximity of residence to work on the number of days that people walk or bike to work, on average.

e. The city planner wants more people to walk and bike to work. She looks at the results in these tables and concludes that, in order to increase the number of bike and pedestrian commuters, building affordable housing close to the business district where many of the city's residents work is a better use of city money than launching a health initiative. Do you agree? Explain your answer.

11. For each of the following examples, answer these questions: Which is the independent variable? Assess the relationship according to the first two criteria for determining causality discussed in this chapter: association, and temporal and logical order.

a. Political scientists have found that people's ideological alignment as liberal or conservative is positively associated with their parents' ideological alignment.

b. People who strongly agree that "the success of the United States is part of God's plan" tend to disagree that "all of the greatest advances of humanity have come from science and technology" (and vice versa).

c. The Pew Research Center found that about 3% of adults ages eighteen to twenty-nine identified as gender non-binary, compared with 1.6% of the overall population.

d. The GSS shows that military veterans are more likely than non-veterans to have a great deal of confidence in the US military.

12. For each of the following examples, answer these questions: What are the independent and dependent variables? Assess the relationship according to the first two criteria for determining causality discussed in this chapter: association, and temporal and logical order.

    a. Majorities of gun owners and non-owners favor making private gun sales and sales at gun shows subject to background checks, according to research from the Pew Research Center. Which is the independent variable?

    b. A 2023 *New York Times* headline reads "Addiction Treatment Medicine Is Vastly Underprescribed, Especially by Race, Study Finds."

    c. There is a relationship between Republican and Democratic party affiliation and knowledge about science on climate change.

    d. The GSS shows that supporters of legalizing marijuana are significantly younger than opponents.

13. For each part below, identify the independent variable, dependent variable, and control variable. Also, specify what kind of control variable is described.

    a. A study examines the effect of the diversity of collaborative teams on the creativity of the teams' ideas. It finds that diversity is positively related to creativity only when trust is high among members of the team.

    b. Data show that boys in a California high school are 20 percentage points more likely than girls in the school to be suspended from school. After accounting for gender differences in the frequency of physical altercations, boys are only about 5 percentage points more likely to be suspended.

    c. A researcher finds that the size of protests is weakly associated with achievement of social movements' policy goals. When they account for media coverage of protests, they find that protests covered by media are more strongly associated with achievement of social movement goals, while those not covered by media are not associated with achievement of goals.

    d. Survey respondents who believe that climate is changing due to human activity favor more government spending on protecting the environment. A policy analyst knows gender affects views on government spending (with women somewhat more likely to favor it). The analyst looks at the relationship between beliefs about climate change and government spending on the environment for women and for men and finds that there is a strong relationship between those variables for both gender categories.

e. There is a significant relationship between gender and whether one's spouse works full-time, works part-time, or keeps house. Men are more likely to have spouses who work part-time or keep house, and women are more likely to have spouses who work full-time. An astute student realizes that this is probably due to traditional gender roles and decides to control for sexual identity. The relationship between gender and spouse's work status remains for respondents who identify as heterosexual but does not exist for those who identify as lesbian, gay, or bisexual.

14. For each part below, identify the independent variable, dependent variable, and control variable. Also, specify what kind of control variable is described.

   a. A study finds that people who ski tend to be more politically liberal than people who don't ski. Controlling for the region of the United States where people live, skiing no longer affects political views.

   b. A study finds higher rates of depression in Northern states than in Southern states. When accounting for average hours of sunlight per week, region of residence no longer affects depression rates.

   c. Evangelical Christians are much more likely than people who are not Evangelical Christians to say they consider themselves pro-life. When controlling for the importance of religion in their life, evangelical Christianity is only associated with considering oneself pro-life for those who say religion is important in their lives.

   d. Respondents who lived in a home their family owned at age sixteen are more likely than those who lived in a rental to live in a different city or state from where they lived at age sixteen. A researcher realizes that this seemingly paradoxical effect is probably due to class. She finds that, when considering class, middle-class respondents whose parents owned their home were still more likely to be geographically mobile, but there was no relationship between the variables for low income, working-class, and upper-class respondents.

   e. Does studying improve test scores? A researcher finds test scores are highest for those who study in a group versus alone, but that the number of hours spent studying improves scores for both groups.

Stata Problems

Here we use data from the 2021 General Social Survey (GSS) to examine the relationship between political ideology and opinions about the government's spending on education. We use the variable *libcons* to measure three categories of respondents' political ideologies: liberal, moderate, and conservative. The variable *nateduc* indicates to what extent respondents think that the government is spending the right amount on education. After examining this initial relationship, we control for age using a variable called *age2*, which measures whether respondents are younger than fifty years of age or older than forty-nine years of age.

1. Open GSS2021.dta and use the "tabulate" command to run a cross-tabulation between *libcons* and *nateduc*, with *libcons* as the independent variable. Be sure to ask Stata to provide the chi-square statistic.

2. What percentage of liberals think that the government spends too much on education? What percentage of conservatives think so?

3. Does the output suggest a relationship between political ideology and opinions about education spending?

4. Use the "bysort" prefix with the "tabulate" command to run a cross-tabulation between *libcons* and *nateduc*, controlling for *age2*. Be sure to ask Stata to provide the chi-square statistics.

5. Is there a relationship between political ideology and opinions about education spending among the younger respondents? How can you tell?

6. Is there a relationship between political ideology and opinions about education spending among the older respondents? How can you tell?

SPSS Problems

Here we use data from the 2021 General Social Survey (GSS) to examine the relationship between political ideology and opinions about the government's spending on education. We use the variable *libcons* to measure three categories of respondents' political ideologies: liberal, moderate, and conservative. The variable *nateduc* indicates to what extent respondents think that the government is spending the right amount on education. After examining this initial relationship, we control for age using a variable called *age2*, which measures whether respondents are younger than fifty years of age or older than forty-nine years of age.

1. Open GSS2021.sav and use the "Crosstabs" procedure to run a cross-tabulation between *libcons* and *nateduc*, with *libcons* as the independent variable. Be sure to ask SPSS to provide the chi-square statistic.

2. What percentage of liberals think that the government spends too much on education? What percentage of conservatives think so?

3. Does the output suggest a relationship between political ideology and opinions about education spending?

4. Use the "Crosstabs" procedure to run a cross-tabulation between *libcons* and *nateduc*, controlling for *age2*. Be sure to ask SPSS to provide the chi-square statistics.

5. Is there a relationship between political ideology and opinions about education spending among the younger respondents? How can you tell?

6. Is there a relationship between political ideology and opinions about education spending among the older respondents? How can you tell?

Notes

[1] Data retrieved from https://www.partsgeek.com/mmparts/car_inspection_requirements_by_state_a_compendium.html.

[2] Edward R. Tufte. 1974. *Data Analysis for Politics and Policy*. Englewood Cliffs, NJ: Prentice-Hall, 5–29.

[3] Population density data retrieved from https://www2.census.gov/programs-surveys/decennial/2020/data/apportionment/population-density-data-table.pdf.

[4] Caroline Newman and Alexandra Angelich. 2017. "The Driving Divide." *UVA Today*. September 1. Retrieved from https://www.news.virginia.edu/content/how-americas-political-divide-reflected-your-daily-commute?utm_source=DailyReport&utm_medium=email&utm_campaign=news.

[5] See, for example, Raymond E. Wolfinger and Steven J. Rosenstone. 1980. *Who Votes?* New Haven, CT: Yale University Press; Ruy A. Teixeira. 1992. *The Disappearing American Voter*. Washington, DC: Brookings.

Chapter 14

Describing Linear Relationships between Variables

Correlation and Regression

Social scientists have observed that when income inequality increases, upward social mobility tends to decrease. For example, as the gap between higher and lower income families grows, it becomes less likely that children will grow up to earn higher incomes than their parents. Researchers have observed this phenomenon, known as the "Great Gatsby Curve," in many countries. Economist Raj Chetty has established a research organization called Opportunity Insights to study social mobility in the contemporary United States.

In one paper, Chetty's team examined the strength of the relationship between children's family incomes and the incomes that those children earn as adults.[1] The stronger the relationship, the fewer opportunities there are for social mobility. The Gatsby Curve would predict that, as income inequality has increased in the United States, the relationship between incomes of families of origin and adult incomes would have increased. What did the researchers find?

First, the team examined the correlations between the children's family incomes and their incomes as adults for people born in three different periods: 1971 to 1974, 1975 to 1978, and 1979 to 1982. Economic inequality increased steadily across these periods. Did that seem to stymie upward mobility? Figure 14.1 presents the scatterplots between these two variables for each of the three time periods. Scatterplots show paired values of two variables for cases in the sample. Here, a point on the scatterplot represents the income of the family of origin (x axis) and the income earned as an adult (y axis).[*] Figure 14.1 shows a positive relationship between these two variables, with increases in family of origin incomes (x axis) associated with increases in adult children's incomes (y axis). So, it is true that adult outcomes are affected by the status of their childhood families. However, the scatterplot also suggests that the strength of the relationship between childhood and adult incomes is similar in the three time periods, casting doubt on the implication of the Gatsby Curve: that the relationship between childhood and adult circumstances would be stronger in periods of increased economic inequality.

---

[*] In this analysis, incomes have been converted from dollars to percentiles, such that the value tells us what percentile the income falls in the distribution of all incomes for the relevant year.

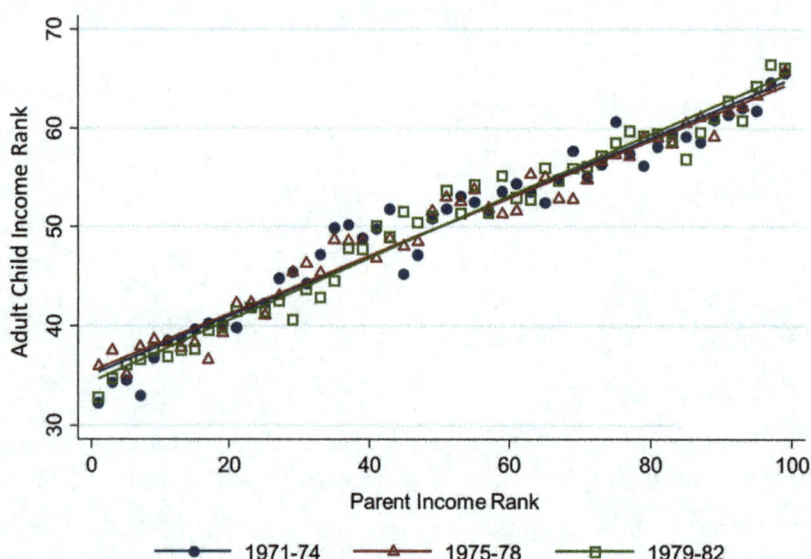

Figure 14.1 Scatterplot of Adult Income Rank against Parents' Income During Childhood (Chetty, et al.)

Source: Opportunity Insights. Retrieved from https://opportunityinsights.org/wp-content/uploads/2018/04/trends_slides.pdf.

The researchers used a technique called regression analysis, a focus of this chapter, which measures the linear relationship between independent and dependent variables. The **regression coefficients** or slopes, yielded by the analyses estimate how much adult income changes for given increases in family of origin income. In one regression model, the coefficient for family of origin income indicates that for every 1 percentile increase in family of origin income, adult income is expected to increase by 0.3 percentiles. By extension, a 30-percentile increase in family of origin income is associated with a 9-percentile increase in adult income (30 x 0.3). In fact, the regression coefficient for each time period is about 0.3, suggesting that social mobility has stayed at roughly the same level in the United States during a period of increasing economic inequality.

This study works with the main statistical technique covered by this chapter, regression, and focuses on one of the key results of regression analysis, the coefficient, or slope, for an independent variable. Before turning to regression analysis, we discuss correlation, an important building block of regression and a useful statistical technique in itself.

⊖ Correlation Coefficients

A **correlation coefficient** measures the **linear relationship** between two interval-ratio variables (x and y), indicating the extent to which the values of the two variables "move together." In other words, the correlation coefficient tells us the degree to which the values of one variable change as the values of the other variable change.

The correlation coefficient communicates two important characteristics of the relationship between two variables: direction and strength. The sign of the correlation coefficient indicates the direction of the relationship. If the correlation is positive, the values of the variables change together in the same direction. With a positive relationship, as the values of the first variable increase (or decrease), the values of the second variable also increase (or decrease). With negative correlations, the values of the two variables move in opposite directions. As the values of the first variable increase (or decrease), the values of the second variable decrease (or increase).

The size of the correlation coefficient communicates the strength of the relationship between two variables. All correlation coefficients range from −1 to +1. The strongest possible relationship between two variables lies at either end of the spectrum (−1 or +1), with either value indicating a "perfect" linear correlation between the two variables. In a perfect positive relationship, every time the value for one variable increases, the value of the other variable also increases at a constant rate. In a perfect negative relationship, every time the value for one variable increases, the value of the other variable decreases at a constant rate. The closer the correlation is to 0, the weaker the relationship, with 0 indicating no linear relationship at all between the two variables. Figure 14.2 presents one general rule for what the absolute value of a correlation indicates about the strength of a relationship between two variables.[2]

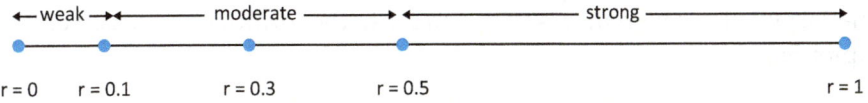

Figure 14.2 Assessing the Strength of Correlation Coefficients by Their Absolute Values

According to Figure 14.2, absolute correlations that fall below 0.1 are generally considered weak, those falling between 0.1 and 0.5 are considered moderate, and those falling above 0.5 are considered strong. With almost any kind of data that social scientists are interested in studying, there are numerous factors that affect the values of any given variable. This means that we never observe a perfect correlation with real social science data. For example, a social scientist studying the factors that influence students' grades will never find one variable that is perfectly associated with grades because they are influenced by so many different variables (e.g., family resources, school resources, friends' attitudes, etc.). It is important to note that assessments of the strength of correlations differ across fields of study. For example, medical researchers consider correlations lower than 0.8 to be of moderate or lower strength.[3]

Correlation coefficients assume that the relationship between two variables is linear, which means that the variables move together in the same way regardless of the value of the variables. In a linear relationship, a change in one variable is always associated with the same size change in the other variable regardless of the value of the first variable. Below, we discuss scatterplots as a way of evaluating the linearity of a relationship between two interval-ratio variables.

Calculating Correlation Coefficients

Although researchers rarely calculate correlation coefficients by hand, examining how they are calculated can facilitate a deeper understanding of this statistic. First, we discuss covariance, an important component of the formula for the correlation coefficient. As its name implies, covariance tells us the extent to which variations around the means for two variables are associated. The formula for the covariance of two variables, x and y, is:

$$COV_{(x,y)} = \frac{\Sigma(x-\bar{x})(y-\bar{y})}{N-1}$$

The quantity in the numerator is called the "sum of cross products." It multiplies the difference between each value for x and the mean of x ($\bar{x}$) by the difference between each value for y and the mean of y ($\bar{y}$). If two values for x and y are both either higher or lower than their respective means, the product of the two differences will be positive. If a value for x is higher than its mean and a value for y is lower than its mean, their product will be negative. The sum of these cross products indicates whether deviations from the mean for x and y both tend to fall above (or below) the variables' respective means (positive sum) or whether values for one variable tend to fall above its mean when values for the other tend to fall below its mean (negative sum). Dividing the sum of cross products by N − 1 yields the covariance between x and y.

The formula for covariance is very similar to the formula for variance, s^2, which we saw in chapter 5:

$$s^2 = \frac{\Sigma(y-\bar{y})^2}{N-1}$$

The difference between variance and covariance is that the numerator for variance is the "sum of squares," $\Sigma(y-\bar{y})^2$, the spread of a variable around its own mean, while the numerator for covariance is the sum of cross products, $\Sigma(x-\bar{x})(y-\bar{y})$.

Whereas variances are always positive because differences from the mean are squared, covariances can be positive or negative, indicating the direction of the relationship between the two variables. However, similar to the variance of an individual variable, the units for a covariance, literally the units of the x variable multiplied by the units of the y variable, are difficult to interpret. There is no way to make intuitive sense of such a unit. Just as we saw in chapter 5 with variance and standard deviation, statisticians convert the covariance into the correlation coefficient to make the units more interpretable.

Like the covariance, the correlation coefficient measures the strength and direction between two variables. The sample correlation between two variables x and y is represented by the letter "r." By formula, the correlation coefficient is:

$$r(x,y) = \frac{COV_{(x,y)}}{s_x s_y}$$

where $COV_{(x,y)}$ is the covariance between x and y, and s_x and s_y are the standard deviations for x and y, respectively. We can think about the correlation coefficient as a standardized version of the covariance. We standardize the covariance of x and y by dividing it by the product of the two variables' standard deviations.

Does this standardization strategy look familiar to you? *Hint*: Look back at chapter 5. There, we saw that dividing the distance of a value from the mean by the variable's standard deviation converts the value into standard deviation units, allowing us to compare values measured in different metrics (e.g., SAT and ACT scores). Similarly, the correlation coefficient reports a standardized measure of association between x and y by dividing the covariance by the product of the standard deviations of x and y. The correlation coefficient converts the original units of the two variables into standard deviation units.

Scatterplots: Visualizing Correlations

Scatterplots show the paired values of two interval-ratio variables for each case in a data set. They are a useful tool for visualizing the correlation between two variables, showing the strength, direction, and linearity (that is, the degree to which the graph of their values forms a straight line) of the relationship between two variables. We make a scatterplot for two variables, x and y, by jointly plotting the values of x and y for each case in the data. The convention in statistics is to place the values of x, the independent variable, on the x axis and the values of y, the dependent variable, on the y axis.

Before turning to scatterplots that use real data, we examine scatterplots using hypothetical data to illustrate how they show the strength and direction of relationships between x and y. Figure 14.3 shows an example of a perfect positive correlation between x and y, on the left, and a perfect negative correlation between x and y, on the right.

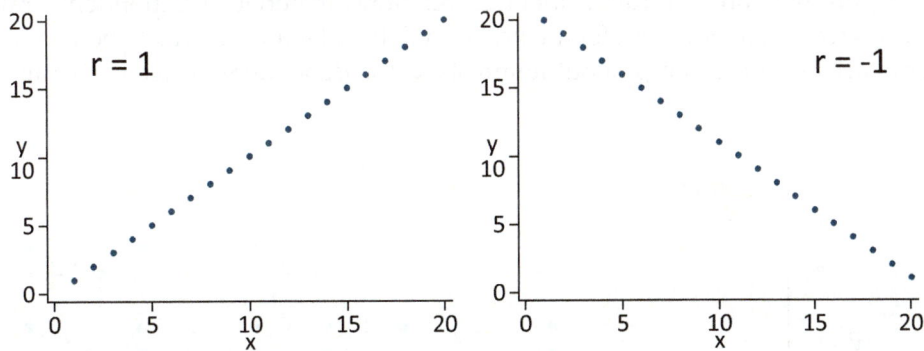

Figure 14.3 Scatterplots of Perfect Positive Correlation (Left) and Perfect Negative Correlation (Right)

The scatterplots on the left and right sides of Figure 14.3 show correlations that are the strongest that they can be in both directions, +1 and −1, respectively. The points in both scatterplots are arranged in diagonal upward (left) and downward (right) trajectories. In both scatterplots, for every change in x, we observe the exact same change in y, indicating a perfectly linear relationship, in which a straight line passes through all points on each plot. For either scatterplot, we can easily envision drawing a single straight line that would pass through all of the points.

Figure 14.4 shows two scatterplots, each with a correlation of 0 between x and y.

In the scatterplot on the left, as the values of x change, there is no change in y. No matter the value of x, the value of y is 5, indicating that the value of y is not at all dependent on the value of x. One can easily imagine drawing a horizontal line through the points in this scatterplot. This represents a correlation of 0, or no association between x and y.

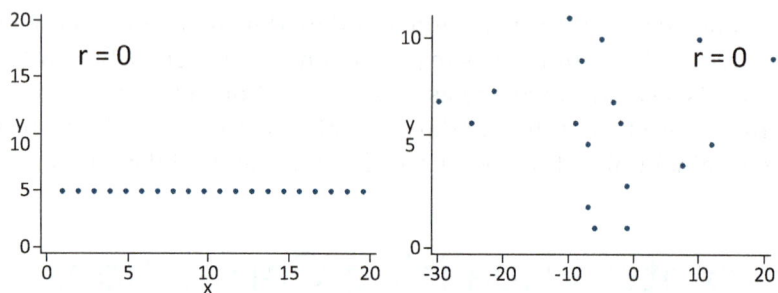

Figure 14.4 Scatterplots of No Correlation between X and Y.

The scatterplot on the right shows the joint values of x and y scattered randomly around the plot area, with no evident trend in how the values of the two variables move together. In this scatterplot, the correlation between x and y is also 0, but the absence of any relationship between the two variables is not as evident as it is in the scatterplot on the left.

As we suggested above, with real social science data, we never observe perfect correlations, nor do we observe the complete absence of correlation (i.e., $r = 0$). Here, we turn to the 2020 wave of the American National Election Study (ANES) to examine the kinds of scatterplots that social scientists encounter in their work with real data.

Figure 14.5 shows a scatterplot of the relationship between respondents' ratings of their feelings toward liberals and feminists. We will treat feelings toward liberals as the independent variable and feelings about feminists as the dependent variable. (Because there is no

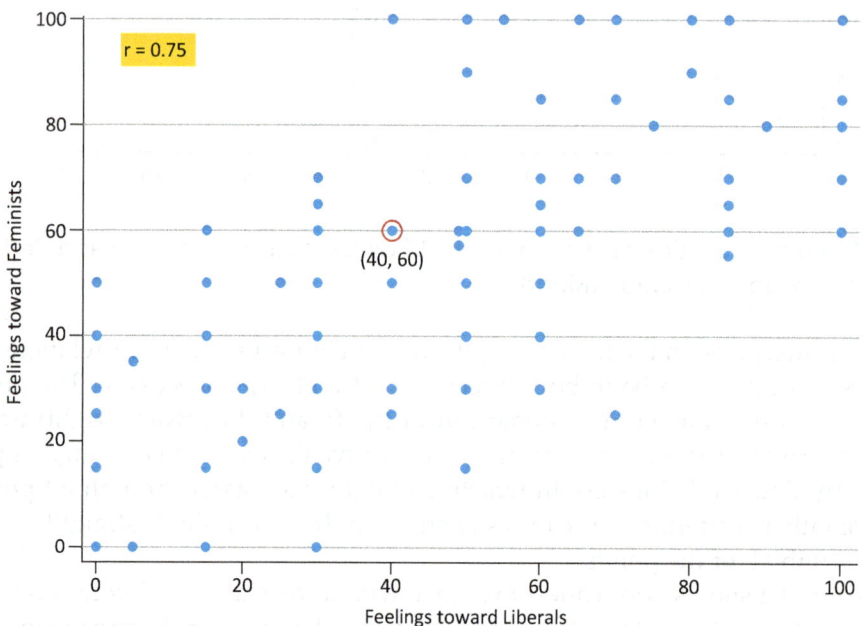

Figure 14.5 Scatterplot of Feelings about Liberals against Feelings about Feminists, 2020 American National Election Study

clear causal relationship between these two variables, we are actually assessing how they are associated with each other.) Following convention, we place feelings toward liberals on the x axis and feelings toward feminists on the y axis.

Each point in the scatterplot represents the paired values of the two variables for one or more respondent in the data.* For example, the circled point represents a respondent who reported feelings of 40 for liberals and 60 for feminists. The coordinates of the point are 40 and 60, written as (40, 60), with the value of the independent variable listed first and the value of the dependent variable listed second.

Note that there are many more points scattered around the area of the scatterplot than there were in Figures 14.3 and 14.4. With a larger sample, the scatterplot is more difficult to interpret. Still, we can use Figure 14.5 to visually assess the relationship between these two variables. To assess the direction of the relationship, we examine the shape of the distribution of points. Does there appear to be an upward or downward trend? In general, as we observe increasingly positive feelings about liberals, we also observe increasingly positive feelings about feminists. This upward trend indicates a positive relationship.

To assess the strength of the relationship, we look at the extent to which the points seem to fall on a straight line. The strongest correlation, +1 or −1, indicates that the points fall on a straight line, either sloping upward or downward, respectively (as in Figure 14.3). It is immediately clear from Figure 14.5 that, although the points are arranged in a general upward direction, they do not fall on a perfectly straight line. Still, you can envision drawing a straight line through the points. The points are not exactly on the line (and some are quite distant from it), telling us that the correlation coefficient will be positive (because the line slopes upward) and somewhat less than 1. A visual inspection of the scatterplot shown in Figure 14.5, then, indicates a positive and somewhat strong relationship between these two variables. In fact, the actual correlation coefficient between feelings about liberals and feminists is 0.75—a strong relationship according to the guidelines that we presented earlier.

Correlation coefficients measure linear relationships between two variables. A linear relationship is represented by a straight line, with every unit change in the independent variable associated with a constant amount of change in the dependent variable. Although we can calculate a correlation coefficient if the relationship between two variables is curvilinear (that is, arranged in a curved line), the result will be misleading. Scatterplots are a useful tool for examining whether the relationship between two variables appears to be linear. Figure 14.6 shows an example of a scatterplot of a **curvilinear relationship**.

The points in Figure 14.6 are not arranged in a straight line. We can see this because there is no consistent upward or downward trend. Instead, we see an upward trend until x reaches about 10, at which point we begin to see a downward trend.** There are many kinds of nonlinear relationships, many of which can be modeled using various statistical techniques, but correlation coefficients are meant only for variables that have a linear relationship with one another.

---

* To produce a less cluttered scatterplot, Figure 14.5 uses a smaller random sample from the ANES data for these two variables. Multiple respondents could have the same pair of values for the variables represented in a scatterplot. The scatterplot does not capture how many respondents have the same pair of values for the two variables.

** The relationship shown in Figure 14.6 is a parabola, which can be modeled by a quadratic equation. More advanced statistics courses cover how to do this.

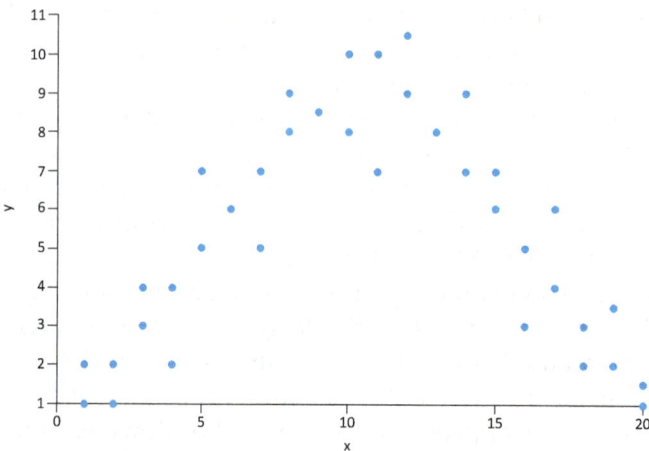

Figure 14.6 Scatterplot of a Curvilinear Relationship

BOX 14.1: IN DEPTH

Correlation Is Not Causation

There are many examples of variables that are highly correlated with each other, but we would be hard pressed to come up with an explanation for these correlations. Edward Tufte finds a statistically robust relationship between the number of mentally ill people in Great Britain between 1924 and 1937 and the number of letters in the first name of American presidents during those same years.[4] The correlation coefficient for these two variables is 0.94. This relationship is nonsensical, unless one is prepared to argue that the transition from Herbert (seven letters) Hoover to Franklin (eight letters) Roosevelt somehow produced an increase in mental illness in Great Britain. This example underscores the importance of having a theory that explains the logic of the causal relationship. Statistical software will calculate slopes and correlation coefficients on any variables that we input. We, the researchers, need to come up with the explanations for those statistical calculations.

There are many examples of statistically strong correlations that are nonsensical. For example:

- per capita cheese consumption and number of people who died by becoming tangled in their bedsheets; $r = 0.95$
- letters in winning words of Scripps National Spelling Bee and number of people killed by venomous spiders; $r = 0.81$
- worldwide noncommercial space launches and sociology doctorates awarded in the United States; $r = 0.79$

Sometimes a relationship may appear to be nonsensical but may reflect the influence of a lurking variable. Tufte again finds a strong statistical relationship, over time, between the number of mentally ill people in Great Britain and the number of radio receiver licenses issued ($r = 0.99$). At first glance, this appears to be nonsensical. But, upon further consideration, this correlation may reflect the influence of scientific progress. New technologies led to the spread of radios and, by extension, radio licenses. And, at the same time, scientific progress led to increased awareness and diagnoses of mental illness. So, what appears on the surface to be a nonsensical relationship actually reflects the influence of scientific progress.

In the next section, we turn to **linear regression** a method of measuring how much our independent variable "influences" our dependent variable.

Regression: Fitting a Line to a Scatterplot

We saw in the previous sections that scatterplots are a useful tool for visualizing correlations between two variables. Scatterplots can also help us to visualize how linear regression works. Linear regression offers us a formal way of drawing the line that best fits the points on a scatterplot. As we will see, the **regression line** allows us to make predictions about the value of the dependent variable at given values of the independent variable.

Returning to the example of the relationship between feelings about liberals and feminists, Figure 14.7 shows the same scatterplot for the two variables as shown in Figure 14.5 with the addition of the regression line.

Earlier, we envisioned drawing a straight line through the points on a scatterplot. In Figure 14.7, we have drawn this line, which is called the regression line, through the scatterplot. We saw earlier that there is a strong positive correlation between these two variables (0.75). The line is consistent with a strong positive correlation because many of the points are clustered closely around it. We can also see that it has a fairly steep upward slope.

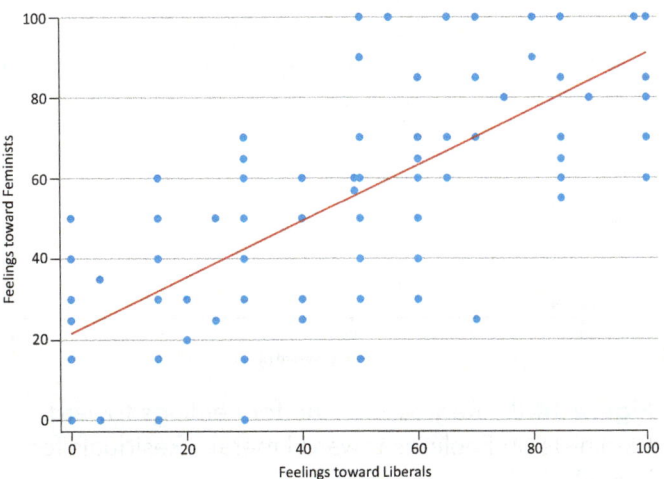

Figure 14.7 Scatterplot of Feelings toward Liberals and Feminists, with Regression Line

The regression line shows the predicted value of the dependent variable for every value of the independent variable. The predicted value of the dependent variable is the value that corresponds with a given value of the independent variable on the line. For a person who rated their feelings about liberals at 30, we look at the point on the line that corresponds to 30 on the x axis; the corresponding value on the y axis tells us the predicted value for that person's rating of feminists: a little higher than 40. We notate a predicted value by placing a "hat" above the variable, like this: $\hat{y}$.

What does it mean that the regression line is the "best-fitting" line, and how do we know exactly where to draw that line through the points on a scatterplot? We address these two questions below, in that order.

The "Best-Fitting" Line

In this section, we discuss conceptually why the regression line is referred to as the "best-fitting" line to represent the linear relationship between two variables.* To begin, we present another version of the scatterplot of feelings toward liberals and feminists, shown in Figure 14.8.

In Figure 14.8, notice that a rectangle encloses a column of points. All of these points represent respondents in the data that have the same value for the independent variable: 30. However, these individuals do not share the same value for the dependent variable. In fact, the values of the dependent variable (feelings about feminists) for people who rated their feelings about liberals at a value of 30 range from very cold (0) to warm (70). We know that the regression line represents the predicted values of the dependent variable for each value of the independent variable. The fact that most of the points that have a value of 30 for the independent variable do not fall on the regression line shows that there is some degree of error in the prediction of the dependent variable for people who rated their

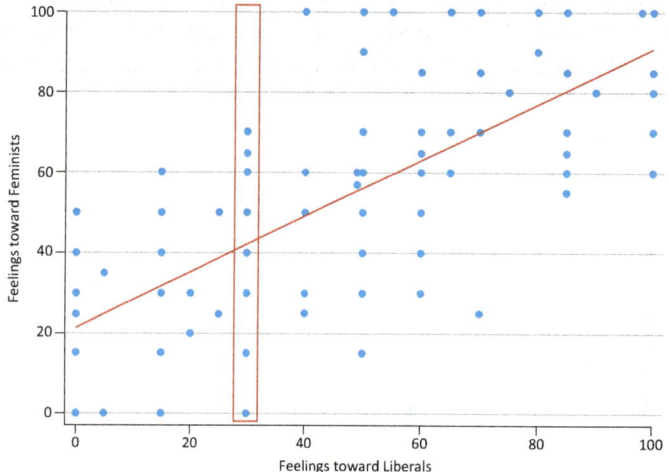

Figure 14.8 Regression Line for Feelings toward Feminists on Feelings toward Liberals, Residuals for X = 30

feelings toward liberals as 30. In fact, we can see that very few points in the scatterplot fall exactly on the regression line, indicating error in the prediction of the value of the dependent variable at every value of the independent variable.

The vertical distance between each data point and the regression line is called a **residual**. We can calculate the size of the residual for each case in the data by subtracting

---

* In this chapter, we focus on regression that models linear relationships between independent and dependent variables. Regression equations can also model curvilinear relationships between variables when the variables have been transformed. (Technically, these also are linear equations; what makes a regression equation linear is not whether the relationship between the variables is linear—that is, can be represented with a straight line—but the fact that the terms in the regression equation are additive.) Regression models with these kinds of transformations are beyond the scope of this book. See Box 14.3 for a brief discussion of one such model.

the value of y predicted by the regression line for a given value of x from the actual value of y for that same value of x:

$$Residual = y - \hat{y}$$

where y is the value of the dependent variable for each case, or point in the scatterplot, and $\hat{y}$ is the predicted value of the dependent variable for each case. The vertical distances between each of the points enclosed in the rectangle in Figure 14.8 and the line are the residuals for those cases in the data. The smaller the vertical distance between the data point and the regression line, the smaller the residual.

The regression line is the "best-fitting" line in the sense that it minimizes the overall residuals, or the difference between the predicted values and actual values of y for given values of x. More specifically, a regression line minimizes the sum of the squared residuals. Thus, the line in Figure 14.8 is drawn such that it minimizes the total squared distance between the predicted feeling thermometer rating of feminists and the actual feeling thermometer rating of feminists at each value of the independent variable.

No regression line will ever fit social science data perfectly. Even the "best-fitting" line will never be a "perfect-fitting" line, like the ones shown in the hypothetical examples of Figure 14.3. A regression line represents the best among less-than-perfect options for drawing the line. Remember, the social world is messy. For social scientists, no independent variable can perfectly predict a dependent variable. Thus, when we model the linear relationship between two variables, we know that in the real world many cases will not have the value for the dependent variable predicted by the independent variable. We can think of a regression line as a simplified representation of the general relationship between two variables, an elegant representation of a complicated reality. This is what we mean when we say it is a *model* of the relationship. Later in the chapter, we present two formal measures for assessing the "fit" of a regression line. First, we discuss two key elements of a regression line: the slope and intercept.

Slope and Intercept

The formula for a regression line formally describes the linear relationship between two variables, x and y. In Figure 14.8, we could estimate the predicted value of y for a given value of x by visually examining the regression line. The equation for a regression line allows us to calculate that value precisely. The equation for a regression line is:

$$\hat{y} = a + bx$$

where the predicted value of y, called "y-hat" and denoted by $\hat{y}$, for a given value of x is equal to the slope of x (b) multiplied by the value of x plus the intercept of the line (a).

BOX 14.2: IN DEPTH

Alternative Terms and Notation for Regression Lines

You may recall studying linear equations in previous math classes. The equation for a line always follows the same basic form and includes the slope of the line, the y-intercept, and the two variables (x and y). Outside of statistics, linear equations are sometimes presented with the terms in a different order or with different letters used to represent the slope and intercept. One common form is y = mx + b. In that case, the slope is represented by "m" and the intercept by "b." Do not be confused! The equation functions in the same way as the equation we are using here (y = a + bx). You may also have learned that the slope is "rise over run"—that is, the amount that the line increases, or "rises," for each unit of horizontal change, or "run."

The **slope** of an independent variable in linear regression indicates how much we can expect the dependent variable to change for a one-unit increase in the independent variable. A positive slope predicts an increase in the dependent variable for a one-unit increase in the independent variable, while a negative slope predicts a decrease in the dependent variable for a one-unit increase in the independent variable. The slope is always measured in units of the dependent variable.

The **intercept** term in the regression equation gives the expected value of the dependent variable when the independent variable is equal to 0. This means that the intercept is the value of the dependent variable where the regression line crosses the y axis. The intercept of a regression equation should be interpreted only when a value of 0 for the independent variable is meaningful. Depending on the independent variable, a value of 0 may or may not be meaningful. For example, if the independent variable is the number of times that children challenge their parents' rules in a given time frame, 0 is a meaningful, if implausible, value. But if the independent variable is the weight in pounds of participants in a study, then a value of 0 would be neither plausible nor meaningful.

We usually rely on statistical software to calculate slopes and intercepts, especially with large data sets. Returning to our example of feelings toward liberals and feminists, we will regress feelings about feminists (y) on feelings about liberals (x). Using statistical software, we find an intercept of 21.68 and a slope of 0.69. Thus, the regression equation for the relationship between those variables is written as:

$$\hat{y} = 21.68 + 0.69x$$

We will use an annotated scatterplot of this relationship, shown in Figure 14.9, to interpret the slope and intercept for this particular equation. Figure 14.9 shows the slope and intercept of the regression line, both circled. The slope for feelings toward liberals is 0.69.

To interpret this slope, we can say: "For every 1-point increase in the feelings about liberals, we expect feelings about feminists to increase by 0.69 degrees." This means that, after we move one unit to the right from the regression line, we must move 0.69 units up to reach the line again. (See inset in Figure 14.9.) We always interpret the slope in units of the dependent variable, given a one-unit change in the independent variable. In this case, both variables are measured in feeling thermometer degrees, so a one-unit increase in the independent variable is an increase of 1 point on the feeling thermometer, and the predicted

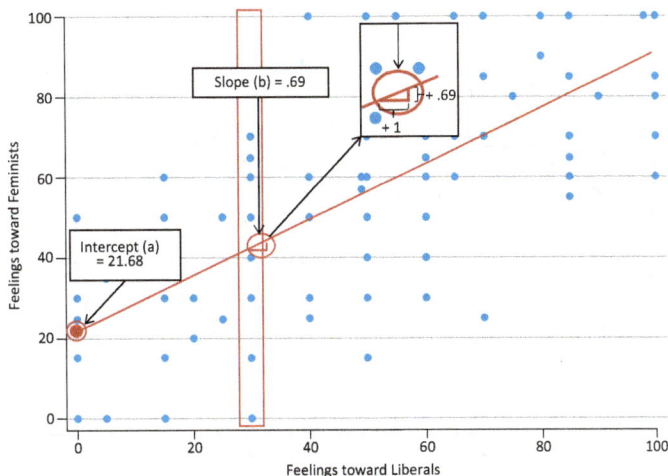

Figure 14.9 Annotated Scatterplot and Regression Line for Feelings toward Feminists against Feelings toward Liberals

increase in the dependent variable is also stated in feeling thermometer degrees. Figure 14.9 depicts the slope at a particular point on the regression line: at the increase from 30 to 31 degrees for the independent variable. Because this regression models a linear relationship, the expected change in the dependent variable is the same regardless of where the one-unit increase in the independent variable occurs. In this case, we expect a 0.69-point increase in the dependent variable no matter where the one-unit increase in the independent variable occurs on the scale of the independent variable. The slope could have been depicted as a one-unit increase in the independent variable from 3 to 4, 10 to 11, or 27 to 28. They would all be associated with the same change in the dependent variable.

Figure 14.9 also shows the intercept for this equation (21.68). This means that for a respondent who rates their feelings about liberals as 0, we would estimate that person's rating of feminists to be 21.68. Taken together, the slope and intercept in a regression equation tell us where and how to draw the regression line: the intercept indicates where the line should cross the y axis, and the slope indicates how steeply upward or downward the line should be drawn.

Regression equations allow us to make precise predictions about the value of a dependent variable for a given value of the independent variable. Take a moment to examine the regression line shown in Figure 14.10, which depicts the same relationship that we have worked with throughout this chapter.

What is your estimate of the predicted value of y when x is 50 degrees? We can "eyeball" a rough estimate by following the red dotted line up from 50 to the regression line and then to the left until our eye meets the y axis. That is the estimated value of y when x equals 50. We can see that the value is close to 60 degrees, but where exactly?

We can find the exact answer by plugging the slope and intercept (given above) and the value of x (50) into the equation for the regression line:

$$\hat{y} = 21.68 + 0.69x$$

$$\hat{y} = 21.68 + 0.69(50) = 56.18$$

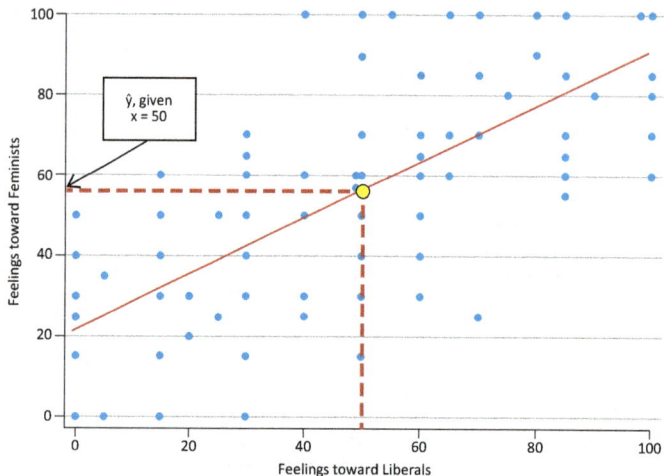

Figure 14.10 Using the Regression Line to Estimate the Predicted Value of Y When X Equals 50

So, for a person who rates their feelings about liberals at 50 degrees, the regression equation predicts that this person would rate their feelings about feminists as 56.18. How close was your eyeball estimate to the one generated by the equation?

Calculating the Slope and Intercept

The equation for the slope, b, is:

$$b = \frac{Sum\ of\ cross\ products}{Sum\ of\ squares_x} = \frac{\Sigma(x-\bar{x})(y-\bar{y})}{\Sigma(x-\bar{x})^2}$$

The slope divides the sum of the cross products of x and y by the sum of squares for x. This shows us how we would go about manually calculating a slope, a task that we almost always delegate to statistical software.

Once we have found the slope, we can find the intercept of the line. The equation for the intercept, a, is:

$$a = \bar{y} - b(\bar{x})$$

where $\bar{y}$ is the mean of y, $\bar{x}$ is the mean of x, and b is the slope.

In the next section, we return to the concept of goodness-of-fit and present two formal measures of the fit of regression lines.

Goodness-of-Fit Measures

Goodness-of-fit measures for a regression model assess the fit of the model—that is, how accurately the regression equation represents the actual data. They do so by assessing the relationship between the predicted values of the dependent variable generated by the regression equation and the actual or mean values of the dependent variable. Since the predicted values are generated by the relationship between the independent and dependent variables, goodness-of-fit measures tell us how well the independent variable(s) perform as predictors of the dependent variable. As we noted earlier, no regression line will ever fit social science data perfectly. In this section, we examine two commonly used goodness-of-fit measures in regression: r-squared and the standard error of the estimate.

R-Squared (r^2)

R-squared (r^2), also referred to as the coefficient of determination, indicates the proportion of the variation in the dependent variable that is accounted for by the independent variable.[*] As its name implies, we can find r^2 by squaring the correlation coefficient for x and y. As noted above, the correlation between feelings about liberals and feminists equals 0.75. We simply square 0.75 to find the r^2 for the regression of these two variables: $0.75 \times 0.75 = 0.56$. We can think of this as a "shortcut" method for determining r^2. If we take a look at another way to calculate r^2, we can get a better sense of what it means. (In practice, analysts almost always use statistical software to calculate r^2.)

R^2 is a proportion, specifically the proportion of variance in the dependent variable explained by the independent variable(s) in the regression equation. To see this, we can examine an alternative formula for r^2 that does not rely on the correlation coefficient:

$$r^2 = \frac{explained\ sum\ of\ squares}{total\ sum\ of\ squares} = \frac{\sum(\hat{y}-\bar{y})^2}{\sum(y-\bar{y})^2}$$

To find r^2, we divide the variation of the predicted values of y ($\hat{y}$) from the mean of y ($\bar{y}$) (i.e., the "explained sum of squares") by the variation of the actual values of y from the mean of y ($\bar{y}$) (i.e., the "total sum of squares"). Since the predicted values of y are generated from the relationship between x and y, r^2 tells us the proportion of the total variation of y from $\bar{y}$ that is accounted for by variation in x. The value of r^2 always ranges from 0 to 1, with 0 indicating that none of the variation in the dependent variable is accounted for by variation in the independent variable and 1 indicating that all of it is accounted for by the independent variable. Because r^2 is a proportion expressed as a decimal, we can multiply it by 100 to express it as a percentage. So we could say that 56% of the variation in feelings about feminists is accounted for by variation in feelings about liberals. The higher the r^2, the greater the explanatory power of the independent variable.

R^2 tells you how well the independent variable you have chosen predicts the values of the dependent variable. It cannot tell you whether the relationship between the two variables is causal and it cannot tell you whether your independent variable or variables are the most important predictors of your dependent variable. For these questions, researchers rely on theory and prior research, as well as the process of assessing causality discussed in chapter 13.

As you might guess, as with correlations, we never see r^2 values of 0 or 1 with real social science data. In the social world, there are too many influencing factors for a single

---

[*] By convention, lowercase r^2 is used where there is one independent variable, and uppercase R^2 is used where there are multiple independent variables.

independent variable (or even a set of them) to fully account for the variation in a dependent variable. In general, r^2 values higher than 0.25, or 25%, are considered to indicate robust explanatory power of the independent variable(s). However, when the values of the independent and dependent variables are aggregates (e.g., the mean GPA for students in a school and the mean extracurricular participation for students in a school), we have higher expectations of the r^2. This is because, when we aggregate values across cases (e.g., average values for all students in a school), residuals will generally be smaller than when we are measuring values for single cases (e.g., individual students). To put it differently, it is easier for social scientific approaches to explain patterns for groups than to explain patterns for individuals.

Standard Error of the Estimate

Another measure of the goodness-of-fit for a regression line is called the **standard error of the estimate**, sometimes referred to as the Root Mean Square Error (RMSE). Whereas r^2 uses the mean of y as the reference point by comparing predicted and actual values of y to the mean, the standard error of the estimate uses the predicted values of y as the reference point. The standard error of the estimate indicates how far, on average, the actual values of the dependent variable fall from the values predicted by the regression line. By formula, the standard error of the estimate ($SE\hat{y}$) is:

$$SE_{\hat{y}} = \sqrt{\frac{sum\ of\ squared\ residuals}{N-2}} = \sqrt{\frac{\Sigma(y-\hat{y})^2}{N-2}}$$

Notice the similarity between this formula and the formula for standard deviation. The major difference is that the standard deviation sums the squared deviations of y from its mean, while the standard error of the estimate sums the squared deviations of y from the predicted values of y yielded by the regression equation. This quantity, $\Sigma(y-\hat{y})^2$, is referred to as the "sum of squared residuals." The sum of squared residuals is divided by $N-2$ because there are two estimated parameters in the regression equation, the slope and intercept. As with r^2, researchers rarely calculate the standard error of prediction by hand.

Smaller standard errors of prediction indicate less error in prediction and, therefore, better-fitting regression lines. Unlike r^2, which is measured as a proportion, the standard error of the estimate is measured in the units of the dependent variable. In the example used throughout this chapter, the standard error of the estimate, as computed by statistical software, is 20.5. This means that the average variation of the actual ratings for feelings about feminists from the predicted value is 20.5 degrees.

Dichotomous ("Dummy") Independent Variables

In addition to using interval-ratio variables, we can also use dichotomous (sometimes called **dummy**) variables—variables with two categories—as independent variables in a regression equation. The two categories of a dichotomous variable can be ordered such that one category indicates the presence of a characteristic (coded as 1) while the other indicates its absence (coded as 0 and called the "reference category"). Because a dichotomous variable has only two categories, the interpretations of the slope and intercept differ from the interpretations when the independent variable is interval-ratio.

Recall that, when the independent variable is interval-ratio, the intercept indicates the predicted value of the dependent variable when the independent variable is equal to 0. When the independent variable is a dichotomous variable, the intercept is equal to the *mean* of the dependent variable when the dichotomous variable is equal to 0.

Recall that the slope for an interval-ratio variable indicates the predicted change in the dependent variable for a one-unit increase in the independent variable. Intuitively, applying this interpretation to the slope for a dichotomous variable does not make sense because the variable has only two categories. The slope for a dichotomous variable indicates the *difference between the means for the dependent variable for the two categories of the independent variable*. We can think of it as adjusting the intercept of the line for the category of the independent variable that has been assigned a value of 0.

We will illustrate how to interpret the slope and intercept terms in a regression equation with a dichotomous independent variable with data from the 2021 General Social Survey (GSS). The dependent variable, attitudes about government assistance, is the mean of individual respondents' opinions about whether it should be the government's responsibility to do the following four things:

- reduce the differences in income between people with high incomes and those with low incomes;
- do everything possible to improve the standard of living of all poor Americans;
- see to it that people have help in paying for doctors and hospital bills; and
- do more to solve our country's problems.

Each item ranges from 1 to 5, with higher values indicating greater agreement that the item should be the government's responsibility. The items are combined into one variable that is the average of each respondent's opinion on the five items; its value ranges from 1 to 5. The independent variable is a dichotomous variable indicating the respondent's marital status, either married (coded as 1) or unmarried (coded as 0). Using statistical software, we find that the regression equation for the relationship is:

$$\hat{y} = 3.34 - 0.31x$$

The intercept, 3.34, indicates the mean score for attitudes about government assistance when the independent variable is equal to 0. Because 0 represents those who are unmarried, the intercept indicates that unmarried people's mean score for attitudes about government assistance is 3.34. The slope is −0.31, indicating that the mean score for the dependent variable is 0.31 points lower for those who are married (i.e., have a value of 1 for marital status) than for those who are unmarried (i.e., have a value of 0 for marital status).

As we saw in the previous example, we can use the regression equation to calculate predicted values of the dependent variable at given values of the independent variable. In this example, since there are only two possible values for the independent variable, there are only two possible predicted y values. For unmarried respondents (i.e., marital status equals 0), the predicted value of y is 3.34 (3.34 − 0.31(0) = 3.34). For married respondents (i.e., marital status equals 1), the predicted value of y is 3.03 (3.34 − .31(1) = 3.03). Note that the predicted values of the dependent variable are equal to the sample means for each group.

Figure 14.11 shows the scatterplot of opinions about government assistance and marital status.

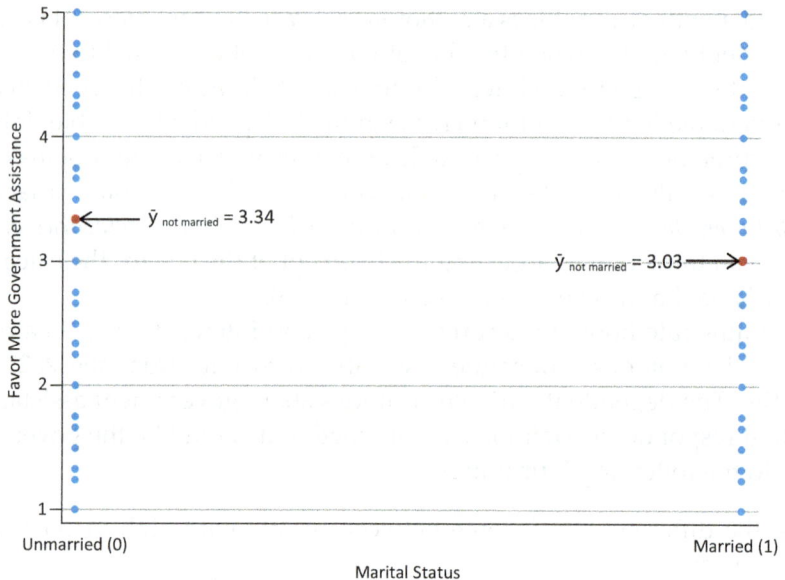

Figure 14.11 Scatterplot of Opinions about Government Assistance against Marital Status, 2021 GSS

The scatterplot in Figure 14.11 looks quite different from the scatterplots that we have seen previously in this chapter. In previous scatterplots, points are scattered about the area of the plot because they are plotted across a range of values for the independent variable. In Figure 14.10, the points are arranged in two columns, one for each category of the independent variable (i.e., not married and married), which are the only two points on the x axis. Unlike in the previous scatterplots, we cannot imagine drawing a best-fitting line through the points on the scatterplot in Figure 14.11. When the sole independent variable is a dichotomous variable, there is no regression line. Instead, there are two separate intercepts, one for each category of the variable. The red dots on the y axis represent the two intercepts. We can see that the intercept for unmarried respondents is 3.34 (the mean of the dependent variable for unmarried respondents) and the intercept for married respondents is 3.03 (the mean of the dependent variable for married respondents—intercept (3.34) plus slope (−0.31) = 3.03). These intercepts are the points where the regression line for each category of the independent variable would cross the y axis if there were also interval-ratio independent variables in the equation. (Regression equations that include more than one independent variable use a technique called multiple regression, the topic of the next section.)

Does the negative slope for marital status indicate that being married "causes" lower support for government assistance to those in need? Here, we must be mindful of the adage "correlation does not imply causation." A regression equation with one independent variable estimates the relationship between just that variable and the dependent variable. As we saw in chapter 13, a relationship between two variables does not necessarily imply a causal connection. We should not conclude that a relationship between two variables is causal without attempting to rule out the influence of other variables that could be accounting for the relationship. In our example, it is possible that there is no causal impact of being married on support for government assistance and other factors actually account for the relationship between the two variables. It also could be that there is a causal relationship, but that the slope for marital status overestimates its direct causal impact on the dependent variable.

In the next section, we will consider multiple regression—regression with more than one independent variable—as a way to address questions of causality in regression analysis.

BOX 14.3: IN DEPTH

Regression Models for Nonlinear Relationships

This chapter focuses on linear regression with interval-ratio dependent variables. However, there are also regression models that can estimate the effects of independent variables on categorical dependent variables (i.e., nominal and ordinal variables). In fact, many outcomes of interest to social scientists are not interval-ratio variables. How likely are voters to choose one candidate over another? What is the probability that someone will be imprisoned at some point in their life? How likely is it that people will get married at least once during their lives? Because these are all examples of categorical outcomes, regular linear regression is an inappropriate technique for modeling the relationship between them and independent variables of interest. An analyst would have to opt instead for a different kind of regression model. There are a variety of linear and nonlinear regression models, depending on the level of measurement of the dependent variable and the nature of the relationship between dependent and independent variables. Here, we focus on **logistic regression**, which is used with binary dependent variables. Logistic regression predicts the probability that a dummy dependent variable (coded as 0 or 1) will be equal to 1 for given values of the independent variables.

Figure 14.12 depicts the curve that shows the predicted probability that a dependent variable's value is 1, at different values of an independent variable. This is an S-shaped curve (called a logistic curve) rather than the straight prediction line seen in linear regression. Unlike in linear regression, where the effect of the independent variable is the same regardless of the value of x, the "effect" of an independent variable in a logistic model is not constant.

Just as with linear regression, logistic regression yields regression coefficients. However, the coefficients in logistic regression equations are not easily interpretable because they represent the rate of change in the "log odds" of y for a one-unit change in x (or, for dummy independent variables, the difference in the "log odds" of y between the two categories of the variable). To make intuitive sense of the coefficients in this model, which are called "logit" coefficients, analysts frequently transform them into "odds ratios" and predicted probabilities when they report their results. (Delving into the mechanics of logistic regression is beyond the scope of this book. *Regression Models for Categorical and Limited Dependent Variables*, by L. Scott Long, is an excellent source.)

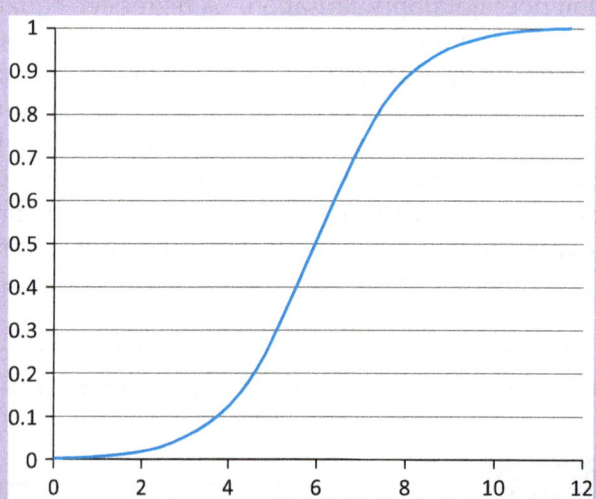

Figure 14.12 Logistic Curve Showing the Predicted Probability that y = 1 for a Binary Dependent Variable

We use data from the 2021 General Social Survey to illustrate how this works. The dependent variable, fear of walking in one's neighborhood at night, has two categories: yes and no. The independent variable is gender (coded 1 for woman, 0 for man). Using logistic regression, we generate the logit coefficient and odds ratio for the gender variable as well as the predicted probabilities of feeling afraid to walk in one's neighborhood at night for women and men.

The logit coefficient for the gender variable is 0.90, indicating that the "log odds" of feeling afraid to walk in one's neighborhood is 0.90 higher for women than men. This gives us some sense that women, on average, are more likely to feel afraid to walk in their neighborhood at night, but we cannot make substantive sense of a difference in "log odds." Our statistical software also reports an odds ratio, in this case, 2.45. This indicates that the odds of feeling afraid to walk in one's neighborhood at night are 2.45 times greater for women than men. We can go one step further and convert the odds ratio into predicted probabilities for women and men. The predicted probability of feeling afraid to walk in one's neighborhood at night is 0.44 for women and 0.24 for men.

Multiple Regression

Analysts who use regression almost always use multiple regression because it allows for the estimation of the relationship between two variables while holding constant other factors. We saw in the previous chapter that we can hold a third variable constant by cross-tabulating the relationship between two variables separately for each category of the third variable. In multiple regression, we hold variables constant by removing from each independent variable the variance that it shares with the other independent variables. A Venn diagram, shown in Figure 14.13, uses circles to represent the variances of three variables: y (the dependent variable), x_1 (a first independent variable), and x_2 (a second independent variable). The three scenarios correspond to three different linear relationships between the variables, with different degrees of correlation, or shared variance, among the independent variables.

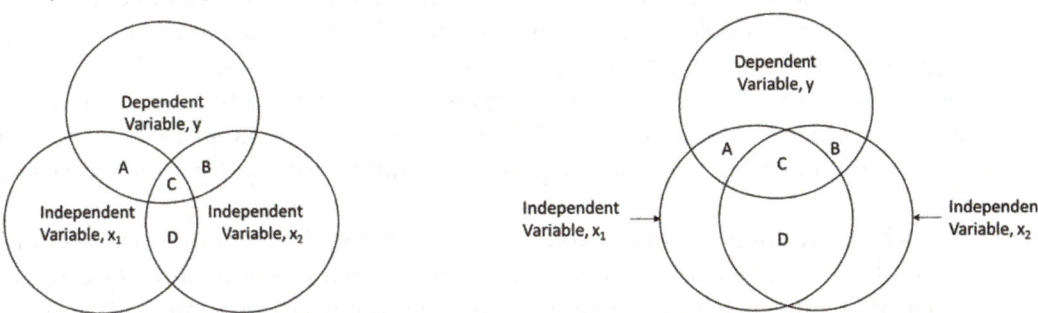

Figure 14.13 Venn Diagrams of Shared Variance for a Dependent Variable and Two Independent Variables with No Correlation, Weak Correlation, and Strong Correlation

In Scenario 1, the two independent variables are uncorrelated, as represented by the fact that the variance that x_1 shares with y (A) and the variance that x_2 shares with y (B) do not overlap. The independent variables share no variance in Scenario 1, which means we could use two separate bivariate regressions for each of them to predict y without it affecting the regression coefficient for either x_1 or x_2.

This is not true for Scenarios 2 and 3, in both of which x_1 and x_2 share variance, represented by the areas of C and D. When independent variables are correlated, multiple regression estimates the unique impact of each independent variable on the dependent variable, represented by the areas of A for x_1 and B for x_2. Under Scenarios 2 and 3, where x_1 and x_2 are correlated, failure to include one of those variables in the regression equation would result in an overestimation of the unique effect of x_1 or x_2 on y. If we failed to control for x_2, then the area of C would be treated as if it were shared only between x_1 and y, resulting in an overestimation of the unique effect of x_1 on y. We can see that the more strongly correlated the independent variables, the larger the area of C. Thus, the overestimation of the effect of an independent variable when a correlated variable is not included in the regression equation is higher the more strongly correlated the variables are.

When we hold a variable constant, or control for it, imagine that we are comparing only the cases that have the same value for the control variable. For example, if we are examining the relationship between how many children people have and time spent on household labor and we control for gender, we would be examining that relationship separately for men and women.

A multiple regression equation looks similar to a bivariate regression equation. Here is the generic form of the regression equation with two independent variables:

$$\hat{y} = a + b_1 x_1 + b_2 x_2,$$

where a is the intercept, b_1 is the slope for the independent variable called x_1, b_2 is the slope for the independent variable called x_2, and $\hat{y}$ is the predicted value of y for given values of x_1 and x_2.

We interpret the intercept in the same way as we do for bivariate regression: it is the predicted value of y when all independent variables are equal to 0. We interpret the slopes for the independent variables in the same general way as we would in bivariate regression for interval-ratio or dichotomous independent variables. The difference is that, in multiple regression, the slope for each independent variable indicates the effect of that variable on the dependent variable, holding constant the other variables in the equation. In multiple regression, the slope for an independent variable is sometimes referred to as a "partial slope," because it "partials out" the variance shared by that independent variable and the others included in the equation. Each slope represents the change in the dependent variable for a one-unit change in that independent variable, *holding the other independent variables constant*.

When the independent variables are measured in different units (for example, number of children and hours of housework), the slopes should not be directly compared to each other. When this is the case, a larger slope does not necessarily mean that the variable has a larger effect on the dependent variable than variables with smaller slopes.[*]

Let's return to the question we posed above regarding the results from the regression of support for government assistance on marital status. Does the negative slope for marital status mean that being married "causes" people to have weaker support for government assistance? While regression results can never prove that an independent variable causes changes in a dependent variable, we can decrease the likelihood that a coefficient reflects a spurious relationship by controlling for factors that previous research or common sense tell us might be accounting for the relationship. There probably are factors that are both correlated with marital status and negatively associated with favorable attitudes toward government assistance. We can use multiple regression to hold such factors constant. Should the relationship between marital status and support for government assistance remain once we have "controlled" for those factors, this increases our confidence that there is a causal relationship between the two variables.

For example, married people might tend to have lower support for government assistance because they tend to have higher family incomes. Higher incomes might be associated with lower support for government assistance because they buffer people from personal experiences of needing assistance. If this were the case, then the relationship between marital status and government support would be spurious. We will use multiple regression to test this idea by adding family income as a second independent variable in the regression equation.

---

[*] Standardized slopes, often called betas, permit comparison of the strength of the effects of different independent variables.

Using statistical software, we find the following multiple regression equation for the effects of marital status (x_1) and family income (x_2) on support for governmental assistance for those in need:

$$\hat{y} = 3.39 - .26x_1 - .01x_2$$

where 3.39 is the intercept, −0.26 is the slope for marital status, and −0.01 is the slope for family income. Take a moment to return to the bivariate regression of support for government assistance on marital status. In that equation, the slope for marital status was −0.31. We see that, once we control for family income, the slope for marital status changes to −0.26. We can say: "Controlling for family income (i.e., comparing individuals with the same income), we expect married respondents' support for government assistance to be 0.26 points lower than unmarried respondents' support." This means that, once the shared variance between marital status and family income is removed, the negative "effect" of being married on support for government assistance declines by 16.1%. We cannot say for sure whether this coefficient reflects a causal impact, but we do know that there is a relationship between marital status and opinions about government assistance even after accounting for family income.

We also see that the slope for family income is −0.01. Since family income is measured in 10,000s (i.e., income is divided by 10,000), one unit is equal to $10,000. This means that, for every $10,000 increase in income, we expect support for government assistance to decline by 0.01 points, holding marital status constant.

Statistical Inference for Regression

In a previous section, we discussed residuals—the difference between predicted and actual values of the dependent variable—as one kind of error in regression models. In this section, we shift to inferential statistics to think about another kind of error that we have seen before in this book: the difference between the relationships among variables estimated for a single sample and the true relationship among those variables in the population. A regression equation for a population is written as:

$$y = \alpha + \beta x + \varepsilon,$$

where the intercept, slope, and error term are written as the Greek letters alpha, beta, and epsilon (α, β, and ε, respectively) to denote that these are unknown population parameters.

Note that we have added the error term, ε, to the population regression equation, representing the amount of variation in the dependent variable that is unaccounted for by the independent variable(s). The error term indicates that, for any given value of the independent variable, x, the actual values of the dependent variable, y, will vary in the population. We saw this previously when we examined sample data, but here we are adding a formal error term to the regression equation.

In this section, we examine two statistics that we can use to test specific hypotheses about the relationship between the sample regression and the population regression: the F-statistic and the standard error of the slope.

The F-Statistic

The F-statistic allows us to test the statistical significance of a regression model overall. Essentially, the F-statistic tells us the probability that a specific regression model fits the data better than an "intercept-only" model. In a regression model, predicted values of the dependent variable are generated using the intercept and the slopes of the independent variables. In an intercept-only model, the predicted value for every case is equal to the sample mean of the dependent variable. So, the F-statistic addresses the question of whether we would be better off using our specific regression equation to predict the dependent variable than using the mean of the dependent variable as the predicted value of the dependent variable for any given value of the independent variable.

To calculate the F-statistic, we use two sums of squares that we have seen previously in this chapter, the explained sum of squares and the sum of squared residuals. By formula, the F-statistic is:*

$$F = \frac{\left(\dfrac{\text{explained sum of squares}}{p-1}\right)}{\left(\dfrac{\text{sum of squared residuals}}{N-p}\right)} = \frac{\left(\dfrac{\Sigma(\hat{y}-\bar{y})^2}{p-1}\right)}{\left(\dfrac{\Sigma(y-\hat{y})^2}{N-p}\right)}$$

where p is the number of parameters estimated in the regression model (number of slopes plus one for the intercept) and N is the sample size. The F-statistic indicates the ratio of variance explained by the model to variance that is not explained by the model. The larger the F-statistic, the better the performance of the regression model relative to the intercept-only model.

We can use the F-statistic to conduct a formal hypothesis test that determines whether the regression model under examination is a better fit for the data than the intercept-only model. We follow the same general steps as for other hypothesis tests:

1. State null and alternative hypotheses. Formally, the null hypothesis is that all of the slopes in the regression equation (1 through j) are equal to 0: $H_0: \beta_1 = \beta_2 = \ldots = \beta_j = 0$; the alternative hypothesis is that at least one of the slopes is unequal to 0: H_a: at least one $\beta_1 - j \neq 0$. Less formally, the null hypothesis is that the regression equation does not explain variation in the dependent variable any better than the intercept-only model. The research hypothesis is that the regression equation does explain variation in the dependent variable better than the intercept-only model.

2. Select the alpha-level. If using decision F, calculate degrees of freedom and find the appropriate decision F in the F-table (appendix C). Just as we saw in chapter 11 with ANOVA, there are two degrees of freedom for the F-table: the degrees of freedom used in the numerator (DF1, located in the columns of the table) and the degrees of freedom used in the denominator (DF2, located in the rows of the table) of the

---

* Another commonly used formula for the F-statistic, which is equivalent to the one provided above, is: $F = \dfrac{\left(\dfrac{R^2}{p-1}\right)}{\left(\dfrac{1-R^2}{N-p}\right)}$.

F-formula. $DF1 = p - 1$; p, the number of unknown parameters, is the number of independent variables plus the intercept. $DF2 = N - p$. Calculate the sample F-statistic (generally using statistical software) and the p-value (if using statistical software).

3. Compare decision F to sample F, OR compare your p-value to alpha. Reject the null hypothesis if sample F > decision F, OR if the p-value < alpha.

We will use our previous example of the regression of support for government assistance on marital status and family income to illustrate how to use the F-statistic. First, we state the null and alternative hypotheses. The null hypothesis is that the slopes for the marital status and family income variables are both equal to 0 ($H_0: \beta_m = \beta_i = 0$). The alternative hypothesis is that at least one of those slopes is unequal to 0 ($H_a: \beta_m$ or $\beta_i \neq 0$).

We will select an alpha-level of 0.05 for this test, meaning that we will reject the null hypothesis only if we would observe our F-statistic in five out of one hundred samples (or fewer) if there were actually no relationship between the independent variables and support for government assistance. The next step is to locate the decision F in the F-table for $\alpha = 0.05$. In this case, $DF1 = p - 1 = 3 - 1 = 2$. (p = 3 because we have two independent variables + one intercept = three predictors.) $DF2 = N - p = 3{,}418 - 3 = 3{,}415$. In the F-table for $\alpha = 0.05$, we move down the column until we reach 3,415 degrees of freedom. The rows skip from 200 to ∞, so we will use the value for F in the ∞ row: 2.99.

Next, we calculate the F-statistic for our regression equation, a task that we would almost never do by hand. We will go through the calculation because it can help you to understand where the statistic comes from. There are more than three thousand cases in this analysis, so we will not calculate the sums of squares by hand. Generated by statistical software, the necessary information for calculating F is:

Explained sum of square: $\Sigma(\hat{y} - \bar{y})^2 = 88.1$

Sum of squared residuals: $\Sigma(y - \hat{y})^2 = 4{,}463$

N: 3,418

We can plug this information into the formula for F:

$$F = \frac{\left(\dfrac{\text{explained sum of squares}}{p-1}\right)}{\left(\dfrac{\text{sum of squared residuals}}{N-p}\right)} = \frac{\left(\dfrac{\Sigma(\hat{y}-\bar{y})^2}{p-1}\right)}{\left(\dfrac{\Sigma(y-\hat{y})^2}{N-p}\right)} = \frac{\left(\dfrac{88.1}{3-1}\right)}{\left(\dfrac{4{,}643}{3{,}418-3}\right)}$$

$$= \frac{\left(\dfrac{88.1}{2}\right)}{\left(\dfrac{4{,}643}{3{,}415}\right)} = \frac{(44.1)}{(1.36)} = 32.4$$

Our sample F is 32.4, showing a high ratio of explained to unexplained variance in our regression model, boding well for our significance test.

Finally, to determine whether to reject the null hypothesis, we can either compare the F we obtained for our sample to our decision F or compare the p-value generated by statistical software to our alpha-level. As with other hypothesis testing, we reject the null hypothesis if the sample F > decision F, OR if the p-value < alpha. Our sample F, 32.4, is much larger than our decision F, 2.99. Thus, we can reject the null hypothesis that the slopes for marital status and family income are both equal to 0 in the population. Similarly, our sample p-value, < 0.001 (generated by statistical software), is much below our alpha of 0.05. This leads us to the same decision: reject the null hypothesis. We conclude that marital status and family income together explain a statistically significant amount of the variation in support for government assistance. Or, to put it differently, collectively, marital status and family income have a statistically significant effect on support for government assistance.

The F-test gives us information about all of the slopes in a regression model at the same time. But what if we want to test the statistical significance of a single slope? Can the effect we find of a single independent variable be generalized to the population? To do that, we must use the standard error of each slope, the topic of the next section.

Standard Error of the Slope

Unlike the F-statistic, which simultaneously tests whether all of the regression coefficients are equal to 0, the **standard error of the slope** allows us to test that hypothesis for a particular slope. The logic of this test is based on the fact that, just as we saw with sampling distributions of proportions and means in chapter 7, there is a sampling distribution of slopes. Any slope, b, from a single sample drawn from a population will likely differ from the true population slope, β.

Just as we saw with means and proportions, if we randomly drew a large number of samples of equal size from a population and measured the slopes for each sample, those sample slopes would arrange themselves in a normal curve. This is called the sampling distribution of slopes. The standard error of a slope is the standard deviation of the sampling distribution of slopes. The standard error of the slope tells us how much, on average, a sample slope varies from the population slope due to sampling error.

Since we do not know the true population standard deviation of slopes, we must estimate it using information from our sample. By formula, the standard error of the slope when there is just one independent variable is:*

$$SE_b = \frac{SE_{\hat{y}}}{\sqrt{\text{sum of squares}_x}} = \frac{\sqrt{\dfrac{\text{sum of squared residuals}}{N-2}}}{\sqrt{\Sigma(x-\bar{x})^2}}$$

$$= \frac{\sqrt{\dfrac{\Sigma(y-\hat{y})^2}{N-2}}}{\sqrt{\Sigma(x-\bar{x})^2}}$$

---

* The formula for the standard error of a slope in a multiple regression equation is more complicated. We do not show it here, because analysts almost always use statistical software to generate standard errors for partial slopes.

The equation divides the standard error of the estimate ($SE_{\hat{y}}$) by the square root of the sum of squares for the independent variable. Thus, it works with two sums that we have seen before: the sum of squared residuals and the sum of squares for the independent variable. N − 2 functions as the degrees of freedom because there are two unknown parameters being estimated: one slope and the intercept. In most cases, we rely on statistical software to calculate the standard error of the slope.

Just as we saw in chapters 9 and 10, we use the standard error to test hypotheses about the population slope, β. This standard error is the standard deviation of the sampling distribution of the slopes, but it is an estimate rather than a precise measure. Because we are relying on an estimate of the standard error, we use the t-distribution to conduct tests of statistical significance about the slope. We follow the same general steps for testing hypotheses about population means and proportions that we have seen in previous chapters. However, since these statistics are rarely calculated by hand in practice, we show how you would follow the steps using statistical software to generate some of the statistics. When using statistical software, your output will include both t and the p-value for your t. You can use these p-values to assess your null hypothesis, or you can find a decision t-value in the t-table and compare it to the t for your sample results. The steps of the hypothesis test are:

1. State null and alternative hypotheses.
2. Select alpha-level. If using decision t, calculate degrees of freedom and find the appropriate decision t in the t-table.
3. Calculate standard error.
4. Calculate sample t and compare it to decision t OR compare the p-value associated with your t-value to alpha.
5. If decision t > sample t OR if the p-value < alpha, reject H_0; if decision t < sample t OR if the p-value is ≥ alpha, fail to reject H_0.

The most commonly tested null hypothesis for slopes is that the population slope, β, is equal to 0: H_0: β = 0. Here, we will return to the example that regresses feelings about feminists on feelings about liberals. Recall that the slope for feelings toward liberals is 0.59, indicating that feeling thermometer ratings of feminists are expected to increase by 0.59 degree for every 1-degree increase in the independent variable, feelings about liberals. We will test the null hypothesis that the population slope is equal to 0. In other words, the hypothesis is that there is no relationship in the population between feelings about liberals and feelings about feminists. The alternative hypothesis is that the slope is not equal to 0 (H_a: β ≠ 0).

We will test this hypothesis at the 0.05 alpha-level (p = 0.05).

Degrees of freedom = N − 2, because there is only one independent variable in the regression equation.* Here, DF = 7,268 − 2 = 7,266. Using the t-table, we find that the decision t for a two-tailed test at an alpha of 0.05, with ∞ degrees of freedom, is 1.96.

With 7,268 cases in the sample, we will not calculate the standard error of the slope completely by hand. We will work from the sum of squared residuals, the sum of squares for x, and the sample size to calculate it. Statistical software yields the following information:

---

* If there is more than one independent variable, DF = N − p, where p is the number of independent variables in the equation plus the intercept.

Sum of squared residuals: 941,113
Sum of squares$_x$: 1,223,567
N: 7,268

We can plug these figures into the formula for the standard error of the slope:

$$SE_b = \frac{\sqrt{\dfrac{\text{sum of squared residuals}}{N-2}}}{\sqrt{\text{sum of squares}_x}} = \frac{\sqrt{\dfrac{\Sigma(\hat{y}-\bar{y})^2}{N-2}}}{\sqrt{\Sigma(x-\bar{x})^2}}$$

$$= \frac{\sqrt{\dfrac{3{,}048{,}318}{7{,}268-2}}}{\sqrt{6{,}225{,}029}} = \frac{20.5}{2495} = 0.008$$

The average distance of the sample slope from the population slope is 0.008 points.

To find the sample t associated with the slope, we divide the sample slope, b, by the standard error of the slope:*

$$t = \frac{b}{SE_b} = \frac{0.59}{0.008} = 73.75$$

The sample t of 73.75 is much larger than the decision t of 1.96. Thus, we can reject the null hypothesis that the population slope is equal to 0. In practice, statistical software will report the t-value associated with the slope for each independent variable and the p-value for that t. In that case, you can avoid all of the hand calculations! The p-value for this slope is < 0.001. Independent variables for which the p-value is less than alpha have statistically significant effects on the dependent variable.

Note that we could have conducted a one-tailed test, in which the null hypothesis stated that the population slope was greater than or less than 0. Technically, we also could have set the population slope equal to a specific value other than 0 in the null hypothesis, but this is a rare circumstance because we usually do not have enough information to hypothesize that the population slope is equal to a specific non-0 value.

● Assumptions of Regression

For both regression and correlation, we assume that the relationship between x and y is linear. We must also be aware of outliers because the presence of even a single outlier can drastically affect the values of the slope and intercept and the goodness-of-fit measures. We can visually inspect the data for linearity and outliers with a scatterplot.

---

* Technically, $t = \dfrac{b-\beta}{SE_b}$. Because in this case we are testing the null hypothesis that $\beta = 0$, we shorten the equation here to $t = \dfrac{b}{SE_b}$.

If we want to use inferential statistics with regression to conduct tests about the true relationship between x and y using a sample, then we must make the following three additional assumptions:

1. *Independence of data*

   This assumption means that observations should not be correlated with each other, such that the residuals in the regression model are independent of each other. To assess this assumption, it is useful to revisit the sampling design to ensure that cases were selected independently of each other.

2. *Homoscedasticity of residuals*

   This assumption means that the variance of the residuals is the same across the regression line (i.e., for every value of the independent variable). Since the regression line is drawn through the predicted values of y for every value of x, we can check this assumption visually by plotting the residuals against their predicted values.

   Figure 14.14 shows the residuals plotted against the predicted feeling thermometer ratings of feminists from our earlier example in this chapter. These are homoscedastic residuals, because the variance of the residuals is about the same regardless of the predicted value of the dependent variable. Data that violated this assumption would show a clear pattern of larger and smaller variances of the residuals across the range of predicted values of the dependent variable.

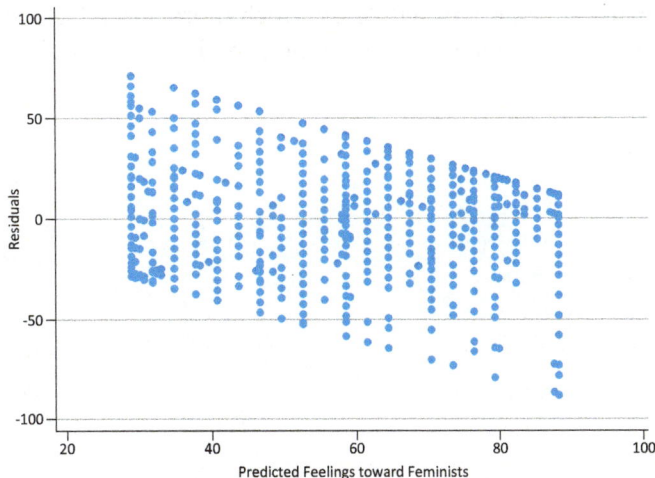

Figure 14.14 Scatterplot of Residuals against Predicted Feelings toward Feminists

3. *Errors are normally distributed*

   The third assumption is that the residuals in the population are normally distributed. We can never know for sure the shape of the actual population distribution, but we can check our sample for normality. The easiest way to do this is to examine a histogram of the residuals to see whether it generally falls into a normal distribution.

Figure 14.15 presents a histogram of the residuals for the regression of feelings toward feminists on feelings toward liberals. The residuals do not follow a perfectly normal distribution, but it is close, meaning that this assumption is not violated.

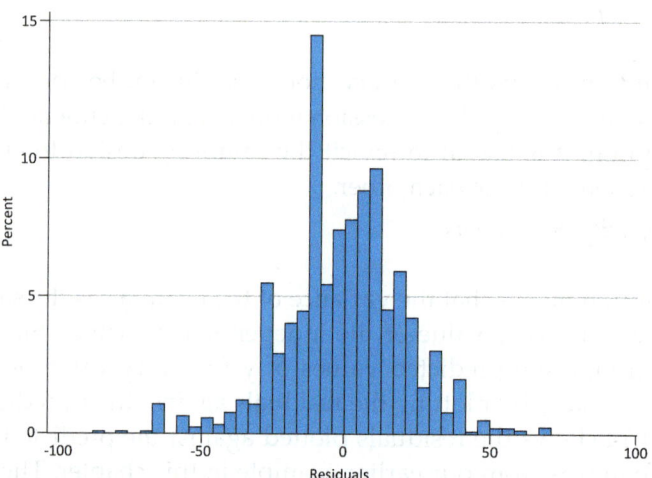

Figure 14.15 Histogram of Residuals from Regression of Feelings toward Feminists on Feelings toward Liberals

If you study statistics further, you will learn much more about how to test the assumptions for regression, what to do when they are not met, and how to model nonlinear relationships.

Chapter Summary

This chapter introduced correlation and regression, two ways of analyzing linear relationships between variables.

- The correlation coefficient, r, measures the linear association between two interval-ratio variables (x and y); by formula:

$$r_{(x,y)} = \frac{COV_{(x,y)}}{s_x s_y} \text{ where } COV_{(x,y)} = \frac{\Sigma(x-\bar{x})(y-\bar{y})}{N-1}$$

and s_x and s_y equal the standard deviations of x and y, respectively.

- The direction of a correlation indicates whether values for x and y move in the same (positive correlation) or opposite (negative or inverse correlation) direction.

- The strength of a correlation ranges from −1 to +1, with stronger correlations closer to −1 or +1 and weaker correlations closer to 0.

- A scatterplot is a graph with points representing the joint values of x and y for each case; it is useful for assessing direction and strength of correlations as well as checking for linearity.

- A linear relationship between two variables can be represented by a straight line.

- Linear regression models the relationship between two variables, x and y; finding the best-fitting line to represent the relationship by minimizing the sum of squared residuals.

- A regression line is the line that best fits the data. It is represented by the equation $\hat{y} = a + bx$, where a is the constant, b is the slope, and $\hat{y}$ is the predicted value of the dependent variable (y) at a given value of the independent variable (x).

- Assumptions of linear regression: (1) linear relationship; (2) no outliers; (3) independence of observations; (4) homoscedasticity of residuals; and (5) errors are normally distributed.

- The slope (regression coefficient) is the expected change in the dependent variable for a one-unit increase in the independent variable. Represented by b in the regression equation. By formula:

$$b = \frac{Sum\,of\,cross\,products}{Sum\,of\,squares_x} = \frac{\Sigma(x-\bar{x})(y-\bar{y})}{\Sigma(x-\bar{x})^2}$$

Interpret the slope by saying: "For every one-unit increase in the independent variable, the dependent variable increases by b units." For multiple regression, interpret the slope by saying: "For every one-unit increase in the independent variable, the dependent variable increases by b units, when holding the other independent variables constant." Use the names of the variables and their units. For example, "For every one year increase in education income increases by $950 a year, when holding gender constant."

- The intercept is the expected value of the dependent variable when the independent variable(s) are equal to 0. Represented by "a" in the equation for the regression line. By formula:

$$a = \bar{y} - b(\bar{x})$$

- R-squared (r^2 or R^2) is a goodness-of-fit measure for regression. It indicates the proportion of variance in the dependent variable accounted for by the variance in the independent variable(s); by formula:

$$r^2 = \frac{\text{explained sum of squares}}{\text{total sum of squares}} = \frac{\sum(\hat{y}-\bar{y})^2}{\sum(y-\bar{y})^2}$$

Interpret r-squared by saying: "The independent variable(s) accounted for r-squared percent of the variance in the dependent variable." Use the name of the dependent variable and independent variable(s) and the value of r-squared. For example, "Years of education and gender accounted for 15% of variation in income."

- The standard error of the estimate is a goodness-of-fit measure for regression that indicates how far, on average, the actual values of the dependent variable fall from the values predicted by the regression line; by formula (with one independent variable):

$$SE_{\hat{y}} = \sqrt{\frac{\text{sum of squared residuals}}{N-2}} = \sqrt{\frac{\sum(y-\hat{y})^2}{N-2}}$$

- The F-statistic tests the statistical significance of a regression model overall by assessing whether the regression model does a better job of generating predicted values of the dependent variable than an "intercept-only model." It also tests the null hypothesis that all of the slopes in the model are equal to 0, with the alternative hypothesis being that at least one of the slopes is unequal to 0; by formula:

$$F = \frac{\left(\dfrac{\text{explained sum of squares}}{p-1}\right)}{\left(\dfrac{\text{sum of squared residuals}}{N-p}\right)} = \frac{\left(\dfrac{\sum(\hat{y}-\bar{y})^2}{p-1}\right)}{\left(\dfrac{\sum(y-\hat{y})^2}{N-p}\right)}$$

where p is the number of parameters estimated in the regression model (number of slopes plus one for the intercept) and N is the sample size.

- To assess statistical significance for the overall regression model, compare the p-value associated with the F-statistic to alpha OR compare sample F to decision F. If p < alpha OR sample F > decision F, reject the null hypothesis.

- The standard error of the slope is used to test the statistical significance of a single slope; by formula:

$$SE_b = \frac{SE_{\hat{y}}}{\sqrt{sum\,of\,squares_x}} \sqrt{\frac{sum\,of\,squared\,residuals}{N-2}} = \frac{\sqrt{\frac{\sum(y-\hat{y})^2}{N-2}}}{\sqrt{\sum(x-\bar{x})^2}}$$

- The t-value associated with the slope by formula:

$$t = \frac{b}{SE_b}$$

- To assess statistical significance of the slope, compare the p-value associated with the t-value for the slope to the alpha-level OR compare the sample t-value with decision t. If p < alpha OR sample t > decision t, reject the null hypothesis.
- A dichotomous, or "dummy," variable is a variable with two categories, coded as 0 and 1. The category coded as 0 is called the "reference category."
- Logistic regression is a regression model that can be used with dichotomous (dummy) dependent variables.

Using Stata

In this section, we will examine the relationship between three variables using data from the World Values Survey (WVS): the immigration status of respondents (*immigrant*), respondents' perceptions of corruption in government and business in their country (*corrtotal*), and respondents' overall trust in a number of key social institutions, such as churches, universities, and the armed forces (*conftotal*).* Immigrant is a dichotomous variable (0 = non-immigrant; 1 = immigrant); *corrtotal* is a scale ranging from 5 to 20, with higher values indicating stronger perceptions of corruption; and *conftotal* is a scale ranging from 15 to 60, with higher values indicating greater trust. The dependent variable is trust in social institutions, and the two independent variables are perceptions of corruption and immigration status.

First, we will inspect the relationship between perceptions of corruption and trust in social institutions for linearity. Next, we will estimate a regression model of the effects of perceptions of corruption and immigration status on having confidence in major social institutions. Finally, we will assess two assumptions of statistical inference for regression using the residuals and predicted values generated by the regression model.

Open `WVSWave7.dta`. To generate a scatterplot of confidence in social institutions against perceptions of corruption, enter the following command into Stata:

---

* *conftotal* and *corrtotal* were generated using the following original variables from the WVS: Q64 through Q78 for *conftotal*; and Q113 through Q117 for *corrtotal*. The values for *conftotal* and *corrtotal* are the sums of the values for each of the respective sets of WVS variables for each respondent.

```
scatter conftotal corrtotal
```

Notice that we enter the variable to be placed on the y axis directly after the "scatter" command, followed by the variable to be entered on the x axis. Stata generates the scatterplot in a new window (Figure 14.16).

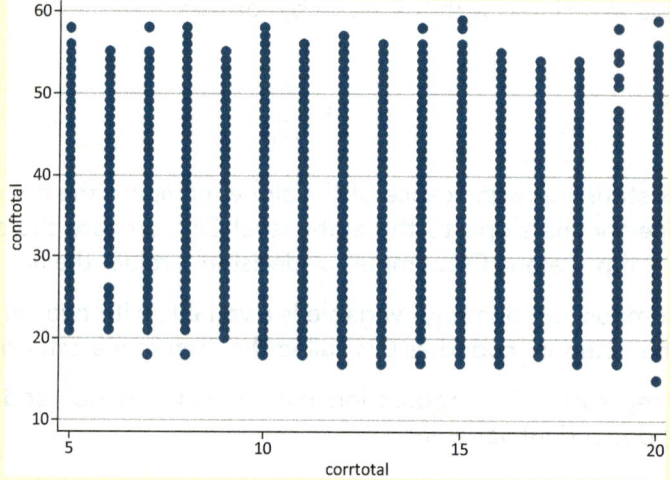

Figure 14.16

As we discussed earlier in this chapter, deciphering patterns in scatterplots with many cases (in this case, 66,003) can be difficult. If we draw a small random sample from the data, we may be able to assess linearity more effectively. Enter the following commands into Stata:

```
sample 1
```

```
scatter conftotal corrtotal
```

With the "sample 1" command, we have told Stata to draw a random sample of 1% of the cases in the data set (i.e., 660 cases). We can now examine a scatterplot of this random subsample of the data, as shown in Figure 14.17.

The scatterplot shows a generally linear relationship that appears to be trending downward. There are no obvious "bends" in the scatterplot, so we will proceed with correlation and regression analysis.

Before proceeding, we must reopen the original `WVSWave7.dta` file so that we are analyzing the full sample. Now we can ask Stata to calculate the correlation coefficient for the two variables by entering the following command:

```
corr conftotal corrtotal
```

The "corr" command, short for "correlate," asks for the correlation between two variables. The order in which the two variables are listed does not matter. The output is shown in Figure 14.18.

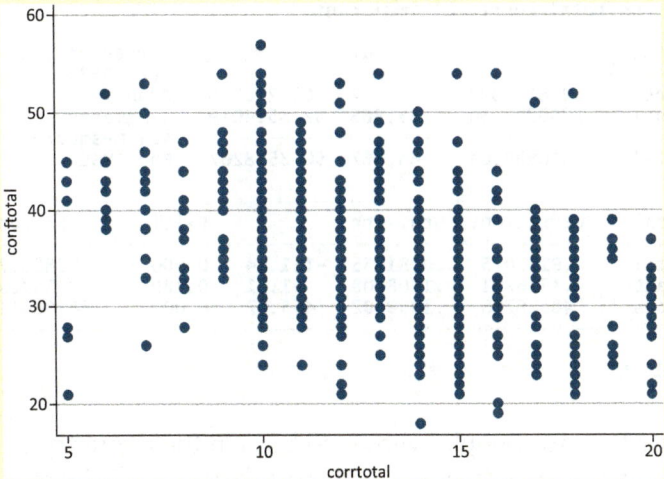

Figure 14.17

```
. corr conftotal corrtotal
(obs=66,003)

             | confto~l  corrto~l
   conftotal |  1.0000
   corrtotal | -0.3985    1.0000
```

Figure 14.18

We see that the correlation between *corrtotal* and *conftotal* is −0.3985, a negative relationship, as suggested by the scatterplot. Although statistical software calculates correlation coefficients to many decimal places, researchers usually present them rounded to two places. We would say that the correlation between these variables is −0.40.

To conduct the regression, type the following command into Stata:

`regress conftotal corrtotal immigrant`

The command tells Stata that it will be performing a regression analysis on the set of variables following "regress." "Regress" is immediately followed by the dependent variable, in this case *conftotal*. The independent variables can be listed in any order following the dependent variable. The results will be the same regardless of the ordering of the independent variables. The output is shown in Figure 14.19.

The output for a regression analysis provides a wealth of information. We will begin with the slopes and intercept, shown in the bottom portion of the output. The "Coefficient" column provides the slopes for the independent variables and the intercept. The partial slope for *corrtotal* is −0.91, indicating that we would expect a 0.91 decrease in confidence for every one-unit increase in perceptions of corruption, controlling for immigrant status. The coefficient for immigrant is −0.15, indicating that the mean confidence for immigrants

```
. regress conftotal corrtotal immigrant

      Source |       SS           df       MS      Number of obs   =    65,868
-------------+----------------------------------   F(2, 65865)     =   6219.41
       Model |  526851.093         2  263425.546   Prob > F        =   0.0000
    Residual |  2789738.54    65,865  42.3554019   R-squared       =   0.1589
-------------+----------------------------------   Adj R-squared   =   0.1588
       Total |  3316589.64    65,867  50.3528267   Root MSE        =   6.5081

------------------------------------------------------------------------------
   conftotal | Coefficient  Std. err.      t    P>|t|     [95% conf. interval]
-------------+----------------------------------------------------------------
    corrtotal |  -.9125025   .0081955  -111.34   0.000    -.9285657   -.8964392
    immigrant |  -.1586701   .1208708    -1.31   0.189    -.3955769    .0782366
        _cons |   48.26775   .1059382   455.62   0.000     48.06011    48.47539
------------------------------------------------------------------------------
```

Figure 14.19

is 0.15 units lower than for people who were born in their country of residence, controlling for perceptions of corruption. (As with correlation coefficients, we round these figures to two decimal places.)

The intercept (located in the row labeled "_cons" because intercepts are sometimes referred to as constants) is 48.3, indicating the predicted level of confidence for those who have values of 0 for *corrtotal* and *immigrant*. In other words, the predicted level of confidence for non-immigrants who have a score of 0 for *corrtotal* is 48.3. In this case, 0 is not a possible value for *corrtotal*, since *corrtotal* is equal to the sum of respondents' values for five variables, such that the smallest possible value for *corrtotal* is five. Here, then, we have a case where the utility of the intercept is limited.

We can see that Stata has performed hypothesis tests for both slopes and the intercept. Specifically, it has tested the null hypothesis for each slope and the intercept that the true population parameter is equal to 0, with the two-tailed alternative hypothesis being that the parameter is unequal to 0. The "Std. err." column provides the standard error for each of these estimates; the "t" column, the test statistic for each estimate; the "P > | t |" column, the right-tail probability associated with the test statistic; and the "95% Conf. interval" column, confidence intervals for each estimate based on their standard errors. We can see that the partial slope for perceptions of corruption has a p-value that is below 0.001, indicating that there is only a very small chance that we would observe a regression coefficient as large as this if the true coefficient in the population were equal to 0. In other words, the effect of perception of corruption and on confidence in social institutions, controlling for immigrant status is statistically significant. The coefficient for immigrant, however, has a p-value greater than 0.05 (0.189), indicating that we cannot reject the null hypothesis of no relationship between immigrant status and confidence in social institutions, controlling for perceptions of corruption.

The top portion of the output presents goodness-of-fit statistics for the regression model. In the far-right column, we see the F-statistic, 6,219, and its associated p-value, < 0.001. This suggests that the regression model fits the data better than does an intercept-only model. We also see the R^2 for the model, 0.16, indicating that the independent variables account for 16% of the variation in the dependent variable. Finally, we see that the standard error of the estimate (called "Root MSE," or Root Mean Square Error, in the output) is 6.5, indicating that the actual values of the dependent variable fall about 6.5 units from the values predicted by the regression line, on average.

Checking Assumptions

Checking the key assumptions of regression for statistical inference must happen after the regression model is run, because we need to use the predicted values and residuals produced by the model to assess the assumptions. We can ask Stata to create new variables that measure the predicted value ($\hat{y}$) of the dependent variable (*conftotal*) and the residual value ($y - \hat{y}$) for each case in the data set. Enter the following commands into Stata to generate predicted values and residuals:

```
regress conftotal corrtotal immigrant
predict conftothat
predict conftotres, residuals
```

The "regress" command is the same one as above, and we begin with it because the subsequent commands will use the results of the regression model. The two "predict" commands are known as post-estimation commands because they are making predictions based on the estimated regression model. The first "predict" command tells Stata to calculate the predicted value of the dependent variable for each case in the data set based on the regression model. This command creates a new variable, which we are calling *conftothat*, containing the predicted values of the dependent variable ($\hat{y}$) for each case. The next "predict" command uses the "residuals" option to signal to Stata that it should create a new variable that measures the residual ($y - \hat{y}$) for each case. We are naming this new variable *conftotres*.

You can always generate residuals and predicted values in this way. With these new variables, we can check two of the assumptions of linear regression when using inferential statistics. First, we can check the assumption of homoscedastic residuals by plotting the residuals against their predicted values. (Refer to page 571 for discussion of homoscedasticity of residuals.) Enter the following command into Stata:

```
scatter conftotres conftothat
```

Here, we are requesting a scatterplot of residuals against predictions of the dependent variable. The scatterplot is shown in Figure 14.20.

Note that Stata uses the label "Fitted values" for the variable that measures predicted values of the dependent variable (*conftothat*). We see that the variance of the residuals is about the same regardless of the predicted value of the dependent variable.

Next, we can check whether the residuals are normally distributed by requesting a histogram of the residuals (*conftotres*):

```
histogram conftotres, percent
```

The histogram is shown in Figure 14.21.

The histogram shows that the residual values are distributed roughly normally.

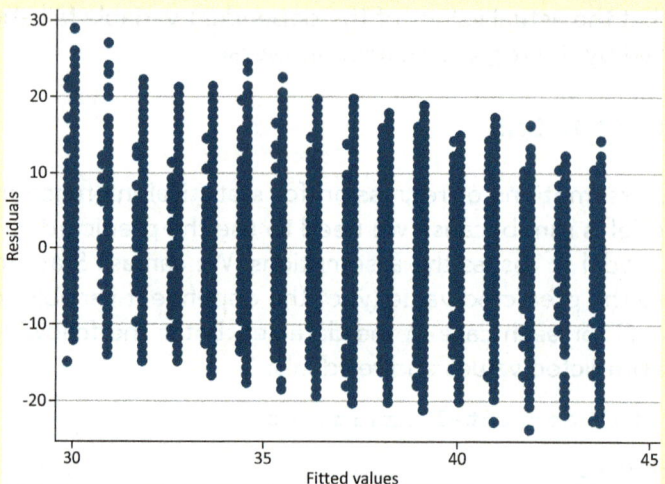

Figure 14.20

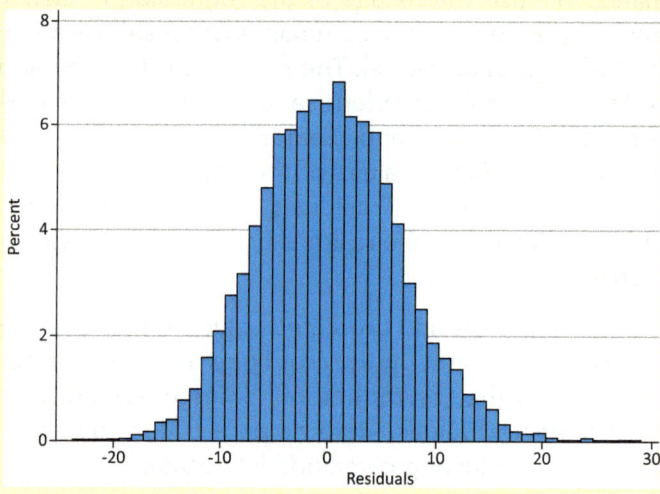

Figure 14.21

Review of Stata Commands

- Generate a scatterplot

    `scatter dependent variable independent variable`

- Request a correlation coefficient

    `corr first variable second variable`

- Linear regression

    `regress dependent variable independent variable(s)`

- Post-estimation commands following regression to be conducted directly after a regression model: generate a new variable containing predicted y-values for each case

    `predict new variable name`

- Post-estimation commands following regression to be conducted directly after a regression model: generate a new variable containing residual values for each case

    `predict new variable name, residuals`

Using SPSS

In this section, we will examine the relationship between three variables using data from the World Values Survey (WVS): the immigration status of respondents (*immigrant*), respondents' perceptions of corruption in government and business in their country (*corrtotal*), and respondents' overall trust in a number of key social institutions, such as churches, universities, and the armed forces (*conftotal*).* Immigrant is a dichotomous variable (0 = non-immigrant; 1 = immigrant); *corrtotal* is a scale ranging from 5 to 20, with higher values indicating stronger perceptions of corruption; and *conftotal* is a scale ranging from 15 to 60, with higher values indicating greater trust. The dependent variable is trust in social institutions, and the two independent variables are perceptions of corruption and immigration status.

First, we will inspect the relationship between perceptions of corruption and trust in social institutions for linearity. Next, we will estimate a regression model of the effects of perceptions of corruption and immigration status on having confidence in major social institutions. Finally, we will assess two assumptions of statistical inference for regression using the residuals and predicted values generated by the regression model.

Open `WVSWave7.sav`. To generate a scatterplot of confidence in social institutions against perceptions of corruption, we use the following SPSS sequence:

`Graphs → Legacy Dialogs → Scatter/Dot`

This opens the "Scatter/Dot" dialog box. We select "Simple Scatter" and then click on "Define." This opens the "Simple Scatterplot" dialog box. We place our dependent variable (*conftotal*) into the "Y Axis" box and our independent variable (*corrtotal*) into the "X Axis" box. We click on "OK," and SPSS produces the output shown in Figure 14.22.

As we discussed earlier in the chapter, deciphering patterns in scatterplots with many cases (in this case, 66,003) can be difficult. If we draw a small random sample from the data, we may be able to assess linearity more effectively. We will regenerate a scatterplot,

---

* *conftotal* and *corrtotal* were generated using the following original variables from the WVS: Q64 through Q78 for *conftotal*; and Q113 through Q117 for *corrtotal*. The values for *conftotal* and *corrtotal* are the sums of the values for each of the respective sets of WVS variables for each respondent.

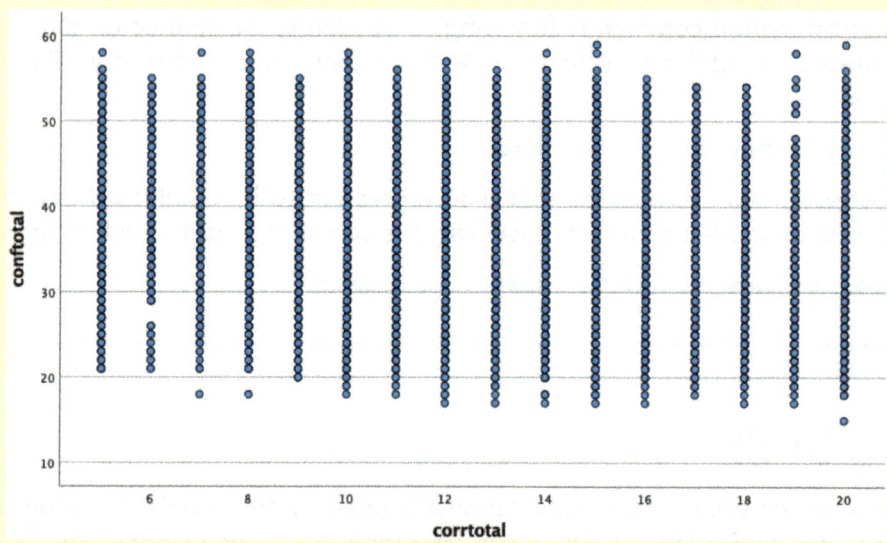

Figure 14.22

but, this time, we will ask SPSS to construct the scatterplot using a random selection of 1% of the cases. Under:

`Data → Select Cases`

select the "Random sample of cases" option and click on "Sample." (See the SPSS section at the end of chapter 7 for a step-by-step review.) The scatterplot of this random subsample of the data is shown in Figure 14.23.

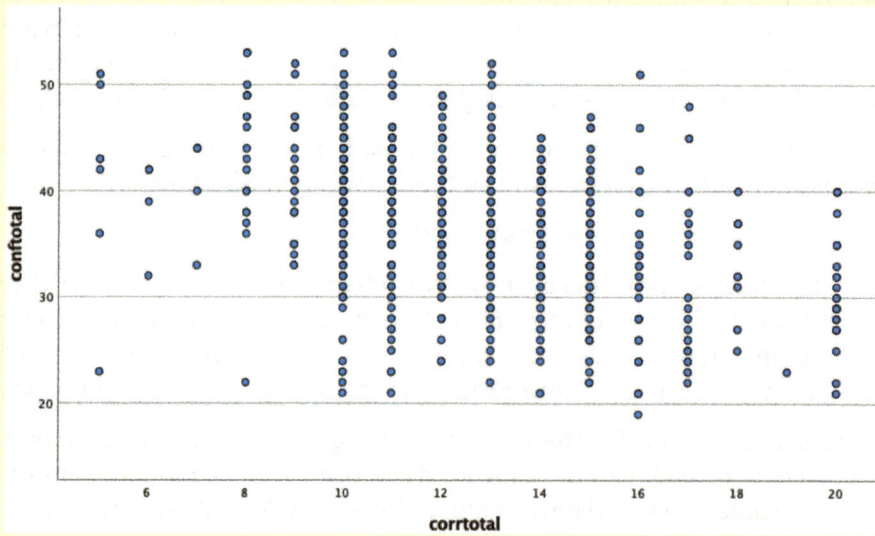

Figure 14.23

The scatterplot shows a generally linear relationship that appears to be trending downward. There are no obvious "bends" in the scatterplot, so we will proceed with correlation and regression analysis. Before doing so, remember to first restore the full sample for analysis by returning to "Data" → "Select Cases" and selecting "all cases."

We can instruct SPSS to calculate the correlation coefficient for the two variables by using the following procedure:

`Analyze → Correlate → Bivariate`

This opens the "Bivariate Correlations" dialog box (Figure 14.24). We move the two variables—*conftotal* and *corrtotal*—into the "Variables" box and click on "OK."

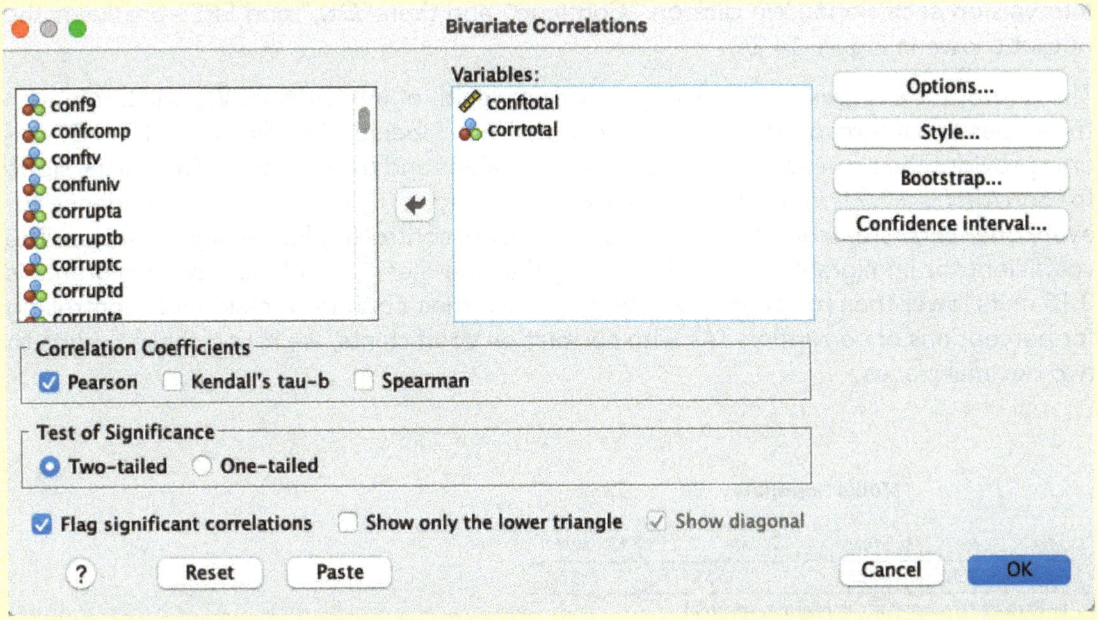

Figure 14.24

The output is shown in Figure 14.25.

Correlations

		conftotal	corrtotal
conftotal	Pearson Correlation	1	-.399**
	Sig. (2-tailed)		.000
	N	72572	66003
corrtotal	Pearson Correlation	-.399**	1
	Sig. (2-tailed)	.000	
	N	66003	75003

**. Correlation is significant at the 0.01 level (2-tailed).

Figure 14.25

We see that the correlation between *corrtotal* and *conftotal* is −0.399, a negative relationship, as suggested by the scatterplot. Although statistical software calculates correlation coefficients to many decimal places, researchers usually present them rounded to two places. We would say that the correlation between these variables is −0.40.

To conduct the regression, we use the following sequence:

`Analyze → Regression → Linear`

This opens the "Linear Regression" dialog box. We place our dependent variable (*conftotal*) into the "Dependent" box and our independent variables (*corrtotal* and *immigrant*) into the "Independent(s)" box. We click on "Statistics" and place a check mark in the small box next to "Confidence Intervals." (This tells SPSS to calculate 95% confidence intervals on each slope.) We click on "Continue" and then "OK," and SPSS produces the output shown in Figure 14.26.

The output for a regression analysis provides a wealth of information. We will begin with the slopes and intercept, shown in the bottom table, labeled "Coefficients." The "B" column provides the slopes for the independent variables and the intercept. The partial slope for *corrtotal* is −0.91, indicating that we would expect a 0.91 decrease in confidence for every one-unit increase in perceptions of corruption, controlling for immigrant status. The coefficient for immigrant is −0.15, indicating that the mean confidence for immigrants is 0.15 units lower than for people who were born in their country of residence, controlling for perceptions of corruption. (As with correlation coefficients, we round these figures to two decimal places.)

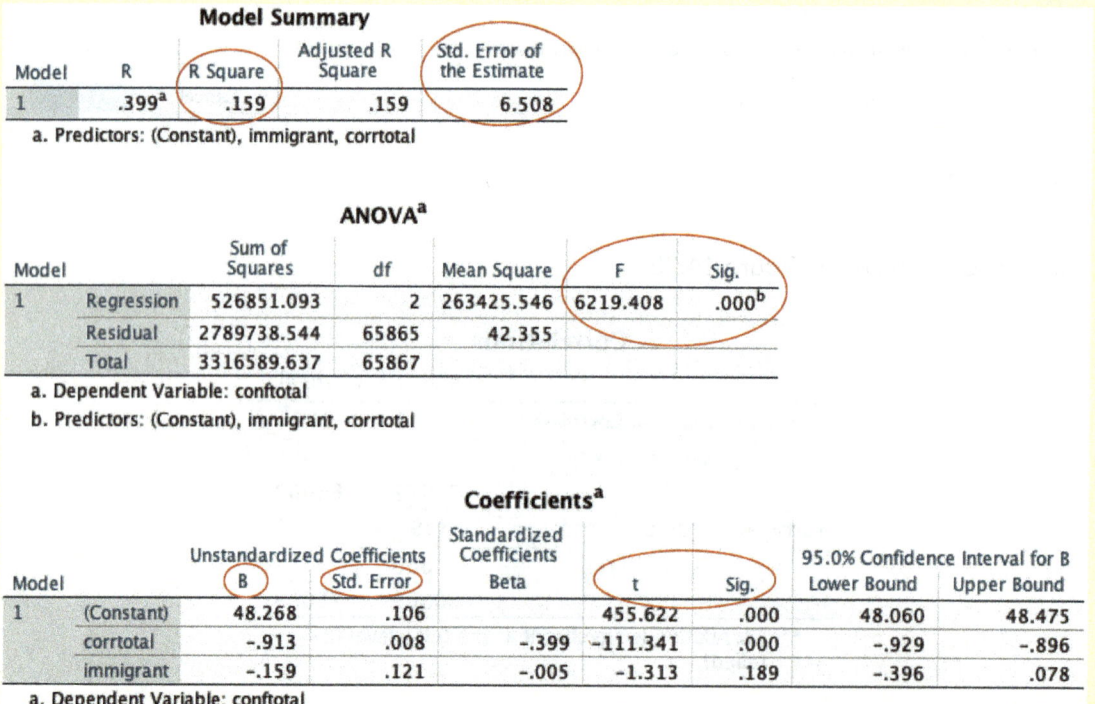

Figure 14.26

The intercept (located in the row labeled "(Constant)" because intercepts are sometimes referred to as constants) is 48.3, indicating the predicted level of confidence for those who have values of 0 for *corrtotal* and *immigrant*. In other words, the predicted level of confidence for non-immigrants who have a score of 0 for *corrtotal* is 48.3. In this case, 0 is not a possible value for *corrtotal*, since *corrtotal* is equal to the sum of respondents' values for five variables, such that the smallest possible value for *corrtotal* is five. Here, then, we have a case where the utility of the intercept is limited.

We can see that SPSS has performed hypothesis tests for both slopes and the intercept. Specifically, it has tested the null hypothesis for each slope and the intercept that the true population parameter is equal to 0, with the two-tailed alternative hypothesis being that the parameter is unequal to 0. The "Std. Error" column provides the standard error for each of these estimates; the "t" column, the test statistic for each estimate; the "Sig." column, the two-tailed probability associated with the test statistic; and the "95% Confidence Interval" column, confidence intervals for each estimate based on their standard errors. We can see that the partial slope for perceptions of corruption has a p-value that is below 0.001, indicating that there is only a very small chance that we would observe a regression coefficient as large as this if the true coefficient in the population were equal to 0. In other words, the effect of perception of corruption and on confidence in social institutions, controlling for immigrant status is statistically significant. The coefficient for immigrant, however, has a p-value greater than 0.05 (0.189), indicating that we cannot reject the null hypothesis of no relationship between immigrant status and confidence in social institutions, controlling for perceptions of corruption.

The top portion of the output, labeled "Model Summary," presents goodness-of-fit statistics for the regression model. We see the R^2 for the model is 0.16, indicating that the independent variables account for about 16% of the variation in the dependent variable. We see that the standard error of the estimate is 6.5, indicating that the actual values of the dependent variable fall about 6.5 units from the values predicted by the regression line, on average. Finally, in the middle table, labeled "ANOVA," we see the F-statistic, 6,219, and its associated p-value, < 0.001. This suggests that the regression model fits the data better than does an intercept-only model.

Checking Assumptions

Checking the key assumptions of regression for statistical inference must happen after the regression model is run, because we need to use the predicted values and residuals produced by the model to assess the assumptions. When we use the regression procedure, SPSS automatically generates a series of variables that can be used to check the key assumptions. To access these variables, we use the same regression procedure we used above.

`Analyze → Regression → Linear`

But before we click on "OK," we click on "Plots." This opens the "Linear Regression: Plots" dialog box (Figure 14.27).

Notice, on the left side of the box, a series of variables that SPSS generates automatically. We will use two of these variables to check the assumption of homoscedastic residuals of

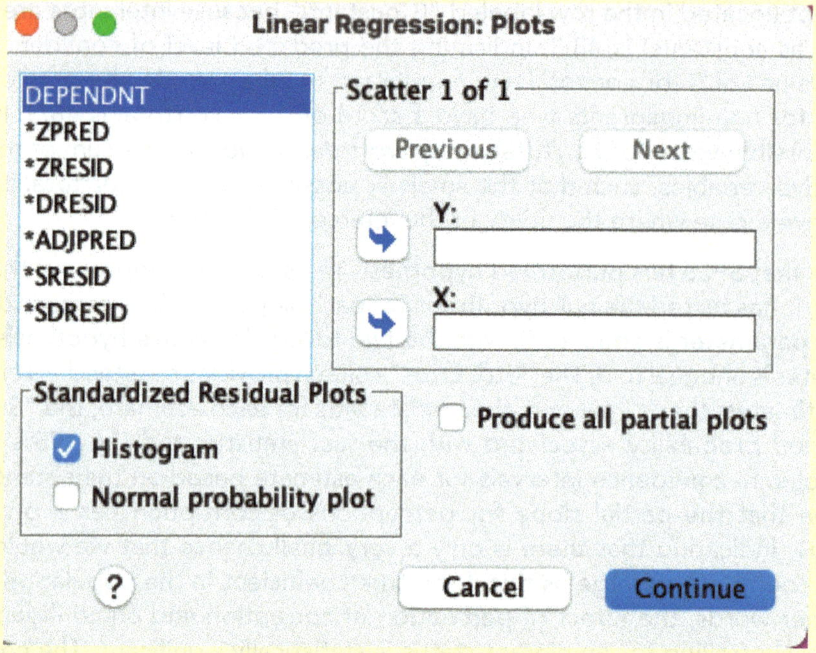

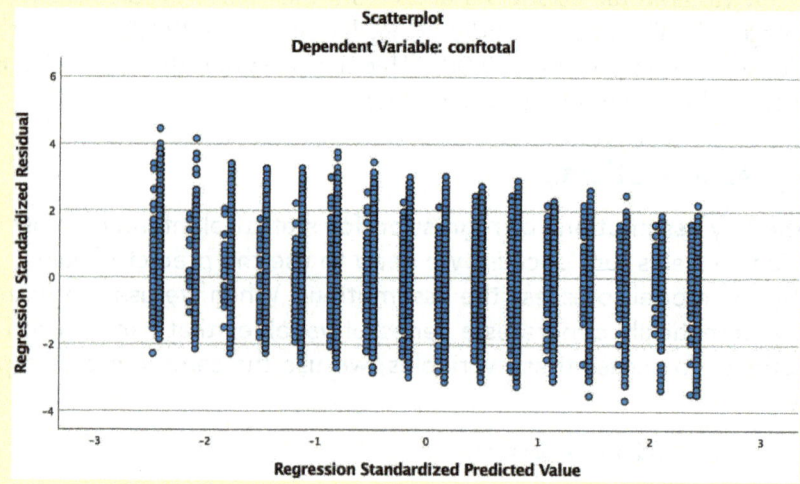

Figure 14.27

the dependent variable. (Refer to page 571 for discussion of homoscedasticity of residuals.) Toward that end, we will generate a scatterplot of the standardized residuals (*zresid* in SPSS) against the standardized predicted values (*zpred*).* We move *zresid* into the "Y" box and *zpred* into the "X" box. Click on "Continue" and then "OK." The scatterplot, shown in Figure 14.28, is included with the SPSS regression output.

Figure 14.28

---

* Earlier in this chapter, we checked the assumption of homoscedasticity by plotting the residuals against the predicted values of the dependent variable. SPSS generates only standardized versions of these variables. One can use either unstandardized or standardized versions to test the assumption of homoscedasticity.

We see that the variance of the residuals is about the same regardless of the predicted value of the dependent variable.

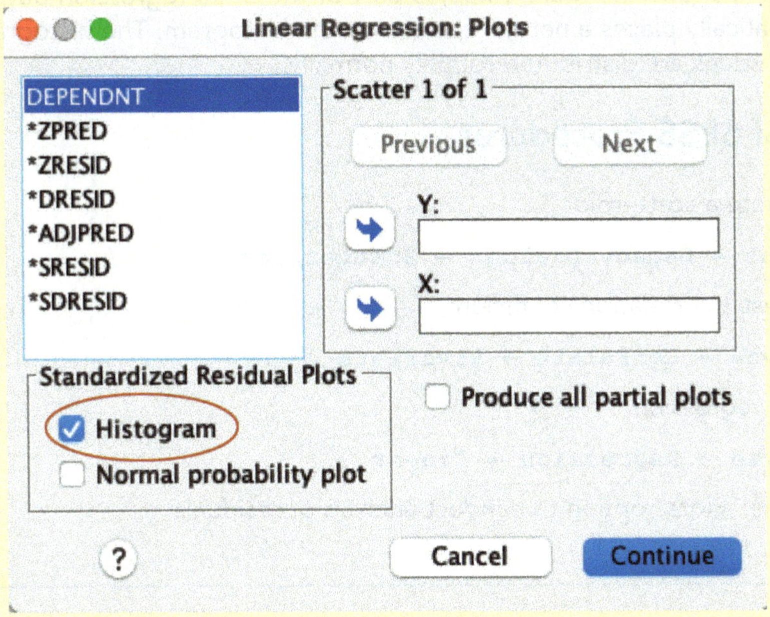

Figure 14.29

Next, we can check whether the residuals are normally distributed by requesting a histogram of the residuals. We use the same SPSS sequence we just used to generate the scatterplot. Using the "Linear Regression: Plots" dialog box (shown in Figure 14.29), we will

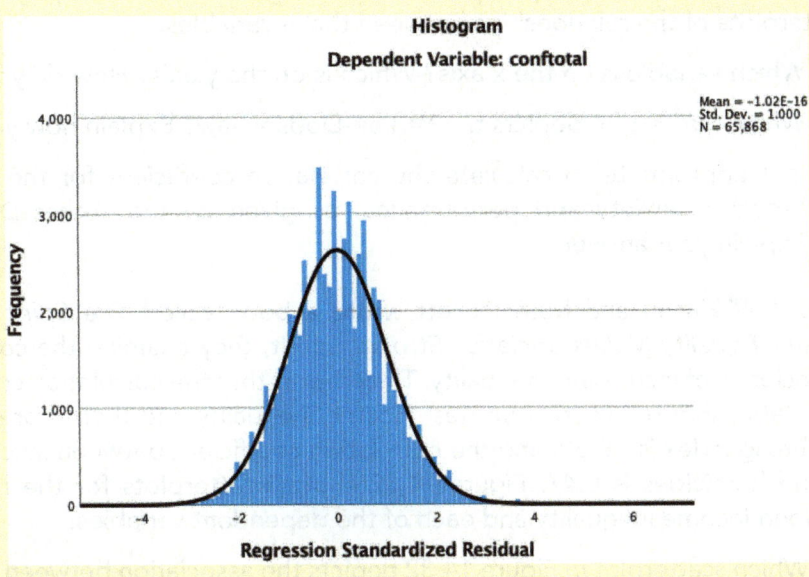

Figure 14.30

tell SPSS to generate a histogram of the standardized residuals. We place a check mark in the small box next to "Histogram," then click on "Continue" and then "OK."

The histogram, shown in Figure 14.30, is part of the SPSS regression output. Note that SPSS automatically places a normal curve over the histogram. The histogram shows that the residual values are distributed roughly normally.

Review of SPSS Procedures

- Generate a scatterplot

    `Graphs → Legacy Dialogs → Scatter/Dot`

- Request a correlation coefficient

    `Analyze → Correlate → Bivariate`

- Linear regression

    `Analyze → Regression → Linear`

- Use the "Plots" option to conduct analysis of residuals

Practice Problems

1. According to the Yerkes–Dodson law, people tend to perform best when they have some anxiety about a task. However, not enough anxiety and too much anxiety have been shown to dampen performance. Figure 14.31 (next page) shows five scatterplots of the relationships between these variables.

    a. Which variable is on the x axis? Which is on the y axis? How do you know?

    b. Which scatterplot depicts the Yerkes–Dodson law? Explain how you know.

    c. Is it appropriate to calculate the correlation coefficient for the relationship between anxiety and performance, as given by the Yerkes–Dodson law? Explain your answer.

2. Richard Wilkinson and Kate Pickett wrote a book called *The Spirit Level: Why Greater Equality Makes Societies Stronger*. In it, they examine the consequences for societies of increasing inequality. They report that the correlation coefficient for the relationship between countries' income inequality and their scores on a child well-being index is −0.62, and the correlation coefficient between income inequality and homicides is 0.47. Figure 14.32 shows scatterplots for the relationships between income inequality and each of the dependent variables.

    a. Which scatterplot in Figure 14.32 depicts the association between income inequality and homicides? Between income inequality and child well-being? How can you tell?

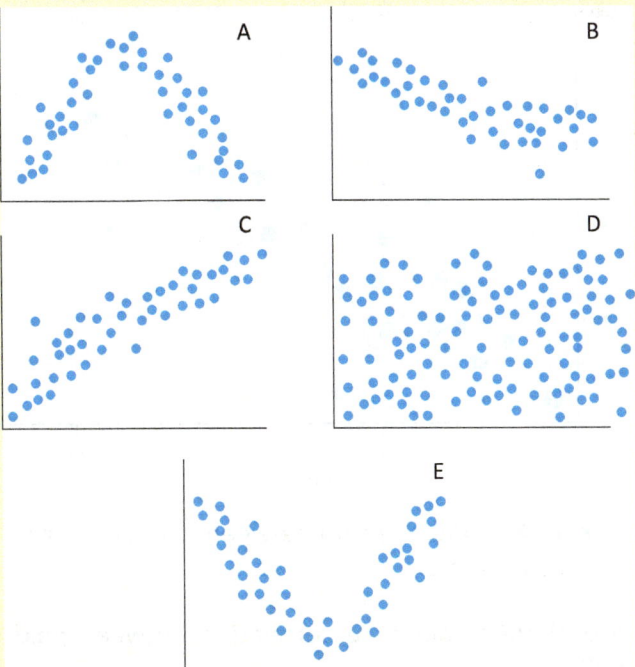

Figure 14.31 Five Scatterplots of the Relationship between Two Variables

b. Which correlation is stronger?

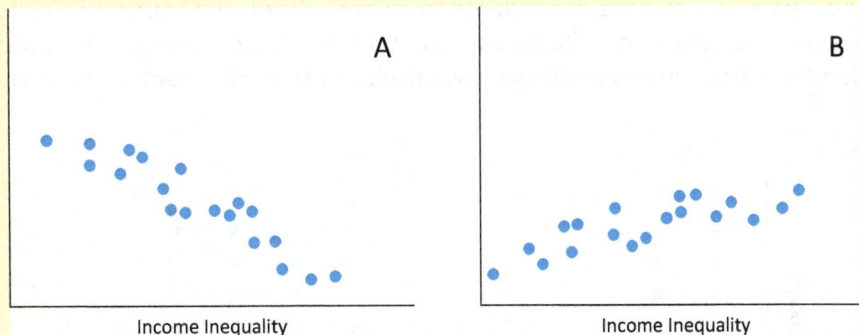

Figure 14.32 Scatterplots of the Relationship between Income Inequality and Two Dependent Variables

3. Figure 14.33 shows a scatterplot of the relationship between the number of paid sick days allowed per year by employers and employees' job satisfaction, measured on a scale from 0 to 100.

   a. Identify the coordinates of the point circled in the scatterplot. How many paid sick days does the person's employer allow per year? What is the employee's job satisfaction rating?

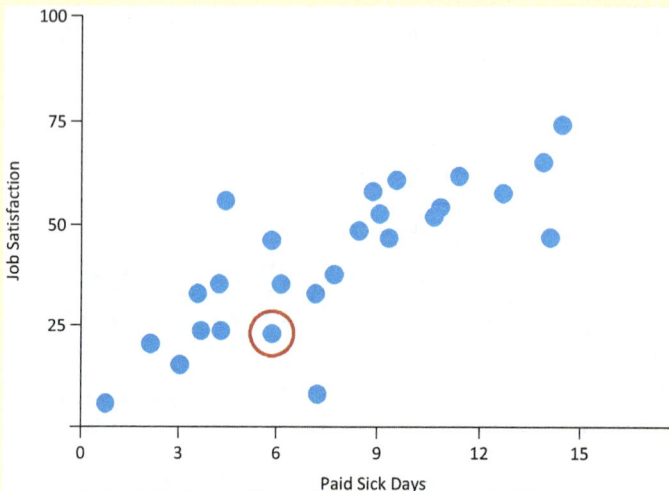

Figure 14.33 Relationship between Paid Sick Days and Job Satisfaction

b. Is there a positive or negative correlation between paid sick days and job satisfaction?

c. If the correlation between paid sick days and job satisfaction is 0.71, how would you describe the strength of the relationship?

4. For her final project in statistics, a student decided to study how excitement about college varied among students, depending on how long they had been pursuing their degree. In a sample of students from her school, she found a 0.04 correlation between time spent in college and excitement about college. The student concluded that time spent in college was unrelated to excitement about college.

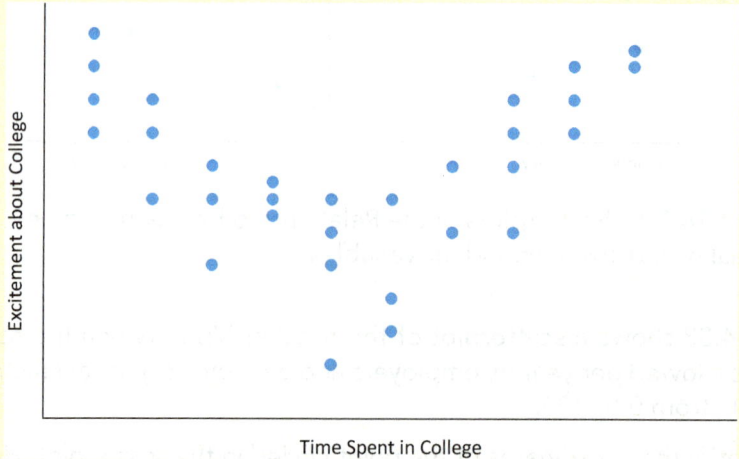

Figure 14.34 Scatterplot of Students' Excitement about College by Time Spent in College

a. Do you agree with the student's conclusion based on the correlation coefficient? Explain your answer.

b. Figure 14.34 shows the scatterplot of excitement about college by time spent at college in the statistics student's sample. Does the scatterplot support the student's conclusion of no relationship between the variables? Explain your answer.

c. What lesson can the student learn from comparing the correlation coefficient for the relationship between time spent in college and excitement about college to the scatterplot?

5. Is attending class regularly positively associated with college students' academic performance? A professor has been telling students for years that the most important thing they can do as college students is to show up to class, whether or not they have completed their work. Noticing that some students continue to disregard her advice, the professor decides to analyze the empirical relationship between class attendance and GPA in a sample of twenty students who are all taking the same courses during their first semester of college. Table 14.1 presents the number of classes that each student attended during their first semester of college and their first-semester GPA. The mean and standard deviation are noted for each variable.

Table 14.1 Number of Classes Attended and GPA for First Semester of College

Student	Classes Attended: $\bar{x} = 94.2$, $s = 21.25$	GPA: $\bar{y} = 3.13$, $s = 0.53$
1	120	3.5
2	90	2.8
3	60	2.2
4	100	3.2
5	110	3.2
6	115	3.7
7	120	3.9
8	70	3.2
9	80	3
10	97	2.9
11	93	3
12	50	2.4
13	56	2.1
14	117	3.75
15	111	3.7
16	105	3.8
17	98	3.3
18	87	3.2
19	105	3.2
20	100	2.5

a. For each student, calculate the difference between the number of classes that the student attended and the mean number of classes that all twenty students attended.

b. For each student, calculate the difference between the student's GPA and the mean GPA for all twenty students.

c. Calculate the cross product for each student.

d. Find the sum of cross products for these twenty students.

e. What is the covariance between number of classes attended and GPA?

f. What is the correlation between number of classes attended and GPA?

g. Does the correlation support the professor's advice to students about the importance of attending class?

6. Figure 14.35 shows two scatterplots, each with the best-fitting line drawn through the points on the plot. Points represent cases in the data with particular values for the x and y variables. In each plot, the "x" represents the point on the best-fitting line when x equals 20.

   a. Which scatterplot has larger residuals? How do you know?

   b. How many negative residuals are there in scatterplot A? How many in scatterplot B? How do you know?

   c. In each scatterplot, a single point in the plot is circled. Calculate the residual for each of the circled points in scatterplots A and B.

   d. The means for the y variable in the two scatterplots shown in Figure 14.35 are both equal to twenty, and the best-fitting lines cross the y axis at the same point and have the same slope. Use the formula for r^2 to explain which scatterplot in Figure 14.35 has a larger r^2. *Do this without actually calculating r^2.*

   e. Which relationship between x and y shown in Figure 14.35 shows a larger proportion of variation in y accounted for by variation in x? Explain how you know.

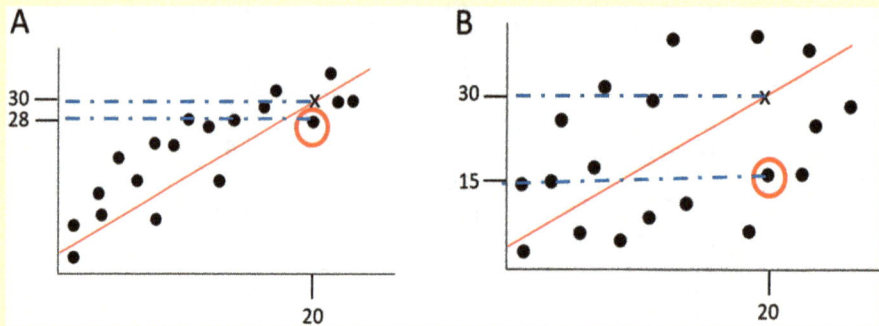

Figure 14.35 Best-Fitting Lines for Two Scatterplots

7. The 2020 American National Election Study (ANES) asked respondents to rate how warmly or coldly they felt about the #MeToo movement on a feeling thermometer scale from 0 to 100 degrees. In the sample of 6,158 respondents, a regression of temperature ratings for the #MeToo movement on age (measured in years) shows that the relationship can be represented by the following equation: $\hat{y} = 69.7 - 0.20x$

   a. Identify the slope and the intercept in the regression equation.
   b. Explain in words what the intercept means. Is this intercept useful?
   c. Explain in words what the slope means. What does the slope tell us about the direction of the relationship?
   d. Calculate the predicted temperature rating for a person who is thirty-two years old. Calculate the predicted rating for a person who is seventy-one years old.
   e. What is the difference between the predicted temperature ratings for a thirty-two-year-old and a seventy-one-year-old?
   f. The standard error for the slope is 0.023. Use this and other relevant information to determine whether we can reject the null hypothesis that the slope for age is equal to 0 in the population.

8. Another researcher working with the 2020 ANES wants to investigate the relationship between age and feeling thermometer ratings of the Supreme Court. He conducts a bivariate regression between the two variables. The regression equation looks like this:

   $$\hat{y} = 45.5 + 0.29x$$

   a. Identify the slope and the intercept in the regression equation.
   b. Explain in words what the intercept means. Is this intercept useful?
   c. Explain in words what the slope means. What does the slope tell us about the direction of the relationship?
   d. Calculate the predicted temperature rating for a person who is thirty-two years old. Calculate the predicted rating for a person who is seventy-one years old.
   e. What is the difference between the predicted temperature ratings for a thirty-two-year-old and a seventy-one-year-old?
   f. The standard error for the slope is 0.015. Use this and other relevant information to determine whether we can reject the null hypothesis that the slope for age is equal to 0 in the population.

9. In Problem 7, we saw the regression equation predicting feeling thermometer ratings of the #MeToo movement based on age. A researcher adds respondents' party affiliation as an additional independent variable. The variable, Democrat, measures whether the respondent identifies as a Democrat (1) or another party (0).

a. The slope for Democrat is 19.7. Explain what the slope for the Democrat variable means here.

b. The researcher's statistical software program produces an F-statistic of 303.8 for the model and standard errors of 0.02 and 0.86 for the slopes of age and party affiliation, respectively. All three of these statistics are associated with p-values less than 0.001. Would we be better off using this regression equation to predict feeling thermometer ratings of the #MeToo movement than using the sample mean ratings as the predicted rating for all respondents? Which statistic answers this question?

c. The researcher wants to know the standard error of the estimate for the regression equation. He knows that the sum of squared residuals—$\Sigma(\hat{y}-y)^2$—equals 4,795,686, and the sample size is 5,922. Use this information to calculate the standard error of the estimate. Explain what it means in words.

d. The researcher wants to know the r^2 value for the regression equation. He knows that the sum of all squared deviations of the mean rating of the #MeToo movement from the predicted value for each respondent $\Sigma(\hat{y}-y)^2$ is 492,196. He knows that the sum of squared deviations of the mean rating of the #MeToo movement from the actual rating for each respondent $\Sigma(\hat{y}-y)^2$ is 5,287,883. Find r^2 and explain what it means in words.

10. In Problem 8, we saw the regression equation predicting feeling thermometer ratings of the Supreme Court based on age. A researcher adds respondents' party affiliation as an additional independent variable. The variable, Democrat, measures whether the respondent identifies as a Democrat (1) or another party (0).

a. The slope for Democrat is –14.6. Explain what the slope for the Democrat variable means here.

b. The researcher's statistical software program produces an F-statistic of 239.3 for the model and standard errors of 0.02 and 0.79 for the slopes of age and party affiliation, respectively. All three of these statistics are associated with p-values less than 0.001. Would we be better off using this regression equation to predict feeling thermometer ratings of the Supreme Court than using the sample mean ratings as the predicted rating for all respondents? Which statistic answers this question?

c. The researcher wants to know the standard error of the estimate for the regression equation. He knows that the sum of squared residuals—$\Sigma(\hat{y}-y)^2$—equals 1,139,422 and the sample size is 2,745. Use this information to calculate the standard error of the estimate. Explain what it means in words.

d. The researcher wants to know the r^2 value for the regression equation. He knows that the sum of all squared deviations of the mean rating of the Supreme Court from the predicted value for each respondent $\Sigma(\hat{y}-y)^2$ is 198,635. He knows that the sum of squared deviations of the mean rating of the Supreme Court from the actual rating for each respondent $\Sigma(\hat{y}-y)^2$ is 1,338,057. Find r^2 and explain what it means in words.

11. Inspired by a study conducted using data from Germany, a pair of political scientists in the United States want to know if the frequency of visits to "alt-right" Facebook pages is associated with racially motivated hate crimes. The researchers create state-level measures of visits to "alt-right" Facebook pages (visitors to these pages per one thousand people) and hate crime incidents for a single year. The slope for the "alt-right" Facebook page visitors variable is 2.3.

   a. Explain what the slope means in words.

   b. Next, the researchers added an additional independent variable to the equation, the state-level percentage of the white male population older than the age of eighteen that is unemployed. The slope for this variable was 3.1. Explain what this slope means in words.

   c. After controlling for the percentage of white men older than eighteen who are unemployed, the slope for the "alt-right" Facebook pages variable declines from 2.3 to 1.5. Based on this fact, which of the Venn diagrams in Figure 14.36 accurately represents correlations among visitors to "alt-right" Facebook pages, unemployed white men older than eighteen, and racially motivated hate crimes? Explain how you know.

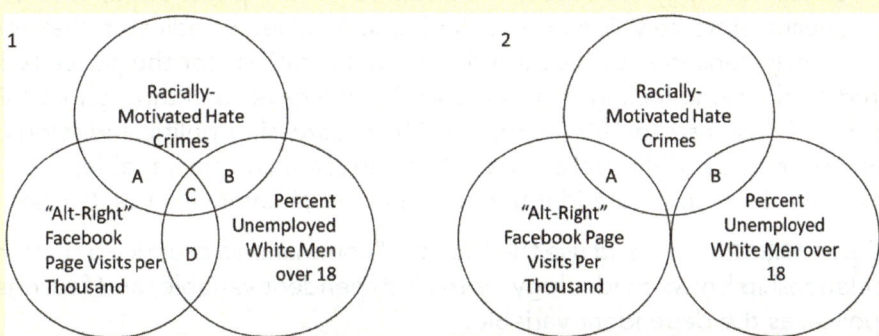

Figure 14.36 Two Venn Diagrams for Correlations among Visitors to "Alt-Right" Facebook Pages, Unemployed White Men Older than Eighteen, and Racially Motivated Hate Crimes

   d. What is the equation for the regression line with both independent variables, given that the intercept is 40, x_1 is "alt-right" Facebook page visitors, and x_2 is percentage of white men older than eighteen who are unemployed?

   e. Calculate the predicted number of racially motivated hate crimes for a state in which 13% of white men older than eighteen are unemployed and "alt-right" Facebook pages had five visitors per one thousand people in the state.

12. Imagine that the following regression equation models the effects of knowing one's neighbors, x_1, and shopping in the community, x_2, on trust in the people who live in one's community, y: $\hat{y} = 13 + 13x_1 + 1.25x_2$. Knowing one's neighbors is measured by two categories (1 = knows at least one neighbor and 0 = knows none of the

neighbors). Shopping in the community is measured on a scale from 0 to 5, with 0 being never shops in the community and 5 being exclusively shops in the community. Trust in the people who live in one's community is measured from 0 (no trust) to 100 (complete trust).

a. What is the predicted trust rating for a person who knows none of their neighbors and never shops in their community? Give the answer using information from the equation, without plugging values into the regression equation. Then, confirm that your answer is correct by plugging values into the regression equation.

b. What is the predicted trust rating for a person who knows at least one of their neighbors and never shops in their community? Find the value without plugging values into the regression equation. Then, demonstrate that you are correct by plugging values into the regression equation.

Stata Problems

Here, we use data from the 2020 American National Election Study (ANES) to investigate the relationship between political ideology and feeling thermometer ratings of the police. Political ideology (V201200) is measured on a 7-point scale, where the lowest value, 1, means extremely liberal and the highest value, 7, indicates that the respondent is extremely conservative. Feeling thermometer ratings for the police (V202171) are measured on the same feeling thermometer that we have seen throughout this book, on a scale from 0 (coldest) to 100 (warmest). After regressing political ideology on feeling thermometer ratings of the police, we will introduce a control variable, poc, which indicates whether the respondent identifies as a person of color (1 = yes, 0 = no).

1. Open `ANES20.dta` and use the "scatter" command to produce a scatterplot of the relationship between ideology, as the independent variable, and feelings about the police, as the dependent variable.
2. Does there appear to be a relationship between these two variables, according to the scatterplot?
3. Use the "corr" command to get the correlation coefficient between these two variables.
4. What is the correlation coefficient? What does it tell us about the strength of this relationship?
5. Use the "regress" command to run a bivariate regression between the variables.
6. What is the slope? What does it mean, in words?
7. What is the intercept? What does it mean, in words?
8. What is the standard error of the slope, and what p-value is associated with the slope? Can we conclude that the population slope is unequal to 0?
9. What is the r^2 value for the model? What does it mean, in words?

10. What is the standard error of the estimate? What does it mean, in words?
11. What are the F-statistic for the model and its associated p-value? What do they mean, in words?
12. Now, use the "regress" command to run the same regression equation from earlier, but add *poc* as a second independent variable.
13. What is the slope for *poc*? What does it mean, in words? Is it "statistically significant"?
14. What is the slope for ideology in this new model, and how does it compare to the first model? What does this comparison suggest about differences in political ideology for people of color and whites? Is it "statistically significant"?
15. What is the intercept in this new model? What does it mean, in words?
16. Give the r^2 and the standard error of the estimate for this new model. Compare these statistics to those from the bivariate model. Does the second model fit the data better?
17. Use two "predict" commands to generate, first, predicted values of the feeling thermometer for every respondent and, second, residuals for every respondent.
18. Use the "scatter" command to request a scatterplot of residuals against predictions of the dependent variable. Is the variance of the residuals about the same regardless of the predicted value of the dependent variable?
19. Use the "histogram" command to generate a histogram of the residuals. Are the residuals normally distributed?

SPSS Problems

Here, we use data from the 2020 American National Election Study (ANES) to investigate the relationship between political ideology and feeling thermometer ratings of the police. Political ideology (*V201200*) is measured on a 7-point scale, where the lowest value, 1, means extremely liberal and the highest value, 7, indicates that the respondent is extremely conservative. Feeling thermometer ratings for the police (*V202171*) are measured on the same feeling thermometer that we have seen throughout the book, on a scale from 0 (coldest) to 100 (warmest). After regressing political ideology on feeling thermometer ratings of the police, we will introduce a control variable, *poc*, which indicates whether the respondent identifies as a person of color (1 = yes, 0 = no).

1. Open `ANES20.sav` and use the "Scatter/Dot" procedure to produce a scatterplot of the relationship between ideology, as the independent variable, and feelings about the police, as the dependent variable.
2. Does there appear to be a relationship between these two variables, according to the scatterplot?
3. Use the "Bivariate" procedure to get the correlation coefficient between these two variables.

4. What is the correlation coefficient? What does it tell us about the strength of this relationship?
5. Use the "Linear" procedure to run a bivariate regression between the variables.
6. What is the slope? What does it mean, in words?
7. What is the intercept? What does it mean, in words?
8. What is the standard error of the slope, and what is its associated p-value? Can we conclude that the population slope is unequal to 0?
9. What is the r^2 value for the model? What does it mean, in words?
10. What is the standard error of the estimate? What does it mean, in words?
11. What are the F-statistic for the model and its associated p-value? What do they mean, in words?
12. Now, use the "Linear" procedure to run the same regression equation from earlier, but add *poc* as a second independent variable.
13. What is the slope for *poc*? What does it mean, in words? Is it "statistically significant"?
14. What is the slope for ideology in this new model, and how does it compare to the first model? What does this comparison suggest about differences in political ideology for people of color and whites? Is it "statistically significant"?
15. What is the intercept in this new model? What does it mean, in words?
16. Give the r^2 and the standard error of the estimate for this new model. Compare these statistics to those from the bivariate model. Does the second model fit the data better?
17. Navigate to the "Plots" dialog box using the "Linear" procedure to request a scatterplot of residuals against predictions of the dependent variable. Is the variance of the residuals about the same regardless of the predicted value of the dependent variable?
18. Navigate to the "Plots" dialog box using the "Linear" procedure to request a histogram of the residuals. Are the residuals normally distributed?

Notes

[1] R. Chetty, N. Hendren, P. Kline, E. Saez, and N. Turner. 2014. "Is the United States Still a Land of Opportunity? Recent Trends in Intergenerational Mobility." *American Economic Review* 104(5): 141–147.

2. J. Cohen. 1988. *Statistical Power Analysis for the Behavior Sciences*, second edition. Hillsdale, NJ: Erlbaum; G. E. Gignac and E. T. Szodorai. 2016. "Effect Size Guidelines for Individual Differences Researchers." *Personality and Individual Differences* 102: 74–78.

3. H. Akoglu. 2018. "User's Guide to Correlation Coefficients." *Turkish Journal of Emergency Medicine* 18(3): 91–93.

4. Edward R. Tufte. 1974. *Data Analysis for Politics and Policy*. Englewood Cliffs, NJ: Prentice-Hall, 88–89; see also www.tylervigen.com.

Solutions to Odd-Numbered Practice Problems

Please note that due to rounding, your answers may differ slightly from the answers presented here.

Chapter 1 Solutions

1a. Index of inequality

1b. Average level of trust

1c. Metropolitan area

1d. The sampling strategy has not been described in the information provided.

3a. It is a closed-ended survey item because response categories are provided for the survey taker.

3b. Frequency of social media posting

3c. Never, Rarely, Often, All the time

3d. The variable is ordinal-level because the categories can be ranked in terms of frequency. The categories are not separated by equal intervals, however, so the variable is not interval-level.

5a. The measure is not reliable because the respondent gave very different answers over the ten weeks, ranging from "never" to "all the time." Since the question was about a typical week, the respondent's answers should not have varied so drastically.

5b. The reliability of the measure could be improved if the categories of the variable gave more precise frequencies of social media posting. The categories could give specific ranges of number of postings. For example, "often" could be replaced with "5 to 7 times."

7a. Simple random sampling

7b. The sampling frame will exclude members of the homeless population who do not stay in shelters, which could be a substantial proportion of the population. Thus, the sample will likely not be representative of the city's homeless population.

7c. It is likely that fewer people seek the services of shelters during the summer months because of the warmer weather. Drawing the sample during the winter might yield a more representative sample of the homeless population because people who normally do not stay at shelters may seek refuge at shelters during the winter.

9a. Ice cream flavor; a nominal variable

9b. I do not agree with the title of the list because the word "Best" in the title of the list indicates that the categories of the variable can be rank-ordered in terms of preference. In fact, managers were asked to indicate whether they would be serving each ice cream flavor, not how much people liked each flavor.

9c. A better title for the list would be "The Summer's Most Frequently Served Ice Cream Flavors." This title is more accurate because it ranks flavors according to how many shops will serve the flavor, not which flavors are better than others.

11a. Do social media campaigns reduce the number of children separated from their families? Do protest events reduce the number of children separated from their families?

11b. For the first question, the variables are number of social media campaigns and number of children separated from their families. The independent variable is social media campaigns, and the dependent variable is number of children. For the second question, the variables are number of protest events and number of children separated from their families. The independent variable is protest events, and the dependent variable is number of children.

11c. All of the variables from both questions are interval-ratio.

13a. Relative self-discipline may look like a ratio-level variable because one of the values in the table is 0 (Child 5). However, the 0 point on this scale is not a true 0. In this case, a value of 0 means that the child's self-discipline score is equal to the mean self-discipline score from the prior study. Thus, relative self-discipline is an interval-level variable.

13b. Child 1: Low; Child 2: High; Child 3: Low; Child 4: High; Child 5: Medium

13c. The new variable is ordinal. The categories can be rank-ordered, but the distances between each of the categories are not equal.

15a. The variable is measuring the gender of respondents' sexual partners over the last 5 years.

15b. The variable has five categories: "Exclusively male" (1), "Both male and female" (2), "Exclusively female" (3), "Don't know" (8), "No answer" (9)

15c. 436

15d. This is a nominal variable because the categories cannot be rank-ordered.

15e. "Don't know" and "No answer" should be categorized as missing data because those responses do not tell us whether the respondent had all male, all female, or both male and female partners.

17a. Nominal

17b. Nominal; or, ordinal because the categories could be rank-ordered if the variable were used as an indicator of the strength of Democratic or Republican affiliation.

17c. Interval-ratio

17d. Ordinal

17e. Interval-ratio

17f. Ordinal

17g. Interval

17h. Nominal

17i. Interval-ratio; the categories girl and not-girl can be rank-ordered in the sense that 0 is the absence of the characteristic and 1 is the presence of the characteristic (i.e., whether or not the person is a girl).

19a. See Unit of Analysis column in the table below.

	Research Project	Unit of analysis	Sampling Method
1	*Research Question*: Are college Republicans or Democrats more likely to engage in campus activism? *Data Collection*: Make a list of all members of the Republican and Democrat student organizations at a large university. Use a random number generator to draw a sample from the list.	Individual	Original data; simple random sample
2	*Research Question*: Does gender affect the number of close friends that high school students have? *Data Collection*: Make a list of all high schools in Texas and draw a random sample of schools. Draw random samples of students from each of the sampled schools.	Individual	Original data; cluster sampling
3	*Research Question*: Are companies' parental leave policies related to the gender composition of the company's leadership? *Data Collection*: Find ten companies that will allow a researcher to collect information about their organizational practices and staff.	Company	Original data; nonprobability sampling
4	*Research Question*: Is the frequency of a city's misdemeanor arrests related to the frequency of felonies committed in the city? *Data Collection*: Use publicly available data from the Bureau of Justice Statistics.	City	Secondary data
5	*Research Question*: Which people have the highest frequency of contact with the police: African Americans, whites, Asians, or Hispanics? *Data Collection*: Create separate sampling frames of all African American, Asian, white, and Hispanic residents of the same geographic area. Draw random samples from each of the four sampling frames.	Individual	Original data; stratified sampling

19b. See Sampling Method column in the table above.

Chapter 1 Stata Problems

1. Respondents have either a numerical value for their age or one of two labels ("iap" or "skipped on web"), which indicate a missing value.
3. `generate agekdbrn2=agekdbrn`
5. Confirm that the transformation proceeded properly by choosing any case (i.e., row) that has a value for *agekdbrn*, scrolling over to the *agekdbrn2* column for that case, and checking whether the value for *agekdbrn* corresponds to the appropriate value for *agekdbrn2*.

Chapter 1 SPSS Problems

1. Respondents have either a numerical value for their age or one of two labels ("iap" or "skipped on web"), which indicate a missing value.
3. `Transform → Recode into Different Variables`
   - Move *agekdbrn* into the "Input Variable" box by clicking on *agekdbrn* and then clicking the arrow to the right of the variable list.
   - Type the new variable name, *agekdbrn2*, into the "Output Variable: Name" box. Then click Change.
   - Click on "Old and New Values." This opens another dialog box where we tell SPSS how to recode the variable, using the specifications in Problem 2.

Chapter 2 Solutions

1a.

	Frequency	Percent	Cumulative Percent
Agree strongly	8,507	19.41	
Agree	17,730	40.46	
Neither agree nor disagree	8,452	19.29	
Disagree	6,324	14.43	
Disagree Strongly	2,810	6.41	
Total	43,823	100.0	

1b. While 19.41% of respondents agree strongly that something makes them feel ashamed, only 6.41% disagree strongly.

1c.

	Frequency	Percent	Cumulative Percent
Agree strongly	8,507	19.41	19.41
Agree	17,730	40.46	59.87
Neither agree nor disagree	8,452	19.29	79.16
Disagree	6,324	14.43	93.59
Disagree Strongly	2,810	6.41	100
Total	43,823	100.0	

1d. The majority of the sample (about 60%) either agrees or agrees strongly that they feel ashamed of something about their country.

3a.

	Frequency	Percent	Cumulative Percent
Agree strongly OR agree	26,237	59.87	59.87
Neither agree nor disagree	8,452	19.29	79.16
Disagree OR disagree strongly	9,134	20.84	100
Total	43,823	100.0	

3b. If 1,474 respondents skipped the question, this means that the total number of respondents was 45,297 (43,823+1,474). Those who agreed that there was something about their country that made them feel ashamed are 57.9% of the total respondents (26,237/45,297). Therefore, the missing data do not challenge the researcher's conclusion.

5a. According to the Cumulative Percent column, a score of 85 is in the 64th percentile, which means that 85 is higher than 64% of the exam scores. I would tell the student to stay in the course based on the percentile—he performed better than 64% of the students in the class.

5b. The variable in the histogram is interval-ratio. The variable in the bar graph is ordinal.

5c. The histogram and the bar graph show the same basic information. If students are focused on the distribution of letter grades, then the bar graph will be more useful.

7a. The promotion rate for men is 0.4 (50/125). The promotion rate for women is 0.24 (72/300). These rates are inconsistent with the claim that men are less likely to be promoted than women. In fact, they show a lower promotion rate for women than men.

7b. The promotion rate for women is 0.48 (72/150). The promotion rate for men is 0.53 (50/94). The rates are different because the denominators for both rates are lower than they were for the rates from part a. The promotion rates for women and men are much closer when they are calculated using the number of employees who were eligible for promotion (i.e., worked at the company for at least one year). Men are still more likely to be promoted but by a much smaller margin.

7c. Thirty-one men left the company during their first year of employment (125 − 94). One hundred fifty women left the company during their first year (300 − 150). The

7d. Fifty percent of women (150/300) and 24.8% of men (31/125) left the company during their first year of employment.

7e. Claims that men are disadvantaged in promotion at the company are not supported by the data. Among employees who are eligible to be promoted, men are slightly more likely to be promoted than women. The company should work on improving retention rates for women because they are much more likely than men to leave during that first year.

9a. Values are shown in the table. For example, San Antonio percent change: 24,208/1,487,738 = .0163.

Rank	City	2016 Population	2017 Population	Numeric Increase	Percent Change
1	San Antonio, TX	1,487,738	1,511,946	24,208	1.63
2	Phoenix, AZ	1,602,042	1,626,078	24,036	1.50
3	Dallas, TX	1,322,140	1,341,075	18,935	1.43
4	Fort Worth, TX	855,504	874,168	18,664	2.18
5	Los Angeles, CA	3,981,116	3,999,759	18,643	0.47
6	Seattle, WA	707,255	724,745	17,490	2.47
7	Charlotte, NC	843,484	859,035	15,551	1.84
8	Columbus, OH	863,741	879,170	15,429	1.79
9	Frisco, TX	163,816	177,286	13,470	8.22
10	Atlanta, GA	472,967	486,290	13,323	2.82
11	San Diego, CA	1,406,682	1,419,516	12,834	0.91
12	Austin, TX	938,200	950,715	12,515	1.33
13	Jacksonville, FL	880,893	892,062	11,169	1.27
14	Irvine, CA	266,385	277,453	11,068	4.15
15	Henderson, NV	292,005	302,539	10,534	3.61

9b.

Rank	City
1	Frisco, TX
2	Irvine, CA
3	Henderson, NV
4	Atlanta, GA
5	Seattle, WA
6	Fort Worth, TX
7	Charlotte, NC

Rank	City
8	Columbus, OH
9	San Antonio, TX
10	Phoenix, AZ
11	Dallas, TX
12	Austin, TX
13	Jacksonville, FL
14	San Diego, CA
15	Los Angeles, CA

Frisco, TX, has the highest annual percentage change, and Los Angeles has the lowest.

9c. The advantage of presenting percentage change is that it adjusts for the total population of the city. Numeric increases will tend to be larger in cities with larger populations.

11a. This is a time series chart.

11b. There are three variables: percentage of people who live below the poverty line, gender, and year.

11c. Older women have been more likely to live below the poverty line during each year from 1966 to 2016. One reason for this could be that women's incomes are lower than men's, on average.

13a. As calculated below, drama had the highest percentage of views, at 34%, while documentary had the lowest, at 14%:

$$\text{Science Fiction: } \frac{(700{,}000 + 300{,}000)}{4{,}180{,}500} \times 100 = 23.9\%;$$

$$\text{Comedy: } \frac{(480{,}000 + 275{,}000 + 450{,}000)}{4{,}180{,}500} \times 100 = 28.8\%;$$

$$\text{Documentary: } \frac{(375{,}500 + 200{,}000)}{4{,}180{,}500} \times 100 = 13.8\%;$$

$$\text{Drama: } \frac{(400{,}00 + 250{,}000 + 750{,}000)}{4{,}180{,}500} \times 100 = 33.5\%$$

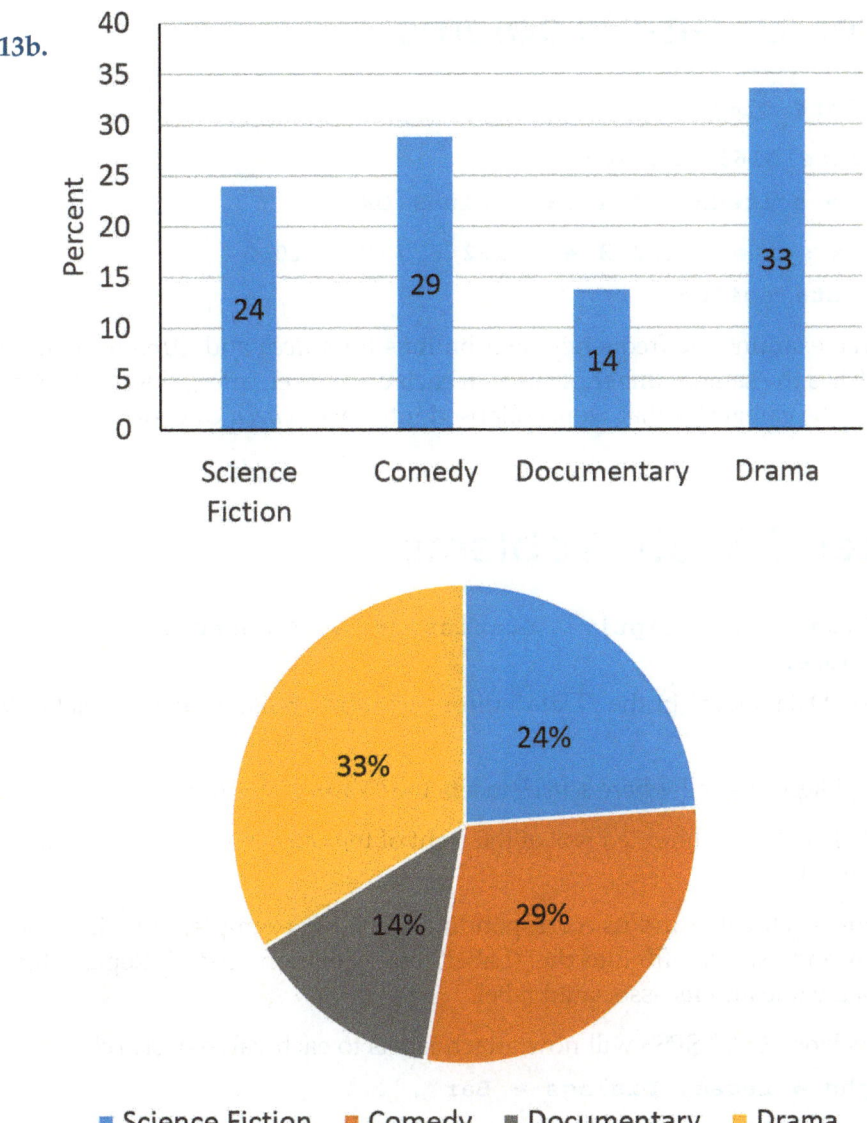

13b.

Since there are only four categories for the genre variable, some people may find that the pie chart shows more clearly the relative size of each category. However, the bar graph shows the same information. In this case, choosing between these two visual representations is really a matter of taste and personal preference.

15a. The charts look different because the y axes have different ranges.

15b. Panel A appears more stable over time because the y axis ranges from −10 to 10 compared to −3 to 3 for Panel B. Larger ranges deemphasize change, while smaller ranges emphasize it.

Chapter 2 Stata Problems

1. `tabulate Q163`

    `tabulate Q163, nolabel`
3. `recode scitech 1/4=1 5=2 6/10=3  OR`

    `recode scitech 1 2 3 4=1 5=2 6 7 8 9 10=3`
5. `tabulate scitech`

    You can examine the frequency distributions for *scitech* and *Q163* to make sure that *scitech* has the same number of cases in each of its three categories as *Q163* has in the sum of the categories that were collapsed into each *scitech* category.

Chapter 2 SPSS Problems

1. `Analyze → Descriptive Statistics → Frequencies Utilities → Variables`
3. Go to "Data View" in the "Data Editor" window. Be sure that "Variable View" is active.

    - Highlight the cell where *scitech* meets the "Values" column.
    - Click on the highlighted box at the right of the cell. This opens the "Value Labels" dialog box.
    - Enter each value and its corresponding label. For example, put 1 into the "Value" box and "worse off" into the "Label" box. Click on "Add." Repeat this step for each value and its associated label.
    - Click on "OK." SPSS will now attach labels to each value of *scitech*.
5. `Graphs → Legacy Dialogs → Bar . . .`

    We see that the majority of respondents around the world (almost 74%) think that the world is better off because of science and technology. That is, about 74% of respondents chose values of 6 through 10, with 10 indicating the respondent thinks the world is "a lot better off" because of science and technology.

Chapter 3 Solutions

1a.

Men and Women Coachella Headliners, by Race, Column, and (Total) Percentages

	Men	Women	Total
White	53.3 (40.0) (N = 8)	80.0 (20.0) (N = 4)	60.0 (N = 12)
Black	46.7 (35.0) (N = 7)	20.0 (5.0) (N = 1)	40.0 (N = 8)
Total	100.0 (75.0) (N = 15)	100.0 (25.0) (N = 5)	100.0 (N = 20)

1b. 40.0% of all individual Coachella headliners have been white men. 53.3% of men who have headlined Coachella are white.

1c. White men have headlined Coachella more often than any other group (about 40% of the time), and Black women have headlined least often (only 1 person, Beyoncé, accounting for 5% of all headlining acts).

1d. Representation of Black performers has been higher among men than women. Among men, 46.7% of headliners are Black, and among women, only 20% of headliners are Black.

1e. Yes, it does make sense to assess the strength. To do that, we compare the differences between column percentages within rows. If there were no relationship, we would expect half of all men and women to be white, and half to be Black. However, the table shows that this is not the case. 26.7% more women headliners are white than men headliners (80 – 53.3). And, 26.7% fewer women are Black than are men (46.7 – 20). So, gender is related to racial identity of Coachella headliners.

It does not make sense to assess the direction of the relationship because both variables are nominal. Because the categories are not rank-ordered, the relationship between the variables has no direction.

1f.

Men and Women Coachella Headliners, by Race, Column Percentages

	Men	Women	Total
White	53.3 (N = 8)	66.7 (N = 4)	57.1 (N = 12)
Black	46.7 (N = 7)	33.3 (N = 2)	42.9 (N = 9)
Total	100.0 (N = 15)	100.0 (N = 6)	100.0 (N = 21)

1g. With the addition of a single case to the table, a Black woman, we see that the percentages of white and Black women changed. Now 33.3% of women headliners have been Black, and 66.7% of them have been white. The relationship between gender and race has weakened with the addition of this case because the percentage differences across the columns and within rows has declined.

1h. It tells us that in small samples, the results of a cross-tabulation are very sensitive to small changes.

3a.

Effect of Gender on Paid Leave

Paid Leave:	Men	Women	Total
Yes	70.3% (N = 263)	56.0% (N = 410)	60.8% (N = 673)
No	29.7% (N = 111)	44.0% (N = 322)	39.2% (N = 433)
Total	100% (N = 374)	100% (N = 732)	100% (N = 1,106)

3b. Gender is the independent variable, and paid leave is the dependent variable.

3c. 410 of the women in the sample (56% of them) received pay while on leave.

3d. Yes, gender influences whether one's parental leave is paid. Men are more likely to receive pay when they take parental leave from work. Specifically, among men who took leave, 70% received pay, compared to only 56% of women who took leave. This difference of 14 percentage points indicates a moderately strong relationship.

5a. There is a negative relationship between the importance of owning a gun to one's identity and supporting women's rights. As the importance of owning a gun to one's identity increases, supporting women's rights decreases. This is a moderately strong relationship, with a 15.6 percentage-point difference between the lowest and highest category of importance of gun ownership to identity agreeing that supporting women's rights describes them "extremely well."

5b. The table does show a negative relationship between the importance of gun ownership to one's identity and the extent to which people identify as supporting women's rights. However, the majority of those who saw gun ownership as very important to their identities still identified strongly as supporting women's rights (60%). And, only a minority of respondents in any category of the importance of gun ownership to identity felt that supporting women's rights didn't describe them well at all. Thus, the evidence in this table suggests that the blogger may be overstating the role that sexism plays in American gun culture. Analysts should examine how additional attitudinal and behavioral measures regarding gender are distributed across gun owners to further examine the blogger's claim.

7a.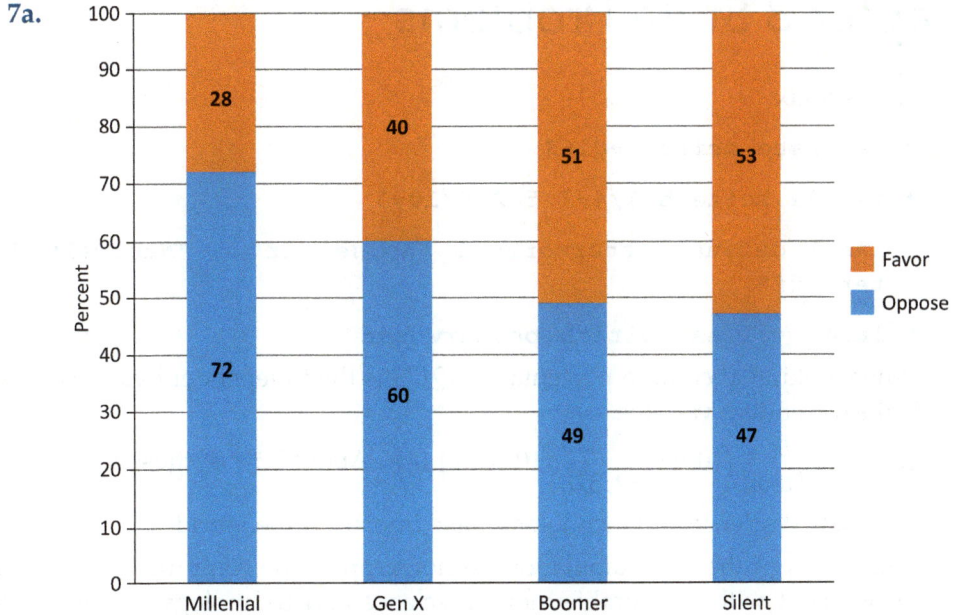

7b. The headline is technically correct if we assume that respondents who stated they "don't know" if they agree that the border wall should be extended are really in favor of it. However, we cannot know this for sure. Also, among the two oldest generations, Boomers and the Silent Generation, people are so evenly split in opposition to the border wall (48% favor versus 49% oppose for Boomers; 46% favor versus 47% oppose for the Silent Generation), that categorizing those who don't know what they think about the issue as either favoring or opposing the wall changes the conclusion. If the reporter had decided to classify those who didn't know what they thought as opposing the wall, then her headline could have read: "Majority of Older Americans Oppose Border Wall." For this reason, it is disingenuous of the reporter to claim that the majority of older Americans support the border wall.

9a. Panel A: (1) whether the respondent is a police officer or a member of the public and (2) opinions about how body cameras would change the way the public interacts with the police. Panel B: (1) whether the respondent is a police officer or a member of the public and (2) opinions about how body cameras would change the way that police officers interact with the public.

9b. 33% of police officers think that body cameras would make the public more likely to cooperate with the police. 59% of civilians agree with that opinion.

9c. 50% of police officers think that body cameras would make police officers act more appropriately. 66% of civilians agree with that opinion.

9d. The relationship is a little stronger in panel A than panel B. For example, the difference between police and the public in agreement that body cameras would make the public more cooperative with police is 26 percentage points; the difference between police officers and the public on whether body cameras would make the police act more appropriately is smaller, at 16 percentage points.

Chapter 3 Stata Problems

1. Stata syntax:
   - `generate scitech=Q163`
   - `recode scitech 1/4=1 5=2 6/10=3`
   - `label define betterworse 1 "Worse off" 2 "Neutral" 3 "Better off"`
   - `label values scitech betterworse`

3a. When looking at column percentages, Q173 is the independent variable, and *scitech* is the dependent variable.

3b. $col\% = \dfrac{f}{col\ total}(100) = \dfrac{37{,}615}{52{,}370}(100) = 71.8\%$; About 72% of those who are religious believe that science and technology have improved the world.

3c. The results of the cross-tabulation do not support the claim that religious belief is inconsistent with a favorable view of science and technology. For example, there is only a 5 percentage-point difference between non-religious and religious people who think that science and technology has improved the world (76.8 – 71.8 = 5).

3d. When looking at row percentages, *scitech* is the independent variable, and Q173 is the dependent variable.

3e. $row\% = \dfrac{f}{row\ total}(100) = \dfrac{37{,}615}{61{,}833}(100) = 60.8\%$; About 61% of those who believe that science and technology have improved the world are religious.

3f. The results of the cross-tabulation do not support the claim that religious belief is inconsistent with a favorable view of science and technology. For example, there is only a 10.6 percentage-point difference between religious people who think that science and technology has damaged the world and religious people who think that it has improved the world (71.4 – 60.8 = 10.6).

Chapter 3 SPSS Problems

1. SPSS procedure:

   `Transform → Recode into Different Variables`

   Using the "Recode into Different Variables: Old and New Values" dialog box, indicate how SPSS should perform the recode. For each value, type the old value in on the left, add the new value on the right, and click "add." (1 through 4 = 1; 5 = 2; 6 through 10 = 3)

   In Variable View, highlight the cell where *scitech* meets the "Values" column. Open the "Value Labels" dialog box. Enter each value and its corresponding label: 1 = "Worse off", 2 = "Neutral," 3 = "Better off." Click on "Add" after entering each value and its associated label. Click on OK.

3a. When looking at column percentages, Q173 is the independent variable, and *scitech* is the dependent variable.

3b. $col\% = \frac{f}{col\ total}(100) = \frac{37,615}{52,370}(100) = 71.8\%$; About 72% of those who are religious believe that science and technology have improved the world.

3c. The results of the cross-tabulation do not support the claim that religious belief is inconsistent with a favorable view of science and technology. For example, there is only a 5 percentage-point difference between non-religious and religious people who think that science and technology has improved the world (76.8 − 71.8 = 5).

3d. When looking at row percentages, *scitech* is the independent variable, and Q173 is the dependent variable.

3e. $row\% = \frac{f}{row\ total}(100) = \frac{37,615}{61,833}(100) = 60.8\%$; About 61% of those who believe that science and technology have improved the world are religious.

3f. The results of the cross-tabulation do not support the claim that religious belief is inconsistent with a favorable view of science and technology. For example, there is only a 10.6 percentage-point difference between religious people who think that science and technology has damaged the world and religious people who think that it has improved the world (71.4 − 60.8 = 10.6).

Chapter 4 Solutions

1a. 1.75: The mean distance from campus is 1.75 miles.

1b. 1.65: The median distance from campus is 1.65 miles. Half the students live more than 1.65 miles away and half live less than 1.65 miles away.

1c. 0: The most common distance from campus was 0 miles.

1d. The mean and median are both acceptable in this case. The median is the most accurate measure because the distribution is slightly positively skewed (because the mean is higher than the median).

3a. mean, median, mode

3b. mode

3c. mean, median, mode

3d. median (if the variables are conceptualized as strength of support), mode

3e. mode

3f. median, mode

5a. Aces: $\bar{y} = \frac{\Sigma y}{N} = \frac{(67+85+76+78)}{4} = 76.5$ Sun: $\bar{y} = \frac{\Sigma y}{N} = \frac{(65+71+105+71)}{4} = 78$

76.5 − 78 = 1.5

5b. Game 1: 2; Game 2: 14; Game 3: 29; Game 4: 7

5c. $\bar{y} = \dfrac{\Sigma y}{N} = \dfrac{(2+14+29+7)}{4} = 13$

7a. The average score is 50, suggesting that the typical response is that people do not feel strongly one way or the other about the death penalty for murder.

7b. Half of the scores lie below 59.5, and half above it.

7c. To find the median as a percent of the mean, divide the median by the mean, and multiply by 100: (59.5/50) x 100 = 119%; The median is 119% of the mean, which suggests low values in the left tail of the distribution are pulling the mean down (left, or negative, skew).

7d. When the distribution is skewed to one side, in this case the left, it is better to use the median to characterize the typical value in the distribution because it is not affected by values in the tail like the mean is.

9a. This is a bimodal distribution.

9b. The best measure of central tendency is the mode because both the median and the mean would misrepresent the typical value for this distribution. There are two typical values, one at the lower end of the distribution and one at the upper end.

11. I do not agree with the student because they have found the mean of a nominal variable. The values assigned to the categories of hair color are arbitrary labels with no numeric meaning. The only measure of central tendency that they can find for this variable is the mode. In this case, the mode is black hair.

13. The Times should use the name "Generation Z."

15a. Mean, median, or mode

15b. Mode

15c. Median

17a. Because of the way the mean is calculated, it can be divided into fractions, even though credit cards do not actually exist in fractions.

17b. The statement about the "average American" is slightly misleading because the mean says something about the distribution of values (here, number of credit cards), not about the individuals themselves. The statement 'An average American has 3.6 credit cards' personalizes the statistic, suggesting a misleading precision since no one can have a fraction of a card, and implies a uniform distribution that is not realistic. In contrast, it could be phrased as 'The average number of credit cards among American adults is 3.6 which more accurately reflects a statistical average from collective data, acknowledging variability among individuals without suggesting everyone has the same number of cards.

17c. If the median is lower than the mean, the distribution is right skewed, toward the higher values.

Chapter 4 Stata Problems

1. `summarize Q106`

   The middle of this scale is 5, which we can think of as feeling neutral on this issue. The mean is 6.3, which indicates that, on average, respondents lean slightly toward agreeing that there should be greater incentives for individual effort (i.e., greater income differences).

3. `tabulate Q106`

   We can see from the Percent column in the frequency distribution that the mode is 10. That is, the most commonly chosen response was 10, or completely agreeing that there should be greater incentives for individual effort.

5. `histogram Q106, percent`

Chapter 4 SPSS Problems

1. `Analyze → Descriptive Statistics → Frequencies`
3. The median is 7, indicating that half of the values fall at or above 7, and half of them fall at or below 7. Half of all respondents agree at least somewhat strongly (values of 7, 8, 9, and 10) that there should be greater incentives for individual effort.
5. The mean (6.3), median (7), and mode (10) suggest that the distribution of values for this variable are not symmetrical. With a median and mode higher than the mean, we would expect larger percentages of cases concentrated at higher values of the scale, potentially indicating a distribution that is skewed to the left.
7. The histogram shows a distribution that is not symmetrical, but it is not dramatically skewed to the left. Instead, large percentages of cases are concentrated across the values of this variable, with the largest percentage of cases (21.4%) falling on the very end of the scale (10). Reporting all three measures of central tendency—the mean, median, and mode—would capture the fact that many people agree at least somewhat that there should be greater incentives for individual effort (median = 7; mode = 10), but significant percentages of people also agree at least somewhat that incomes should be made more equal (e.g., 11.3% of people have value of 1, completely agreeing that incomes should be made more equal).

Chapter 5 Solutions

1a. Range New York = 12,000 − 2,199 = 9,801. We could say, "The smallest and largest rents for two-bedroom apartments in New York differ by $9,801" or "Rents ranged from $2,199 to $12,000."

Range Grinnell = 600 − 250 = 350. We could say, "The smallest and largest rents for two-bedroom apartments in Grinnell differ by $350" or "Rents ranged from $250 to $600."

1b. First rank-order all the rents from the lowest to the highest for each of the two cities.

New York: Quartile 1 case number = N(.25) = 14(.25) = 3.5; Quartile 1 value is the mean of the values for cases 3 and 4: $\frac{2,500 + 2,500}{2} = \$2,500$; Quartile 3 case number = N(.75) = 10.5; Quartile 3 value is the mean of the values for cases 10 and 11: $\frac{5,375 + 7,500}{2} = \$6,438$; IQR = 6,438 − 2,500 = $3,938; We can say, "The middle half of New York rents for two bedroom apartments is between $6,438 and $2,500, a range of $3,938."

Grinnell: Case numbers for Quartiles 1 and 3 are the same as New York because the N is the same. Quartile 1 value is the mean of the values for cases 3 and 4: $\frac{300 + 325}{2} = \$313$; Quartile 3 value is the mean of the values for cases 10 and 11: $\frac{450 + 475}{2} = \$463$; IQR = 463 − 313 = $150; We can say, "The middle half of Grinnell rents for two bedroom apartments is between $463 and $313, a range of $150."

1c. The table shows the calculation of the standard deviation, $s = \sqrt{\frac{\Sigma(y - \bar{y})^2}{N-1}}$, for each city.

In New York, rents differ from the mean by an average of $3,196; in Grinnell, by only $97.

New York	$y - \bar{y}$, $\bar{y}$ = 5145.00	$(y - \bar{y})^2$	Grinnell	$y - \bar{y}$, $\bar{y}$ = 398.00	$(y - \bar{y})^2$
2199.00	−2946.00	8678916.00	250.00	−148.00	21904.00
2200.00	−2945.00	8673025.00	300.00	−98.00	9604.00
2500.00	−2645.00	6996025.00	300.00	−98.00	9604.00
2500.00	−2645.00	6996025.00	320.00	−78.00	6084.00
3300.00	−1845.00	3404025.00	335.00	−63.00	3969.00
3350.00	−1795.00	3222025.00	350.00	−48.00	2304.00
3800.00	−1345.00	1809025.00	375.00	−23.00	529.00
4000.00	−1145.00	1311025.00	400.00	2.00	4.00
4300.00	−845.00	714025.00	445.00	47.00	2209.00
5375.00	230.00	52900.00	450.00	52.00	2704.00
7500.00	2355.00	5546025.00	475.00	77.00	5929.00
9000.00	3855.00	14861025.00	475.00	77.00	5929.00
10000.00	4855.00	23571025.00	500.00	102.00	10404.00
12000.00	6855.00	46991025.00	600.00	202.00	40804.00

New York	$y - \bar{y}$, $\bar{y} = 5145.00$	$(y - \bar{y})^2$	Grinnell	$y - \bar{y}$, $\bar{y} = 398.00$	$(y - \bar{y})^2$
		$\Sigma(y-\bar{y})^2 = 132,826,116.00$			$\Sigma(y-\bar{y})^2 = 121,981.00$
		$\dfrac{\Sigma(y-\bar{y})^2}{N-1} = 10,217,393.54$			$\dfrac{\Sigma(y-\bar{y})^2}{N-1} = 9,383.15$
		$\sqrt{\dfrac{\Sigma(y-\bar{y})^2}{N-1}} = 3,196.47$			$\sqrt{\dfrac{\Sigma(y-\bar{y})^2}{N-1}} = 96.87$

1d. For New York, the IQR is the best measure because its distribution is skewed to the right, with some very high values distorting the range. Also, the standard deviation works better with symmetrical distributions. Grinnell's distributions are not skewed, so all three measures of variability are appropriate to use.

1e. The mean rent for a two-bedroom apartment in New York City is $5,145. There is a large range in rent in NYC, from a low of $2,199 to a high of $12,000. The standard deviation is $3,196, meaning that on average rents differ from the mean by $3,196. However, rental costs are skewed in New York, with a small number of very expensive apartments. The middle half of New York rents for two-bedroom apartments—from the 25th to the 75th percentile—was between $6,438 and $2,500, a range of $3,938.

The mean rent for a two-bedroom apartment in Grinnell is $398, with a range from a low of $250 to a high of $600. The standard deviation is $97, meaning that on average rents differ from the mean by $97. The middle half of Grinnell rents for two-bedroom apartments was between $463 and $313. Unlike NYC, the distribution of Grinnell rents was not skewed.

3a. We see that there is a good deal of variation in responses, with most responses clustered around categories of agreement and disagreement and fewer responses in the middle of the scale.

3b. 3.27 is more likely.

3c. Obtain N by adding the frequency for each bar in the histogram: 40 + 60 + 80 + 50 + 23 + 29 + 45 + 70 + 95 + 90 + 60 = 642. Quartile 1 case number = N(0.25)= 642(0.25)= 160.5. Quartile 1 value is halfway between the values of cases 160 and 161; looking at the figure, both cases 160 and 161 ranked their attitude as 2. Quartile 3 case number = N(0.75) = 642(0.75) = 481.5. Quartile 3 value is halfway between the values of cases 481 and 482; looking at the figure, both cases 160 and 161 ranked their attitude as 8. The middle half of respondents—from the 25th to the 75th percentile—ranked their attitude about whether social media encourages bullying from a low of 2 to a high of 8.

3d. Teenagers' attitude about whether social media encourages bullying averaged 5.5 on a scale from 0 to 10. However, their attitudes varied widely, with most responses clustered around categories of either agreement or disagreement and fewer responses in the middle of the scale. On average, respondents' ranking differed from the mean by 3.27 points. The middle half of respondents—from the 25th to the 75th percentile—ranked their attitude about whether social media encourages bullying from a low of 2 to a high of 8.

3e. Because the distribution is essentially bimodal, the IQR is appropriate; because it is not highly skewed, the standard deviation is also appropriate. The range is appropriate but does not add information since it simply reports that respondents answered both at the high and low end of the scale.

5. Even though the range is the same for boys and girls, ranges can be affected by outlier cases, which may exist among boys or girls. According to the standard deviations, there is more variation for boys than girls. In other words, there is more agreement among girls about the role of social media in bullying behavior than among boys.

7a. The range for both variables is 4 (1–5).

7b. Look at cumulative percent: The 25th percentile is within the cases who chose 2, and the 75th percentile is within the cases who chose 3.

Alternatively: Case number for bottom quartile = N(0.25) = 8,257(0.25) = 2064.25. The value is halfway between the values of cases 2,064 and 2,065. Case number for top quartile = N(0.75) = 8,257(0.75) = 6192.75 The value is halfway between the values of cases 6,192 and 6,193. Add frequencies column to get cumulative frequency; 2,064th and 2,065th cases have a value of 2 and the 6,192nd and 6,193rd cases have a value of 3.

IQR is from 2 to 3. The middle half of respondents rated how hopeful they are about how things are going in the country between 2 and 3 on a 5-point scale, or between "a little" and "somewhat."

7c.

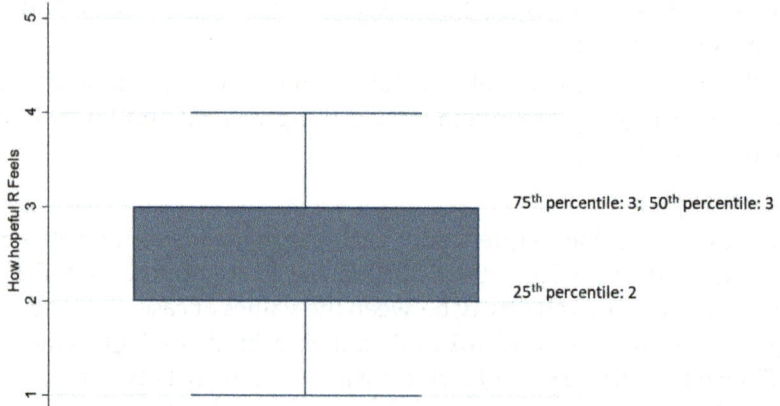

9a. IQR is from 3 to 5.

9b. The IQR for how worried respondents feel about how things are going in the country is both higher and wider than that for how hopeful they feel. In other words, respondents tend to feel more worried about how things are going in the country, with the middle half saying between "somewhat" and "very" and spreading across three response categories. In contrast, respondents feel less hopeful about how things are going in the country, with the middle half saying between "a little" and "somewhat" and spreading across only two response categories.

11a. The dotted distribution is the one after Shiru because students' coffee consumption is more spread out around the mean, indicating more variation. The solid line is the distribution before Shiru because consumption is more tightly clustered around the mean.

11b. The distribution flattened out, with more cases moving to either tail, because many math, science, and technology students are likely consuming more coffee and some social science students would be consuming less coffee as part of the protest.

13a. Subtract the mean from each film's earnings and divide each difference by the respective standard deviation: *Moonlight*: $\frac{65,046,687 - 25,000,000}{10,000,000} = \frac{40,046,687}{10,000,000} = +4.0$;

The Departed: $\frac{291,465,034 - 100,000,000}{45,000,000} = \frac{191,465,034}{45,000,000} = +4.3$

13b. The Departed earned 4.3 standard deviations higher than the mean earnings for its peers, compared to 4.0 standard deviations for *Moonlight*. *The Departed* is the higher earner by this metric, but by only a small margin.

13c. Relative earnings is probably a better metric than absolute earnings because small independent films are released in fewer theaters and have less publicity than Hollywood studio films. Hollywood studio films will almost always out-earn independent films in absolute terms.

Chapter 5 Stata Problems

1. `tabstat V202143 V202144, stat(sd mean iqr p25 p50 p75 range)`
3. The range for both candidates is 100. This means that the lowest value chosen by some number of voters for both candidates is 0, and the highest is 100.
5. `graph box V202143`

   `graph box V202144`

Chapter 5 SPSS Problems

1. *Standard deviation*:

   `Analyze → Descriptive Statistics → Descriptives`

   Under "Options," place a check mark next to "Standard Deviation."

   Mean:

   `Analyze → Descriptive Statistics → Descriptives`

Interquartile Range:

`Analyze → Descriptive Statistics → Explore`

Values for the 25th, 50th, and 75th percentiles:

`Analyze → Descriptive Statistics → Frequencies`

After moving the variables into the "Variable(s)" box, click on "Statistics." This opens the "Frequencies: Statistics" dialog box. Check the small box next to "Quartiles."

Range:

`Analyze → Descriptive Statistics → Descriptives`

Under "Options," place a check mark next to "Range."

3. The range for both candidates is 100. This means that the lowest value chosen by some number of voters for both candidates is 0, and the highest is 100.

5. `Analyze → Descriptive Statistics → Explore`

(Generate each box plot separately with the "Explore" command.)

Chapter 6 Solutions

1a.

	Percent Agree
Always	24.1 (1,798/7,452)
Most of time	40.4 (3,007/7,452)
About half of time	18.9 (1,408/7,452)
Some of time	15.8 (1,176/7,452)
Never	0.8 (63/7,452)

1b. Probability of randomly choosing always = 0.24; most of the time = 0.404; half the time = 0.189; some of the time = 0.158.

1c. $0.189 + 0.404 + 0.241 = 0.834$.

1d. In 2020, 64.5% followed politics most of the time or always ($0.404 + 0.241 = 0.645$). This is considerably higher than the similar number for 2012, which was 51%. This supports the claims that the 2020 electorate was more active than the 2012 electorate.

3a. $0.127 \times 0.311 = 0.039$.

3b. Multiplication rule (with independence).

3c. 0.127 x 0.127 = 0.016.

3d. 1 − (0.873)(0.873) = 0.238.

5a. We must use the Multiplication Rule without independence because each time that the woman goes on a date, that date is not placed back into the pool of potential dates. Therefore, the denominator of the fraction for each subsequent draw declines by one.

5b. $\frac{8}{10} \times \frac{7}{9} \times \frac{6}{8} \times \frac{5}{7} \times \frac{4}{6} = \frac{6720}{30,240} = 0.22$.

7a. The probability that a randomly chosen household with have a net worth between $21,345 and $16,721,760 is: 0.9 − 0.25 = 0.65.

7b. The probability that a household will have a net worth of at least $16,721,760: 1 − 0.90 = 0.10.

9a. First, find the z-score associated with 1.5 hours: $z = \frac{1.5 - 5}{1.5} = \frac{-3.5}{1.5} = -2.33$. The probability associated with a z-score of 2.33 in the normal table is 0.0099. Thus, the probability of a randomly chosen teenager listening to music for less than 1.5 hours per day is 0.0099, or 0.99%.

9b. First find the z-scores for both values: z for 2.25 hours = $\frac{2.25 - 5}{1.5} = \frac{-2.75}{1.5} = -1.83$ and z for 6 hours = $\frac{6 - 5}{1.5} = \frac{1}{1.5} = 0.67$. Sketch a normal curve and mark the z-scores of -1.83 (to the left of the center) and 0.67 (to the right of the center). Shade in the areas between -1.83 and the mean and between 0.67 and the mean. Next, find the probability associated with each z-score in the normal table: probability associated with area greater than 0.67 = 0.2514, probability associated with area less than −1.83 = 0.0336. Area of interest is between z = −1.83 and z = 0.25. Area between −1.83 and 0 = 0.466; area between 0 and 0.67 = 0.249. Desired area = 0.466 + 0.249 = 0.715. The probability that a randomly selected teenager will listen to 2.25-6 hours of music a day is 0.715, or 71.5%.

9c. Find the z-score associated with 8.5 hours: $z = \frac{8.5 - 5}{1.5} = \frac{3.5}{1.5} = 2.33$. The probability associated with 2.33, as shown in the normal table, is 0.0099. Since the mean of the distribution is 5, we know that half of all values lie to the right of 5. Thus, the probability of randomly selecting a teenager with a value between 5 and 8.5 = 0.5 − 0.0099 = 0.4901.

9d. First, use the normal table to find the area in the body of the table that comes as close as possible to 0.25, the area to the right of the value at the 75th percentile. The closest area to 0.25 is 0.2514, which is associated with a z-score of 0.67. Next, solve for y using the formula for z: $z = \frac{y - \mu}{\sigma} \rightarrow y = \mu + \sigma(z) = 5 + (1.5 \times .67) = 6.01$ hours

11a. As shown below, in the veteran's section, 11 students, or 55%, received F grades. In that section, no student received an A. In the new professor's section, none of the students received a grade of F, but 10 (50%) received A grades.

Student	Veteran Professor	New Professor
1	F	B
2	F	A
3	B	B
4	F	B
5	F	A
6	F	A
7	D	A
8	D	A
9	C	C
10	F	B
11	F	B
12	F	A
13	F	A
14	F	B
15	D	B
16	D	B
17	C	B
18	D	A
19	D	A
20	F	A

11b. All of the students in the new professor's section would receive an A or a B because all of their scores exceed the pooled mean of 71.3. Once the raw scores for students in this section are converted to z-scores, they will be at least some fraction of a standard deviation above the mean, which means they would be assigned grades of A or B. Only one student in the veteran's section would receive an A or B because that is the only score that exceeds the pooled mean. It is the only score that, once converted to a z-score would fall some number of standard deviations above the mean. The veteran's proposed solution does not address the problem because all students in the new professor's section would do well, and only one student in the veteran's section would do well.

11c. This solution will effectively address the problem because it would convert each student's raw score into a z-score using the mean score for the student's own section. The mean score for the veteran's section is much lower, 53.5, than the mean score for the veteran's section, 89.1. Eleven students in the veteran's section would have a grade of A or B with this method, and ten in the new professor's section would receive a grade of A or B.

Chapter 6 Stata Problems

1. First, find the z-score associated with 38: $z = \dfrac{y - \mu}{\sigma} = \dfrac{38 - 37.31}{5.50} = 0.13$; next, enter the following command into Stata: `display normal(.13)`

   According to the output, 55.2% of the curve's area lies to the left of $38.00. Thus, the probability that a nurse would make more than this amount is equal to $1 - 0.552 = 0.448$.

3. `display invnormal(.3)`

   The z-score at which 30% of the curve falls to the left is $-.524$. Use the z-score formula to find the wage associated with this z-score: $z = \dfrac{y - \mu}{\sigma} \rightarrow y = \mu + \sigma(z) = 37.31 + 5.5(-.524) = \$34.43$.

Chapter 6 SPSS Problems

1. `Analyze → Descriptive Statistics → Descriptives`

   Make sure to place check mark next to "Save standardized values as variables."

3. $z = \dfrac{y - \mu}{\sigma} = \dfrac{0 - 14.8}{17.4} = \dfrac{-14.8}{17.4} = .85$.

Chapter 7 Solutions

1a. $SE = \dfrac{\sigma}{\sqrt{N}} = \dfrac{0.47}{\sqrt{100}} = \dfrac{0.47}{10} = 0.047$

1b. $z = \dfrac{\bar{y} - \mu}{SE} = \dfrac{3.25 - 3.12}{0.047} = 2.77$

1c. Find the probability associated with your z-score in the Normal Table, appendix A. p=0.0028. There is a 0.28% chance of selecting a random sample with a mean GPA at or above 3.25.

1d.

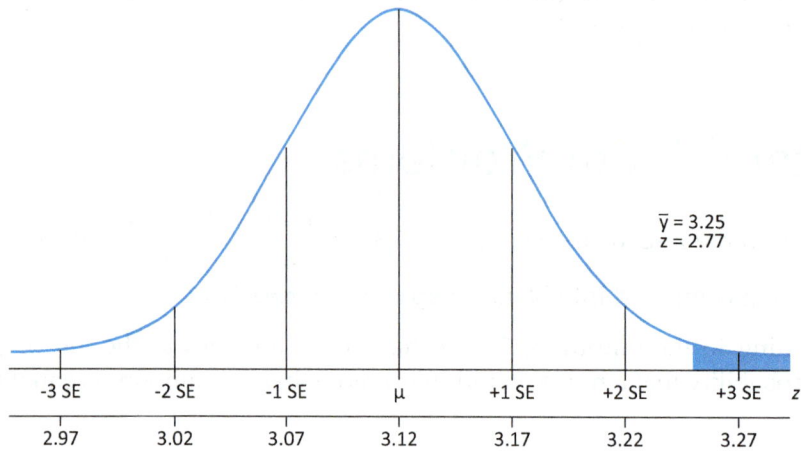

3a. $SE = \sqrt{\dfrac{\pi(1-\pi)}{N}} = \sqrt{\dfrac{0.9(1-0.9)}{100}} = \sqrt{0.001} = .032$

3b. $z = z = \dfrac{p-\pi}{SE} = \dfrac{0.87 - 0.90}{.032} = \dfrac{-0.03}{.032} = -0.94$

3c. Find the probability associated with your z-score in the Normal Table, appendix A. p=0.1736. There is a 17.36% chance of selecting a random sample with a percentage of in-state students at or below 87%.

5a. First, find the standard error: $SE = \sqrt{\dfrac{\pi(1-\pi)}{N}} = \sqrt{\dfrac{0.23(1-0.23)}{200}} = \sqrt{0.0009} = 0.03$. Next, find the z-score: $z = \dfrac{p-\pi}{SE} = \dfrac{0.30 - 0.23}{0.03} = \dfrac{0.07}{0.03} = 2.33$. Next, find the probability associated with the z-score of 2.33 in the Normal Table. The probability associated with z = 2.33 is 0.0099. This is the probability to the right of the z-score. There is a 0.0099 probability that in a random sample of 200 people, at least 30% would be religiously unaffiliated.

5b. Since half of the area of the curve falls to the right of the population proportion, .23, subtract .0099 from 0.5: 0.5 − 0.0099 = 0.49.

5c. First, find the standard error: $SE = \sqrt{\dfrac{\pi(1-\pi)}{N}} = \sqrt{\dfrac{0.23(1-0.23)}{35}} = \sqrt{0.005} = .007$. Next, find the z-score: $z = \dfrac{p-\pi}{SE} = \dfrac{0.30 - 0.23}{0.07} = \dfrac{0.07}{0.07} = 1.0$. Next, find the probability associated with the z-score of 1.0 in the normal table. The probability associated with z = 1.0 is .1587. This is the probability to the right of the z-score. There is a 0.1587 probability that in a random sample of 35 people, at least 30% would be religiously unaffiliated.

5d. The probability for part a is lower because the sample size is larger, which means that the standard error of the sampling distribution is smaller. Given the same numerator in the formula for z-score, the z-score will be larger for larger samples due to the larger standard error. Larger z-scores yield smaller probabilities.

7a. In this case, the standard deviation of 1.8 is also the standard error. This is because the polling firm has drawn repeated random samples, creating a sampling distribution. The standard deviation of a sampling distribution is its standard deviation.

7b. The center of curve is the mean of sample means, 85. To find the value for the first standard error to the right of the mean, add the standard error (see part a), 1.8, to 85: 85 + 1.8 = 86.8. For the second standard error to the right of the mean, add 1.8 to 86.8: 86.8 + 1.8 = 88.6. To find the value for the first standard error to the left of the mean, subtract the standard error from 85: 85 − 1.8 = 83.2. For the second standard error to the left, subtract the standard error from 83.2: 83.2 − 1.8 = 81.4. The result is shown below.

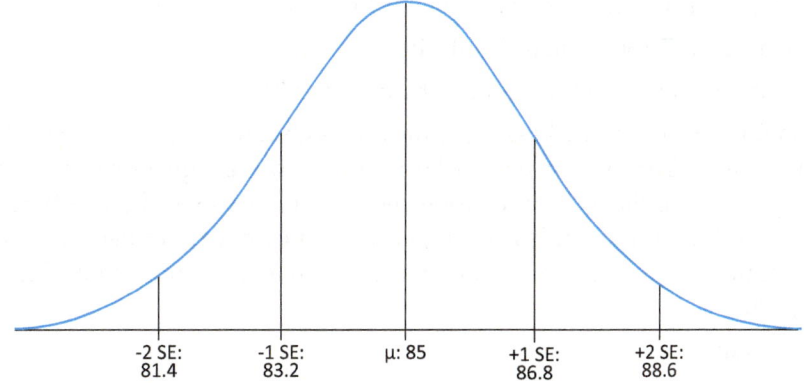

7c. The sampling distribution drawn for part b takes the shape of a normal curve: bell shaped, unimodal, and symmetric, whereas the distribution of values shown in Figure 7.24 is skewed to the left. Per the Central Limit Theorem, a sampling distribution takes the shape of the normal curve once the number of samples is large enough (at least 30 for means). The distribution of the values for a variable from a single sample, as shown in Figure 7.24, can look much different from a normal distribution because we are much more likely to observe very high or low values for a single variable than we are to observe very high or very low sample means.

7d. To find the probability that the sample mean of 82.5 would be randomly drawn from all of the random samples conducted by the polling firm, we find the z-score associated with the sample mean. We found the standard error in part a, 1.8, and plug it into the z-score formula:

$$z = \frac{\bar{y} - \mu}{SE} = \frac{82.5 - 85}{1.8} = -1.39$$

The probability associated with this z-score in the normal table is 0.0823. Because 0.0823 is the probability to the right of the absolute value of the z-score, and the actual z-score falls to the left of the center of the distribution, subtract the probability from 1 to find the area to the right of z = −1.39: 1 − 0.0823 = 0.9177. The probability of drawing a sample mean that is greater than 82.5 is 0.9177.

9a. While there are very large and very small households, average household size in a given county will "average out" large variations.

9b. A sampling distribution.

9c. 30% of the sample means (six) are less than the population mean, and 65% (fourteen) fall at or above it.

9d. We would expect half of the sample means to fall at or above the population mean, and half below it. The twenty samples in this example do not match up with those expectations because twenty is a fairly small number of samples. The greater the number of samples drawn, the more closely the sampling distribution will arrange itself into a normal curve.

11a. Sample proportion: 7/75, or 0.093, which should be denoted by "p"

11b. Population proportion: 0.027, which should be denoted by π

11c. Sample mean: $77,866, which should be denoted by $\bar{y}$

11d. Population mean : 8.3, which should be denoted by μ

13. Pair 1 is B because, with 200 sample means, both distributions take on the shape of a normal curve, but the samples of larger size (on the right) produce a distribution with a smaller standard error, as reflected by the narrower distribution. Pair 2 is A. The distribution on the left is not shaped like a normal curve because there are only 25 means, but the one on the right is shaped normally because it has 4 times as many means, 100.

15a. 50 respondents

15b. $\bar{y} = \dfrac{\Sigma(0 \times 7)+(1 \times 5)+(2 \times 5)+(3 \times 2)+(4 \times 3)+(5 \times 3)+(6 \times 4)+(7 \times 3)+(8 \times 4)+(9 \times 6)+(10 \times 8)}{50}$

$= \dfrac{259}{50} = 5.18$

15c. First, find the standard error of the sampling distribution: $SE = \dfrac{\sigma}{\sqrt{N}} = \dfrac{0.70}{\sqrt{50}} = 0.099$.

Next, find the z-score: $z = \dfrac{\bar{y} - \mu}{SE} = \dfrac{5.18 - 4.95}{0.099} = \dfrac{0.23}{0.099} = 2.32$. Next, find the probability associated with 2.32 in the Normal Table: 0.0102. This means that there is about a 1% chance that the friends would draw a random sample with a mean that falls at or above the original sample mean of 5.18.

15d. 52% of people in the town strongly support or strongly oppose legalized abortion:

$\dfrac{14 + 12}{50} = 0.52$, or 52%

15e. First, find the standard error of the sampling distribution:

$SE = \sqrt{\dfrac{p(1-p)}{N}} = \sqrt{\dfrac{0.52(1-0.52)}{50}} = 0.071$. Second, transform the sample proportion into a

z-score: $z = \dfrac{p - \pi}{SE} = \dfrac{0.52 - 0.515}{0.071} = \dfrac{0.005}{0.071} = 0.07$. Next, find the probability associated with

0.07 in the normal table: 0.4721. There is a 47.21% chance of drawing a random sample within which at least 52% of the sample has strong opinions about legalized abortion.

15f. It is much more likely that the friends would draw a sample in which at least 52% of respondents held strong views about abortion. As shown in part e, the probability of drawing a sample with a proportion greater than 0.52 is 47.21%. In contrast, part c shows that the probability of drawing a sample with a mean above 5.18 is 1%.

15g. The data suggest that each station is telling only part of the story. There is both strong support for and opposition to legalized abortion.

Chapter 7 Stata Problems

1. `summarize Q251`
3. `sample 100, count`
5. There are only 95 values, which means that five of the cases that were included in the sample must have missing values for the variable Q251.
7. `summarize _b[Q251]`

    The mean of the 100 sample means is 6.13, and the standard error is .263.

Chapter 7 SPSS Problems

1. Open the WVS data and use the "Descriptives" procedure to generate summary statistics for Q251 in the full sample.

    `Analyze → Descriptive Statistics → Descriptives`
3. `Data → Select Cases`

    Select the "Random sample of cases" option, and click on "Sample." This opens the "Select Cases: Random Sample" dialog box. Tell SPSS to randomly select 100 cases.
5. There are only 95 values, which means that five of the cases that were included in the sample must have missing values for the variable Q251.
7. The means are different because SPSS has drawn a different set of random cases.

Chapter 8 Solutions

1a. 33.2% of respondents have "hardly any" confidence in organized religion: 879/2650 = 0.332

1b. $SE_p = \sqrt{\frac{p(1-p)}{N}} = \sqrt{\frac{0.332(1-0.332)}{2650}} = \sqrt{\frac{0.222}{2650}} = 0.009$. For a 95% confidence interval, find the z-score associated with the right-tail probability equal to half of alpha, 0.05: $\alpha/2$ = 0.025. The z-score is 1.96. Now calculate the margin of error: Margin of error = $z(SE_p)$ = 1.96(0.009) = 0.0176; Finally, calculate the lower and upper bounds of the confidence interval: CI = p ± margin of error → 0.332 ± 0.0176 → (0.314, 0.350)

1c. The 95% confidence interval for the proportion of all Americans who have "hardly any" confidence in organized religion ranges from 0.314 to 0.350. Over repeated sampling, 95% of the confidence intervals will include the population parameter. This means that there is a 95% chance that the population proportion falls within the current interval: 31.4% to 35.0%.

3a. Step 1: Estimate the standard error: $SE_{\bar{y}} = \frac{s}{\sqrt{N}} = \frac{32.3}{\sqrt{8152}} = 0.358$

Step 2. Find degrees of freedom and the corresponding t-value for a 95% confidence interval: DF = N − 1 = 8152 − 1 = 8151. Use the ∞ row of the t-table. Find the t-value at the intersection of the ∞ row and the 95% confidence interval column: 1.96.

Step 3. Find the margin of error: Margin of error = $t(SE_{\bar{y}})$ = 1.96(0.358) = 0.702

Step 4: Calculate lower and upper bounds of confidence intervals: $\bar{y}$ ± margin of error = 45.3 ± 0.702 = (44.6, 46.0)

3b. The 95% confidence interval for Americans' feelings about the Democratic Party ranges from 44.6 to 46.0. Over repeated sampling, 95% of the confidence intervals will include the population parameter. This means that there is a 95% chance that the population mean falls within the current interval: 44.6 to 46.0.

5a. $SE_{men} = \sqrt{\frac{p(1-p)}{N}} = \sqrt{\frac{0.3(1-0.3)}{951}} = 0.015$; Use the z-score associated with 0.1 in the normal table: 1.29; Margin of error = $z(SE_p)$ = 1.29(0.015) = 0.019; Calculate lower and upper bounds of confidence interval: p ± margin of error = 0.30 ± 0.019 = (0.281, 0.319)

5b. $SE_{women} = \sqrt{\frac{p(1-p)}{N}} = \sqrt{\frac{0.37(1-0.37)}{1072}} = 0.015$; Use the z-score associated with 0.1 in the normal table: 1.29; Margin of error = $z(SE_p)$ = 1.29(0.015) = 0.019; Calculate lower and upper bounds of confidence interval: p ± margin of error = 0.37 ± 0.019 = (0.351, 0.389)

5c. The confidence intervals do not overlap, suggesting that the true population proportions of men and women who volunteer for causes they find important are different, with a higher proportion of women volunteering.

5d. Standard errors: $SE_{women} = \sqrt{\frac{p(1-p)}{N}} = \sqrt{\frac{0.37(1-0.37)}{38}} = 0.078$; $SE_{men} = \sqrt{\frac{p(1-p)}{N}} = \sqrt{\frac{0.3(1-0.3)}{70}} = 0.055$;

Use the z-score associated with .1 in the normal table: 1.29;

Margin of error for women: $1.29(.078) = .101$; Margin of error for men: $1.29(.055) = .071$

CI for women: $0.37 \pm 0.101 = (0.27, 0.47)$; CI for men: $0.30 \pm 0.071 = (0.23, 0.37)$

5e. Here, the confidence intervals overlap, so we cannot conclude that young women are more likely than young men to volunteer for causes that are important to them. The sample sizes here are smaller than in parts a and b. Because sample size is in the denominator in the formula for standard error, smaller samples yield larger standard errors, all else equal. Here, the sample proportions and level of confidence are the same. Since the margin of error is the product of the z-score and the standard error, larger standard errors mean larger margins of error. In turn, larger margins of error mean larger confidence intervals.

7a. The mean age of smokers in general is 33.

7b. The mean age of smokers in the sample is 37.

7c. The bottom of the confidence interval is 34.5, and the top is 39.5. It does not include the actual mean age of smokers in the population because the lower bound is higher than 33.

9. We would expect 5% of the samples to exclude the population parameter and 95% of them to include the parameter. In this case, we would expect 190 of the confidence intervals to include the parameter of interest (200 x 0.95) and ten of them to exclude it (200 x .05).

11a. For a 99% confidence interval, 1% (0.01) of the area of the normal curve will be distributed across both tails. Thus, we find .005 in the normal table (or the closest proportion that does not exceed it) and its associated z-score, which is 2.58.

11b. For a sample of 50, degrees of freedom are 49 (N – 1). Since there is no row for 49 in the t-table, we go to the next lowest row, which is 40. Find the column for the 99% confidence level and scan down until you reach row 40. The t-value at this intersection is 2.704.

11c. The t-value is larger because the width of the t curve depends on sample size, with larger samples associated with narrower curves, and smaller samples with larger curves. The width of the normal curve, which we use for z-scores, does not depend on sample size. T-values will always be larger than z-scores at the same level of confidence, but the difference between them decreases as sample size increases.

13. $N = 0.25\left(\dfrac{Z}{B}\right)^2$, where N is sample size, z is the z-score associated with desired level of confidence, and B is acceptable margin of error. $N = 0.25\left(\dfrac{1.65}{0.03}\right)^2 \approx 757$. $N = \sigma^2\left(\dfrac{Z}{B}\right)^2$, where N is sample size, z is the z-score associated with desired level of confidence, B is acceptable margin of error, and σ is the population standard deviation.

$$N = 15^2\left(\dfrac{1.96}{2.5}\right)^2 = 225(0.615) \approx 139$$

15. The article is discussing survey data, which are usually collected from a sample drawn from a population. In that case, a proportion of 0.51, or 51%, is only a point estimate. Because of sampling variability, the point estimate will vary from the true population percentage. A confidence interval around the point estimate of 0.51 could drop below 0.5, which would mean that "51% might not mean a majority," as the title states.

Chapter 8 Stata Problems

1. `ci means rank, level(90)`
3. The standard error is 0.039.
5. `ci proportions lowrank, level(90)`
7. The 90% confidence interval is (0.184, 0.214). There is a 0.9 probability that the true proportion of the population who see themselves falling on the bottom half of the social scale falls within this confidence interval because 90% of all confidence intervals will include the population mean.

Chapter 8 SPSS Problems

1. `Analyze → Descriptive Statistics → Explore`
3. The standard error is 0.039.
5. `Analyze → Descriptive Statistics → Explore`
7. The 90% confidence interval is (0.184, 0.214). There is a 0.9 probability that the true proportion of the population who see themselves falling on the bottom half of the social scale falls within this confidence interval because 90% of all confidence intervals will include the population mean.

Chapter 9 Solutions

1. $SE = \sqrt{\dfrac{\pi(1-\pi)}{N}} = \sqrt{\dfrac{(0.5)(0.5)}{700}} = 0.019$

Calculate $z = \dfrac{0.53 - 0.50}{0.019} = 1.58$

Look up p-value for z = 1.58 and compare to alpha; p-value is 0.0571, which is greater than alpha = 0.01. We cannot reject the null hypothesis. We are not confident that candidate A has the support to win the gubernatorial election.

3. $SE_{\bar{y}} = \dfrac{s}{\sqrt{N}} = \dfrac{25}{\sqrt{85}} = 2.71$

Sample $t = \dfrac{\bar{y} - \mu}{SE_{\bar{y}}} = \dfrac{54 - 50}{2.71} = 1.48$

Decision t = 1.66. Sample t < Decision t. Thus, we cannot reject the null hypothesis. We are not confident that this candidate is rated positively by her constituents.

5a. H_0: The proportion of people who identify as multiracial in the West is the same as in the US population as a whole. (H_0: $\pi = 0.029$.)

H_a: The proportion of people who identify as multiracial in the West is greater than in the US population as a whole. (H_a: $\pi > 0.029$.)

5b. It is a one-tailed test because the alternative hypothesis is that the population percent in the West is greater than a specific value.

7. 1) Set up hypotheses: H_0: $\pi = 0.029$; H_a: $\pi > 0.029$

2) Set alpha: 0.05

3) Estimate standard error

$SE = \sqrt{\dfrac{\pi(1-\pi)}{N}} = \sqrt{\dfrac{0.029(1-0.029)}{350}} = 0.009$

4) Calculate z

$z = \dfrac{p - \pi}{SE} = \dfrac{0.045 - 0.029}{0.009} = \dfrac{0.016}{0.009} = 1.78$

5) Look up p-value for z = 1.78 and compare p-value to alpha. p-value = 0.0375, less than the chosen alpha-level of 0.05. Reject the null hypothesis. Residents of the West are statistically significantly more likely to identify as multiracial than residents of the US overall. In our sample, 4.5% of residents of the West identified as multiracial, compared to 2.5% of all US residents. But if we use a stricter alpha threshold of 0.01, we cannot reject the null hypothesis; we cannot conclude that residents of the West are more likely than all US residents to identify as multiracial.

9a. 1) State null and alternative hypotheses: H_0: The population mean number of emergency room visits per year for all low-income households is 0.5. H_0: $\mu = 0.5$

H_a: The population mean number of emergency room visits per year for all low-income households is greater than 0.5. $H_a: \mu > 0.5$

2) alpha = 0.05; DF = 104, decision t for one-tailed alpha-level of 0.05, and 100 DF, = 1.660.

3) $SE_{\bar{y}} = \dfrac{s}{\sqrt{N}} = \dfrac{0.41}{\sqrt{105}} = 0.040$

4) Sample $t = \dfrac{\bar{y} - \mu}{SE\bar{y}} = \dfrac{0.65 - 0.5}{0.040} = \dfrac{0.15}{0.040} = 3.75$

5) Sample t > decision t. Reject the null hypothesis. Conclude: Low-income households have a higher number of emergency visits per year.

9b. While there is a statistically significant difference between the mean number of emergency visits per year for low income and overall households, the difference is small, only 0.15 more visits per year, on average, for low-income households. Depending on the cost of these visits to hospitals and to families, the difference may or may not be important in practical terms.

9c. This requires calculating a confidence interval.

$CI = \bar{y} \pm t(SE\bar{y})$ where $SE\bar{y} = \dfrac{s}{\sqrt{N}}$ and t at DF=104 and 95% confidence = 1.984. We calculated SEȳ in part a as 0.040.

$CI = \bar{y} \pm t(SE\bar{y}) = 0.65 \pm 1.984(0.040) = 0.65 \pm 0.079 = 0.57$ to 0.73. In repeated sampling, 95 times out of 100, the true population mean number of visits to emergency rooms for low-income households in Anytown would fall between 0.57 and 0.73 per year.

11a. Because the null hypothesis was rejected, it is possible that we committed a Type I error, which occurs when we reject the null hypothesis when it is actually true. Type I error is a "false positive." Type II error occurs when we fail to reject a null hypothesis that is false.

11b. The probability of committing a Type I error is equal to the alpha-level, which is .01 in this case.

13a. The firm's interpretation of the p-value is incorrect because they are saying that the p-value is the probability (0.1%) that the sample statistic, 75%, is not the true population proportion, the percent of all 18- to 34-year-olds who view the product favorably. What we can say about this p-value is that we would have observed the sample statistic of 75% or higher only 0.1% of the time if the true favorability rating in the population of 18- to 34-year-olds were 50%.

13b. In this case, a two-tailed hypothesis test doesn't make sense because they assume that their marketing campaign increased the product's favorability. When there is a reason to suspect that the population parameter would be greater than or less than the specified value, a one-tailed test is appropriate.

13c. Here, a right-tailed test would be appropriate because they believe that the overall favorability rating among 18- to 34-year-olds is greater than 50%.

13d. We know that two-tailed tests are more conservative than one-tailed tests, meaning that it is more difficult to reject the null hypothesis with a two-tailed test than a one-tailed test. This is because the p-value associated with the test statistic is doubled in a two-tailed test. Here, the p-value is 0.001. If the test were two-tailed, the p-value would be 0.002, the probability of obtaining the sample statistic if the null hypothesis were true. In this case, conducing a two-tailed test would not change the substantive meaning of the results because the p-value of 0.002 would still be lower than any conventional alpha-level adopted by the firm.

15a. It should be a test for means because we are examining the mean percent unemployment cited in a sample of 30 speeches. Tests for proportions can only be conducted on variables with two categories. This variable is an interval-ratio variable, ranging from 0 to 100.

15b. 1) State null and alternative hypotheses: H_0: The mean percent unemployment cited in all of the candidate's speeches is 4%. H_0: $\mu = 4$. H_a: The mean percent unemployment cited in all of the candidate's speeches is greater than 4%. H_a: $\mu > 4$.

2) alpha = 0.05; DF = 29; decision t for one-tailed alpha-level of 0.05, and 29 DF, = 1.699.

3) $SE_{\bar{y}} = \dfrac{\sigma}{\sqrt{N}} = \dfrac{2}{\sqrt{30}} = 0.37$

4) Sample $t = \dfrac{\bar{y} - \mu}{SE_{\bar{y}}} = \dfrac{12 - 4}{0.37} = \dfrac{8}{0.37} = 21.62$

5) Sample t > decision t. Reject the null hypothesis. Conclude: In all of his speeches, the candidate cites unemployment rates that are on average higher than the rate given by the Federal Reserve, 4%.

15c. The practical significance is potentially very large because the candidate's inflation of the unemployment rate could convince voters that the economy is doing worse than it is. They could choose to vote for the candidate based on his promise to fix an economy which may not be as broken as he claimed.

Chapter 9 Stata Problems

1. `ttest Q163 = 7.08 if B_COUNTRY==156`
3. The null hypothesis is that people from China feel the same about science and technology as do people across the world. The two-tailed alternative hypothesis is that there is a difference between how people from China view science and technology and how people across the world view it.
5. The output for the two-tailed test shows that the p-value associated with the sample t is less than 0.001. Under any conventional alpha-level that could be adopted, the null hypothesis would be rejected. There is very strong evidence that people in China have different views (more favorable) of science and technology compared to the rest of the world.

Chapter 9 SPSS Problems

1. Data → Select Cases → If. Enter B_Country = 156 into the box. Click continue and then click OK.

3. The null hypothesis is that people from China feel the same about science and technology as people across the world. The two-tailed alternative hypothesis is that there is a difference between how people from China view science and technology and how people across the world view it.

5. The output for the two-tailed test shows that the p-value associated with the sample t is less than .001. Under any conventional alpha-level that could be adopted, the null hypothesis would be rejected. There is very strong evidence that people in China have different views (more favorable) of science and technology compared to the rest of the world.

Chapter 10 Solutions

1a. **Null hypothesis**: The proportion of same-sex couples who have both earned bachelor's degrees is the same as the proportion of opposite-sex couples who have both earned bachelor's degrees. ($H_0: \pi_{same\text{-}sex} - \pi_{opposite\text{-}sex} = 0$)

Alternative hypothesis: The proportion of same-sex couples who have both earned bachelor's degrees is higher than the proportion of opposite-sex couples who have both earned bachelor's degrees. ($H_a: \pi_{same\text{-}sex} > \pi_{opposite\text{-}sex} = 0$)

1b. $SE_{p_{same\text{-}sex} - p_{opposite\text{-}sex}} = \sqrt{\dfrac{\pi(1-\pi)}{N_{same\text{-}sex}} + \dfrac{\pi(1-\pi)}{N_{opposite\text{-}sex}}}$, where π is the pooled proportion =

$\dfrac{f_{same\text{-}sex} + f_{opposite\text{-}sex}}{N_{total}} = \dfrac{(0.30 \times 200) + (0.18 \times 450)}{650} = \dfrac{60 + 81}{650} = 0.22$

$SE_{p_{same\text{-}sex} - p_{opposite\text{-}sex}} = \sqrt{\dfrac{0.22(1-0.22)}{200} + \dfrac{0.22(1-0.22)}{450}} = 0.035$

1c. $z = \dfrac{(p_{same\text{-}sex} - p_{opposite\text{-}sex})}{SE_{p_{same\text{-}sex} - p_{opposite\text{-}sex}}} = \dfrac{(0.30 - 0.18)}{0.035} = \dfrac{0.12}{0.035} = 3.43$

1d. The p-value associated with the sample z of 3.43 is less than 0.001. This is lower than the designated alpha-level of 0.025. Thus, the null hypothesis should be rejected. We conclude that there is a higher proportion of same-sex couples who both hold bachelor's degrees than opposite-sex couples.

3a. Pooled proportion = $\dfrac{f_{same\text{-}sex} + f_{opposite\text{-}sex}}{N_{total}} = \dfrac{(0.30 \times 75) + (0.18 \times 100)}{175} = \dfrac{22.5 + 18}{175} = 0.23$

$$SE_{p_{opposite\text{-}sex} - p_{same\text{-}sex}} = \sqrt{\frac{0.23(1-0.23)}{75} + \frac{0.23(1-0.23)}{100}} = 0.064$$

$$z = \frac{(p_{same\text{-}sex} - p_{opposite\text{-}sex})}{SE_{p_{same\text{-}sex} - p_{opposite\text{-}sex}}} = \frac{(0.30 - 0.18)}{0.064} = \frac{0.12}{0.064} = 1.88$$

3b. In this case, we should not reject the null hypothesis of no difference because the p-value associated with the sample z of 1.88 is 0.0301, higher than the alpha-threshold of 0.025.

3c. We draw different conclusions because the sample sizes are smaller in this problem than in Problem 1. Since sample sizes are in the denominators in the formula for the standard error of the difference, smaller sample sizes mean larger standard errors. Larger standard errors mean smaller z-scores because the standard error is the denominator in the z-score formula. Turning to the normal table, we can see that the smaller the z-score, the larger the p-value.

5a. The dependent variable is emphasis on appearance, and the independent variable is gender of main character.

5b. The null hypothesis is that there is no difference between books with boy and girl protagonists in emphasis on appearance ($H_0: \mu_g - \mu_b = 0$). The alternative hypothesis is that there is a difference in emphasis on appearance between books with girl and boy protagonists ($H_a: \mu_g - \mu_b \neq 0$).

5c.

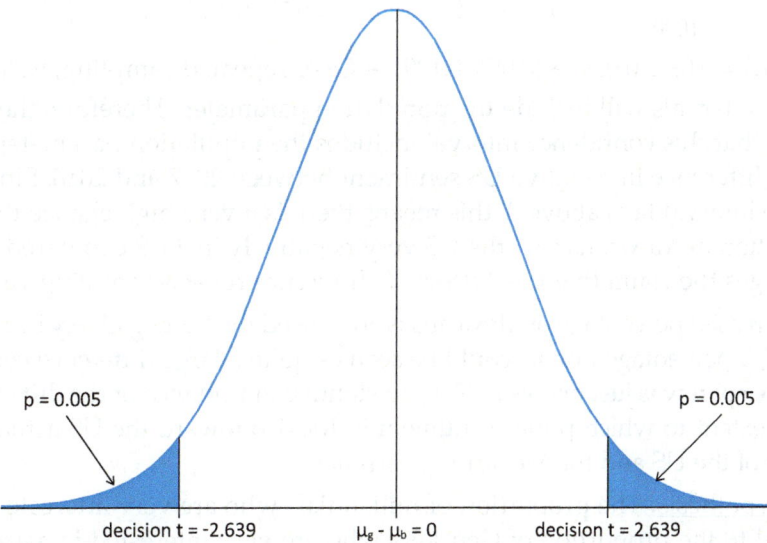

5d. $SE_{\bar{y}_g} = \frac{s_g}{\sqrt{N_g}} = \frac{8}{\sqrt{50}} = 1.13$, $SE_{\bar{y}_b} = \frac{s_b}{\sqrt{N_b}} = \frac{2}{\sqrt{50}} = 0.28 \rightarrow$

$SE_{\bar{y}_g - \bar{y}_b} = \sqrt{SE_{\bar{y}_g}^2 + SE_{\bar{y}_b}^2} = \sqrt{(1.13)^2 + (0.28)^2} = \underline{1.17}$; $t = \frac{(\bar{y}_g - \bar{y}_b)}{SE_{\bar{y}_g - \bar{y}_b}} = \frac{(30-20)}{1.16} = \frac{10}{1.16}$

$= \underline{8.62}$

5e. Sample t, 8.62, is greater than decision t, 2.639. Therefore, we should reject the null hypothesis of no difference between means.

7a. $CI = (\bar{y}_{US} - \bar{y}_B) \pm t\,(SE_{\bar{y}_{US} - \bar{y}_B})$, where $SE_{\bar{y}_{US} - \bar{y}_B} = \sqrt{SE_{\bar{y}_{US}}^2 + SE_{\bar{y}_B}^2}$

$SE_{\bar{y}_{US}} = \frac{s_{US}}{\sqrt{N_{US}}} = \frac{300}{\sqrt{103}} = 29.56$; $SE_{\bar{y}_B} = \frac{s_B}{\sqrt{N_B}} = \frac{250}{\sqrt{175}} = 18.90$

$SE_{\bar{y}_{US} - \bar{y}_B} = \sqrt{SE_{\bar{y}_{US}}^2 + SE_{\bar{y}_B}^2} = \sqrt{(29.56)^2 + (18.90)^2} = \underline{35.08}$

$CI = (\bar{y}_{US} - \bar{y}_B) \pm t\,(SE_{\bar{y}_{US} - \bar{y}_B}) = (1500 - 1000) \pm 1.96(35.08) = 500 \pm 68.76 = (431.24, 568.76)$

7b. Based on the 95% confidence interval, the reporter is right to say that the cost of participation is higher in the US than Brazil. The confidence interval for the difference between mean costs does not include $0.

9a. $CI = (P_{2018} - P_{2017}) \pm z(SE_{p_{2018} - p_{2017}})$, where $SE_{p_{2018} - p_{2017}} = \sqrt{SE_{p_{2018}}^2 + SE_{p_{2017}}^2}$, where SE for

each group $= SE_p = \sqrt{\frac{p(1-p)}{N}}$: $SE_{2018} = \sqrt{\frac{0.31(1-0.31)}{1000}} = 0.015$ and

$SE_{2017} = \sqrt{\frac{0.22(1-0.22)}{1000}} = 0.013 \rightarrow SE_{p_{2018} - p_{2017}} = \sqrt{0.015^2 + 0.013^2} = 0.02 \rightarrow (0.31 - 0.22) \pm$

$1.96(0.02) = 0.09 \pm 0.039 \rightarrow (0.051, 0.129) \rightarrow$ Over repeated sampling, 95% of all confidence intervals will include the population parameter. Therefore, there is a 95% chance that this confidence interval includes the population parameter, or the actual difference in negative US sentiment between 2017 and 2018. Since the confidence interval falls above 0, this means there is a very high chance that a greater proportion of Mexicans saw the US very negatively in 2018 compared to 2017. This challenges the claim that the 2018-2017 difference reflects sampling variation.

9b. The increased percentage of Mexicans who viewed the US negatively in 2018 compared to 2017, 9 percentage points, could be seen as quite a large difference considering that the time span was just one year. The substantive importance of the difference depends on the extent to which public sentiment in Mexico toward the US actually affects the actions of the US and the Mexican governments.

11. <u>Null hypothesis</u>: The proportion of millennials who are very interested in astrology is equal to the proportion of Gen Xers who are very interested in astrology. (H_0: $\pi_M - \pi_{GenX} = 0$)

<u>Alternative hypothesis</u>: The proportion of millennials who are very interested in astrology is greater than the proportion of Gen Xers who are very interested in astrology. (H_a: $\pi_M > \pi_{GenX}$)

Alpha-level is 0.01, located in one tail because this is a one-tailed test, for a decision p-value of 0.01.

$$SE_{p_m - p_{GenX}} = \sqrt{\frac{\pi(1-\pi)}{N_m} + \frac{\pi(1-\pi)}{N_{GenX}}}$$, where π is the pooled proportion =

$$\frac{f_m + f_{GenX}}{N_{total}} = \frac{(0.48 \times 100) + (0.37 \times 100)}{200} = \frac{48 + 37}{200} = 0.425;$$

$$SE_{p_m - p_{GenX}} = \sqrt{\frac{0.425(1-0.425)}{100} + \frac{0.425(1-0.425)}{100}} = 0.070$$

$$z = \frac{(p_m - p_{GenX})}{SE_{p_m - p_{GenX}}} = \frac{(0.48 - 0.37)}{0.07} = \frac{0.11}{0.07} = 1.57$$ The p-value associated with the sample z of 1.57 is 0.0582, higher than our alpha-level of 0.01. Therefore, we cannot reject the null hypothesis of no difference in interest in astrology between millennials and Gen Xers. The results of the hypothesis test suggest that there is not a relationship between these two variables.

13. $CI = (p_M - p_{GenX}) \pm z(SE_{p_m - p_{GenX}})$, where $SE_{p_m - p_{GenX}} = \sqrt{SE_{p_m}^2 + SE_{p_{GenX}}^2}$,

where SE for each group = $SE_p = \sqrt{\frac{p(1-p)}{N}} : SE_m = \sqrt{\frac{0.48(1-0.48)}{100}} = 0.050$ and

$SE_{GenX} = \sqrt{\frac{0.37(1-0.37)}{100}} = 0.048 \rightarrow SE_{p_m - p_{GenX}} = \sqrt{0.050^2 + 0.048^2} = 0.069 \rightarrow (0.48 - 0.37) \pm$

1.282(0.069) = 0.11 ± 0.088 → (0.020, 0.198) → Over repeated sampling, 80% of all confidence intervals will include the population parameter. Therefore, there is an 80% chance that this confidence interval includes the population parameter, or the actual difference in astrology interest between millennials and Gen Xers. Since the confidence interval falls above 0, this means there is a 0.8 probability that a greater proportion of millennials than Gen Xers are very interested in astrology.

15. The null hypothesis is that there is no difference between Black and white respondents in feelings about Vice President Kamala Harris ($H_0: \mu_B - \mu_W = 0$). The alternative hypothesis is that there is a difference between Black and white respondents in feelings about Vice President Kamala Harris ($H_a: \mu_B - \mu_W \neq 0$).

Decision t for a two-tailed hypothesis test at the 0.01 alpha-level and 5,986 DF is 2.576.

$$SE_{\bar{y}_B} = \frac{s_B}{\sqrt{N_B}} = \frac{25.4}{\sqrt{638}} = 1.01, \quad SE_{\bar{y}_W} = \frac{s_W}{\sqrt{N_W}} = \frac{38.2}{\sqrt{5350}} = 0.522 \rightarrow$$

$$SE_{\bar{y}_B - \bar{y}_W} = \sqrt{SE_{\bar{y}_B}^2 + SE_{\bar{y}_W}^2} = \sqrt{(1.01)^2 + (0.522)^2} = \underline{1.13}$$

$$t = \frac{(\bar{y}_B - \bar{y}_W)}{SE_{\bar{y}_B - \bar{y}_W}} = \frac{(80.2 - 47.0)}{1.14} = \frac{33.2}{1.14} = \underline{29.12}$$

Sample t, 29.12, is greater than decision t, 2.576. Therefore, we should reject the null hypothesis of no difference between means.

17. $CI = (\bar{y}_B - \bar{y}_W) \pm t(SE_{\bar{y}_B-\bar{y}_W})$, where $SE_{\bar{y}_B-\bar{y}_W} = \sqrt{SE_{\bar{y}_B}^2 + SE_{\bar{y}_W}^2}$, where SE for each group (also calculated in Problem 15) is $SE_{\bar{y}_B} = \dfrac{s_B}{\sqrt{N_B}} = \dfrac{25.4}{\sqrt{638}} = 1.01$ and $SE_{\bar{y}_W} = \dfrac{s_W}{\sqrt{N_W}} = \dfrac{38.2}{\sqrt{5350}} = 0.522 \rightarrow SE_{\bar{y}_B-\bar{y}_W} = \sqrt{SE_{\bar{y}_B}^2 + SE_{\bar{y}_W}^2} = \sqrt{(1.01)^2 + (0.522)^2} = 1.14$

$CI = (\bar{y}_B - \bar{y}_W) \pm t(SE_{\bar{y}_B-\bar{y}_W}) = (80.2 - 47) \pm 2.576(1.14) = 33.2 \pm 2.94 = (30.3, 36.1)$

Chapter 10 Stata Problems

1. `prtest voted2016, by(endsmeet2) level(95)`

Chapter 10 SPSS Problems

1. `Analyze → Compare Means → Independent-Samples T Test`

    This will open the "Independent-Samples T Test" dialog box. Move *voted16* into the "Test Variable(s)" box. Move *endsmeet2* into the "Grouping Variable" space and click on "Define Groups." Specify value 1 (worried) for Group 1 and value 0 (not worried) for Group 2. Click on "Continue" and then "OK."

Chapter 11 Solutions

1a. Registered Democrats rated the UN the highest, with a mean of 72.3.

1b. Yes, there is variation in every category, as indicated by the fact that all three standard deviations are greater than 0.

1c. If there were no within-group variation, each standard deviation would be equal to 0.

1d. This is the *between* sum of squares because it calculates differences between each group mean and the total mean ($\bar{y}_{group} - \bar{y}_{total}$), not between all values and their respective group means (y-ȳ).

1e.

Between Groups Sums of Squares, Feeling Thermometer Ratings of UN and Political Party Registration

	Between Groups Sum of Squares		
Political Party (N)	$\bar{y}_{group} - \bar{y}_{total}$	$(\bar{y}_{group} - \bar{y}_{total})^2$	$N_{group}(\bar{y}_{group} - \bar{y}_{total})^2$
Democratic (1,084)	64.2 − 54.2 = 10.0	100.0	161,800.0
Republican (1,294)	41.7 − 54.2 = −12.5	156.3	202,187.5
Independent (1,000)	52.5 − 54.2 = −1.7	2.89	2,890
			$\sum N_{group}(\bar{y}_{group} - \bar{y}_{total})^2 = 366,877.5$

3a. The null hypothesis is that there are no differences in average ratings of the UN between the three groups ($H_0: \mu_{Rep} = \mu_{Dem} = \mu_{Ind}$). The alternative hypothesis is that there are differences in average ratings of the UN across these groups ($H_a: \mu_{Rep} \neq \mu_{Dem} \neq \mu_{Ind}$).

3b. Find within and between degrees of freedom: $DF_{within} = N_{total}$ − number of groups = 4098 − 3 = 4095; $DF_{between}$ = number of groups − 1 = 3 − 1 = 2; The F table shows that the decision F is 2.99.

3c. First find mean squares within and between:

$$MS_{within} = \frac{SS\ within}{DF\ within} = \frac{3{,}892{,}974.6}{4095} = 950.7\ ;\ MS_{between} = \frac{SS\ between}{DF\ between} = \frac{366{,}877.5}{2} = 183{,}438.8\ ;$$

next, find sample F: $F = \dfrac{MS\ between}{MS\ within} = \dfrac{183{,}438.8}{950.7} = 192.0$

3d. Sample F, 192.0, is larger than critical F, 2.99. Therefore, we reject the null hypothesis of no differences across groups.

3e. Since we are rejecting the null hypothesis, we conclude that there are significant differences in average ratings of the UN across categories of political party registration.

5a. Tukey's test would show which of the pairs of party registration categories were significantly different from each other in rating of the UN.

5b. The three pairs are Democratic-Republican, Democratic-Independent, and Republican-Independent. The p-value of 0.000 indicates that all three pairs have statistically significant differences in rating of the UN at a p of <0.001.

5c. There is a statistically significant relationship between party registration and mean rating of the United Nations on a feeling thermometer, at an alpha-level of 0.05. Each party registration category is significantly different from the others. Democratic respondents rated the UN significantly more favorably than Independents, and Independents rated the UN significantly more favorably than Republicans.

7a. First calculate the mean for each of the social class groups. Poverty ȳ=167; Working-class ȳ=251; Professional ȳ=382. Remaining calculations are shown in the table below.

Calculation of Within Groups Sums of Squares

Poverty			Working Class			Professional		
y	y − ȳ	(y − ȳ)²	y	y − ȳ	(y − ȳ)²	y	y − ȳ	(y − ȳ)²
150	−17	289	250	−1	1	400	18	324
100	−67	4,489	300	49	2,401	375	−7	49
200	33	1,089	240	−11	121	310	−72	5184
180	13	169	195	−56	3,136	450	68	4624
205	38	1,444	270	19	361	375	−7	49
		Σ(y − ȳ)² = 7,480			Σ(y − ȳ)² = 6,020			Σ(y − ȳ)² = 10,230

7b. Add together the sum of squares for each group: 7,480 + 6,020 + 10,230 = 23,730

7c. Calculation of Between Groups Sums of Squares

	$(\bar{y}_{group} - \bar{y}_{total})$	$(\bar{y}_{group} - \bar{y}_{total})^2$	$N_{group}(\bar{y}_{group} - \bar{y}_{total})^2$
Professional	382 − 267.7 = 114.3	13,064.49	65,322.45
Working Class	251 − 267.7 = −16.7	278.89	1,394.45
Poverty	167 − 267.7 = −100.7	10,140.49	50,702.45
			$\Sigma N(\bar{y}_{total} - \bar{y}_{group})^2 = 117,419.35$

7d. $DF_{within} = N_{total} - k$ (where k is the number of groups) = 15 − 3 = 12; $DF_{between} = k - 1 = 2$. Critical F = 3.89.

7e. $MS_{within} = \dfrac{SS\ within}{DF\ within} = \dfrac{23,730}{12} = 1,977.5$; $MS_{between} = \dfrac{117,419.35}{2} = 58,709.7$

7f. $F = \dfrac{MS\ between}{MS\ within} = \dfrac{58,709.7}{1,977.5} = 29.7$. At the 0.05 alpha level, this sample statistic is larger than critical F (3.89), which means that we can reject the null hypothesis that there is no difference in mean words heard across social class groups. There is a statistically significant relationship between mean words heard and social class.

7g. We cannot tell which differences are statistically different from each other from the ANOVA. We would need to conduct a post-hoc test such as Tukey's.

9a. Panel A is most likely to yield a significant sample F because this panel has the least variation within groups and the most variation between groups.

9b. Panel C is least likely to yield a significant sample F because this panel has the widest variation between groups and the widest variation within groups. Agreement that lesbians and gays should be allowed to adopt appears completely unrelated to these three religious affiliations in Panel C.

11a. The means do suggest that traveling or living abroad is related to support for increasing immigration, with higher means for students who had more experience outside the US. The mean differences suggest that students with more experience outside the US have higher support for increasing immigration, and those with less experience have lower support.

11b. The null hypothesis is that mean support for increasing immigration is the same for all three levels of experience outside the US.

11c. The output shows that the p-value associated with the sample F is 0.00, lower than any tradiational alpha-level. This means that we can reject the null hypothesis that all of the means are equal.

11d. The note to the table indicates that the asterisks next to the differences show p-values that are lower than particular alpha-levels (0.05, 0.01, and 0.001). Thus, every difference marked by any asterisks is "statistically significant" at the alpha-level noted at the bottom of the table. There are statistically significant differences between those who lived abroad and those who traveled outside the US at alpha <0.05 and between those who lived abroad and those who neither lived nor traveled abroad at alpha <0.001. There is no significant difference between those who traveled abroad and those who did neither.

Chapter 11 Stata Problems

1. `oneway V202184 pid3, tabulate`
3. The null hypothesis is that the means for all three groups are equal in the population. The alternative hypothesis is that at least one of the groups has a mean that differs from another in the population.
5. The output provides us with DF 1 (2) and DF 2 (3,335). We use the F table to find critical F: 2.99. We reject the null hypothesis because critical F is smaller than sample F.
7. The p-value for this test is less than 0.001, indicating that the difference in variances in feelings about rural Americans across party affiliation likely reflects true differences in the population.
9. All of the group means are different from each other at the 0.05 alpha-level.

Chapter 11 SPSS Problems

1. Analyze → Compare Means → One-Way ANOVA

   In the "Options" dialog box, select "Descriptive" to generate descriptive statistics; select Homogeneity of variance test to test the assumption of equal variance across groups. In the "Post-Hoc" dialog box, select Tukey to generate the Tukey post-hoc test.

3. The null hypothesis is that the means for all three groups are equal in the population. The alternative hypothesis is that at least one of the groups has a mean that differs from another in the population.

5. The output provides us with DF 1 (2) and DF 2 (3,335). We use the F table to find critical F: 2.99. We reject the null hypothesis because critical F is smaller than sample F.

7. The p-value for this test is less than 0.001, indicating that the difference in variances in feelings about rural Americans across party affiliation likely reflects true differences in the population.

Chapter 12 Solutions

1a. The independent variable is age, and the dependent variable is whether the person gets news from social media.

1b. The null hypothesis is that there is no relationship in the population between age and getting news from social media.

1c. To find the decision chi-square at a 0.05 alpha-level, first find the degrees of freedom: $(r-1)(c-1) = (4-1)(2-1) = 3$. Find the chi-square statistic in the chi-square table for 3 degrees of freedom at the 0.05 alpha-level: 7.815.

1d. $F_e = \left(\dfrac{row\ frequency}{total\ frequency}\right)(column\ frequency)$

The expected values for each cell, rounded to one decimal place, are shown in the table below.

Expected Frequencies for Problem 1d.

	18 to 29	30 to 49	50 to 64	65 and Older	Total
Got news from social media	692.7	1,949.7	1,583.8	1,470.8	5,697
Did not get news from social media	784.3	2,207.4	1,793.2	1,665.2	6,450
Total	1,477	4,157	3,377	3,136	12,147

1e. Follow each of the steps, as shown below:
- Find the difference between the actual frequency and the expected frequency for each cell in the table, and square each difference:
  - People 18–29 who got news from social media: 1,122 − 692.7 = 429.3 →184,298.5
  - People 30–49 who got news from social media: 2,411 − 1,949.7 = 461.3 → 212,797.7
  - People 50–64 who got news from social media: 1,317 − 1,583.8 = −266.8 → 71,182.2
  - People 65 and older who got news from social media: 846 − 1,470.8 = −624.8 →390,375.0
  - People 18–29 who did not get news from social media: 355 − 784.3 = −429.3 →184,298.5
  - People 30–49 who did not get news from social media: 1,746 − 2,207.4 = −461.4 → 212,890.0
  - People 50–64 who did not get news from social media: 2,060 − 1,793.2 = 266.8 → 71,182.2
  - People 65 and older who did not get news from social media: 2,289 − 1,665.2 = 623.8 → 389,126.4
- Divide each squared difference by the expected frequency:
  - People 18–29 who got news from social media: $\frac{184,298.5}{692.7} = 266.1$
  - People 30–49 who got news from social media: $\frac{212,797.7}{1,949.7} = 109.14$
  - People 50–64 who got news from social media: $\frac{71,182.2}{1,583.8} = 45.05$
  - People 65 and older who got news from social media: $\frac{390,375.0}{1,470.8} = 265.76$
  - People 18–29 who did not get news from social media: $\frac{184,298.5}{784.3} = 234.98$
  - People 30–49 who did not get news from social media: $\frac{212,890.0}{2,207.4} = 96.4$
  - People 50–64 who did not get news from social media: $\frac{71,182.2}{1,793.2} = 39.70$
  - People 65 and older who did not get news from social media: $\frac{389,126.4}{1,665.2} = 233.68$
- $\chi^2 = \frac{(Fo - Fe)^2}{Fe} = 266.1 + 109.14 + 45.05 + 265.76 + 234.98 + 96.4 + 39.70 + 233.68 = 1,290.39$

1f. The null hypothesis of no relationship between age and getting news from social media should be rejected because sample chi-square is (much!) larger than decision chi-square.

1g. There is a statistically significant relationship between age and getting news from social media. A smaller percentage of each successively older age group reports getting news from social media, from a high of 76% among 18–29-year-olds, to a low of 27% among those 65 and older.

3a. About 36% of students from low-income families "melt" over the summer, which makes them more likely than students from medium or high-income families to do so.

3b. The observed frequencies can be found by multiplying the proportion of students in each income group who melted or did not melt by the total number of students in that income group. For the first cell: 900 students x 0.356 = 320.4. Each frequency is shown below:

Summer Melt?	Income		
	Low	Medium	High
Yes	320	130	50
No	580	970	750

3c. Yes, the observed frequencies are different enough from the expected frequencies to conclude that there is a statistically significant relationship between the variables. We know this because the larger the differences between observed and expected frequencies, the larger the sample chi-square statistic. The larger the sample chi-square, the greater the chance that there is a statistically significant relationship. In this case, we can see that the sample chi-square, 276.7, is larger than the decision chi-square at 2 degrees of freedom for any alpha-level in the chi-square table. We know, then, that observed frequencies are different enough from expected frequencies to yield a significant sample chi-square.

3d. Since there is a relationship between family income and summer melt in the statewide data, and low-income students are much more likely to melt than higher-income students, the college's most efficient use of resources to decrease overall summer melt would be to target low-income students.

5a. Use the formula for standardized residual, $R = \dfrac{Fo - Fe}{\sqrt{Fe}}$ to find R for each cell. For the first cell: $\dfrac{320 - 160.7}{\sqrt{160.7}} = 12.56$.

Residuals for Practice Problem 5a

Summer Melt?	Income		
	Low	Medium	High
Yes	12.56	−4.74	−7.77
No	−5.85	2.21	5.39

5b. A positive residual means that the observed frequency is less than the expected frequency, and a negative residual indicates that the observed frequency is larger than the expected frequency.

5c. The standardized residual for low-income summer melt is by far the highest, which suggests that higher summer melt for low-income graduates makes the strongest contribution to the chi-square statistic, followed by lower than expected melt for high income graduates. While the residuals are not themselves tests of "statistical significance," this provides further support for the conclusion that focusing on reducing melt for low-income students would be the most effective use of resources.

7a. The test that matches the researcher's question is a chi-square test of independence. The null hypothesis for that test is that there is no relationship between vehicle status and the likelihood that it will cut off other vehicles. The alternative hypothesis is that there is a relationship.

7b. The test that matches this researcher's question is a chi-square goodness-of-fit test. The null hypothesis for this test is that the population rates at which vehicles cut off other vehicles, by status of the vehicle, are the same rates at which people cut off others in conversation, by income quintile. The alternative hypothesis is that these rates do not match in the population.

7c. These are the expected values for the goodness-of-fit test. We can tell because the expected frequencies for the chi-square test of independence are $F_e = \left(\frac{row\ frequency}{total\ frequency}\right)(column\ frequency)$. As an example, the expected number of vehicles in status category 1 that cut off other vehicles would be: $\left(\frac{34}{274}\right)(41) = 5.1$.

This does not match the expected frequency for that group in the figure. The expected frequency for the goodness-of-fit test is the proportion of people in the lowest income quintile who cut off others in conversation multiplied by the number of category 1 status vehicles in the sample: $.08(41) = 3.3$. This matches the expected frequency for this group shown in the figure. We can verify that this is true for every group in the analysis.

7d. The conclusion is inaccurate for the chi-square goodness-of-fit test, which was the test from part c. We do reject the null hypothesis, but the null hypothesis is not that there is no relationship. The null hypothesis is that the relationship in the population between vehicle status and cutting off other vehicles mirrors the relationship between people's income and how often they cut off people in conversation. An accurate conclusion is that the relationship between vehicle status and cutting off other vehicles is not the same as the relationship between income and cutting off people in conversation.

7e. The expected frequency should be at least 5 for each category. Here, one expected frequency is less than 5. Therefore, the analysis should not proceed.

9a. Yes, there does appear to be a relationship, because percentages of vehicles that cut off other vehicles differs across status categories.

9b. The title of the article states specifically that "rich" people drive differently than others. The chi-square statistic is technically a test of the overall relationship between the two variables, not a test of whether one category's distribution is different from all the rest. Technically, then, the test does not demonstrate that category 5 is different from the rest. Additionally, the title of the article equates the status of one's vehicle with their financial status. While this assumption may hold true for many, it is also true that some people drive vehicles that do not correspond to their financial status.

11. No, it is not possible because the dependent variable is not measured at the ordinal or nominal level, which is a requirement for a chi-square test. Instead, a two-sample t-test would be appropriate.

13a. The null hypothesis is that the proportion of people of color who prefer *Get Out* is equal to the proportion of whites who prefer *Get Out*. The alternative hypothesis is that the proportions are unequal.

13b. The cross-tabulation is shown below.

13c. The null hypothesis is that there is no relationship in the population between the racial identity of viewers and preferences for *Get Out* over *Psycho*. The alternative hypothesis is that there is a relationship.

Movie Preference by Race (Column Percentages)

	People of Color	Whites	Total
Prefer *Get Out*	88% (N = 30)	78% (N = 32)	82.7% (N = 62)
Prefer *Psycho*	12% (N = 4)	22% (N = 9)	17.3% (N = 13)
Total	100% (N = 34)	100% (N = 41)	100% (N = 75)

13d. Yes, it is correct to draw that conclusion because the results of a two-sample test for a difference in proportions are equivalent to the results of a chi-square test of independence for a 2 x 2 table.

13e. No, their preferences do not differ to a great enough extent in the sample to conclude that the preferences are actually different in the population.

Chapter 12 Stata Problems

1. `tabulate freemind libcons, column expected cchi2 chi2`
3. Of conservatives in the sample, 4.83% strongly agree that understanding the world requires freeing one's mind of old traditions; about 15% of liberals strongly agree.
5. Yes, there is a relationship in the population. We can see this because the sample chi-square, 398.5, has a p-value of less than 0.001, which is below any traditional alpha threshold.

Chapter 12 SPSS Problems

1. `Analyze → Descriptive Statistics → Crosstabs`

    Put the dependent variable (*freemind*) into the "Row(s)" box and the independent variable (*libcons*) into the "Column(s)" box. Click on "Statistics." In the "Crosstabs: Statistics" dialog box, check the small box next to "Chi-square." Click on "Continue." Now click on "Cells" to request observed frequencies, expected frequencies, column percentages, and standardized residuals in each cell.

3. Of conservatives in the sample, 4.83% strongly agree that understanding the world requires freeing one's mind of old traditions; about 15% of liberals strongly agree.

5. Yes, there is a relationship in the population. We can see this because the sample chi-square, 398.5, has a p-value of less than 0.001, which is below any traditional alpha threshold.

Chapter 13 Solutions

1a. Does social class affect general happiness?

1b. Does the relationship between class and happiness hold for both those born inside the United States and those born outside the US?

1c. Independent variable: class; dependent variable: happiness; control variable: birth location

1d. Among those born inside the US, happiness increases with social class. For example, the percentage saying they are very happy is 14% for the lower or working class and 23% for the middle or upper class; conversely, over 30% of the lower or working class say they are not too happy, compared with only 18% of the middle or upper class. Among those born outside the US, happiness also increases with social class, with 22% of lower or working class people reporting being very happy vs. 28% of middle or upper class respondents, and 26% of lower or working class people saying they are not too happy, compared with only 11% of middle or upper class people. Overall levels of happiness are slightly higher for those born outside the US, but the percentage point gaps between the lower and higher social classes are comparable to those of US-born people.

1e. There is no evidence of a spurious or moderating relationship because the differences in percentages across columns are similar in the cross-tabulations for both those born inside and outside the US. If the relationship were spurious, the relationship would not hold among one of those groups; if it were moderating, the relationship would be notably less or different among one of the groups.

3a. The following figure represents the test of a spurious or moderating relationship between social class and happiness. This model was not empirically supported.

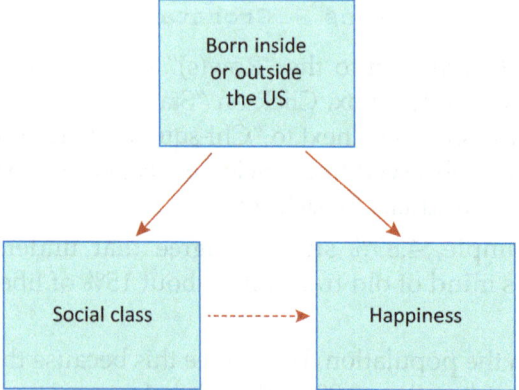

3b. Below are explanations for whether the analysis meets the causal criteria:

- Association: There is a positive association between class and happiness.

- Ruling out alternative explanations: Immigrant status was ruled out as a plausible alternative explanation or moderating variable. There could be other plausible explanations to rule out, such as other demographic factors (gender, race, age) or other characteristics. (*Authors' note*: In fact, the relationship between class and happiness is quite persistent across most other variables.)

- Logical order and temporal order: Class precedes happiness in temporal order, and it is logical that class would lead to happiness rather than the other way around.

5a. There is a positive relationship between age and days without power.

5b. We should not conclude that age causes the amount of time that people go without power. We must rule out alternative explanations first.

5c. The figure shows that age is not related to days without electricity once population density is controlled. Population density, the antecedent variable, is related to both the age of the people in the area and days spent without electricity. Specifically, the results indicate that densely populated areas have younger populations, on average, and densely populated areas spent less time, on average, without electricity.

7a. No, the relationship disappears once we control for parents' political ideology. There are very small differences across columns among those with conservative and liberal parents. For instance, among those with conservative parents, around 80% of millennials were conservative, regardless of how many YA books they read. Additionally, the chi-square statistic is not significant among those with liberal or conservative parents.

7b. It is an antecedent variable because it is a control variable that is temporally prior to both the independent variable and the dependent variable, and it influences both of them. Parents' political ideology comes temporally prior to how many YA romance books their children read.

7c. Since the relationship between reading YA romance and political ideology goes away once parent's political ideology is held constant, it must be that parents' political ideology affects how many YA romance books their children read and that parents' political ideology affects children's political ideology.

9a. The title of the article implies that positive gym class experiences have a causal impact on one's level of physical activity as an adult. The description says that there is a positive association, and gym class experiences do come before what one does as an adult, but no alternative explanations have been ruled out.

9b. They are both modeled as intervening variables.

9c. In the model on the top, positive gym class experiences affect appreciation of team work, which, in turn, affects physical activity as an adult. In the model on the bottom, positive gym class experiences affect self-esteem, which, in turn, affects physical activity as an adult.

9d. The tables show that there is no relationship between appreciation of team work and physical activity for adults, but there is a relationship between self-esteem and physical activity. This means that the top model in the figure cannot be accurate. Appreciation of team work cannot function as an intervening variable if it is not associated with physical activity.

11a. Parents' ideology is the independent variable and individuals' ideology is the dependent variable. There is an association between the variables and parents' ideology is logically and temporally prior.

11b. Either variable could be the independent variable. The variables are associated, but it is difficult to establish which one occurs first in time or logically. They are likely both associated with other explanatory variables, such as religious affiliation or strength of religious belief.

11c. Age is the independent variable and gender identification is the dependent variable. They are associated and age occurs first logically and temporally.

11d. Veteran status is the independent variable and confidence in the military is the dependent variable. They are associated and veteran status most likely precedes confidence in the military in time and logically, although we might also expect those with more confidence in the military would be more likely to enlist.

13a. Independent variable: diversity of team; dependent variable: creativity of ideas; control variable: trust (a moderating variable)

13b. Independent variable: gender; dependent variable: suspension; control variable: physical altercations (a mediating, or intervening variable)

13c. Independent variable: protest size; dependent variable: goal achievement; control variable: media coverage (mediating, or intervening variable)

13d. Independent variable: view on climate change; dependent variable: view on government spending to protect the environment; control variable: gender. This is a nonspurious relationship and the control is not antecedent, intervening, or moderating.

13e. Independent variable: gender; dependent variable: spousal work status; control variable: sexual identity (moderating variable)

Chapter 13 Stata Problems

1. `tabulate nateduc libcons, column chi2`
3. Yes, there is a relationship between political ideology and attitudes about government spending on education. For example, the vast majority of liberals (about 85%) think the government spends too little on education, compared to about 56% of conservatives, for a difference of 29 percentage points. The p-value for the chi-square test is less than 0.001, indicating that we can reject the null hypothesis of no relationship between these variables in the population.
5. The same pattern of the relationship between political ideology and attitudes about government spending on education that we saw in the full sample holds true among the younger respondents. Among those 50 or younger, the more conservative people are, the more they think the government spends too much on education; and the more liberal people are, the more they think the government spends too little. The p-value for the chi-square test is less than 0.001, indicating that this relationship likely holds in the general population.

Chapter 13 SPSS Problems

1. `Analyze → Descriptive Statistics → Crosstabs`

    Place *nateduc* in Row box, *libcons* in Column box, and check chi-square in the "Statistics" box.
3. Yes, there is a relationship between political ideology and attitudes about government spending on education. For example, the vast majority of liberals (about 85%) think the government spends too little on education, compared to about 56% of conservatives, for a difference of 29 percentage points. The p-value for the chi-square test is less than 0.001, indicating that we can reject the null hypothesis of no relationship between these variables in the population.
5. The same pattern of the relationship between political ideology and attitudes about government spending on education that we saw in the full sample holds true among the younger respondents. Among those 50 or younger, the more conservative people are, the more they think the government spends too much on education; and the more liberal people are, the more they think the government spends too little. The p-value for the chi-square test is less than 0.001, indicating that this relationship likely holds in the general population.

Chapter 14 Solutions

1a. The research question posits a causal relationship between anxiety and performance. Therefore, anxiety is on the x-axis because it is the independent variable, causing changes in levels of the dependent variable, performance, which is on the y-axis.

1b. Scatterplot A depicts the Yerkes–Dodson law because, according to the law, performance will be lower at low and high levels of anxiety, but higher at middle levels of anxiety. In other words, as anxiety increases, so does performance, until a tipping point is reached, after which increases in anxiety are associated with decreases in performance.

1c. The correlation coefficient for this relationship should not be calculated because the relationship between the two variables does not follow a straight line. Instead the relationship is curvilinear.

3a. The coordinates of the point are (6, 25). The person gets 6 paid sick days per year, and their job satisfaction rating is 25.

3b. There is a positive correlation between paid sick days and job satisfaction.

3c. This is a strong relationship.

5a. The differences between attendance for each student and mean attendance are shown in the table below.

Calculations for Problem 5a

Student	$(x_i - \bar{x})$
1	25.8
2	−4.2
3	−34.2
4	5.8
5	15.8
6	20.8
7	25.8
8	−24.2
9	−14.2
10	2.8
11	−1.2
12	−44.2
13	−38.2
14	22.8
15	16.8
16	10.8
17	3.8
18	−7.2
19	10.8
20	5.8

5b. The differences between GPA for each student and mean GPA are shown in the table below.

Calculations for Problem 5b

Student	$(y_i - \bar{y})$
1	0.37
2	−0.33
3	−0.93
4	0.07
5	0.07
6	0.57
7	0.77
8	0.07
9	−0.13
10	−0.23
11	−0.13
12	−0.73
13	−1.03
14	0.62
15	0.57
16	0.67
17	0.17
18	0.07
19	0.07
20	−0.63

5c. The cross-products for each student are shown in the table below.

Calculations for Problem 5c

Student	$(x_i - \bar{x})(y_i - \bar{y})$
1	9.55
2	1.39
3	31.81
4	0.41
5	1.11
6	11.86
7	19.87
8	−1.69
9	1.85
10	−0.64
11	0.16
12	32.27
13	39.35
14	14.14
15	9.58
16	7.24
17	0.65
18	−0.50
19	0.76
20	−3.65

5d. The sum of the twenty cross-products is: $\Sigma(x_i - \bar{x})(y_i - \bar{y}) = 175.52$.

5e. The covariance is: $\dfrac{\Sigma(x_i - \bar{x})(y_i - \bar{y})}{N-1} = \dfrac{175.52}{19} = 9.24$

5f. The correlation is: $r_{(x,y)} = \dfrac{COV_{(x,y)}}{s_x s_y} = \dfrac{9.24}{(21.25)(.53)} = 0.82$

5g. The correlation of 0.82 indicates a strong association between attending class and doing well academically. The professor's advice to attend class is supported by the correlation.

7a. The slope is –0.20. The intercept is 69.7.

7b. The intercept of 69.7 means that, on average, a person who is zero years old would give the #MeToo movement a thermometer rating of 69.7 degrees. The intercept is not useful in this case because it is not meaningful to know how a person who is zero years old would rate the #MeToo movement.

7c. The slope of –0.20 means that, on average, for each increase in age by one year, people rate the #MeToo movement 0.20 degrees *lower* on the feeling thermometer. We can see that there is a negative relationship between age and ratings of the #MeToo movement.

7d. For a 32 year-old: $\hat{y} = 69.7 - 0.20x = 69.7 - 0.20(32) = 63.3$;

For a 71 year-old: $\hat{y} = 69.7 - 0.20x = 69.7 - 0.20(71) = 55.5$;

7e. The difference in predicted ratings is 7.8 degrees.

7f. In order to assess whether the slope is different from 0 in the population, we need to find sample t: $t = \dfrac{b}{SE_b}$. In this case, $t = \dfrac{-0.20}{0.023} = -8.7$. The sample size is 6,158. Thus, we use the ∞ row in the t table to find decision t. We see that, no matter what the alpha value, decision t-values are lower than the absolute value of sample t, 8.7. Thus, we can reject the null hypothesis that the population slope is equal to 0.

9a. The slope for Democrat, 19.7, means that, on average, Democrats rate the #MeToo movement 19.7 degrees higher than do non-Democrats, controlling for age.

9b. Yes, the model provides better estimates of ratings of the #MeToo movement than simply assigning everyone the sample mean as a predicted value. The F-statistic tells us this because it evaluates the model globally, whereas standard errors of slopes evaluate those particular slopes. Because the F-statistic is associated with a p-value less than 0.001, we can reject the null hypothesis that the regression model does not fit the data better than an intercept-only model.

9c. $SE_{\hat{y}} = \sqrt{\dfrac{\Sigma(y-\hat{y})^2}{N-2}} = \sqrt{\dfrac{4{,}795{,}686}{5{,}922-2}} = \sqrt{\dfrac{4{,}795{,}686}{5920}} = 28.5$; This means that the average variation of respondents' actual ratings of the #MeToo movement from the values predicted by the equation is 28.5 degrees.

9d. $r^2 = \dfrac{\Sigma(\hat{y}-\bar{y})^2}{\Sigma(y-\bar{y})^2} = \dfrac{492{,}196}{5{,}287{,}883} = 0.09$; 9% of the variation in feeling thermometer ratings of the #MeToo movement is accounted for by variation in age and party affiliation.

11a. The slope of 2.3 means that for every additional visitor to "alt-right" Facebook pages per 1,000 people in the state, we expect racially motivated hate crimes in the state to increase by 2.3 incidents, on average.

11b. This slope means that for every percentage-point increase in unemployed white men over 18 in a state, we expect the state's racially motivated hate crimes to increase by 3.1, on average, controlling for the number of visitors to "alt-right" Facebook pages per 1,000 people.

11c. Venn diagram 1 depicts correlations among the three variables. Since the slope for "alt-right" Facebook page visitors declined once they controlled for the unemployed white male population, this means that the two independent variables are correlated. This is represented by areas C and D in diagram 1. In Venn Diagram 2, there is no correlation between the independent variables. If that were the case, the slope for "alt-right" Facebook page visitors would have remained the same after the introduction of the white male unemployed variable.

11d. The equation for the regression line is $\hat{y} = 1.5x_1 + 3.1x_2 + 40$.

11e. $\hat{y} = 1.5x_1 + 3.1x_2 + 40 = 1.5(5) + 3.1(13) + 40 = \underline{87.8}$

Chapter 14 Stata Problems

1. `scatter V202171 V201200`
3. `corr V202171 V201200`
5. `regress V202171 V201200`
7. The intercept is 40.87, which means that someone with a score of 0 on the independent variable would be expected to rate the police at 40.87 degrees. In this case, 0 is not a possible value of the independent variable.
9. The r^2 is 0.248, which means that about 25% of the variation in ratings of the police is explained by political ideology.
11. The F-statistic is 2,089.92, with a p-value less than 0.001. This means that the regression model overall fits the data better than an intercept only model would.
13. The slope for *poc* is -9.58. This means that, on average, those who identify as people of color rate the police about 9.58 degrees lower than do whites, holding constant political ideology. The slope's standard error, 0.627, is associated with a p-value less than 0.001. This means that the slope is "statistically significant," or that the slope in the population is not equal to 0.
15. The intercept for the new model is 44.19, which is the predicted police feeling thermometer score for a white person (*poc*=0) who scores a 0 on the ideology scale. Again, the ideology scale begins at 1, limiting the utility of the intercept.
17. `predict V202171hat`

    `predict V202171res, residuals`
19. `histogram V202171res, percent`

    Yes, they are roughly normally distributed.

Chapter 14 SPSS Problems

1. **Graphs → Legacy Dialogs → Scatter/Dot**

   Select "Simple Scatter" and then click on "Define." Place the dependent variable into the "Y Axis" box, and the independent variable into the "X Axis" box.

3. **Analyze → Correlate → Bivariate**

5. **Analyze → Regression → Linear**

   Place the dependent variable into the "Dependent" box and the independent variable into the "Independent(s)" box.

7. The intercept is 40.87, which means that someone with a score of 0 on the independent variable would be expected to rate the police at 40.87 degrees. In this case, 0 is not a possible value of the independent variable.

9. The r^2 is 0.248, which means that about 25% of the variation in ratings of the police is explained by political ideology.

11. The F-statistic is 2,089.92, with a p-value less than 0.001. This means that the regression model overall fits the data better than an intercept-only model would.

13. The slope for *poc* is –9.58. This means that, on average, those who identify as people of color rate the police about 9.58 degrees lower than do whites, holding constant political ideology. The slope's standard error, 0.627, is associated with a p-value less than 0.001. This means that the slope is "statistically significant," or that the slope in the population is not equal to 0.

15. The intercept for the new model is 44.2, which is the predicted police feeling thermometer score for a white person (*poc* = 0) who scores a 0 on the ideology scale. Again, the ideology scale begins at 1, limiting the utility of the intercept.

17. **Analyze → Regression → Linear**

    Click on "Plots." This opens the "Linear Regression: Plots" dialog box. Move *zresid* into the "Y" box and *zpred* into the "X" box. Click on "Continue" and then "OK." Yes, the variance is about the same across all values of the dependent variable.

Glossary

Addition Rule: The probability of *either* outcome A *or* outcome B is the sum of the probability of A and B.

aggregate data: Data for which the unit of measurement is the group, not the individual; for example, counties, countries, or organizations.

alpha: The threshold used to determine whether to reject the null hypothesis; the probability of a Type I error.

alternative hypothesis: A hypothesis stating that the true population parameter is less than, not equal to, or greater than a specified value.

analysis of variance (ANOVA): Used to compare means among groups; most commonly used when there are three or more groups.

antecedent variable: A control variable that is temporally prior to both the independent and dependent variables and influences both.

association: A relationship between variables in which changes or differences in the independent variable are associated with changes or differences in the dependent variable.

bar graph: A graphical representation of frequencies or percentages for each category of a nominal or ordinal-level variable.

between groups sum of squares: A measure of how much variation there is between groups; used in analysis of variance.

big data: Data that emerge as a byproduct of the electronic tracking of people's behavior online and in the real world.

bivariate relationship: A relationship between two variables.

case: A single member of a data set; each individual or group under study.

Central Limit Theorem: States that if a sample size is large enough, the sampling distribution (under infinite repeated samples) will be a normal curve, centered around the true population mean or proportion, regardless of whether the actual variable is normally distributed in the population.

chi-square test for goodness-of-fit: Test of significance used with frequency distributions. Assesses whether sample results conform to expected frequencies based on a model.

chi-square test of independence: Test of significance used with cross-tabulations. Assesses whether there is a relationship between two variables by examining the differences between observed frequencies and expected frequencies.

closed-ended survey item: Survey item that provides respondents with pre-defined response categories.

cluster sample: A method of sampling where one randomly samples clusters of cases instead of individual cases and then randomly samples cases from within these clusters.

clustered bar graph: A graphical display of the relationship between two variables;

shows the distribution of the dependent variable with a clustered set of bars for each category of the independent variable.

codebook: Provides essential information about each variable in a data set.

column percentages: The percentages that break down the distribution of the variable presented in the columns of a crosstabulation; calculated by dividing the frequency in a cell by the total number of cases in the column.

Complement Rule: If the probability of outcome A plus the probability of outcome B is 100%, the probability of outcome A is equal to one minus the probability of outcome B.

concept: An abstract factor or idea, not always directly observable, that a researcher wants to study.

confidence interval: An interval estimate of a population parameter that covers a range of values.

confidence level: Under repeated sampling, the proportion of confidence intervals that would contain the population parameter.

continuous variable: A variable with values that can be continually subdivided.

control variable: A variable that we hold constant when we examine the relationship between the independent and dependent variables.

correlation coefficient: Measures the linear relationship between two interval-ratio variables; tells us the degree to which the values of one variable change as the values of the other variable change.

cross-tabulation: A table that presents the frequency distribution for the dependent variable separately for each category of the independent variable.

cumulative percentage: The percentage of cases that are equal to or lower than a particular value for a variable.

cumulative probability: The proportion of cases that are equal to or lower than a particular value for a variable.

curvilinear relationship: A relationship between two variables follows a curved line rather than a straight line.

decision t: The t-value that is the cutoff for making a decision about a null hypothesis for a variable that follows the t-distribution; the t-value that corresponds to the researcher's alpha-level at specified degrees of freedom.

degrees of freedom: Reflects the number of factors that influence the calculation of a statistic.

dependence: Two variables are dependent on each other when one variable influences the second variable.

dependent variable: A variable that is affected by another variable.

descriptive statistics: Statistical techniques for describing the patterns found in a set of data.

direction of a relationship: Used with variables that are ordinal or interval-ratio. Indicates whether values of the dependent variable increase (with a positive direction) or decrease (with a negative direction) as values of the independent variable increase.

discrete variable: A variable measured in whole numbers that cannot be broken down further.

dummy variable: A variable with two categories, coded as 0 and 1. The category coded as 0 is called the "reference category."

ecological fallacy: The error of drawing inferences about individuals based on the groups to which they belong.

Empirical Rule: States that, for any normally distributed variable, a fixed proportion of cases will fall between any given standard deviations from the mean or beyond any given standard deviation from the mean.

expected frequency: For the chi-square test of independence, the number of cases that would be in each cell if there were no relationship between the two variables (i.e., if the column percentages were identical for all rows).

experimental control: The random assignment of research participants to treatment and control groups to ensure that participants in one group are not systematically different from those in the other group.

F-statistic: In regression, used to test the statistical significance of a regression model overall by assessing whether the regression model does a better job of generating predicted values of the dependent variable than an "intercept-only model." In analysis of variance, the ratio of between groups to within groups variation; used to test whether there are statistically significant differences among groups.

frequency: The number of cases in a sample falling into each category of a variable.

frequency distribution: A table that represents the frequencies, percentages, and cumulative percentages for each category of a variable.

growth mindset: An approach that views intelligence as something that develops over time through hard work and effort.

histogram: A graphical representation of the distribution of an interval-ratio-level variable.

hypothesis: A prediction about the value of a parameter or how variables are related in a population.

hypothesis testing: A procedure that allows the researcher to reject or not reject a null hypothesis about a population parameter in light of an observed sample statistic.

independence: Two variables are independent of each other when one variable has no influence on the second variable.

independent effect: The effect of an independent variable on a dependent variable when holding additional explanatory variables constant.

independent variable: A variable that causes a change in another variable.

inferential statistics: Statistics that examine whether information from a sample can be generalized to a population.

intercept: In regression, the expected value of the dependent variable when the independent variable(s) are equal to 0. The value of the dependent variable where the regression line crosses the y axis. Represented by "a" in the equation for the regression line.

interquartile range: The difference between the 25th percentile value of a variable and its 75th percentile value.

interval-level variable: A numerical variable where the distance between each consecutive value of the variable is identical, with no true value of 0.

interval-ratio variable: A numerical or quantitative variable where the distance between each consecutive value of the variable is identical; either an interval-level or ratio-level variable.

intervening variable: A control variable that intervenes between x and y. Also called a **mediating variable**.

level of measurement: Refers to whether a variable's values are nominal, ordinal, or interval-ratio; determines what statistical techniques can be applied to variables.

linear regression: A linear model that measures the effect of an independent variable, x, on a dependent variable, y.

linear relationship: A relationship between two variables that can be represented by a straight line. In a linear relationship, every unit change in the independent variable is associated with a constant amount of change in the dependent variable.

logistic regression: A regression model that can be used with binary (dummy) dependent variables. Logistic regression predicts the probability that a dummy dependent variable (coded as 0 or 1) will be equal to 1 for given values of the independent variables.

margin of error: In a confidence interval, the amount of error above and below the point estimate of the population parameter caused by sampling variability.

mean: The average; the sum of all values of a variable divided by the number of cases.

measure of central tendency: A statistic that gives us information about typical or middle values of a variable; the mean, median, or mode.

measurement: The process of transforming concepts into observable data, or variables; also called **operationalization**.

median: The value that lies in the middle of an ordered distribution of values, such that half of the values lie below the median and half lie above it.

mediating variable: Another term for an **intervening variable**; a control variable that intervenes between the independent and dependent variables.

mixed methods: Methods that employ both qualitative and quantitative data and analysis.

mode: The value of a variable that appears most frequently in a sample or population.

moderating variable: A control variable that affects the relationship between the independent and dependent variable such that the magnitude of the relationship varies depending on the value of the moderating variable.

moderation: A relationship between variables in which the effect of the independent variable on the dependent variable varies, depending on the value of the control, or **moderating**, variable. This is also called **statistical interaction**.

Multiplication Rule: If the probabilities of outcomes A and B are independent of each other, the probability of *both* outcome A and outcome B occurring is found by multiplying the probability of outcome A by the probability of outcome B. If outcomes A and B are not independent of each other, then the probability of *both* outcome A and outcome B occurring is found by multiplying the probability of outcome A by the probability of outcome B, after adjusting the probability of outcome B for its dependence on A.

multistage cluster sampling: A form of cluster sampling in which the random selection of clusters passes through several stages before selecting a random sample of individual cases.

negative relationship: A relationship between two ordinal or interval-ratio variables in which the values of the dependent variable decrease as values of the independent variable increase.

nominal-level variable: A variable that is not numerical and whose categories cannot be rank-ordered.

nonprobability sample: A sample in which cases are self-selected or are not drawn randomly.

nonresponse bias: A form of bias occurring when individuals who are invited to take a survey vary systematically in the likelihood that they will complete the survey.

non-spurious relationship: A relationship between an independent and a dependent variable that persists even after the introduction of another explanatory variable.

normal distribution: A bell-shaped curve that can be used to find probabilities at specified values or ranges of values for variables that are normally distributed. The basis of inferential statistics.

null hypothesis: A hypothesis stating that an unknown population parameter is equal to a specified value or that a difference between the parameters of two groups is equal to 0 or a specified value.

observed frequency: The number of cases in a given cell of a cross-tabulation. Used in calculating chi-square.

one-sample test: A hypothesis test that allows the researcher to reject or not reject a null hypothesis about a population parameter in light of an observed single sample statistic.

one-tailed test: In a hypothesis test, an alternative hypothesis stating that the population parameter is above or below a specified value but not both.

open-ended survey item: A survey item that does not provide respondents with response categories.

operationalization: The process of transforming concepts into observable data, or variables; also called **measurement**.

ordinal-level variable: A variable with values that can be rank-ordered but that are not numerical and where the distance between each value of the variable is not identical.

percent change: A method of understanding the magnitude of a change in percentage over time.

percentage: A standardized version of frequency that divides the number of cases in each category of a variable by the overall number of cases and multiplies by 100.

percentile: The position of any given case relative to the overall distribution for a variable, expressed in a percentage rank.

pie chart: A graphical representation of frequencies or percentages for each category of a nominal or ordinal-level variable.

point estimate: A single value estimate (e.g., mean, proportion) of a population parameter.

population: Every individual or case in a category of interest.

population parameter: A statistic (such as a mean or proportion) for a population.

positive relationship: A relationship between two ordinal or interval-ratio variables in which the values of the dependent variable increase as values of the independent variable increase.

post hoc test: A follow-up test conducted after obtaining results from another analysis, such as **analysis of variance**. In analysis of variance, a post hoc test investigates which differences between groups are statistically significant. One post hoc test is **Tukey's test**.

power of a hypothesis test: The probability of rejecting a false null hypothesis.

precision: The width, or range, of a confidence interval.

probability sample: A sample in which every member of the population has an equal probability of being selected for the sample, because the selection of cases from the population is made randomly.

proportion: Percentage expressed as a decimal between 0 and 1.0.

qualitative method: Analysis of data that are not numerical, such as the text of documents, interviews, or field observations.

quantitative analysis: Analysis that uses statistical techniques to analyze numerical data.

r-squared: Goodness-of-fit measure for regression that indicates the proportion of variation in the dependent variable that is accounted for by the independent variable(s).

random sampling: Sampling method in which every member of a population has an equal probability of being selected for the sample.

range: A statistic calculated by subtracting the lowest value of a variable from its highest value.

rate: The frequency of an event or outcome relative to the number of times that the event or outcome could have occurred in a given group.

ratio: The size of one category relative to another.

ratio-level variable: A numerical variable with identical distances between each value and with a meaningful value of 0 that represents a true value of 0 for the variable being measured.

raw frequency: The number of cases in each category of a variable.

recoding: Combining or collapsing categories of a variable, using statistical software.

regression coefficient: The expected change in the dependent variable for a one-unit increase in the independent variable; represented by "b" in the regression equation. A positive coefficient predicts an increase in the dependent variable for a one-unit increase in the independent variable, while a negative coefficient predicts a decrease in the dependent variable for a one-unit increase in the independent variable. Also called **slope**.

regression line: The line that best fits a scatterplot; represented by the equation $\hat{y} = a + bx$, where a is the intercept, b is the slope (coefficient), and $\hat{y}$ is the predicted value of the dependent variable (y) for a given value of the independent variable (x).

relative frequency: The size of each response category relative to the overall number of cases; expressed as a percentage or proportion.

reliability: The extent to which the values of a variable are unaffected by the measurement process or instrument.

research question: A question, answerable with data, that asks how two or more variables are related.

residual: The vertical distance between each data point and the regression line.

row percentages: Used in cross-tabulation; the percentage of the group in the row that falls into each category of the other variable. Calculated by dividing the frequency in a cell by the total number of cases in the row.

sample: A group of individuals or cases drawn from the larger population of interest.

sample statistic: A statistic (such as a mean or proportion) that describes a sample.

sampling: The process of selecting cases from the population to study.

sampling distribution: A frequency distribution of a statistic (e.g., mean or proportion) obtained through a theoretical repeated sampling process.

sampling frame: A full list of the members of a population.

scale: An ordinal-level variable that asks respondents to place themselves somewhere on a continuum, such as ranging from "strongly agree" to "strongly disagree."

scatterplot: A graph showing the paired values of two interval-ratio variables for each case in a data set.

secondary data: Data that have been collected previously, usually by someone else and often for a purpose that differs from an individual researcher's.

simple random sample: A method of sampling that starts with a list of all members of a population and randomly draws a desired number of them.

slope: The expected change in the dependent variable for a one-unit increase in the independent variable; represented by "b" in the regression equation. A positive slope predicts an increase in the dependent variable for a one-unit increase in the independent variable, while a negative slope predicts a decrease in the dependent variable for a one-unit increase in the independent variable. Also called **regression coefficient**.

spurious relationship: A relationship between independent and dependent variables that disappears or weakens substantially when we introduce a control variable.

stacked bar graph: A graphical display that shows the distribution of the dependent variable with just one bar for each category of the independent variable.

standard deviation: A statistic that measures the average distance of each value of a variable from the variable's mean.

standard error: The standard deviation of the sampling distribution; the average distance between a sample statistic and the population parameter.

standard error of the estimate: Goodness-of-fit measure for regression that indicates how far, on average, the actual values of the dependent variable fall from the values predicted by the regression line.

standard error of the slope: Used to test the statistical significance of a single slope in regression; indicates how much, on average, a sample slope varies from the population slope due to sampling error.

standardized residuals: Indicate the relative contribution of each cell to the overall chi-square statistic; show the size and direction of the difference between the expected and observed frequency for each cell.

statistical control: A data analysis process to ensure that a third variable does not account for the relationship between independent and dependent variables.

statistical independence: Two variables are statistically independent of each other when one variable has no influence on the second variable.

statistical interaction: Occurs when the effect of the independent variable on the dependent variable varies, depending on the value of the control, or **moderating**, variable. Also called **moderation**.

statistical significance: A finding that a difference between a sample statistic and the hypothesized population parameter, or the difference between parameters for two groups, is unlikely to be due to sampling error; it is more likely reflective of a genuine difference between the two.

stratified random sample: A method of sampling that allows the researcher to randomly sample from subgroups in a population to ensure that the sample is representative of population subgroups that are of interest.

strength of relationship: In a cross-tabulation, refers to the sizes of the differences in percentages across categories of the independent variable. In correlation, refers to how closely two variables are related to each other as reflected in the correlation coefficient.

t-distribution: A family of bell-shaped probability distributions that can be used to find probabilities at specified values or ranges of values for variables that follow this distribution.

time series chart: A graphical representation of the change in a variable over time.

total percentage: In a cross-tabulation, the frequency in a cell divided by the total number of cases.

Tukey's post hoc test: A test used after a statistically significant result for an analysis of variance to determine which group means are significantly different from each other.

t-value: Refers to values of the t-distribution; the distance between any value of a sample statistic that follows a t-distribution and the population parameter, expressed in standard errors.

two-sample test: A hypothesis test that allows the researcher to reject or not reject a null hypothesis about the difference between two population parameters in light of an observed difference between two sample statistics.

two-tailed test: In a hypothesis test, an alternative hypothesis that states that the population parameter is either above or below a specified value.

Type I error: An error in which we reject a null hypothesis that is true.

Type II error: An error in which we fail to reject a null hypothesis that is false.

unit of analysis: The object of study, either individuals or groups.

univariate statistics: Statistics that describe one variable.

validity: The extent to which variables actually measure what they claim to measure.

variability: The level of diversity in a sample or a population, measured through range, interquartile range, variance, or standard deviation.

variable: A characteristic that can take on more than one category or value.

variance: A statistic that measures the average distance squared of each value in a variable from the variable's mean; it is the standard deviation squared.

within groups sum of squares: A measure of how much variation there is within groups; used in analysis of variance.

z-score: Refers to values of the normal distribution; the distance between any value of a normally distributed variable and its mean or between any sample statistic and the population parameter, expressed in standard deviations or standard errors.

Appendix A: Normal Table

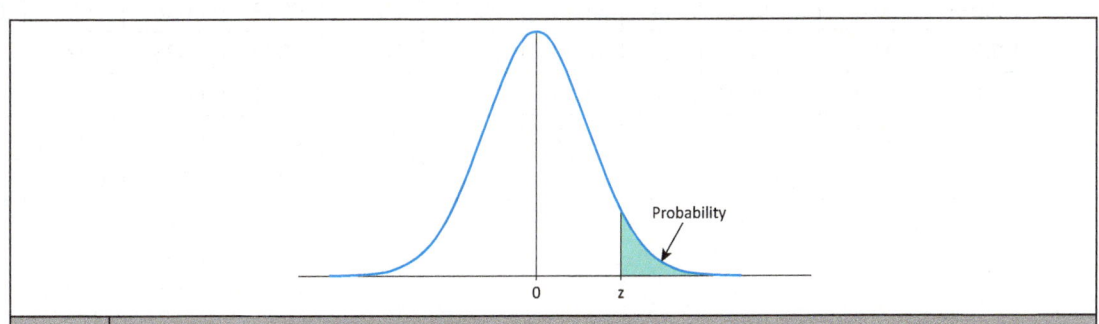

z	.00	.01	.02	.03	.04	.05	.06	.07	.08	.09
0.0	.5000	.4960	.4920	.4880	.4840	.4801	.4761	.4721	.4681	.4641
0.1	.4602	.4562	.4522	.4483	.4443	.4404	.4364	.4325	.4286	.4247
0.2	.4207	.4168	.4129	.4090	.4052	.4013	.3974	.3936	.3897	.3859
0.3	.3821	.3783	.3745	.3707	.3669	.3632	.3594	.3557	.3520	.3483
0.4	.3446	.3409	.3372	.3336	.3300	.3264	.3228	.3192	.3156	.3121
0.5	.3085	.3050	.3015	.2981	.2946	.2912	.2877	.2843	.2810	.2776
0.6	.2743	.2709	.2676	.2643	.2611	.2578	.2546	.2514	.2483	.2451
0.7	.2420	.2389	.2358	.2327	.2296	.2266	.2236	.2206	.2177	.2148
0.8	.2119	.2090	.2061	.2033	.2005	.1977	.1949	.1922	.1894	.1867
0.9	.1841	.1814	.1788	.1762	.1736	.1711	.1685	.1660	.1635	.1611
1.0	.1587	.1562	.1539	.1515	.1492	.1469	.1446	.1423	.1401	.1379
1.1	.1357	.1335	.1314	.1292	.1271	.1251	.1230	.1210	.1190	.1170
1.2	.1151	.1131	.1112	.1093	.1075	.1056	.1038	.1020	.1003	.0985
1.3	.0968	.0951	.0934	.0918	.0901	.0885	.0869	.0853	.0838	.0823
1.4	.0808	.0793	.0778	.0764	.0749	.0735	.0721	.0708	.0694	.0681
1.5	.0668	.0655	.0643	.0630	.0618	.0606	.0594	.0582	.0571	.0559
1.6	.0548	.0537	.0526	.0516	.0505	.0495	.0485	.0475	.0465	.0455
1.7	.0446	.0436	.0427	.0418	.0409	.0401	.0392	.0384	.0375	.0367
1.8	.0359	.0351	.0344	.0336	.0329	.0322	.0314	.0307	.0301	.0294
1.9	.0287	.0281	.0274	.0268	.0262	.0256	.0250	.0244	.0239	.0233

APPENDIX A Normal Table

z	Second Decimal Place of z									
	.00	.01	.02	.03	.04	.05	.06	.07	.08	.09
2.0	.0228	.0222	.0217	.0212	.0207	.0202	.0197	.0192	.0188	.0183
2.1	.0179	.0174	.0170	.0166	.0162	.0158	.0154	.0150	.0146	.0143
2.2	.0139	.0136	.0132	.0129	.0125	.0122	.0119	.0116	.0113	.0110
2.3	.0107	.0104	.0102	.0099	.0096	.0094	.0091	.0089	.0087	.0084
2.4	.0082	.0080	.0078	.0075	.0073	.0071	.0069	.0068	.0066	.0064
2.5	.0062	.0060	.0059	.0057	.0055	.0054	.0052	.0051	.0049	.0048
2.6	.0047	.0045	.0044	.0043	.0041	.0040	.0039	.0038	.0037	.0036
2.7	.0035	.0034	.0033	.0032	.0031	.0030	.0029	.0028	.0027	.0026
2.8	.0026	.0025	.0024	.0023	.0023	.0022	.0021	.0021	.0020	.0019
2.9	.0019	.0018	.0018	.0017	.0016	.0016	.0015	.0015	.0014	.0014
3.0	.0013	.0013	.0013	.0012	.0012	.0011	.0011	.0011	.0010	.0010

Appendix B: Table of t-Values

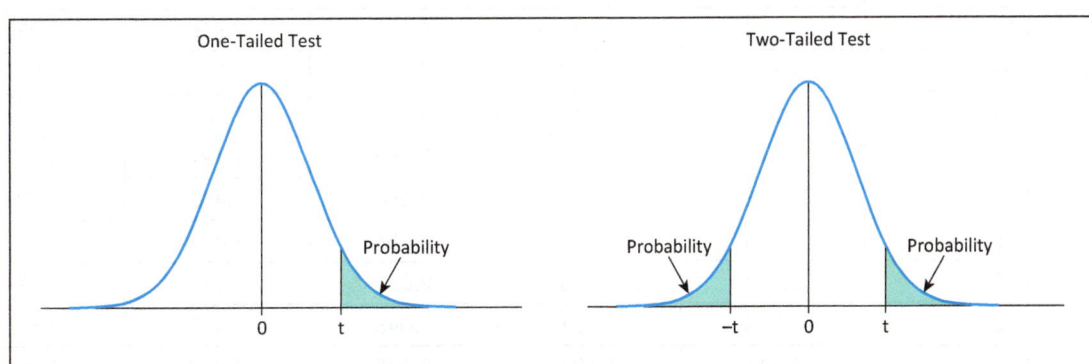

	p-Value for One-Tailed Test (p-Value for Two-Tailed Test)				
	0.1 (0.2)	0.05 (0.10)	0.025 (0.05)	0.01 (0.02)	0.005 (0.01)
	Confidence Level				
DF	80%	90%	95%	98%	99%
1	3.078	6.314	12.706	31.821	63.656
2	1.886	2.920	4.303	6.965	9.925
3	1.638	2.353	3.182	4.541	5.841
4	1.533	2.132	2.776	3.747	4.604
5	1.476	2.015	2.571	3.365	4.032
6	1.440	1.943	2.447	3.143	3.707
7	1.415	1.895	2.365	2.998	3.499
8	1.397	1.860	2.306	2.896	3.355
9	1.383	1.833	2.262	2.821	3.250
10	1.372	1.812	2.228	2.764	3.169
11	1.363	1.796	2.201	2.718	3.106
12	1.356	1.782	2.179	2.681	3.055
13	1.350	1.771	2.160	2.650	3.012
14	1.345	1.761	2.145	2.624	2.977
15	1.341	1.753	2.131	2.602	2.947

APPENDIX B Table of t-Values

	p-Value for One-Tailed Test (p-Value for Two-Tailed Test)				
	0.1 (0.2)	0.05 (0.10)	0.025 (0.05)	0.01 (0.02)	0.005 (0.01)
	Confidence Level				
16	1.337	1.746	2.120	2.583	2.921
17	1.333	1.740	2.110	2.567	2.898
18	1.330	1.734	2.101	2.552	2.878
19	1.328	1.729	2.093	2.539	2.861
20	1.325	1.725	2.086	2.528	2.845
21	1.323	1.721	2.080	2.518	2.831
22	1.321	1.717	2.074	2.508	2.819
23	1.319	1.714	2.069	2.500	2.807
24	1.318	1.711	2.064	2.492	2.797
25	1.316	1.708	2.060	2.485	2.787
26	1.315	1.706	2.056	2.479	2.779
27	1.314	1.703	2.052	2.473	2.771
28	1.313	1.701	2.048	2.467	2.763
29	1.311	1.699	2.045	2.462	2.756
30	1.310	1.697	2.042	2.457	2.750
40	1.303	1.684	2.021	2.423	2.704
50	1.299	1.676	2.009	2.403	2.678
60	1.296	1.671	2.000	2.390	2.660
80	1.292	1.664	1.990	2.374	2.639
100	1.290	1.660	1.984	2.364	2.626
120	1.289	1.658	1.980	2.358	2.617
∞	1.282	1.645	1.960	2.326	2.576

Source: R. A. Fisher and F. Yates. 1949. "Statistical Tables for Biological, Agricultural and Medical Research." *Statistical Tables for Biological, Agricultural and Medical Research* (third edition). Table III, 46.

Appendix C: F Table, for Alpha = 0.05

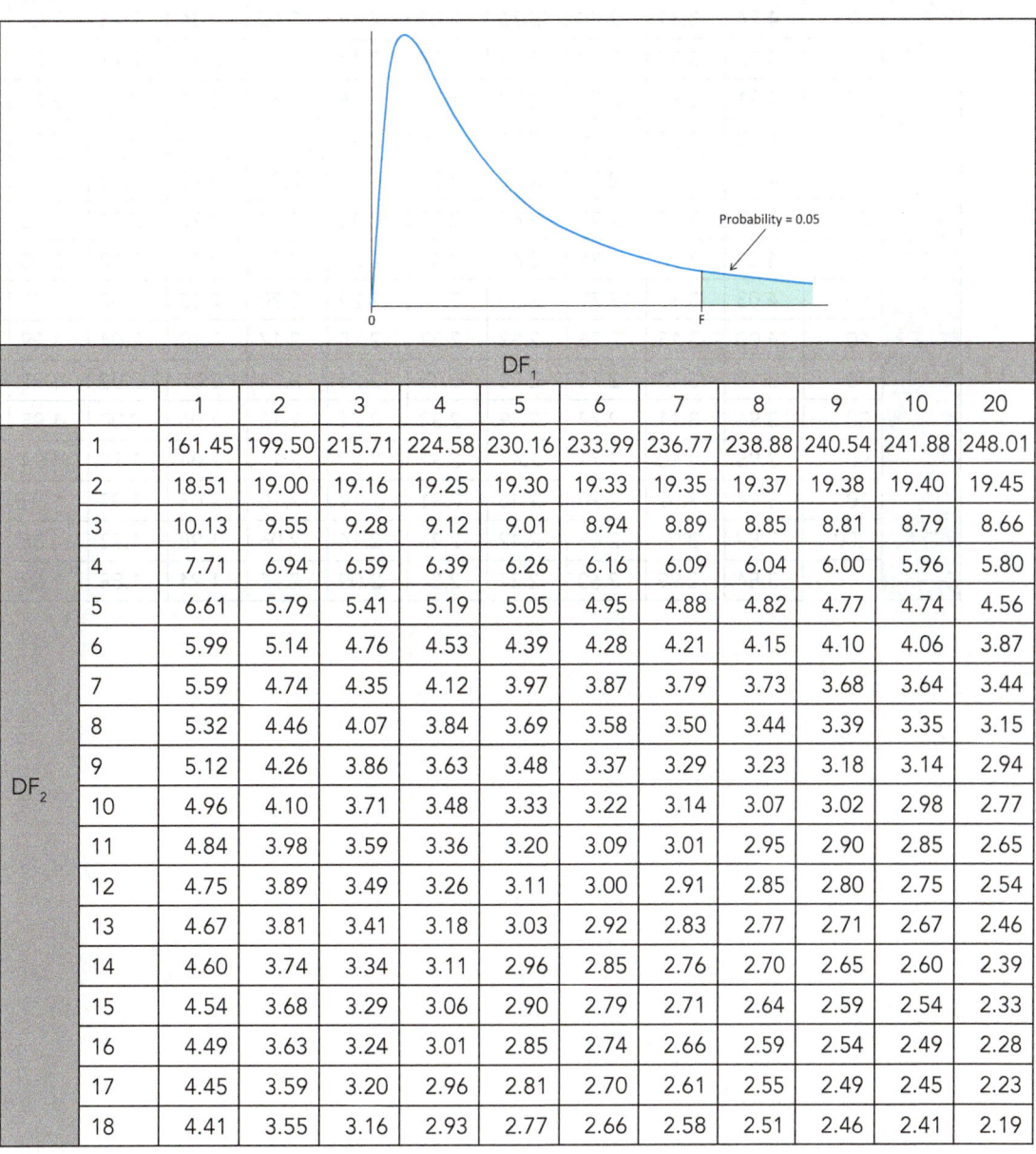

		DF_1										
		1	2	3	4	5	6	7	8	9	10	20
DF_2	1	161.45	199.50	215.71	224.58	230.16	233.99	236.77	238.88	240.54	241.88	248.01
	2	18.51	19.00	19.16	19.25	19.30	19.33	19.35	19.37	19.38	19.40	19.45
	3	10.13	9.55	9.28	9.12	9.01	8.94	8.89	8.85	8.81	8.79	8.66
	4	7.71	6.94	6.59	6.39	6.26	6.16	6.09	6.04	6.00	5.96	5.80
	5	6.61	5.79	5.41	5.19	5.05	4.95	4.88	4.82	4.77	4.74	4.56
	6	5.99	5.14	4.76	4.53	4.39	4.28	4.21	4.15	4.10	4.06	3.87
	7	5.59	4.74	4.35	4.12	3.97	3.87	3.79	3.73	3.68	3.64	3.44
	8	5.32	4.46	4.07	3.84	3.69	3.58	3.50	3.44	3.39	3.35	3.15
	9	5.12	4.26	3.86	3.63	3.48	3.37	3.29	3.23	3.18	3.14	2.94
	10	4.96	4.10	3.71	3.48	3.33	3.22	3.14	3.07	3.02	2.98	2.77
	11	4.84	3.98	3.59	3.36	3.20	3.09	3.01	2.95	2.90	2.85	2.65
	12	4.75	3.89	3.49	3.26	3.11	3.00	2.91	2.85	2.80	2.75	2.54
	13	4.67	3.81	3.41	3.18	3.03	2.92	2.83	2.77	2.71	2.67	2.46
	14	4.60	3.74	3.34	3.11	2.96	2.85	2.76	2.70	2.65	2.60	2.39
	15	4.54	3.68	3.29	3.06	2.90	2.79	2.71	2.64	2.59	2.54	2.33
	16	4.49	3.63	3.24	3.01	2.85	2.74	2.66	2.59	2.54	2.49	2.28
	17	4.45	3.59	3.20	2.96	2.81	2.70	2.61	2.55	2.49	2.45	2.23
	18	4.41	3.55	3.16	2.93	2.77	2.66	2.58	2.51	2.46	2.41	2.19

APPENDIX C F Table, for Alpha = 0.05

		\multicolumn{11}{c}{DF$_1$}
		1	2	3	4	5	6	7	8	9	10	20
DF$_2$	19	4.38	3.52	3.13	2.90	2.74	2.63	2.54	2.48	2.42	2.38	2.16
	20	4.35	3.49	3.10	2.87	2.71	2.60	2.51	2.45	2.39	2.35	2.12
	21	4.32	3.47	3.07	2.84	2.68	2.57	2.49	2.42	2.37	2.32	2.10
	22	4.30	3.44	3.05	2.82	2.66	2.55	2.46	2.40	2.34	2.30	2.07
	23	4.28	3.42	3.03	2.80	2.64	2.53	2.44	2.37	2.32	2.27	2.05
	24	4.26	3.40	3.01	2.78	2.62	2.51	2.42	2.36	2.30	2.25	2.03
	25	4.24	3.39	2.99	2.76	2.60	2.49	2.40	2.34	2.28	2.24	2.01
	26	4.23	3.37	2.98	2.74	2.59	2.47	2.39	2.32	2.27	2.22	1.99
	27	4.21	3.35	2.96	2.73	2.57	2.46	2.37	2.31	2.25	2.20	1.97
	28	4.20	3.34	2.95	2.71	2.56	2.45	2.36	2.29	2.24	2.19	1.96
	29	4.18	3.33	2.93	2.70	2.55	2.43	2.35	2.28	2.22	2.18	1.94
	30	4.17	3.32	2.92	2.69	2.53	2.42	2.33	2.27	2.21	2.16	1.93
	40	4.08	3.23	2.84	2.61	2.45	2.34	2.25	2.18	2.12	2.08	1.84
	50	4.03	3.18	2.79	2.56	2.40	2.29	2.20	2.13	2.07	2.03	1.78
	60	4.00	3.15	2.76	2.53	2.37	2.25	2.17	2.10	2.04	1.99	1.75
	70	3.98	3.13	2.74	2.50	2.35	2.23	2.14	2.07	2.02	1.97	1.72
	80	3.96	3.11	2.72	2.49	2.33	2.21	2.13	2.06	2.00	1.95	1.70
	90	3.95	3.10	2.71	2.47	2.32	2.20	2.11	2.04	1.99	1.94	1.69
	100	3.94	3.09	2.70	2.46	2.31	2.19	2.10	2.03	1.97	1.93	1.68
	200	3.89	3.04	2.65	2.42	2.26	2.14	2.06	1.98	1.93	1.88	1.62
	∞	3.84	2.99	2.60	2.37	2.21	2.09	2.01	1.94	1.88	1.83	1.57

Appendix D: Chi-Square Table

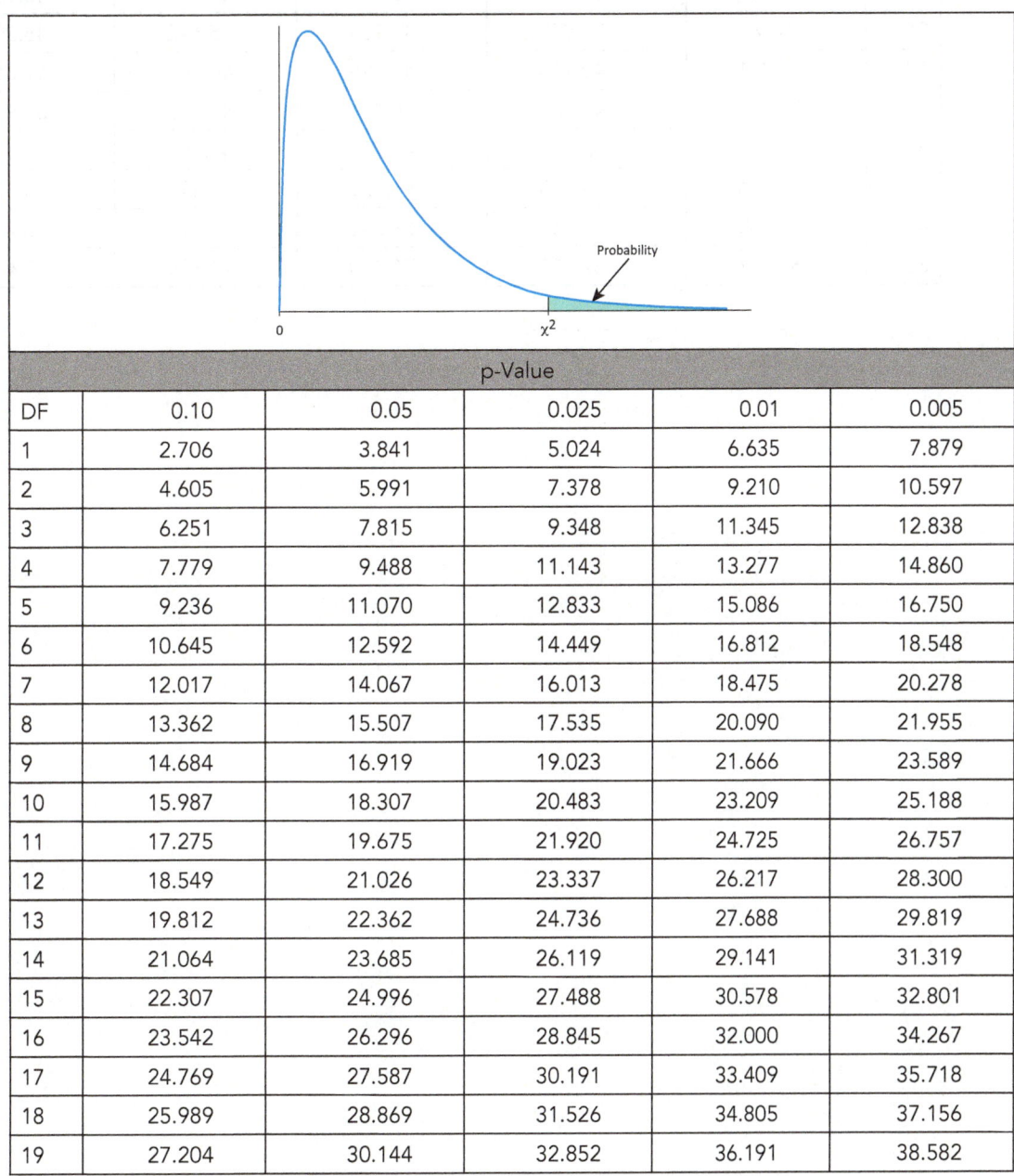

DF	p-Value				
	0.10	0.05	0.025	0.01	0.005
1	2.706	3.841	5.024	6.635	7.879
2	4.605	5.991	7.378	9.210	10.597
3	6.251	7.815	9.348	11.345	12.838
4	7.779	9.488	11.143	13.277	14.860
5	9.236	11.070	12.833	15.086	16.750
6	10.645	12.592	14.449	16.812	18.548
7	12.017	14.067	16.013	18.475	20.278
8	13.362	15.507	17.535	20.090	21.955
9	14.684	16.919	19.023	21.666	23.589
10	15.987	18.307	20.483	23.209	25.188
11	17.275	19.675	21.920	24.725	26.757
12	18.549	21.026	23.337	26.217	28.300
13	19.812	22.362	24.736	27.688	29.819
14	21.064	23.685	26.119	29.141	31.319
15	22.307	24.996	27.488	30.578	32.801
16	23.542	26.296	28.845	32.000	34.267
17	24.769	27.587	30.191	33.409	35.718
18	25.989	28.869	31.526	34.805	37.156
19	27.204	30.144	32.852	36.191	38.582

APPENDIX D Chi-Square Table

	p-Value				
DF	0.10	0.05	0.025	0.01	0.005
20	28.412	31.410	34.170	37.566	39.997
21	29.615	32.671	35.479	38.932	41.401
22	30.813	33.924	36.781	40.289	42.796
23	32.007	35.172	38.076	41.638	44.181
24	33.196	36.415	39.364	42.980	45.559
25	34.382	37.652	40.646	44.314	46.928
26	35.563	38.885	41.923	45.642	48.290
27	36.741	40.113	43.195	46.963	49.645
28	37.916	41.337	44.461	48.278	50.993
29	39.087	42.557	45.722	49.588	52.336
30	40.256	43.773	46.979	50.892	53.672
40	51.805	55.758	59.342	63.691	66.766
50	63.167	67.505	71.420	76.154	79.490

Appendix E: Selected List of Formulas

- ## Chapter 3: Cross-Tabulations

  - $\text{col\%} = \dfrac{f}{\text{col total}}(100)$

  - $\text{row\%} = \dfrac{f}{\text{row total}}(100)$

  - $\text{tot\%} = \dfrac{f}{\text{total } N}(100)$

- ## Chapter 4: Measures of Central Tendency

  - $\bar{y} = \dfrac{\Sigma y}{N}$

  - $\bar{y} = \dfrac{\Sigma fy}{N}$ (for a frequency distribution)

  - Median case number $= \dfrac{N+1}{2}$

- ## Chapter 5: Measures of Variability

  - $s = \sqrt{\dfrac{\Sigma(y-\bar{y})^2}{N-1}}$

  - Range = highest value − lowest value
  - IQR = value at 75th percentile − value at 25th percentile

- ## Chapter 6: Probability and the Normal Distribution

  - $z = \dfrac{y - \mu}{\sigma}$ (for distance between an individual score and population mean)

Chapter 7: Sampling Distributions

- $z = \dfrac{\bar{y} - \mu}{SE}$, where $SE_{\bar{y}} = \dfrac{\sigma}{\sqrt{N}}$ (for distance between a sample mean and population mean)

- $z = \dfrac{p - \pi}{SE}$, where $SE_p = \sqrt{\dfrac{\pi(1-\pi)}{N}}$ (for distance between a sample proportion and population proportion)

Chapters 8–10: Inference for Means and Proportions

Reader should turn to page 715 for formulas.

Chapter 11: Analysis of Variance

- $F = \dfrac{MS_{between}}{MS_{within}}$

- $MS_{within} = \dfrac{SS_{within}}{DF_{within}}$, where $DF_{within} = N_{total} - k$, and k is the number of groups

- $MS_{between} = \dfrac{SS_{between}}{DF_{between}}$, where $DF_{between} = k - 1$, and k is the number of groups

Chapter 12: The Chi-Square Test

- $F_e = \left(\dfrac{\text{row frequency}}{\text{total frequency}}\right)(\text{column frequency})$

- $\chi^2 = \sum \dfrac{(F_o - F_e)^2}{F_e}$ where $DF = (\text{rows} - 1)(\text{columns} - 1)$

Chapter 14: Correlation and Regression

- $r_{(x,y)} = \dfrac{COV_{(x,y)}}{s_x s_y}$, where $COV_{(x,y)} = \dfrac{\Sigma(x-\bar{x})(y-\bar{y})}{N-1}$ and s_x and s_y equal the standard deviations of x and y, respectively

- $b = \dfrac{\text{Sum of cross products}}{\text{Sum of squares}_x} = \dfrac{\Sigma(x-\bar{x})(y-\bar{y})}{\Sigma(x-\bar{x})^2}$

- $a = \bar{y} - b(\bar{x})$, where $\bar{y}$ is the mean of y, $\bar{x}$ is the mean of x, and b is the slope

- $r^2 = \dfrac{\text{explained sum of squares}}{\text{total sum of squares}_x} = \dfrac{\Sigma(\hat{y}-\bar{y})^2}{\Sigma(y-\bar{y})^2}$

APPENDIX E Selected List of Formulas

Inference for Means and Proportions

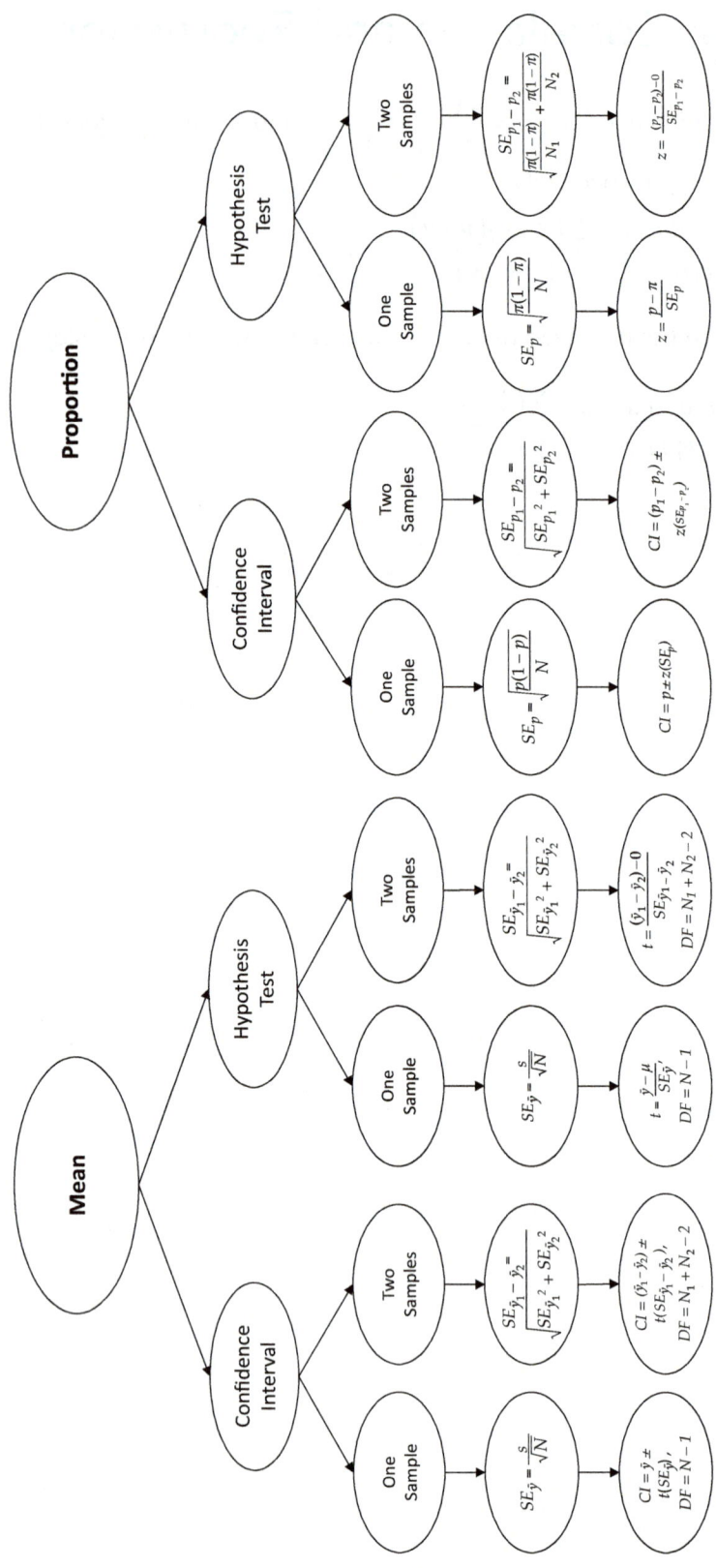

Appendix F: Choosing Tests for Bivariate Relationships

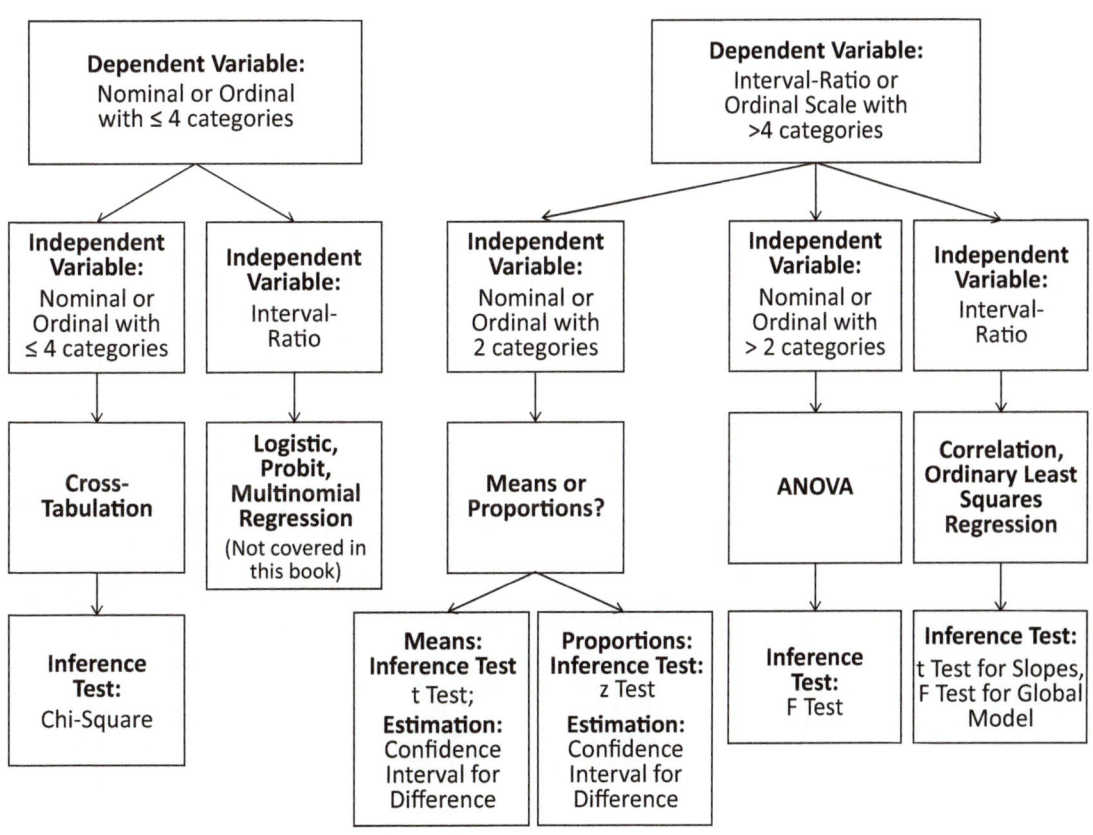

Index

abortion rights: gender and opinions on, 117; generational differences on, 117–20, *118*; race and opinions on, 117, 119, *119*

ACT scores, SAT scores compared to, 208

Adamczyk, Amy, 308

Addition Rule: definition of, 696; examples of, 237–38, 245; form of, 237; z-scores and, 252–53, *253*

address lists, 13

age. *See* generational differences

age2 variable, 540–42

agekdbrn variable problems: for SPSS, 52–53; for Stata, 50–52

age of first childbearing, 198, 429–30, *430*, 437, 437–39, *438*, *439*, 439–40

aggregate data: definition of, 696; of expected and observed frequencies, 467; residual and, 558; units of, 56

aggregate level, definition, 6

alpha level: ANOVA and, 433–34, *434*, 438; chi-square test for goodness-of-fit and, 480; chi-square test of independence and, 466; definition of, 696; F-statistic and, 566–67; in hypothesis testing, 363, 368; standard error of slope and, 569; in two-sample tests, 398, *398*, 405–6, *406*

alternative hypothesis: ANOVA and, 433, 437–38; chi-square test for goodness-of-fit and, *479*, 479–80; chi-square test of independence and, 465–66; definition of, 353, 696; F-statistic and, 566, 567; null hypothesis compared to, 353; null hypothesis rejection and acceptance of, 358; in one-sample tests, 353–55; in one-tailed and two-tailed tests, 354; setting up, 363, 368; standard error of slope and, 569; in two-sample tests, 393–95, 398, 405; in two-tailed tests, 354

American National Election Study. *See* ANES

analysis of variance (ANOVA): alpha level and, 433–34, *434*, 438; assumptions of, 432; between groups sum of squares and, 434–36, *435*; decision f and, 433–34, *434*, 438; definition of, 696; degrees of freedom and, 433–34, *434*, 438; F-statistic and, 430, 434–36, *435*, 438, *438*; hypothesis testing using, *431*, 431–32; independent variables and, 428; mean squares and, 436; null and alternative hypothesis and, 433, 437–38; as omnibus test, 429; post hoc test and, 429, *439*, 439–40; in psychology and social psychology, 428–29; repeated measures of, 432; sample f compared to decision f in, 436–39; SPSS for, 444–47, *445*, *446*; Stata for, 442–44, *443*; steps of, 432–39; t-tests compared to, 430, 440; Tukey's post hoc test and, 439–40, *441*; variation within and between groups and, 429–30, *430*; within groups sum of squares and, 435–36, *436*

"Analyze an Existing Variable" command: SPSS, 41, *42*, *43*; Stata, 28, *28*

ANES (American National Election Study): on Biden feelings, 231–32; on CDC feelings, 338–42; on death penalty support by gender, *460*, 460–61; on democracy satisfaction in United States by gender, *461*, 462; on government spending on health care by gender, *460*, 461; on health insurance attitudes, 244–45, *245*; on Internet news use, 174–84; on party identification, religiousness and race, 516–17, *517*; on party identification and political ideology, *120*, 120–21, *121*; on political ideology by gender, 509–12, *509–12*; political ideology scale of, 66–68, *67*, *68*, 474–76, *474–76*, 596–98; on rural feelings and political parties, 456–58; on Russia as threat to United States opinions by gender, 122, *122*; on social class and happiness, 516, *516*; standard deviation in context of data from, 205, *205*; on Trump feelings, 71, *72*, 231–32; units in, 14

ANOVA. *See* analysis of variance

antecedent variables: control variables as, 507, *508*; definition of, 696; spurious relationships and, *511*, 511–12

anxiety, COVID-19-related, 392–93, *393*, 410

"apples and oranges" comparisons, standard deviation for, 206, 208–11, *208–11*

app ratings, 367–69, *368*

679

association: bivariate, 135; causation and, 501–2; definition of, 696; strength of, 465. *See also* non-spurious relationships; spurious relationships
auto inspections, traffic deaths and, 500–503, *501–3*

bar graphs: clustered, 135–37, *136*, 702; definition of, 81, 696; effective uses of, 70, *70*; frequency distributions compared to, *75*, 75–76, *76*; histograms compared to, 73; pie charts compared to, 76, *77*; in SPSS, 97, *98*; stacked, 135–37, *136*, 702; in Stata, 87, *87*
Base Reference Manual, Stata, 31
B_COUNTRY variable: in SPSS, 293, *293*, 390; in Stata, 287, *287*, 389–90
"best-fitting" line, regression line as, 551–53, *552*
between groups sum of squares: ANOVA and, 434–36, *435*; calculation of, 434; definition of, 696
bias: nonresponse, 13, 699; publication, 359–60; sampling frames and, 16
Biden, Joe, 6, 11; feeling thermometer ratings on, 231–32; frequency of votes for, 57, *57*, 60, *60*; gender and support for, 459–60; NRA evaluations of voters for, *201*, 201–2, *202*
big business, standard deviation in opinion on, *205*, 206, 207
big data: cell phones and, 16; definition of, 15, 696; ethical and privacy issues with, 16–17; sampling frames and, 16; from social media, 16; traditional survey methods compared to, 15–16
bimodal distribution, measures of central tendency and, *170*, 170–71
binary dependent variables, logistic regression and, 561–62, *562*
bivariate relationships: definition of, 696; visual representations of, 135–37, *136*. *See also* cross-tabulations
Black Lives Matter: generational attitudes on, 55; on police contact and time, 377, 380; standard deviation in opinion on, *205*, 205–6, 207
Bonds, Barry, 208, 209, *209*
Boston Public Schools, 65
box plots: interquartile range and, 201–2, *202*; in SPSS, 221, *221*; in Stata, 215–16, *216*
Bureau of Justice Statistics, 74, 377, 380
Bureau of Labor Statistics, 270

case, definition of, 696
categorical variables, 10
causal relationships: control variables and, 502–3; criteria for, 505–7; explanations of, 500–502; between independent variables and dependent variables, 505–6, *506*; spurious relationships and, 502–3. *See also* non-spurious relationships; spurious relationships
CDC. *See* Centers for Disease Control and Prevention
cell phones, big data and, 16
Centers for Disease Control and Prevention (CDC): feeling thermometer ratings on, 338–42; gender and party identification in views of, 521–23, *522*, *523*, 526–29, *527–29*
Central Limit Theorem: definition of, 275, 696; hypothesis testing and, 351–52; sampling distributions and, 275–76, *276*; sampling distributions of the means/proportions and, 284
central tendency. *See* measures of central tendency
certainty: in chi-square test of independence, 464; confidence intervals and, 311–12, 331; definition of, 696
Chetty, Raj, 543
childbearing, age of first, 198, 429–30, *430*, 437, 437–39, *438*, *439*, 439–40
childhood poverty rate, in United States, 327–29, *327–29*, 331–32, *332*
children, number of, *58–60*, 58–62, *62*
China, science views in, 389–90
chi-square statistic: calculation of, 468–69, *469*; chi-square test for goodness-of-fit and, *480*, 481; in controlled cross-tabulation, *516*, 516–17, *517*; hypothesis testing with, *462*, 462–65, *463*; p-value and, 460–61, 470–71, *471*; standardized residuals and, *471*, 471–72
chi-square test for goodness-of-fit: alpha level and degrees of freedom in, 480; chi-square statistic and, *480*, 481; definition of, 696; expected frequencies and, *480*, 480–81; null and alternative hypotheses in, *479*, 479–80
chi-square test of independence: alpha level and, 466; alternative hypothesis and, 465–66; assumptions of, 477; certainty in, 464; chi-square statistic calculation and, 468–69, *469*; decision chi-square compared to sample chi-square in, *469*, 469–70, *470*; decision chi-square value in, 466; definition of, 696; degrees of freedom and, 466; expected frequencies and, 466–67; logic of, 464–65; for military service and political ideology relationship, 474–75, *474–75*; null hypothesis and, 464, 465–66; sample size and, 477–79, *478*; sampling distribution and, 464; SPSS and, 485–88, *486–88*; standardized residuals and, *471*, 471–72; Stata and, 483–85, *484*; statistical independence and, 463; statistical significance and, 477–79, *478*; steps of, 465–71; two-sample tests' relationship with, *476*, 476–77

choice over lives, control and, 214–22
class. *See* social class
closed-ended survey items, 5, 696
clustered bar graphs: cross-tabulations compared to, 137; definition of, 697; examples of, 135–37, *136*
cluster sampling, 13, 696–97
codebooks: definition of, 14, 697; variable information in, 14–15, *15*
coefficient of determination, 557. *See also* r-squared
collapsing/combining categories: in cross-tabulations, 130; frequency distribution tables and, 67–69, *68*, *69*; in statistical software programs, 67
college graduates' mean income, 362–66, *363*, *364*, *366*
column percentages: calculating, *123–25*, 123–26; cross-tabulations and, 121–26; definition of, 122, 697; examples calculations of, 128–29, *129*; independent and dependent variables' relationships and, 125–26; p-value and, 462; raw frequencies and, 123–25, *124*, *125*; row percentages compared to, 127, *128*; total percentages compared to, 128
combining/collapsing categories, frequency distribution tables and, 67–69, *68*, *69*
command-driven use: SPSS, 31; Stata, 23
Command Window, Stata, 24, 30
communist's book acceptance, by gender, *476*, 476–77
Complement Rule, 697; calculating, 238–40; z-scores and, 252, *252*
compressed axis, 78, *79*
concepts: definition of, 4, 697; measurement of, 4–5, *5*; multiple dimensions of, 4
confgov variable, 442–47
confidence intervals: assumptions for, 335; certainty and precision of, 311–12, 331; confidence levels and range of, 330–31; definition of, 308, 309, 697; estimation and, 311; examples for means and, 322–26, *324*, *326*; examples for proportions and, *316–18*, 316–19; hypothesis testing for means and, 365–66, 369; interpreting, 331–32; interval-ratio variables and, 335; lower and upper bounds of, 314–15, 317, *318*, 323, 325–26; margins of error and, 308–9, 311; for mean, 319–26; misinterpretation of, 331; point estimates and, 311, 317, 319, 324; population parameters and, 312, 331, 332; presidential election polls and, 319; probability and, 312; for proportions, 312–19; sample size and range of, 326–29, *327–29*; sample size determination and, 332–35;

sample statistics and, 331; SPSS for means and, 340, *341*; SPSS for proportions and, 340–42, *341*; Stata for means and, *338*, 338–39; Stata for proportions and, 339, *339*; t-distribution and, 319–22, *321*; t-values and, 321, *322*; for two-sample tests and differences between means, 403–4; for two-sample tests and differences between proportions, 408–10, *409*; two-tailed tests compared to, 361
confidence levels: confidence interval range and, 330–31; definition of, 313, 697; degrees of freedom and, 321, 323, 325; ethics and, 331; example selection of, 316, 323, 325; selection of, 313; t-value and, 321, 323, 325; z-scores associated with, 313, *314*, 315, *315*, 316, *318*
conftotal variable, 575–88
conservatives, standard deviation in opinion on, *205*, 206, *207*
continuous variables: definition of, 9, 697; histograms and, 73, 212, 245
control and choice over lives, 214–22
control variables: as antecedent variables, 507, *508*; causal relationships and, 502–3; chi-square statistic in cross-tabulations and, *516*, 516–17, *517*; definition of, 697; as intervening variables, 508, *508*; moderating variables and, 514; multiple regression and, 563–64; non-spurious relationships and, 514, *515*; SPSS effect of independent variable on interval-ratio variable, controlling for third variable, 526–29, *527–29*; SPSS for controlling for third variable in cross-tabulations, 524–26, *525*, *526*; spurious relationships and, 507–8, *508*; Stata controlling for third variables in cross tabulations, 519–20, *520*, *521*; Stata effect of independent variable on interval-ratio variable, controlling for third variable, 521–23, *522*, *523*; statistical interaction and, 514–16, *515*
correlation coefficients: absolute value of, 545, *545*; calculating, 545–47; covariance and, 646; definition of, 544, 697; for direction and strength of relationships, 544–45; linear relationships and, 545, 549; negative relationships and, 545; nonsensical relationships and, 550; positive relationships and, *548*, 549; scatterplots and, *547*, 547–49, *548*; size of, 545; standard deviation and, 546–47; sum of cross products and, 546
corrtotal variable, 575–88
corruption perceptions, institutional trust, immigrant status and, 575–88
covariance, correlation coefficients and, 646

COVID-19: childhood poverty rate during, 327; death toll from, 6; gender and anxiety, stress, insomnia from, 392–93, *393*, 410; vaccine trials, 350–52, *351*, 373–74, *374*

"Create a New Variable" command: SPSS, 34–36, *35–37*; Stata, 26–27

cross-tabulations, xi; chi-square statistic in controlled, *516*, 516–17, *517*; clustered and stacked bar graphs compared to, 137; collapsing categories in, 130; column percentages and, 121–26; definition of, 118, 697; direction of relationships and, 132–34, *133*, *134*; expected frequency calculations for, 466–67; explaining story of, 134–35; independent and dependent variables relationships and, 119–22, *120–22*; measures of central tendency compared to, 154; raw frequencies in, 123–25, *124*, *125*; relationships between variables and, 118–30; row percentages and, 126–27; SPSS for, 141–44, *142–44*, *152*; SPSS for controlling for third variable in, 524–26, *525*, *526*; Stata controlling for third variables in, 519–20, *520*, *521*; Stata for, 139–41, *140*, *141*, 150–51, *151*; statistical significance and, *460*, 460–62, *461*; strength of relationships in, 130–32, *131*, *132*; total percentages and, 127–28. *See also* chi-square statistic

cultural changes, generational differences and, 55–56

cumulative percentage: calculating, *61*, 61–63, *62*; definition of, 60–61, 80, 697; frequency distributions and, 61–62, *62*; interquartile range and, 201; percentile and, 63

cumulative probability: definition of, 253, 697; z-scores and, 253–55, *254*, *255*

Current Population Survey, 64–65

curvilinear relationships: definition of, 697; scatterplots of, 549, *550*

daily total steps, 156–58, *156–58*

Data Editor: SPSS, 32, *32*; Stata, 24–25, *25*

data files, Stata, 29

data gathering, in social sciences, 4

Data View, SPSS, 32, 32–33

death penalty support, by gender, *460*, 460–61

decision chi-square: look up value of, 466; sample chi-square compared to, *469*, 469–70, *470*

decision f: ANOVA and, 433–34, *434*, 438; sample f compared to, 436–39

decision t: definition of, 697; in one-sample tests, 365, *366*, 368–69; in two-sample tests, *398*, 398–401

degrees of freedom: ANOVA and, 433–34, *434*, 438; chi-square test for goodness-of-fit and, 480; chi-square test of independence and, 466; confidence levels and, 321, 323, 325; definition of, 697; in hypothesis testing, 364, 368; in t-distribution, 320–21, *321*; in two-sample test, *398*, 398–99

democracy in United States, gender and satisfaction with, *461*, 462

democratic governance opinions, 306–7

demographics, rate calculations for, 64

denominator, for rate calculation, 64–65

dependence: definition of, 697; probability and, 240

dependent variables: causal relationships between independent variables and, 505–6, *506*; column percentages and relationship of independent variables with, 125–26; definition of, 10–11, 697; independent variables' relationships to, 119–22, *120–22*; intercept and, 554; logistic regression and binary dependent, 561–62, *562*; regression line showing predicted value of, 551; row percentages and relationship of independent variables with, 127; slope and, 554

descriptive statistics, xi; definition of, 12, 697; for "getting to know" data, 80–81

deviation, definition of, 203. *See also* standard deviation

dice roll outcomes, probability in, *235*, 235–36

dichotomous (dummy) variables: categories of, 558–59; definition of, 697; intercept and, 559; logistic regression and, 561; scatterplots and, 560, *560*; slope and, 559

direction of relationships: clear language and examples for explaining, 135; correlation coefficients for, 544–45; cross-tabulations and, 132–34, *133*, *134*; definition of, 697; in scatterplots, 547, *547*; types of, 132–33

directories, 13

discrete variables, definition of, 9, 697

diversity of values, variability and, 197

divorce law opinions, marital status and, 128–29, *129*

divorce rates, 66

Dobbs v. Jackson Women's Health Organization, 117

do-files, Stata, 29, 29–30, *30*

drug effectiveness, opioid addiction relapse and: confidence intervals for, 316–18, *318*; two-sample tests for differences between proportions and, 404–10, *406*, *409*

dummy (dichotomous) variables: categories of, 558–59; definition of, 697; intercept and, 559; logistic regression and, 561; scatterplots and, 560, *560*; slope and, 559

Dweck, Carol, 17

ecological fallacy, definition of, 6, 697
economic security, voting related to, 425–27
edlevel variable, 483–88, *484*, *486–88*
education: government spending on, 540–42; political ideology and, 483–88, *484*, *486–88*; presidential elections influenced by age and, *515*, 515–16; television viewing and, 133–34, *134*, 139–44
elections. *See* presidential elections
Emerson, Ralph Waldo, 158
Empirical Rule: definition of, 247, 697–98; normal distribution and, 247–49, *248*; sampling distributions and, 312
employment status: levels of measurement of, 8; ordinal-level variables in, 9
endsmeet2 variable problems: in SPSS, 426–27; in Stata, 425–26
error messages, Stata, 28–29
estimation, sampling distributions and, 311
ethics: big data and, 16–17; confidence levels and, 331
exam scores, teaching methods and, 432–37, *433–35*
excitement of life, *57*, 57–58, *60*, 60–61, *61*
expected frequencies: chi-square test for goodness-of-fit and, *480*, 480–81; chi-square test of independence and calculation of, 466–67; definition of, 698; difference between observed frequency and, 467, *468*
experimental control, definition of, 11, 698
explanatory variables, non-spurious relationships and, 512, *513*, 515–16
extreme weather events, neighborhood impact of, *316*, 316–17

Facebook, 16
face-to-face surveys, 13
false null hypothesis, 373, 700
false positives/false negatives, in hypothesis testing, 371
Family Strong program, 322–23, 330–31
Felson, Jacob, 308, 311
feminists, feelings towards liberals and, *548*, 548–49, 551–52, *552*, 554–55, *555*, *556*
first childbearing, age of, 198, 429–30, *430*, *437*, 437–39, *438*, *439*, 439–40
flu vaccines, miscarriage and spurious relationship with, 506
food insecurity, measurement of, 6–7
foreign aid opinions, nativity status and, 130–32, *131*, *132*
fraction, probability as, 235
freeminds, political ideology and, 498–99

freemind variable problems, 498–99
free throw shooting, probability in, 236
frequency, definition of, 80, 698
frequency distributions: bar graphs compared to, *75*, 75–76, *76*; cumulative percentage and, 61–62, *62*; definition of, 56, 81, 698; interquartile range in, 199–201, *200*; mean in, 166–67, *167*; measures of central tendency compared to, 154; median in, 165–66; mode in, 165, *165*; nominal-level variables and, 56–57, *57*; ordinal-level variables and, 57–58, *58*; percentages and, *59*, 59–60, *60*; percent change and, 63; raw frequencies in, 56; relative size assessments and, 59; in SPSS, 90–92, *90–92*; standard deviation comparing, *205*, 205–6; in Stata, *82*, 82–83, *83*. *See also* skewed distributions
frequency distribution tables: combining/collapsing categories and, 67–69, *68*, *69*; elements of, 66; interval-ratio variables and, 71–72, *72*, *73*; missing values and, 66–67, *67*
frequency polygon, definition of, 698
F-statistic: alpha level and, 566–67; ANOVA and, 430, 434–36, *435*, 438, *438*; calculating, 566; definition of, 698; hypothesis testing and, 566–67; null and alternative hypothesis and, 566, *567*; p-value and, 567; regression lines and, 566–68; standard error of the slope and, 565

Gallup polls, 308–9
Gates, Bill, 199
Gatsby Curve, 543–44, *544*
gay marriage support: by age, 76–77, *77*, *78*; by political party and education level, 75–76, *75–77*. *See also* LGBTQ people
gender: abortion rights opinions and, 117; Biden support by, 459–60; CDC views by, 521–23, *522*, *523*, 526–29, *527–29*; communist's book acceptance opinions. by, *476*, 476–77; COVID-19-related anxiety, stress, and insomnia by, 410; death penalty support by, *460*, 460–61; democracy in United States satisfaction by, *461*, 462; government spending on health care opinions by, *460*, 461; health perceptions and, 472–73, *473*; job interviews, GPA and, 2–3; LGBTQ people, perceptions by region and, *513*, 513–14, *514*; party identification by, *478*, 478–79, 509–10, *510*, *511*; political ideology by, 509–12, *509–12*; politics and gap in, 459–62, *460*, *461*; religiousness and, *465*, 465–72, *468–72*; Russia as threat to United States opinions by, 122, *122*; third-party candidate support by, *462*, 462–65, *463*

General Social Survey. *See* GSS
generational differences: on abortion rights, 117–20, *118*; cultural changes and, 55–56; government spending on military views by, 519–20, *520*, *521*; Internet usage and, *210*, 210–11, *211*; presidential elections influenced by education and, *515*, 515–16
geographic mobility, age of first childbearing and, *437*, 437–39, *438*, *439*, 439–40
"getting to know" data, 55–56, 80–81
Goldberg, Amir, 16
golden mean rule, 156
goodness-of-fit measures, for regression lines, 556–58
Google, 16
Gould, Stephen Jay, 196
government assistance support, marital status and, 559–61, *560*, 564–65, 567–68
government spending: on education, 540–42; on health care, *460*, 461; on military, 519–20, *520*, *521*
government trust, social class and, 442–47
GPA, gender, job interviews and, 2–3
graduation rates, 63–65
graphical user interface (GUI): SPSS, 31; Stata, 23
graphic displays. *See* visual representations of data
Great Gatsby Curve, 543–44, *544*
Gretzky, Wayne, 208, 209, *209*
growth mindset: definition of, 17, 698; math anxiety and, 17–18; tools for, 18–19
GSS (General Social Survey): on age of first childbearing and geographic mobility, 198, *437*, 437–39, *438*; on communist's book acceptance by gender, *476*, 476–77; on economic security and voting, 425–27; on education and political ideology, 483–88, *484*, *486–88*; on excitement of life, *57*, 57–58, *58*; on extreme weather events and neighborhood impact, *316*, 316–17; on freeminds and political ideology, 498–99; on government assistance support and marital status, 559–61, *560*, 564–65, 567–68; on government spending on education, 540–42; on government spending on military, 519–20, *520*, *521*; on government trust and social class, 442–47; on Internet usage, 270; on Internet usage and generational differences, *210*, 210–11, *211*; on marijuana legalization opinions, 308–9; on marital status, 70, *70*; on meat eating habits, *165*, 165–67, *167*; on nativity status and foreign aid opinions, 130–32, *131*, *132*; probability knowledge tested in, 244; on rooms in homes, 199–201, *200*; on social class and happiness, 133, *133*; on social standing perceptions, 347–48; SPSS using, 90–102; Stata using, 81–89; on television viewing and education, 133–34, *134*; units in, 14
GUI (graphical user interface): SPSS, 31; Stata, 23

happiness: measurement of, 6; social class and, 516, *516*
hard work and success opinions: in Mexico, 272–75, *273–75*, *281*, 281–83; in United States, 287–96, *287–96*
Harris, Kamala, 261–62
health care, government spending on, *460*, 461
health insurance, attitudes toward, 244–45, *245*
health perceptions, gender and, 472–73, *473*
heart disease, satellite dishes and, 503–5, *504*, *505*
histograms: bar graphs compared to, 73; continuous variables and, 73, 212, 245; definition of, 81, 698; interval-ratio variables and, 72–73, *73*, 76–77, *77*; probability and, 244–46, *245*; skewed distributions and, 168–69, *169*; in SPSS, 98, *99–102*, 101–2; in Stata, 87–88, *88*
Hollerith, Herman, 20
homoscedastic residuals, regression lines and, 571, *571*
hourly pay, nurses', 270
hours worked, *160*, 160–63, *162*, 169, *169*
hypothesis, definition of, 3, 698
hypothesis testing: alpha level in, 363, 368; alternative hypothesis in, 353–55; with ANOVA, *431*, 431–32; assumptions with, 367; Central Limit Theorem and, 351–52; with chi-square statistic, *462*, 462–65, *463*; confidence intervals and mean, 365–66, 369; definition of, 698; degrees of freedom in, 364, 368; errors and limitations of, 370–74; false positives/false negatives in, 371; F-statistic and, 566–67; logic of, 352–54; for means, 362–69, *363*, *364*, *366*, *368*; one-tailed and two-tailed test choices for, 354; power of, 373, 700; for proportions, 354–60, *356*, *358*, *359*; p-value and, 356–58, *358*; repeated sampling and, 353; sample statistics in, 350, 352, 358–59, 370–71, *371*; simplicity of, 370; SPSS for mean, 382–84, *383*; SPSS for proportion, 381–82, *382*; standard error and, 357, 364–65, 368; standard error of the slope and, 569–70; Stata for mean, 379–80, *380*; Stata for proportion, 377–79, *378*; statistical significance and, 374–75; steps of, 359; t-values and, 362–66, *363*, *366*, 368–69; Type I and Type II errors in, *371*, 371–74, *374*; uses of, 350–51; z-score and, 356–57. *See also* alternative hypothesis; null hypothesis; one-sample tests; one sample tests; one-tailed tests; two-sample tests; two-tailed tests

ideology. *See* political ideology
immigrants: corruption perceptions, institutional trust and, 575–88; wealth inequality and, 323–25, *324, 325*
incarceration rates, 74, *74–75*, 322–23
income: college graduates' mean, 362–66, *363, 364, 366*; differences as incentives for effort, 193–94; marital status and, 564–65; median compared to mean, *171*, 171–72; probability and, *245*, 246; rural and urban county-level differences in, 396–401, *397, 398, 401*, 403–4; skewed distribution and, 276, *276*
income inequality, social mobility and, 543–44, *544*
independence: definition of, 698; Multiplication Rule with, 240, 242–43; Multiplication Rule without, 240–42, *241, 242*; probability and, 239–40; regression lines and data, 571
independent effect, definition of, 512, 698
independent variables: ANOVA and, 428; causal relationships between dependent variables and, 505–6, *506*; column percentages and relationship of dependent variables with, 125–26; definition of, 10–11, 698; dependent variables' relationships to, 119–22, *120–22*; dichotomous, 558–61; intercept and, 554; row percentages and relationship of dependent variables with, 127; slope and, 554; SPSS for interval-ratio variables and effect of, controlling for third variable, 526–29, *527–29*; Stata for interval-ratio variables and effect of, controlling for third variable, 521–23, *522, 523*; strongly correlated, 563, *563*; uncorrelated, 563, *563*; weakly correlated, 563, *563*
inferential statistics, xi; definition of, 12, 698; population parameter estimations and, 309–11, *310*; probability and, 234; regression lines and, 565–70
insomnia, COVID-19-related, 392–93, *393*, 410
institutional trust, corruption perceptions, immigrant status and, 575–88
instructor resources, xii–xiii
Instructor's Manual with Solutions, xii–xiii
intercept: calculating, 556; definition of, 698; dependent variables and, 554; dichotomous variables and, 559; independent variables and, 554; multiple regression and, 564; of regression line, 553–56, *555, 556*; in scatterplots, 554–55, *555, 556*
Internet news use, 174–84
Internet polling, 332–33
Internet usage: generational differences and, *210*, 210–11, *211*; in Japan, 239, *241*; SPSS problems on, 270

interquartile range (IQR): box plots and, 201–2, *202*; calculating, 198–99, *199*; cumulative percentage and, 201; definition of, 198, 698; in frequency distributions, 199–201, *200*; interval-ratio variables and, 199; in SPSS, 220, *221*; in Stata, 215, *215*
interval-level variables, definition of, 9, 698
interval-ratio variables: confidence intervals and, 335; definition of, 9, 698; frequency distribution tables and, 71–72, *72, 73*; histograms and, 72–73, *73*, 76–77, *77*; interquartile range and, 199; mean and, 155; median and, 160; range and, 198; SPSS and independent variable effect on, controlling for third variable, 526–29, *527–29*; Stata and independent variable effect on, controlling for third variable, 521–23, *522, 523*
intervening variables: control variables as, 508, *508*; definition of, 698; spurious relationships and, 512, *512*
inverse relationships. *See* negative relationships
IQR. *See* interquartile range
IQ scores: Addition Rule and, 252–53, *253*; Complement Rule and, 252, *252*; cumulative probability for, 253–54, *254*; for Mensa, 255; normal distribution and, *248*, 248–49; right-tail probability for, *250*, 250–51; sampling distribution of the means and, 279–81, *280*

Japan, Internet usage in, 239, *241*
job interviews, GPA, gender and, 2–3
jury pools, race in, *479*, 479–81, *480*

Kahneman, Daniel, 233–35, 242–43

left-tail probability, *255*, 255–56
levels of measurement: definition of, 698–99; measures of central tendency and, 167; types of, 8–10. *See also* interval-ratio variables; nominal-level variables; ordinal-level variables
Levene Statistics, 446
LGBTQ people: gay marriage support by age, 76–77, *77, 78*; gay marriage support by political party and education level, 75–76, *75–77*; generational attitudes on, 55; perceptions by region, 512–14, *513, 514*; transgender people, societal acceptance of, 56
libcon variable, 483–88, *484, 486–88*
liberals, feelings towards feminists and, *548*, 548–49, 551–52, *552*, 554–55, *555, 556*
Likert scale, 335

"Linda" problem: Multiplication Rule with independence and, 242–43; probability misinterpretations and, 233–34

linear regression: definition of, 699; regression line to scatterplots and, 551, *552*

linear relationships: correlation coefficients and, 545, 549; definition of, 699

logistic regression: binary dependent variables and, 561–62, *562*; definition of, 699; dichotomous variables and, 561; regression coefficients and, 561

logit coefficients, 561–62

lower bounds, of confidence intervals, 314–15, 317, *318*, 323, 325–26

Major League Baseball (MLB), 208, 209, *209*

marginal distribution, 125, *125*

marginal percentages, 125, *125*

margins of error: calculating, 313–14, *314*; confidence intervals and, 308–9, 311; definition of, 699; example calculation of, 316, *318*, 323, 325; political polls with, 271; sample size and, 333

marijuana legalization: confidence interval for support for, 312; estimation and, 311; hypothesis testing for, 354–59, *356*, *358*, *359*, 370–71, *371*; one-tailed tests for, 360; opinions on, 308–9; two-tailed tests for, 360–61, *361*; Type I and Type II errors in hypothesis testing on, *371*, 371–73

marital status: bar graph showing American, 70, *70*; divorce law opinions and, 128–29, *129*; government assistance support and, 559–61, *560*, 564–65, 567–68; income and, 564–65; pie chart showing American, 70–71, *71*

Martin, Terrance, Jr., 197

Massachusetts Department of Education, 65

math anxiety, growth mindset and, 17–18

mean: bimodal distribution and, *170*, 170–71; calculating, 155–58; confidence intervals and hypothesis testing for, 365–66, 369; confidence intervals and two-sample tests for differences between, 403–4; confidence intervals for, 319–26; definition of, 153, 155, 699; in frequency distributions, 166–67, *167*; golden mean rule, 156; hypothesis testing for, 362–69, *363*, *364*, *366*, *368*; interval-ratio variables and, 155; median income compared to, *171*, 171–72; nominal-level variables and, 155; non-existent value as, 163; notation for, 272; one-tailed tests for, 362–66, *363*; outliers, skewed distributions and, 167–69, *168*, *169*; as redistribution, 158–59, *159*; sample size calculations and, 334–35; sample statistic in two-sample tests as differences in proportion or, *402*, 402–3; in SPSS, 180–82, *181*, *182*; SPSS for confidence intervals of, 340, *341*; SPSS for hypothesis testing for, 382–84, *383*; SPSS for two-sample tests for difference between, 419–20, *420*; in Stata, *175*, 175–76; Stata for confidence intervals of, *338*, 338–39; Stata for hypothesis testing for, 379–80, *380*; Stata for two-sample tests for difference between, 414–16, *415*; two-sample tests for differences between, 396–404, *397*, *398*, *401*; two-tailed tests for, 367–69, *368*. *See also* sampling distribution of the means; standard deviation

mean squares, ANOVA and, 436

measurement: aggregate level in, 6; of concepts, 4–5, *5*; definition of, 699; error, 6–7, *8*; of food insecurity, 6–7; of happiness, 6; levels of, 8–10, 167, 698–99; reliable, 7, *8*; scales and, 10; valid, 6–7, *8*

measures of central tendency: bimodal distribution and, *170*, 170–71; choosing appropriate, 167–71; definition of, 699; frequency distributions, cross-tabulations compared to, 154; in frequency distributions, *165*, 165–67, *167*; income and choosing, *171*, 171–72; levels of measurement and, 167; outliers, skewed distributions and, 167–69, *168*, *169*; SPSS and, 179–84; Stata and, 174–78; typical values and, 154–55; uses of, 153–54; variability compared to, 158, 196–97. *See also* mean; median; mode

measures of variability. *See* variability

meat eating habits, *165*, 165–67, *167*

median: bimodal distribution and, *170*, 170–71; calculating, *160*, 160–63, *162*; definition of, 153, 159, 699; in frequency distributions, 165–66; interval-ratio variables and, 160; mean income compared to, *171*, 171–72; nominal-level variables and, 160; non-existent value as, 163; ordinal-level variables and, 160; outliers, skewed distributions and, 167–69, *168*, *169*; in SPSS, 182–83, *183*; in Stata, 176, *176*

"The Median Isn't the Message" (Gould), 196

mediating variable, definition of, 699. *See also* intervening variables

Mensa, IQ scores for, 255

mental health of prisoners, 394–95

Meta, 16

Mexico: hard work and success opinions in, 272–75, *273–75*, 281, *281–83*, 287, *287–88*, *288*; presidential election voting in, *236*, 236–37

military, government spending on, 519–20, *520*, *521*

military service: government spending on military views by, 519–20, *520*, *521*; political ideology and, 474–75, *474–75*

miscarriage, flu vaccines and spurious relationship with, 506

misconceptions, statistics and, x

misleading graphics, 78, *79*

missing values, frequency distribution tables and, 66–67, *67*

mixed methods, definition of, 4, 699

MLB (Major League Baseball), 208, 209, *209*

mode: bimodal distribution and, *170*, 170–71; calculating, 163–64, *164*; definition of, 153, 163, 699; in frequency distributions, 165, *165*; outliers, skewed distributions and, 167–69, *168*, *169*; in SPSS, 182–83, *183*; in Stata, 176–78, *177*, *178*

moderating variables: control variables and, 514; definition of, 699

moderation: definition of, 699; Emerson on, 158; golden mean rule and, 156

multiple regression: advantages of, 562; control variables and, 563–64; equation for, 564; intercept and, 564; slope and, 564; Venn diagrams and, 562–63, *563*

Multiplication Rule: definition of, 699; with independence, 240, 242–43; without independence, 240–42, *241*, *242*

multistage cluster sampling, definition of, 13, 699

mutually exclusive outcomes, probability of, 237

natarms2 variable, 519–20, *520*, *521*

nateduc variable, 540–42

National Hockey League (NHL), 208, 209, *209*

National Rifle Association (NRA), *201*, 201–2, *202*

nativity status, foreign aid opinions and, 130–32, *131*, *132*

NBC News, *68*, 68–69, *69*

negative relationships: correlation coefficients and, 545; definition of, 132–33, 699; example of, 134, *134*

net effect, definition of, 512, 698

neuroscience, 17

New Immigrant Survey (NIS), *324*, 324–26, *326*

newssites variable: in SPSS, 179–83; in Stata, 174–78

new variable generation: in SPSS, 179–80, *180*; in Stata, 174–75

New York Times, 350

NHL (National Hockey League), 208, 209, *209*

NIS (New Immigrant Survey), *324*, 324–26, *326*

nominal-level variables: bar graphs and, 70, *70*; definition of, 9–10, 61, 699; frequency distributions and, 56–57, *57*; mean and, 155; median and, 160; pie charts and, 70–71, *71*

non-probability samples, definition of, 12, 699

nonresponse bias, definition of, 13, 699

nonsensical relationships, correlation coefficients and, 550

non-spurious relationships: control variables and, 514, *515*; definition of, 700; explanatory variables and, 512, *513*, 515–16; modeling, 512–16, *513–15*

normal distribution: definition of, 246, 700; Empirical Rule and, 247–49, *248*; features of, 246, *247*, 277; for probability calculations, 249; sampling distributions as, *277*, 277–78; standard deviation and, 246–47, *247*; t-distribution compared to, 320; terminology for, *255*, 255–56

normal table, z-scores and, *251*, 251–52

NRA (National Rifle Association), *201*, 201–2, *202*

null hypothesis: alternative hypothesis compared to, 353; ANOVA and, 433, 437–38; chi-square test for goodness-of-fit and, *479*, 479–80; chi-square test of independence and, 464, 465–66; definition of, 351, 700; educated guess for, 371; false null, 373, 700; false positives/false negatives and, 371; F-statistic and, 566, 567; generating, 353–54, 355–56, *356*, 362–63, 367–68; in one-sample tests, 353–55; in one-tailed and two-tailed tests, 354; p-value and, 357–58, 462; random sampling and, 370–71, *371*; rejection of, 358–59, *359*, 365, *366*, 400–401; standard error of slope and, 569; as true, 356; in two-sample tests, 393–95, 398, 400–401, 405

numerator, for rate calculation, 64–65

nurses' hourly pay questions, Stata for, 270

observed frequencies: definition of, 700; difference between expected frequencies and, 467, 468

odds ratios, 561–62

omnibus test, ANOVA as, 429

one-sample tests: decision t in, 365, *366*, 368–69; definition of, 700; example of, 351, *351*; null hypothesis and alternative hypothesis in, 353–55; one-tailed and two-tailed test choices for, 354; population parameter in two-sample tests compared to, 395; sample t in, 365, *366*, 369; t-tests and, 362, 367, 369. *See also* hypothesis testing

one-tailed tests: alternative hypothesis in, 354; choosing, 354; definition of, 700; examples of, 360–61; for mean, 362–66, *363*; null hypothesis in, 354

open-ended survey items, definition of, 5, 700

operationalization: definition of, 4, 700; process of, 4–5, *5*. *See also* measurement

operators, Stata, 29, *29*

opioid addiction relapse, drug effectiveness and: confidence intervals for, 316–18, *318*; two-sample tests for differences between proportions and, 404–10, *406*, *409*

ordinal-level variables: definition of, 9, 700; frequency distributions and, 57–58, *58*; median and, 160; range and, 197–98

outliers, skewed distributions and, 167–69, *168*, *169*

Output window, SPSS, 33

Pakistan, presidential election voting in, *236*, 236–37

paper surveys, 13

party identification, *68*, 68–69, *69*; CDC views by, 521–23, *522*, *523*, 526–29, *527–29*; by gender, *478*, 478–79, 509–10, *510*, *511*; ideological identification and, *120*, 120–21, *121*; misleading graphics on, 78, *79*; race, religiousness and, 516–17, *517*; social class and, 505–6, *506*

Pence, Mike, 261–62

percentages: calculating, 58–59; cumulative, 60–63, *61*, *62*, 80, 697; definition of, 58, 80, 700; frequency distributions and, *59*, 59–60, *60*; marginal, 125, *125*; relationships between variables comparing, *118*, 118–19, *119*; as relative frequencies, 58; relative size assessments and, 59. *See also* column percentages; row percentages; total percentages

percent change: calculating, 63; definition of, 63, 80, 700

percentiles: definition of, 63, 80, 700; in SPSS, 218–19, *219*, *220*; in Stata, 214–15, *215*

Pew Research Center, 55, 117–18, *118*, 163–64, *164*

Pfizer COVID-19 vaccine trials, 350–52, *351*

pid3 variable problems: for SPSS, 457–58; for Stata, 456–57

pie charts: bar graphs compared to, 76, *77*; definition of, 81, 700; effective uses of, 70–71, *71*; in SPSS, 95–97, *96*, *97*; in Stata, 86, *86*

Planned Parenthood v. Casey, 117

point-and-click method: SPSS, 31; Stata, 23

point estimates: confidence intervals and, 311, 317, 319, 324; definition of, 310, 700; population parameters and, 311, 326; sample statistics and, 310

police: contact and time measures, 377–84, 412–21; political ideology and feelings on, 596–98

Police Public Contact Survey. *See* PPCS

political ideology: ANES scale, 66–68, *67*, *68*, 474–76, *474–76*, 596–98; education and, 483–88, *484*, *486–88*; freeminds and, 498–99; by gender, 509–12, *509–12*; military service and, 474–75, *474–75*; party identification and, *120*, 120–21, *121*; police feelings by, 596–98

political parties, rural feeling thermometer ratings of, 456–58

political polls: with margin of error, 271; methods for, 332–33; presidential elections, 11, 319

politics, gender gap and, 459–62, *460*, *461*

polling methods, 332–33

"pooled" standard error, 406–7

populations: definition of, 11–12, 700; sample reflecting, 271; statistical notation for, 272

population density, traffic deaths and, 501–2, *502*

population parameters: confidence intervals and, 312, 331, 332; definition of, 272, 700; inferential statistics and estimation of, 309–11, *310*; in one- compared to two-sample tests, 395; point estimates and, 311, 326; sample statistics' relationship with, 272–75, *273–75*; symbols for, 272

positive relationships: correlation coefficients and, *548*, 549; definition of, 132–33, 700; examples of, 133, *133*

"post-fact/post-truth" era, 2

post hoc test: ANOVA and, 429, *439*, 439–40; definition of, 439, 700; Tukey's, 439–40, *441*, 446–47, 703

power of a hypothesis test, definition of, 373, 700

PowerPoint slides, xiii

PPCS (Police Public Contact Survey): codebook, 15; SPSS using data from, 380–84, 416–21; Stata using data from, 377–80, 412–16; units in, 14

practical significance: statistical significance compared to, 374–75, 477–78; in two-sample tests, 410

prayer and worship, frequency of, *123–26*, 123–28, *128*

precision: compromising, 312; confidence intervals and, 311–12, 331; definition of, 700

presidential elections: age and education influencing, *515*, 515–16; in Mexico, *236*, 236–37; in Pakistan, *236*, 236–37; polls, 11, 319; results, 56–57, *57*, 60, *60*, 164; sustainable behaviors and vote choice in, *507*, 507–8, *508*

prisoners, mental health of, 394–95

privacy, big data and, 16–17

probability: Addition Rule of, 237–38, 245, 252–53, *253*, 696; calculating, 235, 235–37, *236*; Complement Rule of, 238–40, 252, *252*, 697; confidence intervals and, 312; cumulative,

253–55, *254, 255,* 697; dependence and, 240; as fraction, 235; GSS testing knowledge of, 244; histograms and, 244–46, *245*; income and, *245*, 246; independence and, 239–40; inferential statistics and, 234; left-tail, *255*, 255–56; misinterpretations of, 233–34; Multiplication Rule of, 240–43, *241, 242*, 699; of mutually exclusive outcomes, 237; normal distribution for calculating, 249; right-tail, *205*, 250–51, *255*, 255–56; rules of, 234–44; of specific sample statistic, 278; Stata for z-score and, *258–60*, 258–61; statistics and importance of, 234

probability samples: definition of, 12, 700; methods for, 12–13; sampling frames and, 16

proportions: calculating, 58–59; confidence intervals for, 312–19; confidence intervals for differences between, 408–10, *409*; definition of, 58, 700; hypothesis testing for, 354–60, *356, 358, 359*; notation for, 272; as relative frequencies, 58; sample size calculations and, 333–34; sample statistic in two-sample tests as differences in mean or, *402*, 402–3; SPSS for confidence intervals of, 340–42, *341*; SPSS for hypothesis testing for, 381–82, *382*; SPSS for two-sample tests for difference between, 417–19, *418, 419*; Stata for confidence intervals of, 339, *339*; Stata for hypothesis testing for, 377–79, *378*; Stata for two-sample tests for difference between, 413–14, *414*; two-sample tests for differences between, 404–10, *406, 409*. *See also* sampling distribution of the proportions

psychology, ANOVA in, 428–29

publication bias, toward statistical significant results, 359–60

punched cards, US Census and, 20, *20*

p-value: chi-square statistic and, 460–61, 470–71, *471*; column percentages and, 462; F-statistic and, 567; hypothesis testing and, 356–58, *358*; misinterpreting, 357–58; null hypothesis and, 357–58, 462; threshold for, 358–59; two-sample tests and, 407–8; z-scores and, *250*, 251

Q48 variable: in SPSS, 217–22; in Stata, 214–17

Q106 variable problems: for SPSS, 194; for Stata, 193–94

Q110 variable: in SPSS, 292–97; in Stata, 287–91

Q163 variable problems: for SPSS, 115–16, 390; for Stata, 115, 389–90

Q173 variable problems: for SPSS, 152; for Stata, 151, *151*

Q251 variable problems: for SPSS, 306–7; for Stata, 306

Quadlin, Natasha, 2–3

qualitative methods: definition of, 3, 700; in social sciences, 3–4

quantitative analysis: definition of, 3, 700; uses of, 2

race: abortion rights opinions and, 117, 119, *119*; in jury pools, *479*, 479–81, *480*; party identification, religiousness and, 516–17, *517*; stereotypes and, avoiding, ix–x; wealth inequality among immigrants by, 323–25, *324, 325*

random sampling: accuracy with, 271; definition of, 701; sample statistic in hypothesis testing and, 370–71, *371*; sampling distribution of the means and, 283; simple, 12–13, 702; standard error and, 283; Stata and, 287–88, *288*; stratified, 13, 702

range: confidence levels' relationship with confidence interval, 330–31; definition of, 197, 701; interval-ratio variables and, 198; limitations of, 198; ordinal-level variables and, 197–98; sample size and confidence interval, 326–29, *327–29*; in SPSS, 217–18, *218*; in Stata, 214, *214*

rankcat variable, 442–47

rates: calculating, 64–65; childhood poverty, 327–29, *327–29*, 331–32, *332*; definition of, 64, 80–81, 701; divorce, 66; graduation, 63–65; incarceration, 74, 74–75, 322–23; ratios compared to, 66

ratio-level variables, definition of, 9, 701. *See also* interval-ratio variables

ratios: calculating, 65–66; definition of, 65, 81, 701; rates compared to, 66

raw frequencies: in cross-tabulation, 123–25, *124, 125*; definition of, 701; in frequency distributions, 56; row percentages and, 126

recidivism, rehabilitative programs and, 10–11

recoding variables: definition of, 701; SPSS, 92–95, *93–96*; Stata, *83–85*, 83–86. *See also* collapsing/combining categories

redistribution, mean as, 158–59, *159*

regression analysis, definition of, 544

regression coefficients: definition of, 701; logistic regression and, 561. *See also* slope

regression lines: alternative terms and notation for, 554; assumptions of, 570–72; as "best-fitting" line, 551–53, *552*; definition of, 701; dichotomous independent variables and, 558–61; equation for, 553; F-statistic and, 566–68; goodness-of-fit measures for, 556–58; homoscedastic residual, 571, *571*; independence of data and, 571; inferential statistics and, 565–70; normally distributed errors and, 571–72, *572*; predicted value of dependent variables in, 551; residual and, 552–53; r-squared and, 557–58;

to scatterplots, 551, *552*; slope and intercept of, 553–56, *555*, *556*; in SPSS, 581–88, *582–84*, *586–87*; standard error of the estimate and, 558; standard error of the slope and, 568–70; in Stata, 575–79, *576–78*, *580*

rehabilitative programs, recidivism and, 10–11

relative frequencies: definition of, 701; percentages and proportions as, 58

relative size assessments, 59

reliability: definition of, 7, 701; validity and, 7, *8*

religious affiliation, 163–64, *164*

religiousness: gender and, *465*, 465–72, *468–72*; party identification, race and, 516–17, *517*; science views and, 150–52, *151*

repeated measures of ANOVA, 432

repeated sampling: accuracy of, 272; hypothesis testing and, 353; sample statistics, population parameters and, 272–75, *273–75*; in Stata, 288–91, *289–91*

research questions: definition of, 3, 701; in social sciences, 3. *See also specific research*

residuals: aggregate data and, 558; definition of, 552, 701; regression line and, 552–53; regression lines and homoscedastic, 571, *571*; standardized, 471, 471–72, 702

resources, student and instructor, xii–xiii

Results Window, Stata, 25–26, 30

Review Window, Stata, 24, 30

right-tail probability, *205*, 250–51, *255*, 255–56

RMSE (Root Mean Square Error), 558

robots, trustworthiness of, 428–29

"Robust Misinterpretation of Confidence Intervals," 331

Roe v. Wade, 117–18

Root Mean Square Error (RMSE), 558

row percentages: column percentages compared to, 127, *128*; cross-tabulations and, 126–27; definition of, 126, 701; examples calculations of, 128–29, *129*; independent and dependent variables' relationships and, 127; raw frequencies and, 126; total percentages compared to, 128

r-squared: calculating, 557; definition of, 700–701; regression lines and, 557–58

rural county incomes, urban county income differences with, 396–401, *397*, *398*, *401*, 403–4

Russia as threat to United States opinions, 122, *122*

sample chi-square, decision chi-square compared to, *469*, 469–70, *470*

sample f, decision f compared to, 436–39

samples: definition of, 11–12, 701; non-probability, 12, 699; population reflected in, 271; probability, 12, 700; statistical notation for, 272; stratified random, 13, 702

sample size: chi-square test of independence and, 477–79, *478*; confidence interval range and, 326–29, *327–29*; confidence intervals and determining, 332–35; margins of error and, 333; mean and calculations of, 334–35; proportions and calculating, 333–34; standard error, z-scores and, 284–85, *285*; statistical significance and, 375; t-distribution and, 329

sample statistics: confidence intervals and, 331; definition of, 272, 701; in hypothesis testing, 350, 352, 358–59, 370–71, *371*; point estimates and, 310; population parameter's relationship with, 272–75, *273–75*; probability of obtaining specific, 278; as two-sample test for differences in proportions or means, *402*, 402–3; z-scores for, 278

sample t: in one-sample tests, 365, *366*, 369; in two-sample tests, 400–401

sampling: cluster, 13, 696–97; definition of, 11–12, 701; methods, 12–14; multistage cluster, 13, 699; repeated, 272–75; simple random, 12–13, 702; skewed distributions and, *276*, 276

sampling distributions: Central Limit Theorem and, 275–76, *276*; chi-square test of independence and, 464; definition of, 701; Empirical Rule and, 312; estimation and, 311; as normal curve, *277*, 277–78; standard deviation estimation and, 279; standard error and, *277*, 277–78; for two-sample tests, 395–96, *396*

sampling distribution of the means: Central Limit Theorem and, 284; definition of, 275, 701; random sampling and, 283; standard error and, 279–83, *280*, *281*; z-scores for, 279–83, *280*, *281*. *See also* mean

sampling distribution of the proportions: Central Limit Theorem and, 284; definition of, 275, 701; standard error and, 283–84; z-score for, 283–84. *See also* proportions

sampling frames: bias and, 16; big data and, 16; challenges in assembling, 13; definition of, 12, 701

satellite dishes, heart disease and, 503–5, *504*, *505*

SAT scores: ACT scores compared to, 208; interquartile range for critical reading, 198–99, *199*; Stata for probability, z-scores and, *258–60*, 258–61

saving work: in SPSS, 41; in Stata, 30

scales: ANES political ideology, 66–68, *67*, *68*, 474–76, *474–76*, 596–98; definition of, 701; as level of measurement, 10; Likert, 335; standard deviation for comparison across different, 208 scatterplots: correlation coefficients and, 547, 547–49, 548; of curvilinear relationships, 549, 550; definition of, 701; dichotomous variables and, 560, 560; on income inequality, 543–44, 544; regression lines to, 551, 552; slope and intercept in, 554–55, 555, 556; in SPSS, 581–83, 586, 586; in Stata, 575–77, 576, 577, 580; strength and direction of relationships in, 547, 547

science views: in China, 389–90; global attitudes, 115–16; religiousness and, 150–52, 151

scitech variable problems: for SPSS, 152; for Stata, 151, 151

secondary data: definition of, 14, 702; sources of, 14–17, 15

sexual practices of college students, 164

simple random sampling, definition of, 12–13, 702

single variables. *See* variables

SIPP (Survey of Income and Program Participation), 245, 246, 276, 276

skewed distributions: measures of central tendency and, 167–69, 168, 169; sampling and, 276, 276; standard deviation and, 212, 212–13

slope: calculating, 556; definition of, 702; dependent variable and, 554; dichotomous variables and, 559; independent variables and, 554; multiple regression and, 564; of regression line, 553–56, 555, 556; in scatterplots, 554–55, 555, 556; standard error of, 565, 568–70, 702. *See also* regression coefficients

Smith College, 12

SNAP (Supplemental Nutrition Assistance Program), 16

social class: government trust and, 442–47; happiness and, 133, 133, 516, 516; party identification and, 505–6, 506

socialists, standard deviation in opinion on, 205, 206, 207

social media, big data from, 16

social mobility, income inequality and, 543–44, 544

social psychology, ANOVA in, 428–29

social sciences: data gathering in, 4; "getting to know" data in, 55–56, 80–81; importance of studying, 1–3; measurement error in, 6–7, 8; qualitative methods in, 3–4; research questions in, 3; statistics in, viii, ix; units of analysis in, 5–6

social standing perceptions, 347–48

sports records, standard deviation for comparing, 208, 209, 209

SPSS (Statistical Package for Social Sciences), viii, ix; *agekdbrn* variable problems for, 52–53; "Analyze an Existing Variable" command in, 41, 42, 43; for ANOVA, 444–47, 445, 446; bar graphs in, 97, 98; *B_COUNTRY* variable in, 293, 293; box plots in, 221, 221; chi-square test of independence and, 485–88, 486–88; command categories in, 33–41; command-driven use in, 31; for confidence intervals of means, 340, 341; for confidence intervals of proportions, 340–42, 341; for controlling for third variable in cross-tabulations, 524–26, 525, 526; "Create a New Variable" command in, 34–36, 35–37; for cross-tabulations, 141–44, 142–44, 152; Data Editor in, 32, 32; Data View in, 32, 32–33; *endsmeet2* variable problems in, 426–27; *freemind* variable problems for, 498–99; frequency distributions in, 90–92, 90–92; government spending on education problems for, 541–42; graphical user interface in, 31; GSS data for, 90–102; histograms in, 98, 99–102, 101–2; for hypothesis testing for mean, 382–84, 383; for hypothesis testing for proportion, 381–82, 382; independent variable effect on interval-ratio variable, controlling for third variable, 526–29, 527–29; instructors using, x; Internet usage problems for, 270; interquartile range in, 220, 221; launching, 31–32; mean in, 180–82, 181, 182; measures of central tendency and, 179–84; median and mode in, 182–83, 183; new variable generation in, 179–80, 180; Output window in, 33; percentiles in, 218–19, 219, 220; *pid3* variable problems for, 457–58; pie charts in, 95–97, 96, 97; point-and-click method in, 31; police feelings and political ideology problems for, 597–98; popularity of, 20; PPCS data in, 380–84, 416–21; *Q48* variable in, 217–22; *Q106* variable problems for, 194; *Q110* variable in, 292–97; *Q163* variable problems for, 115–16, 390; *Q251* variable problems for, 306–7; range in, 217–18, 218; recoding variables, 92–95, 93–96; regression lines in, 581–88, 582–84, 586–87; saving in, 41; scatterplots in, 581–83, 586, 586; *scitech* and *Q173* variable problems for, 152; social standings perceptions problems for, 347–48; standard deviation in, 222, 222; for standardizing variables, 261–62, 262; student advice for, xii; "Transform an Existing Variable" command in, 36–40, 38–41; Tukey's post hoc test in, 446–47; for two-sample test for difference between means, 419–20, 420; for two-sample test for difference between proportions, 417–19, 418, 419; *V201200* and *V202171* variable problems for, 597–98; *V202143* and *V202144* variable problems for, 232; *V202184* variable problems for, 457–58; *V202187* variable in, 340–42; value labels assigned to recoded variables in, 95, 96; for variability, 217–22; variables in, 32–33; Variable View in, 33, *34*; variance in, 222, *222*; *voted16* variable problems in, 426–27

spurious relationships: antecedent variables and, *511*, 511–12; causal relationships and, 502–3; control variables and, 507–8, *508*; definition of, 702; example of, 503–5, *504*, *505*; between flu vaccines and miscarriage, 506; intervening variables and, 512, *512*; modeling, 507–12, *507–12*

stacked bar graphs: cross-tabulations compared to, 137; definition of, 702; examples of, 135–37, *136*

standard deviation: ANES data and, 205, *205*; for "apples and oranges" comparisons, 206, 208–11, *208–11*; calculating, 204–5; for comparison across different scales, 208; correlation coefficients and, 546–47; definition of, 202, 702; examples of, *203*, 203–4; formula for, 204; frequency distribution comparisons using, *205*, 205–6; normal distribution and, 246–47, *247*; notation for, 272; skewed distributions and, *212*, 212–13; in SPSS, 222, *222*; standard error estimation from, 279; in Stata, *216*, 216–17, *217*; usefulness of, 202–3; variance and, 204

standard error: calculating, 313; definition of, 277, 702; example calculations of, 316, *318*, 323, 325; hypothesis testing and, 357, 364–65, 368; "pooled," 406–7; random sampling and, 283; sample size and, 284–85, *285*; sampling distribution of the means and, 279–83, *280*, *281*; sampling distribution of the proportions and, 283–84; sampling distributions and, *277*, 277–78; standard deviation and estimation of, 279; in two-sample tests, 399–400, 406–9

standard error of the estimate: definition of, 702; regression lines and, 558

standard error of the slope: definition of, 702; F-statistic and, 565; hypothesis testing and, 569–70; regression lines and, 568–70

standardized residuals: chi-square statistic and, *471*, 471–72; definition of, 702

standardized tests, 63

standardizing variables: SPSS for, 261–62, *262*; z-scores and, 249–54. *See also* z-scores

Stata, viii, ix; *agekdbrn* variable problems for, 50–52; "Analyze an Existing Variable" command in, 28, *28*; for ANOVA, 442–44, *443*; bar graphs in, 87, *87*; *Base Reference Manual*, 31; *B_COUNTRY* variable in, 287, *287*; box plots in, 215–16, *216*; chi-square test of independence and, 483–85, *484*; coding logic in, 26; command categories in, 26–28; command-driven use in, 23; Command Window in, 24, 30; for confidence intervals of means, 338, 338–39; for confidence intervals of proportions, 339, *339*; controlling for third variables in cross-tabulation in, 519–20, *520*, *521*; "Create a New Variable" command in, 26–27; for cross-tabulations, 139–41, *140*, *141*, 150–51, *151*; Data Editor in, 24–25, *25*; data files in, 29; do-files in, *29*, 29–30, *30*; *endsmeet2* variable problems in, 425–26; error messages in, 28–29; *freemind* variable problems for, 498; frequency distributions in, *82*, 82–83, *83*; government spending on education problems for, 540–41; graphical user interface in, 23; GSS data for, 81–89; histograms in, 87–88, *88*; for hypothesis testing for mean, 379–80, *380*; for hypothesis testing for proportion, 377–79, *378*; independent variable effect on interval-ratio variable, controlling for third variable, 521–23, *522*, *523*; instructors using, x; interquartile range in, 215, *215*; launching, 23–24, *24*; mean in, 175, 175–76; measures of central tendency and, 174–78; median in, 176, *176*; mode in, 176–78, *177*, *178*; new variable generation in, 174–75; for nurses' hourly pay questions, 270; operators in, 29, *29*; percentiles in, 214–15, *215*; *pid3* variable problems for, 456–57; pie charts in, 86, *86*; point-and-click method in, 23; police feelings and political ideology problems for, 596–97; popularity of, 20; PPCS data in, 377–80, 412–16; for probability and z-scores, *258–60*, 258–61; Q48 variable in, 214–17; Q106 variable problems for, 193–94; Q110 variable in, 287–91; Q163 variable problems for, 115, 389–90; Q251 variable problems for, 306; random sampling and, 287–88, *288*; range in, 214, *214*; recoding variables in, *83–85*, 83–86; regression lines in, 575–79, *576–78*, *580*; repeated sampling in, 288–91, *289–91*; Results Window in, 25–26, 30; Review Window in, 24, 30; saving in, 30; scatterplots in, 575–77, *576*, *577*, *580*; *scitech* and *Q173* variable problems for, 151, *151*; social standings perceptions problems for, 347; standard deviation in, *216*, 216–17, *217*; student advice for, xii; "Transform an Existing Variable" command in, 27; for two-sample test for difference between means, 414–16, *415*; for two-sample test for difference between proportions, 413–14, *414*; *User's Guide*, 31; *V201200* and *V202171* variable problems for, 596–97; *V202143* and *V202144* variable problems for, 231; *V202184* variable problems for, 456–57; *V202187* variable in, 338–39; value labels assigned to recoded variables in, *85*, 85–86; for

variability, 214–17; variables in, 25; variance in, *216*, 216–17, *217*; *voted16* variable problems in, 425–26
statistical control, definition of, 11, 702
statistical independence: chi-square test of independence and, 463; definition of, 702
statistical interaction: control variables and, 514–16, *515*; definition of, 702
Statistical Package for Social Sciences. *See* SPSS
statistical significance: chi-square test of independence and, 477–79, *478*; cross-tabulations and, *460*, 460–62, *461*; definition of, 358, 702; hypothesis testing and, 374–75; practical significance compared to, 374–75, 477–78; publication bias toward, 359–60; sample size and, 375; in two-sample tests, 410
statistical software programs: benefits of, 19; collapsing/combining categories in, 67; historical development of, 19–20. *See also* SPSS; Stata
statistics: confidence for, ix–x; "getting to know" data in, 55–56, 80–81; importance of studying, 1–3; instruction strategies for, xi–xii; math anxiety and, 17–18; misconceptions and, x; probability's importance in, 234; real-world approach to, ix; in social sciences, viii, ix; student advice for, xii; tools for, xi, 18–19; truth and, 2; uses of, 2; Yerkes-Dodson Law and, 18. *See also specific types*
stem-and-leaf plot, definition of, 702
STEM courses, student reading and, *402*, 402–3
stereotypes, avoiding, ix–x
stratified random samples, definition of, 13, 702
strength of association, 465
strength of relationships: categories of, 130; clear language and examples for explaining, 135; correlation coefficients for, 544–45; cross-tabulations and, 130–32, *131*, *132*; definition of, 702; in scatterplots, 547, *547*
stress, COVID-19-related, 392–93, *393*, 410
strongly correlated independent variables, 563, *563*
student loan debt, 153–54
student reading, STEM courses and, *402*, 402–3
students: resources for, xiii; statistics advice for, xii
subscripts, in two-sample tests, 395
sum of cross products, correlation coefficients and, 546
sum of squares. *See* between groups sum of squares; within groups sum of squares
Supplemental Nutrition Assistance Program (SNAP), 16
survey items: closed-ended, 5, 696; open-ended, 5, 700

Survey of Income and Program Participation (SIPP), *245*, 246, 276, *276*
sustainable behaviors, vote choice and, *507*, 507–8, *508*

t-distribution: confidence intervals and, 319–22, *321*; definition of, 703; degrees of freedom in, 320–21, *321*; normal distribution compared to, 320; sample size and, 329
teaching, research on, x
teaching methods, exam scores and, 432–37, *433–35*
telephone polling, 333
television viewing, education and, 133–34, *134*, 139–44
Test Bank, xiii
third-party candidate support, by gender, *462*, 462–65, *463*
Thomas, Christopher, 308
TikTok, 3, 55
time series charts: definition of, 81, 703; misleading, 78, *79*; uses of, 74; variables in, *74*, 74–75
tools, for statistics, xi, 18–19
total percentages: column and row percentages compared to, 128; cross-tabulations and, 127–28; definition of, 127, 703; examples calculations of, 128–29, *129*
traffic deaths, auto inspections and, 500–503, *501–3*
"Transform an Existing Variable" command: SPSS, 36–40, *38–41*; Stata, 27
transgender people, societal acceptance of, 56. *See also* LGBTQ people
trees in cities, 158–59, *159*
Trump, Donald, 6, 11; feeling thermometer ratings on, 71–73, *72*, *73*, 231–32; frequency of votes for, 57, *57*, 60, *60*; NRA evaluations of voters for, *201*, 201–2, *202*
trustworthiness of robots, 428–29
truth, statistics and, 2
t-tests: ANOVA compared to, 430, 440; one-sample, 362, 367, 369; two-sample, 397; Type I errors and, 440
Tufte, Edward, 500–502, 550
Tukey's post hoc test: definition of, 703; example of, 439–40, *441*; in SPSS, 446–47
t-values: confidence intervals and, 321, *322*; confidence levels associated with, 321, 323, 325; definition of, 703; hypothesis testing and, 362–66, *363*, *366*, *368–69*; z-score compared to, 321
Tversky, Amos, 233–35, 242–43
Twitter, 3, 16
two-sample tests: alpha level in, 398, *398*, 405–6, *406*; assumptions for, 396; chi-square test of

independence's relationship with, *476*, 476–77; confidence intervals for differences between means in, 403–4; confidence intervals for differences between proportions in, 408–10, *409*; decision t in, *398*, 398–401; definition of, 703; degrees of freedom in, *398*, 398–99; for differences between means, 396–404, *397*, *398*, *401*; for differences between proportions, 404–10, *406*, *409*; examples of, 392–93, *393*; notation for, 395; null hypothesis and alternative hypothesis in, 393–95, 398, 405; null hypothesis in, 400–401; population parameter in one-sample tests compared to, 395; practical significance in, 410; p-value and, 407–8; sample statistic differences in proportions or means in, *402*, 402–3; sample t in, 400–401; sampling distribution for, 395–96, *396*; SPSS for difference between means, 419–20, *420*; SPSS for difference between proportions, 417–19, *418*, *419*; standard error in, 399–400, 406–9; Stata for difference between means, 414–16, *415*; Stata for difference between proportions, 413–14, *414*; statistical significance in, 410; subscripts in, 395; t-tests and, *397*; z-score in, 407, *476*, 476–77. *See also* hypothesis testing

two-tailed tests: alternative hypothesis in, 354; choosing, 354; confidence intervals compared to, 361; definition of, 703; examples of, 360–61, *361*; for mean, 367–69, *368*; null hypothesis in, 354

two variables. *See* cross-tabulations

Type I errors: definition of, 703; hypothesis testing and, *371*, 371–74, *374*; t-tests and, 440

Type II errors: definition of, 703; hypothesis testing and, *371*, 371–74, *374*

typical values, measures of central tendency and, 154–55

uncorrelated independent variables, 563, *563*

United States: childhood poverty rate in, 327–29, *327–29*, 331–32, *332*; COVID-19 vaccine trials in, 350–52, *351*, 373–74, *374*; gender and satisfaction with democracy in, *461*, 462; hard work and success opinions in, 287–96, *287–96*; incarceration rates in, 74, 74–75, 322–23; income inequality in, 543–44, *544*; Russia as threat to, 122, *122*

units of analysis: definition of, 5, 703; in social sciences, 5–6

univariate statistics, definition of, 56, 703

upper bounds, of confidence intervals, 314–15, 317, *318*, 323, 325–26

urban county incomes, rural county income differences with, 396–401, *397*, *398*, *401*, 403–4

US Census Bureau: on childhood poverty rate, 327–29, *327–29*; on mean and median income, *171*, 171–72; on population density and traffic deaths, 501–2, *502*; punched cards and, 20, *20*; on rural county incomes compared to urban county incomes, 396–401, *397*, *398*, *401*, 403–4

User's Guide, Stata, 31

V344 variable, 15, *15*

V201200 and *V202171* variable problems: for SPSS, 597–98; for Stata, 596–97

V202143 and *V202144* variable problems: for SPSS, 232; for Stata, 231

V202184 variable problems: for SPSS, 457–58; for Stata, 456–57

V202187 variable: in SPSS, 340–42; in Stata, 338–39

vaccine trials, COVID-19, 350–52, *351*, 373–74, *374*

validity: definition of, 6, 703; reliability and, 7, *8*

value labels assigned to recoded variables: SPSS, 95, *96*; Stata, *85*, 85–86

variability: definition of, 197, 703; diversity of values and, 197; measures of central tendency compared to, 158, 196–97; SPSS for, 217–22; Stata for, 214–17; types of, 197. *See also* interquartile range; range; standard deviation; variance

variables: antecedent, 507, *508*, *511*, 511–12, 696; binary dependent, 561–62, *562*; categorical, 10; codebook information on, 14–15, *15*; comparing two groups on same, 75–77, *75–78*; continuous, 9, 73, 212, 245, 697; cross-tabulations and relationships between, 118–30; definition of, 3, 703; dichotomous (dummy), 558–61, *560*, 697; discrete, 9, 697; empirical measures as, 4–5, *5*; explanatory, non-spurious relationships and, 512, *513*, 515–16; graphic displays of single, *70*, 70–73, *71*, *73*; interval-level, 9, 698; intervening, 508, *508*, 512, *512*, 698; moderating, 514, 699; percentages and relationships between, *118*, 118–19, *119*; ratio-level, 9, 701; recoding, 83–85, 83–86, 92–95, *93–96*, 701; in SPSS, 32–33; SPSS generation of new, 179–80, *180*; standardizing, 249–54, 261–62, *262*; in Stata, 25; Stata generation of new, 174–75; in time series charts, 74, 74–75. *See also* control variables; dependent variables; independent variables; interval-ratio variables; nominal-level variables; ordinal-level variables; *specific variables*

Variable View, SPSS, 33, *34*

variance: calculating, 204; definition of, 703; in SPSS, 222, *222*; standard deviation and, 204; in Stata, *216*, 216–17, *217*

variation, within and between groups, 429–30, *430*
Venn diagrams, multiple regression and, 562–63, *563*
veteran variable, 519–20, *520*, *521*
visual representations of data: of bivariate relationships, 135–37, *136*; for comparing two groups on single variable, 75–77, *75–78*; misleading, 78, *79*; of single variables, *70*, 70–73, *71*, *73*; types of, 81. *See also specific visual representations*
voted16 variable problems: in SPSS, 426–27; in Stata, 425–26
voting: economic security related to, 425–27; education and age influencing turnout for, *515*, 515–16; sustainable behaviors and, *507*, 507–8, *508*. *See also* presidential elections

Wall Street Journal, *68*, 68–69, *69*
weakly correlated independent variables, 563, *563*
wealth inequality, immigrants and, 323–25, *324*, *325*
WHO (World Health Organization), 360
within groups sum of squares: ANOVA and, 435–36, *436*; calculation of, 434; definition of, 703
World Health Organization (WHO), 360
World Values Survey. *See* WVS
worship and prayer, frequency of, *123–26*, 123–28, *128*
WVS (World Values Survey): on choice and control over lives, 214–22; democratic governance opinions in, 306–7; on global science attitudes, 115–16; on hard work and success opinions in Mexico, 272–75, *273–75*, *281*, 281–83, *287*, 287–88, *288*; on income differences as incentives for effort, 193–94, 217–22; on institutional trust, corruption perceptions and immigrant status, 575–88; on prayer and worship frequency, 123–26, *124–26*; on presidential election voting in Mexico and Pakistan, *236*, 236–37; on science and religiousness, 150–52; on science views in China, 389–90; on television viewing and education, 139–44; units in, 14

Yerkes-Dodson Law, 18

z-scores: Addition Rule and, 252–53, *253*; calculating, 250; Complement Rule and, 252, *252*; confidence levels and, 313, *314*, 315, *315*, 316, *318*; cumulative probability and, 253–55, *254*, *255*; definition of, 249, 703; hypothesis testing and, 356–57; left-tail probability and, *255*, 255–56; normal table and, *251*, 251–52; p-value and, *250*, 251; right-tail probability and, *205*, 250–51, *255*, 255–56; sample size and, 284–85, *285*; for sample statistics, 278; for sampling distribution of the means, 279–83, *280*, *281*; for sampling distribution of the proportions, 283–84; standardizing variables and, 249–54; Stata for probability and, *258–60*, 258–61; t-values compared to, 321; in two-sample tests, 407, *476*, 476–77